# Lecture Notes in Civil Engineering

Volume 167

**Lecture Notes in Civil Engineering** (LNCE) publishes the latest developments in Civil Engineering—quickly, informally and in top quality. Though original research reported in proceedings and post-proceedings represents the core of LNCE, edited volumes of exceptionally high quality and interest may also be considered for publication. Volumes published in LNCE embrace all aspects and subfields of, as well as new challenges in, Civil Engineering. Topics in the series include:

- Construction and Structural Mechanics
- Building Materials
- Concrete, Steel and Timber Structures
- Geotechnical Engineering
- Earthquake Engineering
- Coastal Engineering
- Ocean and Offshore Engineering; Ships and Floating Structures
- Hydraulics, Hydrology and Water Resources Engineering
- Environmental Engineering and Sustainability
- Structural Health and Monitoring
- Surveying and Geographical Information Systems
- Indoor Environments
- Transportation and Traffic
- Risk Analysis
- Safety and Security

To submit a proposal or request further information, please contact the appropriate Springer Editor:

- Pierpaolo Riva at pierpaolo.riva@springer.com (Europe and Americas);
- Swati Meherishi at swati.meherishi@springer.com (Asia - except China, and Australia, New Zealand);
- Wayne Hu at wayne.hu@springer.com (China).

**All books in the series now indexed by Scopus and EI Compendex database!**

More information about this series at http://www.springer.com/series/15087

C. N. V. Satyanarayana Reddy ·
K. Muthukkumaran · Neelima Satyam ·
Ravikiran Vaidya
Editors

# Ground Characterization and Foundations

## Proceedings of Indian Geotechnical Conference 2020 Volume 1

Springer

*Editors*
C. N. V. Satyanarayana Reddy
Geotechnical Engineering Division
Department of Civil Engineering
Andhra University College of Engineering
Visakhapatnam, Andhra Pradesh, India

K. Muthukkumaran
Geotechnical Engineering Division
Department of Civil Engineering
National Institute of Technology
Tiruchirappalli
Tiruchirappalli, Tamil Nadu, India

Neelima Satyam
Discipline of Civil Engineering
Indian Institute of Technology Indore
Indore, Madhya Pradesh, India

Ravikiran Vaidya
Geo Dynamics, Vadodara, India

ISSN 2366-2557 ISSN 2366-2565 (electronic)
Lecture Notes in Civil Engineering
ISBN 978-981-16-3382-9 ISBN 978-981-16-3383-6 (eBook)
https://doi.org/10.1007/978-981-16-3383-6



This Springer imprint is published by the registered company Springer Nature Singapore Pte Ltd.
The registered company address is: 152 Beach Road, #21-01/04 Gateway East, Singapore 189721, Singapore

# Contents

# About the Editors

**Dr. C. N. V. Satyanarayana Reddy** is a Professor of Civil Engineering at College of Engineering, Andhra University, Visakhapatnam, India, and has 28 years of teaching, research, and consultancy experience. His expertise and research interests are in the areas of reinforced soils, ground improvement, landfills, soil retention in excavations, deep foundations, and forensic geotechnical engineering. He obtained B.Tech. in Civil Engineering from Nagarjuna University, M.Tech. in Geotechnical Engineering from Indian Institute of Technology (IIT) Madras, M.E. (Structures) from Andhra University, and Ph.D. from National Institute of Technology (NIT) Warangal, India. He has guided 8 Ph.D. scholars and 92 M.Tech. dissertations. He has more than 110 publications in various national and international journals and seminars/conference proceedings. He has organized more than 30 seminars, workshops, and training programs. He is fellow of Indian Geotechnical Society, The Institution of Engineers (India) and Life member of Indian Roads Congress, Indian Concrete Institute, Indian Society for Technical Education, Indian Society for Rock Mechanics and tunneling Technology.

He handled several research projects funded by DST, UGC and AICTE. He received the Engineer of the Year 2006 Award from Govt. of Andhra Pradesh and The Institution of Engineers (India) A.P. State Centre, Best Academician Award for the year 2014 from Andhra University and Andhra Pradesh Scientist Award 2020 in Civil Engineering Discipline from A.P. State Council of Science and Technology, Govt. of A.P. He served as National Executive Committee member of Indian Geotechnical Society during the period 2013–2020. He served as a member in H-4 Committee on Embankment, Ground Improvement and Drainage Committee of Indian Roads Congress (IRC) for the term 2018–2020. He is serving as member of TC 213 on "Scour and Erosion" of International Society for Soil mechanics and Geotechnical Engineering (ISSMGE).

**Dr. K. Muthukkumaran** is currently Professor of Civil Engineering at National Institute of Technology (NIT), Tiruchirappalli, India. He obtained his Ph.D. in Marine Geotechnical Engineering from Indian Institute Technology (IIT) Madras. He has published more than 120 papers in international and national journals and conferences. He has completed 5 R&D (including ISRO Chandrayaan-2 Mission project) &

50 major consultancy projects in geotechnical engineering and published two patents. Dr. Muthukkumaran's area of research is in geotechnical engineering, which includes pile foundation, soil–structure interaction, marine geotechnics & foundations, field instrumentation, geotechnical physical modeling, ground improvement, and forensic geotechnical engineering. He has received DST Young Scientist Award, IGS-Smt. Indra Joshi Biennial Award and Keynote Paper Award—GEOMATE Conference 2015 at Osaka, Japan.

He has guided 8 PhD, 5 MS (by research) and more than 40 M.Tech. students in Geotechnical and allied research areas. He is a member of Technical Committee (TC-301 on "Preservation of Historic Sites") of International Society for Soil Mechanics and Geotechnical Engineering. He served as a member of IGS National Executive Committeeduring the period 2017–18. He has received DST Young Scientist Award, IGS-Smt. Indra Joshi Biennial Award and Keynote Paper Award—GEOMATE Conference 2015 at Osaka, Japan. He has received NIT Trichy Achiever Awards for research publications, research projects, maximum citation and consultancy projects. He is a recipient of Bharat Vikas Award–2018.

**Dr. Neelima Satyam** is currently Associate Professor and Head in Discipline of Civil Engineering at Indian Institute Technology (IIT) Indore. She obtained her Ph.D. and M.Tech. from IIT Delhi and B.Tech. from SV University, Tirupati. She is actively engaged in teaching, research, and consultancy in the field of geotechnical engineering, particularly in geotechnical earthquake engineering, microzonation and site response studies, landslide hazard and monitoring, liquefaction modeling, micro-seismic data processing, and rock engineering. She received research grants from DST, MHRD, AICTE, ITRA, DAE, NIOT, NRDMS, ISRO, and MoES. She has 97 papers published in journals and conferences of national and international repute. She was Recipient of IEI Young Engineers Award 2011, AICTE Career Award 2012, JSPS fellowship in 2013, and INWES Young Woman Engineer Award in 2012. She is elected National Executive Member of Indian Geotechnical Society (2018–2020) and Indian Society of Engineering Geology (2020–2021).

**Mr. Ravikiran Vaidya** is Principal Engineer and Founder of Geo Dynamics, a company involved in deep foundation testing. He has been instrumental in popularizing and pioneering various methods for deep foundation testing in India. It is to his credit that methods like high strain dynamic testing, integrity testing, sonic logging, etc., are now routinely used all over India to verify pile foundations and also form a part of all specifications of all major projects. He has recently worked on thermal integrity profiling for bored piles and indigenized the bidirectional load testing technology. He has written several technical papers on the subject, and his work has been cited by several authors in various articles and dissertations. He is also Guest Lecturer at academic and industry forums.

# Effect of Plasticity of Fines on Properties of Uniformly Graded Fine Sand

**M. Akhila, K. Rangaswamy, N. Sankar, and M. R. Sruthy**

## 1 Introduction

Even though researchers separate soils based on particle size as sand, silt and clay, in the field, soil always exists as a combination of all these. There are many studies concentrating on the effect of fines on the shear characteristics of sand [1–3] and liquefaction [4–7] but only a few studies have considered the other properties.

Yang and Wei [8] have analysed the change in critical state friction angle for Fujian and Toyoura sands. For clean sand without fines, the critical state friction angle tends to decrease with increasing roundness of sand particles. When those sands were tested with fines (round shape), the critical state friction angle of the mixture tends to decrease with an increase in fines content. But for fines with an angular shape, the critical state friction angle tends to increase with fines content. Phan et al. [9] have conducted one-dimensional consolidation tests on sand–silt mixtures (with low-plastic fines at a constant void ratio and constant relative density) and indicated that the behaviour of the mixtures were similar to those of loose sand. The effect of fines on void ratios was studied by Cubrinovski and Ishihara [10]. The authors reported that the void ratio initially decreases as the fines content increases from 0–20% and above 40% fines, the maximum and minimum void ratios were seen to increase steadily.

It is clear from the literature that the studies on the effect of plasticity of fines on the properties of sand are limited. Hence, the present study is focused on the effect of the amount of fines and the type of fines (or plasticity index of fines) on various properties of sand like specific gravity, limiting void ratios, grain size characteristics, angle of internal friction and compression index.

M. Akhila (✉) · M. R. Sruthy
SCMS School of Engineering and Technology, Ernakulam, Kerala, India

K. Rangaswamy · N. Sankar
NIT Calicut, Kozhikode, Kerala, India



C. N. V. Satyanarayana Reddy et al. (eds.), *Ground Characterization and Foundations*, Lecture Notes in Civil Engineering 167,
https://doi.org/10.1007/978-981-16-3383-6_1

## 2 Material Tested

The soil materials utilized in the present study are natural sand, M sand, natural clay and kaolinite clay. The natural sand was collected from Turavoor region in Kerala state. The M sand was collected from a local quarry in Calicut, and the non-plastic silt powder was derived after sieving the M sand through 75-µm sieve. The natural clay was collected at a depth of 3 m from Pantheerankavu which is 12 km far from south of Calicut city in Kerala. The index and strength properties of natural clay are listed in Table 1. The commercial kaolinite clay was procured from Sajeev and Co. Ltd. at Calicut district of Kerala state. The physical and chemical properties of kaolinite clay have been provided by the manufacturer as shown in Table 2. The results of tests conducted on natural sand are listed in Table 3.

**Table 1** Index and strength properties of natural clay

| Property | Value |
|---|---|
| *Index property* | |
| Specific gravity | 2.56 |
| Liquid limit (%) | 79 |
| Plastic limit (%) | 48 |
| Shrinkage limit (%) | 27 |
| Plasticity index | 31 |
| Clay size (%) | 50 |
| Soil classification | MH |
| *Strength property* | |
| Maximum dry density (kN/m$^3$) | 17.5 |
| Optimum moisture content (%) | 32 |
| UCS (kPa) | 64 |

**Table 2** Properties of kaolinite clay

| Physical (Mass %) | | Chemical (Mass %) | |
|---|---|---|---|
| Acid soluble | 0.94 | $SiO_2$ | 44 |
| Water soluble | 0.35 | $Al2O_3$ | 38 |
| Oil absorption (mm/100 g) | 35 | $Fe2O_3$ | 0.25 |
| Specific gravity | 2.62 | $TiO_2$ | 0.35 |
| pH | $5 \pm 0.5$ | CaO | 0.05 |
| Moisture percentage | 1.5 ±0.06 0.5 | $Na_2O$ | 0.06 |
| TDS | 100 | $K_2O$ | 0.05 |
| | | MgO | 0.07 |

**Table 3** Basic and index properties of sand

| Property | Value |
|---|---|
| Specific gravity | 2.62 |
| $D_{50}$, mm | 0.28 |
| Uniformity coefficient, $C_u$ | 2.36 |
| Coefficient of curvature, $C_c$ | 0.87 |
| $e_{max}$ | 0.858 |
| $e_{min}$ | 0.578 |

Among the numerous trial combinations of low-plastic soils mixtures processed, the combinations of 50% kaolinite + 50% silt, 100% kaolinite and 20% clay + 80% kaolinite mixtures were found to possess plasticity indices of 5%, 10% and 15%, respectively and hence decided to be used for the current work. A total of 17 soil combinations were prepared by mixing the above-mentioned low-plastic soil combinations to the fine sand. The effect of presence of fines in the sand matrix on its properties was investigated by conducting various test including grain size analysis, relative density tests, specific gravity, direct shear test and one-dimensional consolidation tests. The tests were performed as per IS test procedures at different percentage fines (0, 10, 20, 30 40%) and plasticity index of fines (0, 5, 10, 15%).

## 3 Results and Discussions

Test results for index properties and some engineering properties on all the soil combinations adopted in the study are presented in the following sections.

### 3.1 *Effect of Particle Size Characteristics*

A combined dry sieve and hydrometer analysis was performed on all the soil combinations to obtain the particle size distribution. The gradation curve of fine sand is shown in Fig. 1. The values of average particle size, $D_{50}$, were found from the gradation curves, and its variation with respect to amount and PI of fines added is reported in Fig. 2. $D_{50}$ gives an understanding of physical properties of the soil which in turn affect its strength and load bearing properties. It is clear from Fig. 2a that the $D_{50}$ of sand decreases with the addition of fines at every tested value of PI. But, the plasticity index of fines has no much influence on the gradation of the soil (Fig. 2b).

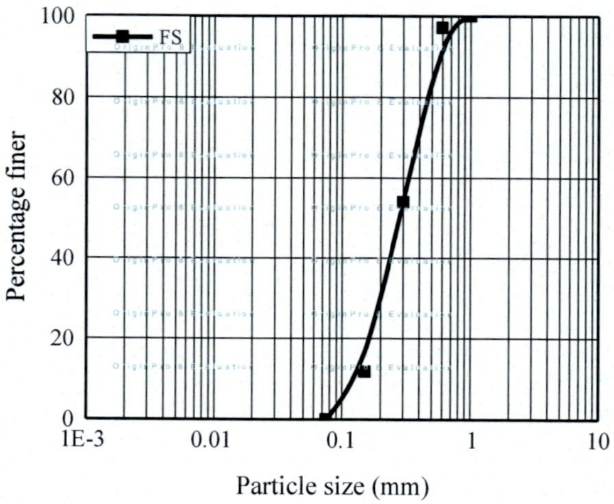


**Fig. 1** Particle size distribution of sand

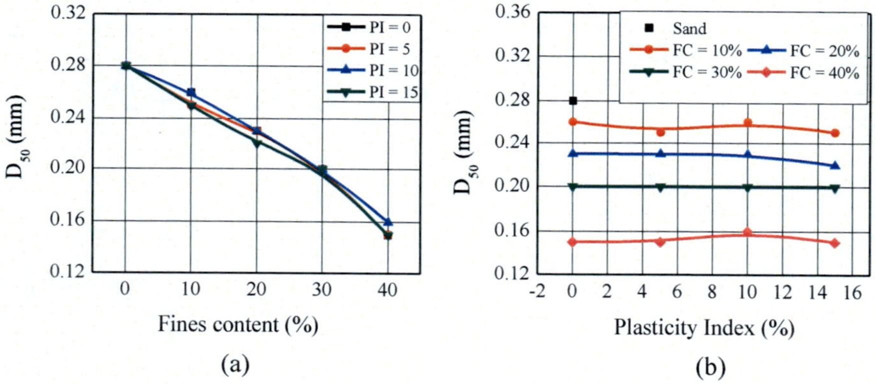


**Fig. 2** Effect of **a** fines content and **b** plasticity index on $D_{50}$

## 3.2 *Effect on Void Ratio*

Relative density tests were performed as per the IS code procedure to arrive at the maximum density of all soil combinations. The minimum densities of soil combinations were attained by pouring it steadily with zero height using paper cone into the CBR mould of 150 mm size. The average values of maximum and minimum densities of soil combinations are reported in tables after repeating the tests thrice. Based on limited densities, the maximum and minimum void ratios are estimated by using the empirical equations, and its variations are shown in Figs. 3 and 4.

It was observed that both the maximum and minimum void ratios decrease as the fines content increases at all tested values of PI of fines. The variation with respect to the PI of fines showed different trends with different fines contents. At low fines

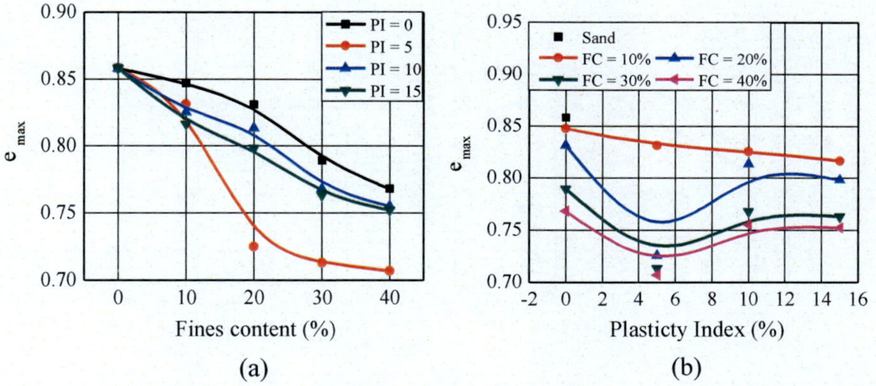


**Fig. 3** Effect of **a** fines content and **b** plasticity index on $e_{max}$

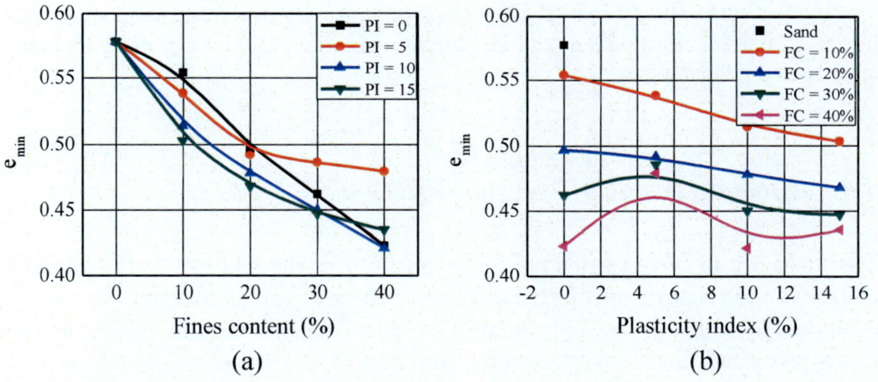


**Fig. 4** Effect of **a** fines content and **b** plasticity index on $e_{min}$

content (10%), both $e_{max}$ and $e_{min}$ decrease with the increase in the plasticity of fines. At higher fines content, a contradiction was observed between the variations in $e_{max}$ and $e_{min}$. At higher fines content, $e_{max}$ shows an initial decrease with the increase in the PI of fines and then shows an increase. But the variation of $e_{min}$ is exactly the opposite.

### *3.3 Effect on Specific Gravity*

The specific gravity of all the soil combinations was found using pycnometer. Figure 5a and b shows the influence of fines on the initial specific gravity of natural fine sand which was found to be 2.62. It is clear from Fig. 5a that specific gravity increases with an increase in fines content. The influence of PI of fines on specific

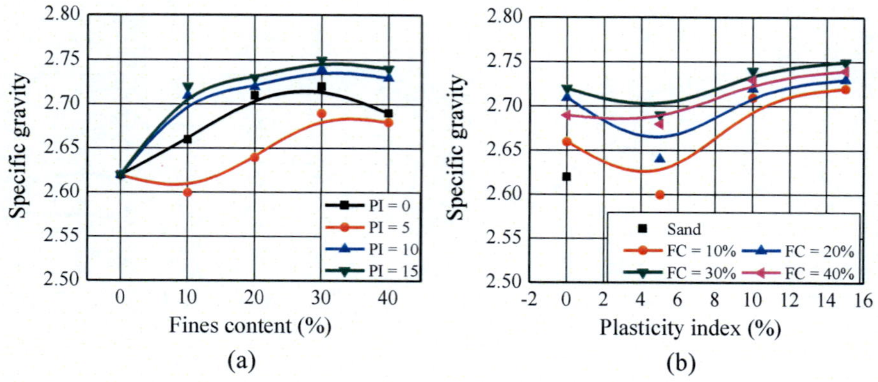


**Fig. 5** Effect of **a** fines content and **b** plasticity index on specific gravity

gravity of sand is shown in Fig. 5b. The specific gravity of natural sand showed an initial decrease followed by a gradual increase with increase in PI of the fines added at all tested fines content.

### 3.4 Effect on Angle of Internal Friction

Direct shear tests were carried out in the small direct shear box (6 × 6 × 5 cm) to understand the changes in shear strength properties of natural sand due to addition of fines. The soil samples were filled in the direct shear box under loosest as well as densest states, and tests were conducted at normal stresses ranging from 100–300 kPa. From the test data, maximum and minimum angle of friction were found and are plotted in Figs. 6 and 7. It can be observed from Figs. 6a and 7a that the value

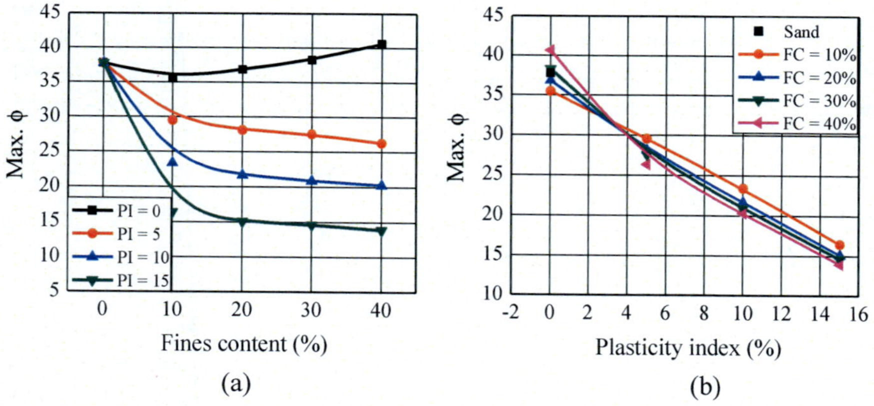


**Fig. 6** Effect of **a** fines content and **b** plasticity index on maximum angle of internal friction

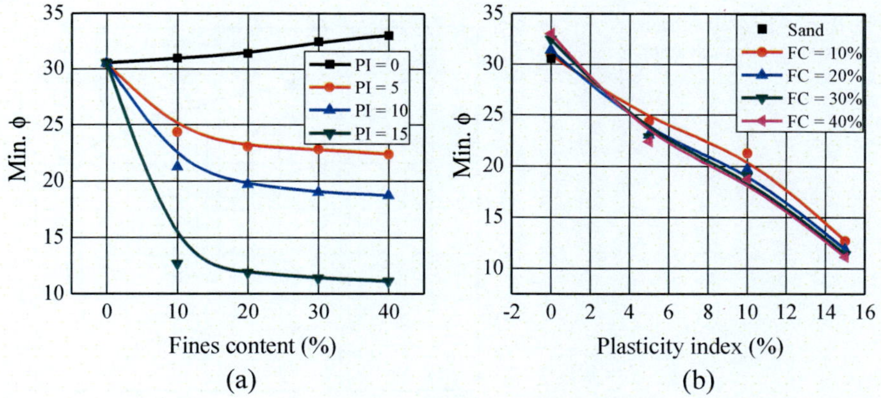


**Fig. 7** Effect of **a** fines content and **b** plasticity index on minimum angle of internal friction

of the angle of internal friction increases with the addition of non-plastic fines (both in loosest and densest state of soil). But on addition of low-plastic fines to the sand, an opposite trend can be seen with a decrease in friction angle corresponding to an increase in fines content.

An average reduction of 62.5% from the initial values was observed in friction angles when low-plastic fines of PI = 15% was added to sand (Figs. 6b and 7b). The reduction trends can be seen overlapping at all fines content. The angle of internal friction decreases significantly with increase in the plasticity of fines.

### *3.5 Effect on Compression Index*

One-dimensional oedometer tests have been conducted on all the soil samples in the loosest and densest possible states. The *e*-log p curve of natural sand is shown in Fig. 8. From this, the compression index is found as 0.04 in the loosest state and 0.03 in the densest state. The effect of fines on compression index of sand is shown in Figs. 9 and 10. It is clear from Figs. 9a and 10a that the addition of low-plastic fines increases the compression index of sand. But the variation in compression index due to the addition of non-plastic fines is negligible. Figures 9b and 10b indicate that at all tested fine content, $C_c$ increases with an increase in PI of fines. Also, if the value of sand is excluded in Figs. 9 and 10, the remaining points will show a linear trend between $C_c$ and PI.

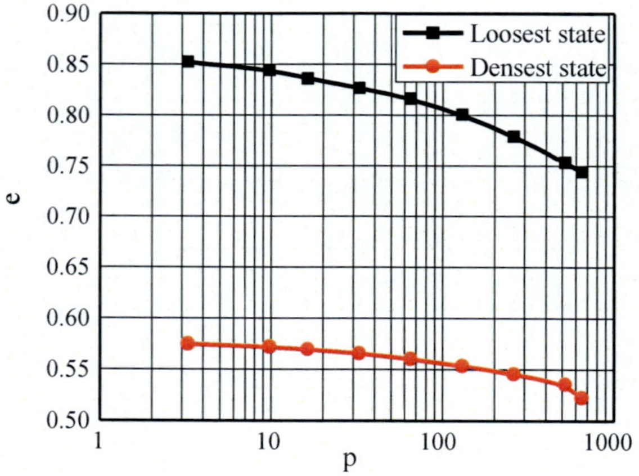


**Fig. 8** *e*-log p plots of sand in loosest and densest state

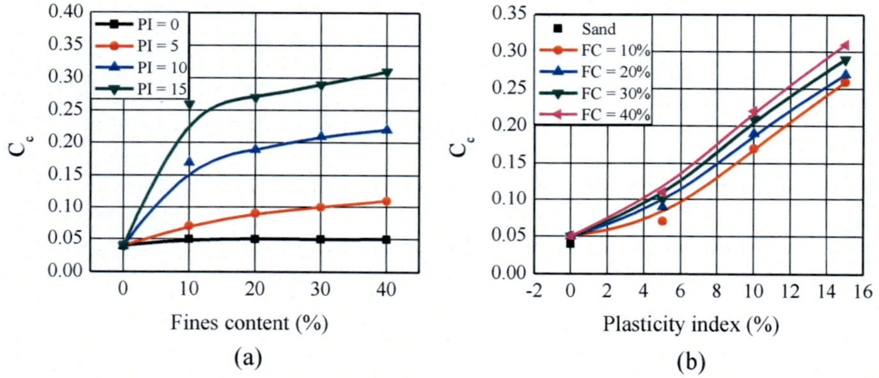


**Fig. 9** Effect of **a** fines content and **b** plasticity index on $C_c$ (loosest possible state)

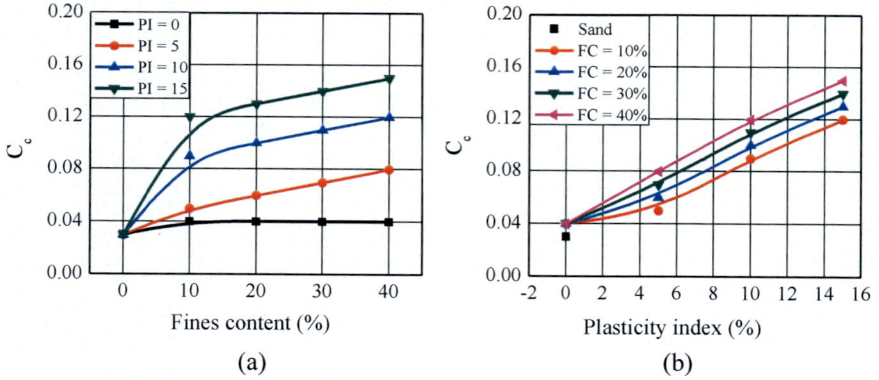


**Fig. 10** Effect of **a** fines content and **b** plasticity index on $C_c$ (densest possible state)

## 4 Summary and Conclusions

The present paper discussed the effect of non-plastic and low-plastic fines on the properties of sand. Effect on grain size characteristics, limiting void ratio, specific gravity, angle of internal friction and compression index is elaborated. The following conclusions are derived:

1. The $D_{50}$ of sand decreases with the addition of fines at every tested value of PI. But, if particular fines content is taken, the plasticity index of fines has no much influence.
2. Both maximum and minimum void ratios decrease as the fines content increases at all testes values of PI of fines. The variation of $e_{max}$ and $e_{min}$ with respect to the PI of fines is contradicting to each other.
3. Value of angle of internal friction is increased with the addition of non-plastic fines (both in loosest and densest state of soil). But an opposite trend is found with the addition of low-plastic fines. The angle of internal friction decreases with increase in the plasticity of fines.
4. The addition of low-plastic fines increases the compression index of sand, but the effect of non-plastic fines is negligible. At all tested fine content, $C_c$ increases with increase in PI of fines.

## References

1. Kuerbis, R., Nagussey, D., Vaid, Y.P.: Effect of gradation and fines content on the undrained response of sand. Proc. Hyd. Fill. Struc. Geotech. Spec. Publ. **21**, 330–345 (1988)
2. Belkhatir, M., Missoum, H., Arab, A., Della, N., Schanz, T.: Undrained shear strength of sand-silt mixture: effect of intergranular void ratio and other parameters. KSCE J. Civ. Eng. **15**(8), 1335–1342 (2011)
3. Naeini, S.A., Baziar, M.H.: Effect of fines content on the steady-state strength of mixed and layered samples of sand. Soil Dyn. Earthq. Eng. **24**, 181–187 (2004)
4. Amini, F., Qi, G.Z.: Liquefaction testing of stratified silty sands. J. Geotech. Geoenviron. Eng. **126**(3), 208–217 (2000)
5. Dash, H.K., Sitharam, T.G.: Undrained cyclic pore pressure response of sand–silt mixtures: effect of non-plastic fines and other parameters. J. Geotech. Geol. Eng. **27**, 501–517 (2009)
6. Shen, C.K., Vrymoed, J.L., Uyeno, C.K.: The effect of fines on liquefaction of sands. In: Proceedings of 9th International Conference on Soil Mechanics and Foundation Engineering, pp. 381–385 (1977)
7. Hernández, Y.A., Towhata, I., Gunji, K., Yamada, S.: Laboratory tests on cyclic undrained behaviour of loose sand with cohesionless silt and its application to assessment of seismic performance of subsoil. Soil Dyn. Earthq. Eng. **79**, 365–378 (2015)
8. Yang, J., Wei, L.M.: Collapse of loose sand with the addition of fines: the role of particle shape. Geotechnique **62**(12), 1111–1125 (2012)
9. Phan, V.-A., Hsiao, D.-H., Nguyen, P.-L.: Effects of fines contents on engineering properties of sand-fines mixtures. Procedia Eng. **142**, 213–220 (2016)
10. Cubrinovski, M., Ishihara, K.: Maximum and minimum void ratio of sand characteristics. Soils Found. **42**(6), 65–78 (2002)

# Comparison of Theoretical and Laboratory Permeability for Coarse-Grained Soil at Different Ground Conditions

**Satyajit Roy, R. K. Bharti, V. K. Jain, Manish Gupta, and R. Chitra**

## 1 Introduction

Permeability is a direct function of average grain size distribution of granular porous media [1]. The inter-relationship is quite effective for preliminary investigation, especially at prefeasibility stage. But proper investigation of soil is required during the designing stage, and it is important to know the actual response of soil towards permeability for structural integrity by laboratory methods. Several researchers made an effort to calculate the co-efficient of permeability and develop several indirect empirical formulae as laboratory testing sometimes takes considerable time in arriving at meaningful conclusion. Empirical correlations are function of grain sizes, porosity/void ratio, $C_u$, $C_c$ and viscosity of pore fluids.

There are various empirical correlations available in the literature such as Hazen, Kozeny-Carman, Breyer, Slitcher, Terzaghi, USBR, Alyamani and Sen. Several investigators have studied these relationships and modified these formulae based on experimental work. The applicability of these formulae depends on the type of soil and compactness of the soil for which co-efficient of permeability is required to be estimated. As per Vukovic and Soro [2], the applications of different empirical formulae to the same porous medium material can yield different values of co-efficient of permeability. Again, soil is not homogeneous, and permeability varies from location to location. Actual ground conditions vary from place to place. Moreover, soil profile is not uniform but varies from one section to other. In some places, it may be dense, partially dense, and loose or submerged in water. Depending upon the condition of ground, permeability will also vary from place to place. Keeping in mind the various ground conditions, attempt has been made to determine the co-efficient of permeability on the soil samples remoulded at different compactness and moisture conditions. In the present study, attempt has been made to correlate permeability

S. Roy (✉) · R. K. Bharti · V. K. Jain · M. Gupta · R. Chitra
Central Soil and Materials Research Station, New Delhi 110016, India



C. N. V. Satyanarayana Reddy et al. (eds.), *Ground Characterization and Foundations*, Lecture Notes in Civil Engineering 167,
https://doi.org/10.1007/978-981-16-3383-6_2

values obtained through different correlations with laboratories values. The study has been carried out with four different types of soils viz. (i) Sind river sand (MP), (ii) Yamuna river sand, Delhi (iii) Rajasthan crushed sand and (iv) Ennore standard sand. The paper discusses the value of co-efficient of permeability determined in the laboratory, and the values are obtained theoretically at different conditions and present the factors affecting the values.

## 2 Established Empirical Formulae

Vukovic and Soro [2] summarized several empirical methods from former studies and presented a general formula:

$$\kappa = g/\nu.C.f(n).d_e^2 \tag{1}$$

where $\kappa$ = co-efficient of permeability; g = acceleration due to gravity; $\nu$ = kinematic viscosity; C = sorting coefficient; $f$(n) = porosity function, and $d_e$ = effective grain diameter. The kinematic viscosity ($\nu$) is related to dynamic viscosity ($\mu$), fluid (water) and density ($\rho$) as follows:

$$\nu = \mu/\rho \tag{2}$$

The values of *C, n* and $d_e$ are dependent on the different methods used in the grain size analysis. According to Vukovic and Soro [2], porosity ($n$) may be derived from the empirical relationship with the co-efficient of grain uniformity $C_u$ as follows:

$$n = 0.255(10.83^{Cu}) \tag{3}$$

where $C_u$ is the co-efficient of grain uniformity and is given by

$$C_u = d_{60}/d_{10} \tag{4}$$

Here, $d_{60}$ and $d_{10}$ in the formula represent the grain diameter in (mm) for which 60% and 10% of the sample, respectively, are finer than $d_{60}$ and $d_{10}$.

Former studies presented the following formulae which took the general form and presented in Eq. (1) but with varying *C, f(n)* and $d_e$ values and their domains of applicability.

### 2.1 Hazen Formula [3]

It was widely used for the estimation of co-efficient of permeability of uniformly graded soils ranges from fine sand to gravel of diameter 0.1 to 3 mm, respectively, and uniformity co-efficient less than 5. This formula only depends on the effective size of grains as.

$$\kappa = (g/\nu) \times 6 \times 10^{-4}[1 + 10(n - 0.26)]d_{10^2}$$

### 2.2 Kozeny-Carman Equation [4]

The KC equation is not appropriate for soil with effective size above 3 mm or clayey soil [5]. The KC equation is widely used and accepted for co-efficient of permeability estimation because it depends on both the effective grain size and porosity (number of pores) of the porous media as given below.

$$\kappa = (g/\nu) \times 8.3 \times 10^{-3}\left[n^3/(1 - n)^2\right]d_{10^2}$$

### 2.3 Breyer

This method does not consider porosity, and therefore, porosity function takes on value 1. Breyer formula is often considered most useful for materials with heterogeneous distributions and poorly sorted grains with uniformity co-efficient between 1 and 20 and effective grain size between 0.06 and 0.6 mm [6].

$$\kappa = (g/\nu) \times 6 \times 10^{-4} \times \log[500/U]d_{10^2}$$

### 2.4 Slitcher [7]

This formula is most applicable for grain size between 0.01 and 5 mm.

$$\kappa = (g/\nu) \times 1.0 \times 10^{-2} \times n^{3.287} \times d_{10^2}$$

### 2.5 Terzaghi [8]

$$\kappa = (g/\nu) \times C_t \times \left[(n - 0.13)/\sqrt[3]{(1-n)}\right]^2 \times d_{10^2}$$

where the $C_t$ = sorting co-efficient and. In this study, an average value of $C_t$ is used. Terzaghi formula is most applicable for coarse grain sand [9]

### 2.6 USBR

$$\kappa = (g/\nu) \times 4.8 \times 10^{-4} \times d_{20}^{0.3} \times d_{10^2}$$

US Bureau of Reclamation (USBR) formula calculates co-efficient of permeability from the $d_{20}$ and does not depend on porosity; hence, porosity function is a unity. The formula is most suitable for medium grain sand with uniformity co-efficient less than 5 [9].

### 2.7 Alyamani and Sen [10]

$$\kappa = 1300 \times [I_o + 0.025(d_{50} - d_{10})]^2$$

where $\kappa$ is the co-efficient of permeability (m/day), Io is the intercept (in mm) of the line formed by $d_{50}$ and $d_{10}$ with the grain size axis, $d_{10}$ is the effective grain diameter (mm), and $d_{50}$ is the median grain diameter (mm). The method considers both sediment grain sizes $d_{10}$ and $d_{50}$ as well as the sorting characteristics. This formula, therefore, is exceptionally different from those that take the general form of Eq. (1) above.

## 3 Materials and Methods

In order to compare results obtained from these correlations with the laboratory test results, 4 sandy soil samples were selected from different sources for comparing the permeability values, and the photographs of samples selected are presented in Fig. 1. The particle size distributions of all 4 samples are performed [11] and presented in Table 1. The grain size distributions indicate that sample 1 and sample 4 have predominance of medium sand. Sample 2 has predominance of fine sand, whereas sample 3 has fine to medium sand in equal proportions. The effective size '$d_{10}$' value of these samples varies from 0.019 to 0.40. $C_u$ for 3 samples which are less than

**Fig. 1** Types of sand selected

5, whereas for sample no. 3, it is 28.95. The different grain sizes of the sands are presented in Table 2.

The maximum density, minimum density and required density at 95, 85 and 75% of relative density for all selected samples as well as relative density achieved due to wet packing are presented in Table 3. Samples were packed in permeability mould in dry loose packing, wet loose packing, packed at 95% of relative density, packed at 85% of relative density and packed at 75% of relative density.

The photographs of the samples are presented in Fig. 1.

The particle size distributions of all 4 samples are presented in Table 1

The different grain sizes of samples are presented in Table 2

The relative density of the soil samples was determined according to IS 2720 (Part 14) [13]

Here, for calculation of required density $\gamma_d$ at which samples are packed, relative density ($I_d$) were considered as 95, 85 and 75%. Moreover, the maximum density,

**Table 1** Particle size distribution

| Mechanical analysis | | | | | | | | |
|---|---|---|---|---|---|---|---|---|
| Sample | 0.002 mm and less | 0.002–0.075 mm | 0.002 to 0.075 mm | 0.425 to 2.0 mm | 2.0 to 4.75 mm | 4.75 mm and above | Soil type (IS:1498) | Remarks |
| | Clay | Silt | Fine sand | Medium sand | Coarse sand | Gravel | | |
| 1 | 0.0 | 1.0 | 10.7 | 74.5 | 11.9 | 1.9 | SP | Medium sand |
| 2 | 0.0 | 7.4 | 75.8 | 7.9 | 7.7 | 1.2 | SP-SM | Fine sand |
| 3 | 0.9 | 17.9 | 35.5 | 37.2 | 8.5 | 0.0 | SM | Fine sand |
| 4 | 0.0 | 1.7 | 26.5 | 70.7 | 1.1 | 0.0 | SP | Medium sand |

**Table 2** Different grain sizes of samples

| Parameters | Sind river sand (MP) | Yamuna river sand | Rajasthan crushed sand | Ennore standard sand |
|---|---|---|---|---|
| $d_{10}$ | 0.40 | 0.09 | 0.019 | 0.19 |
| $d_{20}$ | 0.55 | 0.13 | 0.085 | 0.32 |
| $d_{30}$ | 0.69 | 0.17 | 0.16 | 0.47 |
| $d_{50}$ | 1.00 | 0.22 | 0.35 | 0.72 |
| $d_{60}$ | 1.30 | 0.26 | 0.55 | 0.90 |
| $C_u$ | 3.25 | 2.89 | 28.95 | 4.74 |

**Table 3** Calculated values of minimum, maximum and required density as per IS code [12]

| Type of soil | Sind river sand (MP) | Yamuna river sand | Crushed sand Rajasthan | Ennore standard sand |
|---|---|---|---|---|
| Min density, $\gamma_{min}$ (g/cc) | 1.57 | 1.37 | 1.32 | 1.55 |
| Max density, $\gamma_{max}$ (g/cc) | 1.99 | 1.76 | 1.66 | 1.79 |
| Required density @ 95% of relative density, $\gamma_d$ (g/cc) | 1.96 | 1.74 | 1.64 | 1.78 |
| Required density @ 85% of relative density, $\gamma_d$ (g/cc) | 1.91 | 1.69 | 1.60 | 1.75 |
| Required density @ 85% of relative density,$\gamma_d$ (g/cc) | 1.87 | 1.64 | 1.56 | 1.72 |
| Relative density achieved due to wet packing (%) | 32.3 | 47.2 | 58.6 | 36.2 |

minimum density and required density at 95, 85 and 75% of relative density for all selected samples as well as relative density achieved due to wet packing are presented in Table 3.

## 4 Preparation of Test Sample

In the present study, the theoretical values of co-efficient of permeability's are to be compared with laboratory co-efficient of permeability's compacted at different ground conditions like (i) 95% relative density, (ii) 85% relative density, (iii) 75% relative density (iv) dry loose packing, i.e. at minimum density and (v) wet loose packing. These ground conditions can be achieved by compacting by rodding, dry pouring and placing under water as par Head [14]. The methods of compaction are given below.

## 4.1 Compacting by Rodding

For achieving compactness closer to 95, 85 and 75% of maximum relative densities, compacting by rodding is used.

## 4.2 Dry Pouring

When the sample is to be packed at minimum/low density, dry pouring of sample is used. Here, a funnel fitted with a length of flexible tubing, long enough to reach the bottom of the permeameter cell is used for pouring the sample; the pouring is to be continued until the surface of the sand is at the correct level. The surface is to be levelled carefully with the minimum disturbance. The jolting of the cell or agitating the sample is to be avoided for packing the sample in low density.

## 4.3 Placing Under Water

Here, the valve on the base of the permeameter cell to be connected to the de-aired water supply and then valve to be opened to allow water to enter the cell to about 15 mm above the porous disc. Now, a large funnel fitted with a bung attached to a string or wire is to be supported over the cell, so that tubing reaches to the surface of water in the cell. Sample is now poured into the funnel. Now, the funnel is to be raised so that the end of tubing is just at the water surface. The water surface is to be maintained at about 15 mm above the surface of the placed soil by admitting more water through the base valve. The process is to be continued until the required amount of soil has been deposited in the cell and water added.

The laboratory permeability was determined by constant head method as described in IS 2720 (Part 17) [13], and results are presented in Table 4.

The co-efficient of permeability calculated from grain size analysis using empirical formulae is presented in Table 5.

**Table 4** Result of constant head permeability test

| S. no. | Type of sand | 'k'(cm/s) 95% max. RD | 'k'(cm/s) 85% max. RD | 'k'(cm/s) 75% max. RD | 'k'(cm/s) Dry packing, mini. RD | 'k'(cm/s) Wet packing |
|---|---|---|---|---|---|---|
| 1 | Sind river sand (MP) | $0.40 \times 10^{-04}$ | $0.59 \times 10^{-04}$ | $1.06 \times 10^{-04}$ | $2.2 \times 10^{-04}$ | $0.79 \times 10^{-04}$ |
| 2 | Yamuna river sand | $0.31 \times 10^{-04}$ | $0.53 \times 10^{-04}$ | $0.99 \times 10^{-04}$ | $1.96 \times 10^{-04}$ | $0.57 \times 10^{-04}$ |
| 3 | Rajasthan crushed sand | $0.10 \times 10^{-04}$ | $0.18 \times 10^{-04}$ | $0.37 \times 10^{-04}$ | $0.67 \times 10^{-04}$ | $0.18 \times 10^{-04}$ |
| 4 | Ennore standard sand | $0.78 \times 10^{-04}$ | $1.15 \times 10^{-04}$ | $1.75 \times 10^{-04}$ | $3.77 \times 10^{-04}$ | $1.2 \times 10^{-04}$ |

**Table 5** Co-efficient of permeability based on empirical equations

| Type of sand | Hazen 'k' (cm/sec) | Kozeny-Carman 'k' (cm/sec) | Breyer 'k' (cm/sec) | Sitcher 'k' (cm/sec) | Terzaghi 'k' (cm/sec) | USBR'k' (cm/sec) | Alyamani and Se 'k' (cm/sec) |
|---|---|---|---|---|---|---|---|
| Sind river sand (MP) | $2.19 \times 10^{-01}$ | $9.43 \times 10^{-02}$ | $2.05 \times 10^{-01}$ | $7.32 \times 10^{-02}$ | NA | $1.49 \times 10^{-02}$ | $1.49 \times 10^{-01}$ |
| Yamuna river sand | $1.16 \times 10^{-02}$ | $5.17 \times 10^{-03}$ | $1.06 \times 10^{-02}$ | $4.03 \times 10^{-03}$ | NA | NA | $6.02 \times 10^{-03}$ |
| Rajasthan crushed sand | NA | $5.21 \times 10^{-05}$ | $2.66 \times 10^{-04}$ | $4.05 \times 10^{-05}$ | NA | NA | $2.26 \times 10^{-04}$ |
| Ennore standard sand | $4.23 \times 10^{-02}$ | $1.57 \times 10^{-02}$ | $4.28 \times 10^{-02}$ | $1.23 \times 10^{-02}$ | NA | $4.29 \times 10^{-03}$ | $4.01 \times 10^{-02}$ |

## 5 Result and Discussion

### *5.1 Theoretical Permeability*

The comparison of results of permeability values obtained through existing correlations and laboratory permeability values calculated at different ground conditions is presented in Tables 4 and 5, respectively. The basic reason for this divergence is due to applicability of these formulae under restrictive conditions. Further, depending upon gradations of different type of sand applicability, these formulae are questionable as explained below.

The Hazen and USBR methods are applicable in soil where $C_u < 5$. However, for Rajasthan crushed sand (sample no. 3), the $C_u$ value is 28.7; therefore, this formula is not applicable sample no. 3. Further, Terzaghi method is suitable for coarse sand only; therefore, this empirical correlation cannot be correctly used for medium and fine sand. USBR method is applicable to medium sand, so it cannot be used for coarse and fine sand.

Overall results showed that the co-efficient of permeability calculated by the USBR, and Slitcher methods are giving lower values as compared to other methods as presented in Table 5. These values are in consistent with the conclusions by [2, 9] method. Breyer method is most useful for analysing heterogeneous sample with well-graded grain (Pinder and Ceila 2006). It is the best estimator for sample no.3. However, for less heterogeneous sample, i.e. sample no. 1, 2 and 4, this method underestimates the values. Hazen formula which is based only on the $d_{10}$ particle size is less accurate than Kozney-Carman formula, which is based on entire particle size distribution and particle shape [5]. Therefore, the estimations by Kozney-Carman for sample no. (1, 2 and 4) are more accurate than Hazen, and possibly the best estimation in the study. Alyamani and Sen Method is very sensitive to the shape of the grading curve and more accurate for well-graded sample.

Therefore, the most suitable formulae for estimation of co-efficient of permeability in the studies are as follows:

Sample 1 (Kozney-Carman formula) = $9.43 \times 10^{-02}$ cm/s, Sample 2 (Kozney-Carman formula) = $5.17 \times 10^{-03}$ cm/s, Sample 3 (Breyer formula) = $2.66 \times 10^{-04}$ cm/s, with value of constant head parameter acceptable and Sample 4 (Kozney-Carman formula) = $1.57 \times 10^{-02}$ cm/s.

The results are also compared with Odong [15] and same indicate close resemblance in terms of applicability of correlation for assessment of permeability of different type of sands.

Moreover from empirical equations:

$$k_{\text{Sind river sand (MP)}} > k_{\text{Ennore standard sand}} > k_{\text{Yamuna river sand}} > k_{\text{Rajasthan crushed sand}}$$

## 5.2 *Results of Comparison of Laboratory Co-efficient of Permeability Determined at Different Conditions*

Soil is not homogeneous, and permeability varies from location to location. An actual ground condition varies from place to place. Moreover, soil profile is not uniform but varies from one section to other. In some places, it may be dense, partially dense, and loose or submerged in water. Depending upon the condition of ground, permeability will also vary from place to place. Keeping in mind the various ground conditions, attempt has been made to determine the co-efficient of permeability on the soil samples remoulded at different compactness and moisture conditions. In the present study, sands are compacted at different conditions:

**Dry packing, loose condition and minimum relative density**. In case of dry packing, loose condition, the general sequence of laboratory values of co-efficient of permeability in decreasing order as follows:

$$k_{\text{Ennore standard sand}} > k_{\text{Sind river sand (MP)}} > k_{\text{Yamuna river sand}} > k_{\text{Rajasthan crushed sand}}$$

From the above equation, it is clear that permeability values are more for medium-grained sand; however, sand containing 17.9% silt is having lowest value. Silt helps in filling the voids in between sands, thus reducing permeability value.

In comparison with theoretical permeability values, it is seen that laboratory values of co-efficient of permeability (from Tables 4 and 5) of all the sand compacted in dry and loose condition are far lesser than theoretical values. The basic reason for this divergence is due to applicability of these formulae under restrictive conditions. Further, depending upon gradations of different type of sand applicability, these formulae are questionable as explained above.

**Wet packing, loose condition**. In case of wet packing, loose condition, the general sequence of laboratory values of co-efficient of permeability in decreasing order is as follows:

$$k_{\text{Ennore standard sand}} > k_{\text{Sind river sand (MP)}} > k_{\text{Yamuna river sand}} > k_{\text{Rajasthan crushed sand}}$$

In comparison with theoretical values, laboratory values of co-efficient of permeability of all the sand samples compacted in wet loose condition are much lesser than theoretical values.

When compared with the samples packed in dry and loose condition, it is seen that sand samples packed dry but in the presence of water, the co-efficient of permeability values reduced approximately to (1/3)rd to (1/4)th of the values of co-efficient of permeability of dry packing, loose condition. It is also seen that due to packing in the presence of water, relative density of samples has increased to 32.3% for sand sample of Sind river, MP, 47.2% for Yamuna river sand, 58.6% for Rajasthan crushed sand and 36.2% for Ennore standard sand. Generally, larger grains have higher settling velocity than small grains. So when grains settle through fluids, the larger grains will impact substrate with larger momentum, possibly jolting the grains into tighter packing, therefore with lower porosity and thus increased relative density. Out of all the samples, one with larger silt percentage (17.9%), i.e. Rajasthan crushed sand gained highest relative density and Sind river sand, MP with 1% silt content gained lowest relative density and the sequence is as follows:

$$RD_{\text{Rajasthan crushed sand (17.9 \% silt)}} > RD_{\text{Yamuna river sand (7.4 \% silt)}}$$
$$> RD_{\text{Standard sand (1.7 \% silt)}} > RD_{\text{Sind river MP (1.0 \% silt)}}$$

So, increase in RD values due to wet packing varies from 32. 3 to 58.6%, which indicates loose packing in the presence of water that has improved compactness to medium compactness.

Again, when we compared co-efficient of permeability values due to wet packing with laboratory co-efficient values of dry packing compacted at 75% and 85% of relative density, respectively, it is theoretically understood that RD values should have increased to approximately 80% due to wet packing, but in actual, it increased only up to values that vary from 32.3 to 58.6%. This may be because probability of the presence of any air voids or air bubbles might have replaced with water, thus further reducing the co-efficient of permeability. Packing in the presence of water helped in removing entrapped bubbles of air present between sand particles as well as water acts as lubricating agents for particles to come close to each other and thus reducing permeability values of all four type of sand significantly.

**Dry packing, packed at 95, 85 and 75% of maximum relative density**. In case of dry packing, packed at 95, 85 and 75% of maximum relative density, the general sequence of laboratory values of co-efficient of permeability in decreasing order is as follows:

$$k_{\text{Ennore standard sand}} > k_{\text{Sind river sand (MP)}} > k_{\text{Yamuna river sand}} > k_{\text{Rajasthan crushed sand}}$$

In comparison with theoretical values, laboratory values of co-efficient of permeability of all the sand samples compacted as dry packing packed at 95%, 85% and 75% of maximum relative density are much lesser than theoretical values.

Moreover, as the relative density of sand samples increases, co-efficient of permeability of sand decreases and vice versa.

**Overall comparison**. It is seen that for all the conditions, the general sequence of co-efficient of permeability values are:

$$\begin{aligned} & k_{\text{theoretical values}} > k_{\text{dry packing}} > k_{\text{packed at 75 \% relative density}} > k_{\text{wet packing}} \\ & > k_{\text{packed at 85 \% relative density}} > k_{\text{packed at 95 \% relative density}} \end{aligned}$$

From the above equation, it is generally clear that as the relative density of samples increases, co-efficient of permeability values decreases and vice versa. It is because as relative density increases voids between the sand particles decreases and thus reduces the values of co-efficient of permeability. However, co-efficient of permeability values due to wet packing comes in between the values compacted at 75% & 85% relative density, respectively.

Further, co-efficient of permeability of the samples packed at 95% of maximum relative density reduced to approximately half when compared with samples packed in the presence of water.

## 6 Conclusions

Based on the present study, the following conclusions can be drawn.

1. Empirical equations used for estimating the co-efficient of permeability of soils can relatively lead to underestimation or overestimation unless the appropriate method is used.
2. Compare to empirical values in most of the samples (except sample no. 3), laboratory constant head methods yields very lower values of co-efficient of

permeability. The correlations of grain-size analyses with laboratory constant head methods are generally much weaker, when samples are compacted at 95% of relative density.

3. Kozney-Carman's formula followed by Hazen formula is the best empirical equations and used for wide range of soil sample. However, Breyer formula is the best for estimation of highly heterogeneous soil sample.
4. Slitcher, Terzaghi and USBR formulae grossly underestimated the co-efficient of permeability in comparison with other evaluated formulae.
5. For all the four types of sand, co-efficient of permeability comes out highest for dry packing condition and lowest for samples packed at 95% of maximum relative density.
6. As the relative density of samples increases, co-efficient of permeability values decreases and vice versa.
7. It is also seen that due to packing in the presence of water, relative density of samples has increased to 32.3% for sand sample of Sind river, MP, 47.2% for Yamuna river sand, 58.6% for Rajasthan crushed sand and 36.2% for Ennore standard sand.
8. Co-efficient of permeability of samples packed in loose condition but in the presence of water is approximately from 1/3rd to ¼th of the values of co-efficient of permeability of samples which is dry packed and in loose condition.
9. Co-efficient of permeability of the samples packed at 95% of maximum relative density reduced to approximately half when samples are packed in the presence of water and in loose condition.
10. It is seen that for all the conditions, the general sequence of co-efficient of permeability values is as follows:

$$k_{\text{theoretical values}} > k_{\text{dry packing}} > k_{\text{packed at 75 \% relative density}}$$
$$> k_{\text{wet packing}} > k_{\text{packed at 85 \% relative density}} > k_{\text{packed at 95 \% relative density}}$$

Irrespective of types of packing, the general sequence of co-efficient of permeability values is as follows:

$$k_{\text{Ennore standard sand}} > k_{\text{Sind river sand (MP)}} > k_{\text{Yamuna river sand}} > k_{\text{Rajasthan crushed sand}}$$

## References

1. Freeze, R.A., Cherry, J.A.: Groundwater. Prentice Hall Inc., Englewood Cliffs, New Jersey (1979)
2. Vukovic, M., Soro, A.: Determination of Hydraulic Conductivity of Porous Media from Grain-Size Composition. Water Resources Publications, Littleton, Colorado (1992)
3. Hazen, A.: Some physical properties of sands and gravels, with special reference to their use in filtration. 24th Annual Report of Massachusetts State Board of Health, pp 539–556 (1892)

4. Kozeny, J.: Über Kapillare Leitung Des Wassers in Boden, Sitzungsber, Akad. Wiss. Wien Math, Naturwiss, Kl 2a(136):271–306 (1927)
5. Carrier, W.D.: Goodbye Hazen-Hello. Kozeny-Carman. J. Geotech. Geoenviron. Eng. 1054 (2003)
6. Beyer, W.: To determine the water permeability of gravels and sands from grain distribution. Water Management Water Technology (WWT) 14th year, 165–169 (1964)
7. Slichter, C.S.: Field measurements of the rate of movement of underground waters. U. S. Geological Survey. Water-supply and irrigation paper. Washington, Government Printing Office-122 (1905)
8. Terzaghi, K., Peck, R.B.: Soil Mechanics in Engineering Practice. Wiley, New York (1964)
9. Cheng, C., Chen, X.: Evaluation of methods for determination of hydraulic properties in an aquifer-aquitard system hydrologically connected to River. Hydrogeol. J. **15**, 669–678 (2007)
10. Alyamani, M.S., Sen, Z.: Determination of hydraulic conductivity from complete grain-size distribution curves. Groundwater **31**(4), 551–555 (1993)
11. IS 2720-Part-4: Methods of test for soils-grain size analysis. Bureau of Indian Standards, New Delhi (1985)
12. IS 2720-Part-14: Methods of test for soils-determination of density index (relative density) of cohesion less soil. Bureau of Indian Standards (1983)
13. IS 2720-Part-17: Methods of test for soils-Laboratory determination of permeability. Bureau of Indian Standards, New Delhi (1986)
14. Head, K.H.: Manual of Soil Laboratory Testing, Permeability, Shear Strength and Compressibility Tests, 2 (1994)
15. Odong, J.: Evaluation of empirical formulae for determination of hydraulic conductivity based on grain-size analysis. J. Am. Sci. **3**(3) (2007)

# Prediction of Engineering Properties of Kerala Soil

**Anju Mary Elias and S. Chandrakaran**

## 1 Introduction

Soils are produced as a consequence of denudation of rocks. It will have the same mineralogical composition as that of the parent rock. Some of the main engineering properties of the soils include compressibility, permeability, and shear strength. They are determined either in situ or in the laboratory from the undisturbed samples collected from the site. The test results depend upon the method of testing, loading condition, strain rate, load rate, etc. These properties are of great importance because they are used as design parameters to determine bearing capacity, slope stabilization, etc. To prepare the budget and for preliminary designs, only a rough estimate of the properties is needed. But the tests to determine the properties are time-wasting and costly. There comes the importance of index properties which can be easily determined. The main index properties of the soil are water content, specific gravity, unit weight of soil, particle size distribution, consistency limits, etc. They are obtained from laboratory tests and are used for the classification of soil. Some of the index properties can be correlated with the engineering properties, but it varies for different types of soil. Also, the correlation may not be perfect. So, in the case of small projects where quick determination and a rough estimate of engineering properties are required, a correlation with a factor of safety may be used. But for large projects engineering properties need to be determined. Several predictive and theoretical studies regarding the correlation of index properties with engineering properties are already available in the literature. However, not much study has been done for soils available in Kerala.

A. M. Elias (✉)
Karunya Institute of Technology and Sciences, Coimbatore, Tamil Nadu, India
e-mail: anjumary@karunya.edu.in

S. Chandrakaran
National Institute of Technology, Calicut, Kozhikode, Kerala, India

 
C. N. V. Satyanarayana Reddy et al. (eds.), *Ground Characterization and Foundations*,
Lecture Notes in Civil Engineering 167,
https://doi.org/10.1007/978-981-16-3383-6_3

Kerala state is located in the Southwestern coast of India. Mainly two types of soils are seen here, namely lateritic soil and soft soil. About 40% of the Western Ghats are located in Kerala. They are the mid and high lands of Kerala, and the lateritic soils are seen here. These soils are also known as acid ferruginous soils of Kerala. Studies have been conducted to determine the mineralogy because of its importance in industry and agriculture [1]. The soft soils are seen in the Kuttanad region and Cochin marine region. These are the lowlands of Kerala, and Kuttanad is the place with the lowest altitude in India.

In this project, attempts have been carried out to collect data from literature and geotechnical consulting firms on undisturbed samples that they have used for various infrastructural projects in Kerala. The engineering properties considered are shear strength parameters ($c$ and $\varphi$) and coefficient of consolidation. The relation of these properties with index properties was determined by regression analysis and artificial neural network (ANN). ANN is a statistical technique used to determine the complex relationship of these properties. The main advantage of ANN is that it can be trained, and hence, the compound relation can be determined.

## 2 Scope and Objective

The main scope of this study is to develop predictive equations for engineering properties of soils. Both regression analysis and artificial neural network were used in order to compare the relation of engineering properties, with index properties and other properties of soil. Since many literatures are available for predicting the properties of soils in different places, the applicability of these equations was checked in soils collected from Kerala.

Data for this study was collected from various geotechnical consulting firms in Kerala and was subjected to statistical analysis to check the sufficiency. All the data were combined, and mean, standard deviation, and variance were obtained. The data set was thus prepared, and regression analysis and neural network analysis were performed. The coefficient of determination ($R^2$) value was determined in each case. Predictive equations for shear strength parameters and coefficient of consolidation were suggested. The RMSE values of the developed predictive equation and available equations were compared and studied.

## 3 Methodology

Simple regression analysis and multiple regression analysis of the engineering properties with different index properties were performed. Those index property with a high value of correlation coefficient was selected, and predictive models were developed. A comparative study was also performed using artificial neural network (ANN).

## 3.1 Index Properties

The index properties like bulk density, natural moisture content, percent fines, liquid limit, plastic limit, plasticity index, and other properties like SPT number were considered in this study. Data was collected from the undisturbed samples and used for various infrastructural projects in Kerala, by various geotechnical consulting firms. The tests they performed include grain size distribution, Atterberg limits, tests for determination of shear strength like triaxial tests, direct shear tests, and tests to determine the coefficient of consolidation like oedometer tests. These data were used for the preparation of the data set for this study. Two different data sets, each including 95 samples and 45 samples, were prepared for shear strength parameters and coefficient of consolidation, respectively. Some properties like shrinkage limit, specific gravity, etc., were not determined for every test. Due to their unavailability, these properties were not considered for the regression analysis. The main aim of this study is to determine the best from the available properties, having more correlation to the target engineering properties.

## 3.2 Engineering Properties

In this study, the engineering properties such as shear strength parameters (cohesion ($c$) and angle of internal friction ($\varphi$)) and the coefficient of consolidation ($c_v$) are considered. Shear strength parameters are of great importance in designing the structures and determining the slope stability. In order to estimate the settlement, $c_v$ is also important. But the oedometer test is time-consuming and costly. Here arises the need of a correlative equation for these properties.

For the correlation of shear strength parameters c and φ, the properties considered are bulk density, natural moisture content, SPT value, liquid limit, plastic limit, and plasticity index. For the correlation of coefficient of consolidation, bulk density, moisture content, percent clay and silt, liquid limit, plastic limit, and plasticity index were the properties considered. The relation of the parameters to the engineering property is determined by performing regression analysis. The root mean square error (RMSE), coefficient of determination ($R^2$), and mean square error (MSE) were obtained. The lowest value of $R^2$ implies that the parameters are not related to the engineering properties. Parameters with the highest value of $R^2$ and lowest value of RMSE were selected. Multiple regression analysis was performed on these selected parameters, and correlative equations were formulated. The equation was in the form.

$$Y = b + mx1 + mx2 + mx3 \quad (1)$$

where $b$ is the constant, m is the slope of the regression, and $x1$, $x\ 2$, and $x\ 3$ are the independent variables. The selected parameters were taken as the input parameters for the artificial neural network (ANN) modelling.

**Table 1** RMSE values of available equations

| Parameter | Equation | Author | RMSE |
|---|---|---|---|
| Cohesion (c) | $C = 0.014N'' + 0.18$ | Mahmod Mostafa Abdou Abdel Naiem | 0.2 |
| Angle of internal friction ($\varphi$) | $\Phi = 0.209N'' + 19.68$ | Mahmod Mostafa Abdou Abdel Naiem | 3.28 |
| Coefficient of consolidation ($c_v$) | $c_v = 4258LL^{(-1.75)}$ | Asma Y. Al-Tae'e | 3.99 |

**Table 2** Developed predictive equations

| Parameter | Equation | $R^2$ | RMSE |
|---|---|---|---|
| Cohesion ($c$) | $c = 0.139–0.0004w_n + 0.0012N$<br>$w_n$ = Natural water content<br>N = SPT value | 0.75 | 0.014 |
| Angle of internal friction ($\varphi$) | $\varphi = 16.73–0.08w_n + 0.34N$<br>$w_n$ = Natural water content<br>N = SPT value | 0.8 | 2.63 |
| Coefficient of consolidation ($c_v$) | $c_v = 0.044–0.006\rho + 0.0001f_n–0.0002LL$ $\rho$ = Bulk density(g/cc)<br>$f_n$ = percent fines<br>LL = liquid limit | 0.88 | 1.75 |

### 3.3 Comparison of Other Predictive Techniques

In the present study, the previously predicted equations from various papers were compared. In the case of shear strength, not much literature is available for correlating the shear strength parameters with various properties. Mahmod [2] correlated shear strength parameters with the corrected SPT value. This correlation was applied to the data available. It was found that this equation is not suitable for Kerala soil. In the case of the coefficient of consolidation ($c_v$), less number of correlations are available. Asma Y. Al-Tae'e [3] correlated $c_v$ with the liquid limit (Tables 1 and 2).

### 3.4 Artificial Neural Network Model (ANN)

Artificial neural network is an analytical technique which is nonlinear and is used to solve complex problems. The main feature of ANN is that it consists of neurons that represent the human nervous system. One of the advantages of ANN is that the network can be trained. Its neurons are arranged in layers consisting of the input layer, the hidden layer, and the output layer. The input layer is the most transparent, and it consists of the dependent parameters. All the parameters to be included in the network are given as the input layer. The property to be determined by the network is the output. In the case of hidden layers and the number of neurons in each layer, an optimum number should be provided. They are determined from the trial-and-error process. The available data set is divided into two parts, namely training set and validation set, to provide as feed to the network. The training set consists of 90% of the data and is used to build the network, and validation set consists of 10% of data and is used for the evaluation of the network.

## 4 Results and Discussion

### *4.1 Linear Regression Analysis of Properties*

The linear regression analysis of individual properties with the respective parameters was done to determine the relation. The regression app of MATLAB was used to perform the analysis. From the analysis, it was observed that natural moisture content and SPT value have an $R^2$ value greater than 0.5 which shows its moderate relation with cohesion. Whereas other parameters like bulk density, liquid limit, and plastic limit have an $R^2$ value greater than 0.3 that shows its mild relation, and plasticity index has an $R^2$ value less than 0.3 showing no relation.

The $R^2$ value for parameters that correlates the angle of internal friction ($\varphi$) was obtained. SPT value has an $R^2$ value greater than 0.7 which shows its strong relation with $\varphi$. While properties like bulk density, liquid limit, plastic limit, and plasticity index have a value less than 0.3 which has no relation with $\varphi$. Natural moisture content has an $R^2$ value of 0.38 which shows its mild relation.

In the case of parameters correlating coefficient of consolidation ($c_v$), the $R^2$ value of liquid limit is 0.82 which shows its strong relationship with $c_v$. The values of bulk density, percent fines, and plasticity index are 0.61, 0.59, and 0.5, respectively, showing its moderate relation with $c_v$. The properties like natural moisture content and plastic limit have values 0.49 and 0.43, respectively, which show its mild relationship (Fig. 1).

### *4.2 Development of Predictive Equations*

To develop the predictive equation for cohesion, the parameters which have a moderate and strong relation with cohesion are considered. The parameters are moisture content and SPT value. Multiple regression analysis of these parameters with cohesion is performed, and predictive equations are obtained. For formulating the parameter angle of internal friction, SPT value and moisture content with strong and mild correlation were used.

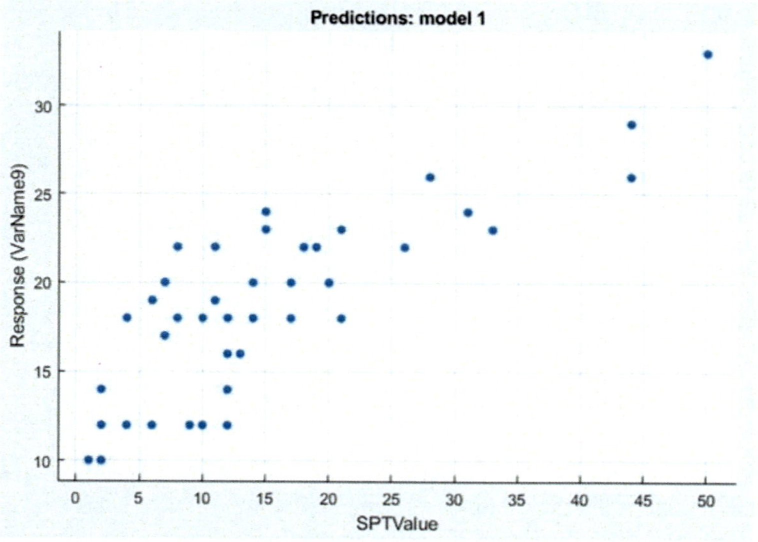


**Fig. 1** Regression analysis of SPT value with angle of internal friction ($\varphi$)

### *4.3 Development of ANN Model*

The model was developed using an artificial neural network [4]. Three models were created each for cohesion, angle of internal friction, and coefficient of consolidation. A total of six iterations were used for training the network for cohesion. Figure 2 shows the performance plot, and the best validation performance was at $4.325e^{-4}$ in the zeroth iteration. The $R^2$ value obtained is 0.90765.

Six iterations were used for training the network for modelling angle of internal friction. The best validation performance was at 6.8943 in the zeroth iteration. The $R^2$ value obtained is 0.8677. A total of eight epochs were used for training the network for the coefficient of consolidation. The best validation performance was at $6.101e^{-7}$ in the second iteration. The $R^2$ value obtained is 0.94615 (Fig. 3).

## 5 Conclusions

In this study, various equations earlier predicted by other authors were compared and concluded. For shear strength, the equation proposed by Mostafa Abdou Abdel Naiem Mahmod for c and $\varphi$ had an RMSE value of 0.2 and 3.28, respectively. The new correlation formulated in this study has an RMSE value of 0.014 and 2.63, respectively. Also, for cohesion and angle of internal friction, parameters like natural water content and SPT value have more relation than others. Hence, these parameters are used for the formulation of correlation.

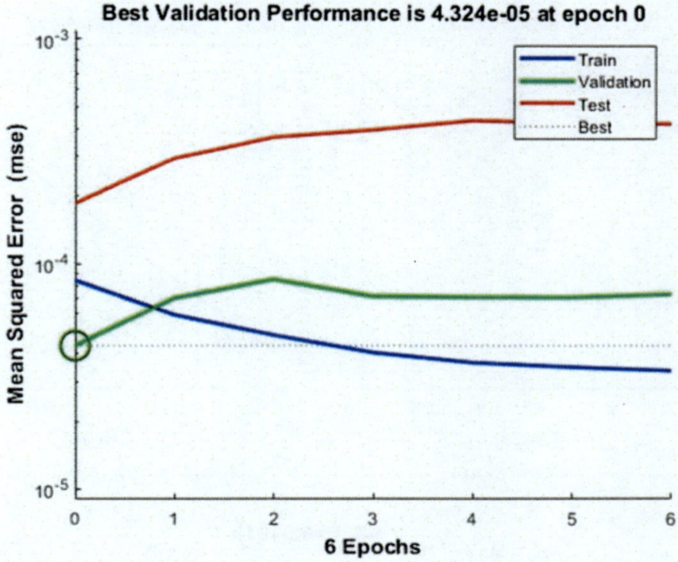


**Fig. 2** Mean squared error plot for cohesion

In the case of the coefficient of consolidation, the equation suggested by Asma Y. Al- Tae'e has an RMSE value of 3.99, whereas the new correlation has an RMSE value of 1.75. For the coefficient of consolidation, the properties like bulk density, percent fines, and the liquid limit have more correlation and hence used for prediction of correlation.

The $R^2$ value of the predicted equation for cohesion, angle of internal friction, and coefficient of consolidation is 0.75, 0.8, and 0.88, respectively. Whereas the $R^2$ value of ANN models for cohesion, angle of internal friction, and coefficient of consolidation is 0.90, 0.86, and 0.94, respectively. It can be concluded that prediction using the ANN model is effective than the conventional regression analysis.

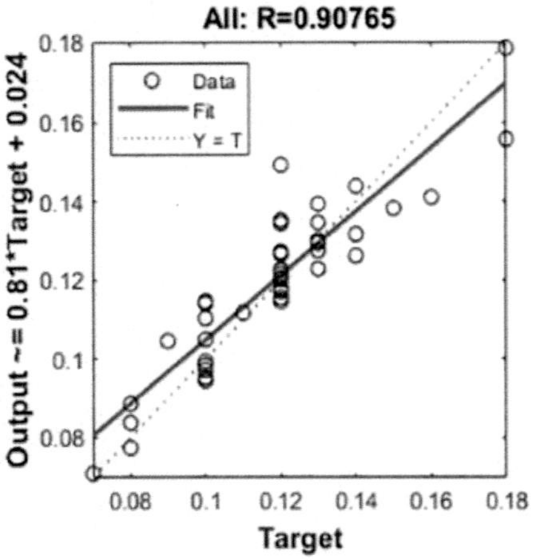


a. cohesion

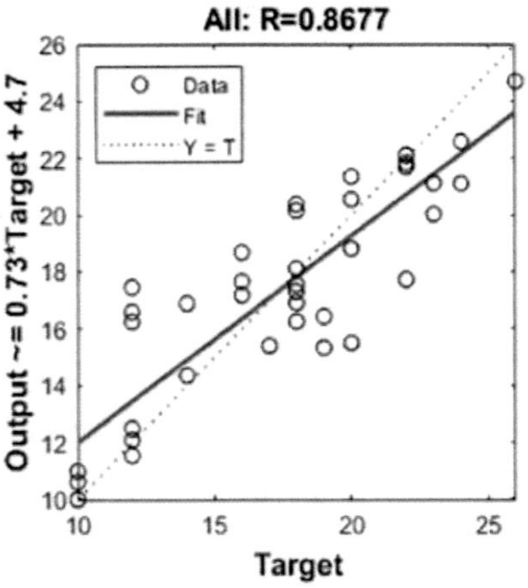


b. angle of internal friction

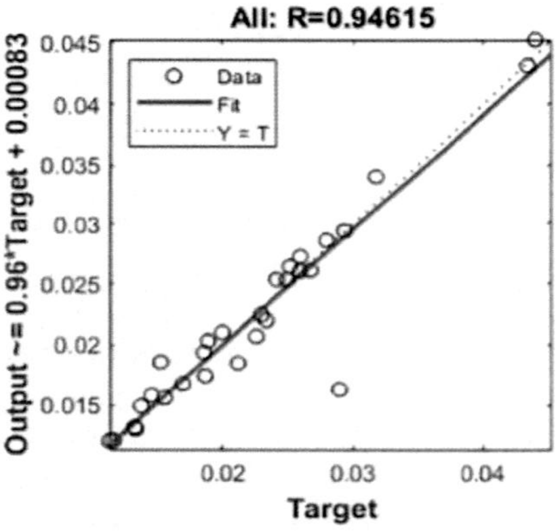


c. coefficient of consolidation

**Fig. 3** Regression plot using ANN

# References


1. Chandran, P., Ray, S.K., Bhattacharya, T., Srivastava, P., Krishnan, P., Pal, D.K.: Lateritic soils of Kerala, India: their minerology, genesis, and taxonomy. Aust. J. Soil Res. **43**, 839–852 (2005)
2. Mahmoud, M.A.A.N.: Reliability of using standard penetration test (SPT) in predicting properties of silty clay with sand soil. Int. J. Civ. Struct. Eng. **3**(3), 545–556 (2013)
3. Al-Tae′e, Y.A., Al-Ameri, A.F.: Estimation of relationship between coefficient of consolidation and liquid limit of middle and south Iraqi soils. J. eng. **3**, 431–440 (2011)
4. Kurnaz, F.T., Dagdeviren, U., Yildiz, M., Ozkan, O.: Prediction of compressibility parameters of the soils using artificial neural network. Springerplus **5**(1), 1–11 (2016)

# Durability Study on Coir Fiber-Reinforced Soil

**Munagala Dhana Teja and M. Muttharam**

## 1 Introduction

Soil is considered by the civil engineer as a complex material. Apart from the testing and classification of various types of soil, in order to determine the stability and physical properties, the knowledge of problems related to foundation design and construction, pavement design, design of embankments and excavation and design of earth dams is necessary. Problems faced before or after construction mainly due to type of soil. Soil stabilization is of major concern in the construction activities due to rapid growth of urbanization and industrialization. The term soil improvement is used for the techniques which improve the index properties and other engineering characteristic of weak (or problematic) soils. Stabilization can increase the shear strength of a soil and/or control the shrink-swell properties of a soil, thus improving the load bearing capacity.

The different types of method used for soil stabilization are like soil stabilization with cement, soil stabilization with lime, soil stabilization using bitumen, chemical stabilization and a new emerging technology of stabilization by geo-textiles and geo-synthetic fibers. Soil reinforcement is an effective and reliable technique for improving strength and stability of the soil.

One of the soil stabilization methods considered as soil stabilization using natural fibers like coir fiber. Coir or coconut fiber belongs to the group of hard structural fibers. It is an important commercial product obtained from the husk of coconut. The coir fiber is elastic enough to twist without breaking, and it holds a curl as though permanently waved. Shorter mattress fibers are separated from the long bristle fibers which are in turn a waste in the coir fiber industry. So this coir fiber waste can be used in stabilization of soil, and thus, it can be effectively disposed of.

M. D. Teja (✉) · M. Muttharam
Division of Soil Mechanics and Foundation Engineering, Civil Engineering Department, College of Engineering Guindy, Anna University, Chennai 600025, India



C. N. V. Satyanarayana Reddy et al. (eds.), *Ground Characterization and Foundations*, Lecture Notes in Civil Engineering 167,
https://doi.org/10.1007/978-981-16-3383-6_4

Durability is one of the major properties to consider in soil stabilization with natural fibers. In order to increase the life of natural fibers, fibers have been successfully treated with bitumen, copper-based chemicals, phenol and some other patented chemical compounds.

## 2 Materials

### 2.1 *Soil*

Natural soil is collected from Perumbakkam village near Cheran Nagar, Changalpattu district, Tamil Nadu state. Then, the soil was sieved through 4.75 mm sieve to remove the gravel fraction. Soil was oven dried for 24 h before execution of geo-technical tests. By conducting laboratory tests, the following index properties of the soil were evaluated (Table 1).

### 2.2 *Stabilizer*

Stabilizers used for this research are raw coconut coir fibers for the soil, and the fiber collected from the industry and after cutting into 20 mm length is shown in Fig. 1.

The properties of the coir fiber used in the present study are shown in Table 2.

**Table 1** Index properties of soil

| Property | Value |
|---|---|
| Liquid limit (%) | 67 |
| Plastic limit (%) | 21 |
| Free swell index (%) | 85 |
| Specific gravity | 2.6 |
| Shrinkage limit (%) | 13 |
| Sand (%) | 3 |
| Silt (%) | 16 |
| Clay (%) | 81 |
| Soil classification | CH |

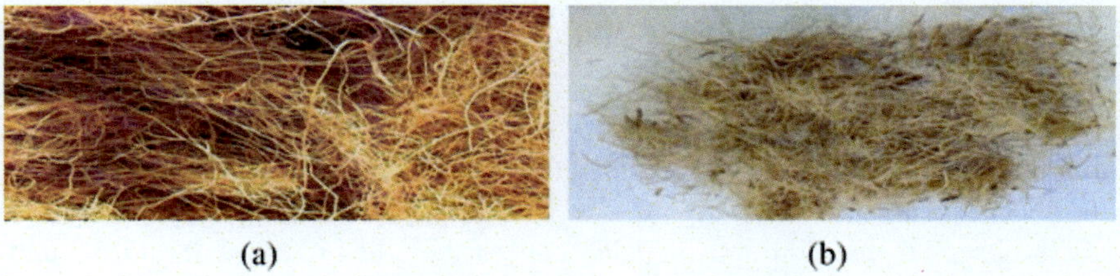

Fig. 1 a Coir fiber and b coir fiber of 20 mm length

**Table 2** Properties of coconut coir fiber

| Property | Value |
|---|---|
| Density | 1.4 g/cc |
| Breaking elongation | 30% |
| Tensile strength | 252 MPa |
| Young's modulus | 0.67 GPa |

### *2.3 Bitumen*

For increasing the life of fiber, the bitumen emulsion is used, which has water content 46%, and brownish-black color. The bitumen emulsion used for this study is collected from industry.

### *2.4 Concentrated HCL*

For finding the durability of coir-reinforced soil, the specimens for UCC test are prepared with optimum fiber content (OFC) by using concentrated hydrochloric acid (HCL) solution as pore fluid. The HCL solution is prepared by mixing the 5 ml of HCL acid with 600 ml of tap water.

## 3 Test Results and Discussion

The interpretations of tests result from the laboratory tests have been discussed in following sections with graphical results.

### 3.1 Compaction Test

The standard Proctor compaction test has been conducted for various samples; the variation of optimum moisture content (OMC) and maximum dry density (MDD) with fiber content is shown in Figs. 2 and 3, respectively.

It is observed that addition of 0.5% coir to the soil increased the OMC and decreased the MDD. Because fibers have tendency to absorb moisture, and this can be attributed to the increase in OMC. And using of materials has weight less than the weight of soil particles at the same volume which leads to decrease MDD, but further increase the fiber content leads to replace the soil particles, and it leads to reduce the soil moisture absorption, so OMC for soil mixture (soil + fiber) decreases. Further increase of fiber content fills up the interlocking spaces within the soil matrix and held the particles together.

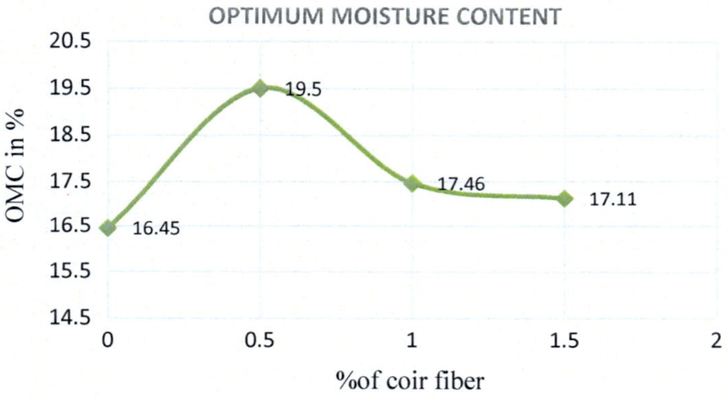


**Fig. 2** Variation optimum moisture content for coir stabilized soil

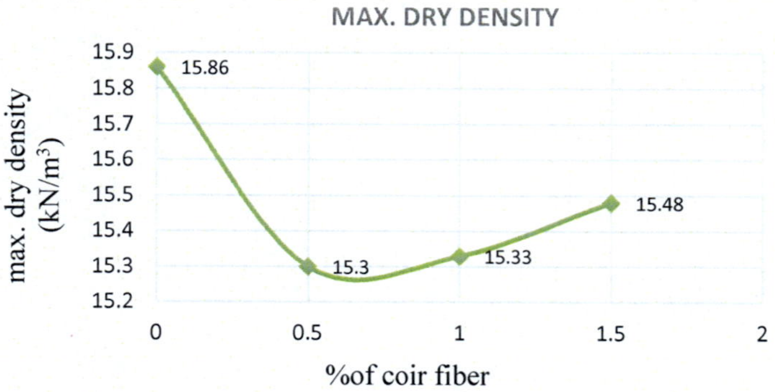


**Fig. 3** Variation of maximum dry density for coir stabilized soil

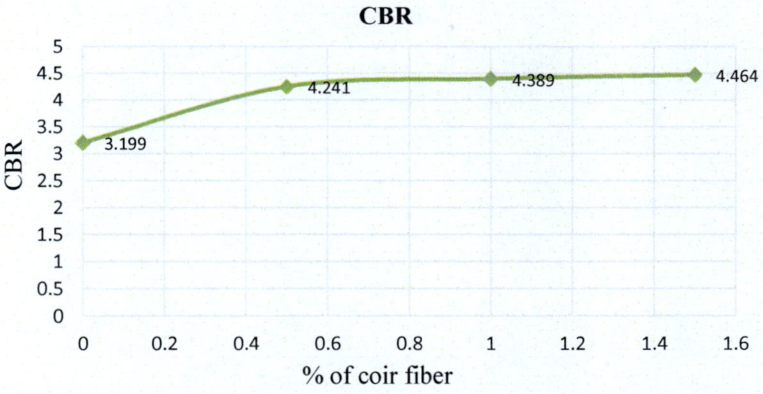


**Fig. 4** CBR for soil stabilized with coir fiber

## 3.2 *California Bearing Ratio Test*

The CBR test has been conducted for various samples. The test results are shown in Fig. 4. It is noticed that the addition of coir fiber increased the CBR value. Addition of 1.5% of coir fiber increased the CBR value by nearly 50%.

## 3.3 *Unconfined Compression Test*

The unconfined compression test has been conducted on various samples of size 3.6 cm diameter and 7.5 cm length. The coir-stabilized soil specimens are prepared at their respective MDD and OMC using water as the pore fluid. The unconfined compressive strength of the soil is increased by the addition of coir fiber. The variation of unconfined compressive strength with the fiber content is shown in Fig. 5. With the addition of 1% fiber, the unconfined compressive strength of the soil is increased by about 17%. The unconfined compressive strength remained nearly the same for further increase in fiber content. Hence, 1% fiber content is taken as optimum fiber content that is the minimum fiber content required to achieve the maximum strength gain.

For finding the durability of coir reinforced soil, the specimens for UCC test are prepared at MDD and OMC with optimum fiber content (OFC), which is 1% in the present study, by using concentrated hydrochloric acid (HCL) solution as pore fluid. The fiber is coated with bitumen. The prepared samples are cured for 7 and 30 days. The unconfined compressive strength of natural soil, soil stabilized with untreated fiber and soil treated with fiber coated with bitumen, is shown in Fig. 6.

From Fig. 6, it is inferred that the inclusion of coir fibers in to the soil as reinforcement shows higher unconfined compressive strength and bitumen-treated coir fiber has more durability compared to untreated coir fiber.

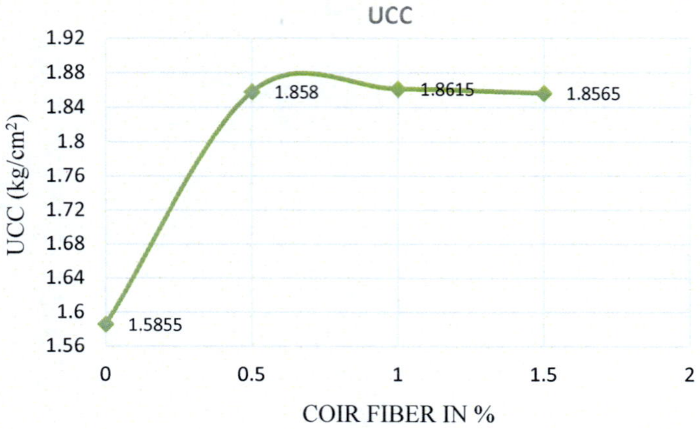


**Fig. 5** UCC for soil stabilized with coir fiber (pore fluid is water)

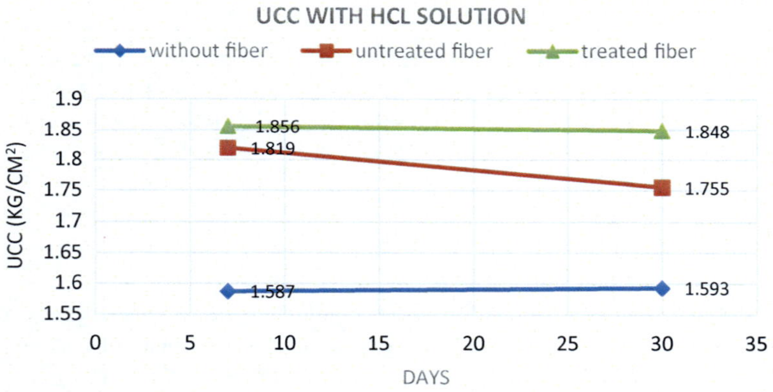


**Fig. 6** UCC for HCL solution as pore fluid

## 4 Conclusions

Based on the laboratory investigations, it can be concluded that the compaction characteristics of soil is influenced by the addition of coir fiber. The CBR value increases with increase of coir fiber content in soil. The unconfined compressive strength of soil is increased with coir fiber addition, and even 0.5% of fiber content is able to increase the strength by about 17%.

As far as the durability of coir fiber stabilized soil is concerned, it is witnessed that the bitumen-coated fiber is more durable than uncoated fiber in the acidic environment.

## References


1. Abhijith, R.P.: Effect of natural coir fibres on CBR strength of soil subgrade. Int. J. Sci. Res. Publ. **5**, 1–4 (2015)
2. Goyal, A., Parkash, V., Kumar, V.: Soil stabilization of clayey soil using jute fibre and gypsum. IJIRSET **5**, 15513–15519 (2016)
3. Chakraborty, A., Sil, A.: Experimental study of unreinforced and reinforced soil retaining wall using shake table facility. J. Civ. Eng. Environ. Technol. **2**(12), 13–21 (2015)
4. Bairagi, H.: Effect of jute fibers on lime treated black cotton soil. Int. J. Eng. Sci. Res. Technol. 705–707 (2014)
5. Singh, H.P.: Strength and stiffness of soil reinforced with jute geo-textile sheets. Int. J. Curr. Eng. Technol. **3**, 1143–1146 (2013)
6. Kulhar, K.S., Raisinghani, M.: Shear strength performance of sandy soil reinforced with jute fiber. J. Basic Appl. Eng. Res. **4**, 624–629 (2017)
7. Upadhyay, P., Singh, Y.: Soil stabilization using natural fiber coir. IRJET **4**, 1808–1812 (2017)
8. Lakshmi, S.M., Sasikala, S., Padmavathi, V., Priya, S., Saranya, V.: Utilization of coconut coir fibre for improving sub-grade strength characteristics of clayey sand. IRJET **5**, 2873–2878 (2018)
9. Soundara, B., Senthil Kumar, K.P.: Effect of fibers on properties of clay. Int. J. Eng. Appl. Sci. (IJEAS) **2**, 123–128 (2015)
10. Chakrabarti, S.K., Saha, S.G., Paul, P., Dewan, A.R., Das, K., Chowdhury, P.K., Gon, D.P., Ray, P.: Specially treated woven jute geo-textiles for river bank protection. Indian J. Fibre Textile Res. **41**, 207–211 (2016)

# Effect of Curing Period on the Geotechnical Properties of Lime-Treated Organic Soils

**Annie Joy, Sruthy Babu, Benny Mathews Abraham, and A. Sridharan**

## 1 Introduction

Organic soils are mostly composed of decayed plant matter and weathered rock material. These soils are known for their inferior engineering behavior such as very high compressibility and low shear strength. In order to improve these properties, organic soils are modified with calcium-based stabilizers like lime, cement and fly ash [1]. Lime is the most common stabilizer used to improve the engineering behavior of weak soils, especially clay. Furthermore, lime is also considered as a low-cost material, making it a popular choice amongst many other effective stabilizers [2]. In contrast with lime stabilization of inorganic clay, reactions between lime and clay are not efficient when organic matter is present in clay minerals. There is evidence in the literature that the presence of high concentrations of organic matter in clay can adversely affect the chemical reaction between lime and clay minerals and can have detrimental effects on the engineering properties of soil [3]. The factors that influence the mechanism in lime-treated organic clay is the moisture content and the insufficient dissolution of the clay minerals during the pozzolanic reaction. Organic matters possess high water retention capacity which limits the quantity of water available for the hydration process. Also, the high water content may produce more spacing between aggregates, thereby reducing the required cementation bond [4]. It has also been observed that organic matter coats the clay particles and thereby behaves as a barrier to lime and clay [5]. Previous studies have shown that the improving effect of lime gets deteriorated in the presence of organic matter with curing period. In this paper, the effect of lime treatment on the engineering properties of highly organic soils for longer curing period (up to 120 days) is investigated and summarized.

A. Joy (✉) · S. Babu · B. M. Abraham
Cochin University of Science and Technology, Kochi, India

A. Sridharan
Indian Institute of Science, Bengaluru, India

 
C. N. V. Satyanarayana Reddy et al. (eds.), *Ground Characterization and Foundations*, Lecture Notes in Civil Engineering 167,
https://doi.org/10.1007/978-981-16-3383-6_5

**Table 1** Physical properties of the bentonite

| Property | Test value |
|---|---|
| Moisture content (%) | 25 |
| Liquid limit (%) | 326 |
| Plastic limit (%) | 45 |
| Plasticity index (%) | 281 |
| Shrinkage limit (%) | 9 |
| Clay size (<0.002 mm) (%) | 81 |
| Silt size (0.002–0.075 mm) (%) | 16 |
| Sand size (>0.075 mm) (%) | 3 |
| Free swell index (cc/g) | 6.25 |
| Specific gravity | 2.7 |

# 2 Materials Used for the Study

## *2.1 Bentonite*

Bentonite is naturally occurring clay with unusual properties like very high swelling capacity, high ion exchange capacity and very low water permeability [6]. For the present study, commercially available sodium bentonite was used. All the tests were conducted on samples prepared using this soil since it contains no organic matter. The physical properties of bentonite used are given in Table 1.

## *2.2 Organic Matter*

Vermicompost collected from College of agriculture, Vellayani, Thiruvananthapuram, was used in the study to convert sodium bentonite into an organic soil. Compost has high cation exchange capacity and hence can increase the cation exchange capacity of soil when added to it. Also, the most essential nutrients in compost are in organic forms which are released slowly. Soil structure can be improved by the binding action between soil organic matter and clay particles via cation bridges and through simulation of microbial activity [6]. The physical properties of organic matter used in the study are summarized in Table 2.

**Table 2** Physical properties of organic matter

| Property | Test value |
|---|---|
| Moisture content (%) | 105 |
| Liquid limit (%) | 165 |
| Plastic limit (%) | 79 |
| Plasticity index (%) | 86 |
| Shrinkage limit (%) | 40 |
| Organic content (%) | 32 |
| Free swell index (cc/g) | 3.85 |
| Specific gravity | 2.05 |

### *2.3 Lime*

Hydrated lime [$Ca(OH)_2$] was used for the treatment. Hydrated lime was prepared by sprinkling water over lime shells, and the crumbled shells were then sieved through IS 425 micron sieve.

## 3 Experimental Investigation

In this study, the effect of curing period on the Atterberg limits, unconfined compressive strength, compressibility and swelling of lime-treated organic soil samples was investigated. In the experimental investigation, artificial organic soil was prepared by mixing bentonite with different percentages of organic matter (0%, 50%, 100%). The organic content in the soil was determined by loss on ignition method. By knowing the amount of organic content induced by the organic matter, the amount required to induce an organic percentage of 50% and 100% was calculated. The identified amount was mixed with bentonite and lime to obtain three different organic soil mixtures as given in Table 3. To study the effect of organic matter on lime stabilized clayey soil, the lime content was fixed from the literature as 6%.

Remolded samples for unconfined compressive strength were prepared with the help of a metallic mold having 3.8 cm inner diameter and 7.6 cm length as per IS

**Table 3** Experimental program for the study

| Sample description | Tests conducted | Curing period (days) |
|---|---|---|
| Bentonite + 6% lime<br>50% bentonite + 50% organic matter + 6% lime<br>100% organic matter + 6% lime | Atterberg limits tests<br>Free swell index test<br>Unconfined compressive strength test<br>Consolidation test | 0<br>7<br>30<br>60<br>90<br>120 |

2720-Part 10 [7]. Predetermined quantity of soil for the required density was filled inside the mold to make the specimen. The prepared samples were cured in labeled airtight polyethene bags. The cured samples were tested to study the effect of curing period. UCC and consolidation samples were made for different curing periods (0, 7, 30, 60, 90 and 120 days) and were kept for curing. All the specimens used in the study were prepared and tested using the standard procedures described by the bureau of Indian standards. Three sets of identical samples were tested to ascertain the repeatability of the tests [8].

# 4 Results and Discussion

## *4.1 Atterberg Limit Test*

Table 4 shows the effect of curing period on liquid limit of lime-treated organic soil. The liquid limit of lime-treated bentonite decreased by 9.7% after 120 days curing period. On the other hand, there was an increase of 4.1% in liquid limit of soil sample containing 100% organic matter when compared to soil sample containing 50% organic matter initially, and it further increased to 8% at 120 days curing period. Liquid limit is directly dependent upon the clay fraction of the soil. It is seen that water holding capacity of organic soils is more and this in turn also increases the liquid limit [9].

The plastic limit of different soil samples at various curing period is given in Table 5. The result shows that plastic limit increases with increase in organic content. This may be due to the high colloidal nature of the organic matter, and also their surface

**Table 4** Effect of curing period on liquid limit

| Curing period (days) | 0 | 7 | 30 | 60 | 90 | 120 |
|---|---|---|---|---|---|---|
| Sample description | Liquid limit (%) | | | | | |
| Organic matter + lime | 152 | 150 | 148 | 154 | 152 | 149 |
| Organic matter + bentonite + lime | 146 | 145 | 148 | 145 | 143 | 138 |
| Bentonite + lime | 310 | 315 | 305 | 299 | 282 | 280 |

**Table 5** Effect of curing period on plastic limit

| Curing period (days) | 0 | 7 | 30 | 60 | 90 | 120 |
|---|---|---|---|---|---|---|
| Sample description | Plastic limit (%) | | | | | |
| Organic matter + lime | 88 | 85 | 88 | 92 | 87 | 85 |
| Organic matter + bentonite + lime | 72 | 63 | 70 | 72 | 73 | 69 |
| Bentonite + lime | 51 | 48 | 50 | 52 | 48 | 49 |

**Table 6** Effect of curing period on shrinkage limit

| Curing period (days) | 0 | 7 | 30 | 60 | 90 | 120 |
|---|---|---|---|---|---|---|
| Sample description | Shrinkage limit (%) | | | | | |
| Organic matter + lime | 40 | 37 | 38 | 36 | 38 | 35 |
| Organic matter + bentonite + lime | 17 | 10 | 7 | 7 | 6 | 11 |
| Bentonite + lime | 9 | 8 | 5 | 10 | 4 | 7 |

area and water absorptive capacity are far in excess of those exhibited by the clays [7].

The shrinkage limit of different soil samples at various curing period is given in Table 6. From the test results, it is noted that shrinkage limit decreases with increase in organic content and curing period for lime-treated organic soil samples. The decrease in shrinkage limit indicates the increase in the expansive nature of the soil with an increase in the percentage of organic content [9].

## *4.2 Free Swell Index*

Free swell index (FSI) represents the ratio of swelled volume of the soil in water per unit weight of the soil. It is the simplest parameter to specify the swelling ability of soils [9]. Table 7 shows the variation of free swell index with curing period. For bentonite + lime soil mixture, the free swell index decreased by 20.7% at the end of 120 days curing period. The decrease in free swell index due to the addition of 6% lime is attributed to the fact that bentonite cations are substituted by calcium, leading to the formation of calcium silicate and aluminate hydrates. The decreased affinity for water of the Ca-saturated bentonite and the formation of a cementitious matrix resists swelling and thus decreases the free swell index [10]. However, when organic matter was introduced, the free swell index increased by 7.8%.

**Table 7** Effect of curing period on free swell index

| Curing period (days) | 0 | 7 | 30 | 60 | 90 | 120 |
|---|---|---|---|---|---|---|
| Sample description | Free swell index (cc/g) | | | | | |
| Organic matter + lime | 4.1 | 3.7 | 4 | 4 | 4.3 | 4 |
| Organic matter + bentonite + lime | 5 | 6.1 | 6.7 | 5.8 | 6.7 | 5.4 |
| Bentonite + lime | 10.9 | 9.1 | 9.1 | 9 | 8.7 | 8.7 |

## 4.3 Unconfined Compressive Strength

The effect of curing period on the unconfined compressive strength of different soil samples is shown in Fig. 1.

The unconfined compressive strength of soil sample containing lime and 100% organic matter was initially 5.95 kPa, and it increased with curing period up to 3 months and then decreased. Same trend was observed in the case of lime-treated organic soil + bentonite samples, i.e., strength decreased after 3 months of curing period. For lime-treated bentonite sample, the strength increased with curing period. In case of organic soil, the initial strength gain occurs since the effect of organic matter do not come into action. After 3 months of curing, the strength decreased because the organic matter inhibits the bond formation. Lime-treated bentonite samples attained more strength even with higher water content which means the presence of organic matter is the only factor that reduced the strength of lime-treated organic samples.

Percentages of strength loss increased with increasing organic matter content and with curing period. It was observed that after 120 days curing period, for lime-treated organic soil + bentonite sample, the unconfined compressive strength decreased by 73% when compared to lime-treated bentonite sample. In the case of lime-treated organic sample, the decrease was 77% when compared to lime-treated bentonite sample. Similarly, when the organic matter was increased from 50 to 100%, the unconfined compressive strength reduced by 14.3% at the end of 120 days curing period. Hence, it can be also concluded that as the presence of organic matter increases, the strength of lime-treated soil decreases drastically. In other words, lime

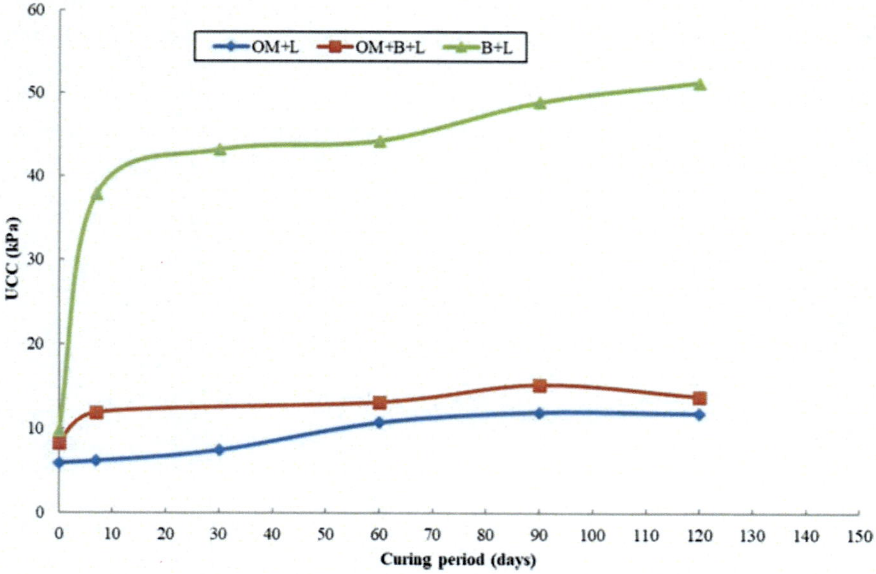


**Fig. 1** Effect of curing period on unconfined compressive strength of different soil samples

treatment is ineffective in strength improvement for highly organic soils. Similar results were reported by Yunus et al. [11] on lime-treated organic soil.

## 4.4 Compressibility Characteristics

The *e*-log *p* curves for different soil samples at 30 days and 120 days curing period are shown in Figs. 2 and 3, respectively. From Figs. 2 and 3, it can be seen that the *e*-log *p* curve of lime-treated soil samples has more slope, i.e., the lime-treated bentonite samples have more compression index, and lime-treated organic matter samples have the least compression index. After 120 days of curing period, it is observed that organic matter samples are more compressible than the bentonite samples. The reason for increasing compressibility of lime-treated organic matter samples may be due to the aggregated structure by molecular complexion involving metallic, organic and clay molecules caused by the added organic substances [8].

Figure 4 depicts the variation of d*e*/d (log *p*) values with pressure range for different soil samples cured for 120 days. From Fig. 4, it can be seen that the d*e*/d (log *p*) value of lime-treated bentonite samples increased with pressure and then decreased, whereas the d*e*/d (log *p*) value of lime-treated organic matter samples increased with pressure.

At the end of 120 days curing period, the d*e*/d (log *p*) value of lime-treated organic soil increased by 6.1% when compared to lime-treated organic soil + bentonite sample. We can conclude that the amount of compression undergone by the lime-treated organic soil sample is significantly higher and goes on increasing with increasing organic matter content.

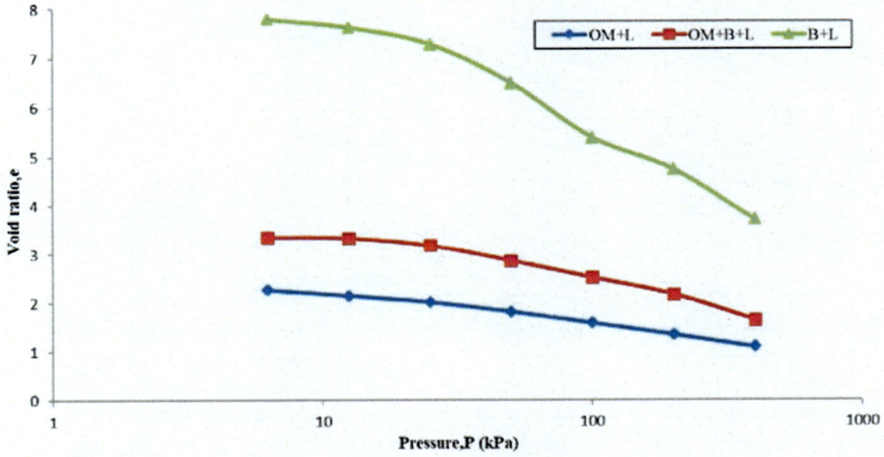


**Fig. 2** *e*-log *p* curve of different soil samples for 30 days curing period

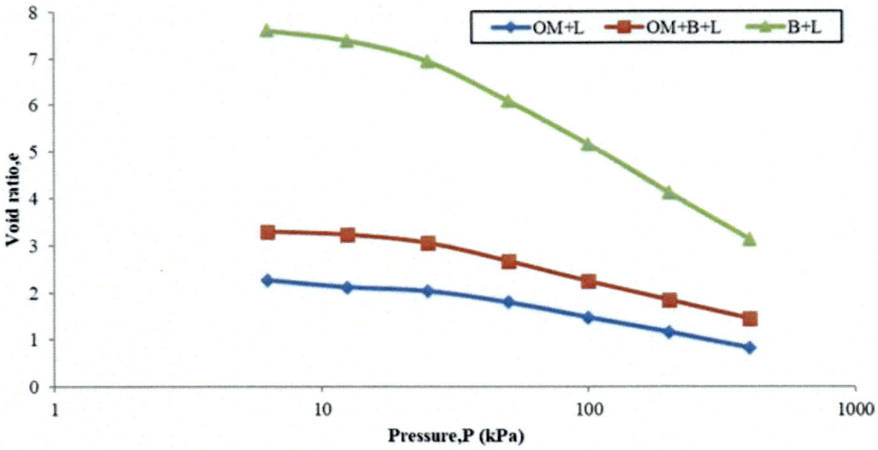


**Fig. 3** *e*-log *p* curve of different soil samples for 120 days curing period

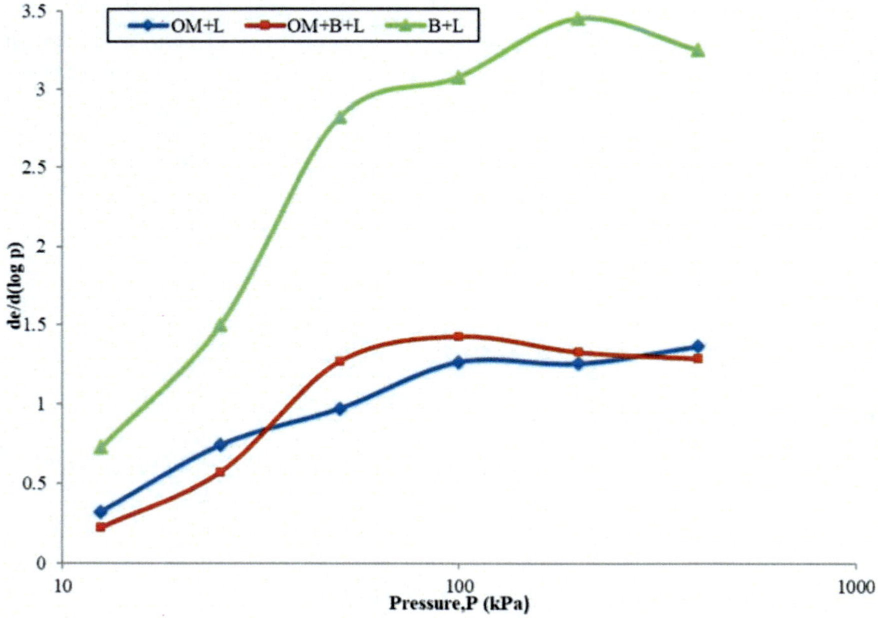


**Fig. 4** Variation of d*e*/d (log *p*) with pressure range for different soil samples cured for 120 days

Compression index is defined as the slope of *e*-log *p* curve. The compression index is extremely useful for the determination of settlement in the field. Figure 5 depicts the variation of compression index (Cc) values of lime-treated soil samples for different curing periods.

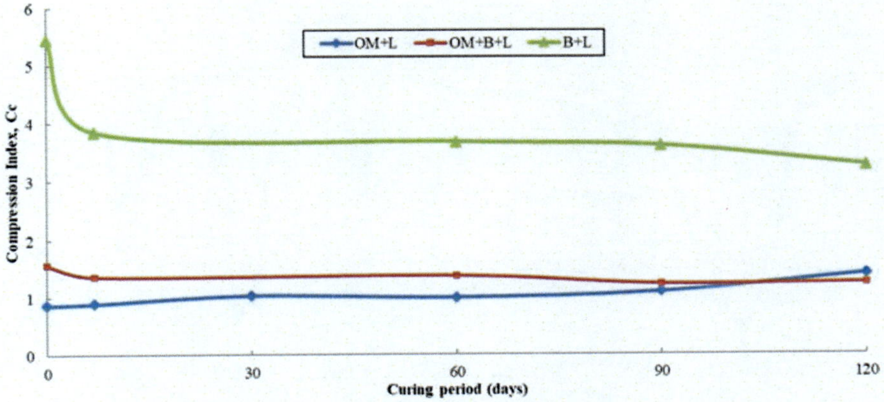


**Fig. 5** Variation of compression index values with curing period of different samples for a pressure range of 200–400 kPa

From Fig. 5, it can be seen that compression index values of lime-treated bentonite samples decrease with curing period and that of lime-treated organic matter samples increase with curing period, i.e., the lime-treated bentonite samples become less compressible with curing, whereas the compressibility of lime-treated organic sample increases with curing period.

At the end of 120 days curing period, the compression index value of lime-treated organic soil increased by 13% when compared to lime-treated organic soil + bentonite sample. We can conclude that the compression index increases with increase in the organic matter content added into the soils. Therefore, the consolidation settlement increases with increase in the organic content added into the soils. The increase in the settlement may be due to the increase in the water content holding capacity of soils with an increase in the percentage of organic content [8].

## 4.5 *Microstructural Analysis*

Microstructure of soil specimens was analyzed by the Fourier transform infrared spectroscopy (FTIR). Fourier transform infrared spectroscopy analysis is a powerful tool to identify the organic functional groups in soil samples. Infrared radiation is passed through a sample. Some of the infrared is absorbed by the sample and few of it is transmitted. The resulting spectrum represents a fingerprint of the sample with absorption peaks corresponding to the frequencies of vibrations between the bonds of the atoms making up the soil structure [12]. High transmittance at a frequency means that there are few bonds to soak up that color light within the sample; low transmittance means that there are a high population of bonds which have vibrational energies like incident light. In the present study, soil samples were characterized using Fourier transform infrared (Thermo Avtar 370 DTGS) in pressed KBr pellets.

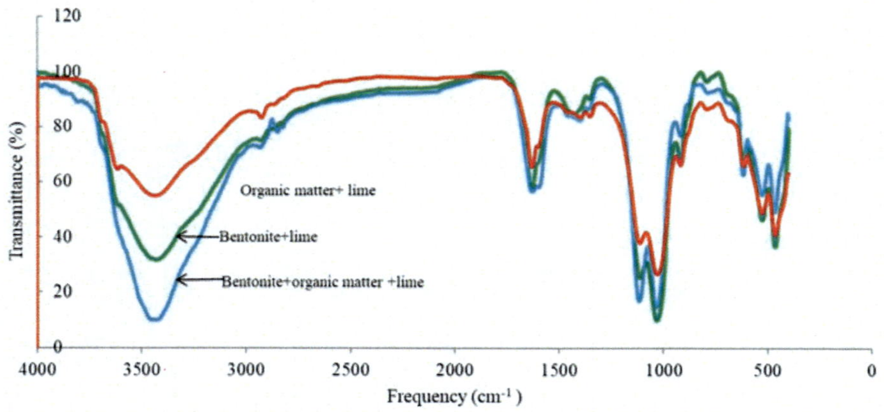


**Fig. 6** FTIR spectrum of soil samples at 120 days curing period

The spectral resolution was set to 4 cm$^{-1}$, and 32 scans were collected for each spectrum over a range of 4000–400 cm$^{-1}$. Figure 6 shows the FTIR spectrum of soil samples at 120 days curing period. With the addition of organic matter, for the same frequency of 3430 cm$^{-1}$, there was a reduction in transmittance from 31.5% in lime-treated bentonite, to 10.12% in lime-treated bentonite + organic matter soil due to aggregation [12]. For lime-treated bentonite + organic matter soil, the formation of aliphatic C–H group was observed at an absorption band of 2900 cm$^{-1}$. For lime-treated organic soil, an increase in transmittance was observed at a frequency of 3430 cm$^{-1}$. This can be attributed to the formation of flock which increased the particle size and pore size [12]. These pores in the microstructure of this soil sample are responsible for the high compressibility and low compressive strength.

## 5 Conclusions

From the results of experimental investigation conducted on lime-treated organic soil samples, the following conclusions are drawn:

1. Liquid limit shows a decreasing trend with curing period. For lime-treated bentonite sample, liquid limit increases after one week and then decreases, and for lime-treated organic matter and organic matter + bentonite samples, liquid limit decreases initially and then increases at two months and one month, respectively, and then again decreases.
2. With the addition of lime, plastic limit increases and the decreases initially for 7 days curing and then increases till certain curing period and then decreases.
3. UCS results show that strength of lime-treated bentonite sample increased with 120 days curing, whereas for lime-treated organic matter samples, strength

decreased after three months of curing (when the deleterious effect of organic matter came into action).

4. The compression index increases with increase in the organic matter content added into the soils. Therefore, the consolidation settlement increases with increase in the organic content added into the soils.
5. The pore size in the microstructure of lime-treated organic matter samples increased after 120 days curing period.
6. The effect of lime stabilization is nullified by the presence of organic matter.

Hence, the results contribute to a better understanding of the impact of curing period on the stabilization of lime-treated organic soils, which can ultimately provide a warning to the geotechnical engineers while choosing ground improvement methods for organic soils.

## References

1. Saride, S., Puppala, A.J., Chikyala, S.R.: Swell-shrink and strength behaviors of lime and cement stabilized expansive organic clays. Appl. Clay Sci. **85**, 39–45 (2013)
2. Mohd Yunus, N.Z., Wanatowski, D., Stace, R., Marto, A., Abdullah, R.A., Mashros, N.: A short review of the factors influenced lime-clay reactions. EJGE **19**, 8305–8318 (2011)
3. Yunus, N.M., Wanatowski, D., Stace, L.R.: Lime stabilisation of organic clay and the effects of humic acid content. Geotech. Eng. J. SEAGS AGSSEA **44**(1), 19–25 (2013)
4. Abbey, S.J., Olubanwo, O., Ngambi, S., Eyo, U., Adeleke, B.O.: Effect of organic matter on swell and undrained shear strength of treated soils. J. Civ. Constr. Environ. Eng. **4**(2), 48–58 (2019)
5. Abood, T.T., Kasa, A.B., Chik, Z.B.: Stabilisation of silty clay soil using chloride compounds. JESTEC **2**(1), 102–110 (2007)
6. Varghese, R., Chandrakaran, S., Rangaswamy, K.: Geo-technical behaviour of different organic matter on clayey soils. Geo-Mech. Geo Eng. 1–10 (2019)
7. IS 2720 Part 10: Methods of test for soils—Determination of unconfined compressive strength. BIS, New Delhi (1991)
8. Varghese, R., Chandrakaran, S., Rangaswamy, K.: Influence of type of organic substances on the strength and consolidation behaviour of inorganic clay soil. Int. J. Geo-Tech. Eng. 1–12 (2019)
9. Varghese, R., Chandrakaran, S., Rangaswamy, K.: Effect of organic content on geotechnical properties of bentonite clay. In: Proceedings of Indian Geotechnical Conference, IGC2016, pp. 1–4, Chennai (2016)
10. Kumar, S., Dutta, R.K., Mohanty, B.: Engineering properties of bentonite stabilized with lime and phosphogypsum. Slovak J. Civ. Eng. **22**(4), 35–44 (2014)
11. Yunus, N.M., Wanatowski, D., Stace, L.R.: Effect of humic acid on physical and engineering properties of lime-treated organic clay. Eng. Technol. **59**(201), 1820–1825 (2011)
12. Hsu, H.T., Lawrence, C.R., Winnick, M.J., Bargar, J.R., Maher, K.: A molecular investigation of soil organic carbon composition across a subalpine catchment. Soil Syst. **2**(1), 6 (2018)

# Influence of Density and Degree of Saturation on the Shear Strength Characteristics of Marine Sands

**K. Natarajan and D. V. Siva Sankara Reddy**

## 1 Introduction

The desk of civil engineers is not merely designing structures from the aesthetic point of view but has to design and built it keeping in mind economy and safety of the structures. To have a stable structure to be placed on sound foundation, a civil engineer ought to have a good idea of the properties of the foundation soil that is confronted with. For a sound foundation design, a thorough investigation of strength and stability characteristics of soil is necessary [1].

Slopes of all kinds including embankments, hills, mountains and manmade cuts and fills. Stay firmly in place because of the shearing strength possessed by the soil or rock of which they are composed of the importance of soil strength analysis was keenly felt during the investigation of the great landslides in the Panama Canal and the failure of fort peck dam [2]. Stability analysis includes checking the structures against required and available shear strength to safeguard it against any possible failure. Thus, it is clear that with all soils  down to the necessity for a clear understanding of the strength characteristics of soils [3].

The shear strength of soil is not an intrinsic property of soil but varies over a considerable range under varying conditions. The shearing strength both in the laboratory specimen and the soil in the natural state is dependent on density, moisture content, state of compaction, grain size, its distribution and particles shape and gradation [4].

Since World War II, fundamental research on shearing strength phenomena has led to the improved understanding of the various phases of the subject, but even today it cannot be claimed that all the factors affecting it have been fully understood [5].

K. Natarajan (✉)
Department of Structural Engineering, Annamalai University, Annamalai Nagar, Chidambaram, Tamilnadu 608002, India

D. V. Siva Sankara Reddy
Department of Civil Engineering, KHIT, Guntur, Andhra Pradesh 522019, India

 
C. N. V. Satyanarayana Reddy et al. (eds.), *Ground Characterization and Foundations*, Lecture Notes in Civil Engineering 167,
https://doi.org/10.1007/978-981-16-3383-6_6

The estimation of shear strength of soil, which exists under natural conditions, from laboratory test data is highly complex in nature because it is difficult to simulate actual conditions existing in site in the laboratory test [6]. Further difficulties arise from non-homogeneity of the representative samples.

An attempt has been made in this work to study the shear strength characteristics of marine sand deposits along the east coast of Tamil Nadu areas of Pichavaram, Pazhayar, Nagapattinam and Samiyarpettai. Borehole sand at different depths was collected by conducting the SPT. The soil samples were tested in loose dry, loose saturated dense dry and dense saturated condition by direct shear test. The variation of the friction angle under different condition has been discussed and presented in this work.

## 2 Experimental Programme

In general, there are two types of direct shear apparatus one the stress-controlled type in which provision is made to increase the shearing force gradually until failure and the other strain-controlled type, wherein shearing displacement is increased, and the force required to produce the increase is measured [7].

Strain-controlled apparatus manufactured by Messrs. Associated Instrument Manufacturers India Private Limited was used for this series of 32 tests. The essential feature of the apparatus is a rectangular box 6 cm $\times$ 6 cm consisting of two separate parts called the upper frame and lower frame [8]. The box is provided with a detachable base plate, one top and another bottom porous plates, one bottom perforated grill plate and another top grill plate. The vertical load is transmitted through a sturdy top cover with ball bearing. The vertical load is applied by means of a loading frame through a lever arm of magnification ratio 1.5. For each test of the series, different normal loads 10 kg, 20 kg, 30 kg, 40 kg, 50 kg, respectively, were applied [9]. To multiply the effect of the weights of tie bar leaver and load hanger, counterweight is provided. By means of the capstan, the height of the leaver fulcrum is adjusted until the leaver is level, and the loading yoke just rests on the loading ball.

The shear failure through the soil at the plane of separation between the two frames is caused by the relative movement between the top and bottom frames. The whole box is placed within a metallic container which is mounted on ball races on a tract over the base plate of the apparatus [9].

The movement of the bottom frame is produced by means of a loading screw worked by a small electric motor through a pulley block. Nine different speeds are possible by suitable combination of pulleys and gears 0.002 mm sensitivity for measuring the shearing stresses. The values of one division of proving ring are 0.2325 kg. The change in thickness is recorded by a gauge. The relative movement of the box is also recorded by a similar gauge [10].

The minimum and maximum density of soil sample were found out for the soils from four different places at Pichavaram, Pazhayar, Nagapattinam and Samiyarpettai. The minimum density is found out through loose filling of soil in the mould. The

**Table 1** Grain size distribution and densities of soil at different depths

| S. No. | Soil collected | Depth of soil occurrence (m) | Minimum density (g/cm$^3$) | Maximum density (g/cm$^3$) |
|---|---|---|---|---|
| 1 | Pichavaram | 3<br>5 | 1.39 | 1.67 |
| 2 | Pazhayar | 3<br>5 | 1.51 | 1.78 |
| 3 | Nagapattinam | 3<br>5 | 1.48 | 1.72 |
| 4 | Samiyarpettai | 3<br>5 | 1.72 | 1.81 |

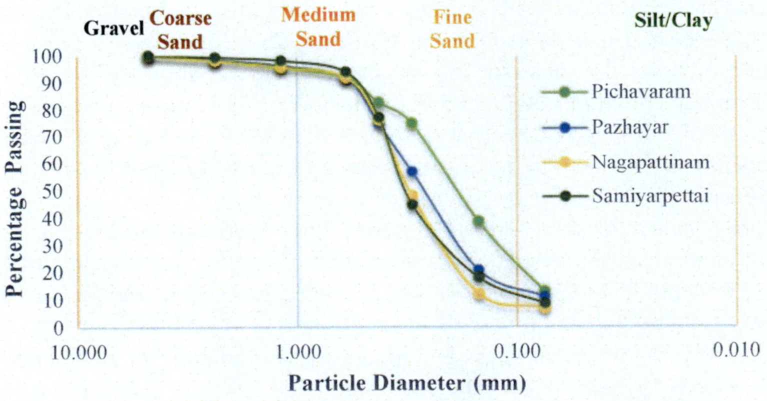


**Fig. 1** Grain size distribution curve of different soils used in the study

maximum density is found out by filling the soil in several layers in through mould a compacting each layer through vibrator. The minimum and maximum densities are shown in Table 1.

Figure 1 shows the grain size distribution of soil, and the results are presented in a graph shown in below with per cent passing versus the sieve size. On the graph, the sieve size scale is logarithmic.

## 3 Results and Discussion

In this study, soil samples were collected from four locations, i.e. Pichavaram, Pazhayar, Nagapattinam and Samiyarpettai. Soil samples were collected at two different depths at each site. The soil samples were tested in loose dry, loose saturated, dense dry and dense saturated condition by direct shear test. The different shear tests have been carried out at a constant strain rate of 0.25 mm per minute for all the

**Table 2** Angle of internal friction for different soil conditions at different depths at Pichavaram location

| S. No. | Test condition | Angle of internal friction for 3 m depth | Angle of internal friction for 5 m depth |
|---|---|---|---|
| 1 | Loose dry | $\varphi = 36°$ | $\varphi = 39°$ |
| 2 | Loose saturated | $\varphi = 33°$ | $\varphi = 34°$ |
| 3 | Densest dry | $\varphi = 47°$<br>$\varphi_{cv} = 36°$ | $\varphi = 49°$<br>$\varphi_{cv} = 38°$ |
| 4 | Dense saturated | $\varphi = 44°$<br>$\varphi_{cv} = 41°$ | $\varphi = 46°$<br>$\varphi_{cv} = 39°$ |

32 tests. The variation of the friction angle under different conditions at Pichavaram location has been discussed and presented in Table 2.

Figure 2 shows the shear stress and normal stress variation for loose dry and loose saturated conditions at 3 and 5 m depths located in Pichavaram. From Fig. 2, it is observed that angle of internal friction is high for loose soil in dry condition at 5 m depth as compared to the angle of internal friction obtained in soil saturated condition.

Figure 3 shows the shear stress and normal stress variation for loose and dense dry conditions at 3 m depth located in Pichavaram. From Fig. 3, it is clearly observed that at 3 m depth of Pichavaram location, the angle of internal friction is more for dense dry soil as compared to loose soil condition.

Figure 4 shows the shear stress and normal stress variation for loose and dense dry conditions at 5 m depth located in Pichavaram. From Fig. 4, it is clearly observed that at 5 m depth of Pichavaram location, the angle of internal friction is more for dense dry soil as compared to loose soil condition.

Figure 5 shows the shear stress and normal stress variation for dense dry and dense saturated conditions at 3 and 5 m depths located in Pichavaram. From Fig. 5,

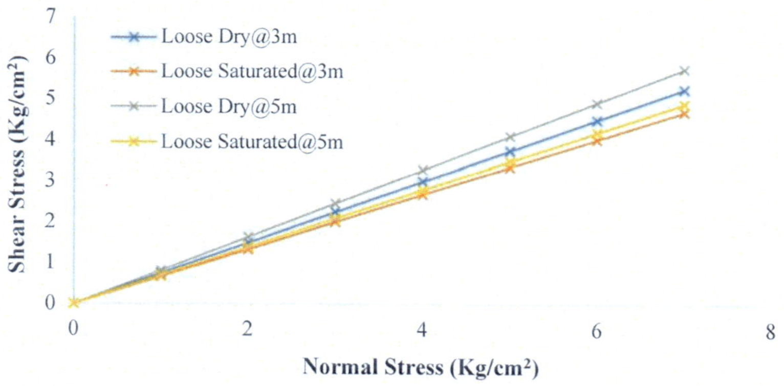


**Fig. 2** Shear stress and normal stress variation of loose soil at 3 and 5 m depths located in Pichavaram

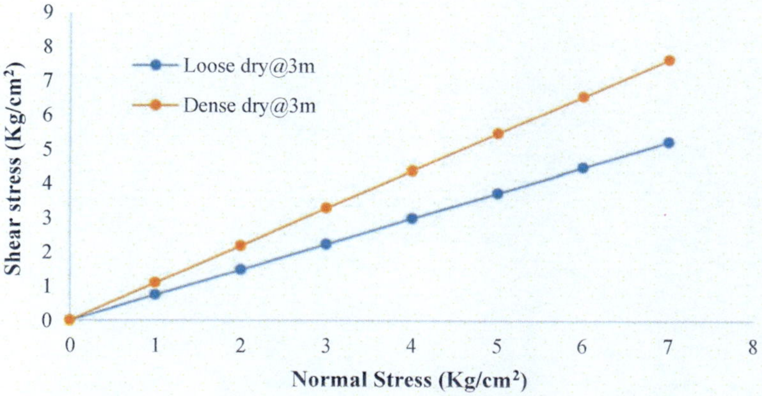


**Fig. 3** Shear stress and normal stress variation of loose and dense dry soil conditions at 3 m depth located in Pichavaram

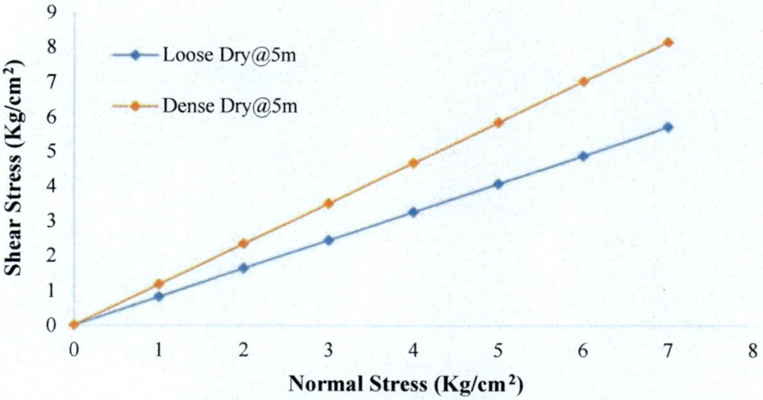


**Fig. 4** Shear stress and normal stress variation of loose and dense dry soil conditions at 5 m depth located in Pichavaram

it is observed that angle of internal friction is high for dense soil in dry condition at 5 m depth as compared to the angle of internal friction obtained in soil saturated condition.

Figure 6 shows the shear stress and horizontal displacement variation for loose dry and loose saturated conditions at 3 and 5 m depths located in Pichavaram. From Fig. 6, it is observed that with increase in the degree of saturation of soil, the horizontal deformation increases for a given shear stress value. This is due to decrease in density as saturation increases. As the depth of soil location increases, the shear stress of the soil increases for a given horizontal deformation.

Figure 7 shows the shear stress and horizontal displacement variation for loose dry and dense dry conditions at 3 m depth located in Pichavaram. From Fig. 7, it is observed that with increase in the degree of saturation of soil, the horizontal

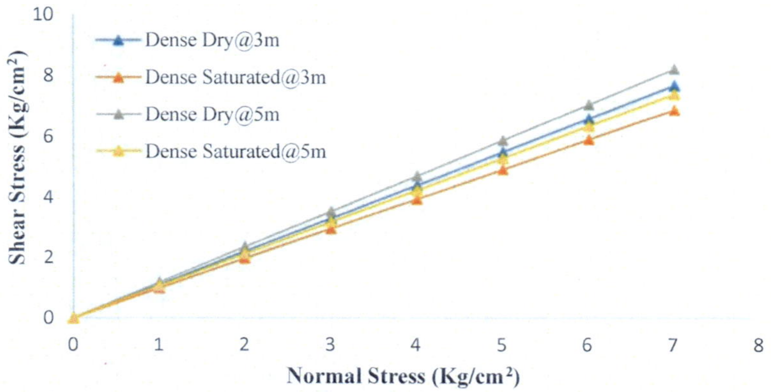


**Fig. 5** Shear stress and normal stress variation for dense dry and dense saturated conditions at 3 and 5 m depths located in Pichavaram

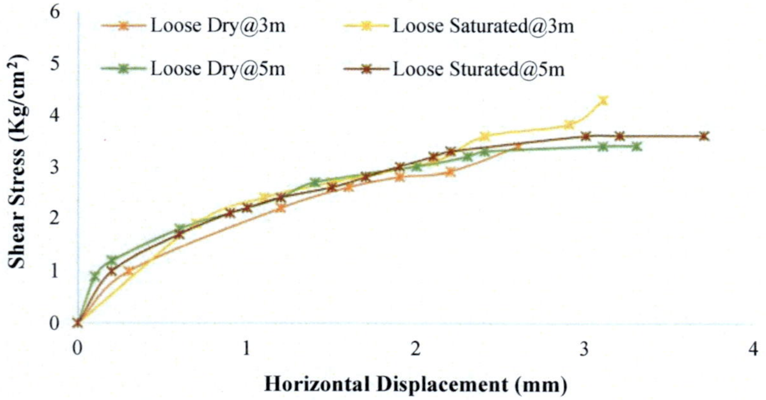


**Fig. 6** Shear stress and horizontal displacement variation for loose dry and loose saturated conditions at 3 and 5 m depths located in Pichavaram

deformation increases for a given shear stress value at same depth. This is due to decrease in density as saturation increases.

Figure 8 shows the shear stress and horizontal displacement variation for loose dry and dense dry conditions at 5 m depth located in Pichavaram. From Fig. 8, it is observed that with increase in the degree of saturation of soil, the horizontal deformation increases for a given shear stress value at same depth. This is due to decrease in density as saturation increases.

Figure 9 shows the shear stress and horizontal displacement variation for dense dry and dense saturated conditions at 3 and 5 m depths located in Pichavaram. From Fig. 9, it is observed that with increase in the degree of saturation of soil, the horizontal deformation increases for a given shear stress value. This is due to decrease in density

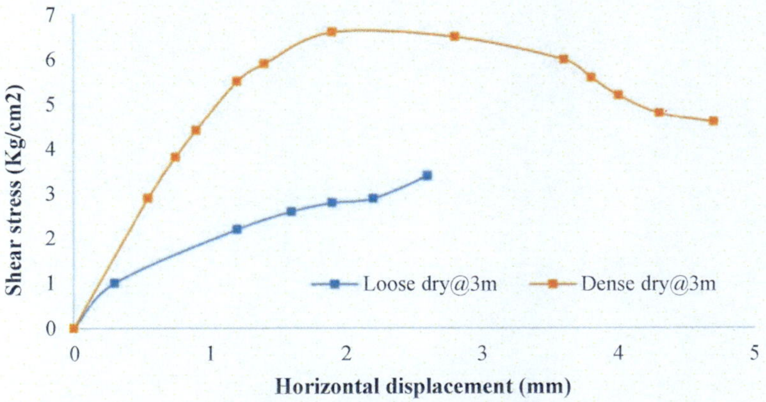


**Fig. 7** Shear stress and horizontal displacement variation for loose dry and dense dry conditions at 3 m depth located in Pichavaram

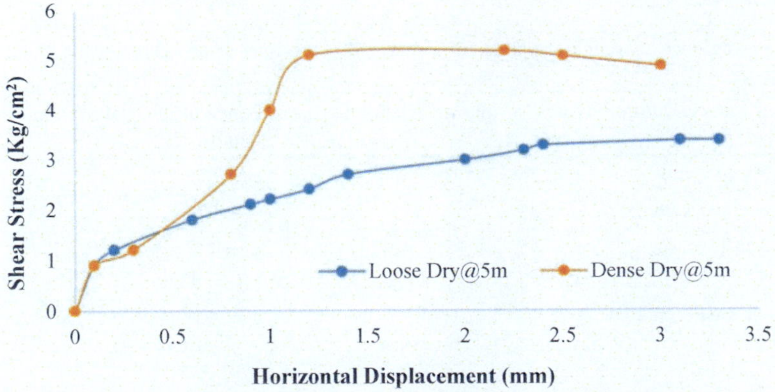


**Fig. 8** Shear stress and horizontal displacement variation for loose dry and dense dry conditions at 5 m depth located in Pichavaram

as saturation increases. As the depth of soil location increases, the shear stress of the soil decreases for a given horizontal deformation.

Angle of internal friction of different soil conditions for different depths at Pazhayar, Nagapattinam and Samiyarpettai locations is given in Tables 3, 4 and 5, respectively.

At Pazhayar location, the normal stress, shear stress and horizontal displacement values are decreasing by 1% as compared to results obtained at Pichavaram location and increasing by 2% than the Nagapattinam location and also observed 1.2% decreasing values than the Samiyarpettai location.

Similarly, at Nagapattinam location, the normal stress, shear stress and horizontal displacement values are increasing by 1.7% as compared to results obtained

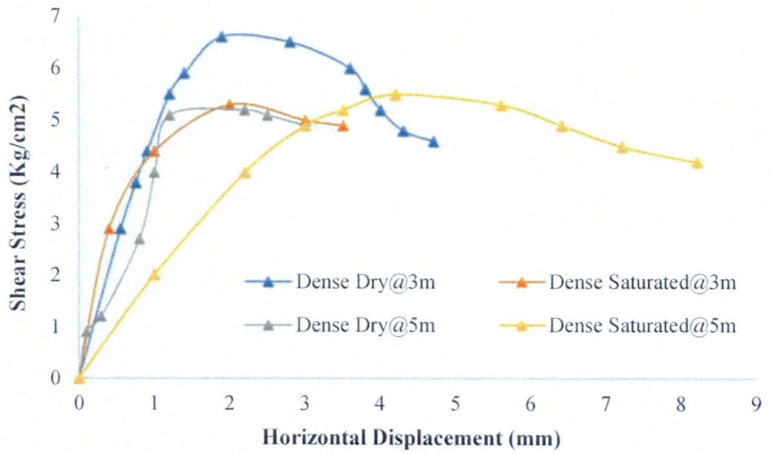


**Fig. 9** Shear stress and horizontal displacement variation for dense dry and dense saturated conditions at 3 and 5 m depths located in Pichavaram

**Table 3** Angle of internal friction for different soil conditions at different depths at Pazhayar location

| S. No. | Test condition | Angle of internal friction for 3 m depth | Angle of internal friction for 5 m depth |
|---|---|---|---|
| 1 | Loose dry | $\varphi = 35°$ | $\varphi = 37°$ |
| 2 | Loose saturated | $\varphi = 32°$ | $\varphi = 34°$ |
| 3 | Densest dry | $\varphi = 45°$<br>$\varphi_{cv} = 41°$ | $\varphi = 47°$<br>$\varphi_{cv} = 43°$ |
| 4 | Dense saturated | $\varphi = 42°$<br>$\varphi_{cv} = 39°$ | $\varphi = 45°$<br>$\varphi_{cv} = 43°$ |

**Table 4** Angle of internal friction for different soil conditions at different depths at Nagapattinam location

| S. No. | Test condition | Angle of internal friction for 3 m depth | Angle of internal friction for 5 m depth |
|---|---|---|---|
| 1 | Loose dry | $\varphi = 33°$ | $\varphi = 36°$ |
| 2 | Loose saturated | $\varphi = 30°$ | $\varphi = 34°$ |
| 3 | Densest dry | $\varphi = 49°$<br>$\varphi_{cv} = 36°$ | $\varphi = 49°$<br>$\varphi_{cv} = 34°$ |
| 4 | Dense saturated | $\varphi = 46°$<br>$\varphi_{cv} = 36°$ | $\varphi = 51°$<br>$\varphi_{cv} = 42°$ |

**Table 5** Angle of internal friction for different soil conditions at different depths at Samiyarpettai location

| S. No. | Test condition | Angle of internal friction for 3 m depth | Angle of internal friction for 5 m depth |
|---|---|---|---|
| 1 | Loose dry | $\varphi = 35°$ | $\varphi = 41°$ |
| 2 | Loose saturated | $\varphi = 36°$ | $\varphi = 39°$ |
| 3 | Densest dry | $\varphi = 53°$<br>$\varphi_{cv} = 44°$ | $\varphi = 47°$<br>$\varphi_{cv} = 43°$ |
| 4 | Dense saturated | $\varphi = 46°$<br>$\varphi_{cv} = 38°$ | $\varphi = 45°$<br>$\varphi_{cv} = 42°$ |

at Pichavaram location and increased by 2.4% than the Pazhayar location and also observed 1.6% decreasing values than the Samiyarpettai location.

Similarly, at Samiyarpettai location, the normal stress, shear stress and horizontal displacement values are decreased by 0.8% as compared to results obtained at Pichavaram location and increased by 3% than the Nagapattinam location and also observed 1.8% increasing values than the Pazhayar location.

## 4 Conclusions

From the results obtained from direct shear test conducted on the soil samples, the following important conclusions may be drawn.

1. There is considerable difference between friction angle using peak shear strength and shear strength corresponding to constant volume.
2. The difference between $\varphi$ and $\varphi_{cv}$ appears to be least for the soils at dry condition as compared to the values when the soils are tested in saturated condition.
3. The difference between $\varphi$ and $\varphi_{cv}$ is more in dense condition for a soil tested than in loosest state.
4. In the case of soil tested in loose condition, it is generally observed that the value obtained from saturated condition is lower than that of the obtained values for the loose sand.
5. The higher difference between the values of $\varphi$ obtained for loose and dense condition indicates better interlocking of soil grains, due to the angularities of the soil grains.

## References

1. Amirtham, T.: The effect of strain rate of the strength characteristic of the river sand. M.E thesis. Soil Mechanics and Engineering, Annamalai University (1969)

2. Das, B.M.: Advanced Soil Mechanics, 5th edn. CRC Press, Taylor & Francis Group, ISBN 9780815379133 (hardback) | ISBN 9781351215183 (ebook)
3. Bishop, A.W., Skempton A.W.: Measurement Strength of Soil, vol. II. Geotech (1950–51)
4. Terzaghi, K., Peck, R.B., Mesri, G.: Soil Mechanics in Engineering Practice, 3rd edn, ISBN: 978-0-471-08658-1, A Wiley Interscience Publication, Wiley, INC, New York
5. Bowles, J.E.: Engineering Properties of Soils and Their Measurement. ISBN : 0071129219, McGraw-Hill Book Company, Inc., New York, New York
6. Dr. Punmia, B.C., Er. Jain, A.K., Er. Jain, A.K.: Soil Mechanics and Foundations, 17th edn, ISBN: 81-7008-791-0, Lakshmi Publications (P) LTD, Golden house, New Delhi-110002, India
7. Turnbull, J.M.: The shearing resistance of soils as determined by direct shear tests at a constant rate of strain. In: Proceedings of the Second International Conference on Soil Mechanics and Foundation Engineering, vol. 1, pp. 211–217. June 21–30, 1948, Rotterdam
8. Rao, V. V. S.: Stress distribution in a shear box. In: Symposium on Pore Pressure and Shear Strength of Sands (1967)
9. Head, K.H.: Manual of Soil Laboratory Testing: Volume III, 3rd New edition, ISBN: 9781482227963, Apple Academic Press Inc., Oakville, Canada
10. IS-2770, (Part XIII): Method of Test for Soils. Part XIII, Direct Shear Test (1965)

# Evaluating Soil Shrinkage Behavior Using Digital Image Analysis Process

**A. G. Sharanya, M. Heeralal, and T. Thyagaraj**

## 1 Introduction

The behavior of soil related to water volume variation can be characterized as swelling or shrinkage. The volume change mechanism associated with the water volume variation has to be understood for determining the hydromechanical behavior of soil. The increase in water content of the compacted soil used in various geotechnical or geo-environmental field application has always been considered as the critical condition for analysis. The response of soil along the drying phase from the extreme saturated state possesses an equal importance to be understood such that the wet–dry moisture variation effect can be explicitly characterized. The reduction in water content and the associated void volume change is generally represented as soil shrinkage curve (SSC) or shrinkage curve (SC). The SSC can be combined with the soil–water characteristic curve (SWCC) or soil–water retention curve (SWRC), and this will aid in understanding the effect of suction potential on the volume change of soil along varying moisture content [1–4]. The characterization of shrinkage curve generally involves four regimes: (i) structural phase, (ii) proportional or normal phase, (iii) residual phase and (iv) zero or no shrinkage phase [2, 3, 5–8]

The contractive volume change is prominently induced by the internal forces such as suction stress existing at or near the particle contact surface [9–12]. The influence of suction stress and the microstructural change relative to the soil moisture movement evidently reveals the mechanism of shrinkage [13, 14]. The contractive volume change characterization necessitates the measurement of water content change and

A. G. Sharanya (✉) · M. Heeralal
National Institute of Technology, Warangal 506004, India
e-mail: sharanyaag@student.nitw.ac.in

T. Thyagaraj
Indian Institute of Technology, Madras, Chennai 600036, India

 
C. N. V. Satyanarayana Reddy et al. (eds.), *Ground Characterization and Foundations*, Lecture Notes in Civil Engineering 167,
https://doi.org/10.1007/978-981-16-3383-6_7

the induced volume change at the same time throughout the drying phase. The moisture content decrease from the initial condition can be measured easily by determining the weight change of soil at regular time period. The accurate determination of volume change to represent the void ratio variation governs the exactitude of shrinkage characterization and the mechanism [8, 15, 16].

This paper aims at characterizing the shrinkage behavior of silty clay soil in Warangal, India, with the use of digital image analysis technique and discusses in detail about the experimental setup framed for the entire process, and the reliability of prediction models is also highlighted. The changes observed in the dimension of the drying soil sample from slurry state to complete dryness are captured and studied to frame the shrinkage behavior as SSC. The cracking of soil during continued drying is eliminated by using a small sample size such that only subsidence is considered in the study. The 3D volume change study is more sophisticated and requires adequate knowledge, whereas the use of micrometer or Vernier caliper for dimension measurement during the entire process of drying results in loss of sample due to handling errors. The vertical subsidence or the thickness can be determined easily with the use of thickness gauge, whereas the horizontal shrinkage needs careful analysis such that volume computation will be reliable. There are notable research works that have used the 2D image analysis process to compute the volume change of samples, and they have been recommended to be efficient in capturing the shrinkage of soil samples [8, 17–21]. In this study, digital imaging was utilized for capturing the radial shrinkage, and the image analysis was carried out using the open-source 'ImageJ' version 1.52a software developed at the University of Wisconsin.

## 2 Materials and Methods

### *2.1 Soil*

The soil obtained from NIT Warangal campus was used in this study. The collected soil samples were air-dried for a week and then oven-dried. The over-dry samples were sieved through 2.36 mm IS sieve and stored in air-tight containers for laboratory study. The liquid limit and plasticity index of the soil were 42% and 11%, respectively. The specific gravity was 2.61, and the soil was classified as silty clay exhibiting low to medium plasticity. The standard proctor compaction test was conducted, and the maximum dry density was 17.12 kN/m$^3$ at an optimum moisture content of 15.5%.

## 2.2 Test Procedures

The sample used for suction measurement was statically compacted to optimum moisture content and allowed for air-induced drying. The suction value was determined at predefined moisture content values. The total suction was measured using the chilled mirror hygrometer technique, and the measured suction values were fitted using the Fredlund and Xing (FX) model as in Eq. (1) [22] and Van Genuchten (VG) model as represented in Eq. (2) [23]

$$(\psi) = w_{\mathrm{s}}\left[1 - \frac{\ln\left(1 + \frac{\psi}{h_{\mathrm{r}}}\right)}{\ln\left(1 + \frac{10^6}{h_r}\right)}\right]\left\{\ln\left[\exp(1) + \left(\frac{\psi}{a_{\mathrm{f}}}\right)^{n_{\mathrm{f}}}\right]\right\}^{m_{\mathrm{f}}-1} \tag{1}$$

where $w_{\mathrm{s}}$ is the saturated water content; $\psi$ is the suction in kPa; $a_{\mathrm{f}}$, $n_{\mathrm{f}}$ and $m_{\mathrm{f}}$ are fitting parameters, and $h_{\mathrm{r}}$ is the residual suction.

$$\theta = \theta_{\mathrm{r}} + \frac{\theta_{\mathrm{s}} - \theta_{\mathrm{r}}}{(1 + |\alpha\psi|^n)^m} \tag{2}$$

$$m = 1 - \frac{1}{n} \tag{3}$$

$\theta_{\mathrm{r}}$ and $\theta_{\mathrm{s}}$ denote the residual and saturated volumetric water content; $\alpha$ and $n$ are independent fitting parameters.

The sample used for shrinkage measurement was prepared at an initial moisture content greater than that of the liquid limit. The sample was allowed to mellow for 24 h sealed in a plastic press bag for moisture equilibrium. The sample was then transferred to an aluminum circular mold of size 42 mm diameter and 12 mm height. The sample container was coated with silicone gel to allow free shrinkage. The sample was placed over a weighing balance and set to allow for air-drying. The detailed experimental setup is as shown in Fig. 1. The measured void ratio and water content was fitted with the parameter fitting model proposed by [7]. Equation (4) was used to fit the experimental data, and the fitting parameters were determined using the solver tool in Microsoft excel. The captured images were processed and analyzed to compute the radial shrinkage for every 2 h from the starting of the test. The experimental setup utilized a 12 MP (Megapixel) hand camera held at a fixed distance to capture the surface area of the sample placed over a weighing balance during the entire process.

$$e(w) = e_{\mathrm{r}} + \frac{e_{\mathrm{s}} - e_{\mathrm{r}}}{\left[1 + (\alpha(w))^{-p}\right]^q} \tag{4}$$

$e_{\mathrm{s}}$ and $e_{\mathrm{r}}$ are the void ratio corresponding to saturated and residual water content; $\alpha, p$ and $q$ are the fitting parameters to be determined using the measured experimental data. The temperature variation during the drying period is shown in Fig. 2, and when

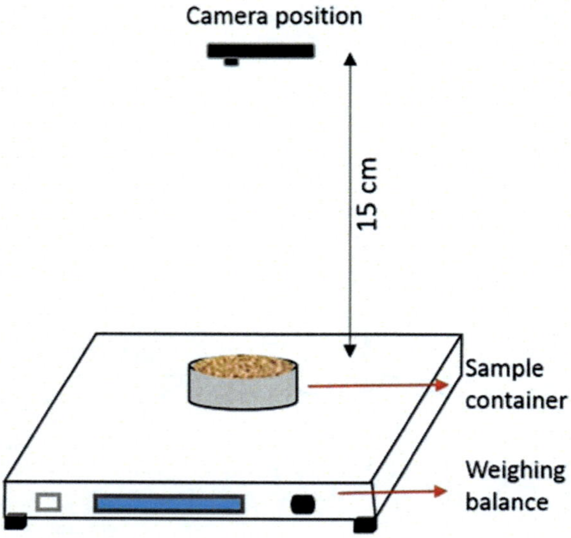


**Fig. 1** Experimental setup for radial shrinkage measurement

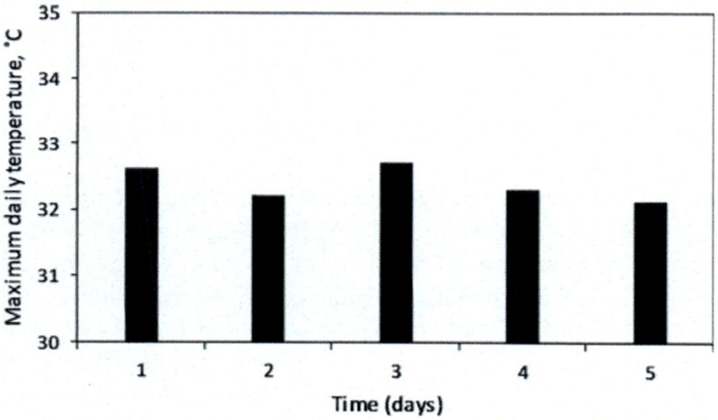


**Fig. 2** Maximum daily variation of temperature

the weight change was negligible, the sample was placed in the oven maintained at 60 °C for 24 h. The final weight measurement and dimensional change were recorded after oven drying the sample.

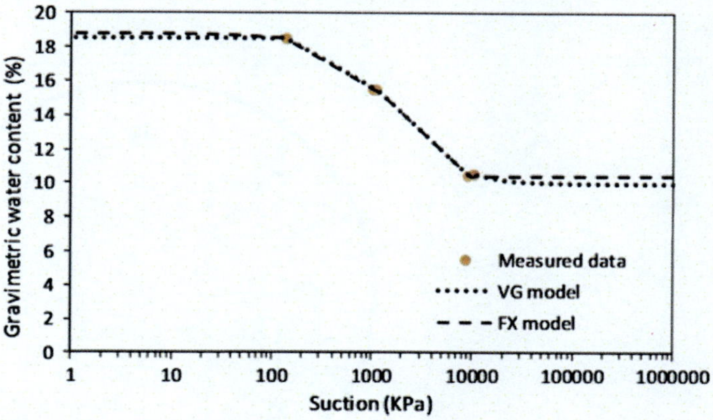


**Fig. 3** Comparison of soil–water retention behavior of silty clay during drying

# 3 Results and Discussions

## 3.1 *Soil–Water Characteristic Curve*

The soil–water retention behavior of the statically compacted soil specimen along the drying phase was fitted using the Van Genuchten model [23] to predict the suction potential up to 1000 MPa (Mega Pascal). The fitting parameters were determined using the SWRCfit software developed by [24], and the unimodal soil–water retention behavior was compared with Fredlund and Xing (FX) model [22]. There was negligible difference in the prediction capability of both the models, and based on the prediction model adopted for shrinkage characterization, Van Genuchten (VG) model was chosen as the appropriate prediction model for the experimental result (Fig. 3).

## 3.2 *Soil Shrinkage Curve*

The shrinkage curve was plotted for the soil prepared at an initial moisture content of 1.5 times the liquid limit. The volume change of sample subjected to air-drying was determined by measuring the dimensions at regular intervals of time until there was negligible change in the weight and the dimension. The weight change was used for the computation of moisture content, and the dimensional change was determined by two approaches. The change in height of sample in saturated state was computed using small pointed pins (Fig. 4), and with further drying, the sample height and diameter were measured using thickness gauge and Vernier caliper, respectively. The sample height was averaged for computing the volume change during the entire process of desaturation. There was a simultaneous capture of digital still images of

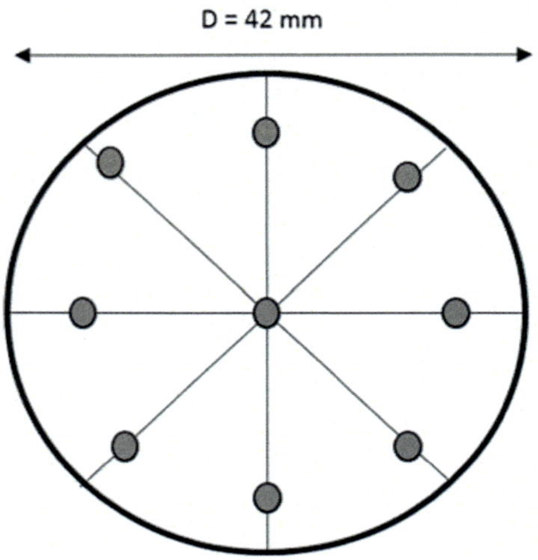


**Fig. 4** Points for thickness measurement in saturated sample

samples, and the diametrical change was computed using the image analysis with a simple and user-friendly open-source 'ImageJ' software.

The image processing of the sample was carried out in four steps as shown in Fig. 5.

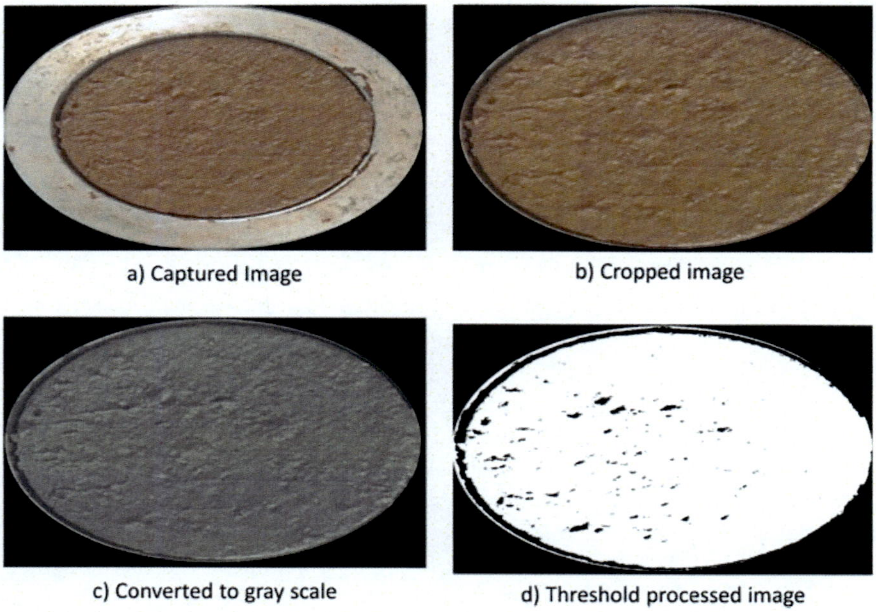


**Fig. 5** Process of image analysis for volume measurement

The image captured was converted into an *8-bit gray scale* image, and then, the *set scale* option was provided with the original dimension of the sample container. This value is the initial size of the sample [19, 21]. The image was further cropped and analyzed to *threshold* the gray scale image. This threshold value will help in demarcating the difference between sample and its background. The threshold process will convert the 8-bit image into a binary image such that the black background and white foreground are visualized. The binary image is then measured using the *measure* option in *analyze tool* to determine its area at various fixed points, and this value is used for volume computation. The volume measured with this image processing is compared with the manual measurement value, and the results show a variation of about 3%. This variation seems to be very less, this is mainly due to the small size of the sample, and the extreme care followed to avoid loss of sample. The testing of large-sized samples will result in high variation due to the chances of occurrence of cracks accompanied with subsidence. The results are obvious that the proposed 2D image processing technique is reliable and provides more accurate result determination compared to the direct measurement practice, especially when the soil is tested in a completely saturated or slurry state. The radial deformation measured with Vernier caliper was higher than that measured from image analysis, and this expected to be the result of loss in sample when handling it for measuring manually. Figure 6 shows the shrinkage curve of the silty clay indicating the presence of structural, proportional and residual shrinkage zone.

The shrinkage curve model proposed by [7] fits the experimental data well, and the entire curve seems to be an inverse of the soil–water characteristic curve. The desaturation of water from the intra-aggregate pores is quantified from the structural shrinkage phase, proportional shrinkage is dominant and followed by the evaporation of water from the inter-aggregate pores, and this represents the entry of air into the soil pore system [25]. The shrinkage in each zone is quantified to be 19.4%, 72.9% and 7.7%, respectively, using the relationship proposed by [7]. Figure 7 shows the

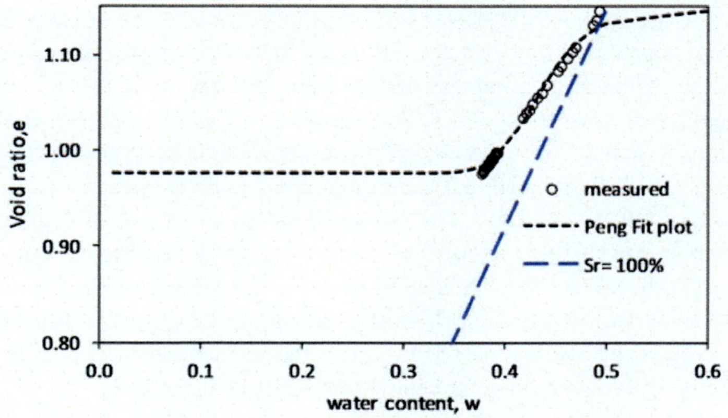


**Fig. 6** Soil shrinkage curve for fully saturated silty clay soil

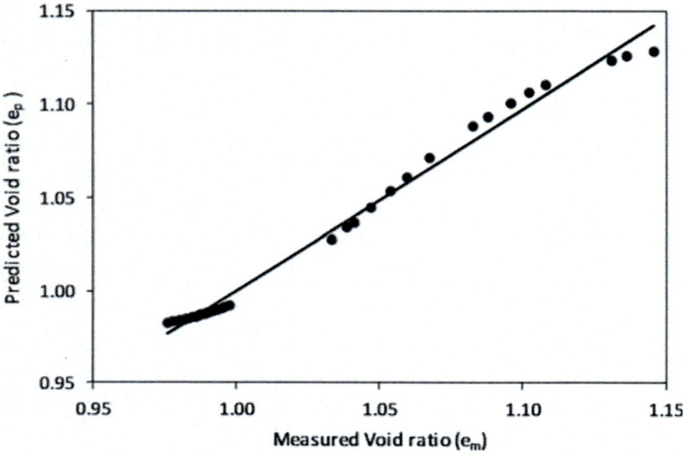


**Fig. 7** Comparison of measured and predicted void ratio

comparison of measured and the predicted void ratio obtained from image analysis results. The line of equity for data points is in high correlation.

## 4 Conclusions

This paper proposes an efficient, inexpensive non-contact image analysis procedure for quantifying soil shrinkage. This procedure can be used to characterize the shrinkage behavior of plastic soils subjected to moisture flux by evaporative dehydration. The simple comparative results show that the image analysis process to compute the dimensional change during shrinkage is advantageous over the conventional practice which usually results in loss of sample crumbs due to manual handling. The single mathematical model proposed by [7] proves to fit the data computed by image analysis with a high goodness-of-fit depicting the water loss from both intra- and inter-aggregate pores of the soil. The characteristic of shrinkage exists in three zones such as structural, proportional and residual. The dominant shrinkage exists for about 79% in the proportional zone indicating that the volume of water lost is equal to the volume of voids. The independent measurement of horizontal and vertical shrinkage is useful in quantifying the anisotropy of contraction due to desiccation. This proposed method can be used for laboratory determination of shrinkage behavior of wide variety of soils which may or may not exhibit the four phases of shrinkage. The shrinkage characterization and the mechanism will help in explicit understanding of hydromechanical behavior of low plasticity soils.

## References


1. Amenuvor, A.C., Li, G., Wu, J., Hou, Y. Chen, W.: An image-based method for quick measurement of the soil shrinkage characteristics curve of soil slurry. Geoderma **363**, 114165 (2020)
2. Haines, W.B.: The volume-changes associated with variations of water content in soil. J. Agric. Sci. **13**, 296–310 (1923)
3. Strik, G.B.: Some aspects of soil shrinkage and the effect of cracking upon water entry into the soil. Aust. J. Soil Res. **15**, 275–277 (1954)
4. Fredlund, M.D., Wilson, G.W., Fredlund, D.G.: Representation and estimation of the shrinkage curve. Unsatuarted Soils 145–149 (2002)
5. Tempany, H.A.: The shrinkage of soils. J. Agric. Sci. **8**, 312–330 (1917)
6. Johnston, J.R.: A study of the shrinking and swelling properties of rendzina soils. Soil Sci. Soc. Am. J. 1–6 (1944)
7. Peng, X., Horn, R.: Modeling soil shrinkage curve across a wide range of soil types. Soil Sci. Soc. Am. J. **69**, 584–592 (2005)
8. Peng, X., Horn, R., Peth, S., Smucker, A.: Quantification of soil shrinkage in 2D by digital image processing of soil surface. Soil Tillage Res. **91**, 173–180 (2006)
9. Lu, N., Dong, Y.: Correlation between soil-shrinkage curve and water-retention characteristics. J. Geotech. Geoenviron. Eng. **143**, 1–11 (2017)
10. Akin, I.D., Likos, W.J.: Suction stress of clay over a wide range of saturation. Geotech. Geol. Eng. **38**, 283–296 (2020)
11. Delage, P., Audiguier, M., Cui, Y.J., Howat, M.D.: Microstructure of a compacted silt. Can. Geotech. J. **33**, 150–158 (1996)
12. Lu, N., Likos, W.J.: Suction stress characteristic curve for unsaturated soil. J. Geotech. Geoenviron. Eng. **132**, 591–602 (2006)
13. Williams, D.J., & Sibley, J.W.: The Behavior at the shrinkage limit of clay undergoing drying. Geotech. Test. J. **15**, 217–222 (1992)
14. Bronswijk, J.J.B.: Relation between vertical soil movements and water-content changes in cracking clays. Soil Sci. Soc. Am. J. **55**, 1220–1226 (1991)
15. Li, L., Zhang, X., Li, P.: Evaluating a new method for simultaneous measurement of soil water retention and shrinkage curves. Acta Geotech. **14**, 1021–1035 (2019)
16. Jain, S., Wang, Y.H., Fredlund, D.G.: Non-contact sensing system to measure specimen volume during shrinkage test. Geotech. Test. J. **38**, 936–949 (2015)
17. Puppala, A.J., Katha, B., Hoyos, L.R.: Volumetric shrinkage strain measurements in expansive soils using digital imaging technology. Geotech. Test. J. **27**, 547–556 (2004)
18. White, D., Take, A., Bolton, M.: Measuring soil deformation in geotechnical models using digital images and PIV analysis. In: 10th International Conference on Computer Methods and Advances in Geomechanics, pp. 997–1002 (2001)
19. Julina, M., Thyagaraj, T.: Determination of volumetric shrinkage of an expansive soil using digital camera images. Int. J. Geotech. Eng. **6362**, 1–9 (2018)
20. Bowman, A.J., Haigh, S.K.: Subsurface deformation mechanisms beneath a flexible pavement using image correlation. Geotechnique **69**, 627–637 (2019)
21. Basson, M.S., Ayothiraman, R.: Effect of human hair fiber reinforcement on shrinkage cracking potential of expansive clay. Bull. Eng. Geol. Environ. **79**, 2159–2168 (2020)
22. Fredlund, D.G., Xing, A.: Equations for the soil-water characteristic curve. Can. Geotech. J. **31**, 521–532 (1994)
23. van Genuchten, M.T.: A closed-form equation for predicting the hydraulic conductivity of unsaturated soils. Soil Sci. Soc. Am. J. **44**, 892–898 (1980)
24. Seki, K.: SWRC fit—a nonlinear fitting program with a water retention curve for soils having unimodal and bimodal pore structure, pp. 407–437 (2007)
25. Tan, Y., Zhang, H., Wang, Y.: Evaporation and shrinkage processes of compacted bentonite-sand mixtures. Soils Found. (2020). https://doi.org/10.1016/j.sandf.2020.03.008

# An Approach for Geotechnical Site Characterization of Brown Field Site of a Steel Plant

Manos De and Shuvranshu Kumar Rout

## 1 Introduction

The drive towards sustainable development in production and manufacturing industries has seen many new concepts in recent times. One of the major challenges to this development has been the availability of land for setting up new plants or retrofitting units that improve the process efficiency and environmental compliance of existing plant units. Availability of new land parcels for the new units have become difficult especially for century-old establishments, further compounded by the growth of densely populated dependent townships that have developed around the industry. Developers have thus turned towards brown field development as opposed to greenfield development for setting up new process units.

The development of brown field sites, though a feasible solution for land availability problem is fraught with other handicaps that require a different type of approach to both engineering design and construction. Greenfield site developments are free of many of the additional challenges experienced in brown field situations especially with reference to the amount of initial exploratory work that can be carried out to characterize the site, particularly the subsoil profile. Since there are no site hindrances, the degree of detailing in investigation depends purely on engineering considerations and a comparative study of the benefits of design optimization versus cost of additional testing. In brown field developments, however, additional constraints are imposed due to lack of clear documentation on existing underground obstructions from past facilities, multiple utility services existing in the area in live

M. De (✉) · S. K. Rout
Tata Consulting Engineers Limited, Jamshedpur, Jharkhand, India
e-mail: mde@tce.co.in

S. K. Rout
e-mail: skrout@tce.co.in



C. N. V. Satyanarayana Reddy et al. (eds.), *Ground Characterization and Foundations*, Lecture Notes in Civil Engineering 167,
https://doi.org/10.1007/978-981-16-3383-6_8

condition, and restricted access to the site for use of exploratory equipment to investigate subsoil conditions. Further, the exploration work had to be coordinated along with the plant operation fully complying with the process and safety requirements of the running plant and its scheduled maintenance program. These challenges call for different approach towards engineering design and execution in brown field sites and are most critical for those parts of the construction that are not directly visible, namely the foundations and substructures.

The project under study in this discussion consists of a new raw material processing plant in an integrated steel plant complex that has been in operation for more than a century and has undergone various stages of development and refurbishment. To increase the plant production, a new plant unit to process one of the basic raw material inputs to the iron-making process was proposed to be constructed within the existing plant area. This area was previously occupied by other manufacturing units which had been taken out of service about 15–20 years back but many of the underground constructions and utility services were in place. There were also some utility facilities in operation in part of the area which had to be mitigated through relocation.

The total area earmarked for the new process plant units is about 45 acres and is spread in two distinct zones, separated by distance of about 500 m and connected by a pipe and cable gallery corridor. Various types of buildings, structures, and equipment foundations were proposed by the process designer. These included main process plant structures, static equipment foundations including tall towers, large storage tanks, material handling system structures, moving equipment foundations, rotary equipment foundations, pipe and cable carrying system structures, service, utility, and office buildings and facilities for maintenance of equipment and storage of spare parts. The type of foundation, sensitivity to settlement and expected bearing pressures for the various types of structures were also specified by the design engineering team at the start of the project and this formed one of the bases for identifying the need for the geotechnical investigation program, together with the expected variation in subsurface conditions across the extent of the project site.

## 2 The Triad Concept

The triad concept was developed by the Environmental Protection Agency (EPA) of USA [1] in dealing with clean-up programs of heavily contaminated and hazardous sites. The basic concept of dealing with the uncertainty is presented in Fig. 1 highlighting the three-prong approach for the decision-making process through (1) systematic project planning, (2) dynamic work plan strategies, and (3) the use of real-time measurement technologies. The triad approach is particularly suitable for brown field project applications since the uncertainties in such sites especially in the subsurface conditions pose serious challenges to the engineering design teams. Application of the triad approach helps in the identification and management of the uncertainties that could have resulted in large errors in decision making. The three

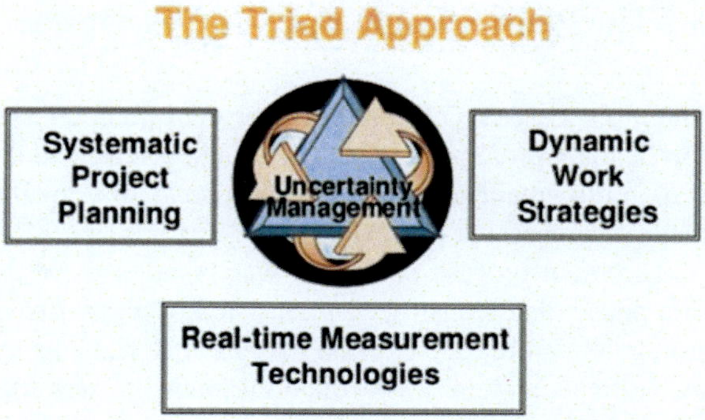


**Fig. 1** Triad approach

elements of the triad approach in the context of geotechnical site characterization are briefly described below.

The systematic project planning element of triad approach involves identification of the variables and factors that can affect the design and project execution activities. This helps in devising cost-effective strategies to anticipate the probable adverse factors and manage them using previous knowledge of the site. The management strategy may consider aspects like level of previous knowledge, budget restrictions, project schedule, resource availability, and regulatory requirements. Common sense approach is used to process this gathered information to arrive at acceptable decisions and the associated uncertainties. The conceptual site model (CSM) is a tool to help organize the available information and identify the need for additional information through site investigation. This plan is dynamic and gets modified to reflect the level and quality of information gathered and the lowering of the associated uncertainties.

The second element of triad planning, a dynamic work plan strategy involves making real-time decision-making in field. This helps in avoiding repeating field investigations to fill gaps in data thus helping to reduce project schedule and cost increases. The field investigation plan is continuously modified during its execution through decision logic updated with accumulated data that is used to collect further information to fill in gaps.

The third element, real-time measurement, is employed in quick time frame to enable real-time decision making and real-time maturation of the CSM. It involves "out-of-box" thinking, deployment of supplementary in situ tests yielding quick and relevant results, and use of software tools for interpretation and mapping of the data for enabling dynamic work plan strategies. Collaborative data that complements the results of the main investigation help provide more detailed image of the site in quicker time. This strategy helps save time and money and limit uncertainty in decision making and the decision achieved by analytical models can be supplemented with further rigorous testing as and when required.

## 3 The Plant Layout and Site Description

The project area identified for the process plant consisted of two land parcels—one for the main process plant of area about 38 acres and the subsidiary gas cleaning and water utility plant area unit of about 7 acres located about 500 m away from the main material processing plant (Fig. 2).

The main process plant unit was planned to be developed in two phases. The first phase plant construction consisted of a raw material conversion unit, material storage structure, flue exhaust system, material conveying system, tracked vehicles for bulk raw and processed material transport, wet cooling system for hot products, dust extraction system structures, various static and rotating equipment foundations, pipe galleries, cable racks, process chemical and waste product storage tanks, storage structures, electrical and control building, and offices along with other auxiliary buildings. The anticipated foundation types for the various types of structures were specified by the designer as part of advance information for proper planning of the subsoil investigation work. In addition, some preliminary information from reconnaissance exploration in the nearby area for relocation of the existing facilities had already been carried out earlier which provided the baseline for the subsurface formation and geology of the area. This was also supplemented by reports from other projects executed in the vicinity in earlier times.

The second phase of the plant involved installation of a dry cooling system for recovery of sensible heat to generate stem both for use in the plant processes and for power generation. The closed dry system would also help establish better control of dust pollution generated during the handling of the product.

The auxiliary unit (Fig. 3) consisted of many processes and chemical plant units for cleaning the exhaust combustion gas for reuse in the plant as fuel and associated water utility system structures. It is comprised of gas coolers, precipitator units, tall tower structures for scrubbing and distillation of gas and wash water, storage tanks,

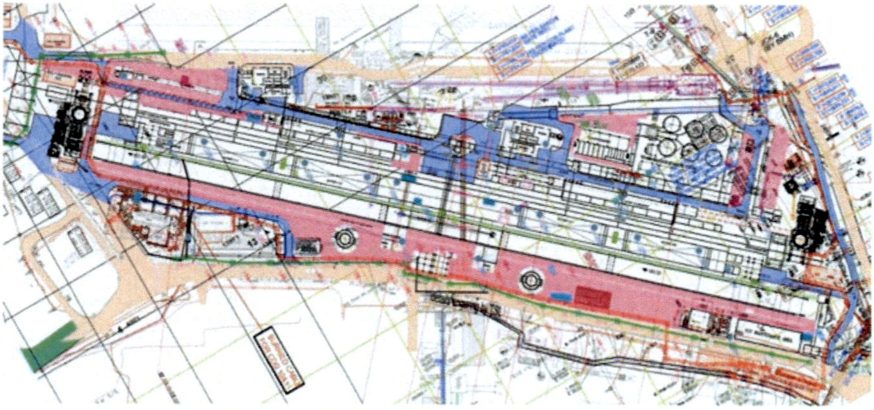

**Fig. 2** Main material processing plant unit—38 acres

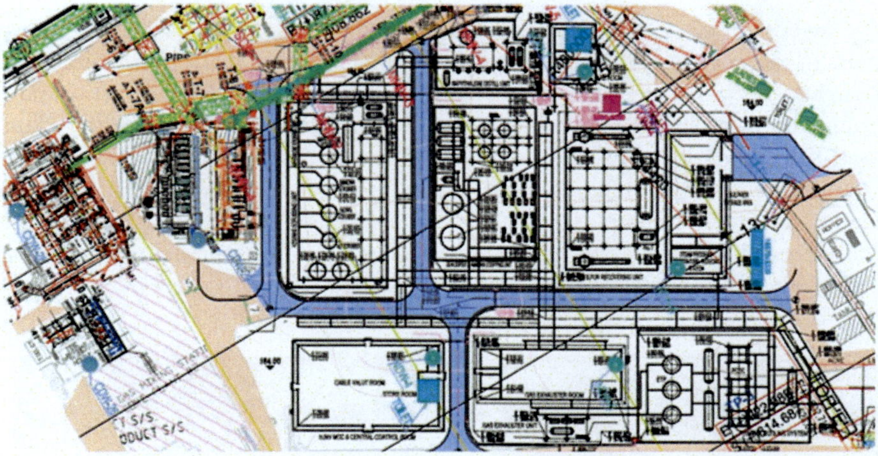

**Fig. 3** Auxiliary gas cleaning and water utility plant unit—7 acres

cooling tower, pump houses, gas compressors and exhauster units, other auxiliary units, and electrical and control building and offices. For these units also anticipated foundation requirements were indicated by the designer which was used along with preliminary subsoil information to form the baseline for further site investigation.

## 4 Application of Triad Concept for Geotechnical Investigation

The first element of the triad concept was utilized for the planning and execution of the geotechnical investigation work.

The systematic project planning stage involved identifying the parameters that would affect the design of the foundation systems and subsequent execution of the works. The important factors selected for the geotechnical characterization included the foundation design parameters specified by designer including structural framing, preferred foundation type and expected bottom of foundation level, expected bearing pressure and foundation size, settlement limits, and the sensitivity of the foundation and structure to differential settlement. Preliminary information on subsoil profiles in vicinity was also available from earlier exploration programs. This data was used to plan the layout and depth of boreholes for the investigation program at the plant site.

The data was used to form the baseline conceptual site model (CSM) for the project. The model consisted of a detailed layout of the site obtained through a total station survey on which the process plant layout was superimposed. The preliminary foundation sizes, levels, and bearing pressure values identified in SPP stage was superimposed on this layout to develop a foundation footprint of the structures. In the next step, the geotechnical properties to meet the foundation design parameters were assessed. The testing plan developed included subsoil stratification through borehole logging, in-situ tests like Standard Penetration Resistance, Plate Load Test, Soil Resistivity Test and laboratory properties of grain size distribution, Atterberg consistency limits, strength, and deformation characteristics. The investigation report [2] would provide foundation design data for the verification of soil bearing capacity/pile capacity and expected foundation settlement for the structural loading from SPP phase. The layout of the investigation plan is shown in Fig. 4 to

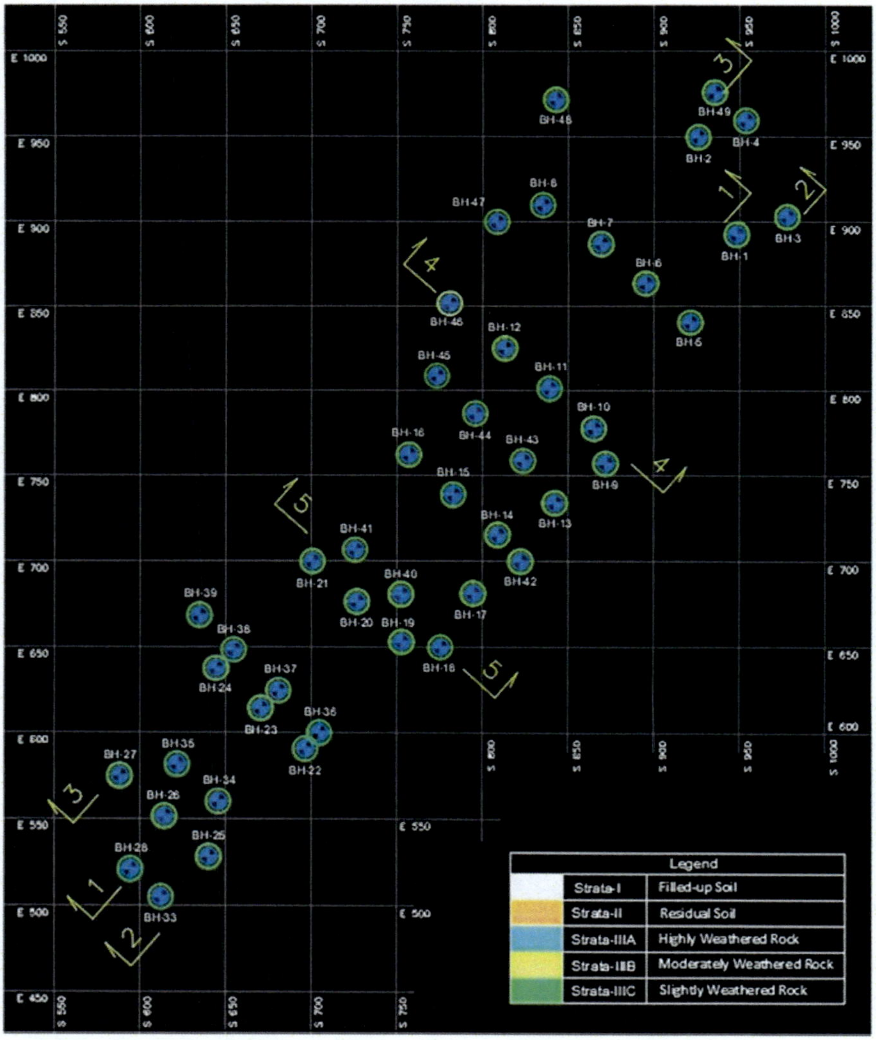


Fig. 4 Map of site showing ground level and rock profile across site in main plant area

cater primarily for the facilities in phase 1 development of the plant, but the test locations planned also covered the areas marked for phase 2 to minimize need for additional test points at later stage. The CSM considered the heightened risk and cost of subsequent investigation for phase 2 while the phase 1 plant would be in operation and the investigation locations were planned accordingly.

In the initial investigation plan, the boreholes were planned based on the guidelines of Indian Standard IS 1892 [3]. Total of 32 boreholes were planned for the site with 28 locations in the main process plant area and 4 locations in the auxiliary plant area. Borehole spacing was adopted at generally 50 m intervals based on previous experience of uniform geologic formation of the site. The depth of exploration was decided to be 30 m for 6 boreholes under critical structures, for other boreholes, termination criteria was set at 3 m inside mica-schist rock with core recovery of 50%. In addition, 5 plate load tests were carried out at the location of the foundations of major structures of the proposed plant using 60 cm square plates at depths of 1.15–2.15 m below ground level, with the procedure conforming to IS 1888 [4] for in-situ verification of bearing capacity of founding soil. Maximum load applied on the plates was 600 kPa and the settlement calculated on the plate varied from 4.3 to 16.0 mm corresponding to safe bearing capacity (SBC) of 300 kPa, with a factor of safety of 2.0 on the applied load. For a prototype footing of size 3.0 $m^2$, the expected settlement under this SBC worked out in the range from 8.0 mm to 80.0 m.

## 5 Applying Other Elements of Triad Concept for Site Characterization

The balance two elements of the triad concept were utilized for modification of the investigation program to meet the requirements of the foundation design and execution works.

Reports of the geotechnical investigation work as per the initial plan were provided as foundation design input to the design team. The soil stratification revealed from this investigation is presented in Table 1.

The geologic formation at the site was typified by metamorphosed mica schist rock. The rock texture was generally foliated and platy having fine to medium-grained deposits. The mineralogy of the rocks showed presence of mostly mica minerals (biotite, chlorite, and muscovite) with occasional deposits of quartz and feldspar. Study of the borehole logs showed that the site had overburden of soil of mainly cohesionless nature, followed by mica-schist rock in varying degrees of weathering occurring at relatively shallow depths.

The recommendation of the report for the bearing capacity at the designated depth and the expected bearing pressure from the structure are summarized in the following Table 2.

The analysis of data from the initial set of investigation showed that for the some of the major plant units, the expected foundation bearing pressures would exceed the

**Table 1** Geotechnical parameters for design

| Strata | Description | Depth (m) below GL | Unit weight (kN/m$^3$) | Cohesion (kN/m$^2$) | Internal friction angle (deg) |
|---|---|---|---|---|---|
| I | Filled up soil | 0.4–2.9 | 18 | 0.5 | 28 |
| II | Residual soil | 1.4–6.6 | 19 | 5 | 30 |
| III A | Highly weathered rock | 0.3–7.9 | 22 | 25 | 35 |
| III B | Moderately weathered rock | 0.75–20.5 | 23 | 35 | 38 |
| III C | Slightly weathered rock | Max 22.6 | 25 | 50 | 40 |

**Table 2** Foundation bearing capacity

| Foundation for structure of | Fdn type | Size (m) | Depth (m) | Settlement sensitivity | Expected load (kPa) | Safe bearing capacity (kPa) |
|---|---|---|---|---|---|---|
| Main process plant | Raft | 25 × 70 | −3.0 | High | 300 | 60 |
| Raw material car | Strip | 3 | −3.0 | High | 200 | 175 |
| Waste flue exhaust | Raft | ϕ 22.5 | −7.0 | High | 250 | 125 |
| Product cooling tower | Isolated | 5 × 6 | −3.5 | High | 200 | 200 |
| Raw material bunker | Isolated | 10 × 12 | −6.0 | High | 350 | 150 |

bearing capacity computed for a total allowable settlement of 20 mm. Discussion with the design team resulted in considering option of placing some of the critical and heavy loaded structures on piled foundations.

Further analysis of the bore logs revealed that in many of the boreholes across the site, weathered rock would be available at quite shallow depths. It was also understood that the suggested bottom of foundation levels was not only guided by geotechnical parameters but also due to functional requirements of the plant units with the presence of cellars and basements accommodating process equipment. The piles that could be reasonably installed from the required bottom of foundation (pile cut-off level) to the termination depth by socketing into rock would be quite short and the short piles would not provide required lateral resistance. Thus, though the piles would be effective in carrying the vertical loads from superstructure loads with desired control of settlement, they would be inadequate in resisting lateral shear from the structure loads.

To eliminate change in design of foundation a different construction technique was planned. The excavation depth was extended beyond design bottom of foundation level till suitable foundation medium corresponding to suitable match of bearing capacity and generated foundation bearing pressure shown in Table 3 was reached. The bearing capacity of the various layers were computed using standards procedures in IS 6403/IS 8009 for foundations in soil and IS 12070 for foundations in rock. The depth between the planned bottom of foundation and the extra depth till suitable founding stratum was found at greater depth was to be filled with mass concrete filling. Detailed study on this proposal was carried out using the information from the soil investigation report, experience of foundation construction in adjacent areas, and techno-economic feasibility.

The second element of triad planning, a dynamic work plan strategy, was thus brought into the picture by changing the initial foundation design and adopting a strategy based on technical merit and project parameters of cost and schedule. The modified foundation design approach was deliberated in detail and agreed to by all stakeholders in the project including the plant owner, the design team, and the execution contractor as the most viable option for the project. The bearing capacity of the various strata revealed in the soil investigation program is shown in Table 3.

Study of the rock profile across the boreholes and the recommended bearing capacity in various strata confirmed that the highly weathered rock of Stratum III A would be able to bear the loads from the superstructure by transferring the loads from the desired bottom of foundation through the intermediate rigid layer of mass concrete fill. This option was analyzed in detail including the cost and time involvement and compared with these parameters of piled foundations.

It was found that the option of removal of the residual soil of stratum II and providing PCC fill from bottom of foundation till the competent bearing layer was reached in weathered rock of stratum III A was an optimum design. Though on

**Table 3** Bearing capacity of open foundations in different strata

| Strata | Description | Foundation type | Depth[a] in layer (m) | Bearing capacity (net, kPa) | Settlement (mm) |
|---|---|---|---|---|---|
| I | Filled up soil | – | – | – | – |
| II | Residual soil | Isolated (B < 5 m) | 2.0 | 200 | 25 |
| III A | Highly weathered rock | Isolated (B < 5 m) | 1.0 | 350 | 25 |
| | | Raft (B < 25 m) | 1.0 | 350 | 50 |
| III B | Moderately weathered rock | Rock bearing[b] | – | 400 | – |
| III C | Slightly weathered rock | – | – | – | – |

[a]The depth of foundation is measured from the top of the layer/strata
[b]Gross bearing on rock calculated as per IS 12070 [5]

purely cost basis, the relatively short piles designed for vertical loads would be less costly than the total cost of removal of the soil and replacement by mass concrete fill there would be significant gain in time for foundation construction. Piling work would involve specialized operations of drilling, pile casting, waiting for 28 days for concrete to gain strength before pile head dismantling before the foundation raft/ pile cap could be constructed. Piles would also have to be load tested—both initial load test and working pile load test would have to be carried out. The alternate foundation proposal involved less specialized and fewer number of activities. The differential cost of construction would be more than offset by the gain in project scheduled completion by about 2 months and the early production from plant after completion would give a much larger beneficial economic impact to the owner.

To implement this effectively, it was necessary to supplement the data from the first phase of soil investigation to get more closely mapped soil stratification data. The weathered rock profile of stratum III A which was found to be adequate for supporting the structure loads needed to be characterized across the site more accurately.

The third element of the triad approach, real-time measurement, was now brought to bear. Along with the initiation of foundation construction activity, additional boreholes were planned at the site—18 boreholes in the main plant area and 5 in the auxiliary plant area. The bore logs were used to map the rock formation in closer grid to trace the depth of the weathered rock layer across the site under the various structures and determine the thickness of the concrete fill required under the foundations. Figures 5, 6, 7 and 8 show the map of the site with the layout of the proposed

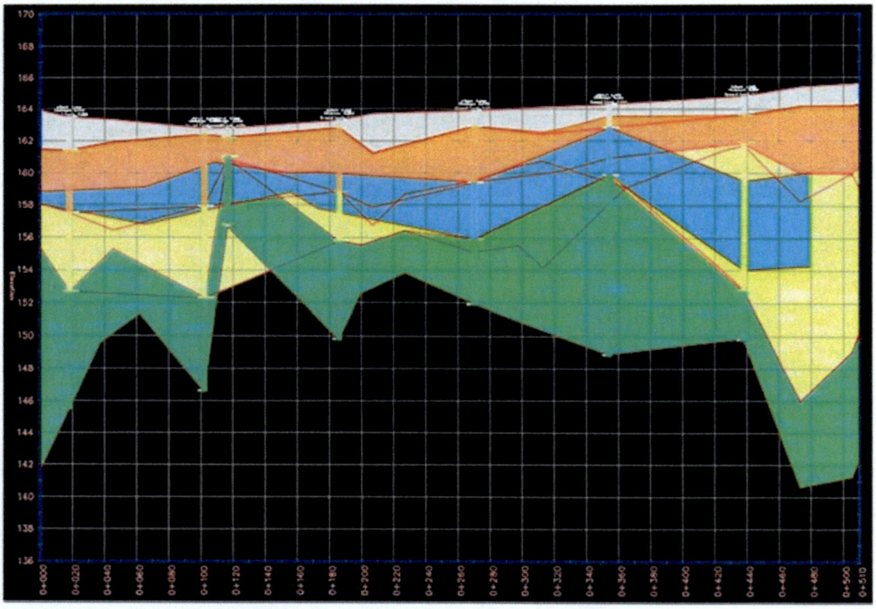


**Fig. 5** Section 1-1—Subsoil profile below main process plant foundation

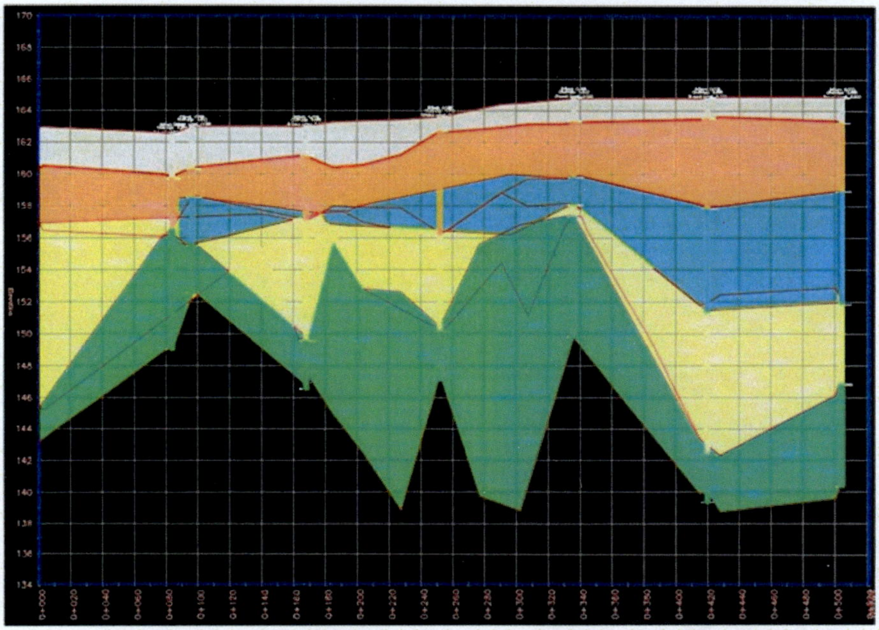


**Fig. 6** Section 2-2—Subsoil profile below foundations of raw material car, raw material bunker, and waste flue exhaust

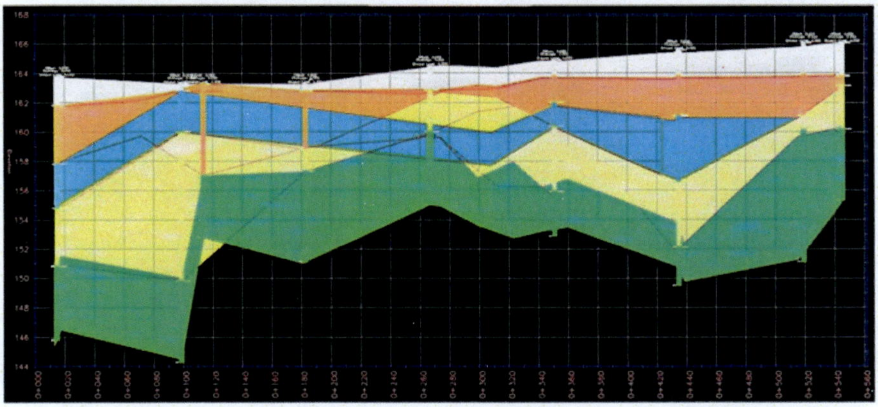


**Fig. 7** Section 3-3—Subsoil profile below product cooling tower and product transfer car track foundation

plant facilities on which is superimposed the layout of boreholes, ground level at the borehole location, and the elevation of top of stratum III A. This map was used to plan the required initial planned depth of cutting for removal of the residual soil of stratum II.

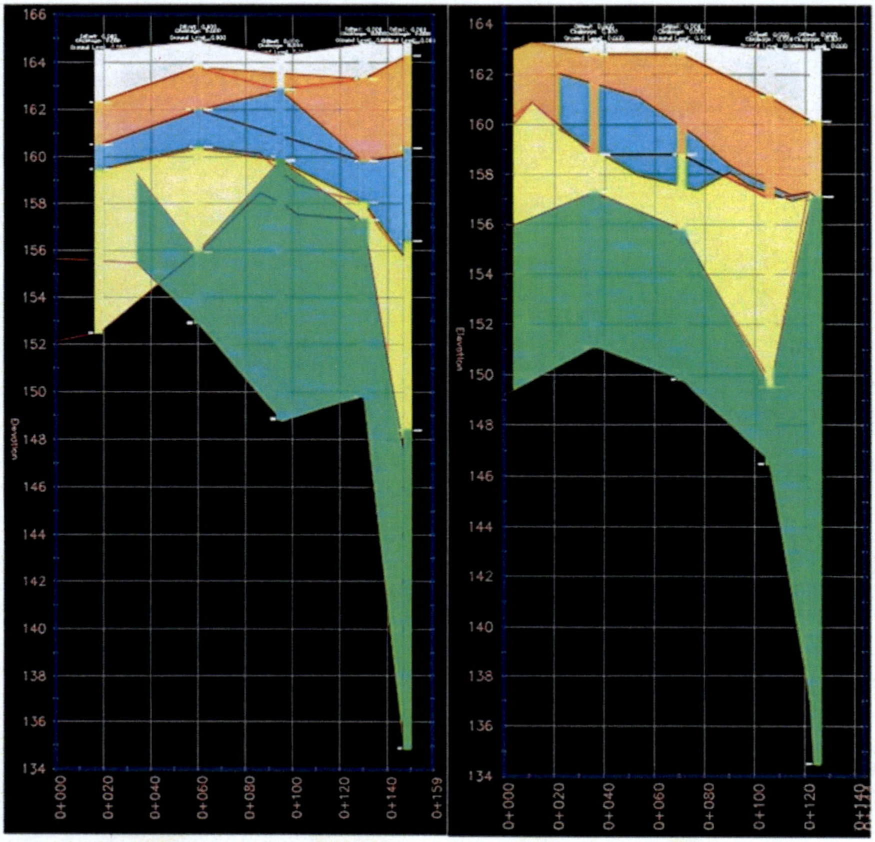


**Fig. 8** Section 4-4—Subsoil profile below product cooling tower and product transfer car track foundation

The RTM approach was deployed during the actual execution of the work at site. The site characterization map is shown in Fig. 9 had mapped the depth of occurrence of suitable foundation strata in terms of bearing capacity for the various structures identified in Tables 2 and 3. For successful execution of the designed foundation solution, the critical activity of correct identification of the stratum III A/III B was necessary during the excavation work at site. To effectively implement this requirement the geotechnical engineers on-site had to devise simple tests for identification of the excavation having reached the weathered rock level before certifying its adequacy for foundation construction. These tests and the acceptable outcome were documented in the form of standard operating procedure (SOP)/site instruction (SI) for the quality assurance activity by the team of geotechnical engineers. The simple tests devised comprised three approaches. The first was a visual inspection of the excavation bottom and comparing it to a pictorial representation of correctly identified formation previously marked by experienced geotechnical engineers. The

Strata - I: Filled-up Soil at Top with all Sub-soil Profiles

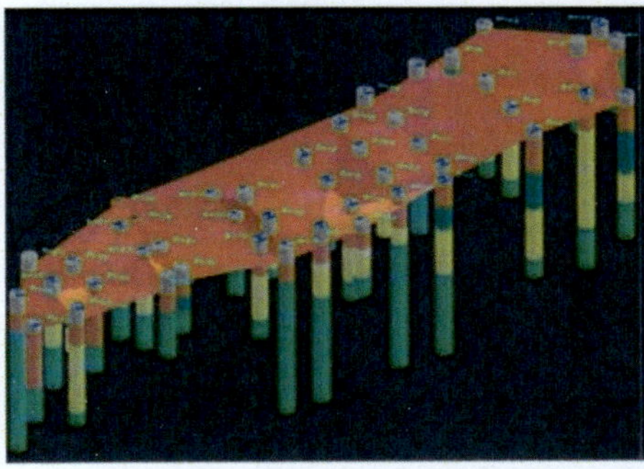

Strata -II : Residual Soil

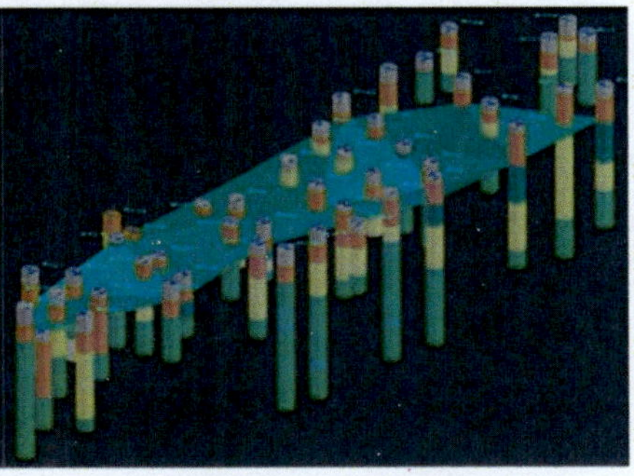

Strata -IIIA: Highly Weathered Rock

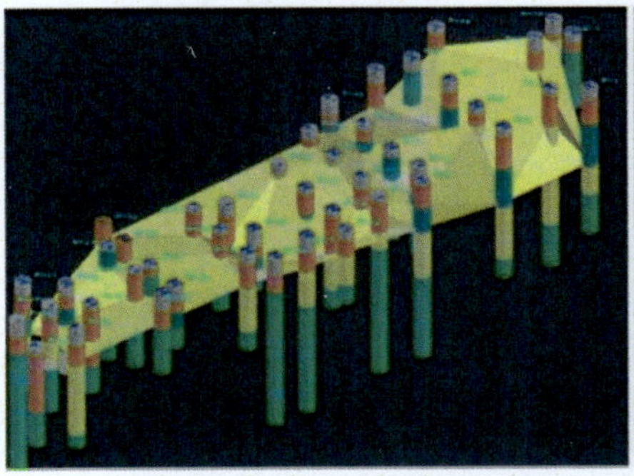

Strata -III B : Moderately Weathered Rock

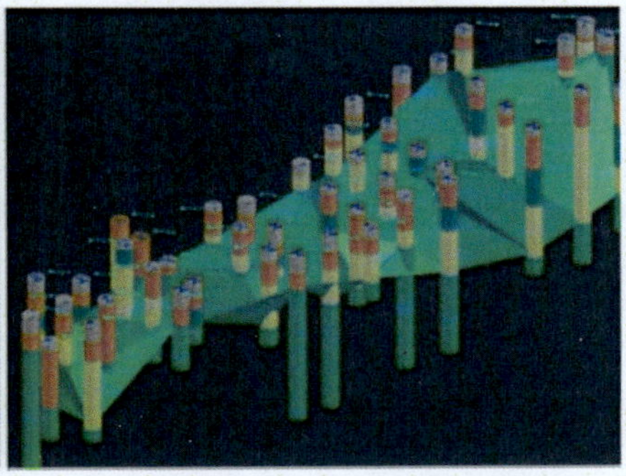

Strata -III C : Slightly Weathered Rock

**Fig. 9** Map of site showing ground level and rock profile across site in main plant area

second test used at site consisted of striking the excavated base layer with a Geological Hammer to check if it formed indentation on the surface. The third test comprised using the teeth of the excavator to scratch the exposed base layer and noticing when the teeth were unable to leave scratch marks on the surface with full hydraulic pressure applied on the cutting tool. The three-step quality check would ensure that the

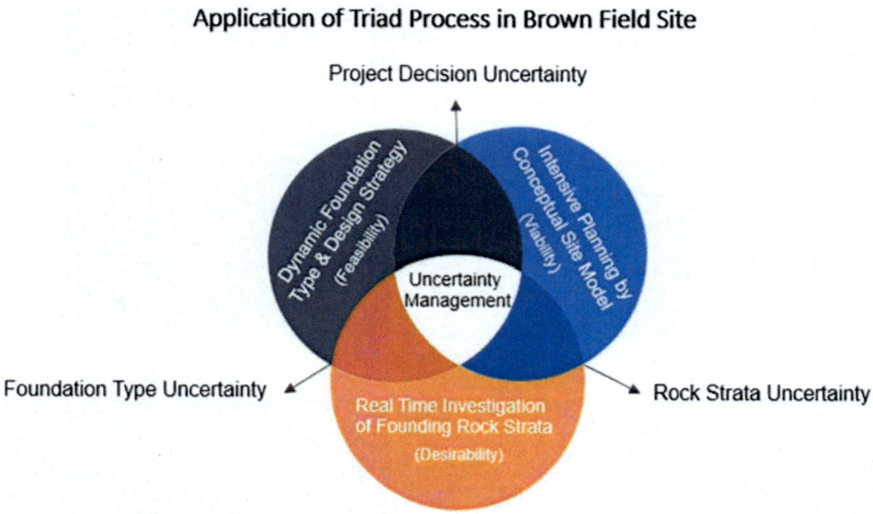


**Fig. 10** TRIAD application parameters for site characterization of brown field site

excavation had reached the desired founding strata and all records of such quality check were maintained for verification.

## 6 Challenges and Mitigations

Despite applying the triad principles for geotechnical characterization [6] of this complex site and project requirements, there were other challenges to be considered during the execution stage. This required continuing with the inter-play between the three tools of the triad approach shown in Fig. 9 for the construction stage activities together with involvement of the design team. An important consideration of great concern with mica-schist rock is that the strength of this type of material decreases rapidly on exposure to atmosphere or when submerged under water. To address this potential risk of degradation of founding medium, the SOP/SI addressed this issue also. Excavation work would be carried out in parts and not over large expanses so that the exposed rock surface is not left unprotected for long time. When the excavation in a zone would reach the mica-schist layer of stratum III A as confirmed by the QA/QC engineer, the base would be leveled off and covered with lean concrete layer of minimum 150 mm thickness within a time gap of not exceeding 2 h. Excavation in other parts would proceed only after the sealing of the rock surface was completed in the exposed zone.

## 7 Conclusions

The execution of the project at this brown field site threw up many complexities due to challenges posed by difficult site. The site location was amidst existing operating plant. The bearing pressure requirement for the different types of structures were widely varying. There was variation in soil layering even within the relatively small zone. There were also project requirements of execution within schedule and budget while addressing safety and quality aspects. To execute the plan for foundation design and construction successfully, proper identification of suitable subsoil stratum and using its geotechnical properties for each specific structure was of great importance. The successful application of the TRIAD approach for subsoil characterization [7] of a brown field site resulted in the following benefits:

1. Development of geotechnical site characterization in brown field site for the process plant
2. Enabled selection of suitable foundation stratum with adequate bearing capacity corresponding to specified superstructure loading and settlement sensitivity of the structure
3. Formulate suitable strategy to deploy verification of adequate founding layer for the structures in varying subsoil conditions
4. Clear presentation of information to all stakeholders to facilitate smooth and consistent decision-making regarding foundation placement
5. Facilitate redeployment of strategies through data generation, integration, and presentation in pictorial representation
6. Even in case of phase-wise plant construction, detailed subsoil characterization at initial stage can lead to significant cost and time saving on overall project from design to construction phases.

**Acknowledgements** Plant Owner (project Client) for engaging Tata Consulting Engineers (TCE) Limited for the detail engineering for the project.

Mr. Abhijit Banik, 3D Modeler, TCE, for creating the subsoil profile mapping from the borehole data that helped the decision-making process for foundation selection, design, and execution. Software used—AutoCAD Civil 3D Geotechnical Module.

## References

1. U.S. Environmental Protection Agency Office of Solid Waste and Emergency Response Brownfields Technology Support Center Washington, DC 20460, Using the Triad Approach to Streamline Brownfields Site Assessment and Cleanup—Brownfields Technology Primer Series (2003)
2. Report on Soil Investigation. S. K. Mitra & Associates, Slat Lake, Kolkata (2010)
3. IS 1892: Code of practice for subsurface investigation for foundations (First Revision). Bureau of Indian Standards, New Delhi (1979)

4. IS 1888: Method of load test on soils (Second Revision). Bureau of Indian Standards, New Delhi (1982)
5. IS 12070: Code of practice for design and construction of shallow foundations on rocks. Bureau of Indian Standards, New Delhi (1987)
6. U.S. Department of Transportation, Geotechnical Site Characterization, FHWA-NHI-16-072: NHI Course No. 132031 (2017)
7. Dr. P. Anbazhagan, Introduction to Site Characterization, NPTEL Course Module 1, 2 & 3, coordinated by IISc, Bangalore, https://nptel.ac.in/courses/105/108/105108078/, available from 2014–05–08

# Effect of Palm Fibres on Lime Blended Sandy Clay

**T. Athira and T. Sini**

## 1 Introduction

Major challenges faced by geotechnical engineers are developing economically feasible and durable methods of ground improvement. Soil stabilization refers to any physical, chemical or biological method, or any combination of such methods, that is employed to improve certain properties of a natural soil to make it adequately serve an intended engineering purpose over the service life of an engineering facility. Through soil stabilization, unbound materials can be stabilized with cementitious materials (cement, lime, fly ash, bitumen or combination of these). For a successful stabilization, laboratory tests followed by field tests may be required in order to determine the engineering and environmental properties. Many stabilizers have been used all over the world to improve the problematic soil in terms of its compressive strength. But in order to improve the tensile strength and enhance the ductility characteristics, fibres have been used nowadays. Both natural and synthetic fibres have been used for stabilization.

T. Athira (✉) · T. Sini
College of Engineering, Trivandrum, Kerala 695016, India

T. Sini
e-mail: sini@cet.ac.in



C. N. V. Satyanarayana Reddy et al. (eds.), *Ground Characterization and Foundations*,
Lecture Notes in Civil Engineering 167,
https://doi.org/10.1007/978-981-16-3383-6_9

**Table 1** Properties of soil

| Property | Value |
|---|---|
| Colour | Light grey |
| Natural water content (%) | 20 |
| Clay (%) | 48 |
| Silt (%) | 14 |
| Sand (%) | 38 |
| Liquid limit (%) | 49 |
| Plastic limit (%) | 25 |
| Plasticity index (%) | 24 |
| Specific gravity | 2.5 |
| MDD ($kN/m^3$) | 18.3 |
| OMC (%) | 15 |
| UCS ($kN/m^2$) | 28.49 |
| pH | 5.72 |

# 2 Experimental Studies

## *2.1 Materials Used*

The soil was collected from Thonnakkal region of Thiruvananthapuram district which belongs to sandy clay category. Laboratory tests were conducted to determine the properties of the soil such as specific gravity, liquid limit, plastic limit, shrinkage limit, unconfined compressive strength, optimum moisture content, maximum dry density and pH. Table 1 shows the engineering properties of the soil. Lime and palm fibres were used as additives.

## *2.2 Methodology*

The soil is first treated with different percentages of lime (5, 10, 12 and 15%) to determine the optimum lime content. With the optimum lime content as constant, the fibre content is varied in order to determine the optimum percentage of fibre by conducting unconfined compression test. The various percentages of fibres used for the treatment are 0.1, 0.2, 0.3 and 0.4%.

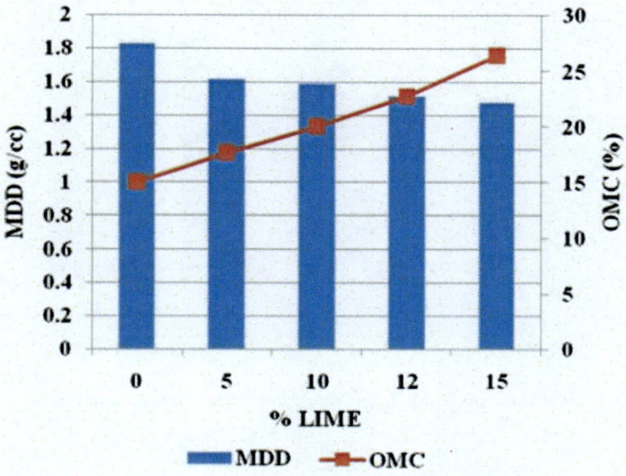


Fig. 1 Variation of MDD and OMC with different lime content

# 3 Results and Discussions

## *3.1 Effect of Lime on Sandy Clay*

**Standard proctor test.** Compaction test was conducted with various percentages of lime such as 5, 10, 12 and 15% to find out the optimum moisture content and maximum dry density. On increasing the proportion of the lime, the optimum moisture content increases, whereas maximum dry density decreases [1, 2]. The pozzolanic reaction between clay and lime is responsible for the increasing the OMC of the mixture. For the same reason, the MDD decreases with increasing the lime content (Fig. 1).

**Unconfined compression test**. The soil sample sieved through 425 μm was taken and was mixed with 0, 5, 10, 12 and 15% of lime and the unconfined compression test was conducted. The results obtained on the zero days were taken and the optimum percentage of lime was determined. From the Fig. 2, it can be inferred that up to 10%, compressive strength increases as the lime content increases. But with further increase in lime content, the compressive strength is found to decrease. Hence, the optimum lime content is taken as 10%. By comparing the UCS values of the sample cured for different time periods, it can be inferred that as the curing period increases the UCS, value also increases. Maximum compressive strength is obtained for the sample after 28 days of curing.

## *3.2 Effect of Lime and Palm Fibre on Sandy Clay*

**Standard proctor test**. Compaction test was conducted with optimum percentage of lime and varying percentage of palm fibre (0.1, 0.2, 0.3 and 0.4%). From Fig. 3,

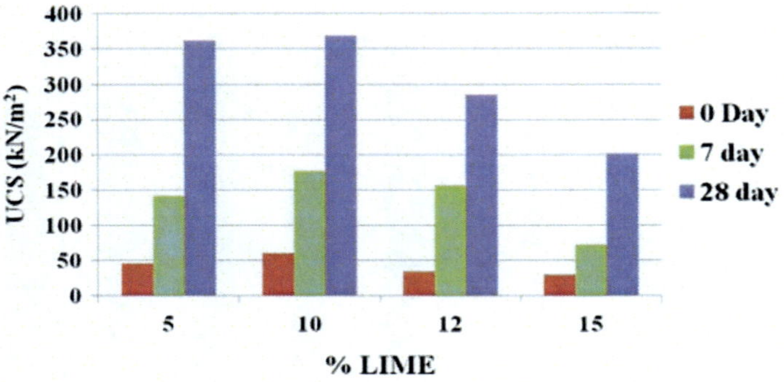


**Fig. 2** Variation of UCS for different lime content and curing period

**Fig. 3** Variation of MDD and OMC with different fibre content

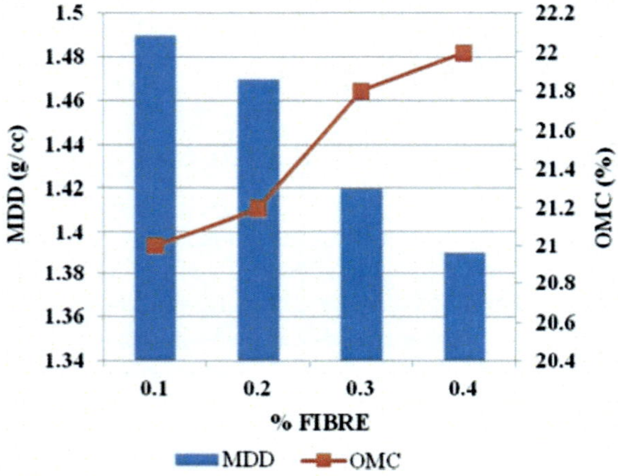


it can be inferred that for constant lime content, addition of nylon fibre decreases MDD whereas increases OMC [3]. Same trend was observed in the work done by M. Arabani and H. Haghsheno. According to them, by adding fibres to the soil–lime mixture, a portion of water is absorbed by the fibres and in order to reach the MDD, a larger amount of water is needed, leading to an increase in OMC. According to Kezdi and brown, the decrease in MDD is due to the growth of micropores which are formed due to the tendency of lime for water absorption and aggregation.

**Unconfined compression test**. The soil sample sieved through 425 μm was taken, and it was mixed with 10% of lime and varying percentage of palm fibre (0.1, 0.2, 0.3 and 0.4%) and the unconfined compression test was conducted. Two different arrangements of palm fibres were experimented, where the fibres were mixed randomly with the soil and also the fibres were mixed in horizontal layers with the soil. The results obtained after 0 and 7 days of curing were recorded.

The variation of UCS with time in the case of random mix is shown in Fig. 4. It can be inferred that as the percentage of palm fibre increases, the UCS increased up

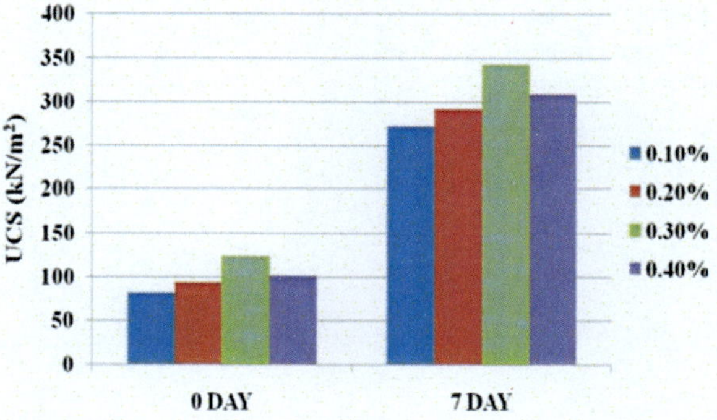


**Fig. 4** Variation of UCS in random mix

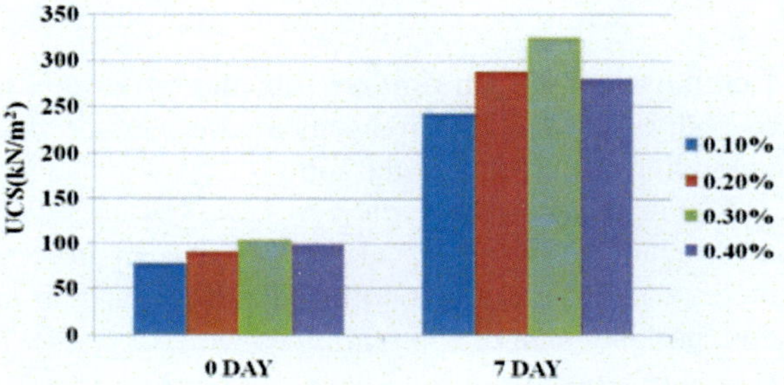


**Fig. 5** Variation of UCS in horizontal mix

to 0.3% and with further increase in fibre content, UCS decreased [4]. Also it can be seen that as the curing period increases, the UCS also increases. Maximum UCS is obtained after 7 days of curing.

In the case of horizontal mix, the fibres were placed in two layers, each layer at a depth of one-third the mould height. From Fig. 5, it can be inferred that the UCS increases with increase in fibre content up to 0.3%. The effectiveness is found to decrease with further addition of fibre [5]. In horizontal mix, also addition of palm fibre improves the compressive strength, but the amount of increase is less compared to that of random mix [6].

According to S. Dhar and M. Hussain [7], the cementitious products formed after lime treatment have rough surfaces and high rigidity, which bind the fibre–soil particles together and provides a compact matrix structure. This helps to increase the effective contact area and interlocking between fibre and modified soil particles, and thus, enables greater mobilization of friction between them with an increase in fibre content. After the peak value, further increment in fibre content forms lumps and adheres to each other; thus, there may be deficiency in the contact between soil and fibre which is responsible for reduction of friction coefficient and hence reduces the stress, and the effectiveness in improvement is reduced.

Because of the time-dependent pozzolanic reactions, the stabilization of the lime-treated soil is a long-term process [8]. Thus, the strength of the stabilized soil increases as the curing time increases.

Fig. 6 Variation of UCS with percentage of palm fibre after 7 day curing

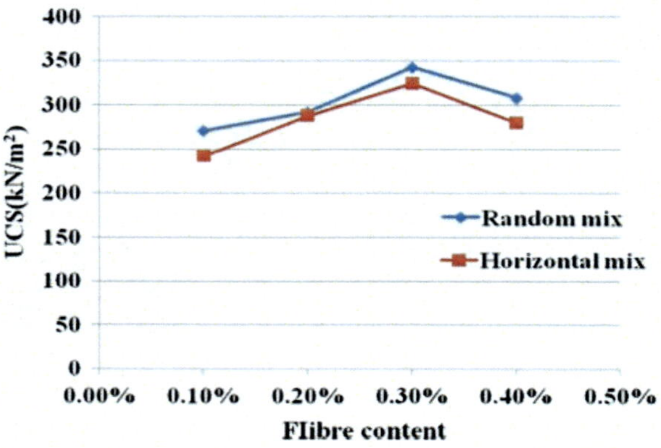


**Comparison between random mix and horizontal mix**. On comparing the results of both types of fibre arrangement, we can see that both cases are effective in improving the strength of the soil.

Figure 6 shows the variation of UCS after 7 day curing for random mix and horizontal mix. Here the peak strength of 0.3% fibre-reinforced soil shows 12 times increase in UCS than unreinforced soil for random mix. But in case of horizontal mix, peak strength of soil reinforced with 0.3% fibre increases by only 11.37 times than unreinforced soil.

## 4 Conclusions

1. With the addition of lime and fibre, optimum moisture content increases
2. Whereas maximum dry density decreases which is due to the pozzolanic reaction between lime and clay.
3. The compressive strength of soil increases with increase in lime up to 10%. After 10%, strength decreases with further increase in lime content. Hence, optimum lime content is obtained as 10%.
4. With 10% lime as constant, effect of fibre reinforcement is studied by varying the fibre content and conducting unconfined compression test.
5. Standard proctor test was conducted for soil with 10% lime and varying percentage of fibre. With the addition of fibre, optimum moisture content increases whereas maximum dry density decreases.
6. As the palm fibre content increases, the compressive strength first increases, but after 0.3%, it is found to decrease.
7. Random mix of palm fibre increases the peak strength by 12 times the unreinforced soil.
8. Horizontal mix of the palm fibre increases the peak strength only by 11.36 times the unreinforced soil.

9. Comparing the random mix and horizontal mix of palm fibres, randomly distributed fibres showed the improvement more effectively that the horizontal mix.

However, in this paper, durability of fibre mixed with soil and lime has not been investigated. There is a chance of decay of fibre which can reversely affect the stabilized soil. Hence, studies regarding durability and ageing of fibres have to be done. Also in this work, curing period up to 7 days has been done. It may not be sufficient to make a conclusion regarding effect of time on compressive strength of soil sample. Hence, curing period extending few months have to be evaluated to reach a conclusion.

## References


1. Anggraini, V., Huat, B.B., Asadi, A., Nahazanan, H.: Effect of coir fibers on the tensile and flexural strength of soft marine clay. J. Nat. Fibers **12**(2), 185–200 (2015)
2. Das, N., Singh, S.K.: Geotechnical behaviour of lateritic soil reinforced with brown waste and synthetic fibre. Int. J. Geotech. Eng. **13**(3), 287–297 (2019)
3. Arabani, M., Haghsheno, H.: The effect of polymeric fibers on the mechanical properties of cement-stabilized clay soils in Northern Iran. Int. J. Geotech. Eng. 1–12 (2019)
4. Cai, Y., Shi, B., Ng, C.W., Tang, C.S.: Effect of polypropylene fibre and lime admixture on engineering properties of clayey soil. Eng. Geol. **87**(3–4), 230–240 (2006)
5. Wei, L., Chai, S.X., Zhang, H.Y., Shi, Q.: Mechanical properties of soil reinforced with both lime and four kinds of fiber. Constr. Build. Mater. **172**, 300–308 (2018)
6. Cicek, E.: The effects of different types of fibres and geotextiles for pavement design. Road Mater. Pavement Design **20**(4), 793–814 (2019)
7. Dhar, S., Hussain, M.: The strength behaviour of lime-stabilised plastic fibre-reinforced clayey soil. Road Mater. Pavement Design **20**(8), 1757–1778 (2019)
8. Rao, S.N., Rajasekaran, G.: Reaction products formed in lime-stabilized marine clays. J. Geotech. Eng.**122**(5), 329–336 (1996)

# Direct Swell Pressure Measurement by Using Newly Designed Proving Ring—A Comparative Study

**Darikandeh Farahnaz, B. V. S. Viswanadham, K. Kayabali, and A. Qureshi**

## 1 Introduction

Expansive soils bear potential for shrinkage or swelling under changing moisture conditions [8]. During the expansion, the upward pressure effects the foundation, and in case the pressure is greater than the foundation pressure, it causes damage. Thus, understanding the amount of this pressure is important in designing the structures. To predict the swelling pressure, several studies have been conducted till date. Sridharan et al. [11] performed a study to investigate the swelling pressures using three different methods. In the first method called swell-consolidation, the sample was permitted to swell at a seating pressure while the sample is getting saturated (full swelling), followed by subsequent loading to bring it back to its initial height. In second method, known as the restricted swell test, three or more specimens were loaded to different pressures (around the estimated swell pressure), followed by inundation with water. In the third method, called as the constant volume, the sample was imbibed in water, and with applying the stress the volume keeping zero. In these methods, the swelling pressure is equal to the stress applied to the specimens, where the deformation curve crosses the stress axis for swelling consolidation test and restrain test. In case of the constant method, stress applied to keep the change in the specimen volume as zero was considered as the swelling pressure. However, a wide discrepancy between the results of these tests was pointed out by the literature [2, 5, 7, 10, 11], who stated the high degree of uncertainty in swelling pressure values obtained using various oedometer tests.

Also, it should be noted that the above investigations carried out on natural expansive soil, whereas the key question is the validity of those procedures for expansive

D. Farahnaz (✉) · B. V. S. Viswanadham · A. Qureshi
Indian Institute of Technology Bombay, Mumbai 400076, India

K. Kayabali
Ankara University, Ankara 06560, Turkey



C. N. V. Satyanarayana Reddy et al. (eds.), *Ground Characterization and Foundations*,
Lecture Notes in Civil Engineering 167,
https://doi.org/10.1007/978-981-16-3383-6_10

soil stabilized with using chemical additives, resulted in modification and pozzolanic reaction that introduces a cementite specimen. The pozzolanic reaction might influenced on actual swell pressure during the loading process in swell-consolidation tests when the sample returns to its initial height. Hence, it is essential to introduce a test method, where the swelling pressure measure in the swelling phase itself.

In our study, the swelling pressure of expansive soil has been measured by a direct method with the help of newly designed proving ring. This method is a kind of constant method test and do not allow vertical displacement; so, the advantage to conventional constant method is not required to control the load in the zero swell test to achieve absolutely zero swell since there will always be some amount of compression after a zero reading [5]. Therefore, a methodology similar to direct measurement of swelling pressure employed by Kayabali and Demir [5] has been applied. The indirect methods used for comparison in this study included the swell-consolidation and restricted swelling tests. It should be noted that all tests have been carried out on natural expansive soil and treated soil using Calcium carbide ($CaC_2$). For this experimental work, a proving ring was designed, fabricated and calibrated which involves the structural and geo-technical knowledge. Since Libii [6] created a successful design based upon the Whittmore-Petrenko proving ring, a similar methodology was applied for our investigation.

## 2 Test Materials

Model soil material is selected in such a way that represents high expansion and high swell pressure. Natural soil was collected from Nanded city which is located in Sinhagad road in Pune city, and Calcium carbide ($CaC_2$) was collected from High Purity Laboratory Chemicals Pvt. Ltd., and stainless steel (SS316). The index properties of the soil, its classification, and chemical composition are shown in Tables 1 and 2, respectively. Table 2 reveals that clay soil had about 70% silica oxide and aluminum oxide (natural pozzolans). In addition, the Calcium carbide had about 65% calcium oxide. Therefore, it was considered that aluminum silicates from the clay together with calcium from Calcium carbide would form cementitious materials of CSH and CASH, in the presence of water. The ring was made in a circular-shape. Two strain gauges (Tokyo sokki kenkyujo Co., Ltd., Japan) were also utilized in the design of the proving ring.

**Table 1** T Basic physical properties of soil and calcium carbide used for the investigation

| Index | Soil | Calcium carbide |
|---|---|---|
| Hygroscopic water content, $w_h$ (%)[a] | 13.35 | – |
| Specific gravity, $G_s$ | 2.685 | 2.24 |
| Liquid limit, $w_L$ (%) | 92 | – |
| Plastic limit, $w_P$ (%) | 45 | – |
| Shrinkage limit (%) | 15.27 | – |
| Free swell index (%) | 135 | – |
| Maximum dry unit weight, $\rho_{dmax}$ (kN/m$^3$) | 12.66 | – |
| Optimum moisture content, $w_{opt}$ (%) | 35.5 | – |
| *Particle size distribution (%)* | | |
| Clay (< 0.002 mm) | 66 | 12 |
| Silt (0.002–0.075 mm) | 26 | 64 |
| Sand (0.074–4.75 mm) | 7 | 24 |
| Gravel (> 4.75 mm) | 1 | – |
| Unified soil classification symbol[b] | CH-MH | – |
| Degree of expansion | High to very high | – |
| pH | 8.14 | 12.6 |

[a]Determined by an air-drying method
[b]*CH* High plasticity clay; *MH* High plasticity silt

# 3 Experimental Procedure

## *3.1 Proving Ring*

In order to measure the swelling pressure during the swelling process, a force sensor was used. This force sensor is a variation of Whittmore-Petrenko proving ring [6]. It was revised based on the design method used by Libii [6] with maximum capacity of 1 kN. Two strain gauges were attached on the inner surface of the ring, and another two were mounted on its outer surface, exactly at the opposite of the ones inside. All four were connected as Wheatstone bridge, which is an electrical circuit. To measure an unknown electrical resistance, the ring becomes a load cell that measures strain. Figure 1 shows the fabricated proving ring. Before starting the tests, proving ring was calibrated for loading. It was calibrated using the oedometer set-up. The proving ring was placed between the rigid solid mass and the supporting beam as shown in Fig. 2. The incremental load applied for the purpose of calibration, in our direct method, was similar to the amount of load used in oedometer tests (0–80 kg). Each step load was kept for two minutes, and the output voltages were read by HBM-MX1615 data logger (Hottinger Baldwin Messtechink, GmbH, Germany).

- Inside diameter: $d_i = 50$ mm;

**Table 2** Major chemical composition of the soil and calcium carbide

| Chemical composition (%) | $SiO_2$ | $Al_2O_3$ | $Fe_2O_3$ | $Na_2O$ | $K_2O$ | CaO | MgO | MnO | $TiO_2$ | $P_2O_5$ | BaO | SrO | SO3 |
|---|---|---|---|---|---|---|---|---|---|---|---|---|---|
| Soil | 49.8 | 17.2 | 13.85 | 2.56 | 2.41 | 7.93 | 2.12 | 0.86 | 1.70 | 0.1 | 0.19 | 0.02 | 0.3 |
| Calcium carbide | 12.77 | 15.48 | 1.28 | 0.35 | 0.47 | 64.97 | 0.26 | 0.01 | 0.36 | 0.06 | 0.47 | 0.01 | 0.11 |

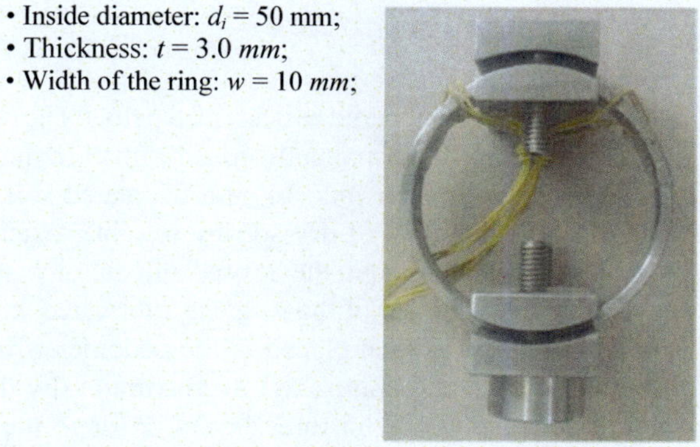


**Fig. 1** View of proving ring, the fabricated ring used in the laboratory

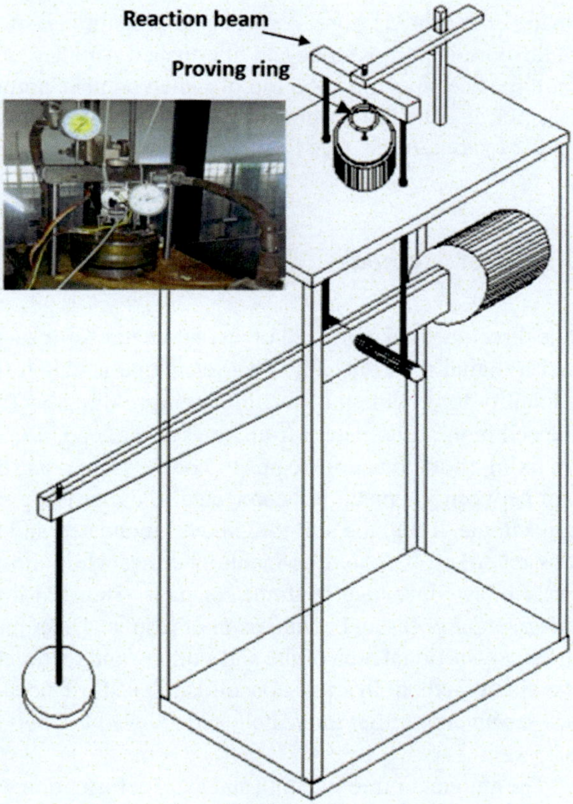


**Fig. 2** The schematic diagram of oedometer test set-up with proving ring

- Thickness: $t = 3.0$ mm;
- Width of the ring: $w = 10$ mm.

## 3.2 Specimen Preparation

The specimens were prepared as described below. The required amount of oven-dried soil (size < 4.75 mm) was manually mixed with the required amount of water to obtain a homogenous soil-water mix. In case of treated soil, soil was mixed by calcium carbide in dry state (6% of dry soil-by weight). After attaining the homogeneous admixture of soil and binder, the desired amount of water was added. Before placing the soil, the inner surface of the ring was lubricated with grease to reduce side-wall friction. The soil was then placed in the oedometer rings (75 mm in diameter and 25 mm in height), and compacted its maximum dry density (MDD) and optimum moisture content (OMC) in three layers. Prior to the tests, several samples were prepared to check the final density. After air-drying of the porous stone, it was positioned at the bottom of a dry oedometer. Subsequently, a filter paper was placed on top of the porous stone. A ring containing the soil specimen was then placed on top of the bottom porous stone and filter paper. Another air-dried filter paper was placed on top of the specimen, on top of which another porous stone and the loading pad was placed. The test specimen was then mounted in the oedometer, and positioned on the loading frame (see Fig. 2).

## 3.3 Test Methods

The direct method consisted of an oedometer set-up, a proving ring, oedometer cell, and a digital read-out unit. This technique is called the direct method because of its ability to provide the swelling pressure directly. To start the tests, after placing the cell in the oedometer set-up, the seating load of 6.25 kPa was applied, followed by fixing of the reaction beam (top of the frame) as shown in Fig. 2. There was no gap between the metal bar connected to the proving ring and the upper cap on the soil sample. Then, the soil sample was inundated and left to swell. The swell force was recorded. It should be mentioned that while most of the swelling takes place in 24 h, swelling may continue for days. Thus, 24 h period was selected only for practical purposes and comparison of results. The remaining net force is divided by the cross-sectional area of the soil sample, and recorded as the swelling pressure for the direct method. In swell-consolidation tests, the cell was placed in the oedometer test set-up and further inundated, and allowed to swell freely under a seating load of 6.25 kPa.

The amount of free swelling and axial deformation was recorded for different time intervals until 7 days. Thereafter, the test specimen was subjected to consolidation under various vertical stress for a period not less than 24 h and/or until the deformation attained a steady state for each load applied over the specimen. The consolidation loads were applied until the specimen attained its original height. The swelling pressure exerted by the soil specimen under zero swelling conditions, and expressed in $kN/m^2$ (kPa). Restricted swelling test was carried out using four samples, and they were prepared in a similar manner as samples for the swell-consolidation tests. The cell was mounted in the oedometer set-up, and the seating load of 6.25 kPa was applied. The dial gauge was further adjusted for the initial value of dial gauge. The range of the vertical stress was carefully selected to cover the range of the expected swelling pressure. The specimen was then inundated with water. The results were plotted as percent swelling versus vertical stress. The point on the horizontal stress axis, where the zero deformation is crossed, was considered as the swelling pressure.

## 4 Results and Discussion

A highly expansive soil was used to compare the swelling pressure, which was measured by different methods. The tests have been repeated four times for each method. Details and data regarding the oedometer swelling pressure test are presented in Tables 3 and 4 for the virgin soil and treated soil, respectively. The results of the direct method test are summarized in Table 5 for the virgin soil and treated soil, respectively. To run the restricted swell test, four specimens were used for four different tests. The results of the swell pressure tests along with applied pressure versus stress are presented in Table 6 for the virgin soil and treated soil, respectively. Data from Tables 3, 5, and 6 revealed the average of the swelling pressures for the virgin soil for swell-consolidation, the direct method, and restricted swell tests measure as 199.7 kPa, 56.4 kPa, and 72.5 kPa, respectively. Data from Tables 4, 5, and 6 depicted the average of the swelling pressures for the soil in treated sate for swell-consolidation, the direct method, and restricted swell tests measure as 65.2 kPa, 14 kPa, and 18.7 kPa, respectively. The results indicated that the highest swelling pressure in case swell-consolidation test, and the lowest for the direct method. Results are in agreement with Basma et al. [3], who mentioned that the most likely reason for the higher swelling pressures, registered during the swell-consolidation test, is due to the high pressure needed to expel the pore water from the voids. However, Kayabali and Demir [5] pointed out that the swell-consolidation test overestimates swelling pressures. In our results, the restricted swell test showed close values observed for the direct swell test. The disadvantage of restricted swell test is that it requires several specimens for a test, if it is to be applied to a field specimen is to get at least three undisturbed specimens at identical initial conditions which becomes difficult. Moreover, as described above, in swell-consolidation tests, the swelling pressure is derived through stress applied over the sample after full swelling. In case of clay, soils treated with using chemical additives (calcium), results in a complex soil matrix behavior. Adding chemical additive to the soil leading to modification and pozzolanic reaction of the soil which has already been reported by Abiodun and Nalbantoglu [1], Rogers et al. [9], and Tonoz et al. [12]. The pozzolanic reaction might be act as a resistance force while bring back the sample height to its initial height. Since in direct method the swell pressure was measured in swelling phase itself, swell pressure might not be affected with a complex soil matrix behavior due to pozzolanic reaction. To argue the existence of cementitious material, typical scanning electron microscopic

**Table 3** Swell pressure (kPa) values as found using the swell-consolidation test on Virgin soil, strain ($\varepsilon$), and consolidation stress ($\sigma$, in kPa)

| Soil | Test No. | Recorded strains (%) | | | | | | | | | | Swell pressure (kPa) |
|---|---|---|---|---|---|---|---|---|---|---|---|---|
| | | $\sigma = 0$ | $\sigma = 10$ | $\sigma = 30$ | $\sigma = 50$ | $\sigma = 70$ | $\sigma = 100$ | $\sigma = 140$ | $\sigma = 190$ | $\sigma = 250$ | $\sigma = 270$ | |
| Virgin soil | T1 | 12.8 | 10.8 | 8.256 | 6.496 | 5.256 | 3.624 | 1.928 | 0.84 | −0.82 | −0.98 | 210 |
| | T2 | 8.56 | 8.20 | 7.00 | 5.88 | 4.76 | 2.80 | 1.24 | 0.01 | −1.00 | −1.20 | 200 |
| | T3 | 10.52 | 9.36 | 8.56 | 7.12 | 5.16 | 3.12 | 1.032 | -0.80 | −2.01 | −2.35 | 189 |
| | T4 | 10.40 | 8.92 | 7.36 | 6.80 | 6.08 | 4.40 | 3.20 | 0.04 | −0.96 | −0.98 | 200 |

**Table 4** Swell pressure (kPa) values as found using the swell-consolidation test on treated soil, strain ($\varepsilon$), and consolidation stress ($\sigma$, in kPa)

| Soil | Test No. | Recorded strains (%) | | | | | | Swell pressure (kPa) |
|---|---|---|---|---|---|---|---|---|
| | | $\sigma = 0$ | $\sigma = 10$ | $\sigma = 30$ | $\sigma = 50$ | $\sigma = 100$ | $\sigma = 200$ | |
| Treated soil | T1 | 6.39 | 4.75 | 2.12 | −1.4 | −6.24 | −11.96 | 46 |
| | T2 | 6.24 | 5.06 | 3.42 | 0.92 | −4.32 | −9.11 | 62 |
| | T3 | 7.50 | 6.23 | 4.56 | 1.91 | −2.90 | −8.98 | 71 |
| | T4 | 8.08 | 6.62 | 5.08 | 2.46 | −2.11 | −15.42 | 82 |

**Table 5** Swell pressure (kPa) values as found using the direct method test on virgin and treated soil

| | Test No. | Swell pressure (kPa) | | Test No. | Swell pressure (kPa) |
|---|---|---|---|---|---|
| Virgin soil | T1 | 62.86 | Treated soil | T1 | 10 |
| | T2 | 42.99 | | T2 | 16 |
| | T3 | 58.86 | | T3 | 18 |
| | T4 | 60.93 | | T4 | 12 |

(SEM) of the virgin soil and treated soil after direct swell test and oedometer test presented in Fig. 3a–d. The micrographs show that the clay platelets flocculated and found flaky pattern after treatment (Fig. 3b and d). The flocculated and flaky patterned of Fig. 3b (after 24 h in direct method) to be further developed in Fig. 3d in oedometer test after about 14 days on treated soil. Figure 3d shows the fine reticulated network of gray patches that might be related to formation of cementitious compounds. These compounds filled the void space and covered the clay clusters. Similar observation reported by Juneja and Shinde [4]. Considering the facts above, the direct method would be more reliable to measure the actual swelling pressure in the swelling phase itself. Moreover, the calculated standard deviation (SD) of swell pressure of untreated soil which was measured by oedometer, restricted, and direct method tests have been obtained 8.57, 8.22, 9.09, respectively. Likewise, the calculated standard deviation (SD) of swell pressure of treated soil measured by oedometer, restricted, and direct method tests have been obtained 15.22, 6.7, 3.65, respectively.

It is known that a low SD indicates that the data points tend to be close to the mean (also called the expected value) of the set, while a high SD indicates that the data points are spread out over a wider range of values. The higher SD value is belonging to swell pressure that is measured by oedometer method on treated soil sample. Discrepancy between the values could be due to existence of cementitious particle in soil matrix. However, more tests are required on different expansive soils, and further on stabilized soils with different admixtures to confirm the applicability of the method.

**Table 6** Swell Pressure (kPa) values as found using the restricted swell test on virgin and treated soil, strain ($\varepsilon$), and consolidation stress ($\sigma$)

| Soil | Test No. | $\sigma$ (kPa) | $\varepsilon$ (%) | Swell pressure (kPa) | Soil | Test No. | $\sigma$ (kPa) | $\varepsilon$ (%) | Swell pressure (kPa) |
|---|---|---|---|---|---|---|---|---|---|
| Virgin soil | T1 | 30 | 1.65 | 80 | Treated soil | T1 | 5 | 0.7 | 17 |
| | | 50 | 0.76 | | | | 10 | 0.2 | |
| | | 70 | 0.2 | | | | 20 | −0.1 | |
| | | 140 | −0.82 | | | | 50 | −0.6 | |
| | T2 | 30 | 0.9 | 70 | | T2 | 5 | 1.4 | 28 |
| | | 50 | 0.2 | | | | 10 | 0.8 | |
| | | 70 | 0.01 | | | | 20 | 0.3 | |
| | | 140 | −2 | | | | 50 | −0.8 | |
| | T3 | 30 | 2.15 | 62 | | T3 | 5 | 0.5 | 12 |
| | | 50 | 1.3 | | | | 10 | 0.1 | |
| | | 70 | −1.2 | | | | 20 | −0.4 | |
| | | 140 | −3.02 | | | | 50 | −0.7 | |
| | T4 | 30 | 1.65 | 78 | | T4 | 5 | 0.6 | 18 |
| | | 50 | 0.76 | | | | 10 | 0.3 | |
| | | 70 | 0.2 | | | | 20 | −0.1 | |
| | | 140 | −1.15 | | | | 50 | −0.3 | |

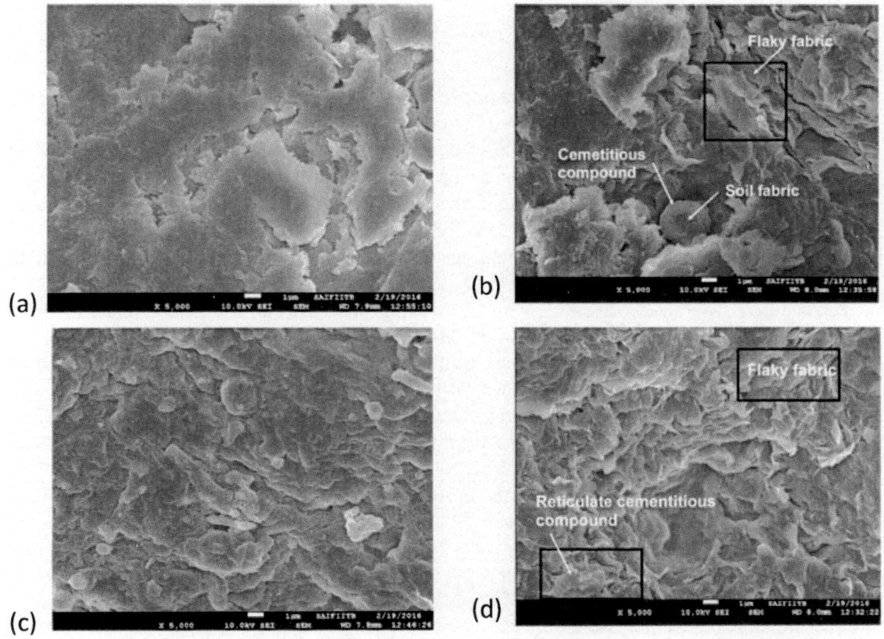


**Fig. 3** Scanning electrons microscope images of **a** untreated soil after direct method test; **b** treated soil after direct method test; **c** untreated soil after oedometer test; **d** treated soil after oedometer test

## 5 Conclusions

In order to measure the swelling pressure directly in the swelling phase, which is more reliable particularly for the stabilized expansive soil, we utilized a direct method in our study. For the purpose of research work, a proving ring was designed, calibrated, and used to measure the amount of swelling pressure. Swell-consolidation and restricted swell tests are called as indirect methods, while measurement with the proving ring is called as the direct method.

Results showed that the highest and the lowest swelling pressures were measured by the swell-consolidation test and direct method, respectively. Stabilization of expansive soil to prevent swelling, and controlling the heave is a quite important consideration for designing infrastructure like roads and railways. Knowing the fact that some of these stabilization methods cause modifications and pozzolanic reactions, which are time dependent and make the soil matrix complex, a direct method for swelling pressure measurement could be more reliable. Further, tests are required using different expansive soils. Subsequently, investigation of swelling pressures for stabilized soils would also be fundamental.

## References


1. Abiodun, A.A., Nalbantoglu, Z.: Lime pile techniques for the improvement of clay oils. Can. Geotech. J. **52**(6), 760–768 (2014)
2. Attom, M.F., Barakat, S.: Investigation of three methods for evaluating swelling pressure of soils. Environ. Eng. Geosci. **6**(3), 293–299 (2000)
3. Basma, A.A., Al-Homoud, A.S., Husein, A.: Laboratory assessment of swelling pressure of expansive soils. Appl. Clay Sci. **9**(5), 355–368 (1995)
4. Juneja, A., Shinde, S.T.: Undrained yielding of black cotton soil treated with calcium carbide residue. Indian Geotech. J. **50**(3), 319–329 (2019)
5. Kayabali, K., Demir, S.: Measurement of swelling pressure: direct method versus indirect methods. Can. Geotech. J. **48**(3), 354–364 (2011)
6. Libii, J.N.: Design, analysis and testing of a force sensor for use in teaching and research. World Trans. Eng. Technol. Educ. **5**(1), 175–178 (2006)
7. Nagaraj, H.B., Munnas, M.M., Sridharan, A.: Critical evaluation of determining swelling pressure by swell-load method and constant volume method. Geotech. Test. J. ASTM **32**(4), 1–10 (2009)
8. Nelson, J., Miller, D.J.: Expansive Soils: Problems and Practice in Foundation and Pavement Engineering. Wiley, New York (1997)
9. Rogers, C.D.F., Glendinning, S., Holt, C.C.: Slop stabilization using lime piles—a case study. Ground Improv. **4**(4), 165–176 (2000)
10. Singhal, S., Houston, S., Houston, W.: Effects of testing procedures on the laboratory determination of swell pressure of expansive soils. Geotech. Test. J. ASTM **34**(5), 1–10 (2011)
11. Sridharan, A., Rao, A.S., Sivapullaiah, P.V.: Swelling pressure of clays. Geotech. Test. J. ASTM **9**(1), 24–33 (1986)
12. Tonoz, M.C., Gokceoglu, C., Ulusay, R.: A laboratory-scale experimental investigation on the performance of lime columns in expansive Ankara (Turkey) clay. Bull. Eng. Geol. Environ. **62**(2), 91–106 (2003)

# Influence of $CaCl_2$ on Compaction and CBR Characteristics of Gypsum ($CaSO_4 \cdot 2H_2O$) Stabilized High Plastic Clay

**Batchu Ramanjaneyulu and Nandyala Darga Kumar**

## 1 Introduction

Soils have either occurred naturally or were made artificially. High plastic clays are found to be the most troublesome soils. These soils are generally clayey, deep, impermeable and are formed by lava basaltic rocks. These soils have high swelling and shrinkage potential, when exposed to moisture changes. Swelling and shrinkage nature of these soils lead to occurrence of cracks in the soil mass. Structures require a stable and steady foundation for their sustainability. The engineering properties of the soil also differ from place to place and depend mainly on soil mineral deposits, water table, and soil–water relationships. The improvements and changes in the soil can be made either by modification or by stabilizing the soil or by using both. Soil modification is done by the compaction and grouting methods. Soil stabilization may be defined as the technique adopted to improve the engineering properties of weaker soil by using different stabilizing agents. Improvement of soil properties by using waste or inexpensive materials is considered to be eco-friendly. Gypsum from the demolished building sites is considered to be waste material, and it can be used in the soil stabilization as stabilizing agent.

Gypsum stabilized clay samples showed an improvement in the stability, strength, and durability when cured in short periods of 3 and 7 days as compared to the 28 days cured samples. The water absorption and soil deterioration are significantly reduced with the admixture content and soaking time. The volume change of clay stabilized with gypsum showed less than 0.15% of soaked sample, and it is noticed as insignificant [1]. Addition of paddy husk did not improve the clay soil strength

B. Ramanjaneyulu (✉)
CVR College of Engineering, JNTUH, Peddapally 501510, India
e-mail: b.ramanjaneyulu@cvr.ac.in

N. D. Kumar
Department of Civil Engineering, JNTUHCE Manthani, Peddapally, Telangana 505212, India

 
C. N. V. Satyanarayana Reddy et al. (eds.), *Ground Characterization and Foundations*,
Lecture Notes in Civil Engineering 167,
https://doi.org/10.1007/978-981-16-3383-6_11

and CBR, but when 2% gypsum is added to clay at its OMC revealed 18% increase in UCS and 33% increase in CBR. Clay soil stabilized with paddy husk even showed reduced values of UCS and CBR as compared to untreated clay soil [2]. The bentonite samples tested corresponding to gypsum mixes such as 2.5, 5, 7.5, and 10% by dry weight revealed a noticeable change in the plasticity, swell percent, and strength parameters. It was mentioned that the gypsum can be used as a stabilizing agent for expansive clay soils [3].

Generally to prevent the solubility of gypsum and to improve the durability of clay samples, the cement or lime is used as admixtures in soil [4]. Good amount of research work had revealed that the use of lime or cement as a stabilizer would enhance the strength of soft soil [5]. Shear strength of cement-treated soil is influenced by the type of cement, amount of cement, physical properties, chemical properties, curing period, and mixing process. Generally, the type of cement and its amount to be used to stabilize the soil are the most considerable parameters [6, 7].

Employing numerical limit analysis and using cement-treated soil characteristics such as shear strength and unit weight of various stabilized soils, the Monte Carlo simulations were developed for the undrained bearing capacity of a surface strip foundation. It is very well valid for practical purposes [8]. The swell percent, swell pressure and UCS of lime, gypsum and lime with gypsum mixtures of compacted soil with curing revealed that the swell percent and swell pressure have reduced with increased additive, while the strength was increased over a period of time. The lowest improvement ratio was reported for gypsum alone added to clay soil. From high plastic clay stabilization point of view, the optimum lime content 6% can be proposed [9].

The variation in strength of soil with curing is due to the cation exchange, flocculation, and binding of particles with cementitious compounds formed after curing. The early accelerated strength is due to the formation of compacted structure with the growth of ettringite needless within the voids. Clay matrix rearrangement and suppression of sulfate effects with the formation of cementitious compounds are the main responsible factors for gaining the strength. Addition of 1% gypsum to soil–fly ash–lime, the strength was seen accelerated after 14 days curing [10]. From the above review, it is noticed that still there is a need to understand the behavior of high plastic clay when stabilized with gypsum and also the influence of $CaCl_2$ on gypsum stabilized soil. The results pertinent to the geotechnical characteristics of high plastic clay which is treated with gypsum are presented and discussed in the following sections.

**Table 1** Basic characteristics of soil

| S. No. | Property | Value |
|---|---|---|
| 1 | % Fine fraction | 70 |
| 2 | % Sand | 27 |
| 3 | % Gravel | 03 |
| 4 | Liquid limit (%) | 68 |
| 5 | Plastic limit (%) | 331 |
| 6 | Plasticity index | 37 |
| 7 | Free welling index (%) | 90 |
| 8 | Specific gravity | 2.69 |
| 9 | Maximum dry density (kN/m$^3$) | 14.35 |
| 10 | Optimum moisture content (%) | 20.8 |
| 11 | California Bearing Ratio (%) | 6.50 |
| 12 | Undrained cohesion (kPa) | 14 |
| 13 | Angle of shearing resistance (deg) | 5.0 |
| 15 | Unconfined compression strength (kPa) | 81 |

# 2 Experimental Investigation

## *2.1 Materials Used*

**Clayey soil**. Clayey soil was collected from Ibrahimpatnam area in Hyderabad, TS, India, from a clear ground at a depth of 0.3 m from the surface. Collected soil was processed and stored in the containers in the laboratory. The basic characteristics of soil are presented in Table 1. The soil has liquid limit and plasticity index are 68% and 37%, respectively.

**Gypsum**. Gypsum is made up of 79% of $CaSO_4$ and 21% of $H_2O$. Gypsum is a by-product of many industrial processes. It was collected from the demolished building sites and processed in the form of fine powder material and stored in the containers under controlled conditions in the laboratory.

**Calcium chloride**. Fused calcium chloride ($CaCl_2$) available in the powder form was purchased from the local market. It was stored in airtight bottles under controlled temperature and moisture conditions.

## *2.2 Tests Conducted*

The laboratory tests conducted are listed in Table 2 along with the Bureau of Indian Standard (BIS).

**Table 2** Tests conducted

| S. No. | Name of the test | BIS code |
|---|---|---|
| 1 | Specific gravity | BIS 2720: Part 3 (1980) [11] |
| 2 | Grain size analysis | BIS 2720: Part 4 (1985) [12] |
| 3 | Atterberg limits | BIS 2720: Part 5 (1985) [13] |
| 4 | Light compaction | BIS 2720: Part 8 (1983) [14] |
| 5 | Direct shear | BIS 2720: Part 13 (1986) [15] |
| 6 | California Bearing Ratio | BIS 2720: Part 16 (1987) [16] |
| 7 | Free swell index | BIS 2720: Part 40 (1977) [17] |

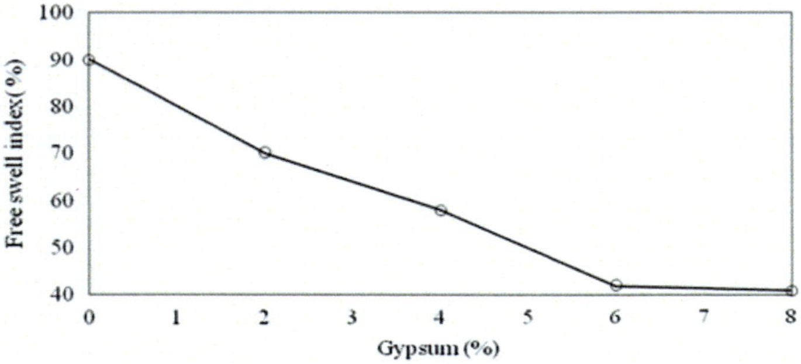


**Fig. 1** Variation of free swell index with % gypsum

# 3 Results and Discussion

## *3.1 Free Swell Index*

Figure 1 shows the variation of free swell index (FSI) with the varied gypsum content. From this figure, it is noticed that as the gypsum increases from 2 to 8%, there is a considerable decrease in the FSI. This decrease is about 50% for the clay which is treated with 8% gypsum. For the gypsum contents 6 and 8%, the FSI has reached a level equal to almost 40%. It can be evidenced from the figure.

## *3.2 Standard Compaction Test Results*

Figure 2 presents the water content–dry density curves obtained from standard compaction test on clay soil treated with gypsum of varied proportions from 2 to 8%. Calcium chloride acts as a soil flocculent and facilitates compaction and also presents the compaction curve of clay soil treated with 6% gypsum + 2% $CaCl_2$. From the figure, it can be noticed that as the % gypsum increases from 0 to 8%,

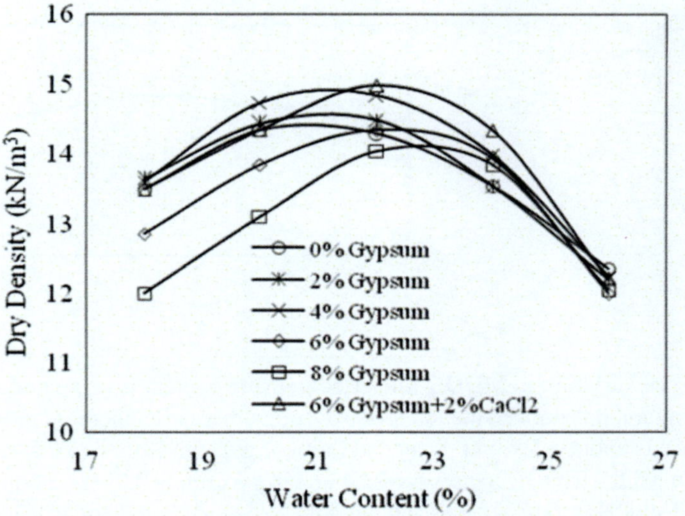


**Fig. 2** Water content–dry density curves for varied gypsum content

there is an increase in the OMC almost linearly. The MDD is increased up to 4% gypsum, and thereafter, it is decreasing. Overall, the range of OMC is between 20.8 and 23.2% for the gypsum content in soil from 0 to 8%. The MDD in clay at 8% gypsum is 14.10 $kN/m^3$, but for untreated clay, its value is 14.35 $kN/m^3$. The clay soil treated at 6% gypsum + 2% $CaCl_2$ is showing the MDD of 15 $kN/m^3$, and it is a good improvement in MDD as compared to the clay soil treated with gypsum alone. The increase in OMC of clay treated with 6% gypsum + 2% $CaCl_2$ is less as compared to the clay treated with 6 and 8% gypsum. It reveals that the addition of 2% of $CaCl_2$ along with gypsum of 6% by dry weight of soil would result in improved MDD.

Figure 3 presents the variation of OMC with the varied content of gypsum. As the gypsum content increases from 0 to 8%, there is an increase in the OMC and this increase is almost linear. The increase in OMC of 8% gypsum treated clay is found to be 11.5% as compared to untreated soil. This increase in OMC can be considered as

**Fig. 3** Variation of OMC for varied gypsum content

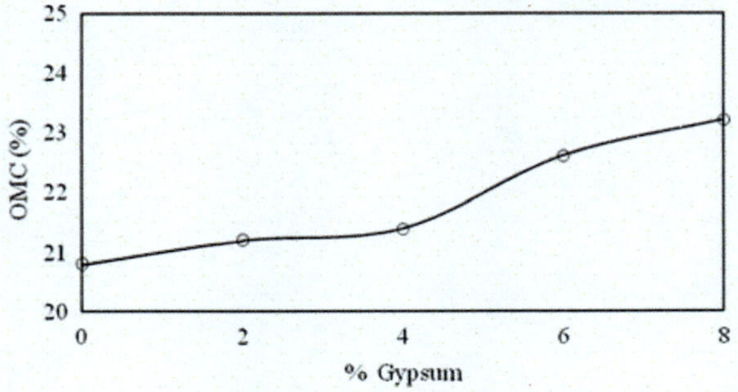

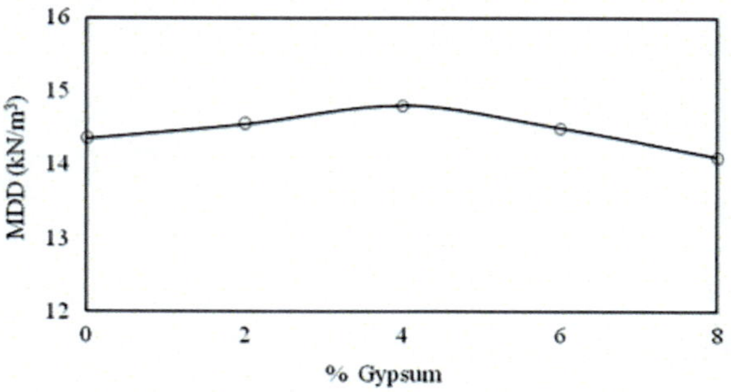


**Fig. 4** Variation of maximum dry density for varied gypsum content

marginal. The variation in MDD with varied gypsum content is presented in Fig. 4. From this figure, it is noticed that up to 4% gypsum, there is an increase in MDD and this increase is about 3.5% and further addition of gypsum from 6 to 8% resulted in decrease in MDD. This decrease in MDD at 8% gypsum is 1.75% as compared to unthreaded soil. However, the clay treated with 6% gypsum + 2% $CaCl_2$ is showing 4.5% increase in MDD as compared to untreated soil. Hence, the effect of $CaCl_2$ addition can be felt in achieving the stabilization process of clay when compared to the stabilization of clay with gypsum alone.

## 3.3 *Un-soaked CBR Variation with Gypsum*

The variation in CBR obtained at 2.5, 5, and 10 mm penetrations is presented in Fig. 5, with varied content of gypsum as well as for clay soil treated with 6% gypsum + 2% $CaCl_2$. From this figure, it is noticed that as the % gypsum increases, there is an increase in CBR and also as penetration increase from 2.5 to 10 mm, there is decrease in CBR value. With the increase in the gypsum content, the increase in CBR is more at 2.5 mm penetration as compared to the 5 and 10 mm penetrations. As compared to the untreated soil, the increase in CBR obtained corresponding to

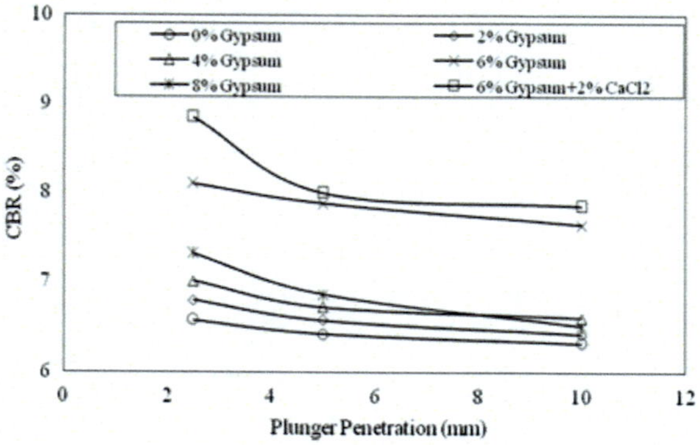


**Fig. 5** Variation of CBR with plunger penetration for varied gypsum content

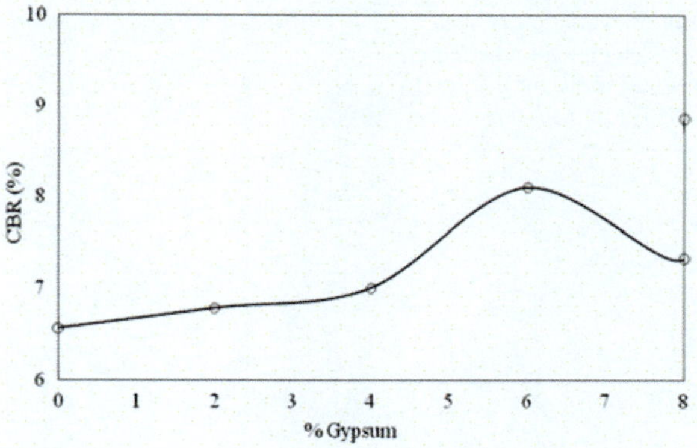


**Fig. 6** Variation of CBR for varied gypsum content

2.5 mm penetration is 23% and 11.5%, respectively, for soil treated with 6% and 8% gypsum. The same soil when treated with 6% gypsum + 2% $CaCl_2$ showing about 35% increase in CBR. In all the cases, the higher values of CBR are noticed corresponding to the 2.5 mm penetration.

Further, the CBR variation with varied content of gypsum is presented in Fig. 6. From this figure, it can be observed that up to 6% gypsum content the CBR is increasing, and for 8% gypsum, it is lower than 6% gypsum content and its value is decreased 10% as compared to the clay treated with 6% gypsum. But, for the same soil when treated with 6% gypsum + 2% $CaCl_2$, the increase in CBR corresponding to 2.5 mm penetration is about 10% as compared to the clay treated with 6% gypsum alone.

## 3.4 *Shear Characteristics of Gypsum-Treated Clay*

Figures 7, 8, and 9 present the variation of shear stress with effective normal stress for different gypsum contents. Figure 7 presents the strength envelops for clay treated with 0, 2, and 4% gypsum content. Figure 8 presents the strength envelops for clay treated with 0, 6, and 8%, and Fig. 9 presents the strength envelops for 6% gypsum + 2% $CaCl_2$ and 6% gypsum for a better comparison of the effect of $CaCl_2$.

From Fig. 7, it is observed that as the effective normal stress increases, the shear stress is almost becoming similar at higher normal stresses. The shear stress variation is only noticed in the initial stages of normal stress for the clay treated with 2 and 4% gypsum. Similar behavior is noticed in shear stress variation for clay soil treated with 6% and 8% gypsum as shown in Fig. 8. From Fig. 9, it is observed that as the normal stress increases, the shear stress also increasing for the cases of clay soil treated with 6% gypsum and 6% gypsum + 2% $CaCl_2$. The strength envelop of clay treated with 6% gypsum + 2% $CaCl_2$ is moving almost parallel and above than the strength envelop of clay treated with 6% gypsum. It is understood that, the addition of 2% $CaCl_2$ to the clay treated with 6% gypsum is causing increase in cohesion

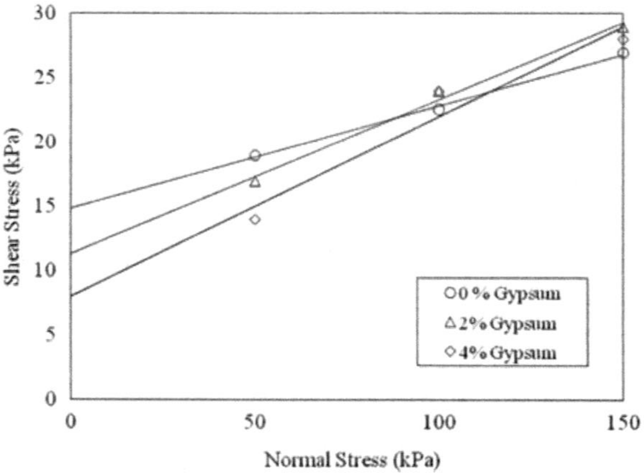


**Fig. 7** Variation of shear stress with applied effective normal stress for 0, 2, and 4% gypsum content

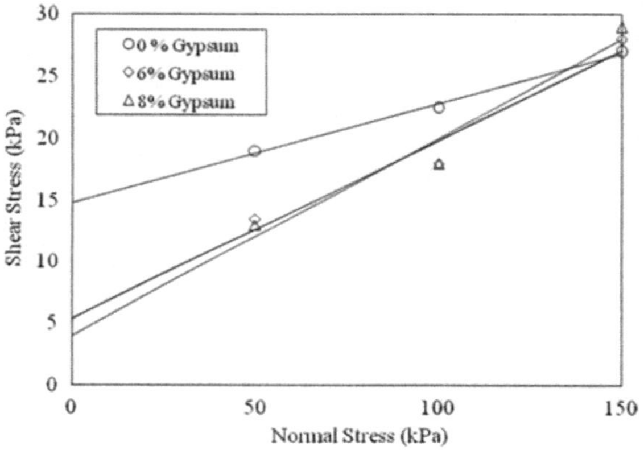


**Fig. 8** Variation of shear stress with applied effective normal stress for 0, 6, and 8% gypsum content

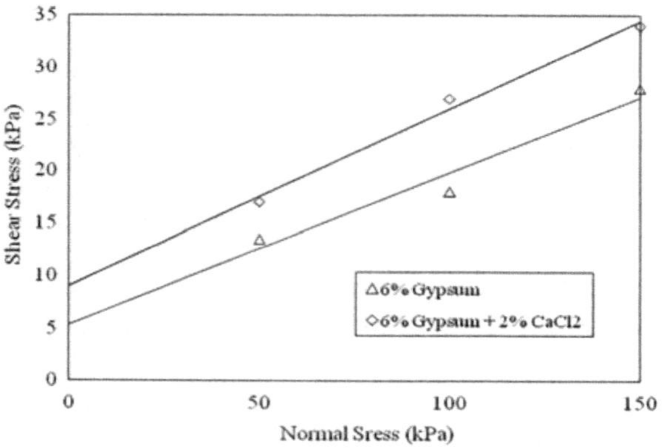


**Fig. 9** Variation of shear stress with applied effective normal stress for 6% gypsum content and 6% gypsum + 2% $CaCl_2$

**Table 3** Shear characteristics of gypsum-treated clay

| S. No. | % Gypsum | Cohesion (kPa) | Angle of shearing resistance (deg) |
|---|---|---|---|
| 1 | 0 | 15 | 5 |
| 2 | 2 | 11 | 7.5 |
| 3 | 4 | 8.0 | 8.85 |
| 4 | 6 | 5.35 | 9.25 |
| 5 | 8 | 4.0 | 10.15 |
| 6 | 6% Gypsum + 2% $CaCl_2$ | 9.0 | 10.75 |

and angle of internal friction. Table 3 presents the variation of cohesion and angle of shearing resistance of gypsum-treated clay. From this table, it is clear that as the % gypsum content increases, the cohesion of clay is decreasing, but the angle of shearing resistance is increasing. The increase in angle of shearing resistance of clay treated with 8% gypsum is 2 times more as compared to the untreated soil and its value is almost similar for both 8% gypsum and 6% gypsum + 2% $CaCl_2$-treated clay. The decrease in cohesion is noticed as 75% in the 8% gypsum treated clay as compared to the untreated clay. Similarly, the decrease in cohesion in the 6% gypsum + 2% $CaCl_2$ treated clay is about 40% as compared to the untreated soil. Overall, it is observed that the clay treated with 6% gypsum + 2% $CaCl_2$ is showing improvement in strength by increased angle of shearing resistance.

## 4 Summary and Conclusions

The test results obtained in the laboratory on gypsum-treated clay are discussed. The geotechnical characteristics such as free swell index (FSI), compaction characteristics such as optimum moisture content (OMC) and maximum dry density (MDD), un-soaked CBR and shear characteristics such as cohesion and angle of shearing resistance of clay treated with gypsum and gypsum + $CaCl_2$ are discussed. From the results discussed above, the following conclusions are drawn.

1. The reduction in FSI is about 50% in the clay soil treated with 8% Gypsum.
2. The OMC is increased with the addition of gypsum content, and MDD is increased initially up to 4% gypsum, thereafter it is decreased.
3. CBR of clay is improved with the gypsum content.
4. Angle of shearing resistance of clay treated with Gypsum has increased with the increased gypsum content, but the cohesion has decreased.
5. Overall, the clay treated with 6% gypsum + 2% $CaCl_2$ has shown good improvement in geotechnical characteristics.

## References


1. Ahmed, A., Issa, U.H.: Stability of soft clay soil stabilized with recycled gypsum in a wet environment. Soils Found. **54**(3), 405–416 (2014)
2. Iskandar, R., Hastuty, I.P., Dianty, W.O.: Clay stabilization by using gypsum and paddy husk ash with reference to UCT and CBR value. In: TALENTA-CEST 2017, IOP Publishing. IOP Conf. Ser.: Mater. Sci. Eng. **309**, 012026 (2018). https://doi.org/10.1088/1757-899X/309/1/012026
3. Yilmaz, I., Civelekoglu, B.: Gypsum: an additive for stabilization of swelling clay soils. Appl. Clay Sci. **44**(1–2), 166–172 (2009)
4. Kamei, T., Ahmed, A., Ugai, K.: The performance of soft clay soil stabilized with recycled gypsum in wet environment. In: Proceedings of the 14th Pan-American Conference on Soil Mechanics and Geotechnical Engineering, 2–6 Oct 2011, Toronto, Ontario, Canada, Paper no. 308, 7 pp. (Proceedings in CD) (2011)
5. Kasama, K., Whittle, A.J., Zen, K.: Effect of spatial variability on the bearing capacity of cement-treated ground. J. Soils Found. **52**(4), 600–619 (2012)
6. Clough, G.W., Sitar, N., Bachus, R.C., Rad, N.S.: Cemented sands under static loading. ASCE J. Geotech. Eng. **107**(6), 799–817 (1981)
7. Consoli, N.C., Rotta, G.V., Prietto, P.D.M.: Influence of curing under stress on the triaxial response of cemented soils. Geotechnique **50**(1), 99–105 (2000)
8. Kasama, K., Whittle, A.J., Zen, K.: Effect of spatial variability on the bearing capacity of cement-treated ground. Soils Found. **52**(4), 600–619 (2012)
9. Kiliç, R., Küçükali, Ö., Ulamiş, K.: Stabilization of high plasticity clay with lime and gypsum (Ankara, Turkey). Bull. Eng. Geol. Environ. **75**, 735–744 (2016)
10. Sivapullaiah, P.V., Jha, A.K.: Gypsum induced strength behaviour of fly ash-lime stabilized expansive soil. Geotech. Geol. Eng. **32**, 1261–1273 (2014)
11. BIS: 2720-Part 3 (1980): Method of testing of specific gravity of soil
12. BIS: 2720-Part 4 (1985): Method of testing of grain size analysis of soil
13. BIS: 2720-Part 5 (1985): Method of testing of Atterberg limits of soil
14. BIS: 2720-Part 8 (1983): Method of testing of light compaction of soil.
15. BIS: 2720-Part 13 (1986): Method of conducting direct shear test of soil
16. BIS: 2720-Part 16 (1987): Method of testing of California Bearing Ratio of soil
17. BIS: 2720-Part 40 (1977): Method of testing of soil for Free Swell Index

# Compaction Characteristics of China Clay–Bentonite–Sand Mix Proportions

**D. N. Jyothi, H. S. Prasanna, B. V. Vidya, and B. S. Pooja**

## 1 Introduction

To know the enhanced engineering results of fine-grained soils, it can be enlightened with the experimental investigations. Compaction as the word itself denotes that the way of increasing the soil density mechanically by lowering the volume of air. Altering the prevailing site foundation soils to yield better results under design and operational loading conditions with the use of vibration compaction methods, vacuum consolidation, soil nailing, grouting methods and dewatering methods. Compaction force has two principle types namely, static force and vibratory force. Static force is simply dead weight of the machine applying the downward force to the soil surface. Vibratory force follows normally engine-driven mechanism, to establish a downward force along with the machine's static weight. Every soil type behaves differently with respect to compaction force. Coarse-grained soils are mainly concerned by rearrangement of seating arrangements of granules. To improve particle stability, frequently fines are introduced throughout the mix. The physio-chemical behavior of fine-grained soils are largely influenced by clay mineralogical compositions. The present experimental study is one such attempt to study the compaction characteristic of natural fine-grained soil through artificial mix proportions of kaolinite-bentonite sand mixtures in varying percentage by the very presence of the least active and most active clay minerals in different proportions.

D. N. Jyothi · H. S. Prasanna (✉) · B. V. Vidya · B. S. Pooja
The National Institute of Engineering, Mysore, India
e-mail: prasanna@nie.ac.in

 
C. N. V. Satyanarayana Reddy et al. (eds.), *Ground Characterization and Foundations*, Lecture Notes in Civil Engineering 167,
https://doi.org/10.1007/978-981-16-3383-6_12

## 2 Literature Review

The study of the compaction of mix proportions having different physical and chemical properties demographically presented in different parts of world has drawn a curious attention of many researchers in the history. Many attempts have been made to correlate the compaction characteristics and index properties of different clay mixtures and sand.

Howell et al. [1] studied the effects of different types of processed clay soils, curing period and the mixing procedure on the laboratory compaction of sand-attapulgite clay, sand-granular bentonite clay, sand–powdery bentonite clay and sand–attapulgite clay–granular bentonite clay mixtures. They conducted compaction test by mixing small amount of attapulgite clay and granular bentonite. It was found that with increase in percentage of attapulgite clay and granular bentonite there is a tendency of increase in maximum density and optimum moisture content.

Sridharan and Nagaraj [2] have conducted a study on the compressibility behavior of remolded fine-grained soils and their correlation with index properties. The results indicate that, in the absence of shrinkage index, the plasticity index can be used to predict the compressibility characteristics with a better correlation than that of the liquid limit. Further in 2005, they brought out the effect of plasticity characteristics on the compaction characteristics for fine-grained soils.

Sivrikaya and Togrol [3] have studied about the compaction behavior of fine-grained soils for varying compaction energy. They have proposed equations correlating the maximum dry density and optimum water content corresponding to the results obtained from standard Proctor and modified Proctor test with the index properties. Chitoori et al. [4] have conducted the experimental investigation on the estimation of quantity of clay mineralogy in fine grained soils. It was found that the results obtained were used to determine the composite chemical characteristics of soils.

Khan and Azam [5] conducted the compressive strength tests on compacted samples of natural clay of high plasticity and clay–sand mixtures containing 20–40% sand. They observed that the compressive strength decreases with an increase in sand content because of increased material heterogeneity and loss of sand grains from the sides during shearing.

Proia et al. [6] has made an attempt to study about the compaction behavior of the sand–bentonite mixtures. It was found that the compaction characteristics were governed by the interaction between the two components each with its peculiar characteristics. With regard of dynamic compaction, it was observed that material exhibits a sand-like behavior for limited percentage of bentonite and shows a bentonite-like behavior for large percentage of bentonite, while for an intermediate percentage of bentonite, a more complex behavior was noticed.

Bose [7] has analyzed the grain size distribution and compaction behavior of sand–bentonite mixtures. The outcome of sieve analysis reveals that with increase in bentonite content, the mean grain size decreases, with an increase in Cu and Cc. To make bentonite qualified to be used as a liner, there should not be any reasonable depletion when the optimum moisture and bentonite content increases along with the decreased dry density in the compacted sand–bentonite mixtures.

Prasanna and Jyothi [8] have studied the compaction characteristics of bentonite–sand, kaolinite–sand mixtures for different mix proportions. They observed that compaction characteristics can be effectively correlated with consistency characteristics such as liquid limit, plastic limit and plasticity index. The shrinkage limit of the mix proportions of both bentonite–sand and kaolinite–sand decreases as the percentage of fines increases. They also developed a plasticity chart which is akin to IS Plasticity Chart.

# 3 Materials and Method

## *3.1 Materials*

Locally available river sand was procured, wet washed and kept in hot air oven for 24 hours at a temperature of $105 \pm 5$ °C. The oven-dried sand was brought to room temperature and sieved through 425 μm IS Sieve to have only fine-sand fraction. The sieved sand was stored in air-tight plastic containers. The grain size distribution of fine sand fraction was done as per IS: 2720 (Part 4) 1985.

China clay is a clay mineral that is part of industrial mineral with a chemical composition $Al_2Si_2O_5(OH)_4$. It is a layered silicate, with a tetrahedral sheet of silica ($SO_4$) linked through oxygen atoms to one octahedral sheet of alumina ($AlO_6$) an octahedra. Kaolinite has a low shrink–swell capacity and a low cation-exchange capacity (1–15 mEq/100 g). It is a soft, earthy, usually white, mineral (dioctahedral phyllosilicate clay), produced by the chemical weathering of aluminum silicate minerals like feldspar.

Bentonite is a clay generated frequently from the alteration of volcanic ash, consisting predominantly of smectite minerals, usually montmorillonite Smectites are clay minerals; i.e., they consist of individual crystallites the majority of which are < 2 μm in largest dimension. Smectite crystallites themselves are three-layer clay minerals. They consist of two tetrahedral layers and one octahedral layer. In montmorillonite, tetrahedral layers consisting of $[SiO_4]$-tetrahedrons enclose the $[M(O_5,OH)]$-octahedron layer (M = and mainly Al, Mg, but Fe is also often found). The silicate layers have a slight negative charge that is compensated by exchangeable ions in the intercrystallite region. The charge is so weak that the cations (in natural form, predominantly $Ca^{2+}$, $Mg^{2+}$ or $Na^+$ ions) can be adsorbed in this region with their hydrate shell. The extent of hydration produces intercrystallite swelling. Depending on the nature of their genesis, bentonites contain a variety of accessory minerals in addition to montmorillonite. These minerals may include quartz, feldspar, calcite and gypsum. The presence of these minerals can impact the industrial value of a deposit, reducing or increasing its value depending on the application. Bentonite presents strong colloidal properties and its volume increases several times when coming into contact with water, creating a gelatinous and viscous fluid. The special properties of bentonite (hydration, swelling, water absorption, viscosity, thixotropy) make it a valuable material for a wide range of uses and applications.

Commercially available bentonite and kaolinite were obtained from Seema chemicals Bengaluru, were stored in air-tight plastic containers. Artificially samples of sand–bentonite–kaolinite were prepared in the laboratory in different mix proportions.

### 3.2 *Method*

**Index tests**. The index property test on various mix proportions was carried out as per BIS Specifications; such as, Specific gravity (IS: 2720-Part 3/sec 2-1980) (ASTM D845-14): Standard test method for specific gravity of soils. Atterberg limits (IS:2720-Part 51985) (ASTM D 4318) Standard test method for liquid limit, plastic limit and plasticity index of soils were conducted. Casagrande cup is used to determine the liquid limit. This test is conducted to determine the soil water content with respect to the number of drops (25) required to bring a 13 mm section of a groove cut into the soil sample together. Plastic limit can be found when a gravimetric water content at which a soil sample can be rolled by hand into a thread of 3.2 mm diameter without breaking.

**Compaction test**. About 3 kg sample of air-dried mix-proportions were prepared based upon the dry weights of the material and predetermined amount of distilled water were added. The samples were packed in an air-tight polythene bag and left for a period of 24 hours to ensure uniform moisture distribution. Standard Proctor test were carried out as per IS:2720 (part 7) 1980 by compacting the mix proportions in three layers with the standard rammer of 2.6 kg giving 25 blows uniformly for each layer from a free fall of 310 mm. Representative sample of it was taken for water content determination. The process is continued till the sample is saturated. Similarly, modified Proctor test was carried out according to IS:2720 (part 8) 1983 with the whole sample compacting it in 5 layers with 25 evenly distributed blows on each layer from 4.9 kg rammer with free fall of 450 mm. In reduced modified and reduced standard Proctor test, procedure and equipment for tests are essentially same as that used in modified Proctor and standard Proctor test, respectively. Each layer received 15 number of blows of a rammer instead of 25 (it is 60% of the respective energy levels).

## 4 Results and Discussions

Figures 1 and 2 show the variations of liquid limit with MDD for constant bentonite and constant kaolinite mix proportions for different energy levels.

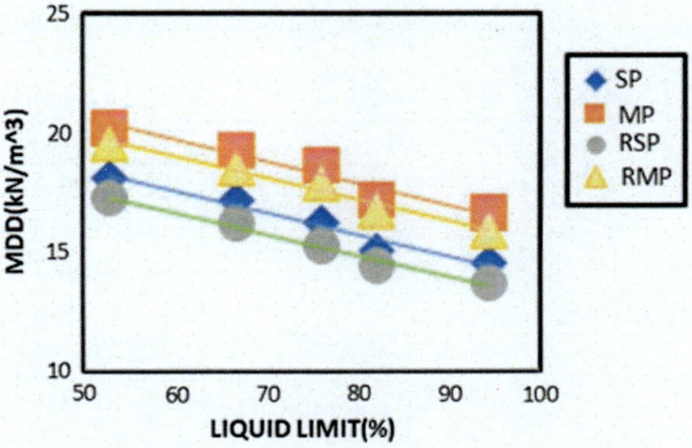


**Fig. 1** Variations of liquid limit, $W_L$ (%) with MDD (kN/m$^3$) for constant bentonite

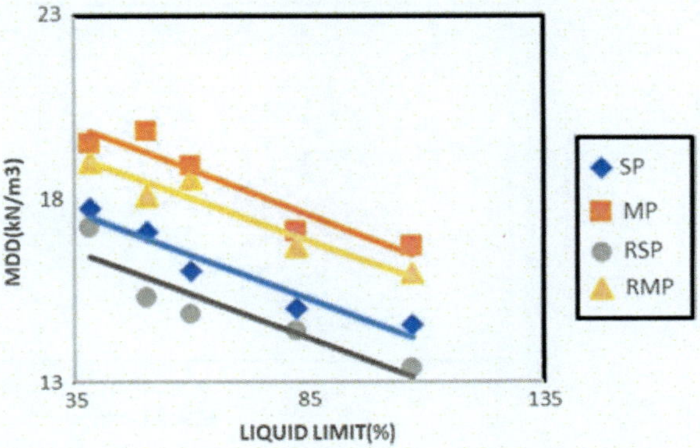


**Fig. 2** Variation of liquid limit, $W_L$ (%) with MDD (kN/m$^3$) for constant kaolinite

From the above figures, it is observed that maximum dry density decreases with increase in the liquid limit. It is because of increasing fine content in the mix proportion which in turn increases the water holding capacity which is also a characteristic feature of a natural fine-grained soils. The mix proportions can be identified by their liquid moisture contents. The corresponding liquid moisture content and maximum dry density are indicated in Table 1 following linear correlations. It is also observed that maximum dry density of mix proportions, which is having high percentage of bentonite is lesser, when compared to maximum dry density of mix proportion which contain higher percentage of kaolinite. This is mainly due to the bentonite clay which is composed of silica and aluminum sheets arranged in such a way that it absorbs large amount of water forming water tight barrier.

Figure 3 and 4 shows the variation of OMC and MDD obtained from different mix proportions irrespective of all energy levels with liquid limit of the mix proportions.

From Fig. 3, it can be observed that MDD decreases with increase in the liquid limit for both constant bentonite and kaolinite group. From Fig. 4, it is observed that OMC increases with increase in the liquid limit for both constant bentonite and kaolinite group. Further MDD and OMC of the mix proportions have a polynomial best fit as given by Eqs. 1 and 2, respectively.

**Table 1** Correlation equations of maximum dry density with liquid limit for both constant bentonite and kaolinite

| Clay minerals | Energy levels | Regression equation | $R^2$ | $R$ |
|---|---|---|---|---|
| Constant bentonite | MP | $\gamma d = -0.091\text{wl} + 25.18$ | 0.95 | 0.97 |
| | RMP | $\gamma d = -0.090\text{wl} + 24.43$ | 0.97 | 0.99 |
| | SP | $\gamma d = -0.091\text{wl} + 22.97$ | 0.96 | 0.98 |
| | RSP | $\gamma d = -0.090\text{wl} + 22.04$ | 0.99 | 0.99 |
| Constant kaolinite | MP | $\gamma d = -0.0487\text{wl} + 21.702$ | 0.89 | 0.94 |
| | RMP | $\gamma d = -0.0448\text{wl} + 20.686$ | 0.93 | 0.96 |
| | SP | $\gamma d = -0.0474\text{wl} + 19.294$ | 0.93 | 0.96 |
| | RSP | $\gamma d = -0.0478\text{wl} + 18.254$ | 0.84 | 0.92 |

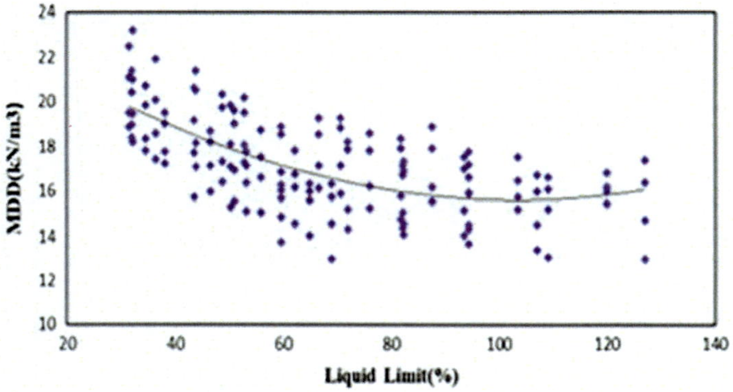


**Fig. 3** Variation of MDD (kN/m$^3$) with liquid limit, $W_L$ (%)

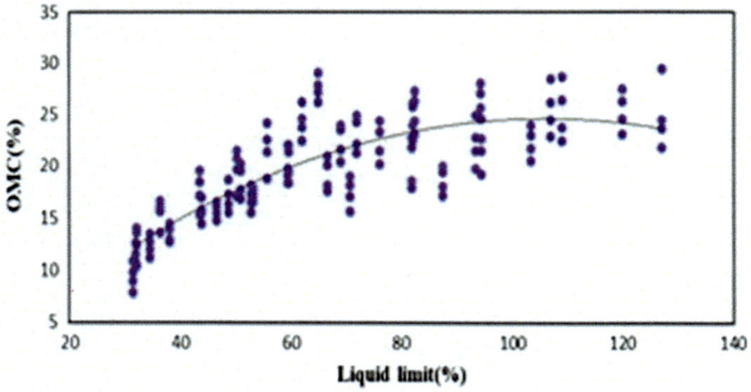


**Fig. 4** Variation of OMC (%) with liquid limit, $W_L$ (%)

$$\text{MDD} = 0.000W_L^2 - 0.168W_L + 24.28 \quad R = 0.67 \tag{1}$$

$$\text{OMC} = -0.002W_L^2 + 0.473W_L - 0.4693 \quad R = 0.83 \tag{2}$$

Figures 5 and 6 show the variation of OMC and MDD obtained from different mix proportions irrespective of all energy levels with plastic limit of the mix proportions.

Figures 7 and 8 show the variation of OMC and MDD obtained from different mix proportions irrespective of all energy levels with plastic index of the mix proportions.

From Figures 5, 6, 7 and 8, compaction characteristics of MDD and OMC can be effectively related with plastic limit and plasticity index with a polynomial best fit having correlation coefficient ranging from 0.63 to 0.84.

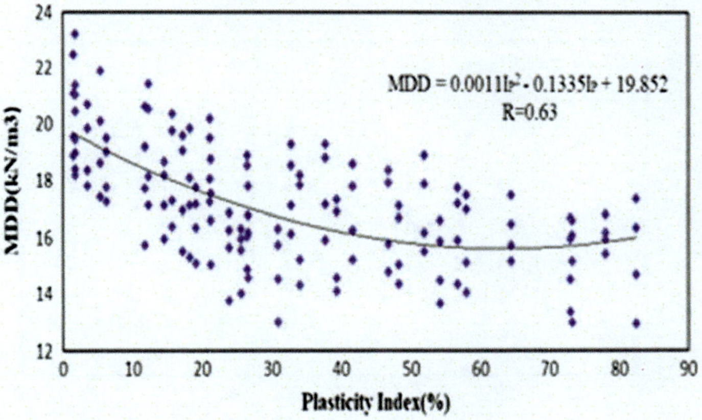


**Fig. 5** Variation of MDD (kN/m$^3$) with plastic limit, $W_P$ (%)

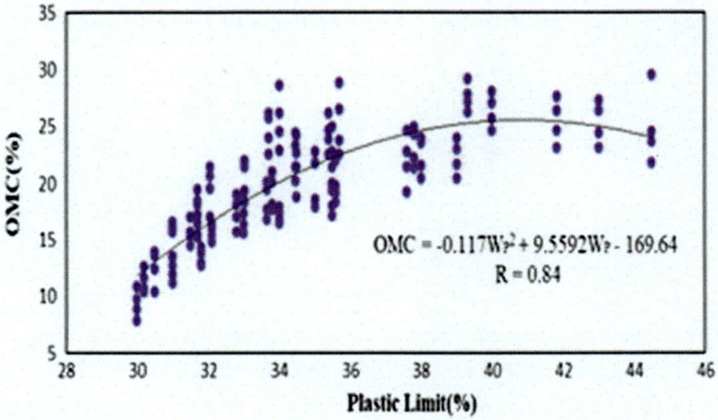


**Fig. 6** Variation of OMC (%) with plastic limit, $W_P$ (%)

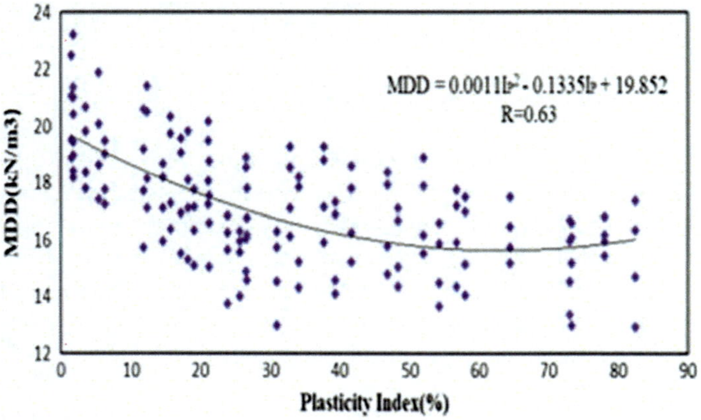


**Fig. 7** Variation of MDD (kN/$m^3$) with plasticity index, Ip (%)

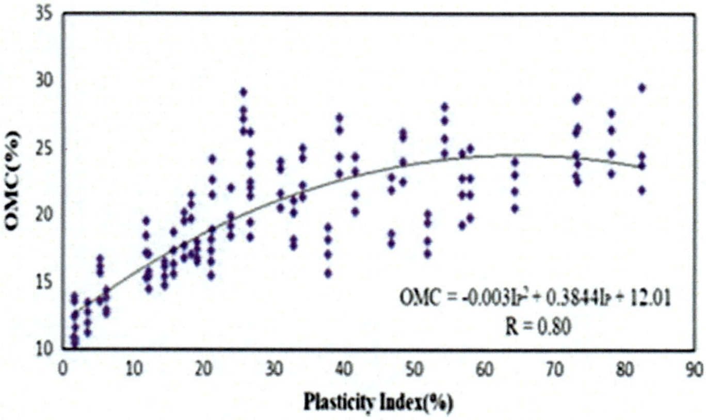


**Fig. 8** Variation of OMC (%) with plasticity index, Ip (%)

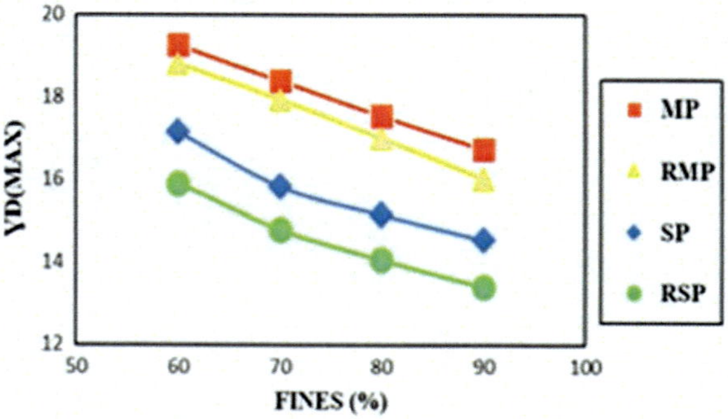


**Fig. 9** Variation of MDD, $\gamma d_{\max}$ (kN/$m^3$) with percent fines, PF (%) for constant bentonite

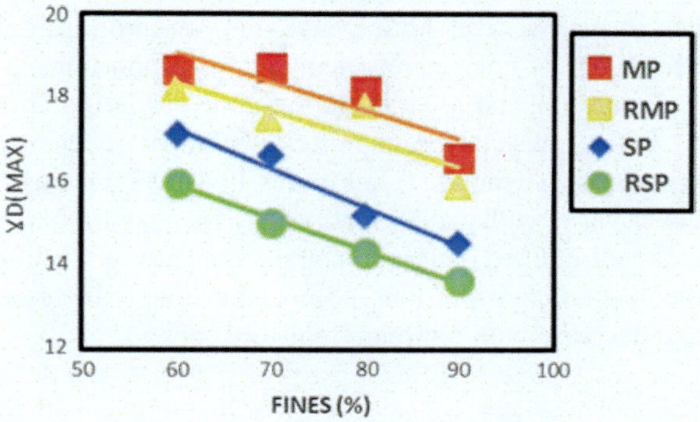


**Fig. 10** Variation of MDD, $\gamma d_{max}$ (kN/m$^3$) with percent fines, PF (%) for constant kaolinite

Figures 9 and 10 show the variation of compaction characteristics -MDD with percent fines of the mix proportions for different compaction energy levels

From Figure 9 and 10, it is observed that as the percentage of fines increases, the value of maximum dry density linearly decreases. The variation of $\gamma d_{max}$ with percent fines of both constant bentonite and kaolinite looks almost similar. The effect of increasing the compaction energy results in an increase in the maximum dry density. However, the increase in the compaction energy does not have a linear relationship with the increase of compactive effort. As the percentage of fines increases, the compaction curve becomes flatter and the maximum dry density will be relatively low because it requires more water for lubrication as the specific surface increases. Mix proportions form a gel due to the increase in moisture content of the mixture called double diffused layer, which causes the enlargement of the size of voids between the particles.

Figure 11 shows the variation of maximum dry density with optimum moisture content irrespective of the mix proportions and energy levels.

From Fig. 11, it can be observed that variation of maximum dry density with optimum moisture content irrespective of the mix proportions and energy levels

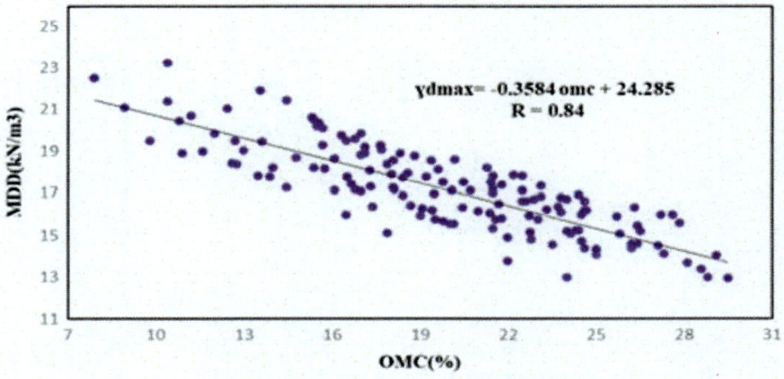


**Fig. 11** Variation of MDD, $\gamma d_{max}$ (kN/m$^3$) with OMC (%) different compactive energy levels

decreases. We can notice marginal variation, and it decreases linearly as the percentage of fines increases in the mix proportions.

Figures 12 and 13 show the variation of MDD and OMC with compactive energy.

Table 2 presents the magnitude of compactive energy imparted to mix-proportions at various compaction test. Figures 12 and 13 represent the compaction energy level, how it varies with maximum dry density and optimum moisture content. It is observed that well-defined pattern of compaction curves for different percentage of fines. For the given mix proportions, compaction energy increases with maximum dry density and decrease with optimum moisture content.

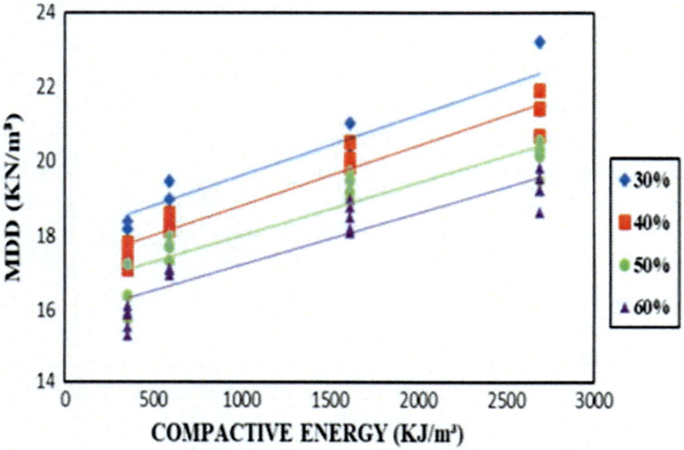


Fig. 12 Variation of MDD (kN/m$^3$) with compactive energy

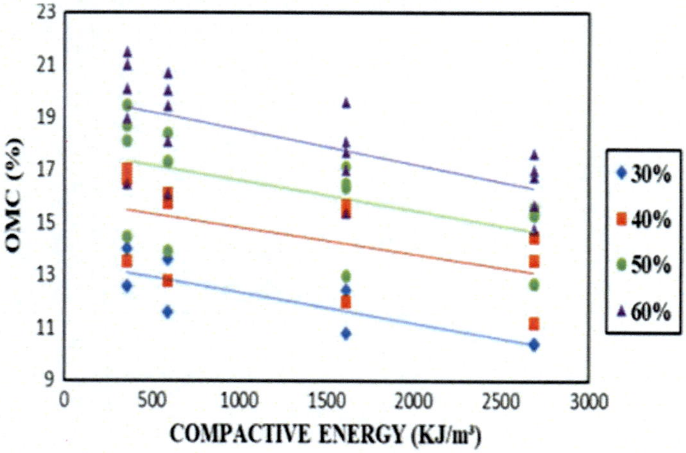


Fig. 13 Variation of OMC (%) with compactive energy

**Table 2** Variation of compactive energy for different energy levels

| Type of compaction test | Compactive energy (kJ/m$^3$) |
|---|---|
| Reduced standard Proctor | 355.25 |
| Standard Proctor | 592.09 |
| Reduced modified Proctor | 1613.94 |
| Modified Proctor | 2689.9 |

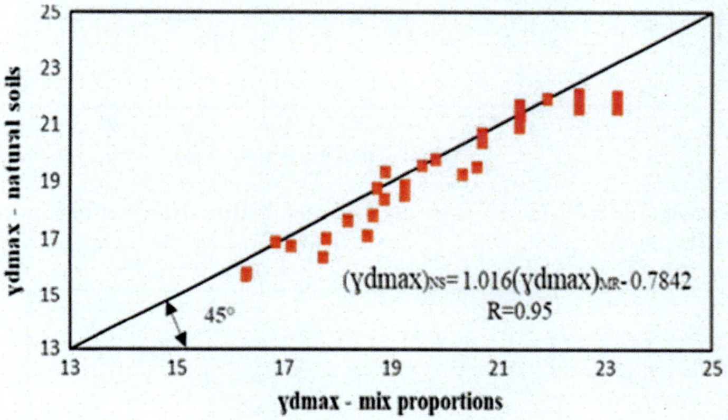


**Fig. 15** Variation of MDD (kN/m$^3$) of natural soils with MDD (kN/m$^3$) of mix proportions of modified Proctor test

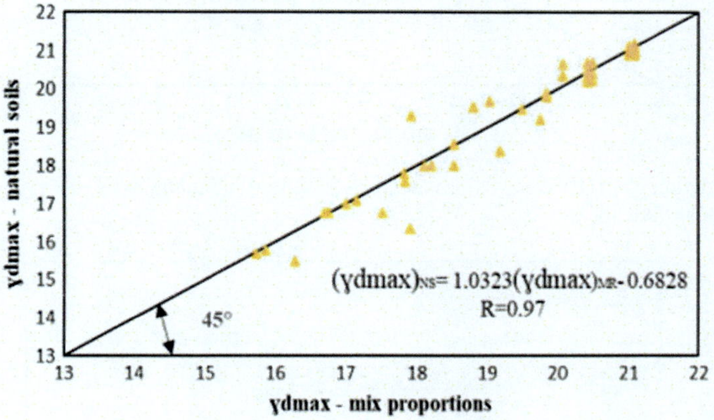


**Fig. 16** Variation of MDD (kN/m$^3$) of natural soils with MDD (kN/m$^3$) of mix proportions of reduce modified Proctor test

Figure 15, 16, 17 and 18 shows the variation of MDD of natural soils with MDD of artificial soils for different energy levels.

Figure 19 shows the variation of OMC of natural soils with OMC of mix proportion of standard Proctor test.

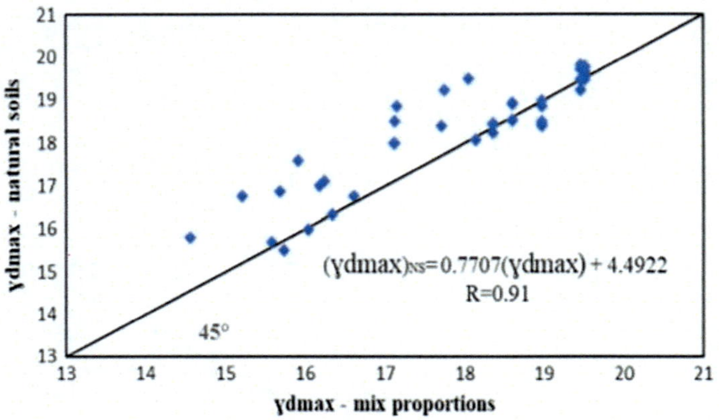


**Fig. 17** Variation of MDD (kN/m$^3$) of natural soils with MDD (kN/m$^3$) of mix proportions of standard Proctor test

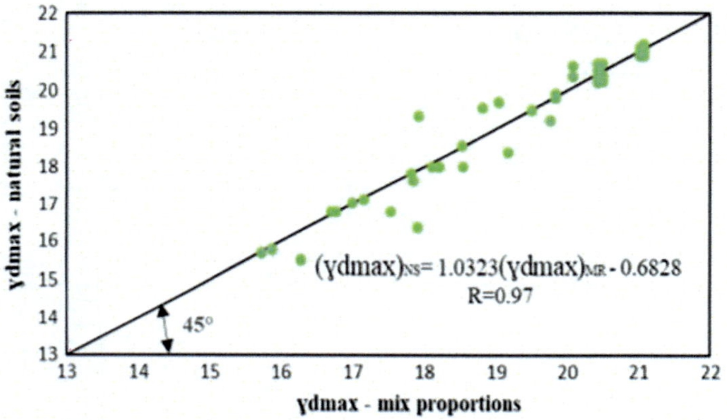


**Fig. 18** Variation of MDD (kN/m$^3$) of natural soils with MDD (kN/m$^3$) of mix proportions of reduced standard Proctor test

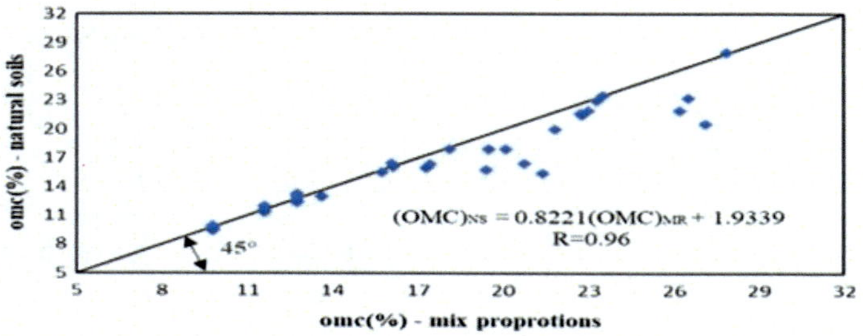


**Fig. 19** Variation of OMC (%) of natural soils with OMC (%) of mix proportions of standard Proctor test

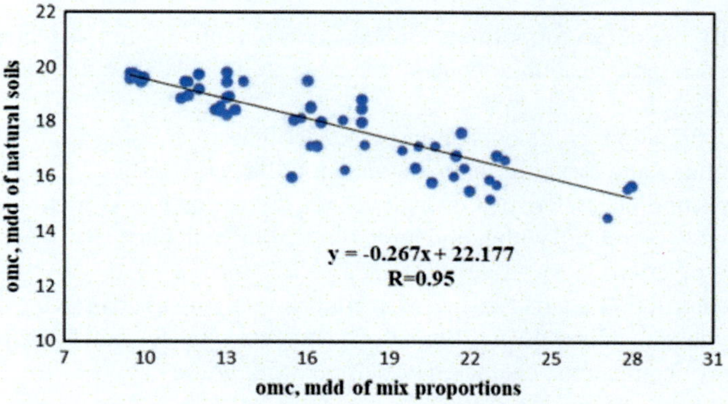


**Fig. 20** Variation of OMC (%), MDD (kN/m$^3$) of natural soils with OMC (%), MDD (kN/m$^3$) of mix proportion of standard Proctor test

$$(\gamma d_{\max})_{\text{Natural soils}} = 1.016(\gamma d_{\max})_{\text{mix proportions}} \text{with a regression } 0.784 \quad (3)$$

Figures 15, 16, 17, 18, 19 and 20 show the variations of maximum dry density of natural soils obtained from author's own experimental work with that of the maximum dry density of mix proportions obtained from the present experimental study from 45-degree line (called equality line). Natural fine-grained soils having both kaolinite and montmorillonite clay minerals in fines component from author's own study were carefully selected to map percent fines in the natural soils with that of the mix proportions and considered for the comparative study. A careful study of these figures depicts that the compaction characteristics of natural fine-grained soils can be accurately estimated with that of maximum dry density and optimum moisture content obtained from artificial mix proportions.

To develop a good and reliable correlation, we must have a high value of coefficient ($R^2$). The correlation is an index of the goodness of fit between the two values used to develop the correlation. It provides a quantitative index to predict values and to incite the accuracy for the future predictions. For models of natural soils and mix proportions, the value of $R^2$ varies between 0.83 and 0.94.

Figure 20 shows the variations of OMC, MDD of natural soils with OMC, MDD of mix proportion of standard Proctor test.

## 5 Conclusions

The experimental study permits to promote the following conclusion which should be valid for bentonite–kaolinite–sand mix proportions.

1. Atterberg limits show that small amounts of bentonite and kaolinite present in the mix proportions are able to influence the liquid limit but not the plastic limit.

2. For all types of mix proportions studied, the maximum dry unit weight increases as the bentonite–kaolinite content increases up to 30% then decreases for all energy levels.
3. The optimum moisture content increases for fines, increasing from 20 to 90% of kaolinite and bentonite sand mixtures for all energy levels.
4. Compaction characteristics of bentonite, kaolinite and sand mixtures can be efficiently correlated with liquid limit, plastic limit and plasticity index as given by the equations.
5. The compaction characteristics of natural fine-grained soils having kaolinite bentonite clay minerals in different percentage can be accurately estimated by means of compaction characteristics of mix proportions.

## References

1. Howell, J.L., Shackelford, C.D., Amer, N.H., Stern, R.T.: Compaction of sand-processed clay soil mixtures. Geotech. Test. J. **20**, 443–458 (1997)
2. Sridharan, N.: Compressibility behavior of remolded fine-grained soils and their correlation with index properties. Can. Geotech. J. **37**, 712–722 (2000)
3. Sivrikaya, T.: Estimating compaction behavior of fine-grained soils for varying compaction energy. Can. Geotech. J. **45**(6), 877–887 (2008)
4. Chitoori, B.: Experimental investigation on the estimation of quantity of clay mineralogy in fine-grained soils. J. Geotech. Geoenviron. Eng. **137**, 997–1008 (2011)
5. Khan, F.S., Azam, S.: Compressive Strength of Compacted Clay-Sand Mixes. Hindawi Publishing Corporation (2014)
6. Proiae, R., et al.: Experimental investigation of compacted sandbentonite mixtures. Procedia Eng. **158**, 51–56 (2016)
7. Bose, B.: Compaction behavior and grain size alteration of sand–clay mix. Int. J. Innov. Technol. Expl. Eng. **9**, 4106–4110 (2019)
8. Prasanna, H.S., Jyothi, D.N.: Compaction characteristics of bentonite, kaolinite and sand mixtures of different mix proportions. Int. J. Recent Technol. Eng. (2020)

# A Lab Study on the Factors Effecting Settlement and Electrical Resistivity of Gypsum Sands

**Raghava A. Bhamidipati and Michael E. Kalinski**

## 1 Introduction

Soils containing gypsum, often termed as gypsiferous or gypseous soils are generally found in the drier parts of the world. Places like Iraq, some middle-eastern states, southern Spain and south western United States have sizeable proportions of gypsiferous soils [1]. These soils generally make stable building material under dry condition but undergo considerable settlement in the presence of water. Gypsum is a moderately soluble calcium–sulfate mineral, belonging to a group of minerals called evaporites. Gypsum soils are often classified as collapsible soils. There has been a lot of damage all over the world because of the settlement of structures constructed on gypsum soils. The hazards associated with these soils and factors causing the collapse settlement have been studied by numerous researchers in the past. Seleam [2], Al-Marsoumi et al. [3] and Fattah et al. [4], were some researchers who had studied the geotechnical and collapsibility properties of gypsum soils from samples obtained from different parts of Iraq. For this research, a laboratory study was conducted on reconstituted specimens poorly graded sand mixed with different percentages of gypsum and the impact of various factors like gypsum content, initial water content, leaching and time of soaking on the settlement of the soil were studied.

For quick and preliminary soil exploration over a large area, electrical resistivity testing is a promising new method that is finding a lot of application these days. Kalinski et al. [5] used this method to estimate the water content of clayey soils. Pandey et al. [6] studied the electrical resistivity of sandy soil using the four electrode

R. A. Bhamidipati (✉)
C V Raman Global University, Bhubaneswar 752054, India

M. E. Kalinski
University of Kentucky, Lexington, KY 40506, USA
e-mail: michael.kalinski@uky.edu

 
C. N. V. Satyanarayana Reddy et al. (eds.), *Ground Characterization and Foundations*, Lecture Notes in Civil Engineering 167,
https://doi.org/10.1007/978-981-16-3383-6_13

**Fig. 1** Sample of sand mixed with gypsum

soil box method. In this investigation, the feasibility of electrical resistivity in investigating gypsum soil horizons was studied in a lab-scale by using a four-electrode soil box method. The electrical resistivity of the soil samples were measured under varying conditions of moisture content, gypsum content and dry density. The results of these studies have been discussed in the following sections.

## 2 Test Setup

Poorly graded Sand was mixed with different percentages of commercial gypsum and samples were prepared for testing (Fig. 1). For studying the factors affecting settlement of gypsum, cylindrical samples of 6.4 cm diameter and 2.8 cm depth were prepared in a shear box. The void ratio of the all samples was maintained close to 0.7. The samples for long-term loading test were prepared in a consolidation load cell of 6.4 cm diameter and 1.8 cm depth.

The electrical resistivity behavior of gypsum–sand mixtures was studied with respect to change in variables like moisture content, gypsum content and dry density of soil samples. The tests were carried out using a lab scale four-electrode soil resistivity box. In this arrangement, the resistivity of the soil is measured by placing the soil sample in a rectangular soil box, and measuring the potential drop between the two inner electrodes of the box caused by passing an electric current through the soil box. Resistivity (Ω m) is calculated as the ratio of voltage and current, multiplied by a box geometry constant ($k$). A schematic of the test setup is shown in Fig. 2. The details of the testing methodology can be found in Bhamidipati and Kalinski [7].

## 3 Factors Influencing the Settlement of Gypsum Sands

The settlement of different gypsum and sand mixtures was studied by conducting several tests as described in this section. Initially, the effect of initial moisture content

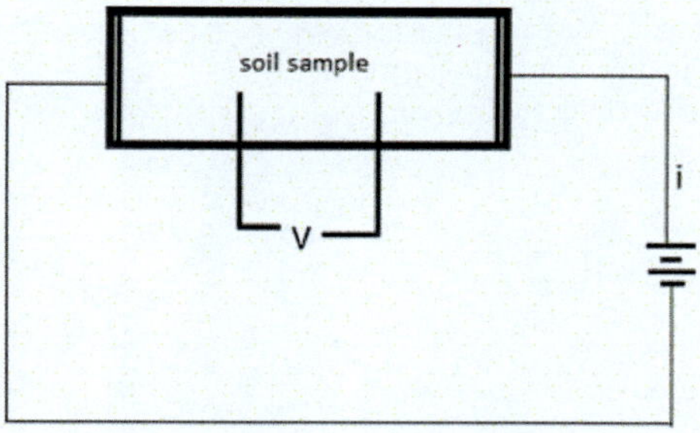


**Fig. 2** Arrangement for a four-electrode soil box resistivity test

on the soil samples was observed. The samples were placed in the shear-trac II loading apparatus and a normal load of 200 kPa was applied on the samples. The axial strain of the samples was recorded for a sufficiently long time until no further strain could be observed. Soil samples were prepared at different initial moisture contents ranging from 0 to 15%. The initial void ratio of all the samples was approximately 0.7. At moisture contents close to 20%, the samples were almost saturated and rapid loading led to the building up of excess pore pressures. This was leading to an incorrect measurement of vertical strain. Hence, those values were not considered for the test. From Fig. 3, it can be seen that the vertical strain increased with increase in initial moisture content of the samples. Also the samples with higher gypsum content (50 and 70%) show a greater vertical strain than the sample with 20% gypsum content.

Leaching of gypsum from soils was studied using a similar setup. In this case, two saturated samples with 30 and 50% gypsum were prepared and a small amount of water (around 2 drops per minute) was continuously percolated through the samples for different time intervals of 10 min, 1100 min (18 h) and 2880 min (48 h). The water coming out from the bottom of the sample was collected in a trough. The settlement of these samples was measured by placing them in the shear-trac II apparatus and applying a normal load of 200 kPa on them. The results from this test are shown in Fig. 4. It was seen that the longer the samples were subjected to leaching, the more

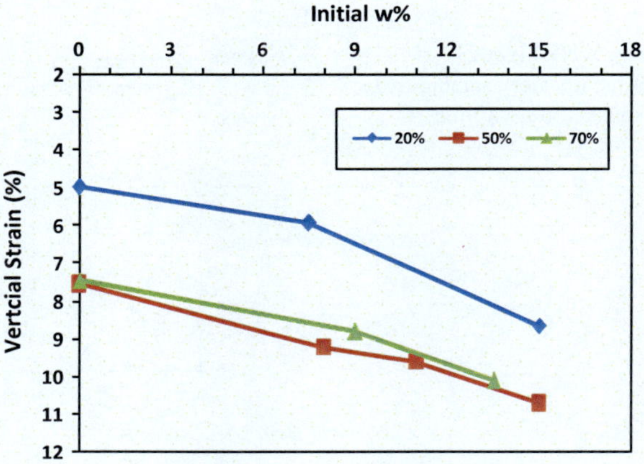


**Fig. 3** Vertical strain versus initial water content for samples subjected to a load of 200 kPa

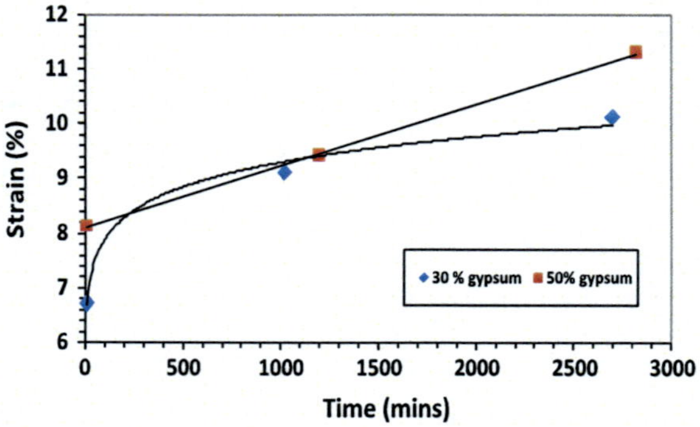


**Fig. 4** Vertical strain at 200 kPa versus time period of leaching for two samples

vertical settlement they were undergoing. Also the sample with 50% gypsum showed greater settlement than the sample with 30% gypsum.

The concentration of gypsum in the effluent water (collected in the trough) was measured at different times using an Extech pocket conductivity meter. For both the 30% gypsum and 50% gypsum sample, the conductivity of the effluent was roughly unchanged with time, indicating a constant rate of dissolution (Fig. 5). The dissolution of 30% gypsum was found to be very slightly higher at the 18 h (1080 min) and 48 h (2880 min) marks. The conductivity of the effluent varied between 3200 and 4000 μS. Tap water at 21 °C was used for the testing. The conductivity of the tap water during the testing was found to be in the range of 670–730 μS.

The long-term settlement of gypsum sand mixtures was studied using three samples with 10%, 30% and 60% gypsum, respectively. The samples were placed in a consolidation load cell, previously inundated under 200 kPa. After completion of settlement, the load was increased to 300 kPa. The decrease in height of the samples was recorded over three days (ranging from 1 to 72 h) (Fig. 6). Greater long-term settlement was observed in samples containing a higher gypsum percentage. Over a 72 h period, the settlement varied from around 0.1 mm for the 10% gypsum sample to 0.6 mm for the 60% gypsum sample.

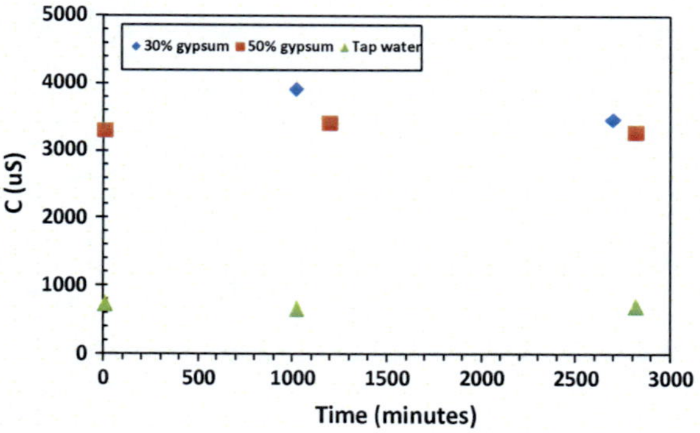


**Fig. 5** Conductivity of effluent water leaching from soils at different times

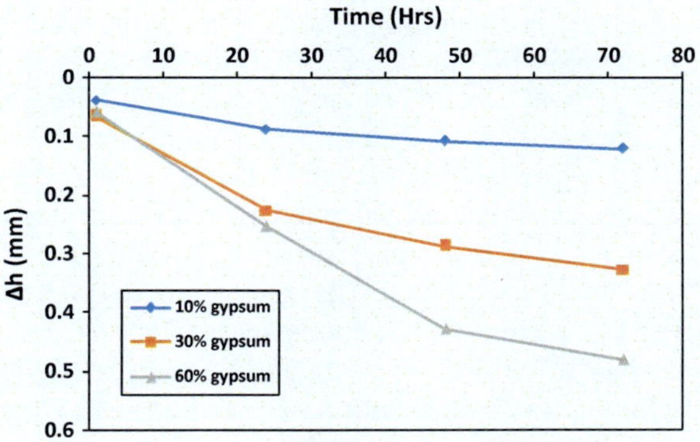


**Fig. 6** Settlement of inundated gypsum sand samples over 72 h under 300 kPa

# 4 Factors Influencing the Electrical Resistivity of Gypsum Sands

The electrical resistivity behavior of gypsum sands was studied using the four-electrode soil box. First, the variation in resistivity with respect to volumetric water content was investigated. Moist sand samples with different proportions of gypsum were prepared and placed in the box at a void ratio of approximately 0.6. The resistivity measurements were taken at different moisture contents as the sample was allowed to air-dry naturally. The results of the test are shown in Fig. 7. It was found that resistivity varied over a wide range with change in moisture content. The resistivity of all samples was typically less than 100 Ω m for volumetric water content more than 0.15. However, resistivity increased very steeply below water content of 0.1 for all the samples, measuring into several hundreds of Ω m. The same trend was found for samples with different gypsum contents.

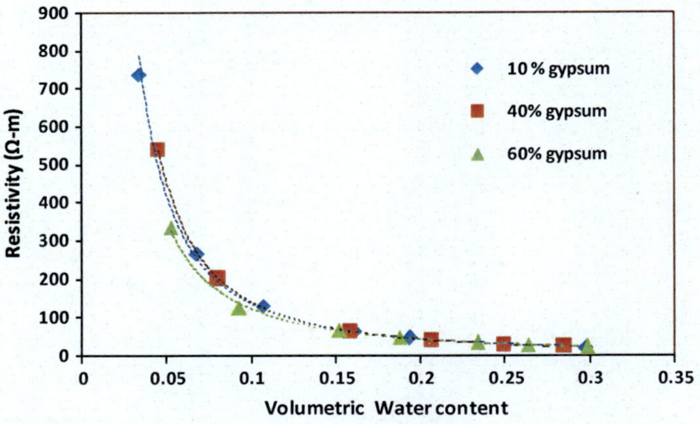


**Fig. 7** Electrical resistivity plotted against water content

Gypsum is a sparingly soluble salt [8] and gets dissociated into $Ca^{2+}$ and $SO_4^{2-}$ ions upon dissolution in water [9]. Electrical resistivity tests were carried out to see if the gypsum content and the presence of ions have any effect on the resistivity of gypsum sands. A detailed investigation for the effect of gypsum content on resistivity of gypsum sands was conducted by studying the resistivity of several saturated gypsum sand mixtures. Saturated mixtures were prepared and placed in the soil box and their resistivity was measured (Fig. 8). Soil samples with gypsum content varying from 0 to 90% were used. The results showed that sand with no gypsum had a resistivity of 27 Ω-m. Resistivity decreased with the addition of gypsum into the soil. Between 10 and 30% gypsum, the resistivity was around 16–17 Ω m and increased slightly up to 20 Ω m with further addition of gypsum. No appreciable change in resistivity was found beyond 50% gypsum.

Further, the influence of dry density of the gypsum sand on the electrical resistivity of the soil was tested. Saturated sand samples of 20 and 50% gypsum were prepared and compacted (by tamping) to different dry densities. The samples showed a slight increase in resistivity with increase in dry density (Fig. 9). For the sample with 20% gypsum, resistivity varied between 15 and 25 Ω m between dry density of 1.5 and 1.86 g/cc. The 50% gypsum sample displayed a resistivity range of 20–30 Ω m, for

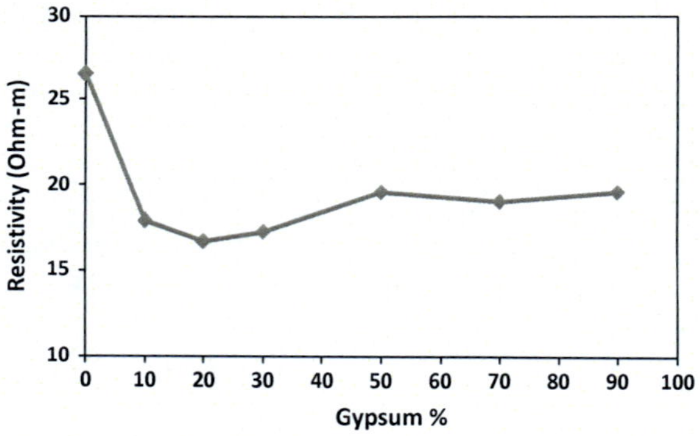


**Fig. 8** Electrical resistivity plotted against gypsum content

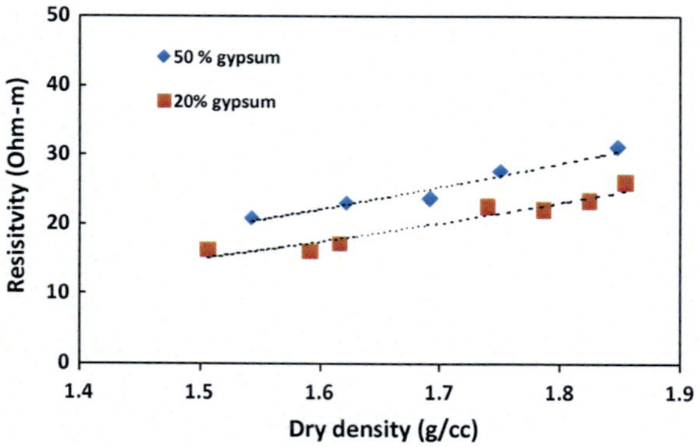


**Fig. 9** Electrical resistivity plotted against dry density

dry density values between 1.55 and 1.85 g/cc. The overall change in resistivity was relatively small, with respect to change in dry density of the soil.

## 5 Results and Discussion

Both initial moisture content and gypsum content effect the settlement of gypsum sands. The addition of water causes an increase in dissolution of gypsum and also causes weakening of any cementing bonds between the sand and gypsum. The long-term settlement of the gypsum soil samples increased with time. It was also found to be dependent on the gypsum content of the soil, with greater settlement taking place with increasing gypsum concentration. This observation was consistent with the findings of Fattah et al. and Mahmood et al. [10]. Percolating water through samples of gypsum sand causes leaching of gypsum. The settlement of soil also increases with leaching. From the tests, it was seen that leaching and settlement was greater when gypsum content was higher. The amount of gypsum dissolved at any given time was roughly constant.

The effect of moisture content, gypsum content and density of the soil sample on the electrical resistivity behavior of gypsum sands were studied and their results were noted. Moisture content had the greatest effect on the resistivity of the soil samples, showing a variation across hundreds of Ohm-m. These results were in agreement with the study conducted by Pandey et al. Resistivity increased very steeply at low moisture contents. In the absence of clay minerals, water is the only main conductor of electricity in otherwise inert sands. Hence, the lower the water content, the greater the resistivity. The presence of gypsum salt in the sand showed only a small change in resistivity values. In general, the resistivity of saturated gypsiferous sands was slightly lower than that of sand without any gypsum. This lowering in resistivity is likely due to the presence of $Ca^{2+}$ and $SO_4^{2-}$ ions. However, the increasing concentration of gypsum shows minimal effect on the resistivity pointing toward the low solubility of gypsum. Lastly, the resistivity of the saturated gypsum soils showed an increase with increasing dry density. The reduction of voids with increasing density causes lower amount of water to be present in the soil matrix. Hence, the lower the water, the greater the resistivity. The results could, however, vary when partially saturated soil is tested. The range of resistivity variation was about 10 Ω-m.

## 6 Conclusions

The study shows that the settlement of gypsum sands is influenced by factors like initial moisture content and gypsum content. Long term settlement (creep) increases with the amount of gypsum in soil. Leaching of gypsum by percolation of water through the soil is also a major factor effecting soil settlement. Thus, these factors should be investigated and considered during the geotechnical investigation stage for any site containing gypsum.

Electrical resistivity testing can also be used a tool to assess soils containing gypsum. Resistivity may not show much variation with density and gypsum content of the sand, but it varies significantly with the moisture content. The results from the testing can be used to make any judgments about the properties of the gypsum soil and any risk of settlement associated with the soil.

## References

1. Cooper, A.H., Calow, R.C.: Avoiding Gypsum Geohazards: Guidance for Planning and Construction. British Geological Survey, Technical Report WC 98/5, Overseas Geology Series (1998)
2. Seleam, S.: Evaluation of collapsibility of gypseous soils in Iraq. J. Eng. **13**(3), 712–726 (2006)
3. Al-Marsoumi, A.-M.H., Kadhum, M.J., Kadhum, S.K.: Some geotechnical properties of sandy gypsiferous soils in Rumaila-Khor Al-Zubair Area, Southern Iraq. Marina Mesopotamica **23**(2), 333–347 (2008)
4. Fattah, M.Y., Yousif, J., Huda, N.: Long-term deformation of some gypseous soils. Eng. Technol. J. Iraq 1461–1483 (2008)
5. Kalinski, R.J., Kelly, W.E.: Estimating the water content of soils from electrical resistivity, Geotech. Test. J. 323–329 (1993)
6. Pandey, L.M.S., Shukla, S.K., Habibi, D.: Electrical resistivity of sandy soil. Geotechn. Lett. **5**, 178–185 (2015)
7. Bhamidipati, R.A., Kalinski, M.E.: Geotechnical and Electrical resistivity properties of gypsum rich sands. Adv. Comput. Methods Geomech. LNCE **56**, 717–729 (2020)
8. Adiku, S., Renger, M., Roth, C.: A simple model for extrapolating the electrical resistivity data of gypsum containing soils from reference soil extract data. Agric. Water Manag. **21**, 235–246 (1992)
9. Bolan, N.S., Syers, J.K., Sumners, M.E.: Dissolution of various sources of gypsum in aqueous solutions and in soil. J. Sci. Food Agric. 527–541 (1991)
10. Mahmood, M.S., Akhtharpour, A., Almahmodi, R., Husain, M.M.A.: Settlement assessment of gypseous sand after time based soaking. IOP Conf. Ser. Mater. Sci. Eng. **737** (2020)

# Evaluation of Liquefaction Susceptibility of Soils in Kerala, India, Based on Equivalent *N* Value and Equivalent Acceleration

**Akshay Vikram, S. M. Alex Abraham, M. R. Greeshma, Iswarya Ani, A. Muhammed Siddik, and S. Adarsh**

## 1 Introduction

Soil liquefaction is one of the most complex phenomena studied in geotechnical earthquake engineering. Darve [1] considered liquefaction as a specific feature of loose and saturated sandy soils. Soil liquefaction occurs when a saturated or partially saturated soil substantially loses strength and stiffness in response to an applied stress such as shaking during an earthquake or other sudden change in stress condition, in which material that is ordinarily a solid behaves like a liquid. The effects of soil liquefaction on the built environment can be extremely damaging. Buildings whose foundations bear directly on sand which liquefies will experience a sudden loss of support, which will result in drastic and irregular settlement of the building causing structural damage, including cracking of foundations and damage to the building structure itself, or may leave the structure unserviceable afterwards, even without structural damage. Soil liquefaction leads to loss of bearing strength, lateral spreading, sand boil, flow failures, ground oscillations, floatation, settlement, overturning of structures. Therefore, the evaluation of liquefaction susceptibility for the design of structures is needed. The major threat recently faced by Kerala was soil piping, which was occurred due to the liquefaction.

Liquefaction prediction and assessment charts developed based on field and laboratory investigations have been widely used for evaluating liquefaction potential [1–11]. The liquefaction charts are characterized by the relationship between the cyclic stress ratio (CSR) versus field measured value such as standard penetration test (SPT) *N* values [12, 13], cone penetration test (CPT) *q*-values [9], and shear wave velocities [8, 14]. The developed liquefaction charts mentioned above are for an earthquake magnitude of 7.5. For other magnitudes, magnitude scaling factors are introduced

A. Vikram (✉) · S. M. Alex Abraham · M. R. Greeshma · I. Ani · A. Muhammed Siddik · S. Adarsh
TKM College of Engineering, Kollam, India



C. N. V. Satyanarayana Reddy et al. (eds.), *Ground Characterization and Foundations*, Lecture Notes in Civil Engineering 167,
https://doi.org/10.1007/978-981-16-3383-6_14

to correct the cyclic stress ratio, based on assumption that there exists an equivalent number of cycles for a given earthquake [15]. But earthquake motions at given sites generally have different waveforms and durations that vary considerably in space and time depending on the characteristics of sites. Therefore, waveforms and duration of earthquake should be considered in liquefaction prediction and assessment charts.

The liquefaction susceptibility of central Kerala (Ernakulam) was evaluated using factor of safety, calculated from CSR and cyclic resistance ratio (CRR) by Akhila et al. [16]. Even though CSR and CRR based estimates are popular for evaluation of liquefaction susceptibility, performing dynamic SPT or CPT becomes inevitable for the field data collection, by simulating the seismic conditions in the field. The liquefaction chart presented in Japanese guidelines by the Ministry of Land, Infrastructure, Transport and Tourism [7] is a simplified approach, which considers the equivalent $N$ value and equivalent acceleration for the assessment of liquefaction potential. This study applies the use of MLIT chart, for evaluating the liquefaction susceptibility of soils in the state of Kerala, India. This method is capable of accounting for the influence of the waveforms and duration of earthquake without performing the real field simulation. In this procedure, the liquefaction susceptibly of any location can be assessed by marking in a second plane between equivalent acceleration and equivalent n value. The equivalent acceleration is modified by applying appropriate correction accounting for the seismic vulnerability of Kerala.

## 2 Methodology

The borehole log details of various sites in Kerala are collected. The SPT $N$ value that corresponds to a depth ranges from 0 to 3.5 m is taken. Collection of geotechnical investigation data for different soils in Kerala involves both the SPT $N$ values and laboratory results. Equivalent $N$ value ($N_{65}$) is the SPT $N$ value ($N$) corresponding to effective vertical stress of 65 kPa [10]. MLIT [7] had introduced Eq. (1) for deriving equivalent $N$ value from SPT $N$ value.

$$N_{65} = \frac{N - 0.019\left(\sigma'_V - 65\right)}{0.0041\left(\sigma'_V - 65\right) + 1} \tag{1}$$

where $\sigma'_V$ = effective overburden pressure of the subsoil.

Effective overburden pressure is calculated with respect to ground at the time of standard penetration test by using Terzaghi's principle. When the fine content is relatively large, equivalent $N$ value should be corrected before applying (Table 1).

**Equivalent accelerations**. MLIT [7] proposed the method (Eq. (2)) of calculating the equivalent acceleration from the effective overburden pressure and maximum shear strength.

**Table 1** Corrections of equivalent $N$ value

| Case 1 | Case 2 | Case 3 |
|---|---|---|
| Plasticity index < 10 or cannot be determined or fine content < 15% then the equivalent $N$ value after correction should be set as $N_{65}/C_N$, $C_N$ is the correction factor of equivalent $N$ value corresponding to fine content | Plasticity index > 10% and < 20% or fine content > 15%, then the equivalent $N$ value after correction should be set as both $N_{65}/0.5$ and $N + \Delta N$ where $\Delta N = 8 + 0.4(\text{Ip} - 10)$ | Plasticity index is 20% or greater, and fine content is > or equal to 15%, then the equivalent n value after correction should set as $N + \Delta N$ the range should be determined with respect to corrected equivalent $N$ value |

*Source* MLIT [7]

$$\alpha_e = 0.7\frac{\tau_{max}}{\sigma_v'}.g \quad (2)$$

Effective overburden pressure ($\sigma_v'$) is calculated using Terzaghi's effective stress principle. Maximum shear stress ($\tau_{max}$) is calculated using software ProShake. Equivalent accelerations are calculated for soil layer using maximum shear stress obtained from the results of the seismic response analysis of ground.

ProShake is a computer program for one-dimensional, equivalent linear ground response analysis. ProShake consists of three managers and they are input manager, solution manager, and output manager. The input manager organizes the input data which include soil model, thickness of soil layer, unit weight of soil, shear wave velocity, shear parameters, and input motions. The solution manager is where the actual site response analyses are executed. The solution manager displays all combinations of soil profiles and ground motions to allow the user to gauge the progress and completeness of the analyses as they are being performed. Shear stress time history can be plotted using the output manager. The liquefaction chart developed by MLIT is for an earthquake magnitude of 7.5 (Fig. 1). The chart is divided into four zones, which are Zone I liquefaction will occur, Zone II high possibility of liquefaction, Zone III low possibility of liquefaction, Zone IV liquefaction will not occur. For other magnitudes, Sassa and Yamazaki [10] had proposed wave correction coefficient (Eq. (3)) to correct equivalent acceleration.

$$C_\alpha = N_r^{-a} \cdot N_{eff}^{a} \quad (3)$$

$N_r$ is effective number of cycles assessed for reference earthquake (1983 Central Japan Sea Earthquake at Akita Port, Japan) which is about 5. Liquefaction prediction and assessments simultaneously consider the influence of the waveforms and durations of earthquakes by plotting the equivalent acceleration which is divided by the wave correction coefficient, $C_\alpha$.

$N_{eff}$ = effective number of cycles which is half the number of half waves above $0.6 \times \tau_{max}$ in the time history of the shear stress variation due to irregular seismic waves. The constant '$a$' depends on relative density of soil. The point corresponds to the corrected equivalent $N$ value, and corrected equivalent acceleration is superimposing

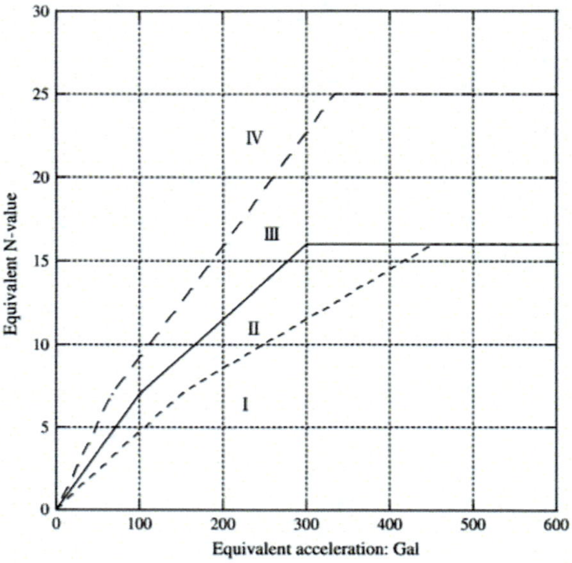


**Fig. 1** Liquefaction chart showing four zones for liquefaction prediction and assessment. [7] *Source* MLIT

with liquefaction chart developed by MLIT Japan to identify the zone of liquefaction susceptibility of the study area. The liquefaction chart proposed by MLIT is characterized by the relationship between the equivalent *N* value versus equivalent acceleration. The MLIT chart has been constructed for reference earthquake of the Central Japan Sea 1983. For other earthquake magnitudes, wave correction coefficients are used to correct equivalent acceleration. The State of Kerala lies in seismic zone 3. As the history of major earthquakes is practically absent in the state, the peak ground acceleration of $0.16 \times g$ of a real case history that occurred at nearby state Tamil Nadu is taken as the input. Possibility of liquefaction is evaluated by the zone in which the point lies (Fig. 1). Flowchart of methodology followed is given in Fig. 2.

## 3 Data Collection and Site Details

Standard penetration test provides samples for the identification purposes and provides a measure of penetration resistance which can be used for geotechnical design purposes. Standard penetration test is conducted by the free fall of safety hammer from a height of 750 mm. The corresponding SPT *N* value is measured and assigned in borehole log. Borehole log is collected along with SPT *N* values. Soil properties like index properties and engineering properties are also collected from different locations in Kerala (Fig. 3).

The data collection on the characteristics, depth of the soil, and the SPT value of 15 sites including Achankovil (Kollam), Kollam District Hospital, Mannar (Alappuzha),

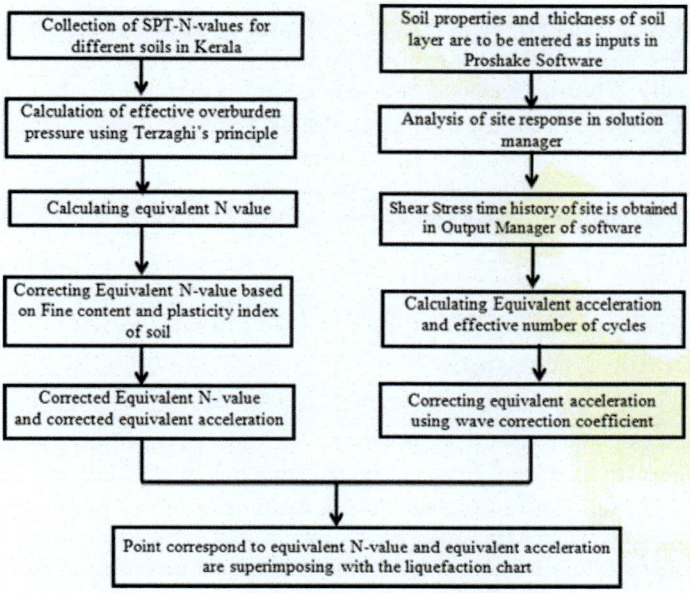


**Fig. 2** Flowchart of methodology

**Fig. 3** Location of borehole sites in Kerala

Kalamassery (Ernakulam), Victoria College (Palakkad), Thekkady (Idukki), Shangumugham (Trivandrum), Vennikulam (Pathanamthitta), Pothencode (Trivandrum), Karunagapally (Kollam), Kalpetta (Wayanad), Ponnani (Malappuram), Peralassery (Kannur), Chavakkad (Thrissur), and Kozhikode Medical College were analysed. The depth of the soil considered is in the range 0–3.5 m, and the soil type observed are reddish-brown clay, brown sand clay, red sandy soil, yellow lateritic sandy clay, brown lateritic sand, grey sand, reddish clayey sand with gravel, brown sand, black fine sand, etc.

## 4 Result and Discussions

Soil profile details and standard penetration test *N* value data of all site have subjected to the analysis for the identification of liquefaction potential. The equivalent acceleration was computed based on the shear stress time history (Fig. 4). The sample calculation results of site Achankovil are shown in Table 2. The points that correspond to corrected equivalent *N* value and corrected equivalent acceleration are plotted in the liquefaction chart (Fig. 5). The same procedure is applied for remaining sites. Liquefaction chart of all sites is plotted here and is analysed. Calculated equivalent *N* values and the corrected equivalent acceleration for different sites are listed in Table 3.

This procedure is applied to all the borehole data of various sites to determine equivalent *N* value and equivalent acceleration. Liquefaction analyses have been done on all sites for a PGA of 0.16 g and an earthquake magnitude of 6. Figure 6 shows the liquefaction susceptibility of all the sites considered in this study.

Result shows that 1.85%, 5.56%, 3.7% and 89% of total boreholes lies in zone 1, zone 2, zone 3 and zone 4, respectively. The results reveal that most of the borehole lies in the 4th zone, which denotes those areas which are not susceptible to liquefaction. Hence the possibility of liquefaction is low in these sites.

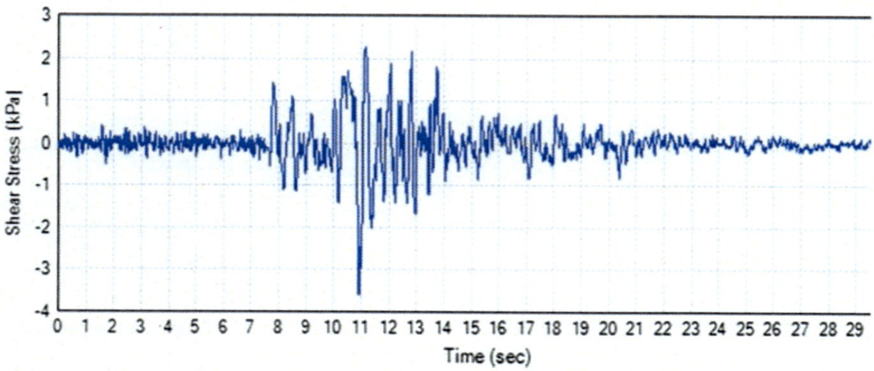


**Fig. 4** Shear stress diagram obtained from ProShake

**Table 2** Calculated results of site Achankovil, Kollam

| Parameter | Value |
|---|---|
| Site | Achankovil, Kollam |
| Depth | 3.5 m |
| Cohesion | 56 kPa |
| Angle of internal friction | 24.3 |
| Relative density, $D_r$ | 20.28% |
| Shear wave velocity | 74.37 m/s |
| SPT $N$-value of soil layer | 9 |
| Top layer soil type | Reddish-brown clay for a depth of 1.6 m |
| Unit weight of top layer | 19 kN/m$^3$ |
| Bottom layer soil type | Brown sandy clay for a depth of 1.9 m |
| Unit weight of bottom layer | 18 kN/m$^3$ |
| Equivalent $N$ value | 9.02 |
| Fine content | 62.80% |
| Plasticity index, $I_p$ | 25 |
| Correction, $N + \Delta N$ | 14 |
| Corrected equivalent $N$ value | 23 |
| Equivalent acceleration, $\alpha_e$ | 41.138 Gal (Galileo) |
| Maximum shear stress | 3.87 kPa (Fig. 4) |
| Correction, $C_\alpha$ | 0.948 |
| Corrected equivalent acceleration | 43.38 Gal (Galileo) |

# 5 Discussion

The result of the study area shows that the majority of the site considered for the study are falling in the Zone-4, which shows that the study regions are not susceptible to the liquefaction. Generally, SPT values are designed to geotechnical design purposes. The unavailability of geotechnical investigation data of soil from various spectra is a barrier in the evaluation of liquefaction potential of all sites in Kerala. For prediction and preparedness against the liquefaction hazard, the government should take initiative in developing a geotechnical investigation data of soil where construction is not possible. Thereby, the prediction of liquefaction possibility of more sites can be done and liquefaction can be prevented by taking remedial measures. It can reduce the damages due to liquefaction in terms of life and assets.

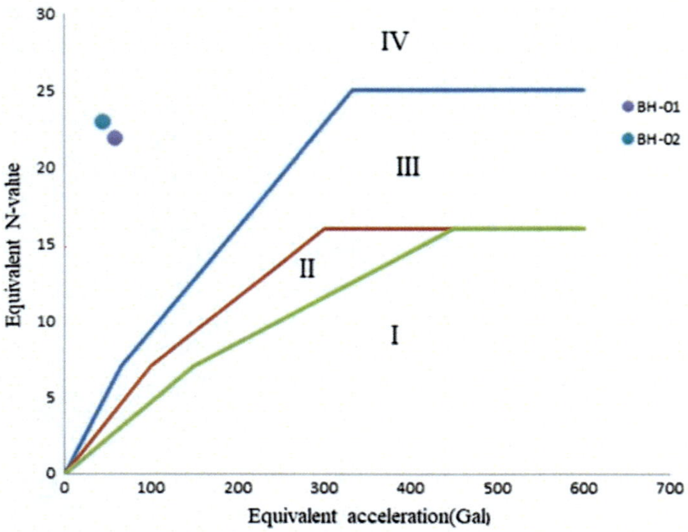


**Fig. 5** Liquefaction susceptibility of site Achankovil, Kollam, marked in liquefaction chart

## 6 Conclusions

Liquefaction prediction and assessment is a vital part of the earthquake-resistant design of structures on liquefiable soils. Liquefaction prediction and assessment methods have been widely used for such design in practice, as well as for disaster prevention and mitigation. In this paper, the liquefaction susceptibility of soils in Kerala is evaluated using liquefaction chart proposed by the Ministry of Land, Infrastructure, Transport and Tourism (MLIT), Japan, which is a simplified liquefaction prediction and assessment method. Apart from the other, this method is capable of considering the influence of the waveforms and duration of earthquake. The SPT $N$ values collected from the various construction sites are subjected to a correction corresponding to the fine content and plasticity index of the soil, to get the equivalent $N$ value. Then, the equivalent acceleration was calculated using maximum shear stress obtained from ProShake 2.0 software for a peak ground acceleration of $0.16 \times g$. The obtained results were subjected to a wave correction corresponding to the earthquake magnitude of the study area. Subsequently, the point corresponds to the equivalent $N$ value and equivalent acceleration was plotted in the reference liquefaction chart proposed by MLIT, Japan, to determine the zone of liquefaction susceptibility of sites. The result shows that majority of the study area lies in Zone 4, which is free from liquefaction.

**Table 3** Corrected equivalent N values and the corrected equivalent acceleration for different sites

| Place | Bore hole | SPT *N* value | Corrected equivalent *N* value | Corrected equivalent acceleration (Gal) |
|---|---|---|---|---|
| Achankovil, Kollam | BH 01 | 9 | 23 | 43.38 |
| | BH 02 | 8 | 22 | 54.00 |
| Kollam District Hospital | BH 01 | 3 | 33 | 54.97 |
| | BH 02 | 7 | 8 | 0.29 |
| Mannar, Alappuzha | BH 01 | 2 | 21 | 50.98 |
| | BH 02 | 4 | 22 | 47.99 |
| | BH 03 | 2 | 22 | 40.45 |
| | BH 04 | 1 | 19 | 71.55 |
| Kalamassery, Ernakulam | BH 01 | 51 | 63 | 2.23 |
| | BH 02 | 45 | 57 | 32.85 |
| | BH 03 | 51 | 65 | 56.03 |
| | BH 04 | 39 | 53 | 2.26 |
| Ponnani, Malappuram | BH 01 | 6 | 8 | 150.83 |
| | BH 02 | 7 | 10 | 69.26 |
| | BH 03 | 4 | 6 | 180.19 |
| Victoria College, Palakkad | BH01 | 8 | 14 | 38.14 |
| | BH 02 | 6 | 11 | 28.50 |
| | BH 03 | 10 | 14 | 35.52 |
| | BH 04 | 22 | 38 | 2.19 |
| | BH 05 | 15 | 33 | 61.06 |
| Thekkady, Idukki | BH 01 | 7 | 17 | 35.97 |
| | BH 02 | 1 | 2 | 2.17 |
| Shangumugham, Trivandrum | BH 01 | 9 | 8 | 1.92 |
| | BH 02 | 11 | 19 | 1.88 |
| Vennikulam, Pathanamthitta | BH 01 | 20 | 22 | 60.57 |
| | BH 02 | 51 | 51 | 2.62 |
| | BH 03 | 32 | 69 | 2.57 |
| | BH 04 | 38 | 79 | 43.98 |
| Karunagapally, Kollam | BH 01 | 2 | 2 | 2.01 |
| | BH 02 | 4 | 4 | 47.94 |
| | BH 03 | 4 | 4 | 45.72 |
| | BH 04 | 7 | 7 | 44.56 |
| Kalpetta, Wayanad | BH 01 | 7 | 8 | 67.17 |
| | BH 02 | 15 | 29 | 2.24 |
| | BH 03 | 20 | 22 | 2.66 |

(continued)

**Table 3** (continued)

| Place | Bore hole | SPT *N* value | Corrected equivalent *N* value | Corrected equivalent acceleration (Gal) |
|---|---|---|---|---|
| Peralassery, Kannur | BH 01 | 6 | 12 | 75.80 |
| | BH 02 | 23 | 23 | 62.49 |
| | BH 03 | 10 | 21 | 3.21 |
| | BH 04 | 17 | 35 | 2.68 |
| | BH 05 | 18 | 37 | 2.72 |
| Chavakkad, Thrissur | BH 01 | 3 | 4 | 72.58 |
| | BH 02 | 3 | 7 | 3.48 |
| | BH 03 | 3 | 4 | 84.01 |
| Pothencode, Trivandrum | BH 01 | 20 | 40 | 39.83 |
| | BH 02 | 15 | 30 | 51.30 |
| | BH 03 | 30 | 48 | 57.55 |
| | BH 04 | 30 | 59 | 6.12 |
| | BH 05 | 26 | 41 | 55.70 |
| Kozhikode Medical College | BH 01 | 36 | 40 | 66.62 |
| | BH 02 | 50 | 54 | 2.16 |
| | BH 03 | 4 | 8 | 2.72 |
| | BH 04 | 4 | 16 | 2.70 |
| | BH 05 | 4 | 17 | 2.78 |
| | BH 06 | 17 | 18 | 2.97 |

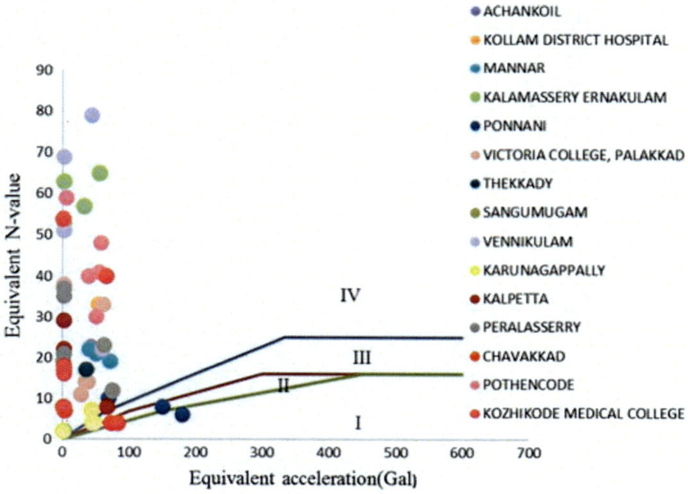


**Fig. 6** Liquefaction susceptibility of various sites in Kerala State

## References

1. Darve, F.: Liquefaction phenomenon of granular materials and constitutive stability. Eng. Comput. **13**(7), 5–28 (1996)
2. Bray, J.D., Sancio, R.B.: Assessment of the liquefaction susceptibility of fine-grained soils. J. Geotech. Geoenviron. Eng. **132**(9), 1165–1177 (2006)
3. Hakam, A., Ismail, F.A., Fauzan: Liquefaction potential assessment based on laboratory test. Int. J. Geomate **11**(26), 2553–2557 (2016)
4. Ishihara, K., Yasuda, S.: Sand liquefaction in hollow cylinder torsion under irregular excitation. Soils Found. **15**(1), 45–59 (1975)
5. Iwasaki, T.: A practical method for assessing soil liquefaction potential based on case studies at various sites in Japan. In: Proceedings of the Second International Conference on Microzonation for Safer Construction Research Application, vol. 2, pp. 885–896 (1978)
6. Maurer, B.W., Green, R.A., Cubrinovski, M., Bradley, B.A.: Evaluation of the liquefaction potential index for assessing liquefaction hazard in Christ Church, New Zealand. J. Geotech. Eng. **140**(7), 04014032 (2014)
7. MLIT (Ministry of Land, Infrastructure, Transport and Tourism, Japan): Ground liquefaction. Technical Standards and Commentaries for Port and Harbour Facilities in Japan, pp. 282–288 (2007)
8. Shelley, E.O., Mussio, V., Rodríguez, M., Acosta Chang, J.G.: Evaluation of soil liquefaction from surface analysis. Geofísica Int. **54**(1), 95–109 (2015)
9. Robertson, P.K., Wride, C.E.: Evaluating cyclic liquefaction potential using the cone penetration test. Can. Geotech. J. **35**(3), 442–459 (1998)
10. Sassa, S., Yamazaki, H.: Simplified liquefaction prediction and assessment method considering waveforms and durations of earthquakes. J. Geotech. Geoenviron. Eng. **143**(2), 04016091 (2017)
11. Seed, H.B., Idriss, I.M.: Simplified procedure for evaluating soil liquefaction potential. J. Soil Mech. Found. Div. (1971)
12. Seed, H.B., Idriss, I.M., Arango, I.: Evaluation of liquefaction potential using field performance data. J. Geotech. Eng. **109**(3), 458–482 (1983)
13. Seed, H.B., Tokimatsu, K., Harder, L.F., Chung, R.M.: The influence of SPT procedures in soil liquefaction resistance evaluations. J. Geotech. Eng. **111**(12), 1425–1445 (1985)
14. Andrus, R.D., Stokoe, K.H., II.: Liquefaction resistance of soils from shear-wave velocity. J. Geotech. Geoenviron. Eng. **126**(11), 1015–1025 (2000)
15. Youd, T.L., Idriss, I.M.: Liquefaction resistance of soils: summary report from the 1996 NCEER and 1998 NCEER/NSF workshops on evaluation of liquefaction resistance of soils. J. Geotech. Geoenviron. Eng. **127**(4), 297–313 (2001)
16. Akhila, M., Rangaswamy, K., Sankar, N.: Liquefaction susceptibility of central Kerala. SN Appl. Sci. **1**(6), 583 (2019)

# Plasticity Characteristics of China Clay-Bentonite–Sand Mix Proportions

**D. N. Jyothi, H. S. Prasanna, H. S. Pavithra, and K. A. Yashaswini**

## 1 Introduction

Kaolinite and Montmorillonite are the two clay minerals present in the natural soils. In kaolinite due to attractive forces dominance, flocculent structure is seen in soil, whereas in montmorillonite due to repulsive forces dispersive structure is seen. The behaviour of fine-grained soil is distinct because of its physico-chemical in nature which is influenced by the soil clay mineralogy. A lot of financial requirements and time is required to study the Plasticity characteristics and activity of fine grained soils having different clay mineralogy around the world. In order to get over this problem, in the present experimental study the plasticity characteristics, activity of artificially prepared China. Clay–Bentonite sand mix proportions which is likely to represent the natural fine-grained soil behaviour are studied. The new plasticity chart in conjunction with IS plasticity chart is also developed. This chart is also used in accessing the physical properties of clay. Acitivity of artificially prepared sand mix proportions chart is also developed and compared with nature soils. In this Experimental study data analysis of artificially prepared soils in laboratory representing the natural fine grained soils.

## 2 Literature Review

Bolton seed et al. [4] in their experimental work on clay mineralogy observed that mixtures of illite and bentonite are less active than mixtures of kaolinite and bentonite in the same properties and have lower swelling characteristics.

D. N. Jyothi · H. S. Prasanna (✉) · H. S. Pavithra · K. A. Yashaswini
The National Institute of Engineering, Mysore, India
e-mail: prasanna@nie.ac.in



C. N. V. Satyanarayana Reddy et al. (eds.), *Ground Characterization and Foundations*, Lecture Notes in Civil Engineering 167,
https://doi.org/10.1007/978-981-16-3383-6_15

Dumbleton and West [6] made an attempt to study the factors affecting the relation between the clay minerals in soil and their plasticity and the relationships between clay content, plastic limit and liquid limit of natural montmorillonite and kaolinitic soils and of artificial mixtures have been examined and compared.

Bain [5] has shown that a plasticity chart as aid to the identification and assessment of industrial clays. This identification can be done by using their Atterberg 'Plastic limit' and 'Plasticity index' values as parameters for an identification chart.

Sridharan et al. [8] studied on shrinkage limit of soil mixtures. The study gives that shrinkage limits of the natural soil will not depend on plasticity characteristics. This study confirms this mechanism with the result obtained using clay-clay, clay-non cohesive soil, and non-cohesive soil mix system.

Polidori [2] investigates on the liquid limit and plastic limit of six inorganic soils with fine silica sand. The study gives that knowing the two out of three parameters, the value of other parameters can be obtained.

Polidori [3] study on activity of clays and activity chart aims to classify soils. The activity chart was adapted to predict residual shear behaviour of cohesive soils.

Bera [1] made detailed experimental work on effect of sand content on engineering properties of fine-grained soil mixed with sand. The study highlights the effect of mixing sand (%)on liquid limit, plastic limit of fine-grained soil.

Uprety (2016) conducted test on index properties of the soils using cyclic triaxialshear test. They observed that the state of knowledge on the effects of plasticity on liquefaction characteristics of the fine-grained soil.

Prasanna et.al [7] made a study on index properties of kaolinite and bentonite sand mixtures. Index properties have an induce on the shrinkage characteristics, compaction, swelling, shear strength, bearing capacity of the kaolinite-sand mixtures and bentonite–sand mixtures. In their study, an effort was also made to match up the index properties of natural soil with that of artificial mix proportions.

Detailed literature review of Plasticity characteristics of kaolinite–bentonite–sand mixture is very scanty.

## 3 Materials and Method

Soil mix proportion contains commercially available clay minerals like china clay and bentonite, along with natural river sand. China clay (kaolin) is a clay mineral that is part of industrial mineral with chemical composition Al2Si2O5(OH)4, it is chemically inert and prepared as a white powder. Bentonite is physically and chemically reactive. It shrinks or swells and exhibits cation exchange. The river sand was initially wet washed to remove the dirt and silts, Later it is oven dried for 24 h at a temperature of $105 \pm 5$ °C and it is brought to a normal temperature and sieved through 425 $\mu$m. Commercially available bentonite and kaolin were obtained from Seema chemicals Bengaluru, were stored in air tight plastic container are mixed with well-prepared sand in varied proportions.

By keeping the bentonite constant the sand and china clay proportions were varied i.e., 10%B constant for mix ratio-1 with China clay varying from 10 to 80% and sand varying from 80 to 10%. Correspondingly 20%B, 30%B, keeping bentonite constant with sand and kaolinite varying proportions were prepared up to mix ratio-8. Similarly keeping kaolin constant, with sand and bentonite varying mix proportions prepared. The collected data has been correlated with plasticity characteristics of natural fine-grained soils and made an effort to develop new plasticity chart which is cognate to the Casagrande's plasticity chart.

The index property test on the mix proportions was carried out as per BIS specifications. In the present experimental investigation plasticity characteristics of a soil mix proportions are evaluated from the Atterberg limits (IS2720-PART 5 1985) to determine liquid limit, plastic limit and shrinkage.

## 4 Experiment Re and Discussion

Figures 1 and 2 shows Variation of Liquid Limit with Percent Fines.

Figure 1 shows the liquid limit with percent fines of MR-1 to MR-8. It can be observed that liquid limit increases from 31.5 to 127% (4 Folds) with the increase in percent fines in the artificial mix proportion. From Fig. 2, for MR-1 to MR-8 liquid limit increase from 31.5 to 69% (2.2 Folds) with the increase in percent fines in the mix proportion. This is due to the fact that increase in fines content results in increase in the water holding capacity, which is also a characteristic feature of a fine-grained soil. From Figs. 1 and 2 it can be also observed that liquid limit of mix proportions which is having higher percentage of bentonite is greater compare to the liquid limit of artificially prepared mix proportions which contain higher percentage of kaolinite which is mainly due to bentonite clay is composed of silica and alumina sheets arranged in such a way that it adsorbs large amount of water forming water tight barrier. It is observed that an effective correlation is seen for both constant bentonite and constant kaolinite mix proportions.

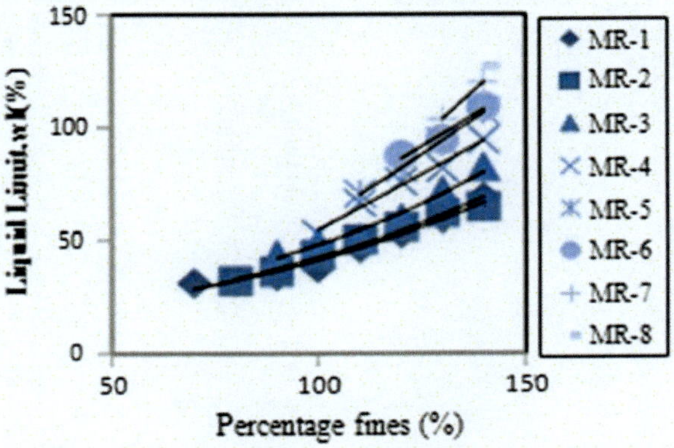


**Fig. 1** Variation of liquid limit, WL (%) with percent fines, PF (%) for constant bentonite

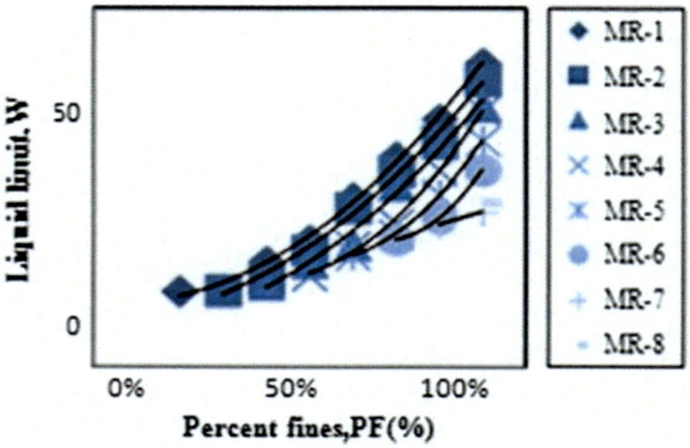


**Fig. 2** Variation of liquid limit, WL (%) with percent fines, PF (%) for constant kaolinite

Combined variation of liquid limit with percent fines is presented in Fig. 3 and the relation is found to be is linear. The relationship between liquid limit and percent fines is given in Eq. 1

$$WL = 1.61PF - 30.1 \tag{1}$$

Using this equation we can predict the value of liquid limit for any other value of percent fines without performing.

Figures 4 and 5 shows Variation of Plastic Limit with Percent Fines.

From Fig. 4, Plastic limit of MR-1 to MR-8 increases from 30 to 44.50%. In Fig. 1 the bandwidth of variation of liquid limit with percent fines is very high compare to that of plastic limit (Fig. 4). Hence, plastic limit which is having small range of bandwidth is considered to have one of the important plasticity characteristics rather than the liquid limit value alone.

This small range of bandwidth can be plotted (Fig. 5) as a combined variation of plastic limit with percent fines which gives the best average polynomial equation shown in Eq. 2. The plastic limit equation for a soil mix proportion is given by Eq. (2).

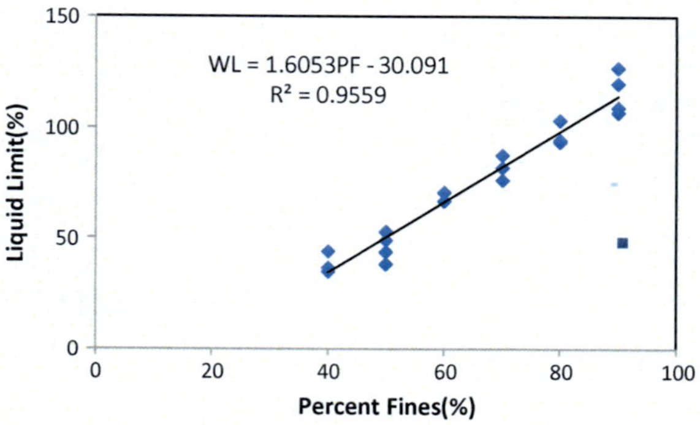


**Fig. 3** Combined variation of liquid limit with percent fines

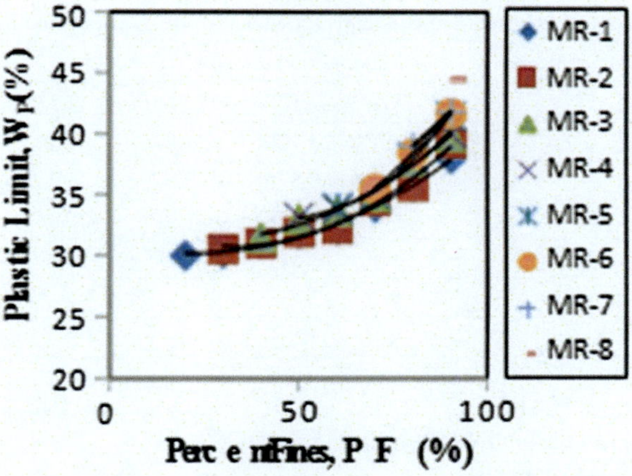


**Fig. 4** Variation of plastic limit $W_P$ with percentage fines

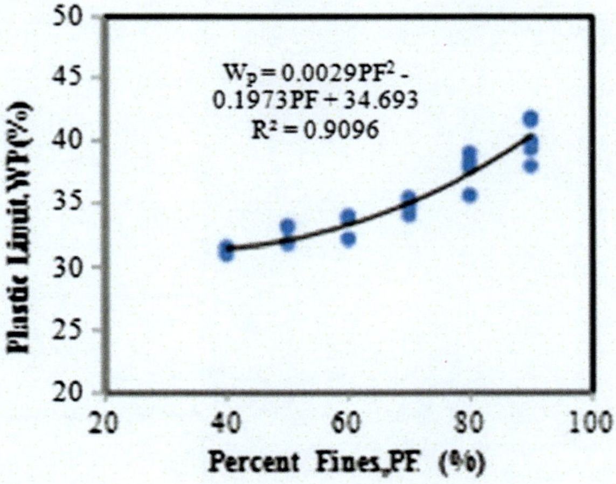


**Fig. 5** Combined Variation (MR-1 to MR-8) of Plastic Limit with Percent Fines

$$WP = K1\ (PF) + K2(PF)2 + 34.59 \tag{2}$$

The Average value of intercept WP = 34.59%, where the slopes K1 and K2 depend on the clay mine.

The regression equations of plotted mix proportion in Fig. 4 and its correlation coefficients are tabulated in Table 1

Figures 6 and 7 shows Variation of Shrinkage Limit with Percent Fines.

From Fig. 6, it is observed that for MR-1 to MR-8, Shrinkage limit decreases from 29.68 to 13.2% (2.24 folds). It can be observed that shrinkage limit is the opposite process of liquid and plastic limit of artificially prepared mix proportions.

Sridharan et al. (1998) shown that shrinkage limit is a function of grain size distribution of the natural soil, irrespective of the clay minerals of the soil and it does not depend on plasticity characteristics and it is for the different natural soil.

**Table 1** Correlation equations for MR-1 to MR-8 of plastic limit with percent fines

| Mix ratio No. | Correlation equation | $R^2$ | $R$ |
|---|---|---|---|
| 1 | $WP = 32.10PF^2 - 16.65PF + 32.53$ | 0.975 | 0.987 |
| 2 | $WP = 24.90PF^2 - 11.42PF + 31.64$ | 0.984 | 0.992 |
| 3 | $WP = 33.57PF^2 - 22.57PF + 34.75$ | 0.996 | 0.998 |
| 4 | $WP = 47.85PF^2 - 41.9PF + 40.58$ | 0.994 | 0.997 |
| 5 | $WP = -5PF^2 + 34.5PF + 13.1$ | 0.994 | 0.997 |
| 6 | $WP = 120PF^2 - 163PF + 89.3$ | 1 | 1 |
| 7 | $WP = 37PF + 6$ | 1 | 1 |

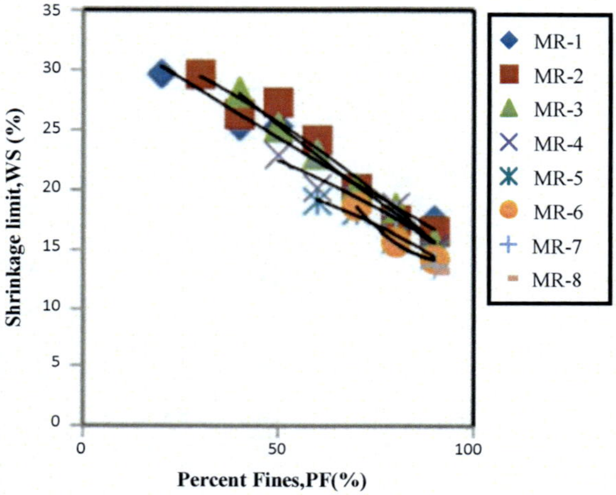


**Fig. 6** Variation of shrinkage limit WS (%) with percent fines

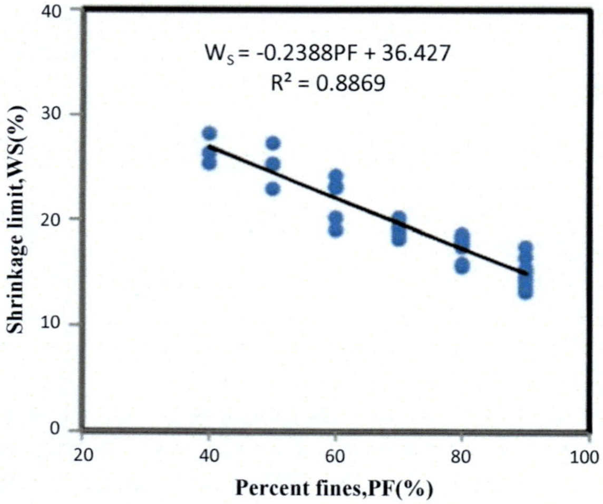


**Fig. 7** Combined variation of shrinkage limit with percent fines

Artificial mix proportion is the true representative of the natural soil behaviour. The best relation between the shrinkage limit and percent fines is the linear relationship. The equation remarked in Fig. 6, refer to the average value of intercept, WS = 36.42%., The general equation representing the relation between shrinkage limit and percent fines is given by Eq. 3.

$$\text{WS} = \text{K3(PF)} + 36.42 \tag{3}$$

where the WS-slope K3 depends on the clay minerals present.

Correlation equations and its coefficients of MR-1 to MR-8 (Fig. 6) are tabulated in Table 2.

Figure 8 shows Plasticity Chart of Artificial Mix Proportion.

Figure 8 shows the plasticity chart of the artificial mix proportion which is cognate to Casagrande's chart. In the above chart mix proportions are classified into silt (M) and clay(C), based on their liquid limit (WL) and plasticity index (IP). The relation between plasticity index (IP) and liquid limit (WL) is remarked in Eq-4 which is akin to A-line equation.

**Table 2** Correlation Equations for MR-1 to MR-8 of Shrinkage limit with Percent fines

| Mix ratio No. | Correlation equation | $R^2$ | $R$ |
|---|---|---|---|
| 1 | $\text{WS} = 1\text{E} - 04\text{PF}^2 - 0.20\text{PF} + 34.384$ | 0.95 | 0.97 |
| 2 | $\text{WS} = -0.001\text{PF}^2 - 0.11\text{PF} + 33.703$ | 0.95 | 0.97 |
| 3 | $\text{WS} = 0.0004\text{PF}^2 - 0.29\text{PF} + 39.299$ | 0.99 | 0.99 |
| 4 | $\text{WS} = -0.001\text{PF}^2 - 0.03\text{PF} + 26.516$ | 0.92 | 0.96 |
| 5 | $\text{WS} = -0.0012\text{PF}^2 + 0.01\text{PF} + 22.253$ | 0.97 | 0.98 |
| 6 | $\text{WS} = 0.0078\text{PF}^2 - 1.47\text{PF} + 83.5$ | 1 | 1 |
| 7 | $\text{WS} = -0.431\text{PF} + 52.47$ | 1 | 1 |

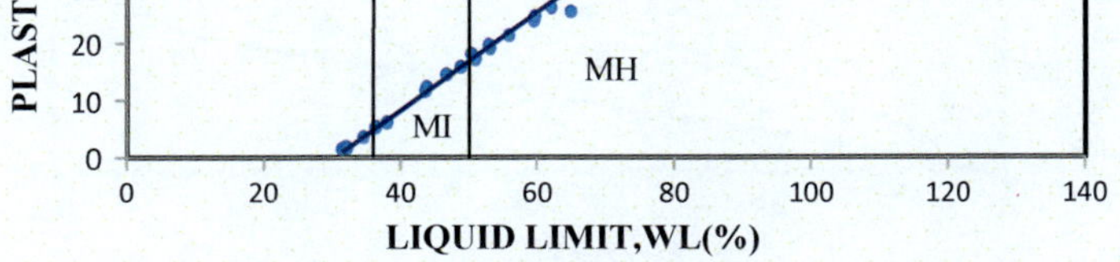


**Fig. 8** Plasticity chart of the artificial mix proportion

$$IP = 0.86(WL-30.5) \quad (4)$$

This Artificially prepared mix proportions line normally spilt the clay and silt materials. This mix proportions are further subdivided into low (L) or high (H) plasticity based on the liquid limit.

An attempt has been made to compare the Casagrande's plasticity chart with plasticity chart of artificial mix proportion. From Fig. 9, it is observed that they are in close agreement with one another.

Casagrande's A-line equation and artificially prepared mix proportion equation of soil mixtures are indicated in Table 3.

Figure 10 shows Activity of artificially prepared mix proportions.

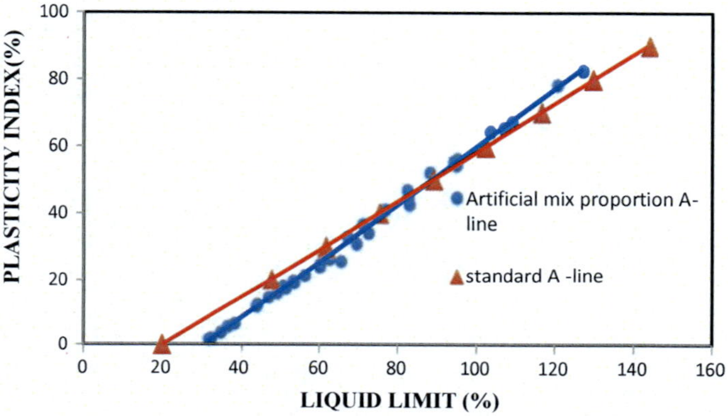


**Fig. 9** Variation of plasticity index, IP with liquid limit, WL of artificial mix proportion, and natural soils

**Table 3** Correlation equation

| | | |
|---|---|---|
| Mix proportion equation | IP = 0.87(WL − 30.5) | $R = 0.998$ |
| A-line equation | IP = 0.73(WL − 20) | $R = 1$ |

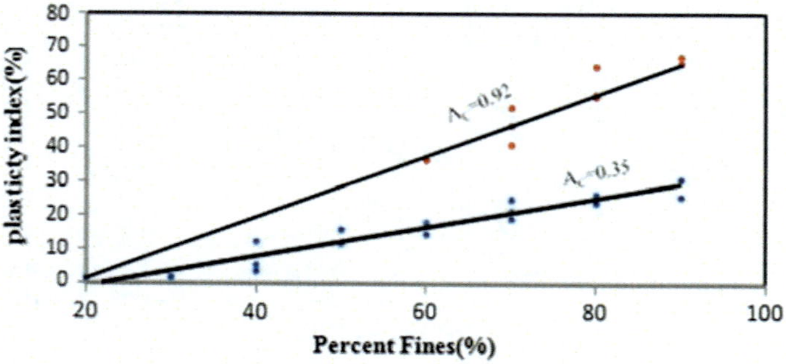


**Fig. 10** Activity chart of artificial mix proportions

**Table 4** Activity of artificial mix proportion

| Mineral | Activity |
|---|---|
| Montmorillonite | 0.5–0.92 |
| Kaolinite | 0.1–0.35 |

The properties of clays and their behaviour are mainly due to presence of certain clay minerals. Activity of a soil is equal to the ratio between plasticity index value and its percent fines, least and highest activity values belong to pure clay minerals kaolinite and bentonite, respectively.

In Fig. 10 activity chart is given for an artificial mix proportions, kaolinite have activity values ranging from 0.1–0.35 and montmorillonite have 0.5–0.92 indicated in Table 4. The experimental study shows that artificial mix proportion activity values match with that of the natural soils as proposed by Skempton in 1953.

Activity of the natural soil with artificial mix proportion is compared and tabulated in Table 4 and 5.

Figure 11 shows Variation of Liquid Limit with Plastic Limit.

Figure 11 shows the variation of plastic limit as a function of liquid limit in the mix proportions for different percentage of fines. The trend lines are parallel with a common slope of 0.079, if all the lines are extended, they pass through the axis of ordinate at $Z = 0.058\text{PF} + 28.95$, which is a function of percent fines. The relationship of liquid limit, plastic limit and percent fines is given by Eq. 5

$$\text{WP} = 0.079\text{WL} + (0.058\text{PF} + 28.95) \tag{5}$$

**Table 5** Activity of natural clay minerals

| Mineral | Activity |
|---|---|
| Montmorillonite | 0.7–1.0 |
| Kaolinite | 0.4–0.5 |

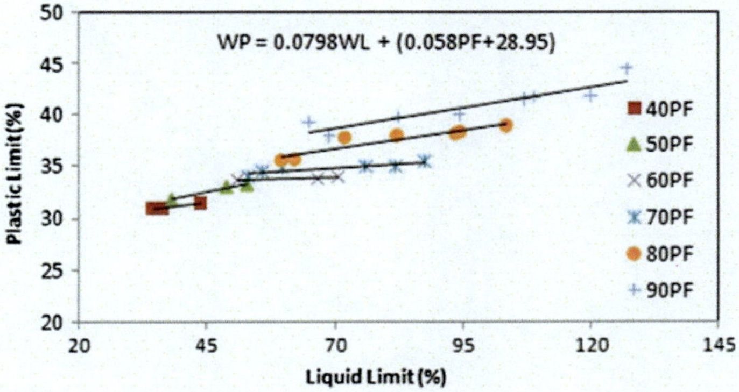


**Fig. 11** Variation of liquid limit with plastic limit

### Plasticity Index

In Fig. 8 plasticity index is defined by difference between liquid limit and plastic limit. Plasticity index can be obtained by liquid limit and percent fines without the value of plastic limit. The relationship between plasticity index (IP), liquid limit (WL) and percent fines is given by Eq. 6

$$IP = 0.86WL - (0.058PF + 28.9) \tag{6}$$

From Eq. 6 for known value of WL and PF, the value of plasticity index is estimated. Using Eq. 1 and 5 for different value of percent fines, the value of WL and IP can be estimated respectively without performing the test.

Figure 12 shows the relation between the calculated and measured data of plasticity index. Calculated plasticity index can be obtained by Eq. 5 (which is graphically given) and made an attempt to compare the IP measured and calculated data of the artificially prepared mix proportions. It can be observed that the values are in close agreement with one another.

Figures 8 and 13 shows the plasticity chart for measured and calculated value of plasticity index with liquid limit for an artificially mix proportions. Both the correlations are in close agreement with the Casagrande's plasticity chart.

Activity of the soil depends on the percent fines. Variation of activity with percent fines shown in Fig. 14, for constant bentonite and constant kaolinite. The correlation between activity and percent fines is given in Eq. 7 with correlation coefficient

$$\begin{aligned} A &= 0.013PF - 0.319 \\ R^2 &= 0.9094 \end{aligned} \tag{7}$$

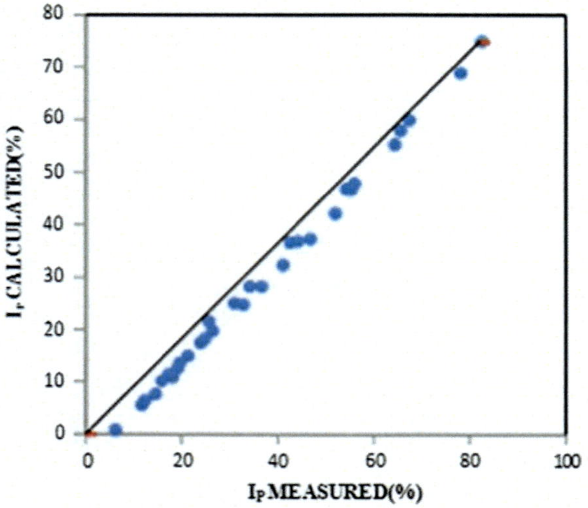


**Fig. 12** Comparison between measured and calculated plasticity index

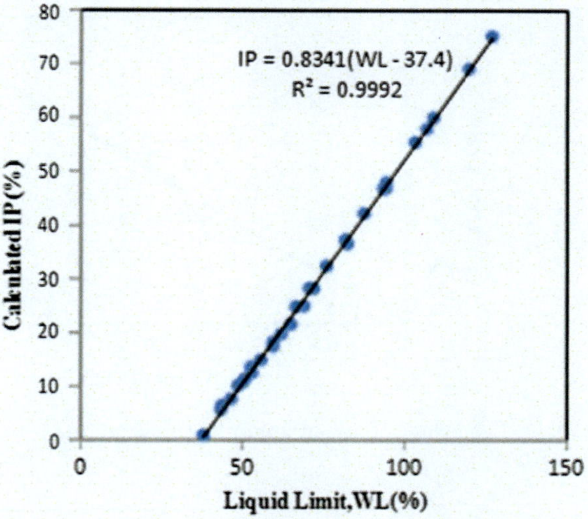


**Fig. 13** Plasticity chart of calculated plasticity index with liquid limit

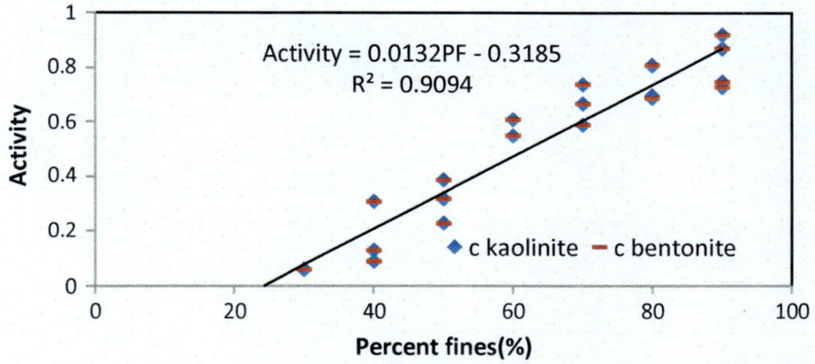


**Fig. 14** Variation of activity with percent fines

Figures 15 and 16 shows Relationship of plasticity characteristics of natural with that of artificial soil mixtures.

In Fig. 15, the liquid limit of artificial mix proportion with that of natural having percent fines which is having same percentage as that of mix proportions and it is observed that from 20 to 80% fines, lies on the line of equality. Hence it can be concluded that artificial mix proportion prepared in the laboratory has similar plasticity characteristics of natural soils.

Figure 16 shows the variation of plastic limit of natural soils obtained from the author's experimental work with that of the plastic limit of mix proportions obtained from present experimental study. It is observed that the range of plastic limit values vary from 30 to 40%.

It can be concluded that the plasticity characteristic of natural fine grained soils (WL, WP) can be estimated with the help of artificial mix proportions with a fair degree of accuracy.

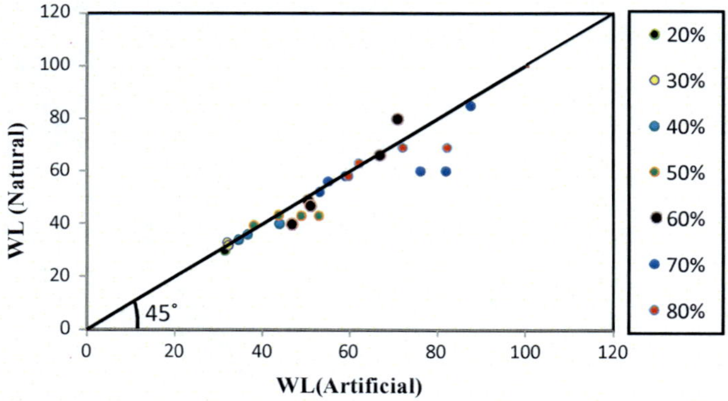


**Fig. 15** Liquid limit of natural with artificial soil mixture

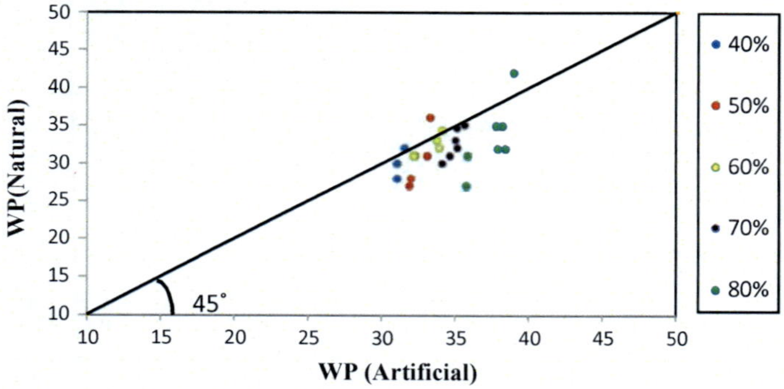


**Fig. 16** Plastic limit of natural with artificial soil mixture

## 5 Conclusions

On detailed study, the following conclusions can be made.

1. Liquid limit and plastic limit increases with increase in percent fines in the mix proportion, which is due to increase in fine content results in the water holding capacity. Liquid limit of mix proportion having higher percentage of Bentonite is greater compare to having higher percentage of Kaolinite this is due to formation of water tight barriers.
2. Shrinkage limit decreases due to grain size distributions of the natural soil and doesn't depend on plasticity characteristics. Shrinkage limit and percentage of fines show linear relationships, which shows that, artificial mix proportion is the true representative of the nature soil behaviour.
3. Plasticity chart of artificial mix proportion is compatible to the Casagrande's plasticity chart and activity values of artificial mix proportions match with the natural soil.

4. A graphical equation is defined to estimate the value of plasticity index for any other value of percent fines and an effort has been made to contrast the measured and calculated data.
5. The liquid limit and plastic limit of artificial mix proportion and natural soils are compared and it is observed that both the soils position on the line of equality. This concludes that artificial mix proportion prepared in the laboratory has similar plasticity characteristics natural soils.

## References


1. Bera, A.K.: Effect of sand content on engineering properties of fine grained soil. EJGE **16**, 1275–1286 (2011)
2. Polidori, E.: Relationship between the Atterberg limits and clay content. Soils Found. **47**(5), 887–896 (2007)
3. Polidori, E.: reappraisal of the activity of clays, activity chart. Soils Found. **49**(3), 431–441 (2009)
4. Seed, H.B., et al.: Clay mineralogical aspects of the atterberg limits. J. Soil Mech. Foundations Div. **90**(4), 107–131 (1964)
5. Bain, J.A.: A plasticity chart as an aid to the identification and assessment of industrial clays. Clay Minerals **9**, 1 (1971)
6. Dumbleton, M.J., West, G.: relation between clay minerals in soils and their plasticity. Clay Minerals **6**, 179 (1966)
7. Prasanna, H.S., et al.: A study on index properties of kaolinite and bentonite sand mixtures. AIP Conf. Proc. **2204**(1), 020009 (2020)
8. Sridharan, A., Prakash, K.: Shrinkage limit of soil mixtures. Geotech. Test. J. **23**(1) (2000)

# Estimation of Pre-consolidation Stress of Compacted Fine-Grained Soils—By User-Friendly Methods

**H. S. Prasanna and Basavaraju**

## 1 Introduction

Over consolidated natural soils exhibit a characteristic stress, known as preconsolidation stress ($\sigma_p'$), which represents the maximum stress to which the soil has been subjected in the past. The soils used in various geotechnical mass applications like construction of earthen embankments, earth dams, etc., will be subjected to some specified compactive effort, which is akin to over consolidation. Hence, compacted soils are also expected to possess characteristic stress similar to preconsolidation stress of over consolidated natural soils. Literature review indicates that very limited such documentations of the studies related with preconsolidation stress of compacted soils is available [1]. Among different methods of estimating preconsolidation stress of fine grained soils, four user friendly methods were selected for comparative study, i.e. Casagrande Method, Log–log Method, N-plot Method and Pacheo- silva's method.

## 2 Literature Review

Over consolidated (OC) and normally consolidated (NC) soils exhibit widely varying compressibility behaviour. Compression index (Cc) is a parameter which helps in the settlement calculations while coefficient of consolidation is an useful parameter in analyzing the time-rate of consolidation behaviour of fine-grained soils. $\sigma_p'$ is the important consolidation characteristics of OC soils. The over consolidation of soil mass can be attributed to many causes such as erosion/removal of previously

H. S. Prasanna (✉) · Basavaraju
The National Institute of Engineering, Mysuru, India
e-mail: prasanna@nie.ac.in

 
C. N. V. Satyanarayana Reddy et al. (eds.), *Ground Characterization and Foundations*, Lecture Notes in Civil Engineering 167,
https://doi.org/10.1007/978-981-16-3383-6_16

existed overburden, desiccation of soil mass, change in the structure due to aging, chemical behaviour of the deposits and the internal pressures due to pore water. Various methods have been mentioned in the literature for the evaluation of $\sigma_p{}'$ of OC soils. A brief description of each of these methods is given below.

1. *Schmertmann method*
   Schemertmann [2] proposed a method for obtaining $\sigma_p{}'$, which also involved a trial and error process.
2. *Janbu method*
   Janbu [3] proposed a procedure for determining the pre-consolidation stress based on constrained modulus ($M$) versus $\sigma'$ plot, where constrained modulus is the reciprocal of coefficient of volume compressibility (i.e., $M = 1/\text{mv}$). According to Janbu, the consolidation stress at which there is a drop marked in the modulus or the stress beyond which the constrained modulus levels out.
3. *The void Index method*
   Burland [4] proposed a method, known as the void index method, to assess the preconsolidation stress for soils for which $\sigma_p{}'$ is not well defined in the conventional e-log $\sigma'$ plot.
   The $e_{100}$ and $C_c$ are to be determined from the laboratory tests, where $e_{100}$ is the void ratio at $\sigma' = 100$ kPa and $C_c$ which is corresponding to virgin portion of the curve. However, for soils which plot above the A-line of the plasticity chart, Burland suggested that $e_{100}$ and $C_c$ could be calculated from the following empirical correlations.

$$e_{100} = 0.109 + 0.679\, e_L - 0.089\, e_L^2 + 0.016\, e_L^3 \quad (1)$$

$$C_c = 0.256\, e_L - 0.04 \quad (2)$$

   where $e_L$ is the void ratio (The percentage of water content at liquid limit of the soil).
   The value of void Index ($I_v$) is determined from the following equation

$$I_v = (e - e_{100})/C_c \quad (3)$$

4. *Jacobsen's Method*
   Jacobsen [5], His evaluation of the stress-strain curves from consolidation tests on Danish over-consolidated clays combined with the aspect of Casagrande and Terzaghi, projected an preconsolidation stress with reference to Eq. 4.

$$\sigma_p' = 2.5\, \sigma_k' \quad (4)$$

   where $\sigma_k{}'$ is the consolidation stress corresponding to the point of maximum curvature on the e-log $\sigma'$ curve defined by [6]. This method appears to be very simple. Hence, the value of $\sigma_p{}'$ depends upon the point of maximum curvature on the e-log $\sigma'$ curve, that was judged by manual identification and in addition to this Jacobsen's method also not checked and that is limitation except for the Danish clays.

5. *Pacheco Silva's method*
   This method is based on empirical constructions done on e-log $\sigma'$ plot [7]
   The advantage climbed by this method is that the method does not give scope for any personal judgement.
6. *n*-$\log_{10} \sigma'$ *method*
   Allam and Robinson [8] proposed the use of *nv/s* $\log_{10} \sigma'$ plot, where $n$ is the porosity of the soil sample corresponding to $\sigma'$, to determine the value of preconsolidation stress, instead of conventional e-$\log_{10} \sigma'$ plot. They validated their method with experiments on soils of known stress-history and found that the results obtained were in agreement with those determined by Casagrande method. This method is very similar to the log–log method proposed by Sridharan et al. [9]. However, it is to be noted that n-$\log_{10} \sigma'$ method requires additional calculations to obtain the values of porosities at different consolidation stresses.
7. *Onituska Method*

Apart from these methods, the literature also documents the bilogarithmic approach by [10]. They suggested the use of $\log_n(1 + e)$ with $\log_{10} \sigma'$ plot instead of conventional e-log $\sigma'$ plot. It is to be noted here that this method is nothing but the log–log method proposed by [9] with the only difference that the logarithm to base 10 is replaced by natural logarithm.

The review of literature on procedures of determining the preconsolidation stress indicates that different methods discussed have their own merits and limitations. It has also been noted that limited work has been reported by the researchers on the preconsolidation stress of compacted fine-grained soil having different clay mineralogy subjected to variation of different compaction energy levels for different placement conditions.

## 3 Materials and Methods

The engineering behaviour of fine-grained soils is largely dominated by the soil clay mineralogy. The soils included two types of clay minerals like montmorillonite and kaolinite in different fraction, in addition to other clay and non-clay minerals. These clay minerals were responsible for the geotechnical engineering behaviour, especially in the fine-grained soils. The proportion of these minerals in natural soil concludes the relative activeness. In this context, it is preferred to conduct the present experimental work on natural soils procured from the field, containing extreme clay minerals having same liquid limit and different plasticity characteristics.

### 3.1 Materials

**Selection of natural soils.** Nearly thirty soils from different locations in Mysore and Chamarajanagar districts were subjected to preliminary laboratory investigation involving liquid limit and free swell tests. Their liquid limits were determined using Casagrande percussion method (IS: 2720—Part 5, 1985), and the nature of their clay mineralogical composition was judged by the free swell index method. Finally, the following soils were identified, for the experimental investigation purpose.

It has been decided to conduct the experimental investigation on two field soils-one kaolinitic and the other montmorillonitic, having lower liquid limit range i.e. $35 < W_L < 50$ and on two field soils-one kaolinitic and the other montmorillonitic, having higher liquid limit range i.e. > 50.

Group of soils having low liquid limit range ($35 < W_L < 50$).

1. Field soil from Bogadi, Mysuru (passing 425 μm sieve), Chamarajanagar district, which contains kaolinite as the predominant clay mineral.
2. Field soil from Nanjangud, Mysuru District (passing 425 μm sieve), Chamarajanagar district, which is a montmorillonitic soil.

Group of soils having high liquid limit range (> 50).

1. Field soil from Kollegala (passing 425 μm sieve), Chamarajanagar District, which contains Kaolinite as the predominant clay mineral.
2. Field soil from Kuderu (passing 425 μm sieve), Chamarajanagar District, which appears to contain Montmorillonitic as the predominant clay mineral.

**Preparations of natural soils for investigation.** The field soils from Bogadi and Nanjungud were wet analysis passing through 425 μm IS Sieve to remove the coarser. They were then oven-dried and powdered to have soil dry analysis for determining the fraction finer than 425 μm size. Processed soils were batching separate in the bin. Similarly, the process for field soils from Kollegala and Kuderu were wet sieved through 425 μm IS sieve. They were also oven-dried, powdered and stored in separate plastic bins.

The below experiments were done on the prepared soil state as per Indian standards.

1. Specific gravity test: Specific gravity of the soils was done by the density bottle test with kerosene as the test liquid. (IS: 2720—Part-3/Sec 1, 1980).
2. Particle size analysis: The particle size distributions of the bogadi, nanjanagud, kollegal and kuderu soil were determined by wet sieve analysis (retained on 425 μm sieve) and Hydrometer analysis (IS: 2720, Part 4, 1985) (passing 425 micron sieve).
3. Free swell ratio: (IS: 2720, Part 40, 1977) suggests that the free swell index of the soil be calculated as

$$\text{FSR} = \left(\frac{V_d}{V_k}\right)$$

Determine the equilibrium sediment volumes of fine-grained oven dry soils passing 425 μm sieve using two 100 ml measuring jars with distilled water and carbon tetra chloride as the test liquids, the initial volumes of soil–liquid suspensions in the two jars being 100 ml.
Calculate the FSR as the ratio of $V_d$ to $V_k$.
Identify the soil as kaolinitic if FSR is less than unity or as montmorillonitic if FSR is more than 1.5.
If FSR is in between 1.0 and 1.5, the soil under consideration is of mixed clay mineral type.

4. Liquid limit test: The liquid limits of the soils were determined by the Casagrande percussion method (IS: 2720—Part 5, 1985). In addition, liquid limit tests by fall cone penetration method (IS: 2720—part 5, 1985) were also conducted on soils from kollegal, with water and kerosene as test liquids in order to know the clay mineralogical dominance.
5. Plastic limit test: It was obtained by rolling the soil into 3 mm thread conventional 3 mm thread rolling method. (IS: 2720—Part 5, 1985)
6. Shrinkage limit test: It was determined by the mercury displacement method (IS: 2720—Part 6, 1972).

**Classification of soil**. The soils used in this experimental investigation have been classified according to Unified Soil Classification System as specified by IS: 1498-1970.

Figures 1 and 2 illustrate the grain size distribution curves of the soils under study. Table 1 represents the physical properties of soils obtained as per the procedures indicated above, including the IS soil classification.

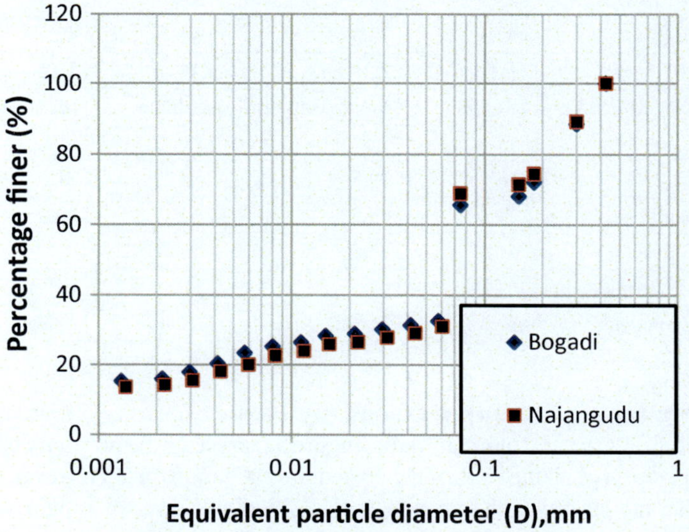


**Fig. 1** Grain size distribution curves for Bogadi and Nanjangud soils

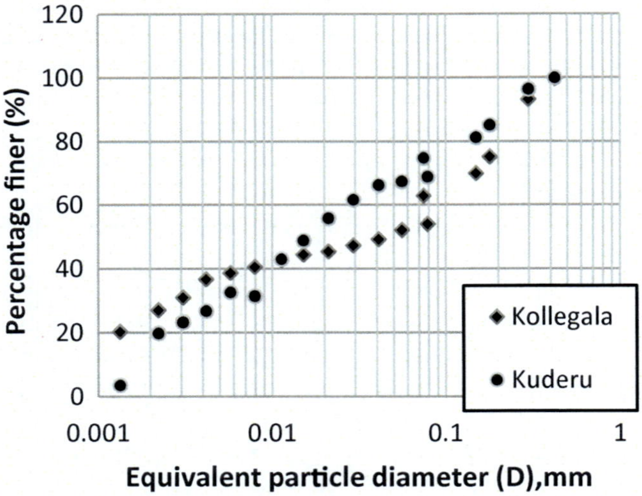


**Fig. 2** Grain size distribution curves for Kollegala and Kuderu soils

**Table 1** Physical properties of the soils under study

| S. No. | Soil property | Bogadi soil | $N'$ gud soil | Kollegal soil | Kuderu soil |
|---|---|---|---|---|---|
| 1 | G | 2.6 | 2.65 | 2.74 | 2.85 |
| 2 | $W_L$ (%) | 46 | 46 | 55 | 54 |
| 3 | $W_P$ (%) | 22 | 23 | 26 | 26 |
| 4 | $I_P$ (%) | 24 | 23 | 29 | 28 |
| 5 | $W_S$ (%) | 13.7 | 18.7 | 15.9 | 11.5 |
| 6 | $I_S$ (%) | 32.3 | 37.3 | 39.1 | 42.5 |
| 7 | FSR | 1.3 | 1.3 | 1.11 | 1.42 |
| 8 | Clay Minerology | Kaolinite | Montmorillonite | kaolinite | Montmorillonite |
| 9 | Clay (%) | 13 | 7.5 | 37.0 | 39.0 |
| 10 | Silt (%) | 16 | 19.5 | 34.5 | 21.0 |
| 11 | Sand (%) | 71 | 60.5 | 28.5 | 40.0 |
| 12 | Gravel (%) | – | 12.5 | – | – |
| 13 | IS classification | CI | CI | CH | CH |
| 14 | Comments | K Soil | M Soil | K Soil | M Soil |

**Clay mineralogy of soils and soil expansivity**. Plasticity chart is used in differentiating inorganic clays from inorganic silts. Figure 3 represents the plasticity chart indicating the position of all the soils of the present investigation on it. However, it should be noted that the plasticity chart cannot be classified by degree of expansiveness.

All geotechnical laboratories cannot afford to have sophisticated instruments such as X-ray diffractometer and the like for the qualitative identification of clay minerals

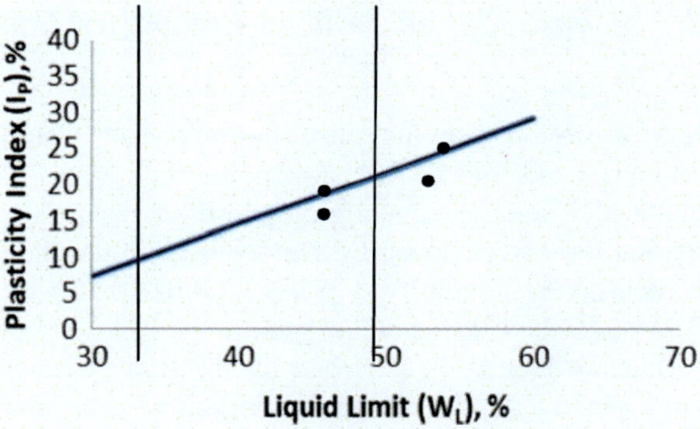


**Fig. 3** Position of the soils under study on the plasticity chart

present. In such cases, identification of clay mineral type by a simple method, which can serve the required purpose with a fair degree of accuracy is a welcome move. In this context, FSR has been shown to serve this purpose admirably well. Hence, following procedure is suggested IGC 2011 paper by Sridharan Prakash and Prasanna to classify the soils based on their degree of expansivity. In order to determine the predominant clay mineral present in the soil containing more than one clay mineral, Prakash and Sridharan suggested that the liquid limit percentage of soils determined by cone penetration/fall cone method with distilled water and $CCl_4$ as the pore liquids. If the fall cone liquid limit of a soil in distilled water is more than that in $CCl_4$, it indicates that the dominate clay mineral in the soil is montmorillonite. On the other hand, if the fall cone liquid limit of a soil in $CCl_4$ is more than that in distilled water, it indicates the dominance of kaolinite clay mineral in the soil.

## *3.2 Compaction Tests*

Standard and modified proctor compaction energy experiments were conducted on all soils under study (IS: 2720, Part-7, 1980; IS: 2720, Part-8, 1983). For each of these tests namely light and heavy compaction tests, about six to eight samples of mass 2.5 kg each, were mixed thoroughly with different percentages of moisture content and those samples were kept inside plastic or polythene bags, it is due to fact that the soil moisture equilibrium will be maintained. The duration of moisture equilibrium period depends upon types of soil, i.e. five to ten days. After this, the compaction tests were conducted on prepared soil samples. From these tests, the values of optimum water content and maximum dry density were determined.

## 3.3 Consolidation Tests on Compacted Soils

**Sample preparation for consolidation testing.** Consolidation tests were conducted on compacted soils of same low liquid limit group (i.e. K-soil and M-soil) and on compacted soils of same high liquid limit group (i.e. K-soil, M-soil). The consolidation tests were done at three levels of initial moulding water contents—water contents corresponding to $\gamma_{dmax}$ (i.e. OMC), 0.95 $\gamma_{dmax}$ on dry side of optimum and 0.95 $\gamma_{dmax}$ on wet side of optimum.

For all these soils identified for the study were mixed with required moisture content and left for attaining equilibrium moisture content. After achieving equilibrium state, the matured soil sample was compacted into consolidation ring to achieve required dry density. This consolidation ring was then assembled in its position on the consolidation cell.

The consolidation cell used for the experimental work is of fixed ring type with double drainage and has the facility to conduct the variable head permeability test also on the soil sample. The dimensions of consolidation ring were 60 mm (dia) × 20 mm (height).

**Load—deformation—time measurements for compacted soils.** Consolidation tests were conducted according to IS: 2720, Part 15 (1986). Next, the consolidation cell was located in its location on the loading cell well equipped with minimum least count of 0.002 mm to measure the vertical deflection of the soil sample, a initial consolidation stress of 6.25 kPa was loaded, and distilled water was added into the consolidation cell.

The soil samples were allowed to equilibrate under the initial stress. Some soil samples revealed swelling on addition of water into the consolidation cell. In such cases, time-deformation readings were noted till the equilibrium was achieved. The samples were loaded from 6.25 to 1600 kPa with a stress increment ratio of unity. Under each consolidation stress increment, time-compression readings were recorded till the near-equilibrium state was reached.

After achieving the near-equilibrium state under a consolidation pressure of 1600 kPa, the samples were unloaded in stages to the seating stress of 6.25 kPa.

(IS: 2720, Part15, 1986).Then the samples were unloaded and weighed and their final heights were recorded.

**Pre consolidation stress of compacted soils**

**General**

Over consolidated (OC) natural soils exhibit a characteristic stress, known as preconsolidation stress ($\sigma_p'$), which represents the higher stress level to which the soil had experienced in the past. The soils used in various geotechnical mass applications like construction of earthen embankments, earth dams, etc., will be subjected to some specified compactive effort, which is akin to over consolidation. Hence, compacted soils are also expected to possess a characteristic stress similar to preconsolidation stress of over-consolidated natural soils. Literature review indicates that very limited such documentations of the studies related with preconsolidation stress of compacted soils is available (Prakash et al. 2014).

## 4 Results and Discussions

Tables 2, 3 and 4 shows the values of preconsolidation pressure for soils under study with different placement conductions by Casagrande, $\log_{10}(1 + e)$, n-plot and Pacheo Silva's methods for both light and heavy compaction energy levels and placement condition respectively (Table 5).

**Low liquid limit group soils (for both energy levels)**
The variation of Pre-consolidation pressure values increases from dry of optimum to optimum state and decreases from optimum to wet of optimum state for both K&M soils. This trend was observed more in M-soils.

**High liquid limit group soils (for both energy levels)**
The variation of Pre-consolidation pressure values increases from dry of optimum towards the wet of optimum through the optimum state. This trend was observed more in K-soils than M-soils.

**The above observations can be explained through the following proposed hypothesis**
Along the compaction curve, as the percentage of water content increases, the dry density increases up to OMC state and decreases on the wet of optimum side and this contribution towards the pre-consolidation pressure also increases along the

**Table 2** Values of pre-consolidation pressure for soils having different placement conditions and energy levels (Casagrande method)

| Sample No. | Type of soil | | Preconsolidation pressure ($\sigma_p'$) kPa | | | | | |
|---|---|---|---|---|---|---|---|---|
| | | | 95% of $\gamma_{dmax}$ (dry of optimum) | | OMC | | 95% of $\gamma_{dmax}$ (wet of optimum) | |
| | | | Light | Heavy | Light | Heavy | Light | Heavy |
| 1 | Low liquid limit group soils | K Soil | 45 | 46 | 125 | 90 | 120 | 86 |
| 2 | | M Soil | 80 | 58 | 187 | 85 | 109 | 95 |
| 3 | High liquid limit group soils | K Soil | 61 | 84 | 200 | 360 | 380 | 150 |
| 4 | | M Soil | 84 | 130 | 170 | 192 | 300 | 230 |

**Table 3** Values of pre-consolidation pressure for soils having different placement conditions and energy levels ($\log 10_{(1+e)}$ method)

| Sample No | Type of Soil | | Preconsolidation pressure ($\sigma_p'$) kPa | | | | | |
|---|---|---|---|---|---|---|---|---|
| | | | 95% of $\gamma_{dmax}$ (dry of optimum) | | OMC | | 95% of $\gamma_{dmax}$ (wet of optimum) | |
| | | | Light | Heavy | Light | Heavy | Light | Heavy |
| 1 | Low liquid limit group soils | K Soil | 43 | 49 | 88 | 95 | 66 | 39 |
| 2 | | M Soil | 76 | 85 | 132 | 88 | 94 | 90 |
| 3 | High liquid limit group soils | K Soil | 51 | 74 | 205 | 250 | 330 | 500 |
| 4 | | M Soil | 70 | 76 | 150 | 160 | 160 | 245 |

**Table 4** Values of pre-consolidation pressure for soils having different placement conditions and energy levels (n-plot method)

| Sample No. | Type of soil | | Preconsolidation pressure ($\sigma_p$') kPa | | | | | |
|---|---|---|---|---|---|---|---|---|
| | | | 95% of $\gamma_{dmax}$ (dry of optimum) | | OMC | | 95% of $\gamma_{dmax}$ (wet of optimum) | |
| | | | Light | Heavy | Light | Heavy | Light | Heavy |
| 1 | Low liquid limit group soils | K soil | 53 | 52 | 76 | 110 | 130 | 75 |
| 2 | | M soil | 78 | 75 | 160 | 92 | 105 | 76 |
| 3 | High liquid limit group soils | K soil | 68 | 82 | 160 | 265 | 300 | 320 |
| 4 | | M soil | 90 | 125 | 160 | 170 | 240 | 245 |

compaction curve up to optimum moisture content state and then tends to achieve an equilibrium value of and may tend to decrease as well beyond the optimum compacted state.

The decrease in the value of preconsolidation stress value on the wet of optimum side can be attributed to the compacted soil fabric which results in the increase of water pressure in pores of soil (+ve) in case of K-soils and in case of M-soils, having combined effect of double layer repulsion and pore pressure.

All soils under study having different clay mineralogical compositions were subjected to same compactive energy levels (light compaction and heavy compaction)

**Table 5** Values of pre-consolidation pressure for soils having different placement conditions and energy levels (Pacheo silva's method)

| Sample No. | Type of soil | | Preconsolidation pressure ($\sigma_p'$) kPa | | | | | |
|---|---|---|---|---|---|---|---|---|
| | | | 95% of $\gamma_{dmax}$ (dry of optimum) | | OMC | | 95% of $\gamma_{dmax}$ (wet of optimum) | |
| | | | Light | Heavy | Light | Heavy | Light | Heavy |
| 1 | Low liquid limit group soils | K soil | 32 | 30 | 105 | 140 | 105 | 40 |
| 2 | | M soil | 70 | 44 | 230 | 88 | 105 | 86 |
| 3 | High liquid limit group soils | K soil | 54 | 76 | 190 | 300 | 440 | 130 |
| 4 | | M soil | 81 | 98 | 168 | 170 | 280 | 250 |

and it is reasonable to expect all of them to have in them the same value of preconolidation stress. On the contrary, the soils under study exhibit different values of preconsolidation stress. This unexpected behaviour can be essentially attributed to the net (R-A) forces that are operative in the placement compacted soil–water system, which in turn depend upon clay mineralogical composition of the soil.

From the Tables 2, 3, 4 and 5, it is observed that,

Figures 4, 5, 6, 7, 8, 9, 10, 11, 12, 13, 14 and 15, shows the variation of preconsolidation stress with consolidation stress for Light and Heavy compaction energy level for different placement conditions for different methods (Tables 6, 7, 8 and 9).

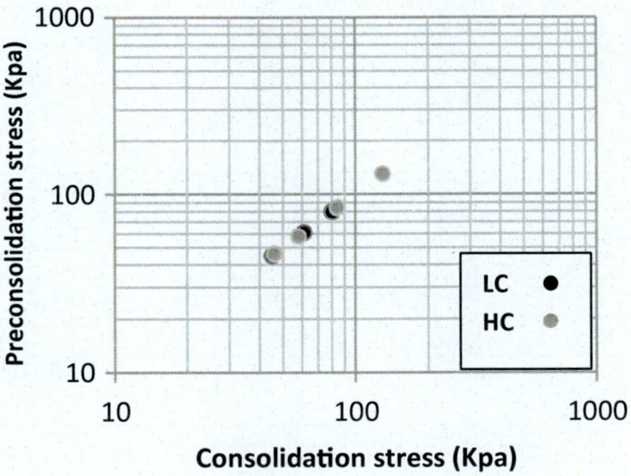


**Fig. 4** Variation of preconsolidation stress with consolidation stress (dry side)

**Fig. 5** Variation of preconsolidation stress with consolidation stress (OMC)

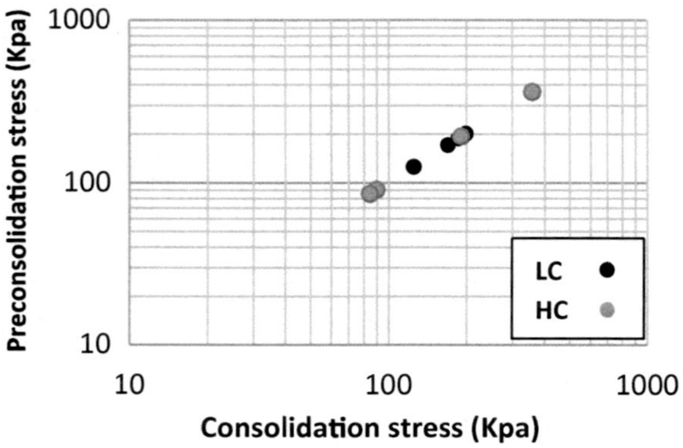


**Fig. 6** Variation of preconsolidation stress with consolidation stress (Wet Side)

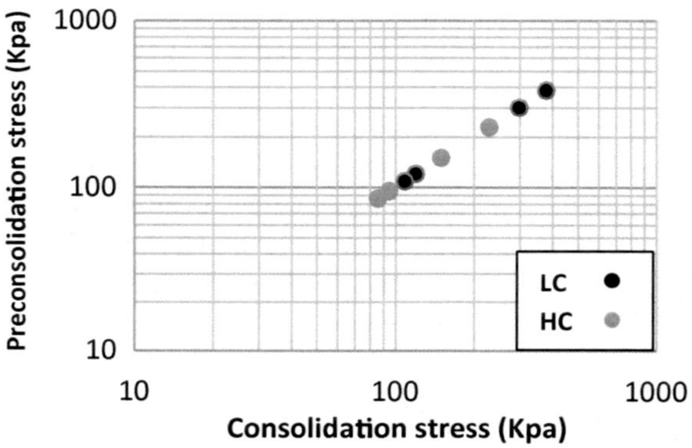


**Fig. 7** Variation of preconsolidation stress with consolidation stress (Dry Side)

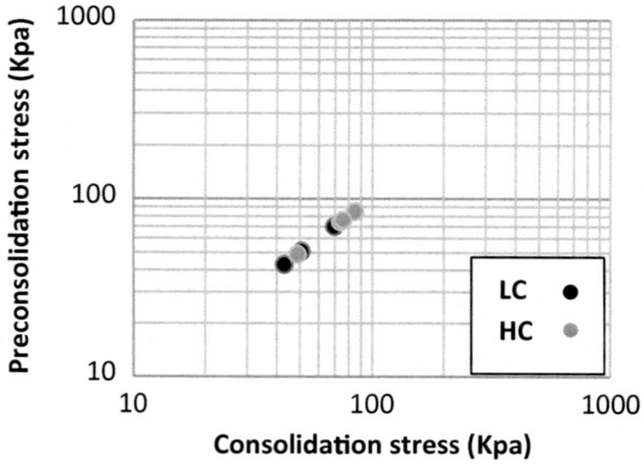

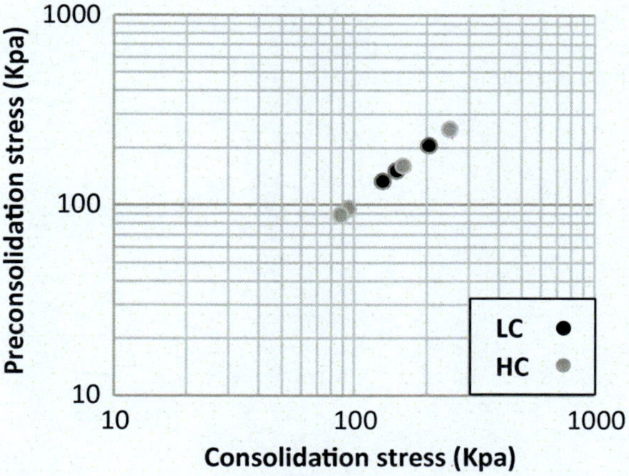


**Fig. 8** Variation of preconsolidation stress with consolidation stress (OMC)

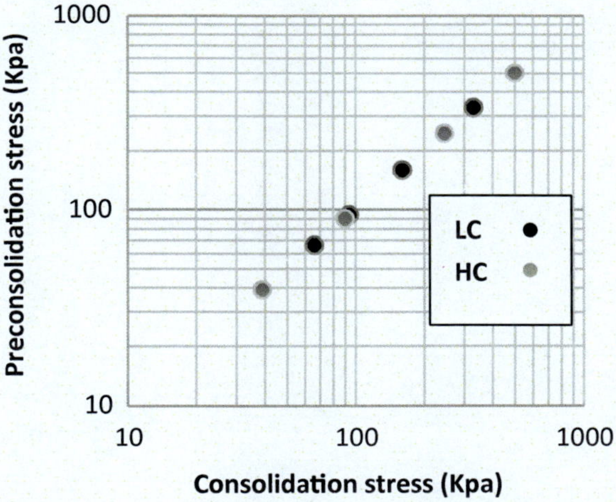


**Fig. 9** Variation of Preconsolidation stress with consolidation stress (wet side)

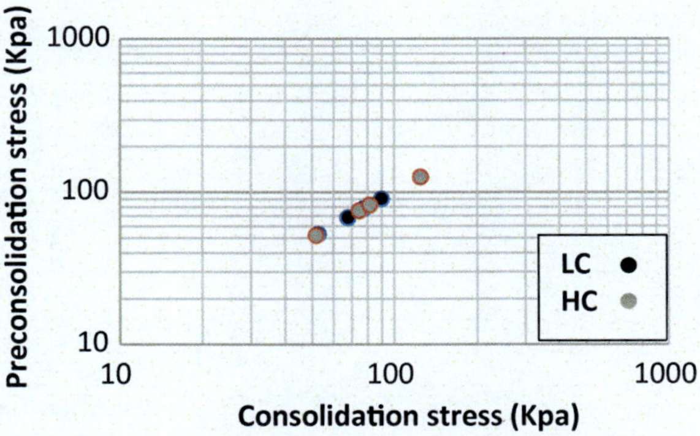


**Fig. 10** Variation of preconsolidation stress with consolidation stress (dry side)

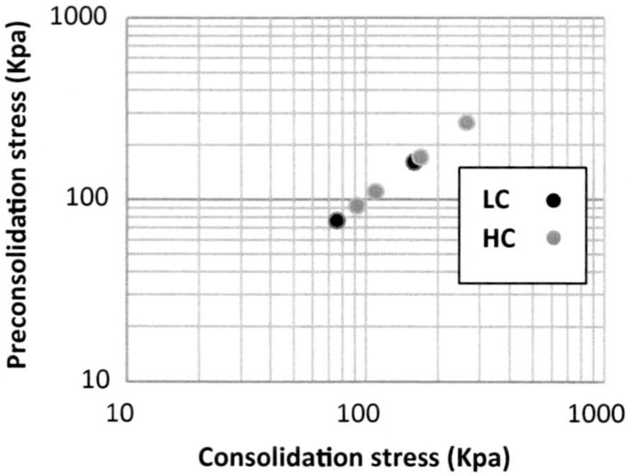


**Fig. 11** Variation of preconsolidation stress with consolidation stress (OMC)

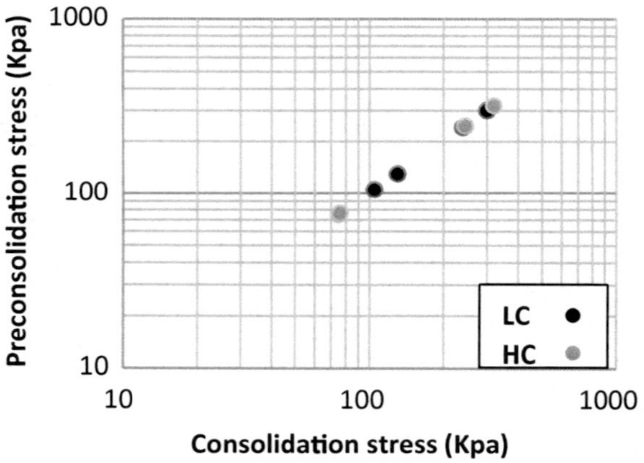


**Fig. 12** Variation of preconsolidation stress with consolidation stress (wet side)

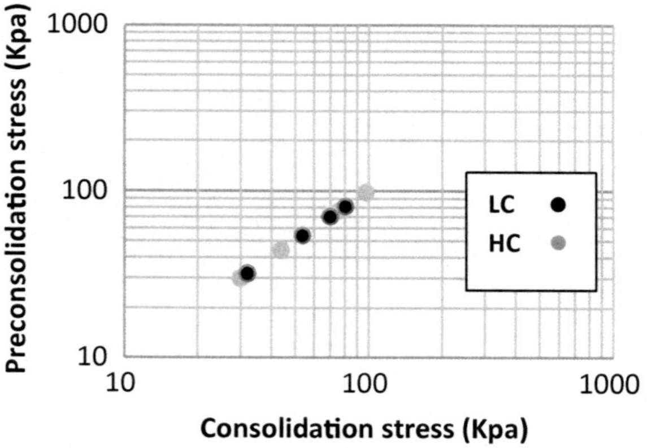


**Fig. 13** Variation of preconsolidation stress with consolidation stress (dry side)

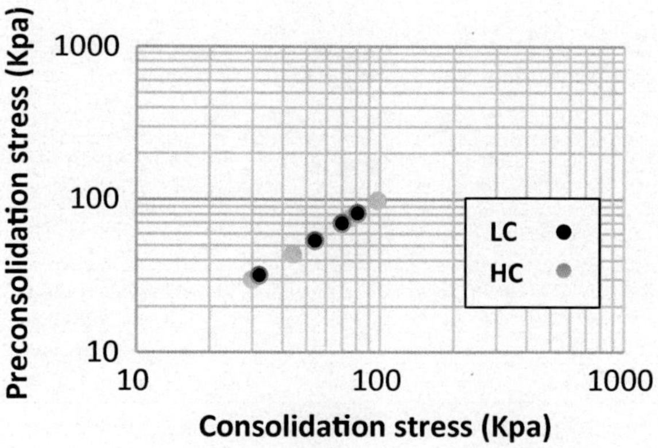


**Fig. 14** Variation of preconsolidation stress with consolidation stress (OMC)

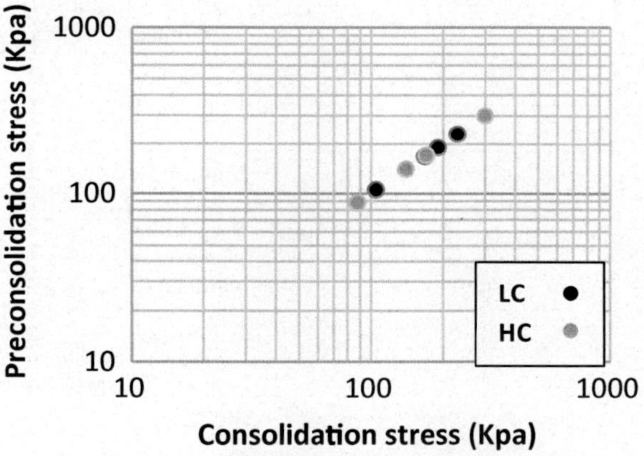


**Fig. 15** Variation of preconsolidation stress with consolidation stress (wet side)

**Table 6** Casagrande method

| Soil No. | Soil | Tendency | |
|---|---|---|---|
| | | LC | HC |
| S1 | Soils of Low liquid limit group | Dry of optimum to OMC (increases) and OMC to wet of optimum (decreases) | Dry of optimum to OMC (increases) and OMC to wet of optimum (decreases) |
| S2 | | Dry of optimum to wet of optimum (increases) | Dry of optimum to wet of optimum (increases) |
| S3 | Soils of high liquid limit group | Dry of optimum to wet of optimum (increases) | Dry of optimum to OMC (increases) and OMC to wet of optimum (decreases) |
| S4 | | Dry of optimum to Wet of optimum (increases) | Dry of optimum to wet of optimum (increases) |

**Table 7** $\mathrm{Log}_{(1+e)}$ method

| Soil No. | Soil | Tendency | |
|---|---|---|---|
| | | LC | HC |
| S1 | Soils of low liquid limit group | Dry of optimum to wet of optimum (increases) | Dry of optimum to OMC (increases) and OMC to wet of optimum (decreases) |
| S2 | | Dry of optimum to OMC (increases) and OMC to wet of optimum (decreases) | Dry of optimum to wet of optimum (increases) |
| S3 | Soils of high liquid limit group | Dry of optimum to wet of optimum (increases) | Dry of optimum to wet of optimum (Increases) |
| S4 | | Dry of optimum to wet of optimum (increases) | Dry of optimum to wet of optimum (Increases) |

**Table 8** N-plot method

| Soil No. | Soil | Tendency | |
|---|---|---|---|
| | | LC | HC |
| S1 | Soils of low liquid limit group | Increases from dry of optimum to wet of optimum | Increases from dry of optimum to OMC and then decreases |
| S2 | | Increases from dry of optimum to OMC and then decreases | Increases from dry of optimum to OMC and then decreases |
| S3 | Soils of high liquid limit group | Increases from dry of optimum to wet of optimum | Increases from dry of optimum to wet of optimum |
| S4 | | Increases from dry of optimum to OMC and then decreases | Increases from dry of optimum to wet of optimum |

## 5 Conclusions

1. The Preconsolidation stress value in Casagrande and Pacheosilva's method increases from dry of optimum to OMC and decreases. This tendency is observed in soils having low liquid limit having different Clay mineralogy subjected to light and heavy compaction energy levels.
2. The preconsolidation stress value in all methods increases from dry of optimum to wet side of optimum through OMC for soils having different clay mineralogy and compaction energy levels.
3. Compaction process induces stress history effect in the compacted fine-grained soil. Compacted soils exhibit preconsolidation stresses, the values of which are strongly dependent upon the clay mineralogical composition of soils and also

**Table 9** Pacheo Silva's method

| Soil No. | Soil | Tendency | |
|---|---|---|---|
| | | LC | HC |
| S1 | Soils of Low liquid limit group | Increases from dry of optimum to OMC and becomes constant | Increases from dry of optimum to OMC and then decreases |
| S2 | | Increases from dry of optimum to OMC and then decreases | Increases from dry of optimum to OMC and then decreases |
| S3 | Soils of high liquid limit group | Increases from dry of optimum to wet of optimum | Increases from dry of optimum to OMC and then decreases |
| S4 | | Increases from dry of optimum to wet of optimum | Increases from dry of optimum to wet of optimum |

on the placement conditions, the compactive effort applied on the soils being the same.

4. Equilibrium void ratio at the seating consolidation stress of 6.25 kPa of Kaolinitic and Montmorilloniticsoils for light compaction energy level is more for Kaolinitic soils in relative comparison to Montmorillonitic soils.
5. Higher equilibrium void ratio for M-soils at lower effective consolidation stresses are more than that of K-soils indicating dominance of double-layer repulsion over the effect of flocculent fabric.
6. Higher value of cumulative change in void ratio at seating stress if 6.25 kPa for M-soils are generally being observed than K-soils for both light and heavy compaction energy levels for soils of same low and high liquid limit group indicating dominance of double-layer repulsion over the flocculent fabric exhibited by K-soil.
7. Pre consolidation stress values increases from dry of optimum to OMC and then decreases for OMC to Wet of optimum for M-soils whereas has a tendency to increase beyond optimum state to wet side of optimum for K-soils which signifies the importance of placement condition and clay mineralogy.
8. Compression index values increase more rapidly with effective consolidation stress in a narrow band in the pre-yield zone and show a tendency to stabilize in the post yield region for soils under study for both light and heavy compaction energy levels which is a characteristic of over-consolidated soil.

## References

1. Prakash, K., et al.: "Yield stresses of compacted soils" A comparative study. In: Proceeding of National Conference "GEPSID". Ludhiana (Oct 11-12 2014)

2. Schemertmann, J.M.: The undisturbed consolidation of clay. Truncation ASCE **120**, 1201–1211 (1955)
3. Janbu, N.: The resistance concept applied to deformation of soils. In: International Proceeding of the 7th International Soil Mechanics and Foundation Engineering Conference, vol. 1, pp. 191–196 (1969)
4. Burland, J.B.: On the compressibility and shear strength of natural clays. Geotechnique **29**, 329–378 (1990)
5. Jacobsen, H.M.: 'Bestemmelse of forbelastingstryklaboratorict, IT in nordiskegeotekikermode', NGM-92. Aalborg **2**, 455–460 (1992)
6. Casagrande, A.: The determination of the preconsolidation load and its significance. In: International Proceedings of 1st International Soil Mechanics and Foundation Engineering Conference, Cambridge, MA, vol. 3, pp. 60–64 (1936)
7. Clementino, R.V.: Discussion of an oedometer test study on the preconsolidation stress of glaciomarine clays. Can. Geotech. J. **40**, 857–872 (2005)
8. Allam, Robinson: Estimation of preconsolidation pressure using n-log p plots. Proc. Ind. Geotech. J. **1**, 95–98 (1997)
9. Sridharan, A., Abraham, B.M., Jose, B.T.: Improved technique for estimation of preconsolidationpressure. Geotechniqe **41**(2), 263–268 (1991)
10. Onitsuka, K., Hong, Z., Hara, Y., Yoshitake, S.: Interpretation of oedometer test data for natural clays. Soils Foundations **35**:61–70 (1995)
11. Burmister, D.M.: Laboratory investigations of soils at flushing meadow park. Trans ASCE **107**, 187 (1942)
12. Burmister, D.M.: The application of controlled test methods in consolidation testing. In: Symposium on Consolidation Testing of Soils, ASTM STP, vol. 126, pp. 83–89 (1951)
13. International Proceedings of the 7th International Soil Mechanics and Foundation Engineering Conference, Vol. 1 pp. 191–196. Transactions of ASCE, vol. 120, pp. 1201–1211

# Relationship Between Various Consolidation Parameters of Compressible Soils

**Siri Ande, Ch. Nageshwar Rao, and Madhav Madhira**

## 1 Introduction

Soft soils are often encountered along coastal regions, swampy areas, and deltas all over the world, because of which construction of any infrastructure project on such soils is indeed a challenge. Consolidation of soft soils is mostly achieved through vertical drains which are coupled with preloading. Based on the one-dimensional consolidation theory, Asaoka [1] proposed a new approach to evaluate the coefficient of consolidation for vertical flow, $c_v$ along with final settlement, $S_f$. Case history of Changi East Reclamation, Singapore comprising of vertical drains coupled with preloading was studied and presented by Arulrajah et al. [2]. Chung et al. [3] presented various case studies which are associated with vertical drains with preloading like Chek Lap Kok airport, Busan Airport, and Changi airport in their research work. Adverse effects such as smear and permeability ratio affect the consolidation in an unfavorable manner by delaying the settlement rate. So, it is important to estimate these parameters that affect the consolidation rate and time in soft soils. The objective of the paper is to analyze and estimate the in-situ consolidation parameters smear ratio ($s$), permeability ratio ($k_h/k_s$) and coefficient of consolidation ($C_h$) from the available time-settlement data. Monitored data of time-settlement data from Indraratna et al. [4] are analyzed and presented.

S. Ande (✉) · Ch. Nageshwar Rao · M. Madhira
VNR Vignana Jyothi Institute of Engineering and Technology, Hyderabad 500090, India

Ch. Nageshwar Rao
e-mail: nageshwarrao_ch@vnrvjiet.in

 
C. N. V. Satyanarayana Reddy et al. (eds.), *Ground Characterization and Foundations*, Lecture Notes in Civil Engineering 167,
https://doi.org/10.1007/978-981-16-3383-6_17

## 2 Methodology

Theoretical values of degree of consolidation, $U_{tc}$, at the end of construction are obtained from Olson [5] for the consolidation with flow in the radial direction for ramp loading as

$$U_{tc} = T_c - \left[(1 - \exp(-AT_r)/A)\right]/T_c \tag{1}$$

where

$$F(n) = \left(n^2/n^2 - s^2\right)\ln(n/s) - 3/4 + \left(s^2/4n^2\right) + k_h/k_s\left(n^2 - s^2/n^2\right)\ln(s) \tag{2}$$

diameter ratio, $n = d_e/d_w$, $d_e$ and $d_w$ are the diameters of the drain well and influence zone respectively, smear ratio, $s = r_s/r_w$, $r_s$ and $r_w$ are the radii of smear zone and drain well, respectively, $T_r = c_r \cdot t/d_e{}^2$—dimensionless time factor, $T_c = c_r \cdot t_c/d_e{}^2$—dimensionless time factor for time, $t_c$, at the end of construction, $F(n)$—function of $n$ and $s$, $k_h/k_s$—permeability ratio and $A = 20/F(n)$. $T_c$ is calculated from the Eq. 1 with $T_r = T_c$ for different diameter ratios and different smear ratios. Knowing $T_c$, the coefficient of consolidation, $c_r$ for flow in radial direction is determined as

$$c_r = \left(T_c * d_e^2\right)/t_c \tag{3}$$

Time-settlement plots are shown in Fig. 1 of different sections from Indraratna et al. [4] who demonstrated the effectiveness of vacuum coupled surcharge loading system over conventional surcharge loading. All the curves from time versus settlement are digitized and analyzed for different time intervals.

In the new method, Final settlements, $S_{fA}$ are estimated from the digitized data of time-settlement plots for different sections based on Asaoka [1]. Based on the settlement, $S_c$ corresponding to the time at the end of first stage of ramp loading, $t_c$, the degree of consolidation, $U_{tc}$, at the end of construction is estimated for different diameter ratios, $n$. $U_{tc}$ is plotted with respect to '$s$', and smear ratio is interpolated corresponding to '$n$' $U_{tc}$ is also plotted against $T_c$ for different smear ratios, $s$. For a given '$n$', $T_c$ is obtained through interpolation. Same procedure is repeated for determining the permeability ratio, $k_h/k_s$ and the corresponding $T_c$. The estimated coefficients of consolidation are compared with those of Indraratna et al. [4]. Details such as drain type, diameter of drain well ($d_w$), equivalent diameter of the influence zone ($d_e$), spacing between the drains ($S$), diameter ratio ($n$), time at the end of construction ($t_c$) for the sections from Indraratna et al. [4] are extracted from Fig. 1.

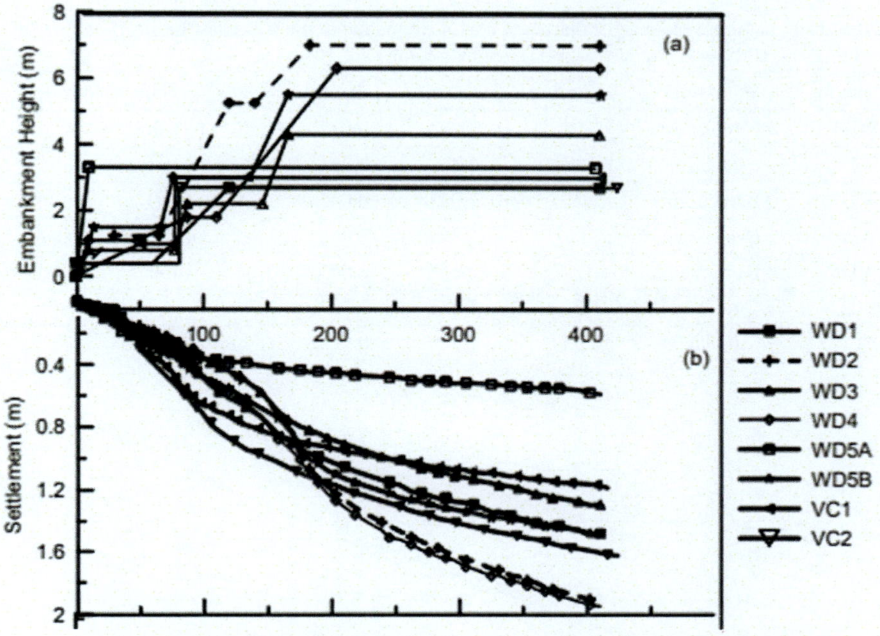


**Fig. 1** Time-settlement plot of staged construction (after Indraratna et al. [4])

# 3 Case Study

Indraratna et al. [4] compared time-settlement responses between consolidation with vacuum surcharge preloading and conventional surcharge loading at seven different sections and found that consolidation due radial flow is faster and lateral displacements are less in the former (vacuum surcharge loading) than the latter. In this study, typically, two types of drains i.e., circular and band-shaped (100 × 4)mm with spacing of the drains ranging between 1.1 and 1.3 m were used. Drains were laid in square pattern and the diameter of equivalent influence zone/unit cell, $d_e$, is obtained. The parameters are listed in Table 1.

## *3.1 Validation*

The proposed method is applied and validated for different sections mentioned in Indraratna et al. [4]. A typical Asaoka plot for section VC1 is shown in Fig. 2 with diameter ratio, *n* of 36 with spacing, *S* as 1.1 m. The final settlement, $S_{fA}$ is obtained as 1.2 m.

Figures 3, 4, 5 and 6 show plots were drawn between $U_{tc}$ and *s* for diameter ratio (*n*) varying between 10 and 40 for different time factors, $T_c$, at the end of

**Table 1** Drain type, spacing, $S$, diameter, $d_w$, of drain, and diameter, $d_e$, of unit cell, $n = d_e/d_w$ and time, $t_c$, at the end of construction (after Indraratna et al. [4])

| Section | Drain type | Spacing | $d_e$ (m) | $d_w$ (m) | $n$ | Time, $t_c$ (days) |
|---|---|---|---|---|---|---|
| WD1 | Circular 34 | 1.1 | 1.243 | 0.034 | 36 | 120 |
| WD2 | Circular 34 | 1.3 | 1.469 | 0.034 | 43 | 180 |
| WD3 | Band drains | 1.1 | 1.243 | 0.065 | 19 | 160 |
| WD4 | Band drains | 1.3 | 1.469 | 0.065 | 22 | 200 |
| WD5B | Band drains | 1.1 | 1.243 | 0.065 | 19 | 160 |
| VC1 | Circular 34 | 1.2 | 1.356 | 0.034 | 40 | 80 |
| VC2 | Circular 34 | 1.23 | 1.389 | 0.034 | 40 | 80 |

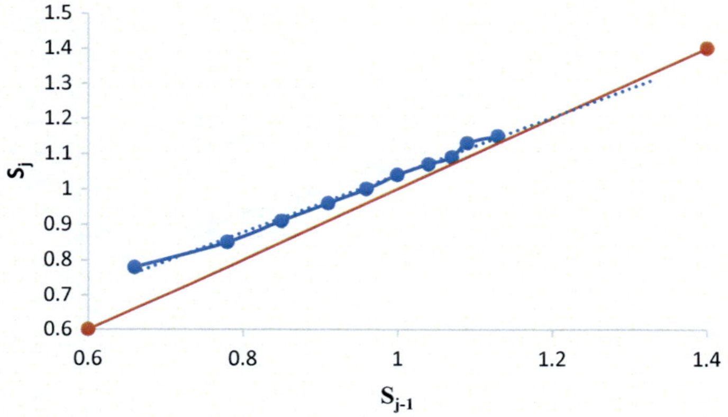


**Fig. 2** Asaoka plot for section VC1

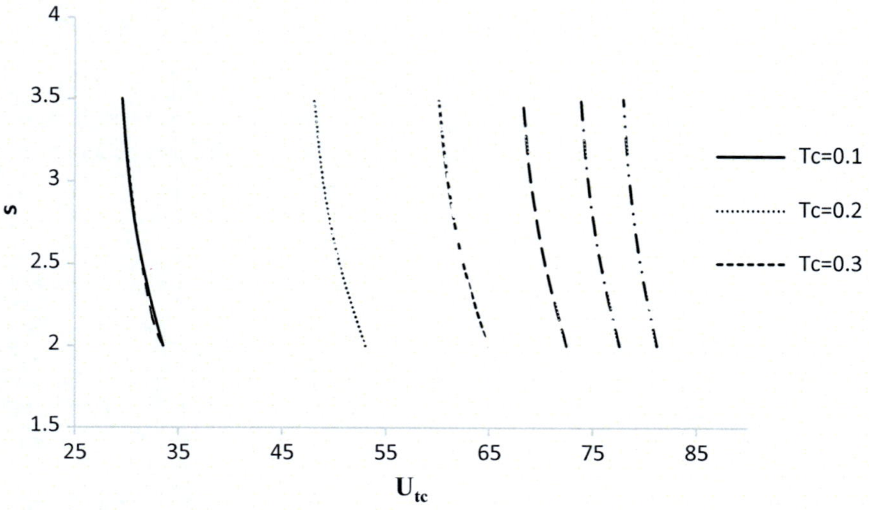


**Fig. 3** $U_{tc}$ versus $s$ for $n = 10$

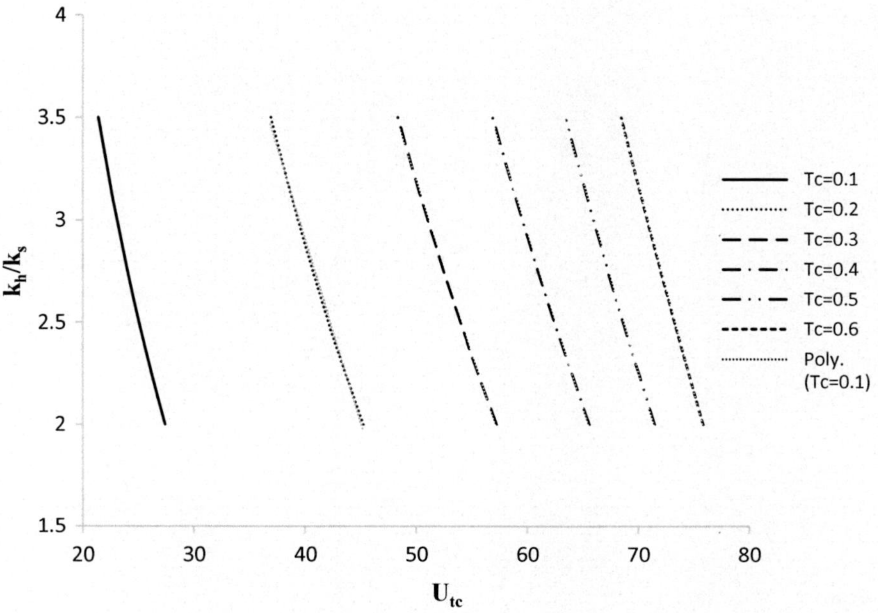


**Fig. 4** $U_{tc}$ versus $s$ for $n = 20$

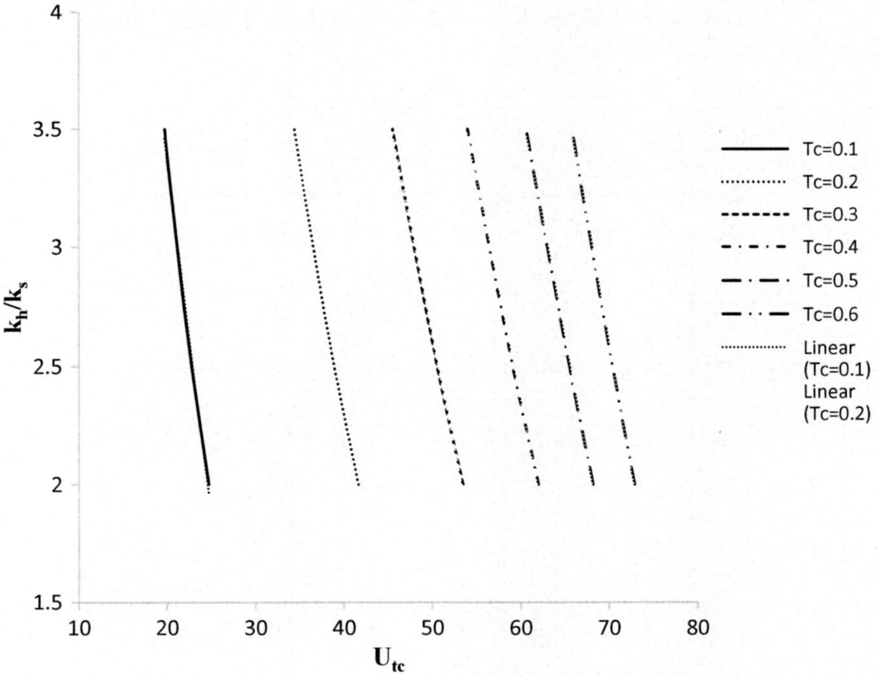


**Fig. 5** $U_{tc}$ versus $s$ for $n = 30$

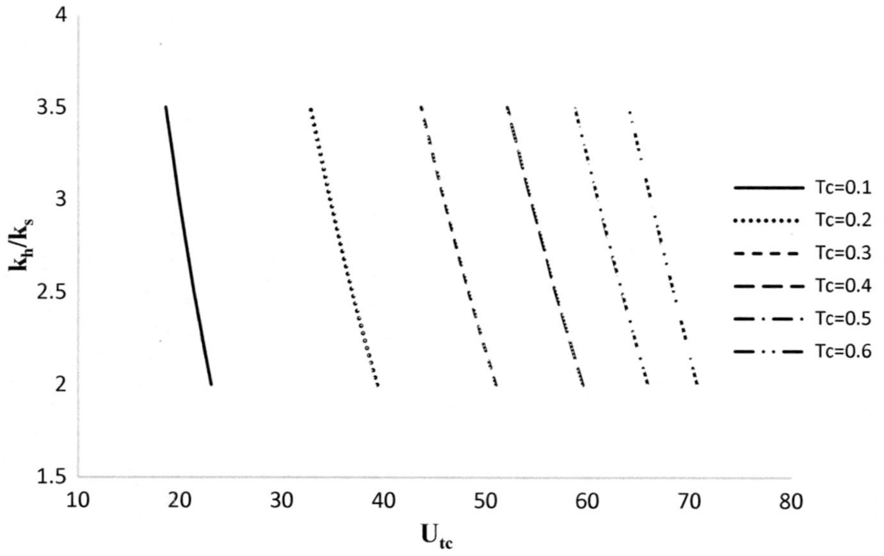


**Fig. 6** $U_{tc}$ versus $s$ for $n = 40$

construction. In a similar way, plots for different '$T_c$' values for varying '$n$' are drawn. '$s$' is estimated through interpolation with '$n$' corresponding to the obtained $U_{tc}$ value.

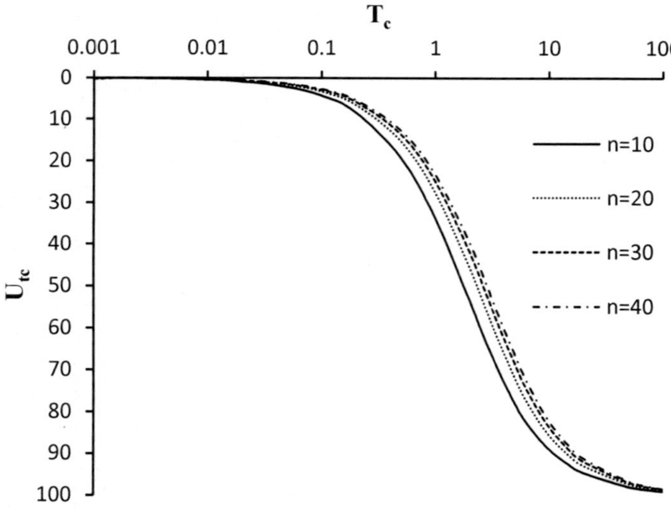


**Fig. 7** $T_c$ versus $U_{tc}$ for different '$n$' under $s = 2$

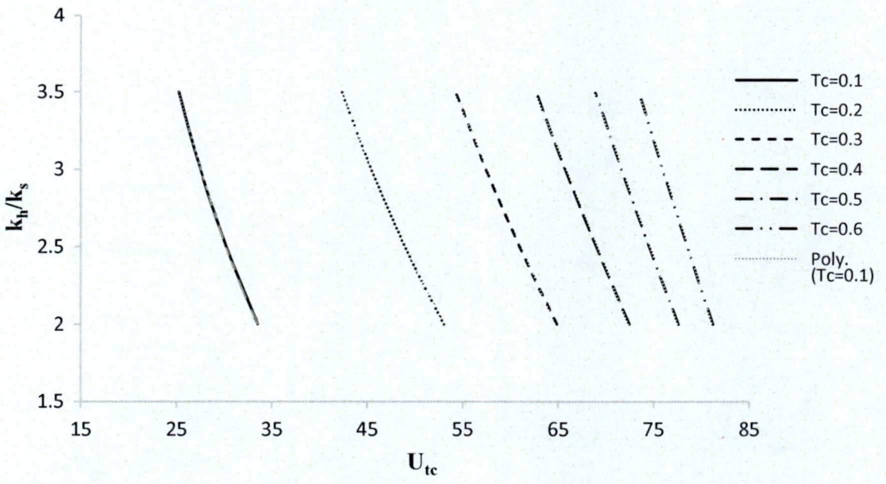


**Fig. 8** $U_{tc}$ versus $k_h/k_s$ for $n = 10$

Figure 7 is a typical plot between time factor, $T_c$, and degree of consolidation, $U_{tc}$, at the end of construction for diameter ratios ($n$) ranging from 10 to 40 for smear ratio, $s = 2$. For different diameter ratios, similar plots are drawn for '$s$' varying from 2.5 to 3.5. $T_c$ is obtained through interpolation for a given '$s$', corresponding to the obtained $U_{tc}$ and for the known '$n$' value.

Figures 8, 9, 10 and 11 show typical plots drawn between $U_{tc}$ and $k_h/k_s$ for different diameter ratios ($n$) varying between 10 and 40 for different time factors, $T_c$, at the end of construction, Similarly, plots for different '$T_c$' for varying '$n$' are drawn. '$k_h/k_s$' value is estimated through interpolation of '$k_h/k_s$' with '$n$' corresponding to the obtained $U_{tc}$ value.

Figure 12 is a typical plot drawn between time factor, $T_c$, at the end of construction and degree of consolidation, $U_{tc}$, at the end of construction for diameter ratios ($n$) ranging from 10 to 40 for permeability ratio, $k_h/k_s = 2$. Similar plots are drawn for '$k_h/k_s$' values varying from 2.5 to 3.5 for varying diameter ratios. $T_c$ is estimated for a given '$k_h/k_s$', through interpolation of $T_c$ and $n$ corresponding to $U_{tc}$.

Results obtained are tabulated in Tables 2 and 3. $U_{tc}$ is obtained from settlement, $S_c$ corresponding to the time, $t_c$, at the end of first construction loading and final settlement. Smear ratio and time factor at the end of construction are estimated or determined from Figs. 3, 4, 5, 6 and 7. The coefficient of consolidation, $c_r$ $T_c$ is determined from Eq. 3. Similarly, permeability ratio, $k_h/k_s$, and the corresponding time factor, $T_c$ at the end of construction are from Figs. 8, 9, 10, 11 and 12. The coefficient of consolidation, $c_r$ $T_c$, is calculated from Eq. 3.

Ratio of coefficient of consolidation ($c_r$ $T_c$) to the coefficient of consolidation, $c_r$ (from Indraratna et al. [4]) ranges between 0.8–1.4 and 0.7–1.4 for different permeability and smear ratios (Table 3).

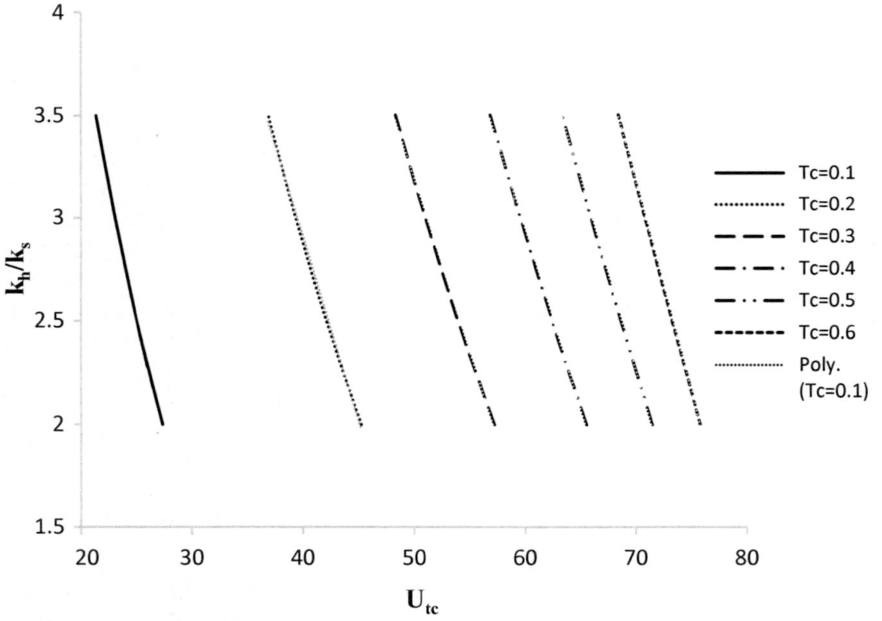


**Fig. 9** $U_{tc}$ versus $k_h/k_s$for $n = 20$

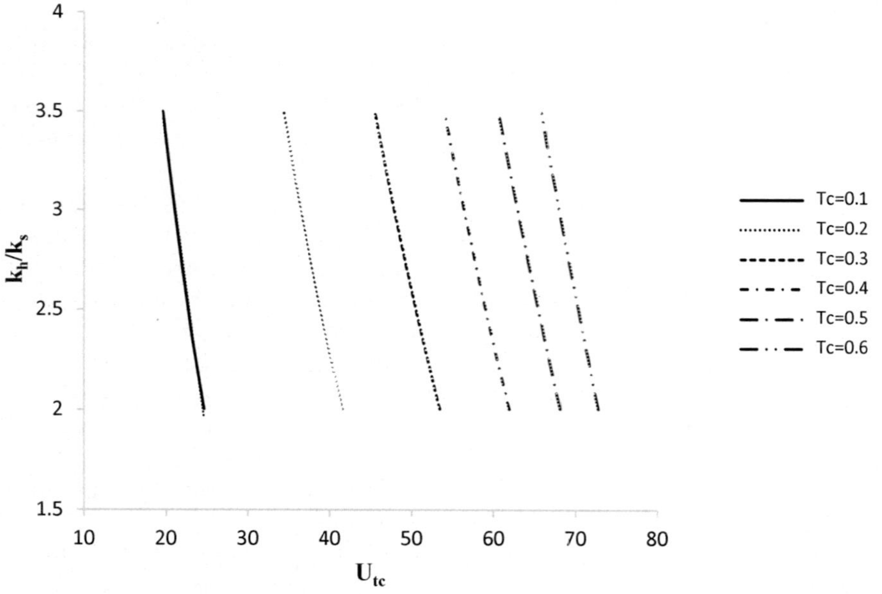


**Fig. 10** $U_{tc}$ versus $k_h/k_s$ for $n = 30$

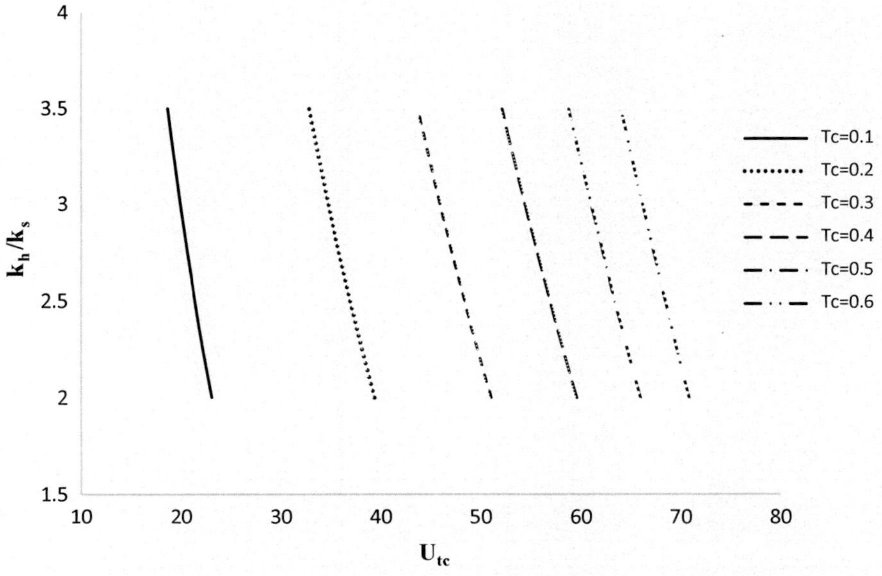


**Fig. 11** $U_{tc}$ versus $k_h/k_s$ for $n = 40$

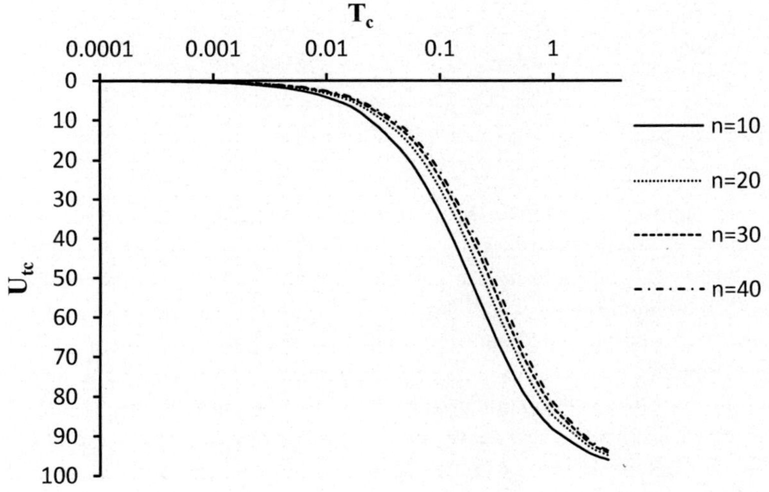


**Fig. 12** $T_c$ versus $U_{tc}$ for different '$n$' under $k_h/k_s = 2$

## 4 Conclusions

As a well-known fact, vertical drains coupled with preloading is an efficacious method to expedite the consolidation by promoting radial flow. In this paper,

**Table 2** Final settlement, $S_{fA}$, degree of consolidation, $U_{tc}$, at the end of construction, smear ratio ($s$), permeability ratio ($k_h/k_s$) and time factor, $T_c$, at the end of construction

| Section | $S_c$ (m) | $S_{fA}$ (m) | $U_{tc}$ (%) ($S_c/S_{fA}$) | $S$ | $T_c$ | $k_h/ks$ | $T_c$ |
|---|---|---|---|---|---|---|---|
| WD1 | 0.6 | 1.6 | 38 | 2.1 | 0.18 | 2.1 | 0.19 |
| WD2 | 1.1 | | 50 | 2.3 | 0.32 | 2.2 | 0.33 |
| WD3 | 0.9 | 1.4 | 64 | 2.1 | 0.38 | 1.9 | 0.37 |
| WD4 | 1.3 | 2.4 | 54 | 2.0 | 0.27 | 2.3 | 0.28 |
| WD5B | 0.9 | 1.6 | 56 | 1.8 | 0.265 | 1.9 | 0.265 |
| VC1 | 0.5 | 1.2 | 42 | 3.0 | 0.25 | 2.5 | 0.24 |
| VC2 | 0.6 | 1.7 | 34 | 3.5 | 0.22 | 3.0 | 0.21 |

**Table 3** Coefficients of consolidation ($c_r$)

| Section | $C_r$ ($\times 10^{-3}$) ($m^2$/year) (Indraratna) | $C_r T_c$ ($\times 10^{-3}$) ($m^2$/year) (for s) | $C_r T_c/C_r$ (for s) | $C_r T_c$ ($\times 10^{-3}$) ($m^2$/year) (for $k_h/k_s$) | $C_r T_c/C_r$ ($k_h/k_s$) |
|---|---|---|---|---|---|
| WD1 | 1.16 | 0.84 | 0.80 | 0.80 | 0.70 |
| WD2 | 1.31 | 1.38 | 1.05 | 1.42 | 1.08 |
| WD3 | 0.89 | 1.33 | 1.40 | 1.30 | 1.40 |
| WD4 | 1.02 | 1.04 | 1.02 | 1.08 | 1.06 |
| WD5B | 0.90 | 0.92 | 1.03 | 0.92 | 1.03 |
| VC1 | 1.70 | 2.07 | 1.20 | 2.02 | 1.10 |
| VC2 | 1.41 | 1.90 | 1.30 | 1.82 | 1.30 |

the crucial factors that control the performance of vertical drains including the phenomenon of smear zone along with the permeability ratio were discussed. A new method to estimate the in-situ parameters, viz., smear ratio ($s$), permeability ratio ($k_h/k_s$) and coefficient of radial consolidation ($c_r$) from the time versus settlement plots is proposed. The proposed method is analyzed and illustrated through a well-documented case history, reported by Indraratna et al. [4]. Time factors, $T_c$, at the end of construction obtained from the plots corresponding to smear ($s$) and permeability ($k_h/k_s$) ratios are nearly the same. The coefficients of radial consolidation ($c_r$ $T_c$) obtained from both the approaches are close and compare well with those of Indraratna et al. [4].

## References

1. Asaoka, A.: Observational procedure of settlement prediction. Soils Found. **18**(4), 87–101 (1978)
2. Arulrajah, A., Bo, M.W., Chu, J., Nikraz, H.: Prefabricated vertical drains to the Changi land reclamation project, Singapore. In: Proceedings of 4th Asian Regional Conference on Geosynthetics, pp. 651–655 (2008)

3. Chu, J., Bo, M.W., Chao, V.: Practical considerations for using vertical drains in soil improvement project. Geotext. Geomembr. **22**(1), 101–117 (2003)
4. Indraratna, B., Rujikiatkamjorn, C., Xueyu, G., Ameratunga, J., Peter.: Performance and prediction of vacuum combined surcharge consolidation at port of Brisbane. In: Sydney Chapter Symposium Australian Geomechanics Society, pp. 45–60 (2011)
5. Olson, R.E.: Consolidation under time dependent loading. J. Geotech. Eng. Div. Proc. ASCE **103**(GT1), 55–60 (1977)

# Implementation of Wavelet Algorithm and Maximum Change-Point Method for the Detection of Ballast Substructure Using GPR

**S. J. Savita, P. Anbazhagan, and Andhe Pallavi**

## 1 Introduction

Ground-penetrating radar is a non-destructive geotechnical tool for detecting the subsurface targets. It consists of transmitting antenna, receiving antenna and display unit. The transmitting antenna transmits an electromagnetic pulse into the ground to detect and classify the various subsurface objects like metal, pipe and air cavity, and the reflected pulses are picked up by the receiving antenna. The reflections from the target depend on the different dielectric value of the object. According to the general principle, if the antenna frequency increases, then depth of penetration decreases with increase in resolution. The depth of penetration increases with decrease in the antenna frequency and resolution also decreases. To detect various patterns of different objects like metal, pipe and air cavity etc., various image and signal processing techniques can be utilized. GPR is widely used in various applications in landmine investigation, geotechnical, underground utilities, archaeology, etc. The interpretation of raw GPR radargram is a challenging task as it needs a better algorithm for improved visualization of the targets.

In this work, two contributions are carried out to map the subsurface targets. The wavelet transform with higher-order statistics method is implemented to map the subsurface targets. The change-point method [1] is also used to detect the abrupt

S. J. Savita · A. Pallavi
EIE Department, RNSIT, Bengaluru, India
e-mail: savita.s.j@rnsit.ac.in

A. Pallavi
e-mail: andhepallavi@rnsit.ac.in

P. Anbazhagan (✉)
Department of Soil Mechanics, IISc, Bengaluru, India
e-mail: anbazhagan@iisc.ac.in



C. N. V. Satyanarayana Reddy et al. (eds.), *Ground Characterization and Foundations*, Lecture Notes in Civil Engineering 167,
https://doi.org/10.1007/978-981-16-3383-6_18

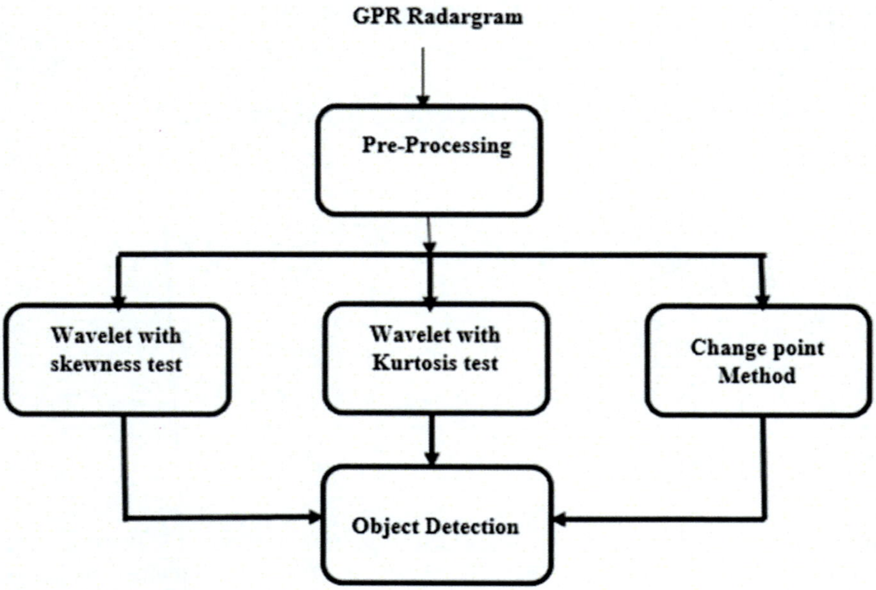


**Fig. 1** GPR processing step

changes in the signal. By change-point algorithm, it is easy to detect the size of the object.

## 2 Methodology

Processing of raw GPR images [2] to map the subsurface targets is one of the difficult tasks. The subsurface targets are obscured with signal noise, echo, reflections from soil, etc. Hence, the interpretation of GPR image is one of the difficult techniques to differentiate between the target and the false target. An algorithm is implemented to detect the different signatures of the target from GPR images. The processing steps are firstly pre-processing of the GPR data, secondly applying the wavelets to the pre-processed signal and next is the detection of abrupt changes in the signal using a maximum change-point method. The block diagram of the processing steps is as shown in Fig. 1.

## 3 GPR Data Collection

The data sets are acquired from Mala GPR Machine at IISc Bangalore. The ground-coupled antenna with 800 MHz has been used to acquire the information of the subsurface targets. In this work, a model railway track is built at IISc without the rails and sleepers. The underground objects used are metal (iron rod), M S steel and clean ballast. The length and width of the track are 4.8, 1.6 m. Ballast is used to construct a railway track. Ballast is irregular shaped stones, and the height of the ballast is 0.6 m. For all these targets, the reflections are different due to the dielectric value of the material.

## 4 Post-Processing Using Wavelets and Change-Point Method

### *4.1 Wavelet Transform*

The algorithm used to filter out the noise in GPR data is discrete wavelet transform, after that a change-point method is applied to detect the abrupt changes in the signal. The wavelet transform is a multiresolution technique to extract the information content in the signal [3]. It is used to denoise the incoming signal. By choosing the wavelet coefficients at a particular value, we can convert the data into the wavelet domain. The coefficients chosen are from higher-order statistics method. In this method, we considered two tests, skewness test and kurtosis test. A signal is decomposed into 10 levels. The wavelet coefficients of a 2D wavelet transform are given by

$$WP^{m}_{x,s}(i) = WP^{n}_{x,s}(i) + WP^{o}_{x,s}(i) \tag{1}$$

where $WP^{m}_{x,s}(i)$, $WP^{n}_{x,s}(i)$ and $WP^{o}_{x,s}(i)$ are wavelet packet coefficients of $m$, $n$ and number of decomposition levels $x = 1, 2, \ldots, x$. number of scales are $s = n\ 1, 2, \ldots$, $i = 1, 2, \ldots, K$, where $K = L/2x$. Where $L$ is the length of the input signal.

The information content in the signal can be extracted by third and fourth-order moments, i.e. skewness and kurtosis test.

**Skewness test**. It is a distribution of asymmetry of the data. If the data are perfectly symmetrical, then its skewness value is zero. The value is positive if it is skewed to the left and negative for right side. It is given by the equation

$$S_3 = \sum_{i=1}^{n} \left(\frac{z_i - \sigma}{S}\right)^3 \tag{2}$$

where $n$ = size of the sample, and $S$ is the standard deviation. The third-order moment of a probability model is referred to as a skewness, so it is 3. The practical formula used for the skewness is,

$$\text{Skewness} = \frac{n}{(n-1)(n-2)} \sum_{i=1}^{n} \left(\frac{z_i - \sigma}{s}\right)^3 \tag{3}$$

For large samples, the skewness distribution is normal with an error of $\sqrt{\frac{6}{n}}$. Hence,

$$\text{Variance (Skewness)} = \frac{6}{n} \tag{4}$$

The Chebyshev inequality for any random variable y is given by

$$\text{Prob}(|S - E(S)| \geq t) \leq \frac{\text{Var}(S)}{t^2} \tag{5}$$

Using inequality of Eq. (5) the skewness test of Eq. (3) can be given as, a $\sqrt{\frac{6}{n}}$. where $a = 1/\sqrt{1-\delta}$.

$$-\frac{1}{\sqrt{1-\delta}} \leq s \geq +\frac{1}{\sqrt{1-\delta}} \tag{6}$$

We have chosen a percentage of $\Phi = 90\%$.

**Kurtosis test**. Kurtosis is also called as a fourth-order moment, which is used to measure the peak reflections of the target. It is given by the equation

$$K_a = \sum_{i=1}^{n} \left(\frac{z_i - \sigma}{S}\right)^4 \tag{7}$$

If the distribution is flat and long tailed, then it has a higher kurtosis value and low kurtosis value for short-tailed distribution. The practical equation is given by

$$K_a = \frac{n(n+1)}{(n-1)(n-2)(n-3)} \sum_{i=1}^{n} \left(\frac{z_i - \sigma}{S}\right)^4 \tag{8}$$

For large samples, the distribution is normal with an error of $\sqrt{24/n}$.
Using the inequality, the kurtosis with $a = \frac{1}{\sqrt{1-\delta}}$.
The kurtosis varies between $\pm \frac{1}{\sqrt{1-\delta}}$ Var (Ka).

## 4.2 Change-Point Method

It is a method used for the detection of the maximum number of abrupt changes in the GPR signal to calculate the size of a target [4]. In this, the GPR signal has been analysed by implementing a change-point method by following these steps.

1. It identifies the abrupt changes in the signal.
2. To find a signal change point, first, it chooses a point and divides the signal into segments [5].
3. Next, it computes an empirical estimate for the maximum number of changes for each section.
4. Within a section at each point, it calculates how much it deviates from empirical value and then adds the deviations for all the points.
5. Region of interest (ROI) of detected target is identified.
6. By this, the starting trace (position) and end trace (position) of the target are easily calculated.

Let us consider a signal $Y_1, Y_2, Y_3, \ldots, Y_{N,}$ the function finds '$m$' number of change points such that,

$$K = \sum_{i=1}^{m-1} \left(Yi - (Y)_1^{m-1}\right)^2 + \sum_{i=m}^{N} \left(Yi - (Y)_m^N\right)^2$$
$$= \sum_{i=1}^{m-1} \left(Yi - \frac{1}{m-1} \sum_{j=1}^{m-1} Yj\right)^2 + \sum_{i=1}^{m} \left(Yi - \frac{1}{N-m+1} \sum_{j=m}^{N} Yj\right)^2$$
$$= m - 1\mathrm{var}([Y_1, \ldots, Y_m - 1]) + N - m + 1\mathrm{Var}([Y_m, \ldots, Y_N]) \tag{9}$$

So findchangepts finds 'm' such that,

$$K(M) = \sum_{i=1}^{m-1} \Delta(Yi; X([Y_1, \ldots, Y_m - 1]))$$
$$+ \sum_{i=1}^{m} \Delta(Yi; X([Y_m, \ldots, Y_N])) \tag{10}$$

$X$ is the empirical estimate, and $\Delta$ is the deviation from empirical value.

As the number of change points increases, residual error decreases. In this algorithm, a MATLAB function called findchangepts and MaxNumChanges is used to detect the abrupt changes in the GPR signal. MaxNumChanges uses an automatic threshold to detect a value of change point [6].

## 5 Results and Discussion

In this work, a GPR technique operates electromagnetic (EM) waves to provide a better resolution and non-destructive measurements of different dielectric contrasts in geological targets. When there is a target like metal or big sized ballast, it develops a strong reflection of electromagnetic waves because of the sharp objects present between the soil and the surrounding rock. The proposed algorithm is verified for different targets. The GPR data consists of a total number of 135 traces with a distance interval of 0.009712 m. The length of the metal piece buried underground is 1.2 m. The model track is constructed at IISc Bangalore as shown in Fig. 2. The length and breadth of the track are 4.8 and 1.6 m. The ballast filled at a height of 0.6 m. Ballast is a collection of big-sized stone which plays a very important role in maintaining a good condition of the track [7].

The signals reflected from the target are picked up by the receiving antenna. Further, it is processed using MATLAB. Figure 3 is the output after applying an average subtraction trace on the GPR raw data [8]. The pre-processed image is represented as a signal as shown in Fig. 4.

The comparison of wavelet with kurtosis and skewness test values for metal (iron rod), clean ballast and metal circular disc is analysed and tabulated.

Table 1 shows the comparison of various targets with skewness and kurtosis tests. It is clearly understood that dB3 performs better compared to the other mother wavelets. Kurtosis test performs better with respect to the higher values than skewness test.

In this Fig. 5, the GPR signal is partitioned into two segments. In each segment, the points are at a minimum distance from maximum number of changes. For metal targets, it accurately detected the target by considering two change points between the trace 25 and 45. By this, it is easy to calculate the size of the target which is discussed in Table 2. The targets used are metal 1 (iron rod), metal 2 (MS steel-large circular disc) and metal 3 (MS steel -small circular disc).

**Fig. 2** Model track

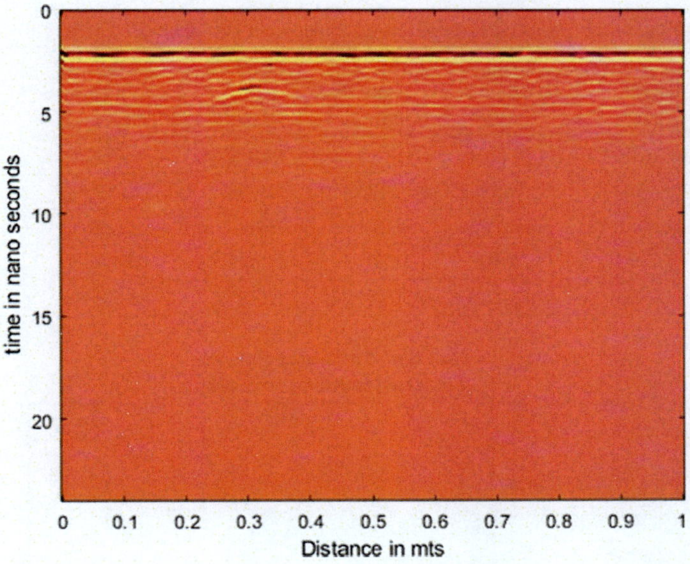


**Fig. 3** GPR image

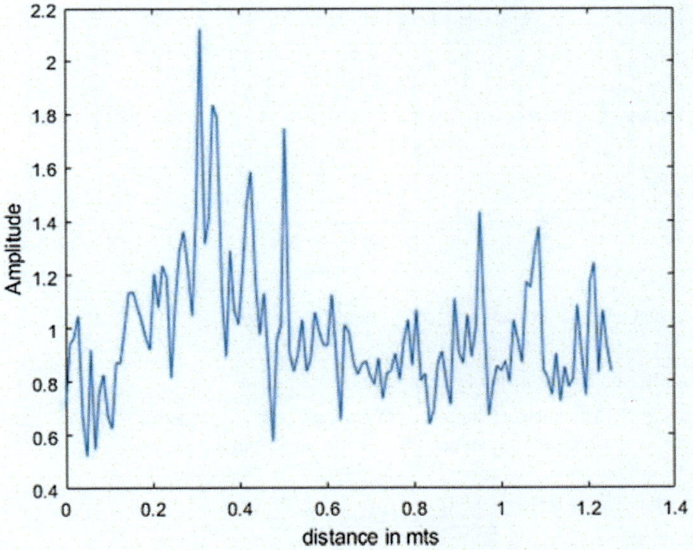


**Fig. 4** Processed GPR signal in MATLAB

**Table 1** Evaluation of results using wavelets (skewness and kurtosis test)

| Wavelets | Metal track target | No target (ballast) | Metal circular disc (large) | Metal track target | Metal circular disc (large) | No target |
|---|---|---|---|---|---|---|
| | Skewness | | | Kurtosis | | |
| Haar | 1.1954 | 0.055802 | 1.32125 | 5.3774 | 3.6417 | 1.7909 |
| dB2 | 1.2128 | 0.20548 | 1.30761 | 4.288 | 4.1728 | 1.6484 |
| dB3 | 1.3041 | 0.26402 | 1.18864 | 4.7827 | 3.5847 | 1.6569 |
| Sym2 | 1.2128 | 0.20548 | 1.30761 | 4.288 | 4.1728 | 1.6484 |

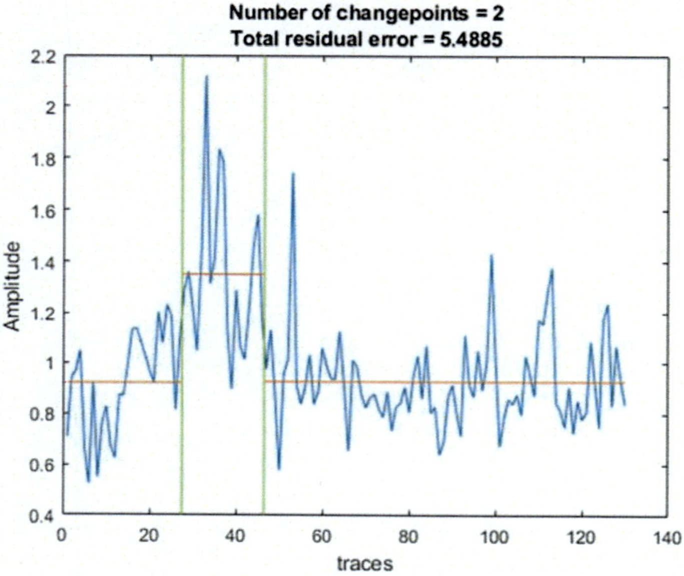


**Fig. 5** Applied change-point method (target–metal 1)

**Table 2** Comparison of results using change-point method

| Object | Traces | Distance interval | Actual value | Obtained value-(change-point method) | Error in % | SNR | Resolution |
|---|---|---|---|---|---|---|---|
| Metal 1 | 135 | 0.009712 | 0.2 m | 0.19 m | 5 | 47 | 1.0296 |
| Metal 2 | 42 | 0.009712 | 0.12 m (Dia) | 0.11 m | 8.3 | 38 | 1.0296 |
| Metal 3 | 37 | 0.009712 | 0.08 m (Dia) | 0.072 m | 10.05 | 36 | 0.8261 |

By implementation of maximum change-point algorithm, it is accurately identified the size of the target in all the three cases with error of 5% and resolution of approximately 1.0.

## 6 Conclusion

The application of GPR in the field of geotechnical engineering gives a better result in interpretation of radargrams by implementing a wavelet algorithm with higher-order statistics and change-point method for detection of the various target. Wavelet transform is applied on raw GPR data to remove the unwanted noise. Haar, dB2, dB3, Sym2 and Sym8 mother wavelets are applied on GPR data to extract the useful information. The two tests called skewness and kurtosis are carried out. In this, kurtosis performs better compared to the skewness test. Later, a change-point method is implemented to identify the size of the target. A MATLAB function called findchangept is used accurately to find the total number of traces in which the object is buried underground. The change-point method works better to estimate the size of the target.

## References


1. Chakar, S., Lebarbier, E., Levy-Leduc, C., Robin, S.: A robust approach for estimating change-points in the mean of an AR (1) process. Bernoulli Soc. Math. Stat. Prob. **23**(2), 1408–1447 (2017)
2. Jol, H.: Ground Penetrating Radar: Theory and Applications, 2nd edn. ISBN: 9780128159774, Elsevier (2019)
3. Javadi, M., Ghasemzadeh, H.: Wavelet analysis for ground penetrating radar applications: a case study. J. Geophys. Eng. 1189–1202 (2017)
4. Lavielle, M.: Using penalized contrasts for the change-point problem. Sig Process **85**(8), 1501–1510 (2005)
5. Nilsen, M.: A Study on Change Point Detection Methods Applied to Beam Offset Detection in Laser Welding. Elsevier Science Direct, (NOLAMP17) (2019)
6. Killick, R., Fearnhead, P., Eckley, I.A.: Optimal detection of changepoints with a linear computational cost. J. Am. Stat. Assoc. **107**(500), 1590–1598 (2012)
7. Vidyaranya, B., Anbazhagan, P., Divyesh, R., Athul.: Identification of heterogeneities in lateritic soils. In: Proceedings of Fifth International Conference on Forensic Geotechnical Engineering (2016)
8. Daniels, D.J.: Ground Penetrating Radar, 2nd edn. London Institution of Electrical Engineers. ISBN 978-0-86341-360-5 (2004)

# Control of Heave Action Using Micropile with Geotextile Layer in Expansive Soil

**Prashant G. Sudani, Sanjay Rajpara, and Mayur G. Vanja**

## 1 Introduction

Some of the partially saturated clayey soil is very sensitive to variation in water content and shows excessive volume changes; such soil is classified as expansive soil and exists in many parts of the world. Various places of India, Australia, Africa, Israel and South America are covered with such soil. Geotechnical engineers around the world face problems associated with expansive soils. It is extended to nearly one-fifth of our country, mostly in the states of Maharashtra, Gujarat, Madhya Pradesh, Uttar Pradesh, Rajasthan, Karnataka, Andhra Pradesh and Tamil Nadu. Expansive soil derives their swelling potential mainly from the properties of the minerals which is present in this type of soil. Any structure located on expansive soil may be subjected to large magnitudes of pressures due to the development of swelling pressure when the moisture content of clay increases.

On the other hand, the decrement in the moisture content of such soil lead to shrinking the volume, which would make an uneven settlement called differential settlement of the foundation. Such a differential settlement can put the negative imprints on the structures such as cracks and deformation sometimes may structure lose their stability and got collapsed. We can reduce economic loss by evaluating a swelling characteristic of the soil before constructing any meaningful structure.

Sometimes the minute to medium soil pressure of expansive soil is adjusted by a dead load of the superstructure, but in many cases, such pressure seems too high which could not adjust by dead load only. In such a case, we must reduce the heave by engineering means. Several methods of heave control were well-explored and tested by several researchers including pre-wetting [1, 2], replacement of the soil

P. G. Sudani (✉)
Department of Civil Engineering, College of Engineering Pune, Pune, Maharashtra, India

S. Rajpara · M. G. Vanja
Department of Civil Engineering, LD College of Engineering, Ahmedabad, Gujarat, India



C. N. V. Satyanarayana Reddy et al. (eds.), *Ground Characterization and Foundations*, Lecture Notes in Civil Engineering 167,
https://doi.org/10.1007/978-981-16-3383-6_19

[3], stabilization [4, 5], drilled pier foundation [6] and many more[7–12]. Outcomes of these methods show positive results, but it seems expansive to some extent. At the same time, the immerging trend of geotextile has the great advantage of imparting the strength along with heave control.

In the present research, an extensive effort was made to control the heave using the application of the geotextile. Series of the laboratory experiment was performed on a model tank with and without reinforcement layer. Observation for the only footing plate, footing plate with micropile, micropile with one reinforcement layer and micropile with two reinforcement layers was carried out. Micropile used in this study has its geometry as 200 mm of length and 16 mm of the diameter. Piles were used to put into action in the pattern of the triangular. Piles were used to place in such a way so that each pile has a 140 mm distance from the other two. As the soil expands, micropile with footing plate has a tendency to rise mainly due to heaving. However, it has been observed that micropile reduces the potential of heaving. In the second stage, the non-woven geotextile of 136 GSM is attached to the micropile with nut and bolt arrangement. This geotextile layer with micropile acts as reinforcement in the soil, pushes the soil downward, and acts as a heave controller. Outcomes of the research show that the geotextile does not control only heave but also imparting the significant strength to the foundation soil.

## 2 Aim and Objectives of the Research

The ultimate aim of the research is to control the heave action of the expansive soil. The in-depth objective of the current research includes the (i) investigation of the expansive soil and their heave action, (ii) investigation of swelling behaviour of the soil under the pile group, (iii) effect of the geotextile layer on the heave control, and (iv) comparison of the outcomes.

## 3 Material Procurement and Test Set-Up

### *3.1 Expansive Soil*

In this research, the soil under investigation is collected from village Rautela, of Dholera City, Gujarat region of India. Figure 1 shows the location of the Google Map from where the soil was procured along with site and oven-dried soil sample. To get an unaltered soil sample, we were used to collecting the soil sample from the depth of 1.5 m from the ground level. The obtained soil was air-dried and pulverized manually.

**Fig. 1** Location map of procurement site and own dried soil sample

## 3.2 *Non-woven Geotextile*

Non-woven geotextile used for control of heaving is procured from the manufacturing company. Specifications of the used geotextile in the research are listed in Table 1. The visual images of the non-woven geotextile are as seen in Fig. 2.

**Table 1** Specification of the non-woven geotextile used in the study

| S. No. | Properties | Unit | Value |
|---|---|---|---|
| 1 | Mass per unit area | GSM | 136 |
| 2 | Tensile strength | KN/m | 24 |
| 3 | Elongation on specified tensile strength | % | 22 |
| 4 | Trapezoidal tearing strength | N | 250 |
| 5 | Puncture strength | N | 200 |
| 6 | Apparent opening size | Micron | 75 |
| 7 | Water flow rate | $L/m^2/s$ | 10 |

**Fig. 2** Visual image of non-woven geotextile used in the research

**Fig. 3** Fabricated micropile system includes a footing plate with micropile (left) and footing plate with three micropiles (right)

## 3.3 Fabrication of Pile

Fabrications of micropile foundation system were done from the steel plate and bars of required diameter and length. Fabricated system of micropile foundation is shown in Fig. 3.

## 3.4 Fabrication of Model Tank

The tank of size 75 cm × 75 cm × 75 cm is fabricated for the purpose of heave control experimental work. Tank with the arrangement of the frame is shown in Fig. 4.

**Fig. 4** Fabricated model tank along with loading frame

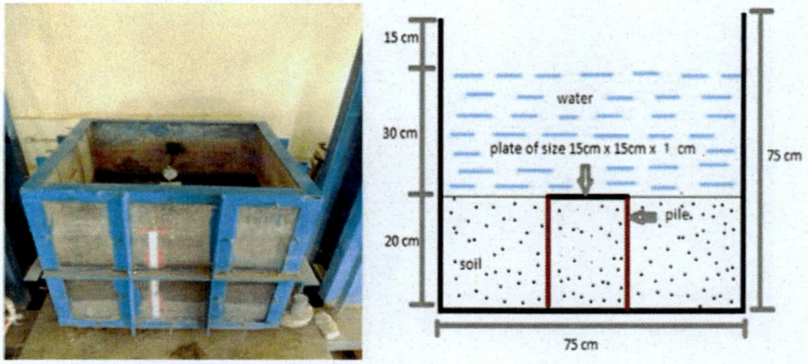


**Fig. 5** Experimental tank (left) and graphical visualization of the tank along with dimensions (right)

# 4 Methodology

## *4.1 Experimental Set-Up*

Experimental set-ups of the tank and pile system along with graphical visualization are shown in Fig. 5.

## *4.2 Experimental Procedure*

For experiment purpose, the tank of size 75 cm × 75 cm × 75 cm is used as shown in Fig. 5. Soil which was procured from Dholera region is the first oven-dried for a period of 24 h. Then it is passed through an IS sieve of size 4.75 mm, the required amount of water is added to the soil to obtain optimum moisture content and mixed thoroughly, then it is stored in a plastic bag up to a period of 5 h. After this, the soil is placed into the tank to the depth of 20 cm in four equal layers of 5 cm. Each layer is compacted with the hammer to obtain maximum dry density. Then the group of three micropiles of diameter 16 mm with footing plate of size 15 cm ($L$) × 15 cm ($B$) × 1 ($H$) cm is installed in the soil. Now the remaining 30 cm portion of the tank is filled with water. And with the provision of the dial gauge, results are observed at different time intervals up to 4 days. Dial gauge arrangement in the model tank during testing and heave action of expansive soil is as well-visualized in Fig. 6.

## *4.3 Experimental Procedure for Micro-Pile with Geotextile*

Non-woven geotextile precisely the size of footing is fitted to the micropile with nut and bolt arrangement. First, the experiment is carried out with a single geotextile

**Fig. 6** Set-up for the heave test with dial gauge arrangement (left), and heave action of soil after four days (right)

**Fig. 7** Three micropile arrangement with single and double layers of geotextile

layer. The layer is fitted to the micropile at $0.33B$, and in the second stage, for double geotextile later it is fitted at a depth of $0.33B$ and $0.66B$, where $B$ is the width of the footing (Fig. 7).

# 5 Result and Discussion

## *5.1 Index and Engineering Properties of Soil*

See Table 2.

**Table 2** Index and engineering properties of expansive virgin soil

| Properties | Measured values | IS code referred |
|---|---|---|
| Specific gravity | 2.54 | IS; 2720(Part-3)-1980 |
| Liquid limit | 70% | IS; 2720(Part 5)-1985 |
| Plastic limit | 33% | IS; 2720(Part 5)-1985 |
| Plasticity index | 37% | IS; 2720(Part 5)-1985 |
| Shrinkage limit | 14.57% | IS; 2720(Part 6)-1978 |
| OMC | 26% | IS; 2720(Part 7)-1980 |
| MDD | 14.9 kN/m$^3$ | IS; 2720(Part 7)-1980 |
| Free swell index | 81% | IS; 2720(Part 40)-1977 |
| Swell pressure | 54.5 kPa | IS; 2720 (Part 41)-1980 |
| Grain size distribution | Clay—52% and Silt—48% | IS; 2720 (Part 4)-1985 |

## 5.2 *Results of Model Test*

Test series for heave control was carried out in the laboratory on the model tank, which was fabricated as discussed previously, and results for various conditions of a group of micropile and geotextile layer are as described in an upcoming section.

**Analysing heaving action on footing plate without any micropile**. Here in the first case, only the footing plate is used to know the initial heave action on the footing. Further readings of the experimental work will compare with this initial value of heave. Observation of the test here displayed on heave versus time plot in Fig. 8. Observation shows the 32.4 mm heave on the foundation. The trend of the plot shows that the heave has a tendency to increase gradually and attains its peak at 32.47 mm after 96 h of time. The reduction in heave can be observed by Formula 1, in percentage reduction in heave.

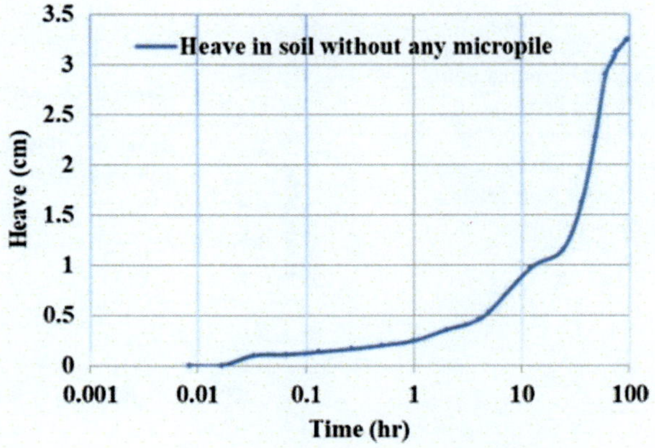


**Fig. 8** Heave-log time graph for soil without any micropile

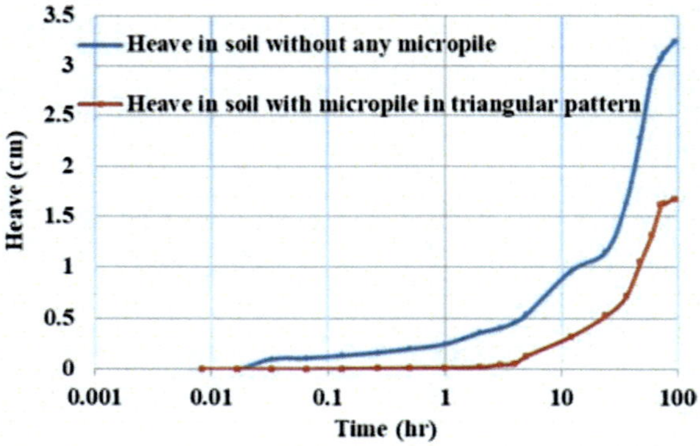


**Fig. 9** Comparison between heave in soil without any micropile and micropile in triangulation pattern

$$\text{RH}(\%) = \frac{\text{Hu} - \text{Hr}}{\text{Hu}} \times 100 \tag{1}$$

Here, Rh (%) stands for percentage reduction in heave; Hu stands for heave of unreinforced soil, and Hr stands for heave of reinforced soil.

**Analysing heave action on footing plate with three micropiles in a triangular pattern**. Findings of the heave control through micropile in triangulation pattern without attaching any geotextile layer are shown in Fig. 9. It is observed that there is a 48% reduction in heave by using three micropiles as compared to footing without any micropile.

**Analysing heave action on footing plate with three micropiles in a triangular pattern and single geotextile layer**. Figure 10 shows the comparison between the heave of soil without any micropile and the single geotextile layer fitted with micropile in a triangular pattern; it is observed that there is a 63% reduction in heave by using single geotextile layer with micropile.

**Analysing heaving action on footing plate with three micropiles in a triangular pattern and double geotextile layer**. Figure 11 shows the comparison between the heaving of soil without any micropile and the double geotextile layer fitted with

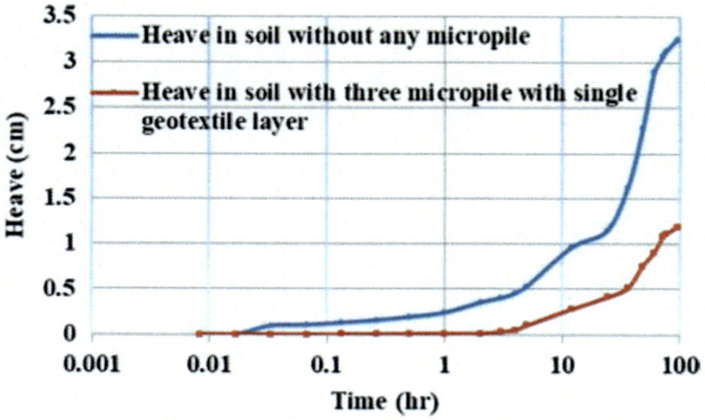


**Fig. 10** Comparison between heave in soil without any micropile and micropile in triangulation pattern with a single geotextile layer

**Fig. 11** Comparison between heave in soil without any micropile and micropile in triangulation pattern with a double geotextile layer

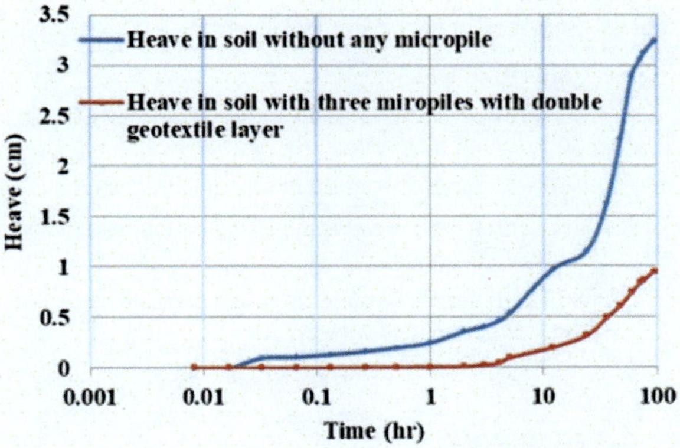


micropile in a triangular pattern; it is observed that there is a 71% reduction in heave by using double geotextile layer with micropile.

## 6 Summary of Findings

In this section, the summary of the complete laboratory experimental series is displayed along with their comparison. Comparisons of the results are shown in Fig. 12. Figure 12 shows that as micropile inserted in the foundation system, it reduces the heave significantly. Besides the micropile, the layer of the geotextiles is also contributing to heave control in an excellent manner (Table 3).

**Fig. 12** Comparison between all the cases of heave control

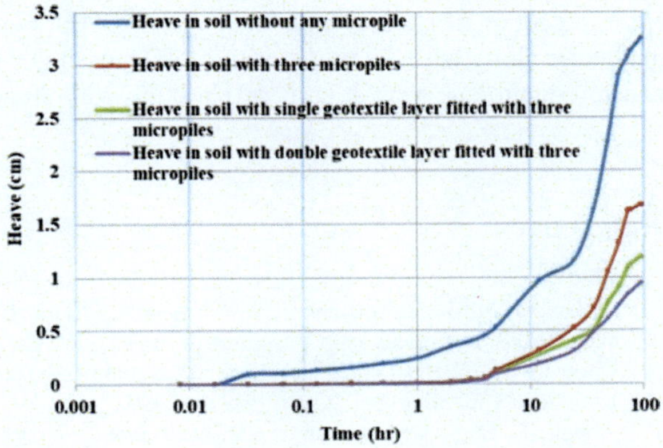

**Table 3** Summary of the experimental work

| S. No. | Types of footing | Reduction in heave (%) |
|---|---|---|
| 1 | Soil without micropile | 0 |
| 2 | Soil with three micropiles in triangulation pattern | 48 |
| 3 | Soil with three micropiles in triangulation pattern and a single layer of geotextile | 63 |
| 4 | Soil with three micropiles in triangulation pattern and double layer of geotextile | 71 |

# 7 Conclusions

Extensive model-based laboratory test series were conducted to evaluate the effectiveness and performance of micropile and geotextile in reducing the heaving of expansive soil. Based on experimental results, the following conclusions have been drawn.

1. Maximum reduction in heave is observed to be 71% by using three micropiles with double geotextile layer.
2. The small increments 8% are observed in heave control with double geotextile layer as compared to single geotextile layer.
3. Percentage reduction in heave increases with an increase in the number of geotextile layer.

A finding of the research shows that the heave action can be reduced with the help of micropile and geotextile layers. Results of the test series show that the micropile group can reduce 48% of the heave action on expansive soil; this value has a significant impact on heave control in the expansive soil. Further with adding the layer of geotextile along with the group of micropile its shows that it gives up to 71% of the heave reduction as compared to only footing without any micropile.

# References

1. McOmber, R.M., Thompson, R.W.:Verification of depth of wetting for potential heave calculations. In: Advances in Unsaturated Geotechnics, pp. 409–422 (2000)
2. Dhowian, A.W.: Heave prediction techniques and design consideration on expansive soils. J. King Saud Univ. Eng. Sci. **2**(2), 355–376 (1990)
3. Arafat, H., Ebid, A.: Optimum replacement depth to control heave of swelling clays. Int. J. Eng. Innov. Technol. (IJEIT) **4**(9) (2015)
4. Kettle, R.J., McCabe, E.Y.: Mechanical stabilization for the control of frost heave. Can. J. Civ. Eng. **12**(4), 899–905 (1985)
5. Sun, N., Fang, Y., Chen, H., Fu, Y., Lu, B.: Nonlinear stabilizing control for ship-mounted cranes with ship roll and heave movements: design, analysis, and experiments. IEEE Trans. Syst., Man, Cybern. Syst. **48**(10), 1781–1793 (2017)

6. Chen, F.H.: Foundations on Expansive Soils vol. 12. Elsevier (2012)
7. Nusier, O.K., Alawneh, A.S., Abdullatit, B.M.: Small-scale micropilesto control heave on expansive clays. Proc. Inst. Civ. Eng. Ground Improve. **162**(1), 27–35 (2009)
8. Nusier, O.K., Alawneh, A.S., Rabadi, R.: Micropiles reinforcement for expansive soils: large-scale experimental investigation. Proc. Inst. Civ. Eng. Ground Improve. **11**(2), 55–60 (2007)
9. Ali, M., Ahmed, S.M.: Micropile technique to control heave on expansive soils. In; Proceedings of Indian Geotechnical Conference, Kochi (paper No. D311) (2011)
10. Nusier, O.K., Alawneh, A.S.: Micropile technique to control upward movement of lightweight structures over expansive soils. Geotech. Geol. Eng. **22**(1), 89 (2004)
11. Ranjan, G., Rao, A.S.R.: Basic and applied soil mechanics. New Age Int. (2007)
12. Shah, M.V., Sudani, P.: Strength and deformation characteristics of laterite rock with different rock matrix's using triaxial system. In: International Congress and Exhibition Sustainable Civil Infrastructures, pp. 85–102. Springer, Cham (2019)

# Assessment of Deformability Characteristics of Sandstone by Direct and Indirect Methods—A Case Study

D. V. Sarwade, P. Senthil, Pankaj Kumar, and Hari Dev

## 1 Introduction

Rock masses are discontinuous and often have heterogeneous and anisotropic properties. Since the rock mass cannot be fabricated according to the project requirements, the properties have to be established. Moreover, there is no single parameter or index, which can fully describe the properties of jointed rock mass. Various parameters have different significance, and only if combined, they can describe a rock mass satisfactorily [1]. According to Lama and Vutukuri [2], the engineering properties of a rock mass depend far more on the system of geological discontinuities within the rock mass than on the strength of the rock itself. Further, the strength of a rock mass is often governed by the interlocking bonds of the unit "elements" forming the rock mass.

Knowledge of the rock mass deformation modulus is important in any rock-engineering project involving tunnel lining design or dam foundations. The deformability of rock mass is dependent mainly upon the compressive strength of the intact rock materials, joints, bedding planes, groundwater condition, applied stress and its condition and softer beds in the system. Different methods have been proposed for measuring or estimating the deformation modulus. These vary from in-situ tests [3] to modulus estimation using rock mass classification systems [4]. The accuracy and reliability of in-situ tests depend on the quality of test execution and consistency of the theoretical assumptions with the real rock mass conditions. Plate load tests, dilatometer tests and flat jack tests are often used in rock engineering projects.

Engineering properties of rock mass can be determined by different methods, which are broadly classified into two general categories, viz. direct and indirect

D. V. Sarwade (✉) · P. Senthil · P. Kumar · H. Dev
Central Soil and Materials Research Station, Olof Palme Marg, HauzKhas, New Delhi 110016, India



C. N. V. Satyanarayana Reddy et al. (eds.), *Ground Characterization and Foundations*, Lecture Notes in Civil Engineering 167,
https://doi.org/10.1007/978-981-16-3383-6_20

methods. The direct methods include laboratory and in-situ tests are somewhat time-consuming and expensive. The indirect methods include empirical or theoretical correlations, combination of intact rock and discontinuity properties using analytical and numerical methods and back analysis using filed observations.

The direct methods include laboratory and in-situ tests. The direct methods have different limitations. To obtain realistic results of rock mass properties, rock of different volumes having a number of different known discontinuity configurations should be tested at relevant stress levels. Such an experimental program, however, would be much time-consuming and expensive. ISRM, ASTM and BIS standards provide guidance related to the specific procedures for performing the actual laboratory and in-situ tests. The indirect methods include empirical or theoretical correlations, combination of intact rock, and discontinuity properties using analytical or numerical methods and back analysis using field observations of the prototype. The indirect methods such as the empirical or theoretical correlations can be used to analyse the data from tests and investigate the reasons for the variation.

Rock mass classification plays an important role in estimating the strength and deformability of rock masses and in assessing the stability of rock slopes. They also serve as an index to rock rippability, dredgeability, excavibility, cuttability and cavibility. During the past 50 years, around the world there have been numerous efforts to create a suitable engineering rock mass classification system so that the preliminary evaluation of feasibility, development and stability/service of engineering structures/projects is fairly reliable.

Hari Dev [5] highlighted the factors affecting deformability characteristics of rock mass, viz. stress level, weathering, repetitive loading and anisotropic behaviour. In the present paper, an effort is made to compare the results of in-situ tests with the results estimated from empirical relations derived from different rock mass classification systems.

## 2 Direct Estimate

Deformation modulus of rock mass can be estimated by conducting Plate Jacking Tests or Plate Loading Test or Goodman Jack tests. Among the three tests, conducting plate load tests are simpler and deformations are measured at the surface. These methods are explained in detail by Palmström and Singh [6]; IS: 7317 [7]; ISRM [8]. These tests provide design parameters which representative of rock mass at the site. In-situ tests are more reliable than laboratory tests or from empirical methods. However, conducting in-situ tests are expensive and time taking. Exploratory drift or open foundation is to be excavated in controlled manner in order to reduce blast disturbed zone. Further muck or loose rocks are to be removed, and fresh rock surface is to be exposed for making site preparation.

Plate load tests conducted at two different dam sites, viz. Thana Plaun HEP and Ujh Multipurpose Project are considered for the study. A total of 28 plate load tests were conducted to study deformability characteristic of rock mass. Number

**Table 1** Details of plate load tests

| S. No. | Project name and location | Nos. of plate load tests | Stress Level (MPa) | Average deformation modulus ($E_d$) GPa | Intact rock UCS (MPa) | Geological description |
|---|---|---|---|---|---|---|
| 1 | Thana Plaun HEP, Himachal Pradesh | 18 | 3.5 | 1.11 | 30 | Medium-grained Sandstone Q: 3.4 to 10.0 |
| 2 | Ujh Multipurpose Project, J&K | 10 | 5.0 | 1.88 | 13 | Hard, Compact Sandstone Q: 5.0 to 9.0 |

of test, deformation modulus, uniaxial compressive strength (UCS) and geological description are given in Table 1.

All the tests values have been considered for averaging the $E_d$ values. It is to be remembered that deformation modulus values vary with stress level at which test is being conducted.

Q values at the test locations considered in this paper have $Q$ in the range of 3–10. The in-situ results have been compared with indirect estimates by using empirical equations. As suggested by Palmström and Singh [6], plate load tests' results were also multiplied by a factor of 2.5 in order to account for blast disturbed zone effects and compared with indirect estimates.

## 3 Indirect Estimates

Indirect estimates are derived from correlations developed between in-situ data set and rock mass classification. Numerous equations were developed using rock mass classification system such as RMR, $Q$ and GSI. Initially, Bieniawski [9] given linear equation for prediction of the deformation modulus of rock masses for RMR > 50 and for RMR < 50, Serafim and Pereira [10] developed power equation. The field data used in this study is available in the form of $Q$-system, thereby RMR-based equations have been converted into $Q$-based by applying the relation RMR $= 15\text{Log}Q + 50$ given by Barton [11, 12].

Barton et al. [13] proposed logarithmic relation based on $Q$-system for hard rocks, i.e. $Q > 1$. This equation is similar to Bieniawski [9] as RMR and $Q$ are correlated logarithmically. Later, Barton [14] modified into a power equation incorporating uniaxial compressive strength and Barton [14] proposed equation for estimating modulus in excavation damage zone. Singh and Bhasin [15] proposed equation for rock mass with $Q > 1$ based on the field data collected in India. Geological Strength Index (GSI)-based equations also proposed by Hoek and Brown [16] similar to Serafim and Pereira [10].

**Table 2** Empirical equation for indirect estimation of $E_d$ values

| | Author | Equation ($E_d$ in MPa) | Remarks |
|---|---|---|---|
| 1 | Bieniawski [9] | $E_d = 2RMR-100$ or<br>$E_d = 30\log_{10} Q$ | RMR > 50<br>Converted into $Q$ using Barton [11] |
| 2 | Barton et al. [13] | $E_d = 25\log_{10} Q$ | Range from 10 to 40LogQ |
| 3 | Serafim and Pereira [10] | $E_d = 10^{(RMR-10)/40}$ or<br>$E_d = 10^{(15\log_{10} Q+40)/40}$ | RMR < 50<br>Converted into $Q$ using Barton [11] |
| 4 | Barton [11] | $E_d = 10Q_c^{1/3}$ | where $Q_c = Q \times UCS/100$ |
| 5 | Singh and Bhasin [15] | $E_d = 12.5\log_{10} Q$ | Applicable for $Q > 1$ |
| 6 | Barton [14] | $E_d = 3Q_c^{1/2}$ | where $Q_c = Q \times UCS/100$<br>(Excavation disturbed zone) |

Equations given in Table 2 were considered for the study in evaluating the equations based on CSMRS in-situ test data.

## 4 Results and Discussions

Deformation modulus of rock mass estimated from direct and indirect estimates are given in Tables 3 and 4 for Thana Plaun HEP, Himachal Pradesh and Ujh Multipurpose Project, J&K, respectively. Ratio between direct estimate and indirect estimate has also been worked. It could be seen that indirect estimates provide modulus values ranging from 2 to 25 times the direct estimate values from plate load tests. It could be seen from Table 3, that all the empirical equations except Barton [14] predict very high modulus of rock mass in comparison with in-situ modulus. The ratio of average values of $E_{d\,(\text{in-situ})}$ to $E_{d\,(\text{empirical})}$ is ranging from 3.9 to 22.4; when plate load test results are multiplied by a factor of 2.5, Barton [14] gives closely matching results.

Similarly, in Table 4, all the empirical equations except Barton [14] predict very high modulus of rock mass in comparison with in-situ modulus. The ratio of average values of $E_{d\,(\text{in-situ})}$ to $E_{d\,(\text{empirical})}$ is ranging from 1.5 to 13.7; when plate load test results are multiplied by a factor of 2.5, Barton [14] gives closely matching results.

From laboratory test, modulus of intact rock reported in the DPR of the project varies from 7.2 to 25 MPa, whereas equations by Barton et al. [13], Bieniawski [9], Serafim and Perira [10] predict modulus of rock mass ($E_d$) as higher than the intact rock modulus.

CSMRS in-situ plate load test data and adjusted data by multiplication with factor of 2.5 are plotted in Fig. 1 along with envelope of empirical equations as given in Table 1. It is revealed that Barton [14] equation which is proposed for excavation damage zone provides nearly matching results to in-situ test results. On applying multiplication factor of 2.5 to plate load test results, the adjusted data set matches

**Table 3** Modulus values from direct and indirect estimates—Thana PlaunHEP, Himachal Pradesh

| Test ID | $Q$ | $E_d$ (GPa) | Barton et al. [13] | Bieniawski [9] | Serafim and Pereira [10] | Barton [11] | Barton [14] | Singh and Bhasin [15] | Sandstone × 2.5 |
|---|---|---|---|---|---|---|---|---|---|
| PLT-1 | 6.0 | 0.49 | 19.45 | 23.34 | 19.58 | 12.14 | 4.02 | 9.73 | 1.22 |
| PLT-2 | 6.0 | 0.39 | 19.45 | 23.34 | 19.58 | 12.14 | 4.02 | 9.73 | 0.98 |
| PLT-3 | 6.0 | 0.27 | 19.45 | 23.34 | 19.58 | 12.14 | 4.02 | 9.73 | 0.68 |
| PLT-4 | 7.0 | 0.41 | 21.13 | 25.35 | 20.74 | 12.77 | 4.35 | 10.56 | 1.02 |
| PLT-5 | 8.0 | 1.72 | 22.58 | 27.09 | 21.81 | 13.35 | 4.65 | 11.29 | 4.30 |
| PLT-6 | 10.0 | 0.58 | 25.00 | 30.00 | 23.71 | 14.37 | 5.20 | 12.50 | 1.46 |
| PLT-7 | 10.0 | 1.73 | 25.00 | 30.00 | 23.71 | 14.37 | 5.20 | 12.50 | 4.33 |
| PLT-8 | 8.0 | 0.82 | 22.58 | 27.09 | 21.81 | 13.35 | 4.65 | 11.29 | 2.04 |
| PLT-9 | 7.3 | 1.38 | 21.58 | 25.90 | 21.07 | 12.95 | 4.44 | 10.79 | 3.46 |
| PLT-10 | 6.0 | 0.31 | 19.45 | 23.34 | 19.58 | 12.14 | 4.02 | 9.73 | 0.77 |
| PLT-11 | 6.0 | 0.65 | 19.45 | 23.34 | 19.58 | 12.14 | 4.02 | 9.73 | 1.62 |
| PLT-12 | 7.0 | 1.10 | 21.13 | 25.35 | 20.74 | 12.77 | 4.35 | 10.56 | 2.76 |
| PLT-13 | 4.0 | 0.76 | 15.05 | 18.06 | 16.82 | 10.62 | 3.29 | 7.53 | 1.89 |
| PLT-14 | 7.1 | 1.61 | 21.28 | 25.54 | 20.86 | 12.83 | 4.38 | 10.64 | 4.03 |
| PLT-15 | 7.1 | 1.59 | 21.28 | 25.54 | 20.86 | 12.83 | 4.38 | 10.64 | 3.98 |
| PLT-16 | 7.5 | 2.01 | 21.88 | 26.25 | 21.29 | 13.07 | 4.50 | 10.94 | 5.03 |
| PLT-17 | 8.0 | 2.18 | 22.58 | 27.09 | 21.81 | 13.35 | 4.65 | 11.29 | 5.45 |
| PLT-18 | 3.4 | 1.95 | 13.29 | 15.94 | 15.82 | 10.07 | 3.03 | 6.64 | 4.88 |
| Average in-situ tests ($E_d$) | | 1.11 | 20.65 | 24.77 | 20.50 | 12.63 | 4.29 | 10.32 | 2.77 |
| Ratio $E_{d(\text{In-situ})}/E_{d(\text{empirical})}$ | | | 18.64 | 22.37 | 18.51 | 11.41 | 3.87 | 9.32 | |
| Ratio $E_{d(\text{PJT arrived})}/E_{d(\text{empirical})}$ | | 2.77 | 7.46 | 8.95 | 7.40 | 4.56 | 1.55 | 3.73 | 1.00 |

**Table 4** Modulus values from direct and indirect estimates—Ujh Multipurpose Project, J&K

| Test ID | $Q$ | Ed (GPa) | Barton et al. [13] | Bieniawski [9] | Serafim and Pereira [10] | Barton [11] | Barton [14] | Singh and Bhasin [15] | Sandstone × 2.5 |
|---|---|---|---|---|---|---|---|---|---|
| PLT-1 | 5.0 | 0.814 | 17.47 | 20.97 | 18.29 | 8.67 | 2.42 | 8.74 | 2.04 |
| PLT-2 | 6.0 | 0.966 | 19.45 | 23.34 | 19.58 | 9.21 | 2.65 | 9.73 | 2.42 |
| PLT-3 | 7.0 | 1.637 | 21.13 | 25.35 | 20.74 | 9.69 | 2.86 | 10.56 | 4.09 |
| PLT-4 | 8.0 | 2.208 | 22.58 | 27.09 | 21.81 | 10.13 | 3.06 | 11.29 | 5.52 |
| PLT-5 | 7.0 | 0.868 | 21.13 | 25.35 | 20.74 | 9.69 | 2.86 | 10.56 | 2.17 |
| PLT-6 | 8.0 | 2.451 | 22.58 | 27.09 | 21.81 | 10.13 | 3.06 | 11.29 | 6.13 |
| PLT-7 | 7.0 | 1.390 | 21.13 | 25.35 | 20.74 | 9.69 | 2.86 | 10.56 | 3.48 |
| PLT-8 | 8.0 | 2.042 | 22.58 | 27.09 | 21.81 | 10.13 | 3.06 | 11.29 | 5.11 |
| PLT-9 | 9.0 | 3.871 | 23.86 | 28.63 | 22.80 | 10.53 | 3.24 | 11.93 | 9.68 |
| PLT-10 | 8.0 | 2.527 | 22.58 | 27.09 | 21.81 | 10.13 | 3.06 | 11.29 | 6.32 |
| Average in-situ tests ($E_d$) | | 1.88 | 21.45 | 25.74 | 21.01 | 9.80 | 2.91 | 10.72 | 4.69 |
| Ratio $E_{d(\text{In-situ})}/E_{d(\text{empirical})}$ | | | 11.42 | 13.71 | 11.19 | 5.22 | 1.55 | 5.71 | |
| Ratio $E_{d(\text{PJT arrived})}/E_{d(\text{empirical})}$ | | 4.69 | 4.57 | 5.48 | 4.48 | 2.09 | 0.62 | 2.28 | 1.00 |

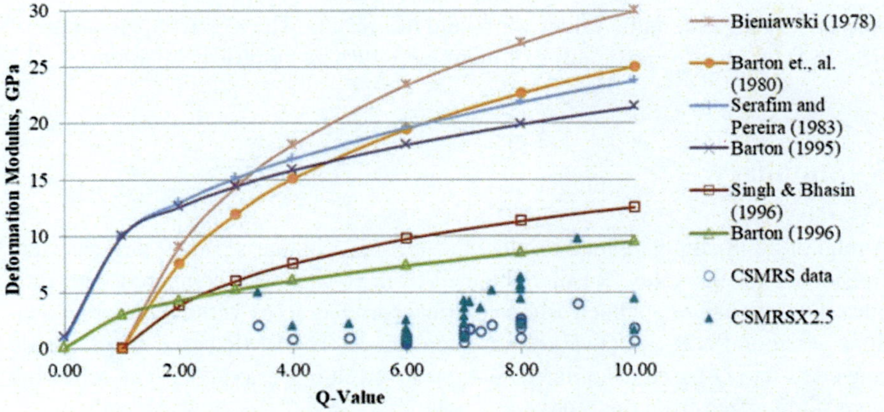


**Fig. 1** In-situ plate load test data along with envelope of empirical equations

closely with Barton [14] equation. Equation proposed by Singh and Bhasin [15] acts as upper bound to adjusted in-situ test data.

Error prediction of empirical methods in comparison with in-situ test tests is shown in Fig. 2. In the figure, it can be seen clearly that all empirical equations

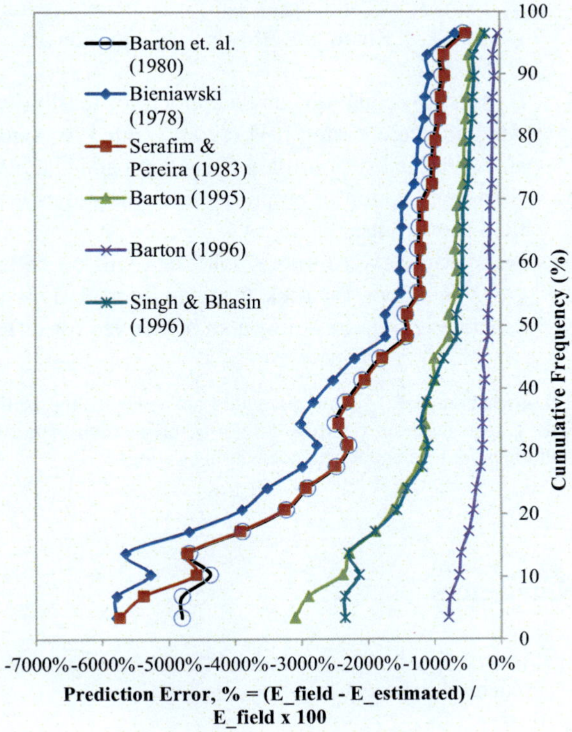


**Fig. 2** Error prediction of empirical methods

predict very high modulus values w.r.t. field test results. The negative percentage of error indicates the overestimation of modulus values by empirical equations.

## 5 Summary

A total of 28 in-situ plate load results of sandstone rock mass from 2 projects were considered for the study. Results obtained from in-situ were compared with rock mass modulus values obtained using empirical correlations by various authors. Equations based on RMR and *Q* values were considered. The RMR-based equations are converted into *Q*-based using inter-relations as given in Barton [11]. The following were summarized from this study:

1. All empirical equations are valid for fair to good rock mass with $Q > 1$ or RMR $> 50$. Even, the present paper studied based on CSMRS data has data for *Q* value in the range of 3–10. At present, no empirical equations available for estimating the deformation modulus for poor rock mass.
2. Equations developed by Bieniawaski [9], Barton et al. [13] and Barton 11 estimates very high modulus for rock mass as compared to in-situ test results.
3. Equations developed by Barton [14] predict value nearer to in-situ test results. Plotting of in-situ plate load test results multiplied by a factor of 2.5 as suggested by Palmström, and Singh [6] gives closely matching with equations given by Barton [14]. Singh and Bhasin [15] equation acts as upper bound to the studied CSMRS data.
4. Barton [14] equation can be applied for feasibility stage designs. However, for detailed stage design it is recommended to conduct in-situ tests by adopting stringent quality control in excavation and site preparation.
5. Error predication of empirical methods shows very high modulus values w.r.t. field test results.
6. As a way forward, more field data will be collected with different range of *Q*/RMR values for rock mass and correlations will be validated. Moreover, separate correlations may also be derived for different rock types.

**Acknowledgements** Authors gratefully acknowledge the motivation and encouragement of Sh. S. L. Gupta, Director, CSMRS during the preparation of manuscript. Sincere thanks to project authorities for the support extended during testing.

## References

1. Bieniawski, Z.T.: Rock Mechanics Design in Mining and Tunneling, 272 p. A.A. Balkema, Rotterdam (1984)

2. Lama, R.D., Vutukuri, V.S.: Handbook on Mechanical Properties of Rocks, 2nd edn. Trans Tech Publications, Switzerland (1978)
3. Heuze, F.E., Salem, A.: Rock deformability measured in-situ—problems and solutions. In: Proceedings of International Congress on Field Measurements in Rock Mechanics (FMRM 77), pp. 375–387, Zurich, Switzerland, April (1977)
4. Hoek, Diederichs: Empirical estimation of rock mass modulus. Int. J. Rock. Mech. Min. Sci. **43**(2), 203–215 (2006)
5. Hari Dev: Stress-deformation properties of rock material and rock mass – An overview. J. Rock. Mech. Tunn. Technol. (JRMTT) **26**(1), 41–51 (2020)
6. Palmström, A., Singh, R.: The deformation modulus of rock masses—comparisons between in-situ tests and indirect estimates. Int J Tunnel. Underground Space Technol. **16**(3), 115–131 (2001)
7. IS-7317: Code of practice for uniaxial jacking test for deformation modulus of rock mass. Bureau of Indian Standards, New Delhi (2015)
8. ISRM: The complete ISRM suggested methods for rock characterization, testing and monitoring: 1974–2006, (Blue Book) ISRM, Turkey
9. Bieniawski, Z.T.: Determining rock mass deformability: experience from case histories. Int. J. Rock. Mech. Min. Sci. Geomech. Abst. **15**(5), 237–247 (1978). https://doi.org/10.1016/0148-9062(78)90956-7
10. Serafim, J.L., Pereira, J.P.: Consideration of the geomechanics classification of Bieniawski. In: Proceedings of International Symposium on Engineering Geological Underground Openings, Lisbon, vol. 1. pp. II33–II42 (1983)
11. Barton, N.: The influence of joint properties in modelling jointed rock masses. In: Keynote Lecture, 8th Congress of ISRM, Tokyo, vol. 3. Balkema, Rotterdam (1995)
12. Barton, N.: Some new Q value correlations to assist in site characterisation and tunnel design. Int. J. Rock. Mech. Min. Sci. **39**, 185–216 (2002)
13. Barton, N., Loset, F., Lien, R., Lunde, J.: Application of the Q-system in design decisions concerning dimensions and appropriate support for underground installations. In: International Conference on Subsurface Space, Rockstore, Stockholm, Subsurface Space, vol. 2, pp. 553–561 (1980)
14. Barton, N.: Estimating rock mass deformation modulus for excavation disturbed zone studies. In: International Conference on Deep Geological Disposal of Radioactive waste, pp. 133–144, Winnipeg, Manitoba, Canada (1996)
15. Singh, R., Bhasin, R.: Q-system and deformability of rock mass. In: Proceedings of Conference on Recent Advances in Tunnelling Technology, New Delhi, pp. 57–67 (1996)
16. Hoek, E., Brown, E.T.: Practical estimates of rock mass strength. Int. J. Rock. Mech. Min. Sci. **34**(8), 1165–1186 (1997)

# Interpretation of Static Cone Penetration Test with Triaxial Test to Determine Undrained Shear Strength of Clayey Soil

Sachin K. More

## 1 Introduction

### *1.1 A Subsection Sample*

In geotechnical engineering practice, most importance is given for determination of index properties and engineering properties of the soil sample. There are several laboratories, and field methods are available for determination of these properties. In laboratory, many of the tests are performed on disturbed and undisturbed soil sample. But these are the representative soil samples which give the approximate values of the test results. There are various laboratory and field tests available for the determination of index and engineering properties of the soil. For major construction work, it is very necessary to have a representative test data for further design process. Clay soil sample is very sensitive from construction point of view, and it is very necessary to take care during construction.

In clay soil, undrained shear strength parameter is very important from design perspective. So, it is very important to have representative values of undrained shear strength of clay soil sample [3]. Laboratory triaxial test is one of the important test to determine undrained shear strength of clay soil, and it gives representative results. But if we can able to find the undrained shear strength of soil directly on site, then achieved values will be based on site conditions [4]. In this research, the correlation has been made between triaxial test and static cone penetration test to achieve the undrained shear strength values of clay soil directly on the site using static cone penetration test. Static cone penetration test can be used for wide range of geotechnical applications in clay soil [5].

S. K. More (✉)
Department of Civil Engineering, Rajarambapu Institute of Technology, Rajaramnagar, Maharashtra 415414, India
e-mail: sachin.more@ritindia.edu

 
C. N. V. Satyanarayana Reddy et al. (eds.), *Ground Characterization and Foundations*, Lecture Notes in Civil Engineering 167,
https://doi.org/10.1007/978-981-16-3383-6_21

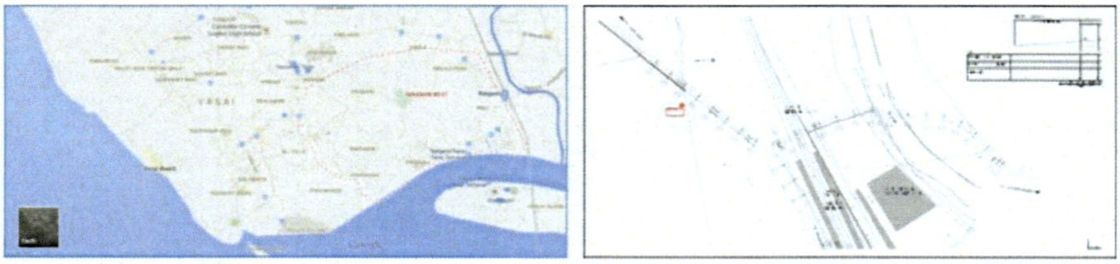

**Fig. 1** Site location and map

## 2 Site Selection

The static cone penetration test can be performed on soft soil, so site is selected in such way that static cone penetration test can be carries out easily. The area around Naigaon West, Mumbai, is deposited with marine clay. So, we have selected Naigaon West, Mumbai, to carry out static cone penetration test. The site is located near Naigaon railway station, Mumbai, which is 5 km away from west coast of Arabian Sea.

The site map is shown below in Fig. 1 (Ref. Genstru Consultants Pune). On the same site, Genstru Consultants were also performing geotechnical investigation for railway over bridge.

## 3 Field Testing and Sample Collection

**Static cone penetration test**: The static cone penetration test used to know soil strata and also to identify soil layers that may be problematic and require additional testing during investigation. The results of static cone penetration tests are used for geotechnical design process and various correlations between soil parameters.

Static cone penetration is internationally recognized as a standard field test to collect data about bearing capacity and frictional resistance of soil. The equipment meets essential requirements of IS: 4968 (Part III). The probing part consists of a cone which has an apex angle of 60°, and overall base diameter of 35.7 mm equivalent to is of 10 cm$^2$. The cone is connected with the sounding rods and mantle tubes. The assembly of sounding rod and mantle tubes is pushed into the soil, by means of hydraulic pumping unit and ram. Basically, the test procedure consists of first pushing the cone and friction jacket and finally the whole assembly in sequence through specified depth and noting the resistance. The readings are taken at every 20 cm of cone penetration, and continuous data of end resistance and sleeve friction is recorded.

**Static cone penetration test results**: The procedure of conducting static cone penetration test is as per IS: 4968 (Part III) and ASTM D 5778. Total four number of static cone penetration test was conducted at 50 m interval named as A, B, C, and D locations, and continuous profile of cone resistance and sleeve friction is recorded. The graph shows the cone resistance and sleeve friction with respect to depth (Fig. 2).

From trends of the all graph, it seems that cone resistance and frictional resistance go on increasing with increase in depth. The range of cone resistance varies from 0 to 3000 kPa, and range of frictional resistance varies from 0 to 1000 kPa. The resistances are recorded up to a depth of 6 m for each 0.2 m interval and are then converted into layers of 1.5 m interval (Table 1).

**Corrections for static cone penetration test results**: The combine cone and frictional resistance shall be corrected for the dead weight of the cone, frictional jacket, and sounding rods. These values shall be corrected for the ratio of ram area to the base area of the cone.

I. Correction for cone resistance

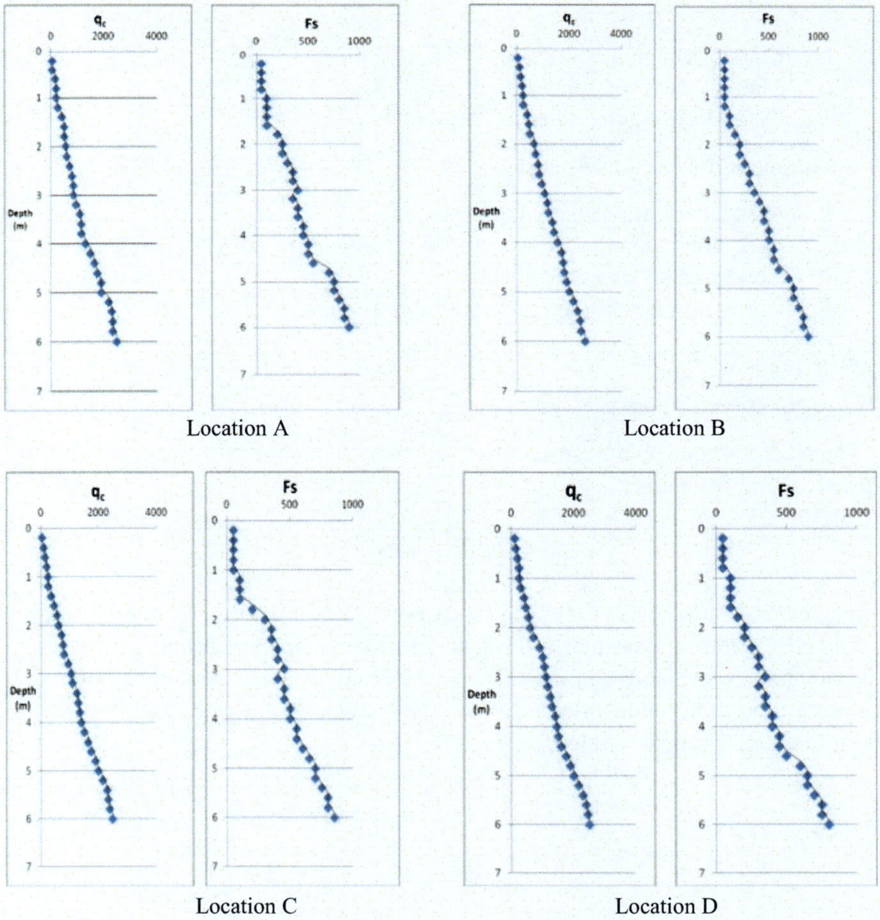


**Fig. 2** Static cone penetration test results

**Table 1** Cone resistance and frictional resistance at every 1.5 m interval

| Depth in m | $q_c$ (kPa) | $f_s$ (kPa) | Depth in m | $q_c$ (kPa) | $f_s$ (kPa) |
|---|---|---|---|---|---|
| *Location A* | | | *Location B* | | |
| 0.00–1.50 | 225.00 | 75 | 0.00–1.50 | 281.25 | 62.5 |
| 1.50–3.00 | 692.85 | 300 | 1.50–3.00 | 785.71 | 250 |
| 3.00–4.50 | 1425.00 | 450 | 3.00–4.50 | 1487.50 | 500 |
| 4.50–6.00 | 2214.28 | 800 | 4.50–6.00 | 2257.14 | 800 |
| *Location C* | | | *Location D* | | |
| 0.00–1.50 | 225.00 | 68.75 | 0.00–1.50 | 262.50 | 75 |
| 1.50–3.00 | 771.42 | 350 | 1.50–3.00 | 842.85 | 250 |
| 3.00–4.50 | 1412.50 | 500 | 3.00–4.50 | 1425.00 | 400 |
| 4.50–6.00 | 2228.57 | 750 | 4.50–6.00 | 2242.85 | 700 |

1. Mass of cone, $m = 0.8$ kg
2. Mass of each sounding rod, $m_1 = 1.5$ kg
3. Correction = $(m + nm_1) \times 10$ kN/m$^2$

II. Correction for frictional resistance

1. Mass of friction jacket $m_f$ kg = 1.3 kg
2. Area of friction jacket, $a = \pi dh = 113.04$ cm$^2$
3. Cone area at base, $b = 10$cm$^2$
4. Correction factor to be added = $\mathrm{m_f/a + z}$ …….

Where $z = (f_s \times b)/a$.

III. Correction for pore pressure:

The recorded cone resistance and frictional resistance should be corrected for pore pressure as shown below,

$$q_t = q_c + u_2(1 - a)$$

a is the area ratio equals to 0.7, and $u_2$ is pore water pressure (Table 2).

Sample Collection. The undisturbed and disturbed samples were collected at every 1.5 m interval. The disturbed samples were collected using standard penetration test sampling tubes, and undisturbed samples were collected through Shelby tubes of 100 mm diameter and 0.9 m length (Fig. 3).

**Table 2** Corrected values of cone resistance and frictional resistance at every 1.5 m interval

| Depth in m | $q_c$ (kPa) | $f_s$ (kPa) | Depth in m | $q_c$ (kPa) | $f_s$ (kPa) |
|---|---|---|---|---|---|
| *Location A* | | | *Location B* | | |
| 0.00–1.50 | 256.27 | 7.73 | 0.00–1.50 | 261.91 | 6.63 |
| 1.50–3.00 | 748.62 | 27.64 | 1.50–3.00 | 840.9 | 23.22 |
| 3.00–4.50 | 1403.85 | 40.92 | 3.00–4.50 | 1572.01 | 45.34 |
| 4.50–6.00 | 2323.90 | 71.89 | 4.50–6.00 | 2366.15 | 71.89 |
| *Location C* | | | *Location D* | | |
| 0.00–1.50 | 255.38 | 7.18 | 0.00–1.50 | 293.48 | 7.74 |
| 1.50–3.00 | 826.32 | 27.64 | 1.50–3.00 | 898.33 | 23.22 |
| 3.00–4.50 | 1496.72 | 45.34 | 3.00–4.50 | 1504.18 | 36.49 |
| 4.50–6.00 | 2343.85 | 67.47 | 4.50–6.00 | 2352.17 | 63 |

**Fig. 3** Collected disturbed and undisturbed samples

## 4 Laboratory Testing

In order to determine index and engineering properties of the soil, various laboratory tests were conducted. The laboratory test program includes determination of specific gravity, grain size analysis, Atterberg's limits, and undrained triaxial tests.

### *4.1 Grain Size Analysis*

The grain size analysis and hydrometer test were conducted on collected soil samples of each location. Mostly for clay soil sample *t,* sieve analysis is preferred because it contains maximum amount of clay and silt particles. Hydrometer test was conducted on the soil samples which are less than 75 μm. After conduction of tests, the results of grain size analysis and hydrometer tests are club together in soil classification curve. Below graph shows the particle size distribution (Fig. 4).

The particle size distribution results indicate the type of soil is silty clay. The content of clay in the soil is more than 80%.

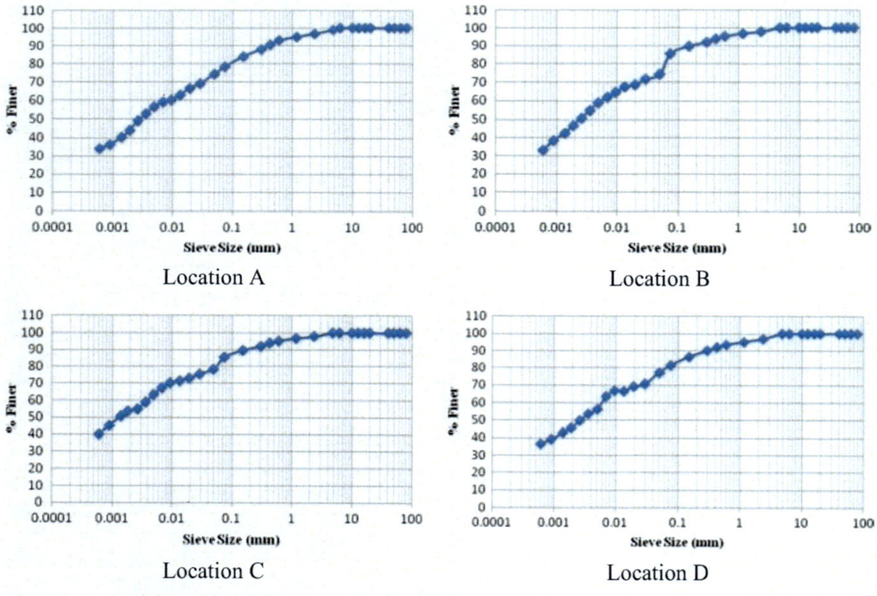


**Fig. 4** Particle size distribution curve for location A, B, C, and D

**Table 3** Results of Atterberg's limit test

| Depth in m | Liquid limit | Plastic limit | Shrinkage limit | Plasticity index |
|---|---|---|---|---|
| *Location A* | | | | |
| 0.00–1.50 | 66 | 36 | 12.67 | 30 |
| 1.50–3.00 | 68 | 37.14 | 13.65 | 30.86 |
| 3.00–4.50 | 69.81 | 39.38 | 14.65 | 30.43 |
| 4.50–6.00 | 72 | 36 | 13.82 | 36 |
| *Location B* | | | | |
| 0.00–1.50 | 67.5 | 33.82 | 12.67 | 33.68 |
| 1.50–3.00 | 69.25 | 36.02 | 13.65 | 33.23 |
| 3.00–4.50 | 70.25 | 37.67 | 14.65 | 32.58 |
| 4.50–6.00 | 71 | 35.65 | 13.82 | 35.35 |
| *Location C* | | | | |
| 0.00–1.50 | 68 | 34.42 | 13.75 | 33.58 |
| 1.50–3.00 | 70 | 36.3 | 16.32 | 33.7 |
| 3.00–4.50 | 70.5 | 39 | 14.55 | 31.5 |
| 4.50–6.00 | 69 | 33.2 | 16.21 | 35.8 |
| *Location D* | | | | |
| 0.00–1.50 | 67 | 33.8 | 14.3 | 33.2 |
| 1.50–3.00 | 69.5 | 31.92 | 14.9 | 37.58 |
| 3.00–4.50 | 68.5 | 36.3 | 15.25 | 32.2 |
| 4.50–6.00 | 70 | 33.97 | 16.2 | 36.03 |

## 4.2 Atterberg's Limits

Atterberg's limits are basic tests to identify the behavior of the clay soil, which measures liquid limit, plastic limit, and shriKage limit. As the water content in dry clayey soil increases, it undergoes dramatic and distinct changes in behavior and consistency. Depending upon the presence of water content, it has four states like solid, semi-solid, plastic, and liquid state. The consistency of soil changes as there is a change in water content. The results of consistency limits are tabulated below. The Atterberg's limits are calculated at every 1.5 m interval at each location (Table 3).

## 4.3 Specific Gravity

Specific gravity of the soil samples was determined using density bottle as per IS: 2720 (Part II) 1980. The tests were performed on collected soil samples from various layers. The values of specific gravity are shown in table below. The specific gravity of soil sample varies in between 1.58 and 1.64 (Table 4).

**Table 4** Results of specific gravity test

| Depth in (m) | Specific gravity Location A | Specific gravity Location B | Specific gravity Location C | Specific gravity Location D |
|---|---|---|---|---|
| 0.00–1.50 | 2.62 | 2.57 | 2.61 | 2.60 |
| 1.50–3.00 | 2.64 | 2.63 | 2.62 | 2.59 |
| 3.00–4.50 | 2.59 | 2.62 | 2.59 | 2.60 |
| 4.50–6.00 | 2.58 | 2.60 | 2.60 | |

## 4.4 Triaxial Test

The triaxial test on cylindrical sample is most widely used to determine undrained shear strength, bearing capacity parameters, and stress–strain relationship. In this research, a strain-controlled triaxial test are conducted as per procedure given in the IS: 2720 (Part 10, 11, and 12). Electrical load cell is used to measure the pore water pressure, displacement, and volume change.

The triaxial tests were conducted on undisturbed soil samples collected from each location to determine undrained shear strength of soil. The strain rate was 0.002 mm/min with different cell pressure of 0.5, 1.0, and 1.5 kg/cm$^2$. The whole assembly of triaxial test is shown in Fig. 5.

The results of triaxial test are tabulated below which shows that undrained shear strength of soil goes on increasing with increase in depth. The determined values of undrained shear strength of clay soil samples are then used for correlation with static cone penetration test data (Table 5).

**Fig. 5** Collected undisturbed and disturbed samples

**Table 5** Results of specific gravity test

| Depth in (m) | Undrained shear strength (Su) kPa | Depth in (m) | Undrained shear strength (Su) kPa |
|---|---|---|---|
| *Location A* | | *Location B* | |
| 0.00–2.00 | 18.49 | 0.00–2.00 | 16.67 |
| 1.60–3.20 | 52.34 | 1.60–3.20 | 54.96 |
| 2.60–4.60 | 99.43 | 2.60–4.60 | 104.94 |
| 4.20–6.50 | 140.47 | 4.20–6.50 | 147.08 |
| *Location C* | | *Location D* | |
| 0.00–2.00 | 16.67 | 0.00–2.00 | 16.67 |
| 1.60–3.20 | 54.96 | 1.60–3.20 | 54.96 |
| 2.60–4.60 | 104.94 | 2.60–4.60 | 104.94 |
| 4.20–6.50 | 147.08 | 4.20–6.50 | 147.08 |

# 5 Determination of Cone Factor Values

It is very reliable estimation of undrained shear strength of clay from cone factor determined from the results of static cone penetration test [6]. The value of cone factor is influenced by type of soil and test methods for determination of undrained shear strength. Many attempts have been made to fine cone factor values. From the studies, it is observed that it is very necessary to calculate the cone factor values suitable for localized, specific type of soil considering all those factors. In this research, we tried to calculate the cone factor values for the marine clay situated at Naigaon West, Mumbai. These cone factor values can be calculated using empirical relationships mentioned below.

## 5.1 *Empirical Cone Factor Values*

There are many empirical correlations that have been proposed to calculate the cone factor value. The researcher Terzaghi 1948 proposes an equation to calculate cone factor value based on recorded values of cone resistance and sleeve friction of static cone penetration test are

$$S_u = \frac{q_c - \sigma_{VO}}{K}$$

$K$ = Empirical cone factor.

$\sigma_{VO}$ = The total overburden stress.

The researcher Lunne proposed an empirical relationship between cone factor and tip resistance of static cone penetration test.

$$S_u = \frac{q_t - \sigma_{VO}}{K_t}$$

Another equation has been proposed to calculate cone factor that is given below,

$$S_u = \frac{q_t - u_2}{K_e}$$

From the empirical relationships mentioned above, the values of cone factor are calculated. These values of cone factor are ranging between 12 and 16 and an average of 14.7. The empirical cone factor values of every layer in each location are calculated, and empirical relationship between index properties, engineering properties, and cone factor has been made.

## 6 Correlation Between Cone Factor, Index Properties, and Engineering Properties

### *6.1 Correlation Between Cone Factor and Plasticity Index*

We have tried to plot the graph between all these three cone factor values $K$, $K_t$, and $K_e$ with plasticity index of the soil, and we observed linear relationship between cone factors and plasticity index. Using correlation between cone factor and plasticity index of soil, a localized equation $K_t = 0.384\text{Ip} + 1.726$ with a correlation coefficient equal to 0.926 is suggested for estimating the cone factor. The analysis is meaningful as it indicated increasing trends of cone factor with plasticity index of clayey soil. The relationship between cone factor and plasticity index is shown in graph below (Figs. 6, 7, and 8).

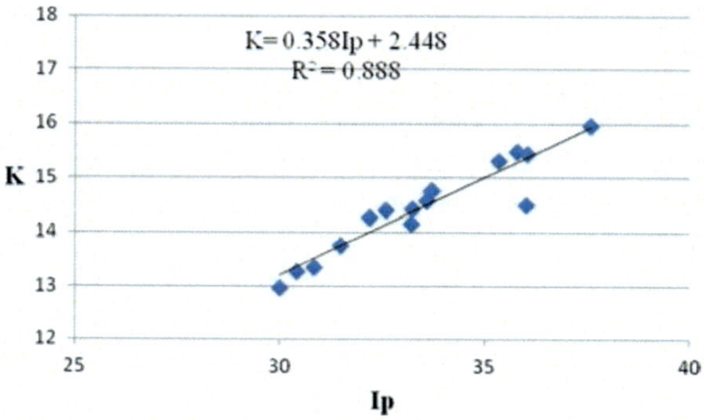


**Fig. 6** Correlation between *K* and plasticity index

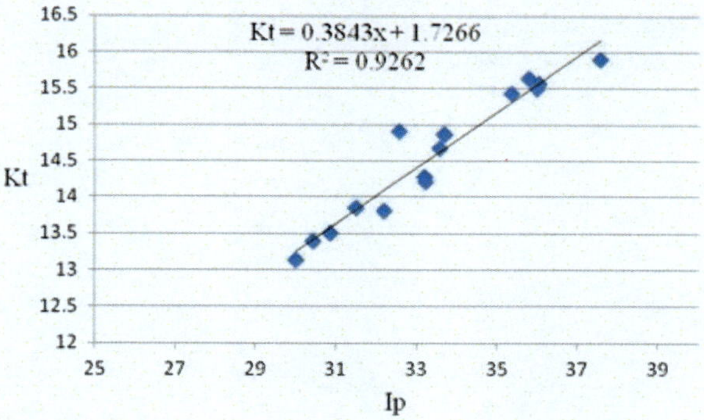


**Fig. 7** Correlation between $K_t$ and plasticity index

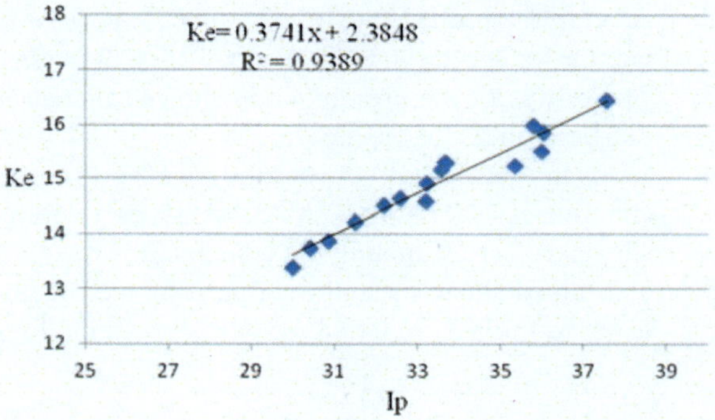


**Fig. 8** Correlation between $K_e$ and plasticity index

### 6.2 *Correlation Between Cone Factor and Undrained Shear Strength*

To calculate in situ undrained shear strength of soil using static cone penetration test, they must correlate with each other. So, undrained shear strength of soil calculated using undrained triaxial test and cone resistance from cone penetration test on similar soil in Mumbai is plotted against each other as shown in graph. From this graph, it seems that undrained shear strength of soil goes on increasing with increase in cone resistance of soil; so there is linear relationship between these two parameters.

Using the correlation between these two parameters, a localized equation for undrained shear strength $S_u = 0.064\,(q_t - \sigma_{vo}) + 4.058$ is suggested with correlation coefficient of 0.989, for the cone factor value of 14.87 (Fig. 9).

## 7 Conclusions

1. From SCPT test results, it is observed that cone resistance and frictional resistance go on increasing with an increase in depth.

**Fig. 9** Correlation between $K_e$ and plasticity index

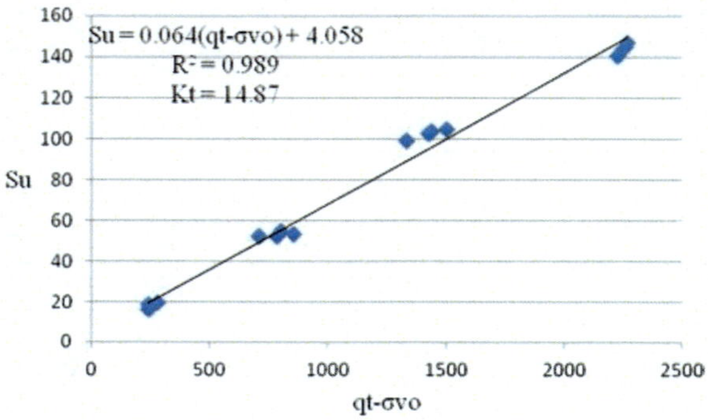


2. An empirical cone factor, $K$, $K_t$, and $K_e$ is derived from the empirical correlation, and the values are in the range of 12–16 or average of 14.87.
3. The results of the present study showed increasing trends of cone factor with plasticity index of clayey soil.
4. Using the correlation between the cone factor and the plasticity index, a localized equation $K = 0.384\text{Ip} + 1.726$ with a correlation coefficient equal to 0.926 is suggested for estimating the cone factor.
5. Undrained shear strength can be estimated using static cone penetration test using formula $S_u = 0.064\ (q_t - \sigma_{vo}) + 4.058$ for cone factor value 14.87.

## References


1. Naeini, S.A., Ziaie Moayed, R.: Evaluation of undrained shear strength of loose silty sands using CPT results. Int. J. Civil Eng. **5**(2) (2007)
2. Nwobasi, P.A.: Estimation of undrained shear strength of soil using cone penetration test. Int. J. Sci. Eng. Res. **4**(9), 409 (2013). ISSN 2229-5518
3. Chen, C.S.: Evaluating undrained shear strength of klang clay from cone penetration test. In: International Conference on In-situ Measurement of Soil Properties and Case Histories, Bali, Indonesia (2001)
4. Karakouzian, M., Burcin Avar, B.: Field measurements of shear strength of an under consolidated marine clay. Eng. Geol. **67**, 233–242 (2002)
5. Rémai, Z.: Correlation of undrained shear strength and CPT resistance. Periodica Polytechnica Civil Eng. **57**(1), 39–44 (2012)
6. Shin, Y.J., Kim, D.: Assessment of undrained shear strength based on cone penetration test (CPT) for clayey soils. KSCE J. Civil Eng. **15**(7), 1161–1166 (2011)

# Stress-Deformation Behaviour of Feldspathic Gneisses as Foundation Medium for a 278 m High-Concrete Gravity Dam in Eastern Himalayas

**Hari Dev and S. L. Gupta**

## 1 Introduction

Rock mass comprises of intact blocks separated by discontinuities namely joints, fractures etc. Pattern and orientations of these fractures mainly govern the properties of the rock mass. Deformability characteristics of rock mass are primarily needed for the design of any structure in or on the rock mass. All the stress analysis theories have been derived assuming the rock mass as continuous, homogeneous, isotropic and linearly elastic (CHILE) material. However, in practise, the actual rock mass is discontinuous, inhomogeneous, anisotropic and nonlinearly elastic (DIANE) due to the effect of structural features, thus, necessitating the actual measurement of parameters [1]. In order to assess the degree of rock mass anisotropy, field tests are conducted in different directions with respect to the apparent rock mass fabric. A number of methods for measurement of deformation modulus [2] of rock mass are available, the important ones are plate jacking test with borehole deformation measurements, plate loading tests with surface deformation measurements, borehole jack with radial deformations inside a drillhole, flat jack tests etc. All these methods have their own benefits and limitations. Borehole jacks or commonly known as Goodman jack tests (GJTs) are conducted inside NX (76-mm diameter) drillholes. The test involves around 0.15 $m^3$ volume of rock mass and takes into account the effect of a few discontinuities, whereas volume of rock mass affected in in plate loading tests (PLT) is much larger. Due to instable nature of drillholes and inability to fix anchors for LVDTs for measurement of deformations in weak and highly jointed rocks, plate jacking tests (PJT) with borehole deformations may not be feasible. Hence, PLT has to be conducted to determine the in situ deformation modulus. Plate loading tests with deformation at surface of loading plate gives lower modulus values due to the effect of weathering, the presence of open joints near the loading surface and ground

H. Dev (✉) · S. L. Gupta
Central Soil and Materials Research Station, New Delhi 110016, India

 
C. N. V. Satyanarayana Reddy et al. (eds.), *Ground Characterization and Foundations*, Lecture Notes in Civil Engineering 167,
https://doi.org/10.1007/978-981-16-3383-6_22

water conditions. Flat jack tests having similar limitations also give lower modulus values. Therefore, all these methods are bound to provide different modulus values [3]. Dev and Singh [4] compared the modulus values obtained from Goodman jack and plate load tests in pyroclastic rocks. The author's experience shows results of plate load and Goodman jack tests in fresh rocks are quite comparable.

The modulus of deformation primarily depends on sample size; testing conditions; nature and persistence of discontinuities; joint properties and infilling material; water conditions; strength of intact material etc. [5]. Due to the influence of so many factors, variation in modulus values is observed when even using a single method and that too within close vicinity. Hence, test results need to be interpreted and correlated with structural features of rock and judgement. Pathak et al. [6] suggested statistical approach to interpret results of PLT for providing values of deformation modulus with desired level of confidence.

The method of excavation also affects the modulus. Drill and blast method of excavation lead to fracturing of rock mass around the opening and widening/opening of existing joints. Weathering of rock mass due to water, air and chemical actions etc. are some of the other prominent factors affecting the modulus values. Dev and Gupta [7] studied the effect of weathering on the deformation modulus of garnetiferous quartzo-feldspathic gneiss rock mass and 40% to 60% decrease in modulus values was reported with alteration from fresh to slightly/moderately weathering.

Anisotropy is a characteristic of intact foliated metamorphic rocks (slates, gneisses, phyllites, schists). These rocks tend to split into planes due to parallel orientation of microscopic grains of mica, chlorite or other platy minerals. Foliation can also be expressed by alternating layers of different mineral composition such as in gneisses. Non-foliated metamorphic rocks such as marble also show some anisotropy due to preferred orientation of calcite grains. Anisotropy is also the characteristic of intact laminated, stratified or bedded sedimentary rocks such as shales, sandstones, siltstones, limestones, coal, etc. The term transverse isotropy is generally used to indicate that a foliated rock has isotropic geomechanics properties in the foliation plane, i.e. transverse to the axis of rotational symmetry, but has varying geomechanics properties perpendicular to the foliation, i.e. along the axis of rotational symmetry [8]. The properties of the intact rock are different with stress application normal or parallel to the foliation.

Since, analysis is carried out considering the rock mass as completely isotropic, therefore, as per the theory of linear elasticity, rock mass can be characterised by a single value of modulus and Poisson's ratio. But in actual practise, it is almost impossible to assign a unique modulus value to any rock or rock mass.

## 2 Geology of the Project Area

The main geological formations around the project are meta sedimentary rocks of Ithun and Hunli formations of Palaeozoic to Precambrian age. These rocks are

**Table 1** The main discontinuities around right bank dam site

| Set No. | Average orientation | Spacing (cm) | Persistence | Condition | Remarks |
|---|---|---|---|---|---|
| $S_1$ | 047°/66° | Closely spaced (5–15 cm) | High | Tight to partly open, rough planer to rough undulatory, Fe-stained | There is reversal of dips from NE to NW direction at higher platform above EL 550 m |
| $S_2$ | 181°/46° | Moderately spaced (20–50 cm) | Medium to high | Tight to partly open, rough, irregular, Fe-stained | |
| $S_3$ | 318°/45° | Moderately spaced (25–60 cm) | Low to Medium | Tight to partly open, rough irregular to smooth planer | |
| $S_4$ | 270°/73° | Moderately to widely spaced (50–80 cm) | Low | Tight, rough irregular, Fe-stained | |

intruded by younger granite and pegmatitic rocks having both concordant and discordant relationships with the host rocks. The rocks exposed in the project are massive to jointed quartzo-feldspathic gneiss, amphibolites gneiss/amphibolites, granitic gneiss with bands of phyllites, volcano-sedimentaries and pegmatites. The different litho units dip towards north, north-east and northwest with varying dips. Light grey-coloured feldspathic gneisses with occasional felsic intrusives classified as poor to fair rock class as per RMR have been encountered. The main discontinuities around right bank dam site are given below in Table 1.

## 3 Tests Conducted

A total of 28 tests were carried out in NX size drillholes using Goodman jack in vertical as well as horizontal drillholes [9]. The transferred stress in the Goodman jack model used in the investigation work is 55% of the applied stress. In view of the expected loading on the foundation due to the proposed dam, the tests were required to be conducted at maximum stress level of 10 MPa. Peak applied stress was kept such that the transferred stress works out to be 1.0, 2.0, 4.0, 6.0, 8.0 and 10.0 MPa. Results from first cycle of 1.0 MPa stress level may not be true representative of the actual rock mass and hence, omitted. Modulus values have been determined for the successive five cycles with transferred stress of 2.0–10.0 MPa as per the design requirements following suggested methods by International Society of Rock Mechanics [10, 11].

### *3.1 Anisotropy*

To assess the anisotropy in rock mass, orientation of applied stress was varied in different directions. Stress was applied in three mutually perpendicular directions as follows:

1. Along the dam axis (*x*-axis)
2. Perpendicular to dam axis (*y*-axis)
3. Normal (Vertical direction) to dam axis (*z* axis)

Therefore, deformation modulus was determined in all the above orientations with respect to dam axis as depicted in the sketch (Fig. 1).

The GJT tests were conducted in fresh rocks at varying depths from 11.40 to 42.40 m in five drillholes (2 horizontal and 3 vertical). Results in different loading directions indicated anisotropic behaviour of rock mass. Results were derived in the light of three-dimensional data. The deformation modulus was denoted by $E_{m(x)}$, $E_{m(y)}$ and $E_{m(z)}$ in three directions as explained above. Direction of loading was kept oblique to the dam axis (almost 450) also and modulus value has been denoted by $E_{m(o)}$. Similarly, unloading modulus or modulus of elasticity calculated using recovery of deformation during unloading was denoted by $E_{e(x)}$, $E_{e(y)}$, $E_{e(z)}$ and $E_{e(o)}$, in respective directions. All the tests were conducted in fully saturated conditions.

Basic rock type in all the drillholes was found to be feldspathic gneiss. However, the minor geological variations like joint infillings, minor shear seams were noticed. As a result of these geological variations along the drillhole, variation in moduli values was also observed.

The arithmetic mean values of deformation modulus with stress applications in different directions are presented in Table 2. The variation in deformation modulus by GJT has also been graphically shown in Figs. 2 and 3 including correlations between deformation modulus ($E_m$) and applied stress ($\sigma_n$) by drawing trendlines. The anisotropy in rock mass in different loading directions was observed.

**Fig. 1** Sketch showing the orientation of applied stress

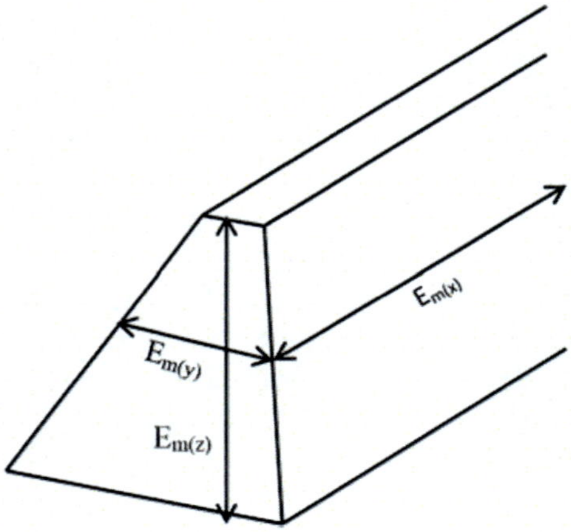

**Table 2** Modulus values from Goodman jack tests

| Applied stress, MPa | 2.0 | 4.0 | 6.0 | 8.0 | 10.0 | Correlation | |
|---|---|---|---|---|---|---|---|
| orientation of stress application | Deformation modulus, GPa | | | | | Equation | $R^2$ |
| Along dam axis ($E_{m(x)}$) | 1.23 | 1.98 | 2.58 | 3.27 | 3.63 | $E_{m(x)} = 0.7691\sigma_n^{0.6814}$ | 1.00 |
| Perpendicular to dam axis ($E_{m(y)}$) | 1.78 | 2.54 | 3.14 | 3.47 | 3.73 | $E_{m(y)} = 1.314\sigma_n^{0.4661}$ | 0.99 |
| Vertical direction ($E_{m(z)}$) | 1.41 | 1.88 | 2.47 | 2.90 | 3.04 | $E_{m(z)} = 0.9902\sigma_n^{0.4980}$ | 0.99 |
| Oblique direction ($E_{m(o)}$) | 1.00 | 1.72 | 2.11 | 2.86 | 3.32 | $E_{m(o)} = 0.597\sigma_n^{0.7412}$ | 0.99 |

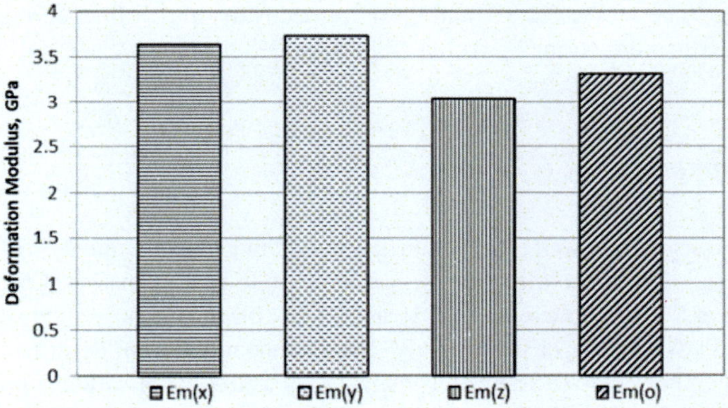


**Fig. 2** Anisotropy in deformation modulus at applied stress 10 MPa

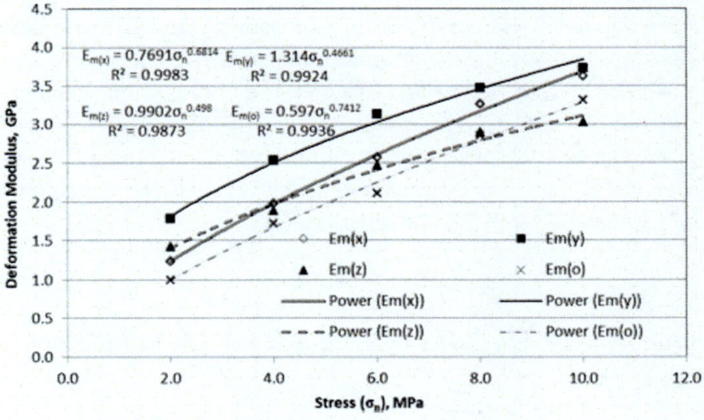


**Fig. 3** Variation in deformation modulus with variation in stress

**Table 3** Results of plate load tests

| Applied stress, MPa | 2.0 | 3.0 | 4.0 | 5.0 | 6.0 | 10.0 (extrapolated) | Correlation equation |
|---|---|---|---|---|---|---|---|
| Modulus of deformation ($E_m$), GPa | 0.50 | 0.80 | 1.06 | 1.38 | 1.50 | 3.5 | $E_m = 0.2703\ {\sigma_n}^{1.0357}$ |

Anisotropic study helps in selecting the critical values of parameters for safe design of structures. Deformation modulus values obtained in the present study using Goodman Jack tests with loading in different directions showed anisotropic behaviour of rock mass. Deformation modulus was observed to be the lowest with loading in vertical direction. Deformation modulus with stress application in horizontal stress directions viz. along ($E_{m(x)}$), perpendicular ($E_{m(y)}$) and oblique ($E_{m(o)}$) to dam axis were found to be 19.7%, 22.8% and 9.4%, respectively, higher than corresponding values in vertical direction ($E_{m(z)}$).

## 3.2 Comparison of GJT and PLT

Plate load tests (PLTs) were also conducted in left and right abutment drifts (5 tests in left and right abutment drifts). Deformation modulus of rock mass was found to be 1.535 GPa at 5 MPa applied stress [12] and 2.140 GPa at 6 MPa applied stress [13] in right and left bank drifts, respectively. The average modulus of deformation from PLT has been presented in Table 3. Though 10 MPa stress application is possible in GJT, stress levels of 5–6 MPa could be applied in PLT due to limitations of testing equipment. Unless tested at the desired stress level, the stress-deformation behaviour cannot be assumed to be similar. However, for the sake of comparison, deformation modulus versus applied stress curve for all data from all the PLTs was drawn and a trend line was drawn to estimate the deformation modulus at stress levels equivalent to GJT (Fig. 4). For this, the upper and lower bound curves were drawn to exclude the outliers or the odd results. The correlation $E_m = 0.2703\ {\sigma_n}^{1.0357}$ was derived by omitting the outliers. The extrapolated deformation modulus of rock mass by PLT at 10 MPa stress level was estimated to be 3.5 GPa which is in coherence with corresponding values of 3–3.7 GPa obtained by GJT.

## 3.3 Comparison of Modulus of Intact Rock ($E_r$) and Rock Mass ($E_m$)

Bieniawski [14] wrote that "Unfortunately, few projects to date have featured a sufficient number of different tests to allow a meaningful comparison of in situ test data. Very different in situ results may be obtained depending on the test method. Under

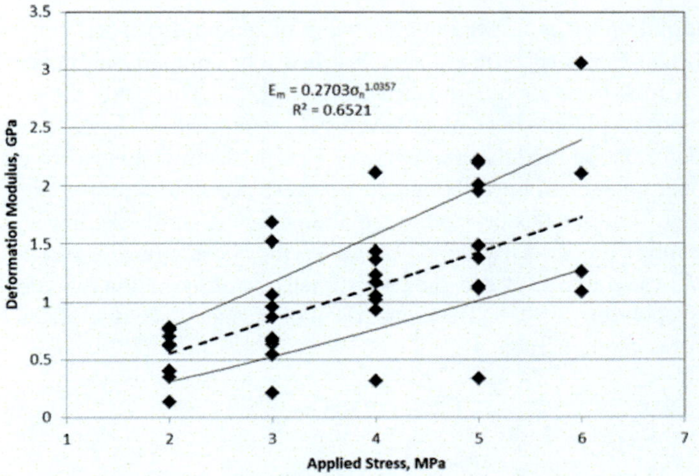


**Fig. 4** Variation of deformation modulus by PLT

these circumstances, it is not helpful to discuss the precision of in situ methods. Even in an extensive in situ test programme in fairly uniform and good quality rock mass conditions, deformability data may feature a deviation of 25% or as much as 10 GPa for an average in situ modulus of 40 GPa. The tests involving full scale prototype behaviour (tunnel relaxation) give different results by comparison with other in situ tests. The choice of the design value for the in situ modulus of deformation thus becomes a matter of engineering judgement. This means that it is difficult to rely on any one in situ method alone, two or more methods should be used to crosscheck the results."

Therefore, variations in test results are obvious and has to be interpreted using own wisdom and experience. Further, no in-situ method is capable of providing unique characteristic value of in situ modulus of deformation due to limitations of capacity of testing equipment and the wide ranging geological variations [5]. Deformations of rock mass in field tests are mainly the result of closing of joint spaces, compaction of joint infilling material and sliding along the discontinuities. For assigning deformation modulus ($E_m$), a unique or specified value, the rock mass has to be tested up to failure. But due to restricted capacity of testing system and to involve all the discontinuities, it is impractical.

At the project site, in situ modulus of deformation was evaluated using two different methods and an attempt has been made to compare $E_r$ and $E_m$ based on extrapolation of observed field test data. Average modulus ($E_r$) of intact feldspathic gneiss rock tested in laboratory was observed to be 34 GPa with variation from 24 to 46 GPa and average uniaxial compressive strength (UCS) has been reported as 33.1 GPa [15]. The in situ modulus ($E_m$) at applied stress of 10 MPa was of the order of 3–4 GPa which is about 1/10th of value of $E_r$. Since, the measured values of modulus of deformation of rock mass ($E_m$ corresponds to certain stress levels actually tested and it is not the unique value of modulus of rock mass; comparison of $E_r$ and $E_m$ as such may not be justified. For comparison, the deformation modulus of rock mass should be corresponds to UCS. Therefore, the deformation modulus versus applied stress plots have to be extrapolated. Hence, likewise PLT (Fig. 4), deformation modulus values by GJT were also plotted against the applied stress. Upper bound and lower bound curves were drawn to filter the exceptionally low and high values. Considering the remaining results within the lower and upper bound, trendline was drawn (Fig. 5). The following correlations for PLT (Eq. 1) and GJT (Eq. 2) were arrived at and used to estimate the deformation modulus at stress levels corresponding to UCS. The comparison of $E_r$ and $E_m$ (Extrapolated using the correlation) is given in Table 4.

$$E_m = 0.2703\ \sigma_n^{1.0357} \tag{1}$$

$$E_m = 0.9416\sigma_n^{0.5788} \tag{2}$$

Therefore, for the sake of comparison, the $E_m$ was estimated using the trendline equations for PLT and GJT (given above) to the stress levels corresponding to UCS. Modulus of rock mass using the correlations was estimated to be of the order of 10

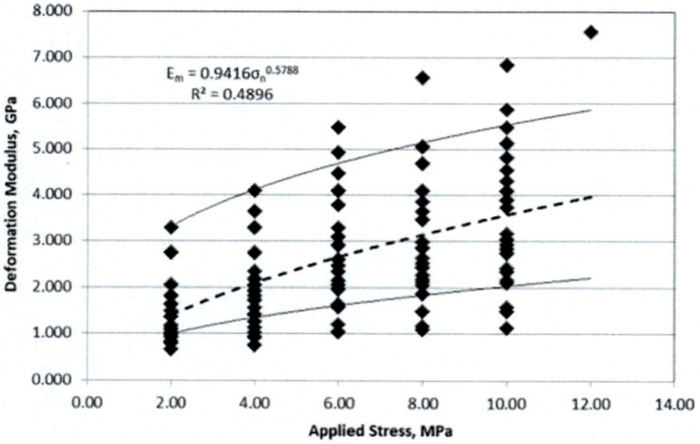


**Fig. 5** Variation of deformation modulus by GJT

**Table 4** Comparison of $E_r$ and $E_m$

| Uniaxial compressive strength of intact rock (UCS), MPa | Deformation modulus of intact rock, $E_r$ (GPa) | Extrapolated deformation modulus of rock mass $E_m$ (GPa) | | Ratio $E_r/E_m$ | |
|---|---|---|---|---|---|
| | | PLT | GJT | PLT | GJT |
| 33.1 | 34 | 10.1 | 7.1 | 3.4 | 4.8 |

and 7 GPa for PLT and GJT at stress level equivalent to UCS. Therefore, ratio of $E_r$ to $E_{m(\text{estimated})}$ using the correlations worked out to be 3.4 and 4.8 for PLT and GJT, respectively. These estimates authenticate the remarks by Bieniawski [14]. Volume of rock mass affected in in situ tests is much large as compared to intact rock. Further, the extent of rock mass affected varies with the test method employed.

In the absence of field tests, $E_m$ is generally estimated by dividing $E_r$ with a factor of 2.5. In situ deformation modulus is primarily dependent on the discontinuities present in the rock mass. Hence, it is advisable to assess $E_m$ due to different testing methods as well as influence of various geological and environmental factors.

## 3.4 Elasto-Plastic Nature of Rock

Similarly, modulus of unloading also denoted by modulus of elasticity ($E_e$) was calculated from the stress-deformation curve using recovery of deformation during unloading. Modulus ratio ($E_e/E_m$) was plotted against the varying applied stress (Fig. 6). During loading, the rock mass deforms, and some of this deformation becomes irrecoverable, generally denoted by plastic or permanent deformation, whereas a part of this total deformation recovers, whilst unloading which is usually known as elastic rebound. The magnitude of plastic and elastic deformation depends mainly on the properties of joints and joint volume. Ratio of modulus of elasticity to modulus of deformation ($E_e/E_m$) in GJT was observed to be 1.19 at 10 MPa applied stress. whereas in PLT, modulus ratio was observed to be 2.43 and 1.57 in right and left bank at stress level of 5 and 6 „ respectively. Stress-deformation curves for one of the GJT and PLT are shown in Figs. 7 and 8 to compare the behaviour of rock under stress in different testing methods. From results of GJT (Fig. 7), the rock mass look to be elastic. In fact, it is due scale effect. Elasticity or plasticity in rock mass is

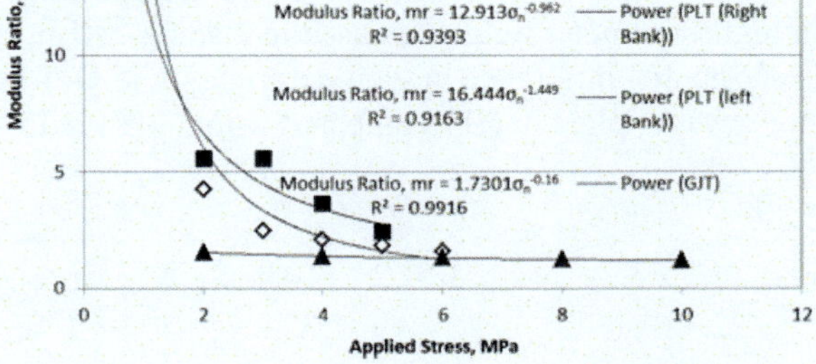


**Fig. 6** Modulus ratio versus applied stress curve

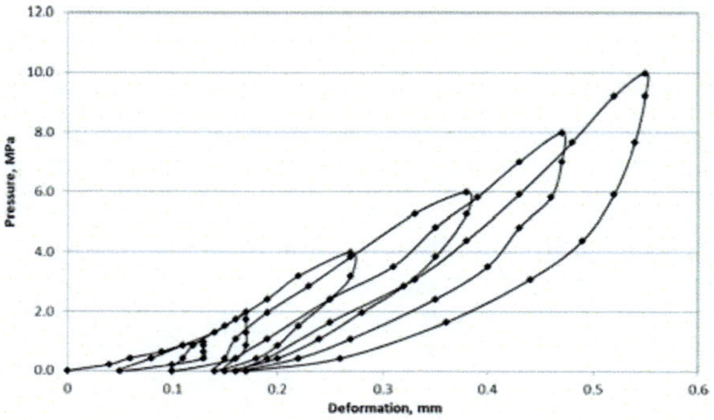


**Fig. 7** Stress-deformation plot for GJT

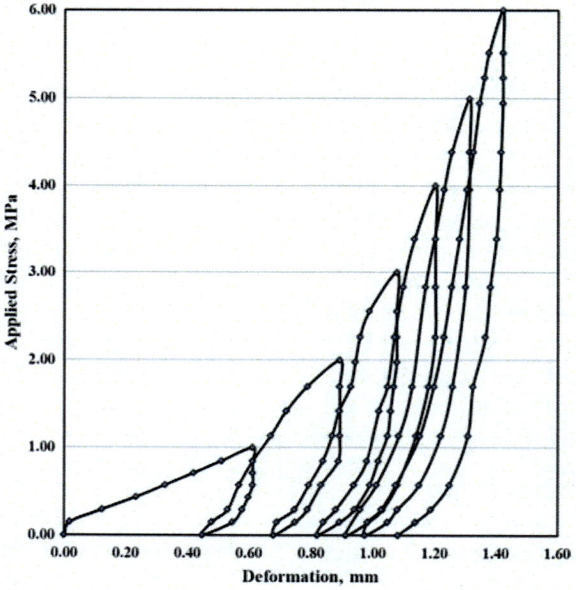


**Fig. 8** Stress-deformation plot for PLT

directly proportional to aperture size and number of joints/discontinuities and properties of joints under the influence of loading platens. GJT affects lesser loading area (two rigid curved plates 20.3 cm × 6 cm); thereby involving lesser number of discontinuities, whereas PLT influences large volume of rock mass (usually 60 cm or higher diameter rigid plate used in loading). Also, as seen in Figs. 7 and 8, there is substantial difference in total deformation in GJT and PLT even though 10 MPa stress was applied in GJT compared with 5 MPa in PLT.

## 4 Conclusions

Based on this study, the following conclusions can be drawn:

For reliable assessment of deformation modulus of rock, the tests should be carried out in fresh rock as weathering leads in significant reduction in modulus values.

Deformation modulus increases with applied stress. However, due to excessive deformations or fracturing of weak rock mass fabric, higher applied stress may not necessarily yield the increase in modulus. Repeated or sustained loading tends to strengthen the rock mass and improves the deformability behaviour given the condition that repeated loading does not cause formation of additional fractures or crushing of rock material.

Rock mass being anisotropic material, deformation modulus varies with application of stress orientation. In the present study, feldspathic gneisses showed anisotropy. Modulus of deformation by GJT at 10 MPa stress level with loading directions along, perpendicular, vertical and oblique directions with respect to dam axis was found to be 3.630 GPa ($E_{m(x)}$), 3.725 GPa ($E_{m(y)}$), 3.033 GPa ($E_{m(z)}$) and 3.317 ($E_{m(o)}$) GPa, respectively. Thus, the variation in deformation modulus was observed to be of the order of 9.4–22.8% when compared with vertical loading direction. Deformation modulus in vertical loading direction ($E_{m(z)}$) was found to be 3.033 GPa at 10 MPa stress level and observed to be the lowest. Hence, the anisotropic studies can be helpful in safe design of structures using critical modulus values.

The estimated modulus by PLT at 10 MPa stress through extrapolation of PLT data worked out to be 3.5 GPa which is in well coherence with modulus values of the order of 3–3.7 GPa measured using GJT at same stress level.

Comparison of intact rock modulus ($E_r$) with rock mass modulus ($E_m$) can be done if the latter is tested up to failure. However, no in situ test method is capable of providing unique value of $E_m$, but an attempt was made to estimate the ultimate modulus of rock mass by extrapolating the field test data up to the stress level corresponding to UCS. $E_r$ was found to be 3.4 and 4.8 times the in situ modulus $E_m$ by PLT and GJT, respectively. Therefore, it is always advisable to measure the in situ modulus rather than assuming it using the correlations in view of influence of geology, environmental factors and test method.

Behaviour of rock mass whether elastic, plastic or elasto-plastic depends mainly on the number of joints, their spacing, orientations, joint properties and properties of infilling material. In cyclic tests, recovery of deformations during unloading increases in stress in successive cycles. This may be attributed to the closing of joint spaces in successive loading cycles and strain hardening of the material. With increased number of loading cycles, behaviour of rock mass shifts towards elastic nature. Due to the involvement of more number of discontinuities under the loading platens, plasticity/elasticity of rock mass is represented better in PLT in comparison with GJT.

## References

1. Harrison, J.P., Hudson, J.A.: Engineering on rock mechanics. In Sarkka, P., Eloranta, P., (eds.) Part 2: Illustrative Workable Examples. Pergamon, Oxford (2000)
2. Bieniawski, Z.T., Bernede, M.J. (coordinators): Suggested methods for determining the uniaxial compressive strength and deformability of rock materials (1979)
3. Palmström, A., Singh, R.: The deformation modulus of rock masses—comparisons between in situ tests and indirect estimates. Tunnel. Underground Space Technol. **16**(3), 115–131(2001)
4. Dev, H., Singh, R.: An analysis of in-situ test data for deformation modulus of weak pyroclastic rock mass. J. Rock Mech. Tunnel. Technol. **21**(1) (2015)
5. Dev, H.: Stress—deformation properties of rock material and rock mass—an Overview. J. Rock Mech. Tunnel. Technol. (JRMTT) **26**(1), 2020 (2020)
6. Pathak, S., Ramana, G.V., Gupta, V.K., Dev, H., Singh, R.: Statistical approach for rock mass deformability characterization. In: Proceedings of Indian Geotechnical Conference, 22–24 Dec 2013, Roorkee, India (2013)
7. Hari Dev and S.L. Gupta (2017). Effect of weathering on modulus of deformation of Gneisses of Peninsular India, Journal of Rock Mechanics and Tunnelling technology (JRMTT), 23 (2) 2017 pp. 113–122.
8. Wittke, W.: Rock mechanics based on an anisotropic jointed rock model rock mechanics based on an anisotropic jointed rock model. In: AJRM, pp. 451 (2014)
9. CSMRS: Report on deformability characteristics of rock mass by Goodman Jack Tests at Right Bank of Dam Axis at Dibang Multipurpose Project, Arunachal Pradesh, CSMRS, New Delhi, May 2016 (2016a)
10. Goodman, R.E., Tran, V.K.: The Measurement of Deformability in Boreholes. Department of Geological Engineering, University of California (1968)
11. ISRM: The complete ISRM suggested methods for rock characterization, testing and monitoring: 1974–2006, Edited by Reşat Ulusay and John A. Hudson, pp. 321–327 (2007)
12. CSMRS: Report on in-situ deformability characteristics of rock mass by uniaxial jacking tests in right bank drift RDR-8, on dam axis of Dibang Multipurpose Project, Arunachal Pradesh, CSMRS, New Delhi, May 2015 (2015)
13. CSMRS: Report on in-situ deformability characteristics of rock mass by uniaxial jacking tests in left bank drift LDR-5, on dam axis of Dibang Multipurpose Project, Arunachal Pradesh, CSMRS, New Delhi, Mar 2016 (2016b)
14. Bieniawski, Z.T.: Engineering Rock Mass Classifications, p. 251. Wiley, New York (1989)
15. ISM: Final report on determination of mechanical properties of rocks from Dibang Hydel Project, NHPC Ltd., Department of Mining Engineering, Indian School of Mines, Dhanbad, Aug 2011
16. IS-12955: 1990 (Part-II). In-situ determination of rock mass deformability using a Flexible Dilatometer—with radial displacement

# Load–Penetration Behaviour of Composite Soil with Nano-Alumina Material Under Soaked and Unsoaked Condition

**S. V. Sivapriya, T. R. Madhu, and Sajid Ali**

## 1 Introduction

Stabilisation of soft soil ground through chemical treatment is a traditional method. For structures such as runway, the time allotted for stabilisation (renovation) will be diminutive and to increase the rate of stabilisation within the stipulated time; innovative additional admixture should be required. Considering this scenario, the nanomaterial came in utilisation to increase the strength of a material, i.e. concrete or soil [1].

The usage of nanoparticles in concrete is to increase its compressive strength [2]. Nano metakaolin along with carbon nanotubes improves the interaction and increases the compressive strength by 18% [3]. Also, Carbon nanotube is widely used in the crack—bridging, filling the hole, modification of soil's microstructure and nucleation effects [4, 5].

The main problem in road system is the formation of crack which may be due to inadequate strength found in the subgrade soil which can be assessed from the CBR value. However, the usage of nanomaterials can mitigates cracks formation due to its functional characteristics [6, 7]. For sandy soil, with the inclusion of nano silicate and artificial pozzolan, the maximum dry density increases and strength increases [9] with the increase in cement content [8]. However, for a clayey soil, addition of nano-alumina reduces the swelling index of the soil in addition to strength improvement [10].

Considering the advantages of nano in increasing the strength of the soil, nano-alumina is added to the soil with lime as an activator as a column. The influence of water content and admixture influence ratio is studied in the laboratory and discussed in this paper.

S. V. Sivapriya (✉) · T. R. Madhu · S. Ali
Department of Civil Engineering, Sri Sivasubramaniya Nadar College of Engineering, Chennai 603110, India



C. N. V. Satyanarayana Reddy et al. (eds.), *Ground Characterization and Foundations*, Lecture Notes in Civil Engineering 167,
https://doi.org/10.1007/978-981-16-3383-6_23

**Table 1** Properties of soil

| Properties | Indian Standard Code | Value |
|---|---|---|
| Specific gravity | IS 2720 (Part III/I) Determination of Specific Gravity of Fine grained Soil [11] | 2.76 |
| Particle percentage, % | IS 2720 (Part IV) Methods of test for soil - Grain size analysis [12] | |
| Sand | | 2.88 |
| Silt | | 28.41 |
| Clay | | 68.71 |
| Liquid limit, % | IS 2720 (Part V) Determination of Liquid and Plastic Limit [13] | 75.8 |
| Plastic limit, % | | 23.49 |
| Differential free swell index, % | Determination of free swell index of soils [14] | 76.48 |
| Soil classification | IS 1498–1970 (Reaffirmed 2002): Classification and identification of soil [15] | CH |

## 2 Materials

The clay needed for experimenting is obtained from Thaiyur, Chennai. The soil was then pulverised, and the properties of the same are listed in Table 1.

Hydrated lime which is used as an activator was obtained from M/s. Shiyal Chemicals and the Nano-sized particle alumina chosen for the test series was obtained from M/s. Astraa Chemicals, India.

## 3 Methodology

The load – penetration characteristics of soil are evaluated by conducting unit cell study with lime and lime + nano -alumina and then compared with virgin soil (Fig. 1). The unit cell model comprises a single column and its equivalent circular influence zone. It is used to represent a column located on the interior of an infinitely large group of columns. A similar concept is used for the penetration analysis on lime and lime + nano-alumina columns. The soil is prepared at a soft consistency of 0.42 under soft consistency. Specimens are prepared and tested after soaked and unsoaked curing conditions [16].

Lime used as an activator is mixed at initial consumption limit (ICL) as suggested by Eades and Grim [17], and the value was found as 4.5%. The sample preparation for untreated ground involves pouring of soil mixture having consistency index of 0.42 into the CBR mould layer-by-layer for achieving uniformity in sample preparation. For the treated ground, the column was prepared by mixing water and lime at a proportion of 1:2 and poured into a predrilled hole inside the soil using a pipe of inner diameter 38 mm. The pipe was first placed in the centre of the mould, and the soil is then placed at various layers. Once the soil is filled, the lime + water is

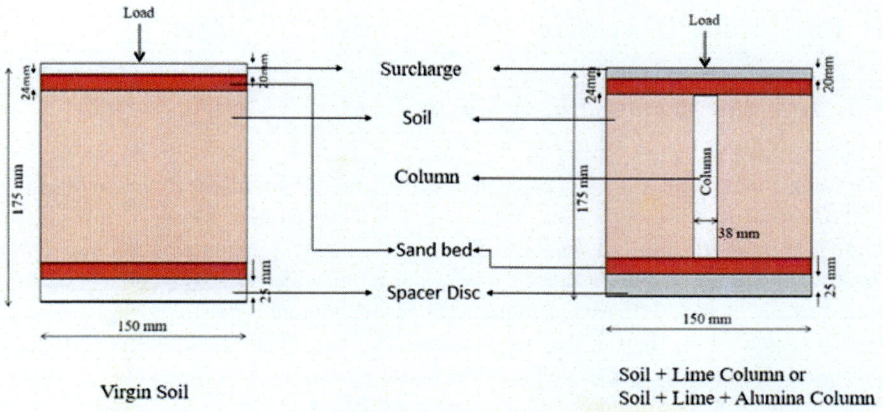


**Fig. 1** Schematic representation of Test sample

poured into the column and then the pipe is removed and allowed for two days for column formation and tested for unsoaked condition (Fig. 2). The preparation of the untreated and treated soil sample is the same for the soaked condition, but the sample was immersed in a bucket of water for 4 days.

An arbitrary amount of 0.5% of nano is added to a lime column for studying its influence on load-penetration behaviour. Upon preparation of the sample, the soil specimen is placed on compression testing machine and a 50-mm diameter plunger is kept on the top surface of the specimen. The load was applied into the prepared soil at a rate of 1.25 mm per minute. The load was applied through a plunger to penetrate the specimen and the load which causes 2.5 mm and 5 mm penetration was recorded.

a. Sample with Column

b. Under Soaked Condition

**Fig. 2** Prepared Sample

# 4 Results and Discussion

## 4.1 Effect of Water Content Versus Load

The soil is mixed with water at a consistency index of 0.42 for all the three conditions. With the known Atterberg's limit of the soil, the water content for the virgin soil is 44.38%. The liquid limit of soil + ICL and soil + ICL + Nano was 54 and 53%, respectively with non-plastic nature. The water required to possess a consistency of 0.42 were 31.74 and 31.16%. The reduction in water content is due to the replacement of ions present in the material/ water with calcium ions, which reduces the plasticity index and shrinkage properties called flocculation and agglomeration [18]. The load at 5 mm penetration is used for interpretation of data (Fig. 3).

For unsoaked condition, there is an increase in load by 58.01 and 88.25% for 5 mm penetration, whereas for soaked condition, the increase is by 6.89 and 20.47% for soil + ICL and soil + ICL + Nano conditions, respectively (Fig. 4). With a reduction in water content due to emission of heat by the addition of admixtures,

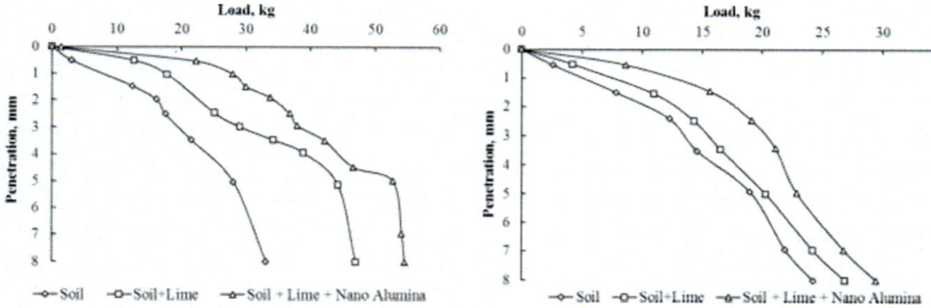


a. Unsoaked Condition b. Soaked Condition

**Fig. 3** Load- penetration characteristics

**Fig. 4** Water content versus load

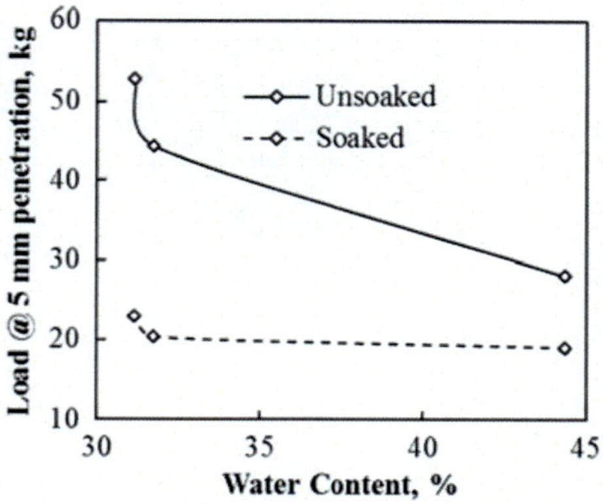

**Table 2** CBR values

| Combinations | Unsoaked | Soaked |
|---|---|---|
| Soil | 1.28 | 0.92 |
| Soil + Lime | 1.83 | 1.09 |
| Soil + Lime + Nano | 2.67 | 1.42 |

due to agglomeration of Nano-Alumina, weak zones were formed with a decrease in density and increase in load in case of unsoaked conditions.

Table 2 shows the CBR values for different proportions. It is observed that CBR value increases with increase in admixture. The chemical reaction occurs due to hydrated lime reduces water holding capacity thereby inducing an increase in CBR value. From the test results, it clearly shows that under unsoaked condition, the CBR value increases by 42. 89% and 108.59% for Soil + ICL and soil + ICL + Nano combination; the same is observed for the soaked condition with an increase in CBR value by 18.48 and 54.35%.

Addition of nano-Alumina enhances the bonding between the soil particles; leading to an increase in CBR value in case of lime stabilised soil. Calcium from the hydrated lime combine with the silica and alumina produces calcium-aluminate a-hydrate, a compound leading to cementitious action which helps in increasing the strength. Weak soil stabilised with nanoparticles enrich the impermeable and load-bearing characteristics of soil, making an unsuitable land into a suitable ground.

**Admixture influencing factor (AIF)** [19], a ratio of CBR (2.5 mm penetration) value of treated soil to untreated soil is referred for further understanding. Soil stabilised with lime increases the CBR value, because of the bonding created through their chemical reaction. From Fig. 5, the ratio increases almost linearly and shows a greater increase with the addition of nano-alumina. To speed up the rate of increase in strength, nano is added. With the addition of nano, the ratio increases about double the time than the initial condition.

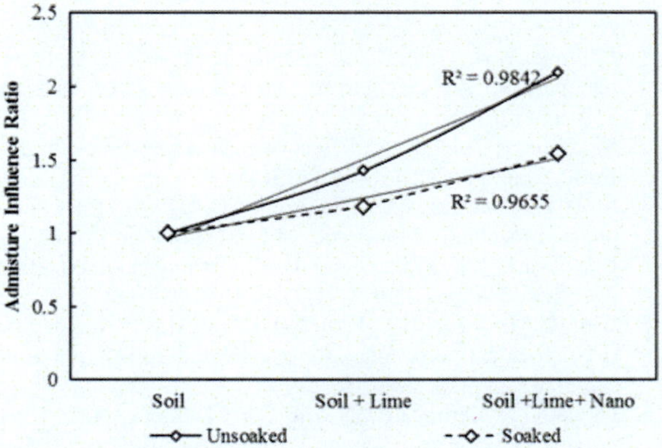


**Fig. 5** Influence of admixtures in *AIF*

## 4.2 Limitations

Tests were conducted till 8 mm penetration. The soil sample was removed, and the condition of the column was observed. Small bulging was observed near the top of the column due to its load distribution mechanism. Due to damage in the ground during sample ejection, the depth of location of bulging was not estimated. In addition to that the dial gauge fixed in the set-up in soaked condition remained constant suggesting that the addition of chemical additives inhibit the swelling potential of the soil.

# 5 Conclusions

Laboratory tests were conducted to understand the influence of nano-alumina with lime in weak soil in this study. A set of three combinations was used; (i) soil, (ii) Soil + ICL and (iii) soil + ICL + nano-alumina. The lime (ICL) and lime (ICL) + nanomaterial were injected into the soil as a column with an inner diameter as 38 mm. The following observations were inferred,

1. The water content reduces with the inclusion of lime and nano into the soil and also contributed to improving the load-bearing strength of the soil.
2. The CBR value increases drastically with the addition of admixture, the increase is about 42. 89% and 108.59% for Soil + ICL and soil + ICL + Nano combination, respectively. The similar increment is also observed for the soaked condition. The increment is about 18.48 and 54.35%, respectively.
3. The AIF increases linearly with the addition of admixtures.

# References

1. Bahmani, S.H., Huat, B.B.K., Asadi, A., Farzadnia, N.: Stabilization of residual soil using $SiO_2$ nanoparticles and cement International College of Auckland. Constr. Build. Mater. **64**(July), 350–359 (2014)
2. Siddique, R., Mehta, A.: Effect of carbon nanotubes on properties of cement mortars Constr. Build. Mater. **50**(1), 116–129 (2014)
3. Morsy, M.S., Alsayed, S.H., Aqel, M.: Hybrid effect of carbon nanotube and nano-clay on physico-mechanical properties of cement mortar. Constr. Build. Mater. **25**(1), 145–149 (2011)
4. Liew, K.M., Kai, M.F., Zhang, L.W.: Carbon nanotube reinforced cementitious composites: an overview. Compos. Part A **91**(1), 301–323 (2016)
5. Kim, G.M., Yang, B.J., Ryu, G.U., Lee, H.K.: The electrically conductive carbon nanotube (CNT)/cement composites for accelerated curing and thermal cracking reduction. Compos. Struct. **158**(1), 20–29 (2016)
6. Ugwu, O.O., Arop, J.B., Nwoji, C.U., Osadebe, N.N.: Nanotechnology as a preventive engineering solution to highway infrastructure failures. J. Constr. Eng. Manage. **139**(August), 987–993 (2013)
7. Steyn ,W.J.: Research and application of nanotechnology in transportation. In: 27th Southern African Transport Conference, pp. 345–353 (2008)

8. Choobbasti, A.J., Vafaei, A., Kutanaei, S.S.: Mechanical properties of sandy soil improved with cement and nanosilica. Open Eng. J. **5**(1), 111–116 (2015)
9. Changizi, F., Haddad, A.: Strength properties of soft clay treated with mixture of nano-$SiO_2$ and recycled polyester fiber. J. Rock Mech. Geotech. Eng. **7**(4), 367–378 (2015)
10. Taha, M.R., Muhie, O., Taha, E.: Influence of nano-material on the expansive and shrinkage soil behavior. J. Nanoparticles **1**(May), 0–13 (2012)
11. Bureau of Indian Standard: IS 2720 (Part III/I) Determination of Specific Gravity of Fine granied Soil, pp. 1–10 (1997)
12. Bureau of Indian Standard: IS 2720 (Part IV) Methods of test for soil - Grain size analysis, pp. 1–40 (1995)
13. Bureau of Indian Standard: IS 2720 (Part V) Determination of Liquid and Plastic Limit, pp. 1–17 (1995)
14. Central Bureau of Indian Standards: IS: 2720 (Part XL) - Determination of free swell index of soils, pp. 1–10 (1977)
15. Bureau of Indian Standard: IS 1498–1970 (Reaffirmed 2002) : Classification and indentification of soil, pp. 1–28 (2002)
16. Bureau of Indian Standard: IS 2720 (Part 16) Laboratory Determination of CBR., pp. 1–17 (1997)
17. Eades, J.L., Grim, R.E.: A quick test to determine lime requirements for lime stabilization. Highw. Res. Board. **1**(1), 61–72 (1966)
18. National Lime Association: Lime treated soil construction manual. January. Virginia, pp. 1–41 (2004)
19. Sivapriya, S.V.: Stress-strain and penetration characteristics of clay modified with crumb rubber. Rev. Fac. Ing. **28**(49), 65–75 (2018)

# Engineering Behavior of Alluvial Rockfill Material

**Uday Bhanu Chakraborty and N. P. Honkanadavar**

## 1 Introduction

River valley projects are being designed and constructed in India/abroad to store the natural water flowing in the rivers and use it latter for different purposes viz. power generation, irrigation and flood control. Rockfill material is being used in the Earth core rockfill dam (ECRD) and concrete faced rockfill dams (CFRD) because of its inherent flexibility, ability to absorb large seismic energy and adoptability to various foundation conditions. In laboratory, large size rockfill materials cannot be tested directly. Various kinds of modeling technique are often used to reduce the particles size so that the specimens prepared with smaller size particles which can be tested in laboratory.

Engineering behavior of rockfill materials has been reported by many researchers [1–5] have carried out laboratory tests on various rockfill materials and concluded that stress–strain behavior is non-linear and stress level dependent. They had also observed that for alluvial (riverbed) rockfill material, and the angle of internal friction increases with increase in particle size ($d_{max}$) [6–8].

This paper deals with the testing of the alluvial rockfill material obtained from a project site in Jammu & Kashmir.

U. B. Chakraborty (✉) · N. P. Honkanadavar
Central Soil and Materials Research Station, New Delhi, India



C. N. V. Satyanarayana Reddy et al. (eds.), *Ground Characterization and Foundations*, Lecture Notes in Civil Engineering 167,
https://doi.org/10.1007/978-981-16-3383-6_24

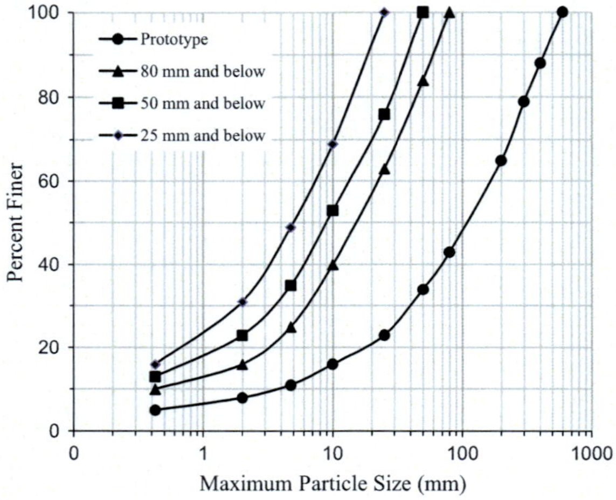


**Fig. 1** Prototype and modeled grain size distribution curves

# 2 Experimental Investigations and Discussion

## 2.1 Material Used

For the present research work, rockfill material collected from a hydropower Project in Jammu & Kashmir. The $d_{max}$ of the material is 600 mm and gradation of the material is shown in Fig. 1. For testing in a large size triaxial cell of 381-mm diameter and 813-mm height, the material has been modeled to maximum particle sizes of 80, 50 and 25 mm by parallel gradation technique as shown in Fig. 1 [9].

## 2.2 Experimental Program

**Drained triaxial tests**: Consolidated drained triaxial tests have been conducted on the modeled materials at various confining pressures (0.6, 1.2 and 1.8 MPa) at Central Soil & Materials Research Station, New Delhi. The stress–strain-volume change response for $d_{max}$ of 80 mm tested with 87% relative density has been presented in Figs. 2 and 3.

The behavior of rockfill material for a relative density is observed to be non-linear and stress dependent. The volume change response shows compression in the initial part of shearing and dilation is noted on further shearing of the specimen.

The value of initial tangent modulus is determined from the stress–strain response using Kondner's (1963) hyperbolic relationship,

$$\sigma_1 - \sigma_3 = \frac{\varepsilon_1}{a + b\varepsilon_1} \tag{1}$$

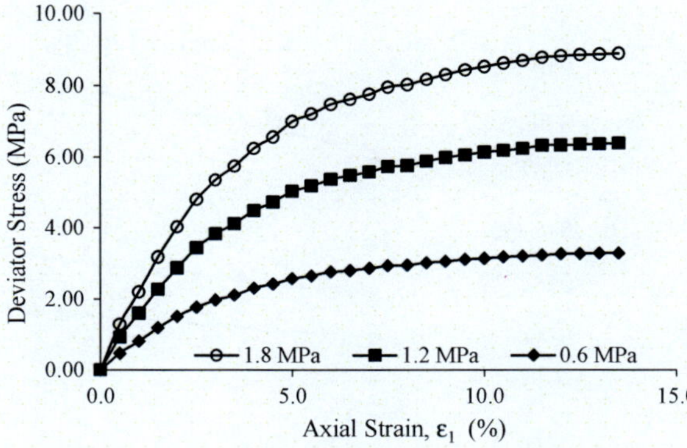


**Fig. 2** Stress–strain relationship of 80 mm maximum modeled rockfill material

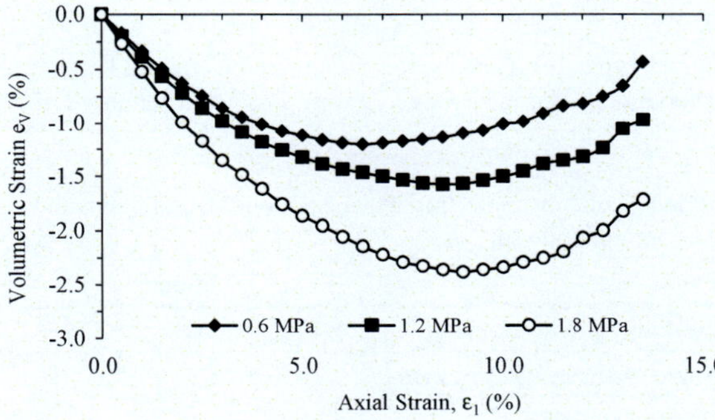


**Fig. 3** Volumetric and axial strain relationship of 80 mm maximum modeled rockfill material

where,

| | |
|---|---|
| $(\sigma_1 - \sigma_3)$ | deviatoric stress. |
| $\varepsilon_1$ | axial strain. |
| $a$ | constant, inverse of initial tangent modulus, $E_i$. |
| $b$ | constant, inverse of ultimate strength, $(\sigma_1 - \sigma_3)_{\text{ult}}$. |

The values of $[\varepsilon_1/(\sigma_1 - \sigma_3)]$ are calculated from the initial part (Initial straight line points) of the stress–strain curves and are plotted against $\varepsilon_1$ (Fig. 4). The intercept of the best fit line in the transformed plot is obtained as the value of the constant a. The reciprocal of the constant a gives the initial tangent modulus. The values of $E_i$ for the riverbed rockfill materials are given in Table 1.

Typical variation of the initial tangent modulus with the maximum particle size at different confining pressures are also represented in Fig. 5 for riverbed rockfill material collected from project site in Jammu & Kashmir. Similar relationships have been obtained for other riverbed rockfill materials [10, 11, 5, 12].

The dilation behaviors, Initial Tangent Modulus and Shear Modulus of materials were also determined different confining pressure and presented in Tables 1, 2 and 3.

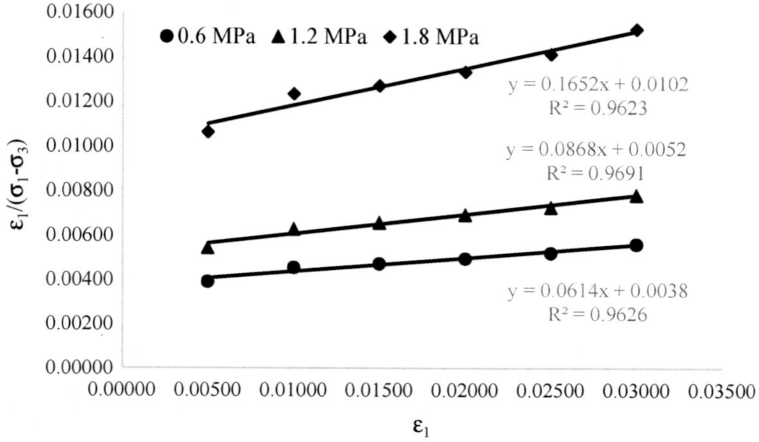


**Fig. 4** Typical Hyperbolic representation by straight line in transformed axis for riverbed rockfill material collected from a hydropower. Project in Jammu & Kashmir ($d_{max} = 80$ mm)

**Table 1** Dilation angle, initial tangent modulus and shear modulus for the riverbed rockfill materials of 25 mm particle ($d_{max}$)

| Properties | 25 mm | | |
|---|---|---|---|
| Confining pressure (MPa) | 0.6 | 1.2 | 1.8 |
| Angle of shearing resistance (°) | 45.2 | | |
| Dilation angle (°) | 14.79 | 14.36 | 13.98 |
| Initial tangent modulus (MPa) | 93.46 | 175.44 | 204.08 |
| Shear modulus (MPa) | 36.20 | 69.41 | 85.66 |
| Poisson ratio | 0.291 | 0.264 | 0.191 |

## 3 Conclusions

From the study, it is observed that the stress–strain behavior of tested rockfill material found non-uniform, non-elastic and stress path dependent. The axial strain and deviator stress increases with increase in confining pressure for all the tested materials ($d_{max}$). From axial strain-volumetric strain behavior, it is observed that the material compress during initial shearing and later shows dilation phenomenon. The effect of dilation decreases with increase in $d_{max}$ and confining pressures for all the materials. Initial tangent modulus increases with increase in confining pressure and $d_{max}$. Angle of shearing resistance ($\phi$) of the tested alluvial rockfill material, increases with increase in $d_{max}$, decrease with increase in confining pressure.

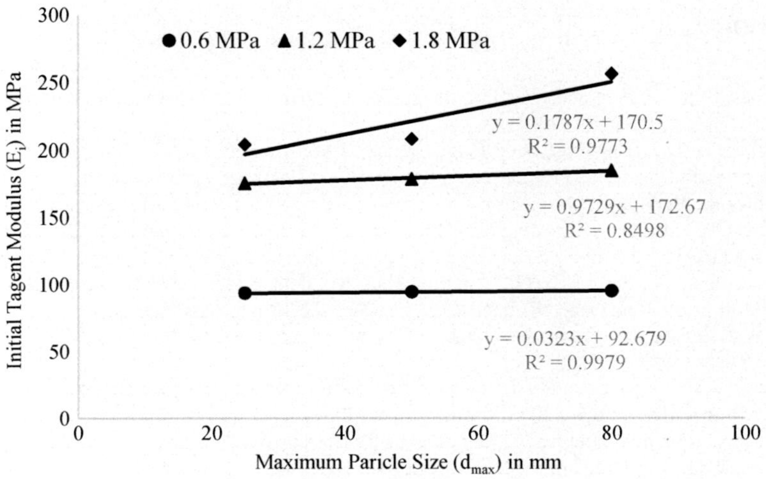


**Fig. 5** Variation of initial tangent modulus with maximum particle size for riverbed rockfill material collected from a hydropower. Project in Jammu & Kashmir

**Table 2** Dilation angle, initial tangent modulus and shear modulus for the riverbed rockfill materials of 50 mm particle ($d_{max}$)

| Properties | 50 mm | | |
|---|---|---|---|
| Confining pressure (MPa) | 0.6 | 1.2 | 1.8 |
| Angle of shearing resistance (°) | 45.4 | | |
| Dilation angle (°) | 14.63 | 14.41 | 14.04 |
| Initial tangent modulus (MPa) | 94.34 | 178.57 | 208.33 |
| Shear modulus (MPa) | 38.23 | 68.47 | 78.94 |
| Poisson ratio | 0.319 | 0.304 | 0.234 |

**Table 3** Dilation angle, initial tangent modulus and shear modulus for the riverbed rockfill materials of 80 mm particle ($d_{max}$)

| Properties | 80 mm | | |
|---|---|---|---|
| Confining pressure (MPa) | 0.6 | 1.2 | 1.8 |
| Angle of shearing resistance (°) | 45.9 | | |
| Dilation angle (°) | 14.31 | 13.87 | 13.77 |
| Initial tangent modulus (MPa) | 95.24 | 185.19 | 256.41 |
| Shear modulus (MPa) | 35.51 | 70.29 | 102.44 |
| Poisson ratio | 0.341 | 0.317 | 0.252 |

## References

1. Abbas, S.M.: Testing and modeling thebehaviourof riverbed and quarried rockfill materials. Ph.D. Thesis, IIT, Delhi (2003)
2. Mirachi, N.D., Chan C.K., Seed H.B., Duncan, J.M.: Strength and deformation characteristics of rockfill materials. Report No. TE 69(5), Civil Engineering Department, University of California, Berkeley, USA (1969)
3. Marsal, R.J.: Large scale testing of rockfill materials. J. Soil Mech. Found. Div., ASCE **93**(2), 27–43 (1993)
4. Varadarajan, A., Sharma, K.G., Venkatachalam, K., Gupta, A.K.: Testing and modeling two rockfill materials. J. Geotech. Geoenv. Eng., ASCE **129**(3), 206–218 (2003)
5. Venkatachalam, K.: Prediction of mechanicalbehaviour of rockfill materials. Ph.D. Thesis, I.I.T.Delhi (1993)
6. Abbas, S.M., Varadarajan A., Sharma, K.G.: Prediction of shear strength parameter of prototype rockfill material. In: IGC, 2003, vol. I, pp. 5–8 (2003)
7. Gupta, A.K.: Constitutive modelling of rockfill materials. Ph.D. Thesis, IIT, Delhi (2000)
8. Honkanadavar, N.P., Sharma, K.G.: Testing and modeling the behavior of riverbed and blasted quarried rockfill material. Int. J. Geomech., ASCE (2014)
9. Lowe, J.: Shear strength of coarse embankment dam materials. In: Proceedings of the 8th International Congress on Large Dams, vol. 3, pp. 745–761 (1964)
10. Okamoto, T.: Evaluation of in-situ strength of rockfill material taking into account of in-situ density and strength by laboratory. In: Wieland Ren & Tan (eds.) New Developments in Dams Engineering. Taylor & Francis Group, London (2004). ISBN 04 1536 2407
11. Varadarajan, A., Sharma, K.G., Abbas, S.M., Dhawan, A.K.: Constitutive model for rockfill materials and determination of material constants. J. Geomech., ASCE, 226–237 (2006)
12. Honkanadavar, N.P.: Testing and modeling the behavior of modeled and prototype rockfill material. Ph.D. Thesis, IIT Delhi (2010)

# Prediction of Shear Strength Parameter Using Basic Index Properties and Modelling the Behaviour of Prototype Riverbed Rockfill Material

**N. P. Honkanadavar**

## 1 Introduction

Rockfill material is being used extensively in the construction of earth core rockfill dam (ECRD) and concrete faced rockfill dam (CFRD) because of its inherent flexibility, capacity to absorb large seismic energy and adoptability to various foundation conditions. The use of modern earth and rock moving equipments and use of locally available rockfill material make such dams economical.

Rockfill material consists of maximum particle size ($d_{max}$) up to a metre in diameter. Rockfill material with such a large particle size is not feasible to test in the laboratory. Therefore, modelling techniques are being used to down size the particles so that the specimens prepared with smaller size particles can be prepared and tested in the laboratory. Among all existing modelling techniques, the parallel gradation technique [21] is most commonly used. The behaviour of riverbed rockfill material has been studied by number of researchers. Marsal [22], Marachi et al. [23], Gupta [5, 9], Abbas [1], Abbas et al. [2], Honkanadavar [10], Honkanadavar and Sharma [11–18] have performed laboratory tests on riverbed rockfill material collected from different river valley projects from India and abroad. They concluded that stress–strain behaviour is nonlinear, inelastic, and stress level dependent. The volume change increases with increase in confining pressure ($\sigma_3$) and $d_{max}$ at failure.

Stress–strain-volume change behaviour of riverbed rockfill material has been characterized by many researchers using hierarchical single surface (HISS) models [1, 25, 26]. From the laboratory test results, they determined the material parameters and back predicted the stress–strain-volume change behaviour using HISS model and compared with the observed behaviour. From the predicted and observed results, they found that both observed and predicted results match closely. Their prediction

N. P. Honkanadavar (✉)
Central Soil and Materials Research Station, New Delhi 110016, India



C. N. V. Satyanarayana Reddy et al. (eds.), *Ground Characterization and Foundations*, Lecture Notes in Civil Engineering 167,
https://doi.org/10.1007/978-981-16-3383-6_25

is based on only two index properties, i.e., uniaxial compressive strength (UCS) and uncompacted void content (UVC).

This paper deals with the testing of riverbed rockfill material obtained from Tehri dam site, Uttarakhand and study its stress–strain-volume change behaviour tested with 87% relative density. Tests were also conducted to determine the index properties of viz. UCS and UVC. In the present study, the third index property, i.e., relative density (RD) is also considered for analyses. Procedures were developed to predict the material parameters using the basic index properties of the rockfill material where UCS represents the strength of the rock from which rockfill materials are derived, and it is independent of $d_{\max}$. UVC includes the effect of gradation, shape, size and surface texture of the rockfill materials, and it is dependent on $d_{\max}$. RD represents the relative compactness of the rockfill materials. Total nine projects materials were considered to develop procedures and predict the strength and elastic material parameters [10]. Using the developed procedures, material parameters were predicted and compared with the laboratory test results. Simulated the triaxial test specimen and back predicted the stress–strain-volume change behaviour for all the $d_{\max}$ of Tehri dam using HISS model. The predicted stress–strain-volume change behaviour of modelled rockfill material is compared with observed behaviour of laboratory tests. Developed procedures were also used to predict the material parameters for the prototype rockfill material and back predicted its stress–strain-volume change behaviour using HISS model.

## 2 Experimental Investigations and Discussion

### 2.1 Material Used

In the present study, rockfill material from Tehri dam site (old dobatta borrow area), Uttarakhand has been used. The rock type is quartzite, tabular grains, equigranular, granoblastic in texture, leucocratic grey in colour and metamorphosed from sedimentary rock. The $d_{\max}$ proposed in the construction of shell portion of Tehri dam is 600 mm. Field prototype grain size distribution tests were conducted, and an average prototype gradation curve with $d_{\max}$ of 600 mm was obtained and presented. Sizes of different per cent finer, i.e., for $d_{10}$, $d_{30}$, $d_{60}$, and average particle size $d_{50}$ were calculated from the prototype gradation curve, and determined coefficient of uniformity ($C_u$) and coefficient of curvature ($C_c$) were determined and presented in Table 1.

The prototype rockfill material has been modelled into $d_{\max}$ of 25, 50, and 80 mm using parallel gradation technique as shown in Fig. 1 for testing in the large size triaxial specimen of size 381 mm diameter and 813 mm height.

**Table 1** $C_u$ and $C_c$ for prototype gradation curve

| Sizes for different per cent finer (mm) | $C_u$ ($d_{60}/d_{10}$) | $C_c$ ($d_{30}{}^2/(d_{60} \times d_{10})$) | $d_{50}$ (mm) |
|---|---|---|---|
| $d_{10} = 1.5$ | 2.67 | 1.77 | 85 |
| $d_{30} = 20$ | | | |
| $d_{60} = 150$ | | | |

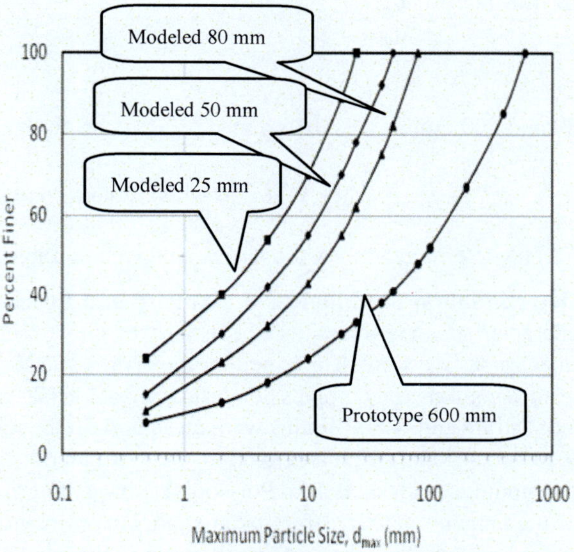


**Fig. 1** Prototype and modelled grain size distribution curves

## 2.2 Experimental Programme

**Determination of index properties**: From the literature, it is known that the behaviour of rockfill material is dependent on RD, confining pressure ($\sigma_3$), individual particle strength, $d_{max}$, shape, surface texture and mineralogy. The individual rockfill particle strength can be represented by UCS of the rock from which rockfill material is derived. Three cylindrical NX (54 mm diameter) size rock core specimens were tested from Tehri dam site as per IS: 1943-1979 [19], and average value of UCS is obtained as 125.36 MPa.

Shape, size, surface texture and gradation of aggregates are represented by a basic characteristic known as UVC for coarse material [3, 4]. The apparatus has been modified and fabricated to determine UVC for rockfill material [10].

The UVC apparatus is designed to test the modelled rockfill material of $d_{max}$ = 4.75, 10 and 19 mm. To determine the UVC for $d_{max}$ of 25, 50, 80 and prototype (600 mm) rockfill material, following procedure has been adopted.

Three modelled rockfill materials of $d_{max}$ = 4.75, 10 and 19 mm were obtained using parallel gradation technique, and they were tested to determine the UVC. The $d_{max}$ versus UVC has been plotted on semi-log graph, and then the UVC for 25, 50,

**Table 2** Index properties of rockfill material

| Properties | RD (%) | $d_{max}$ (mm) | | | | | |
|---|---|---|---|---|---|---|---|
| | | 4.75 | 10 | 19 | 25 | 50 | 80 |
| UVC (%) | | 38.9 | 36.0 | 33.6 | 32.5 | 29.9 | 28.1 |
| UCS (MPa) | | 125.36 | | | | | |
| $\phi$ (°) | 87 | | | | 36.2 | 37.6 | 39.2 |
| $E_{ir}$ (MPa) | 67,800 | | | | | | |
| $\nu_{ir}$ | 0.31 | | | | | | |

80 and 600 mm $d_{max}$ is determined using a best fit linear extrapolation as

$$UVC = -3.83 \ln(d_{max}) + 44.85 \quad (1)$$

The determined index properties are given in Table 2.

**Determination of modulus of elasticity and Poisson's ratio**: The deformability characteristics viz. modulus of elasticity of intact rock ($E_{ir}$) and Poisson's ratio of intact rock ($\nu_{ir}$) were determined for rock cores of NX size for conducting laboratory tests using uniaxial compression testing machine [7]. During the test, the axial load, axial strain and lateral strains were measured till the rock core specimen fails. Axial load versus axial strain and axial load versus lateral strain graphs were plotted, and the modulus of elasticity and Poisson's ratio was determined [10]. The $E_{ir}$ is the slope of the tangent on axial stress-axial strain curve drawn at 50% failure stress, and $\nu_{ir}$ is determined as the ratio of lateral strain to the axial strain at 50% failure stress [6, 20]. The modulus of elasticity and Poisson's ratio of intact rock are given in Table 2.

**Drained triaxial test**: Consolidated drained triaxial tests have been conducted on the modelled rockfill material with $\sigma_3$ varying from 0.4 to 1.2 MPa at Central Soil and Materials Research Station (CSMRS), New Delhi.

Stress–strain-volume change behaviour for all the $d_{max}$ is studied. From the stress–strain plots, it is observed that the behaviour is nonlinear, inelastic and stress level dependent. The volume change behaviour shows compression during the initial part of shearing and dilation with further shearing which decreases with increase in $d_{max}$ and $\sigma_3$. Typical stress–strain-volume change behaviour of 80 mm modelled rockfill material tested with 87% RD has been presented in Fig. 2.

Mean stress versus deviatoric stress was plotted and shear strength parameter, $\phi$ is determined for all the $d_{max}$ tested and presented in Table 2. Using the standard procedures, elastic parameters viz. modulus of elasticity, $E$ and Poisson's ratio, $\nu$ are determined from the laboratory test results and presented in Table 3.

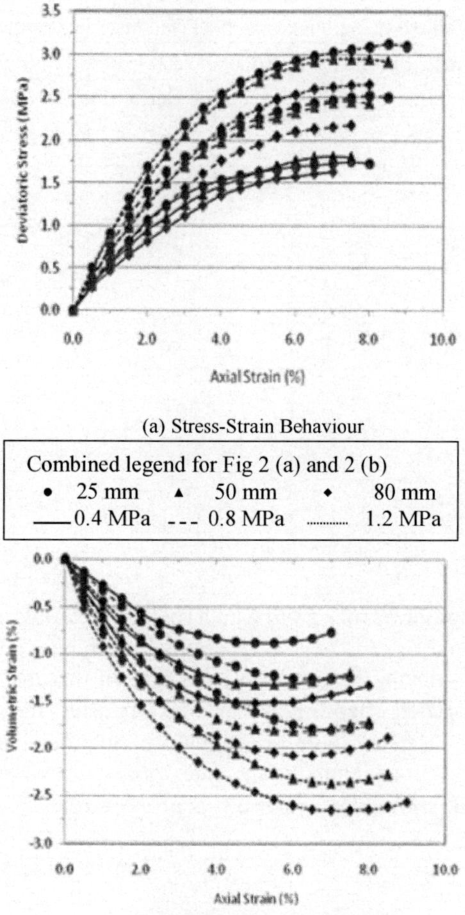


(a) Stress-Strain Behaviour

(b) Volume Change Behaviour

**Fig. 2** Stress–strain-volume change behaviour for 25, 50 and 80 mm $d_{max}$

## 3 Prediction of Material Parameters Using Index Properties

Following procedures have been developed to predict the $E$ and $\nu$ and the non-dimensional parameter $B'$ of the modelled rockfill materials using index property.

$$E/E_{ir} = C(\text{UVC})^{T_1}(\sigma_3/P_a)^{T_2} \tag{2}$$

$$\nu/\nu_{ir} = C'(\text{UVC})^{T_3}(\sigma_3/P_a)^{T_4} \tag{3}$$

where $E$ and $\nu$ are the modulus of elasticity and Poisson's ratio, respectively, for rockfill and $E_{ir}$ and $\nu_{ir}$ are the modulus of elasticity and Poisson's ratio, respectively,

**Table 3** Determined and predicted material parameters

| $d_{max}$ (mm) | $\sigma_3$ (MPa) | $\phi$ (°) | | $E$ (MPa) | | $\nu$ (Poisson's ratio) | |
|---|---|---|---|---|---|---|---|
| | | Det | Pred | Det | Pred | Det | Pred |
| 25 | 0.4 | 36.2 | 36.1 | 69.83 | 76.82 | 0.303 | 0.282 |
| | 0.8 | | | 105.64 | 110.15 | 0.303 | 0.276 |
| | 1.2 | | | 128.36 | 136.00 | 0.303 | 0.273 |
| 50 | 0.4 | 37.6 | 37.9 | 79.52 | 83.57 | 0.298 | 0.279 |
| | 0.8 | | | 114.78 | 119.83 | 0.298 | 0.273 |
| | 1.2 | | | 136.97 | 147.96 | 0.298 | 0.270 |
| 80 | 0.4 | 39.2 | 40.1 | 82.37 | 88.88 | 0.287 | 0.277 |
| | 0.8 | | | 116.92 | 127.44 | 0.287 | 0.271 |
| | 1.2 | | | 155.3 | 157.36 | 0.287 | 0.268 |
| 600 | 0.4 | | 45.4 | – | 122.24 | – | 0.268 |
| | 0.8 | | | – | 175.28 | – | 0.262 |
| | 1.2 | | | – | 216.42 | – | 0.259 |

for intact rock from which rockfill materials are derived. $C$ and $C'$ are the coefficients, and $T_1$, $T_2$, $T_3$ and $T_4$ are the exponents.

From the literature, it is known that the factors viz. RD, UCS, gradation, $d_{max}$, particle shape and surface texture affect the shear strength parameter of the granular materials. Therefore, to incorporate these factors, a non-dimensional parameter B´ has been related to the index properties viz. UCS, UVC and RD. The relationship of $B'$ with index properties is proposed as

$$B' = D(P)^{p_1}\,(\mathrm{UVC})^{p_2}\,(\mathrm{RD})^{p_3} \tag{4}$$

where $P$ is the normalized UCS (Ratio of UCS of the material to the maximum UCS of the material among all the rockfill materials, i.e., UCS/UCS$_{max}$) and D is the coefficient.

Using the developed FORTRAN computer programme, the coefficients and exponents in Eqs. 2–4 are determined adopting a least squares fitting technique. For determining the coefficients and exponents, total nine projects riverbed rockfill materials from India and abroad have been considered [10]. Substituting the values of coefficients and exponents in Eqs. 2–4 becomes

$$E/E_{\mathrm{ir}} = 1.839 \times 10^{-4}(\mathrm{UVC})^{-0.991}(\sigma_3/P_a)^{0.52} \tag{5}$$

$$\nu/\nu_{\mathrm{ir}} = 2.04(\mathrm{UVC})^{0.615}(\sigma_3/P_a)^{-0.031} \tag{6}$$

$$B' = 0.995(\mathrm{UCS/UCS_{max}})^{0.218}(\mathrm{UVC})^{0.164}(\mathrm{RD})^{0.351} \tag{7}$$

Substituting the values of $E_{ir}$, UVC, $\nu_{ir}$ and $\sigma_3$, $E$ and $\nu$ of rockfill material can be determined for any $d_{max}$ using Eqs. 5 and 6. Substituting the normalized UCS ($UCS_{max} = 125.36$ MPa), UVC and RD in Eq. 7, $B'$ value can be determined for any $d_{max}$. Substituting $B'$ in the proposed strength law [10], the major principal stress ($\sigma_1$) at failure is determined for the corresponding minor principal stress ($\sigma_3$). Plotting mean stress $(\sigma_1 + 2\sigma_3)/3$ versus deviator stress ($\sigma_1$–$\sigma_3$), the angle of shearing resistance, $\phi$ is determined for all the $d_{max}$ tested. Using Eqs. 5 and 6, elastic parameters were predicted for all the $d_{max}$ tested. Determined and predicted $\phi$, $E$ and $\nu$ values for all the $d_{max}$ tested are presented in Table 3. From the comparison, it is observed that both determined and predicted results match closely. Therefore, these procedures have been adopted to determine the material parameters of modelled and prototype (600 mm) rockfill material of Tehri dam.

## *3.1 Prediction of Stress–Strain-Volume Change Behaviour Using HISS Model*

**Hierarchical Single Surface (HISS) model**: In this model, a unique and continuous yield function is used that leads to the failure when an ultimate condition is reached. The model is based on associative plasticity and isotropic hardening. The yield function for the model is given in Eq. 8.

$$F = \left[\frac{J_{2D}}{P_a^2}\right] - \left[-\alpha\left[\frac{J_1}{P_a}\right]^n + \gamma\left[\frac{J_1}{P_a}\right]^2\right](1 - \beta S_r)^m \tag{8}$$

where

$$S_r = \frac{\sqrt{27}}{2}\frac{J_{3D}}{J_{2D}^{1.5}} \tag{9}$$

$P_a$ is the atmospheric pressure; $\gamma$, $\beta$ and $n$ are material parameters; $\alpha$ is the hardening function; $J_1$ is the first invariant of stress tensor; $J_{2D}$ and $J_{3D}$ are the second and third invariants of deviatoric stress tensors, respectively. Elastic parameters ($E$, $\nu$), ultimate parameters ($m$, $\gamma$, $\beta$), phase change parameter ($n$), hardening parameters ($a_1$, $\eta_1$) and non-associative parameter ($\kappa$) are the input parameters for the HISS model and are determined using the standard procedures [1, 8, 10, 24].

**Modelling of triaxial testing specimen**. In the present study, the triaxial specimen of size 381 mm diameter and 813 mm height has been tested in the laboratory for Tehri dam modelled rockfill material. A quarter of the triaxial specimen with axisymmetric geometry (Fig. 3) has been modelled using disturbed state concept-soil structure two-dimensional analysis (DSC-SST2D) computer software for considered riverbed rockfill material.

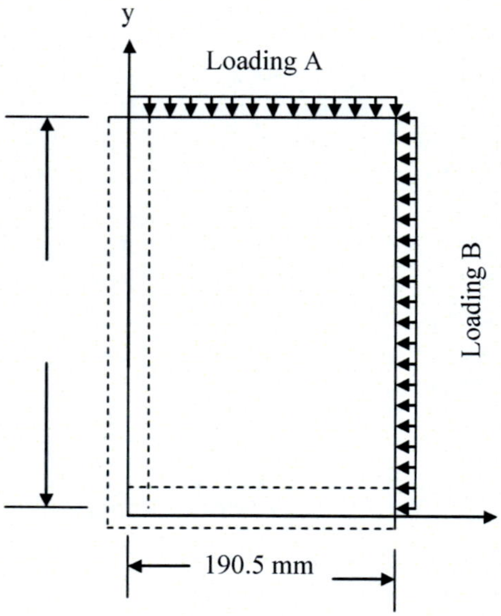


**Fig. 3** Simplified configuration of a triaxial test

The triaxial specimen has been modelled by means of single 8-noded isoparametric solid element for DSC-SST2D. The stresses and strains are assumed to be uniformly distributed over this geometry. The bottom and left hand side of the geometry are axis of symmetry (Fig. 3). At these boundaries, the displacements normal to the boundary are restrained, and the remaining boundaries are fully free to move.

A triaxial test is simulated in two phases; consolidation and shearing. The consolidation phase is simulated by stress controlled, and shearing phase is simulated by strain controlled method. In the first phase, the confining pressure is applied by activating load A and load B by equal amount as shown in Fig. 3.

In the second phase, displacements are reset to zero, and the specimen is sheared by strain controlled test up to desired axial strain level, while the horizontal load B (confining pressure) is kept constant.

Stress–strain-volume change behaviour was predicted and compared with the experimental results for all the tested modelled riverbed rockfill materials considered in the present study. A typical observed and predicted stress–strain-volume change behaviour for 25, 50 and 80 mm tested with 1.2 MPa confining pressure is presented in Fig. 4.

Developed procedures have been used to predict strength and elastic material parameters. Comparing predicted and test results, it is observed that strength and elastic material parameters match closely. Using the predicted material parameters, stress–strain-volume change behaviour of modelled and prototype riverbed rockfill material of Tehri dam (Old Dobatta borrow area) has been back predicted using HISS model for all the $d_{max}$ tested. The predicted behaviour of modelled rockfill material matches closely with the observed behaviour of modelled rockfill material. It is also observed that the predicted behaviour of prototype rockfill material follows

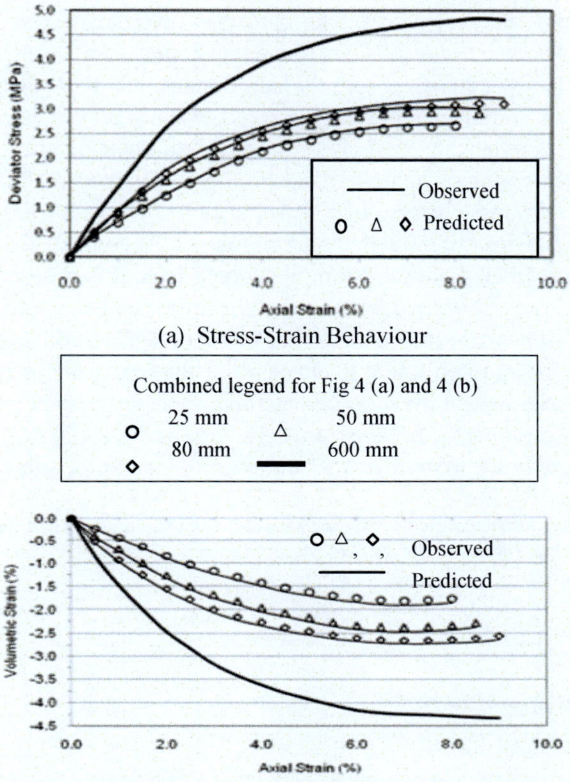


(b) Volume Change Behaviour

**Fig. 4** Observed and predicted stress–strain-volume change behaviour of different $d_{max}$ tested with 87% RD ($\sigma_3 =$ 1.2 MPa)

similar trend as that of modelled rockfill material. The typical observed and predicted behaviour of modelled rockfill material for different $d_{max}$ and predicted prototype ($d_{max} = 600$ mm) of Tehri dam project tested with 87% RD and $\sigma_3 = 1.2$ MPa is shown in Fig. 4.

## 4 Conclusions

The riverbed rockfill material from Tehri dam site, Uttarakhand has been considered in the present study. The material has been modelled into $d_{max}$ of 25, 50 and 80 mm and tested in the laboratory under consolidated drained triaxial test conditions for different confining pressures ranging from 0.4 to 1.2 MPa and 87% RD. The index properties viz. UCS and UVC have been determined.

Procedures have been developed using the index properties viz. UCS, UVC and RD to predict the strength, elastic and non-dimensional parameters of the riverbed rockfill material. Developed procedures have been used to predict strength and elastic material parameters. The predicted material parameters were compared with the observed test results and found that both values match closely. Other material parameters were related with the non-dimensional parameter B´ and predicted the values for modelled and prototype material using best fit extrapolation technique.

Using the predicted material parameters, the stress–strain-volume change behaviour of modelled and prototype riverbed rockfill material is predicted using HISS model based on elasto-plasticity. The predicted stress–strain-volume change behaviour of modelled rockfill material is compared with the observed behaviour of modelled material. From the comparison, it is observed that both the results match closely. It is also observed that the predicted behaviour of prototype rockfill material follows similar trend as that of modelled rockfill material. Therefore, this model appears to be suited to characterize the behaviour of riverbed rockfill material. The advantage of the proposed methods is that the developed methods are more realistic, economical, can be used where large size triaxial testing facilities are not available and quick to predict $\phi$, $E$ and $\nu$ using index properties.

**Acknowledgements** Author wish to thank Director, CSMRS for his support to carry out this work. Also Thanks to the staff of rockfill division, CSMRS for their help in conducting the laboratory tests.

## References

1. Abbas, S.M.: Testing and Modeling the Behaviour of Riverbed and Quarried Rockfill Materials. Ph.D. Thesis, I.I.T. Delhi (2003)
2. Abbas, S.M., Varadarajan, A., Sharma, K.G.: Prediction of shear strength parameter of prototype rockfill material. IGC-2003, vol. I, pp. 5–8. Roorkee (2003)
3. Ahlrich, R.C.: Influence of aggregate properties on performance of heavy-duty hot mix asphalt pavements. Transportation Research Record 1547, Transportation Research Board, National Research Council, Washington, DC (1996)
4. ASTM C1252: Standard Test Method for Uncompacted Void Content. ASTM Standard (1998)
5. Gupta A.K.: Constitutive Modeling of Rockfill Materials. Ph.D. Thesis, I.I.T. Delhi (2000)
6. ASTM D7012: Standard test method for compressive stress and elastic moduli of intact rock core specimens under varying states of stress and temperature (2004)
7. Brown, E.T.: Rock Characterisation Testing and Monitoring, ISRM Suggested Methods. International Society for Rock Mechanics by Pergamon Press (1981)
8. Desai, C.S.: Hierarchical single surface and the disturbed state constitutive models with emphasis on geotechnical applications. In: Saxena, K.R. (ed.) Geotechnical Engineering: Emerging Trends in Design and Practice, Chap. 5. Oxford & IBH Pub. Co. Pvt. Ltd., New Delhi, India (1994)
9. Gupta, A.K.: Constitutive Modelling of Rockfill Material. Ph.D. Thesis, I.I.T. Delhi (2000)
10. Honkanadavar, N.P.: Testing and Modelling the Behaviour of Modeled and Prototype Rockfill Materials. Ph.D. Thesis, I.I.T. Delhi (2010)
11. Honkanadavar, N.P.: Prediction of shear strength parameter for prototype riverbed rockfill material. In: Electronic Version, IGC-2016, 15–17 Dec 2016, IIT Chennai, India, pp. 120–121 (2016)
12. Honkanadavar, N.P.: Prediction of material parameters using index properties and modeling the behaviour of prototype quarried rockfill material. In: Proceedings of Indian Geotechnical Conference, IIT Guwahati, vol. I, p. 75 (2017)

13. Honkanadavar, N.P., Sharma, K.G.: Prediction of shear strength parameter for prototype riverbed rockfill materials using Index properties. In: Proceedings of Indian Geotechnical Conference, India, vol. II, pp. 335–338 (2010)
14. Honkanadavar, N.P., Sharma, K.G.: Prediction of shear strength parameter for prototype quarried rockfill materials using index properties. In: Proceedings of Indian Geotechnical Conference, India, vol. I, pp. 245–248 (2011)
15. Honkanadavar, N.P., Sharma, K.G.: Testing and modeling the behaviour of riverbed rockfill material. In: Proceedings of Indian Geotechnical Conference, India, vol. II, pp. 131–134 (2012)
16. Honkanadavar, N.P., Sharma, K.G.: Effect of particle breakage on stress-strain-volume change behavior of quarried rockfill material. In: Proceedings of Indian Geotechnical Conference, India, vol. I, p. 47 (2013)
17. Honkanadavar, N.P., Sharma, K.G.: Testing and modeling the behaviour of riverbed and blasted quarried rockfill materials. Int. J. Geomech. ASCE **14**, 1–11 (2014)
18. Honkanadavar, N.P., Sharma, K.G.: Testing and modeling the behaviour of blasted quarried rockfill material. In: Proceedings of Indian Geotechnical Conference, India II, pp. 140–141 (2015)
19. IS 1943: 1979 Method for the Determination of Unconfined Compressive Strength of Rock Material
20. IS 9221: 2010 Method For The Determination of Modulus of Elasticity and Poisson's Ratio of Rock Material in Uniaxial Compression
21. Lowe, J.: Shear strength of coarse embankment dam materials. In: Proceedings of the 8th International Congress on Large Dams, vol. 3, 745–761 (1964)
22. Marachi, N.D., Chan, C.K., Seed, H.B., Duncan, J.M.: Strength and Deformation Characteristics of Rockfill Materials. Report No. TE 69(5), Civil Engineering Department, University of California, Berkeley, USA (1969)
23. Marsal, R.J.: Large scale testing of rockfill materials. J. Soil Mech. Found. Div. ASCE **93**(2), 27–43 (1967)
24. Varadarajan, A., Desai, C.S.: Material constants of a constitutive model determination and use. Indian Geotech. J. **23**(3), 291–313 (1993)
25. Varadarajan, A., Sharma, K.G., Venkatachalam, K., Abbas, S.M.: Constitutive modeling of rockfill materials from Tehri Dam. In: Proceedings of Indian Geotechnical Conference, India, vol. I, pp. 592–595 (2002)
26. Varadarajan, A., Sharma, K. G., Venkatachalam, K., Gupta, A.K.: Testing and modeling two rockfill materials. J. Geotech. Geoenviron. Eng. ASCE **129**(3), 206–218 (2003)

# Modeling Using ANN and RNN Approach for Shearing Behavior of Residual Soil

**Manas Bhoi and Milind Amin**

## 1 Introduction

Recently, geotechnical engineers are mainly concern about the strength parameters of the soil and the settlements of different types of foundation, and it has a crucial role to support different types of structures like buildings, dams power plants, etc. Therefore, to avoid the superstructure and foundation failures, the shear strength properties must be well understood [1]. It is the property of the soil that enables the soil to keep it in its state when the surface is not even or when there is a shear force acting on the specimen. These shear strength parameters can be calculated either in the field or in the laboratory. The tests done in the laboratory may include an unconfined compression test, triaxial test, vane shear test, a direct shear test. A direct shear test is used for calculating the soil strength parameters which includes the angle of internal friction ($\phi$) and cohesion $c$ (kPa) shown in Eq. 1.

The shear strength, $\tau$ of soil in terms of effective stress ($\sigma'$) is:

$$\tau = c + \sigma' \tan \phi \quad (1)$$

M. Bhoi (✉) · M. Amin
Pandit Deendayal Petroleum University, Gandhinagar 382007, India
e-mail: manas.bhoi@sot.pdpu.ac.in



C. N. V. Satyanarayana Reddy et al. (eds.), *Ground Characterization and Foundations*, Lecture Notes in Civil Engineering 167,
https://doi.org/10.1007/978-981-16-3383-6_26

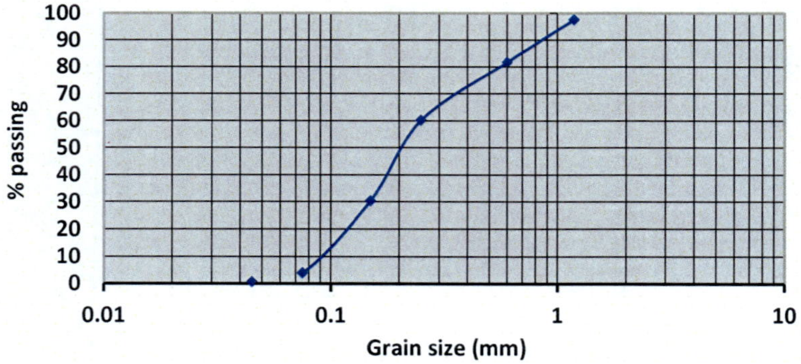


**Fig. 1** Sieve size distribution

The slope expressed in degrees is the angle of shearing resistance or the internal angle of friction ($\phi$), and the intercept is its cohesion $c$ (kPa) [2-4]. The angle of internal friction represents the interlocking between the soil particles, whereas cohesion is mainly due to the intermolecular bond between the adsorbed water surrounding each grain, especially in fine-grained soils [3, 5]. Soils with high plasticity have a lower angle of internal friction and higher cohesion value. Conversely, as the soil grain size increases, the soil internal friction angle increases and its cohesion decreases. Grain size distribution is shown in Fig. 1.

However, experimental determination of the strength parameters is extensive, cumbersome, and expensive. Further, it is not always possible to manage and do the tests in every new situation. In order to successfully deal with such problems, artificial intelligence-based methods have been developed to estimate shear strength parameters [6]. As per [7], empirical relations are widely used in geotechnical engineering practice as a tool to estimate the engineering properties of soils. Considering these, an experimental study is done to generate a data set of geotechnical properties of soil. This paper deals with the relationship of geotechnical properties of soil with shear strength parameters and to develop neural network-based model for its prediction.

## 2 Artificial Neural Network (ANN)

Artificial neural network (ANN) can be described as a computational tool whose design is motivated by the designs and functioning of human brains and components. An ANN is a system composed of a number of artificial neurons. These neurons have multiple numbers of input parameters and single or multiple output parameters to perform elementary calculations.

The particular specification of ANN is that the unit learns from an example in a similar manner to the biological neurons. Biological neurons receive input from sources and merge them in a way to perform a nonlinear operation as result. ANN is similar to the traditional statistical models in which model parameters (i.e., connection weights) are adjusted to calibrate a model called *learning* or *training* [8]. If the units are organized into multiple layers, then all units of each layer are connected with subsequent layers and feed-forward network is developed.

Each layer in a network contains a satisfactory number of neurons depending on its application. The neurons in a layer are linked to the neurons in the next succeeding layer, and each link carries a weight value. The number of neurons in the input layer equals to the number of input data, while the number of neurons in the output layer equals to the number of outputs linked with each of the input. The number of hidden layers and their neurons is often decided by trial-and-error method. The number of neurons in output layers is decided based on the application. Each hidden neuron responds to the weighted inputs; it collects from the linked neurons from the previous input layer.

Once the combined outcome on each hidden neuron is determined, the activation at this neuron is determined by a transfer function. For transfer function, any nonlinear differentiable function can be used, but a sigmoid function is broadly used; nevertheless, there are many other functions [9]. The sigmoid function restrains the outputs of a network between 0 and 1. The input vector ($X_m^n$) is related output vector ($X_j^{n+1}$) by the below equation and can be described as shown in Fig. 2.

$$x_j^{n+1} = F\left(\sum\nolimits_i W_{jm}^n X_m^n\right)$$

where $F(x) = \frac{1}{1+e^{-x}}$ log sigmoid function.

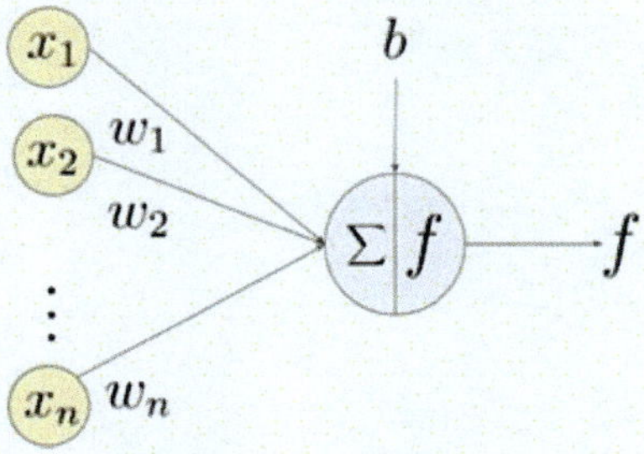


**Fig. 2** The relationship between the input and output vector in the artificial neuron

## 3 Recurrent Neural Network (RNN)

In this case, a recurrent neural network (RNN) model is found to be more effective than standard backpropagation network in simulating and predicting nonlinear shear behavior of residual soil. The direct shear test performed on residual soil is used to train the models developed in this study. The good simulation and prediction of stress–strain behavior prove that the RNN approach can be effectively used to model complex soil behavior [10]. A fair agreement between experimental and the RNN model is observed. The significant variations inherent in the soil behavior are successfully captured by using an appropriate algorithm function and architecture of the neural network.

## 4 Experimental Setup

The soil sample is taken from F-block, Pandit Deendayal Petroleum University, Gandhinagar, which is under construction. Sieve analysis is performed to find the gradation of soil. For direct shear test (shown in Figs. 3 and 4), the soil mass which is used for the experiment is passing 425-μm sieve. The experiment is performed in 6 batches of soil starting with 0% bentonite to 50% bentonite addition with soil, and the water content was taken as 10% for each batch. For each batch, 3 direct shear test is performed with 0.5, 1 and 1.5 kg/cm$^2$ normal stresses. Moisture content is evaluated for each set of soil samples. A total of 18 shear tests are performed with

**Fig. 3** Direct shear test apparatus

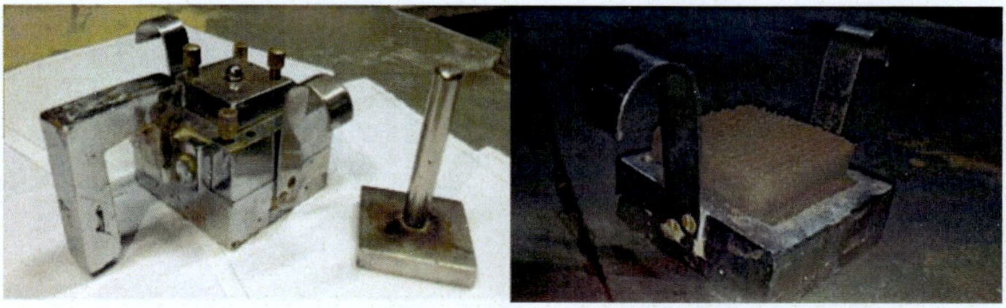

**Fig. 4** Direct shear mold

the help of the direct shear test apparatus. The angle of internal friction Φ (°) and cohesion *c* (kPa) is calculated for each set of test samples (Fig. 5).

## *4.1 ANN*

The MATLAB-neural network tool is used to do the necessary computing work. All computations have been carried out on a windows 10 server with Intel Xeon E5 CPU (10 core) 16 GB of RAM and 4 GB Dual Nvidia Quadro K1200.

In this project, we have selected 6 input parameters from direct shear test, that is, moisture content, clay (Bentonite) content, cohesion (*c*), angle of internal friction (Φ), normal stress ($\sigma'$) and shear strain, whereas output parameter selected is shear stress ($\tau$); an interdependency between the input and the output is established keeping in mind the contribution of all input which contributed for good training of the data. Out of the 662 direct shear results, 0, 10, 20, 30, 50% clay content soil sample batches are chosen for training sets and 40% is used for predicting the data relation in the training network. The mean square error (*R*) value obtained is equal to 0.96775, for developing the network shown in Fig. 6. Various permutations were tried. Partial data sets are given in Tables 1 and 2 used for training and testing data sets, respectively.

While choosing a number of neurons in particular hidden layer and number of hidden layers in activation function, the output varies significantly.

## *4.2 RNN*

For RNN, again 6 input parameters from direct shear test were chosen, that is, moisture content, clay content, cohesion (*c*), angle of internal friction (Φ), normal stress and shear strain, whereas output parameter as shear stress, the training network include 0, 10, 20% of clay content soil samples batches as input whereas selected data sets of 30, 40, and 50% batch were used as simulation shown in Fig. 7. That is 30% at 0.5 kg/cm$^2$ normal stress; 40% at 0.5 and 1.5 kg/cm$^2$ normal stress; and 50%

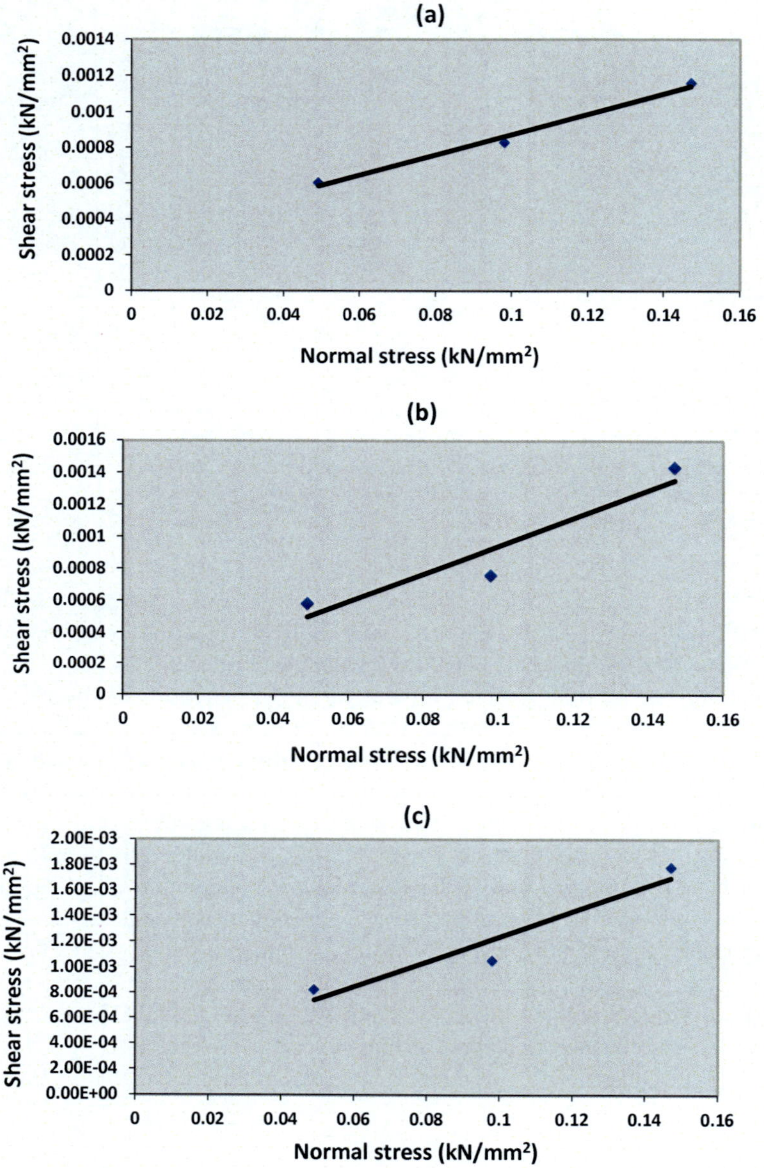


**Fig. 5** Relation between normal stress and shear stress for: **a** for 0% bentonite, **b** for 10% bentonite, **c** for 20% bentonite, **d** for 30% bentonite, **e** for 40% bentonite, **f** for 50% bentonite

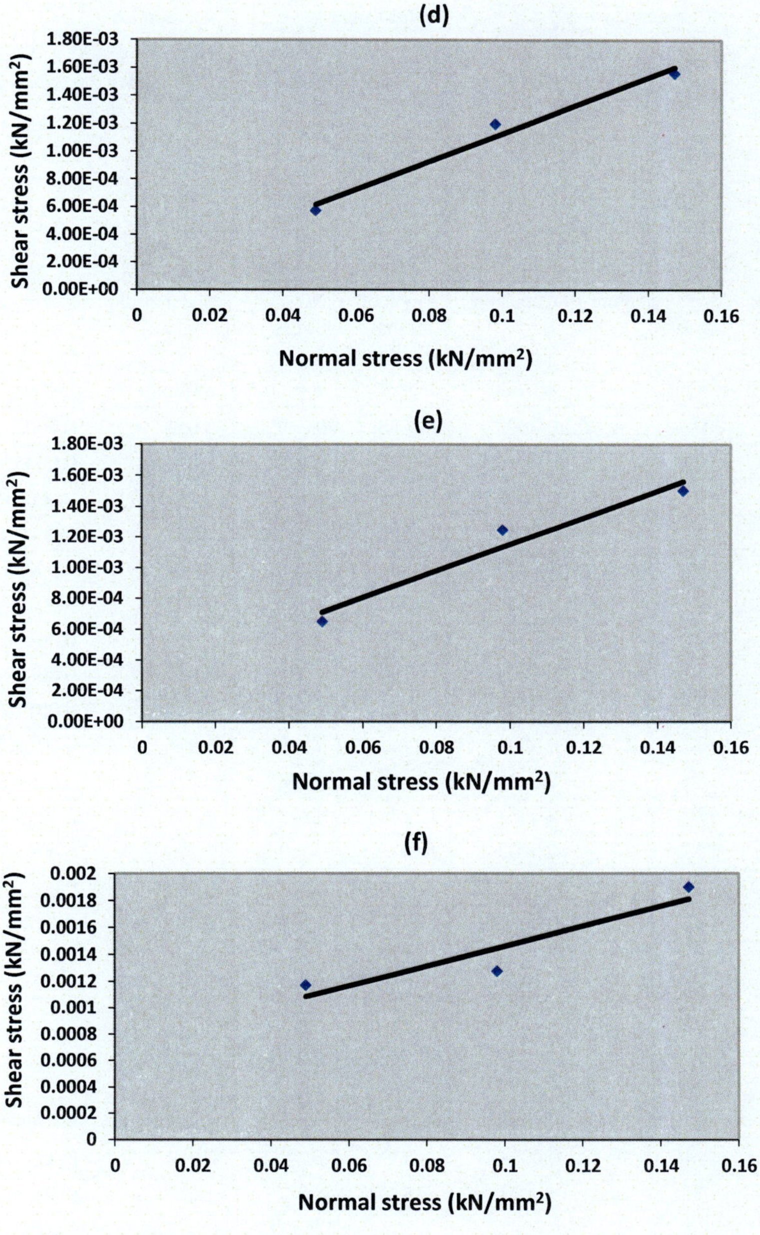


**Fig. 5** (continued)

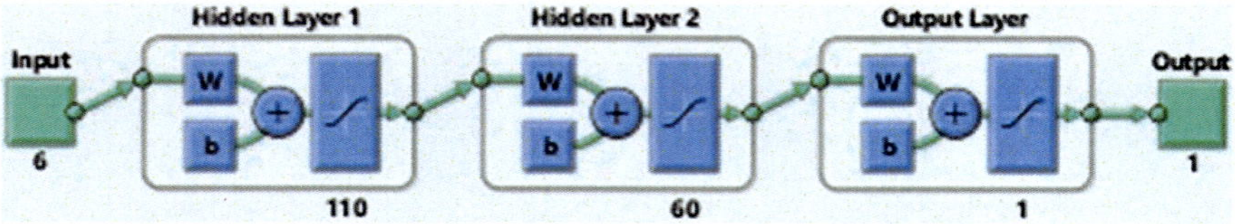


**Fig. 6** ANN model

**Table 1** Training data sets for ANN model in summarized form

| Moisture content | Clay content | Cohesion | Angle of internal friction | Normal Stress | Shear strain | Shear stress |
|---|---|---|---|---|---|---|
| % | % | kN/mm$^2$ | degree | kN/mm$^2$ | mm/mm | kN/mm$^2$ |
| 10 | 0 | 0.0003 | 18.678 | 0.04905 | 0.011 | 0.0002174 |
| 10 | 0 | 0.0003 | 18.678 | 0.0981 | 0.014 | 0.0005184 |
| 10 | 0 | 0.0003 | 18.678 | 0.14715 | 0.017 | 0.000719 |
| 10 | 10 | 0.00007 | 28.533 | 0.04905 | 0.008 | 0.000223 |
| 10 | 10 | 0.00007 | 28.533 | 0.0981 | 0.022 | 0.0006577 |
| 10 | 10 | 0.00007 | 28.533 | 0.14715 | 0.036 | 0.0010368 |
| 10 | 20 | 0.0003 | 31.799 | 0.04905 | 0.011 | 0.00068 |
| 10 | 20 | 0.0003 | 31.799 | 0.0981 | 0.025 | 0.0009977 |
| 10 | 20 | 0.0003 | 31.799 | 0.14715 | 0.031 | 0.0017837 |
| 10 | 30 | 0.0001 | 33.154 | 0.04905 | 0.014 | 0.0005017 |
| 10 | 30 | 0.0001 | 33.154 | 0.0981 | 0.021 | 0.0010591 |
| 10 | 30 | 0.0001 | 33.154 | 0.14715 | 0.037 | 0.0015161 |
| 10 | 50 | 0.0007 | 24.236 | 0.04905 | 0.018 | 0.0010089 |
| 10 | 50 | 0.0007 | 24.236 | 0.0981 | 0.029 | 0.0011092 |
| 10 | 50 | 0.0007 | 24.236 | 0.14715 | 0.045 | 0.0018952 |

**Table 2** Testing data sets for ANN model in summarized form

| Moisture content | Clay content | Cohesion | Angle of internal friction | Normal Stress | Shear strain | Shear stress |
|---|---|---|---|---|---|---|
| % | % | kN/mm$^2$ | degree | kN/mm$^2$ | mm/mm | kN/mm$^2$ |
| 10 | 40 | 0.0003 | 28.533 | 0.04905 | 0.017 | 0.000474 |
| 10 | 40 | 0.0003 | 28.533 | 0.0981 | 0.028 | 0.00112 |
| 10 | 40 | 0.0003 | 28.533 | 0.14715 | 0.043 | 0.001332 |

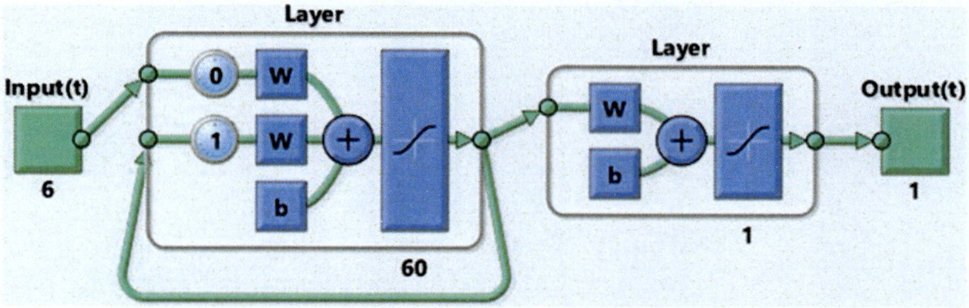


**Fig. 7** RNN model

at 1 and 1.5 kg/cm$^2$ normal stress. The data sets for training and testing stages are given in Tables 3 and 4, respectively, in summarized manner.

**Table 3** Training data sets for RNN model in summarized form

| Moisture content | Clay content | Cohesion | Angle of internal friction | Normal Stress | Shear strain | Shear stress |
|---|---|---|---|---|---|---|
| % | % | kN/mm$^2$ | degree | kN/mm$^2$ | mm/mm | kN/mm$^2$ |
| 10 | 0 | 0.0003 | 18.678 | 0.04905 | 0.015 | 0.000345588 |
| 10 | 0 | 0.0003 | 18.678 | 0.0981 | 0.017 | 0.000629862 |
| 10 | 0 | 0.0003 | 18.678 | 0.14715 | 0.026 | 0.000897414 |
| 10 | 10 | 0.00007 | 28.533 | 0.04905 | 0.011 | 0.0002787 |
| 10 | 10 | 0.00007 | 28.533 | 0.0981 | 0.027 | 0.000719046 |
| 10 | 10 | 0.00007 | 28.533 | 0.14715 | 0.039 | 0.00105906 |
| 10 | 20 | 0.0003 | 31.799 | 0.04905 | 0.016 | 0.000785934 |
| 10 | 20 | 0.0003 | 31.799 | 0.0981 | 0.027 | 0.00103119 |
| 10 | 20 | 0.0003 | 31.799 | 0.14715 | 0.030 | 0.001772532 |
| 10 | 30 | 0.0001 | 33.154 | 0.0981 | 0.026 | 0.001159392 |
| 10 | 30 | 0.0001 | 33.154 | 0.14715 | 0.039 | 0.001571868 |
| 10 | 40 | 0.0003 | 28.533 | 0.0981 | 0.024 | 0.001047912 |
| 10 | 50 | 0.0007 | 24.236 | 0.04905 | 0.016 | 0.00097545 |

**Table 4** Testing data sets for RNN model in summarized form

| Moisture content | Clay content | Cohesion | Angle of internal friction | Normal Stress | Shear strain | Shear stress |
|---|---|---|---|---|---|---|
| % | % | kN/mm$^2$ | degree | kN/mm$^2$ | mm/mm | kN/mm$^2$ |
| 10 | 30 | 0.0001 | 33.154 | 0.04905 | 0.019 | 0.000552 |
| 10 | 40 | 0.0003 | 28.533 | 0.04905 | 0.013 | 0.000435 |
| 10 | 40 | 0.0003 | 28.533 | 0.1471 | 0.051 | 0.00146 |
| 10 | 50 | 0.0007 | 24.236 | 0.0981 | 0.031 | 0.001171 |
| 10 | 50 | 0.0007 | 24.236 | 0.14715 | 0.044 | 0.00189 |

To develop the network, training method is changed to recurrent neural network to compare the change in the predicted values of data sets with ANN prediction. In this network, the data points were divided into two equal layers, one for training and another for simulating. The 1st layer was trained to have an initial delay layer as zero, after training the network; the final delay layer obtained was used as an initial delay layer for repeating the training process. The 2nd layer is used as simulation which included 30, 40 and 50% clay content soil sample batches. It has been observed that the predicted results were quite similar to the original one.

## 5 Simulation Results

The proposed solutions have been designed using the ANN and RNN model, to predict the shear stress for different batches of soil. The calibrated RNN model considered experimental results of 30% at 0.5 kg/cm$^2$ normal stress; 40% at 0.5 and 1.5 kg/cm$^2$ normal stress; and 50% at 1 and 1.5 kg/cm$^2$ normal stress as testing data, and remaining experimental data are used as training data. Figure 8 shows the comparison of experimental data and predicted data using RNN model.

Similarly, ANN model considered experimental results of all test conducted at 0, 10, 20, 30, and 50% bentonite batches as an input, and the result for all test conducted at 40% bentonite batch is predicted. Figure 9 shows the comparison of experimental data and predicted data using ANN model (Fig. 10).

## 6 Conclusions

Based on the experimental results and subsequent simulation of the same using neural network method, the following conclusions are made:

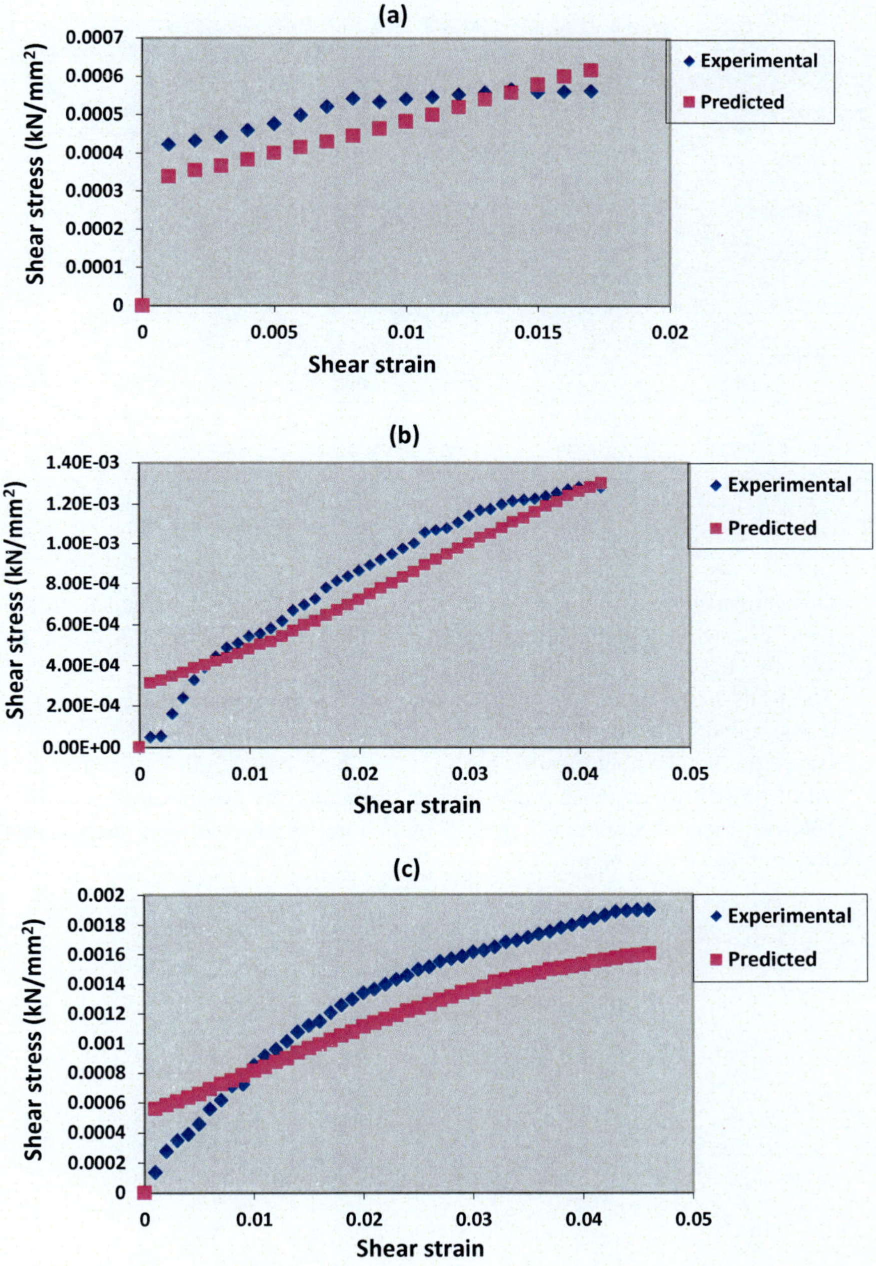


**Fig. 8** Simulated results of RNN model; **a** 30% bentonite @ 0.5 kg/cm$^2$ normal stress, **b** 50% bentonite @ 1 kg/cm$^2$ normal stress, **c** 50% bentonite @ 1.5 kg/cm$^2$ normal stress

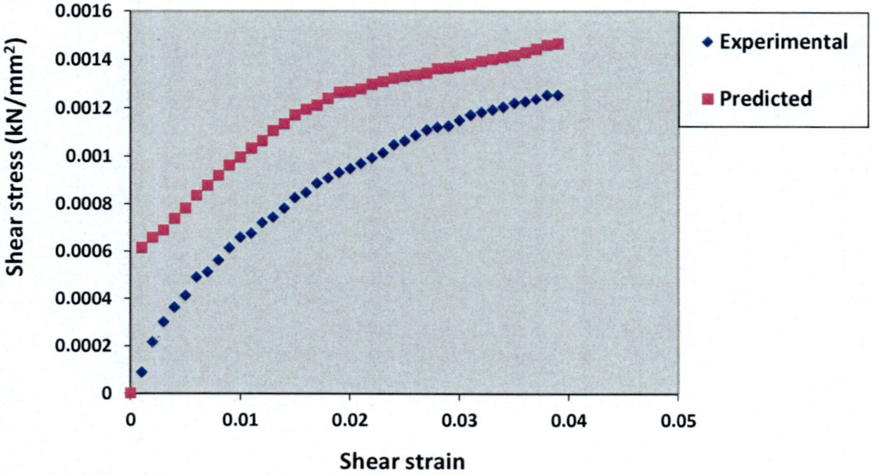


**Fig. 9** Simulated results of ANN model for 40% bentonite @ 1 kg/cm$^2$ normal stress

1. On comparing experimental results with predicted results obtained from the feed-forward backpropagation-based ANN model and RNN model, it can be seen that the RNN model gives better results.
2. The RNN-based prediction model can be used in the actual field cases also as it is simulating the nonlinear behavior with a good accuracy.
3. Based on the simulated results obtain, it can be concluded that there is still a lot of scope for further improvement in predicting the nonlinear stress–strain behavior more accurately using RNN by increasing the no of input parameters and number of data points.

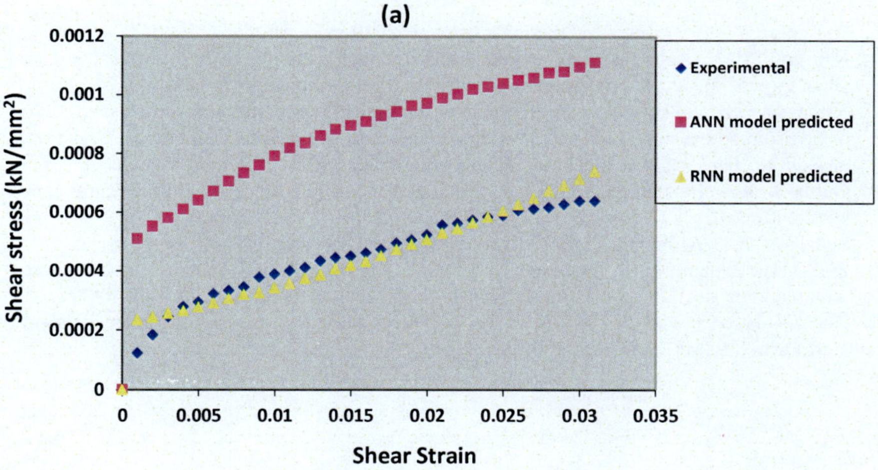


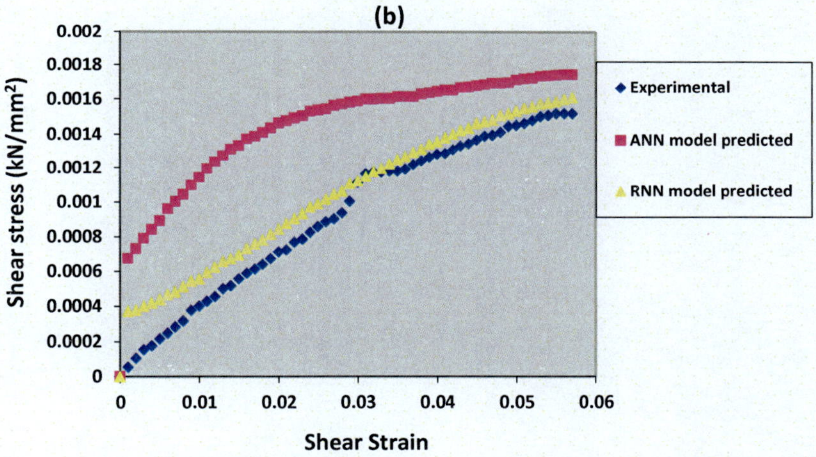


**Fig. 10** Simulated results of ANN and RNN model compared with experimental data: **a** 40% bentonite @ 0.5 kg/cm$^2$ normal stress, **b** 40% bentonite @ 1.5 kg/cm$^2$ normal stress

# References


1. Omotoso, O.A., Mamodu, M.O., Ojo, O.J.: Evaluation of geotechnical properties of laterite soils in Asa-Dam are, Ilorin, southwest Nigeria. World J. Appl. Sci. Technol. **3**(2), 1–9 (2011)
2. Arora, K.R.: Introductory Soil Engineering: Text Book. Nem Chand Jain (Prop), Standard Publishers Distributors, Nai Sarak, Delhi (1988)
3. Murthy, S.: Geotechnical Engineering: Principles and Practices of Soil Mechanics, 2nd edn. Taylor & Francis, CRC Press, UK (2008)
4. Mollahasani, A., Alavi, A.H., Gandomi, A.H., Rashed, A.: Nonlinear neural-based modeling of soil cohesion intercept. KSCE J. Civ. Eng. **15**(5), 831–840 (2010)
5. El-Maksoud, M.A.F.: Laboratory determining of soil strength parameters in calcareous soils and their effect on chiseling draft prediction. In: Proceedings of the Energy Efficiency and

Agricultural Engineering—International Conference, Rousse, Bulgaria (2006)
6. Mousavi, S.M., Alavi, A.H., Gandomi, A.H., Mollahasani, A.: Nonlinear genetic-based simulation of soil shear strength parameters. J. Earth Syst. Sci. **120**(6), 1001–1022 (2011)
7. Sorensen, K.K., Okkels, N.: Correlation between drained shear strength and plasticity index of undisturbed over consolidated clays. In: Proceedings of the 18th International Conference on Soil Mechanics and Geotechnical Engineering, Paris, pp. 1–6 (2013)
8. Gupta, R., Kewalramani, M.A., Goel, A.: Prediction of concrete strength using the neural-expert system (2006).
9. Schalkoff, R.J.: Artificial Neural Networks. McGraw-Hill, New York (1997)
10. Zhu, J.-H., Zaman, M.M., Anderson, S.A.: Modelling of shearing behavior of a residual soil with recurrent neural network. Int. J. Numer. Anal. Meth. Geomech. **22**, 671–687 (1997)
11. IS: 2720-Part 13: Methods of Test of Soils—Direct shear test, Bureau of Indian Standards, New Delhi (1986) (Reaffirmed 2002)

# Assessment of Local Seismic Hazard of Agartala Based on Nonlinear Site Response Analysis

**Rima Das, Rajib Saha, and Rajat Debnath**

## 1 Introduction

The response of any structure depends on its regional seismicity, geology and local soil conditions, and the nature of source mechanism. The local soil geology plays an important role in governing the ground response. The evaluation of ground response is an important task, as it governs the safety of structures located in seismically prone areas. Site response estimation from geotechnical investigations involves a combination of wave propagation theory with the material properties and the expected ground motion computed at the site of interest. Soil material modeling can be broadly classified into three types, linear, equivalent linear (EQL), and nonlinear (NL) models for one horizontal direction of shaking, and nonlinear models for multiple directions of shaking. 1D ground response analysis (GRA) can be performed either in the time domain (nonlinear total and effective stress approaches) or in the frequency domain (linear and equivalent linear total stress approaches) [21]. However, in the sake of the inherent nonlinearity in soil behavior, a time-domain nonlinear GRA can model the soil response during an actual earthquake, more accurately than any frequency domain GRA method [21].

### *1.1 Geological Significance of Tripura*

Agartala city is the capital of Tripura state, in the northeast of India. This is one of the fastest-growing cities in northeast India, with a population of 400,000 residents and a high population density of 10.119 persons per km$^2$ compared to other cities

R. Das · R. Saha (✉) · R. Debnath
Department of Civil Engineering, National Institute of Technology Agartala, Tripura 799046, India



C. N. V. Satyanarayana Reddy et al. (eds.), *Ground Characterization and Foundations*, Lecture Notes in Civil Engineering 167,
https://doi.org/10.1007/978-981-16-3383-6_27

in northeast India. The city is located within the NW portion of the state having latitude of 23.78–23.88 N and longitude of 91.26–91.33 E within the western district of Tripura. The city covers an area of about 62.0 $km^2$ in its present form and functions by the authority of the Agartala Municipal Council (AMC). However, geologically the whole of the area in Tripura forms a part of the Tertiary Naga-Arakan-Yoma basin, a younger terrain that is located to the southwest of the Paleogene fold belt. It is likely that Neogene sediments were deposited on the folded, but not uplifted, Paleogene sediments, and were subsequently co-folded in the latter stage. The tectonic cycle ended with a weak deformation of Dupitila sediments (Fig. 1). However, Dasgupta (1908) first classified the folded sediments of Tripura into 'coal measures' and 'Tipam' groups. The water table is very close to the ground surface and the central part of the city, which is close to the Haora river/Katakhal, which has soft clay, silty-sand alluvium deposits, and mostly sedimentary soils. These sediments, according to the geological survey of India (GSI), were laid down in the

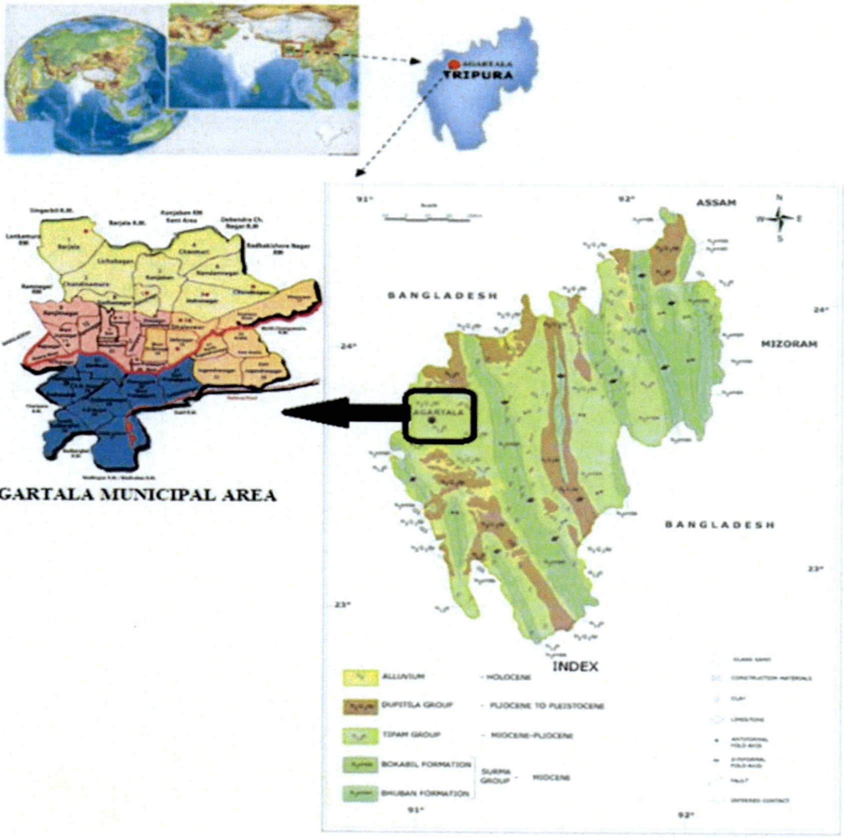


**Fig. 1** Geological map of the study area

Surma basin during the tertiary age (which lasted for 65 million years) in a wide range of environmental conditions governed by localized tectonic movement.

### *1.2 Seismological Significance of Tripura*

Seismically, northeast India is active, and this region is highly populated. The region falls under the highest zone V as per the seismic zoning map of India (BIS 2002). The seismotectonics of this area is complicated because interaction between the active north–south convergence along the Himalaya and the east–west convergence and folding within the Indo-Burma ranges, which attribute deformation known as subduction [10]. These two mobile belts meet at the eastern syntaxes zone, which was the source zone of the 1950 Assam earthquake (Mw = 8.1), the 1897 Shillong earthquake (Mw = 8.7), and two large earthquakes (Mw > 7.0), such as 1930 Dhubri (Mw > 7.0) and 1918 Srimangal earthquakes (Mw > 7.5), which significantly affected the adjoining areas in this region. Further, it is also reported by some other researchers (Mukhopadhyay and Dasgupta [13]; Kayal [9]; Anbazhagan et al. [1], Basu et al. [2]) that the Indian plate actively subducts below the Burmese arc. Based on the previous earthquakes and geomorphological survey, this zone was subdivided into five parts [18]. As per the tectonic framework, it is observed that the study area of Tripura lies within the Tripura fold belt zone, which is in the proximity of the Bengal basin and Indo-Burmese arcs in the western and eastern sides, respectively. It is observed that there are three active faults near Tulashekhar, Chamanu which are relatively near to the study area. The major fault lines are passing in the south through Rupaichari, Matabari, Hrishyamukh, Killa, Jampuijala, Jirania, and Manu.

## 2 Sample Borehole

Primarily, the analysis is carried out from sub-soil data retrieved from twenty boreholes (BH). BH were selected in such a manner that they are uniformly distributed across the central zone of Agartala municipality area (AMC).The SPT is the most important and common in situ tests used in all geotechnical investigations around the world. Shear wave velocity is one of the key parameters in seismic hazard evaluation, which plays an important role during an earthquake as it is directly responsible for the amplification or de-amplification of incoming ground motion at the ground surface which is effected by local soil characteristics [19]. A typical bore log profile along with SPT-N value and shear wave velocity variation is presented in Fig. 2. Several correlations are proposed by various researchers; however, for the present study, correlation proposed by [20] was taken into consideration. Correlations proposed by [20] were based on an extensive study on Agartala soil which included in situ tests and are deemed most suitable for the present study. 20 borehole profiles considered for the present study were classified as Site Class D ($V_{s30}$ 180–360 m/s) or Class E

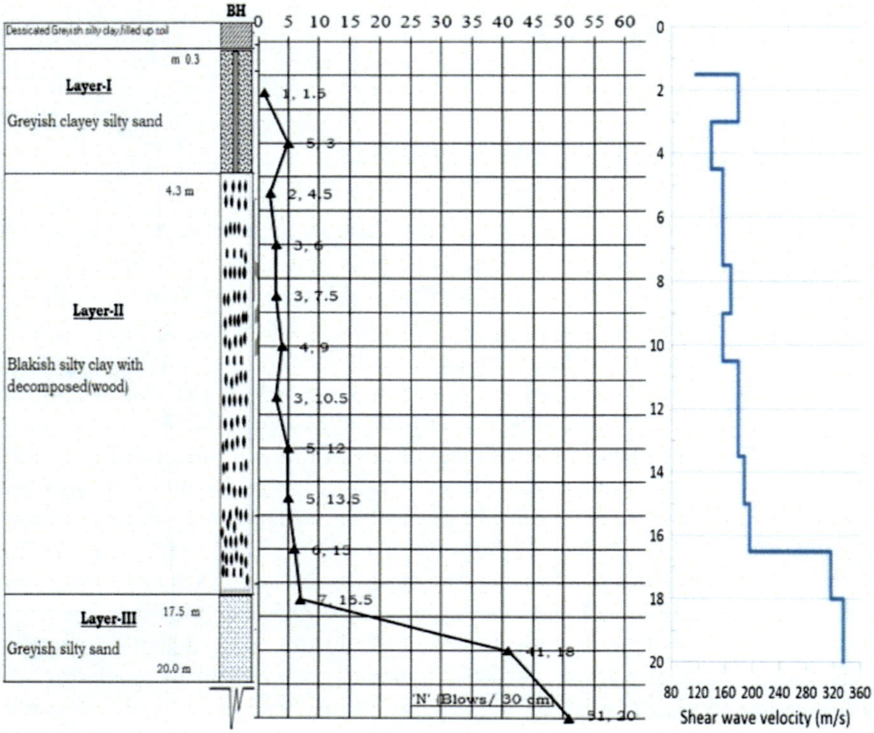


**Fig. 2** Typical bore log profile along with SPT-N value and shear wave velocity variation

($V_{s30}$ < 180 m/s). The $V_{s30}$ of each soil column classified according to the [17] site classification scheme (Tables 1 and 2).

**Table 1** Site classification system as per NEHRP [17] design manual

| NEHRP site class | Description | $V_{s30}$ |
|---|---|---|
| A | Hard rock | > 1500 m/s |
| B | Firm and hard rock | 760–1500 m/s |
| C | Dense soil, soft rock | 360–760 m/s |
| D | Stiff soil | 180–360 m/s |
| E | Soft clays | < 180 m/s |

**Table 2** Name and locations of the salient five sites used for this study

| Borehole profile No. | Name of site | Location of site |
|---|---|---|
| BH1 | Gandhi School Road | 23.8825662, 91.291366 |
| BH2 | Dashamighat, Hariganga Road | 23.828000, 91.265533 |
| BH3 | North Banamalipur | 23.839450, 91.289542 |
| BH4 | Ujan Abhoynagar | 23.847881, 91.288744 |
| BH5 | South Badharghat Railway Station | 23.893869, 91.277146 |

## 3 Methodology

This paper focusses on a non-Masing criteria-based nonlinear 1D GRA carried out using DEEPSOIL v6.0 and liquefaction potential assessment of a soil profile in Agartala. Nonlinear time history analysis is done in DEEPSOIL considering soil profile which is converted into a mass-spring—dashpot system using MDOF lumped mass parameter model. The dynamic equation as follows is solved at each time step using the numerical integration process by Newmark β method (Newmark 1959).

$$[M]\{\ddot{u}\} + [C]\{\mathring{u}\} + [K]\{u\} = -[M]\{I\}\ddot{u}_g \tag{1}$$

where $[M]$ is the mass matrix, $[C]$ is the viscous damping matrix, $[K]$ is the stiffness matrix, $u$ is the vector of relative nodal displacement, $\ddot{u}_g$ is the acceleration at the base of the soil column, and $\{I\}$ is a unit vector. The nonlinear model developed by Hashash and Park [6], Hashash et al. [7], Hashash [8] with hysteretic damping reduction factor, referred to as MRDF procedure, has been employed in the DEEPSOIL code, for performing nonlinear non-Masing GRA. The model used in DEEPSOIL is the pressure-dependent hyperbolic model which defines the inter-relationship between the stress and strains developed in the soil subjected to a cyclic loading and unloading phenomenon. Generally, the development of cyclic shear stress due to the cyclic strains or vice versa is governed by Masing and Extended Masing rules (Park 2003). The hyperbolic model is a modification of the original strain-dependent hyperbolic MKZ (Matasovik Kondner and Zelasko) model Equation

$$\tau = \frac{\gamma G_{\mathrm{mo}}}{1 + \beta\left(\gamma \frac{G_{\mathrm{mo}}}{\tau_{\mathrm{mo}}}\right)^s} = \frac{\gamma G_{\mathrm{mo}}}{1 + \beta\left(\frac{\gamma}{\gamma_r}\right)^s} \tag{2}$$

where $G_{\mathrm{mo}}$ is the initial shear modulus, $\tau_{\mathrm{mo}}$ is the shear stress at 1% shear strain, $\gamma$ is the shear strain, and $\gamma_r$ is the reference shear strain. The model parameters $\beta$ and $s$ are used to define the shape to the backbone curve. The pore water pressure generation models for sands [12], for clays [12], and the pore water pressure dissipation model are implemented in DEEPSOIL code.

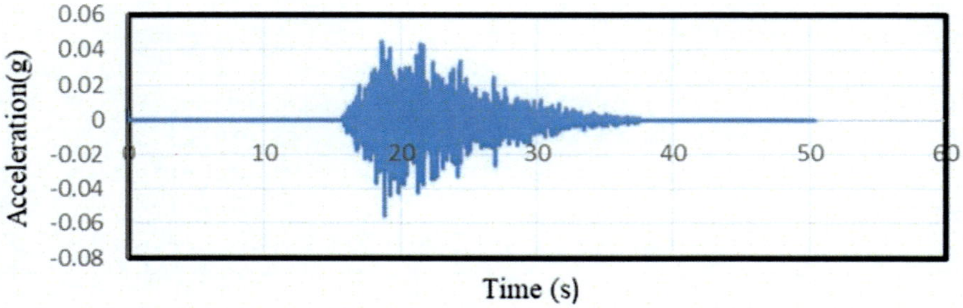


**Fig. 3** Synthetic bedrock motion DHALAI 2017 with PGA $0.056 \times g$

## *3.1 Description of Seismic Loading*

In the present analysis, recent moderate January 3, 2017, Ambassa earthquake (Mw = 5.7) epicenter located at Dhalai district of Tripura is used as input motion. Due to the unavailability of recorded data Dhalai2017, synthetic ground motions considering past local earthquakes using Boore's point source model (Boore 1983, 2003) are used with peak ground acceleration of $0.056 \times g$ (Fig. 3; Table 3).

# 4 Results and Discussion from GRA Studies

## *4.1 Peak Ground Acceleration (PGA) with Depth*

Ground response analyses are carried out on 20 BH at Central Agartala using synthetic motion. Salient 5 boreholes from five important locations are selected from a total sample of 20 boreholes to present peak ground acceleration (PGA) at the top surface for five different site locations at Agartala (Table 4; Fig. 4). The results of acceleration time history and stress time history are obtained for all 20 BH. It has been observed that the PGA of input motions gets amplified at surface.

**Peak ground displacement (PGD) and maximum strain (%) with depth**

Maximum displacement and maximum strain at a depth of every layer are calculated for 20 boreholes subjected to the ground motion from DEEPSOIL analysis. Salient 5 boreholes selected from a total sample of 20 boreholes to present peak ground displacement (PGD) and maximum strain (%) at the top surface for five different site locations at Agartala (Table 5; Fig. 5). Max displacements occur due to loss of strength of soil strata subjected to numerous cyclic loading.

**Amplification factor ($A_F$)**

The amplification ratio is defined as the ratio of amplification at the surface to that of the bed rock. Amplification factor gives an idea about the amplification/attenuation

**Table 3** Presented details of earthquake around the study area used for this

| Long (°) | Lat (°) | Name | Year | Month | Day | Hypocentral distance (km) | Focal depth (km) | Magnitude | Max Acc ($g$)$g$ |
|---|---|---|---|---|---|---|---|---|---|
| 91.9 | 24.1 | Dhalai Tripura | 2017 | 1 | 17 | 61.71 | 28 | 5.7 | 0.0556 |

**Table 4** Salient 5 boreholes selected from a total sample of 20 boreholes presenting peak ground acceleration (PGA) at the top surface for five different site locations at Agartala

| Borehole profile No. | Peak ground acceleration (*g*) |
|---|---|
| BH1 | 0.06080 |
| BH2 | 0.07650 |
| BH3 | 0.06862 |
| BH4 | 0.07620 |
| BH5 | 0.05919 |

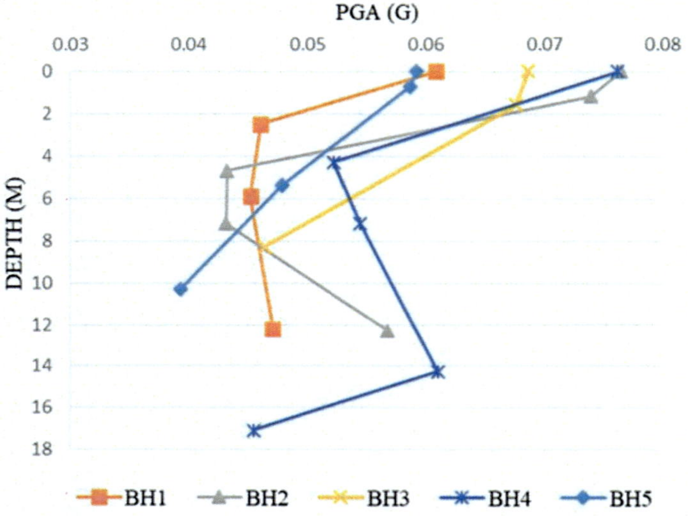


**Fig. 4** PGA profiles in central region of Agartala for EQ Dhalai (2017) EQ

**Table 5** Salient 5 boreholes selected from a total sample of 20 boreholes presenting peak ground displacement (PGD) and maximum strain (%) at the top surface for five different site locations at Agartala

| Borehole profile No. | Peak ground displacement (PGD) (m) | Max strain (%) |
|---|---|---|
| BH1 | 0.001084 | 0.00943 |
| BH2 | 0.001734 | 0.01945 |
| BH3 | 0.001627 | 0.01452 |
| BH4 | 0.00115 | 0.00760 |
| BH5 | 0.00118 | 0.00806 |

of ground accelerations as the stress waves propagate through the soil medium. This is indicative of the damage expected at a particular location. However, higher amplification has been observed in a range between 1.06 and 1.37. Salient 5 boreholes

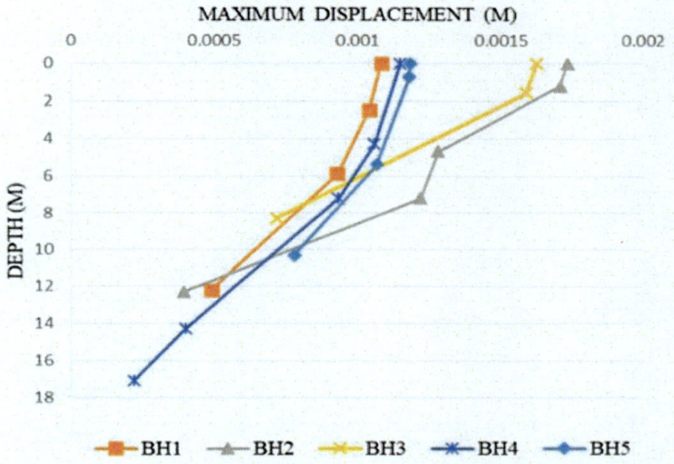


(a)

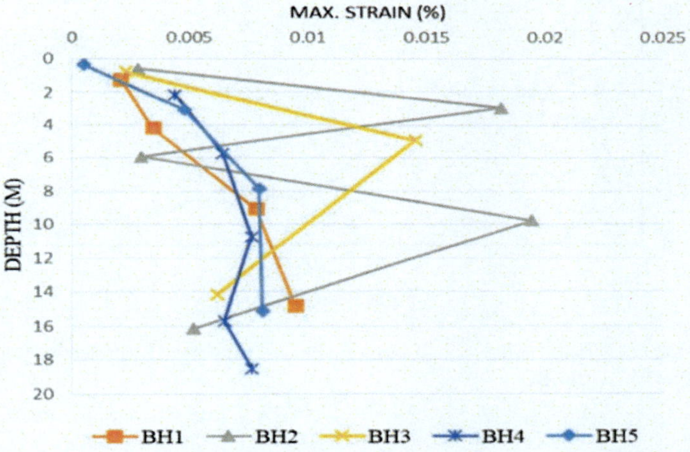


(b)

**Fig. 5** **a** PGD and **b** Max strain (%) profiles in central region of Agartala for Dhalai (2017) EQ

selected from a total sample of 20 boreholes to present amplification factor at the top surface for five different site locations at Agartala at Table 6.

## Response spectra

Response spectrum describes the maximum response of a single-degree-of-freedom (SDOF) system, subjected to a particular input motion, as a function of the natural period (or natural frequency) and the damping ratio of the system [11]. For any particular site, the 5% damped surface spectral acceleration plot can be obtained based on the GRA study. Spectral acceleration at the ground surface, i.e., the response

**Table 6** Salient 5 boreholes selected from a total sample of 20 boreholes presenting amplification factor at the top surface for five different site locations at Agartala

| Borehole profile No. | Amplification factor |
|---|---|
| BH1 | 1.09 |
| BH2 | 1.37 |
| BH3 | 1.23 |
| BH4 | 1.37 |
| BH5 | 1.06 |

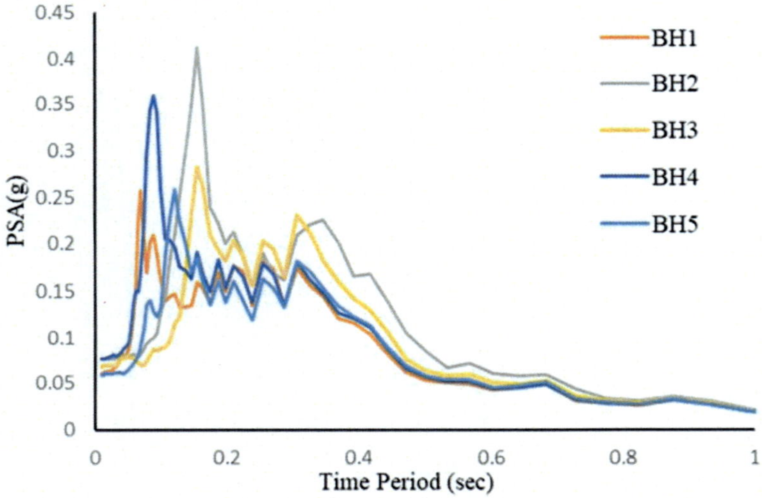


**Fig. 6** Spectral acceleration at the ground surface with 5% damping at five selective BH locations for the motion Dhalai2017 EQ

spectra with 5% damping, is shown in Fig. 6 at five selective BH locations for the motion Dhalai2017. The high value of spectral acceleration has been observed in BH2. It serves as an essential guideline for seismic design of structures. It can be seen that highest responses are expected for structures having low natural period in the range of 0.1–0.2 s.

## 5 Summary and Conclusions

The site response analysis of the central region of Agartala and the detailed site effect analysis of soil profiles are presented in this study. A total of 20 SPT-N boreholes were analyzed with the aim to study the dynamic properties of the soil and to evaluate the behavior of soils upon application of seismic load. The SPT-N values were used to evaluate the Vs30 for all borehole locations. In the NEHRP site classification scheme, site classes D and E are prevalent in the study area. Peak ground acceleration (PGA) at surface obtained for twenty boreholes, located at central zone of Agartala ranges from 0.059 to 0.076 g. From results obtained, it can be stated that the maximum increase in PGA is found to be at ground surface which indicates amplification of the motions due to the presence of soft soil deposit. Peak ground displacement (PGD) was evaluated from twenty boreholes distributed uniformly at central zone of Agartala. It is observed that PGD at the surface level is on the higher side. The range of PGD is found to be in the order of 0.001–0.017. Results indicated wide range along with higher value which gives rise to possibility of ground cracks or failure in future earthquakes. It is observed from the figure that higher strain range is noted for motion Dhalai EQ 2017. Higher strain indicates more dissipation of energy during the cyclic loading phenomenon and hinting at further softening of the soil stratum with the number of loading cycles. Amplification has been observed for the motion Dhalai EQ 2017 with amplification factor range between 1.06 and 1.37. High value of spectral acceleration has been observed for Dhalai EQ 2017 motion. The plots also reflect that higher spectral acceleration is obtained as a result of a higher amplitude of input motion.

## References


1. Anbazhagan, P., Sitharam, T.: Estimation of ground response parameters and comparison with field measurements. Indian Geotech. J. **39**(3) (2009)
2. Basu, D., Madhulatha, B., Dey, A.: A time-domain nonlinear effective-stress non-Masing approach of ground response analysis of Guwahati city, India. Earthq. Eng. Eng. Vib. **18**(1), 61–75 (2017). https://doi.org/10.1007/s11803-019-0490-0
3. Boore, D.M.: Stochastic simulation of high-frequency ground motions based on seismological models of the radiated spectra. Bull. Seismol. Soc. Am. **73**(6A), 1865–1894 (1983)
4. Boore, D.M.: Simulation of ground motion using the stochastic method. Pure Appl. Geophy. **160**(3–4), 635–676 (2003)
5. Dasgupta, S.: Echoes from Old Calcutta. Kolkata: London: Thacker (1908)
6. Hashash, Y.M.A., Park, D.: Nonlinear one-dimensional seismic ground motion propagation in the 433 Mississippi embayment. Eng. Geol. **62**(1–2), 185–206 (2001)
7. Hashash, Y.M., Phillips, C., Groholski, D.R.: Recent advances in non-linear site response analysis. Paper Presented at Fifth International Conference on Recent Advances in Geotechnical (2010)
8. Hashash, Y.M.: DEEPSOIL 7.0. User Manual, p. 1 (2018). www.illinois.edu/~deepsoil
9. Kayal, J.R.: Seismicity of Northeast India and surroundings – development over the past 100 years. J. Geophy. **19**, 9–34 (1998)
10. Kayal, J.R., Arefiev, S.S., Baruah, S., Hazarika, D., Gogoi, N., Kumar, A., Chowdhury, S.N., Kalita, S.: Shillong Plateau earthquakes in northeast India region: complex tectonic model. Current. Sci. **91**, 109–114 (2006)
11. Kramer, S.: Geotechnical Earthquake Engineering, p. 653. Prentice Hall, New Jersey, NJ (1996)

12. Matasovic, N., Vucetic, M.: Seismic response of soil deposits composed of fully-saturated clay and sand layers. In: Proceedings 1st International Conference on Geotechnical Earthquake Engineering (1995)
13. Mukhopadhyay, M., Dasgupta, S.: Deep structure and tectonics of the burmese arc: constraints from earthquake and gravity data. Tectonophysics **149**(3-4), 299–322 (1988)
14. Newmark, M., N.: A method of computation for structural dynamics. J. Eng. Mech. Div. **85**(3), (1959). https://doi.org/10.1061/JMCEA3.0000098
15. Park, Y.M.: Viscous damping formulation and high frequency motion propagation in non-linear site response analysis. Soil Dyn. Earthq. Eng. **22**, 611–624 (2002)
16. Park, D.: Estimation of non-linear seismic site effects for deep deposits of the Mississippi Embayment. Ph.D. Thesis. Department of Civil and Environmental Engineering. Urbana: University of Illinois, p. 311 (2003)
17. Provisions, N.P.: NEHRP Recommended Provisions for Seismic Regulations for New Buildings and Other Structures (FEMA 450). Building Seismic Safety Council, National Institute of Building Sciences, Washington, p. 338 (2003)
18. Saha, R., Debnath, R., Dash, S., Haldar, S.: Engineering Reconnaissance Following the Magnitude 5.7 Tripura Earthquake on January 3, 2017. American Society of Civil Engineers (2020). https://doi.org/10.1061/(ASCE)CF.1943-5509.0001446
19. Sana, H., Nath, S.K., Gujral, K.S.: Site response analysis of the Kashmir valley during the 8 October 2005 Kashmir earthquake (Mw 7.6) using a geotechnical dataset. Bull. Eng. Geol. Environ. **78**, 2551–2563 (2018). https://doi.org/10.1007/s10064-018-1254-1
20. Sil, A., Sitharam, T.: Dynamic site characterization and correlation of shear wave velocity with standard penetration test 'N' values for the city of Agartala, Tripura State, India. Pure Appl. Geophys. (2013). https://doi.org/10.1007/s00024-013-0754-y
21. Tsai, C.-C., Chen, C.W.: Comparison study of 1D site response analysis methods. Earthq. Spectra **32** (2015)
22. Vedatri, A., Bashir, K., Saha, R.: Assessment of seismic local hazard of Agartala town based on nonlinear site response analysis. Disaster Adv. **11**(1) (2018)

# Effect of Alkali-Activated Fly Ash on Shrinkage Characteristics of Expansive Soil

Vamsi N. K. Mypati and Sireesh Saride

## 1 Introduction

The expansive soils are well known as problematic soils due to their swell-shrinkage behavior due to seasonal moisture fluctuation. The structures built on these soils will experience severe distress due to differential settlements [1, 2]. Many highways, highway embankments, and airfield pavements were damaged due to shrinkage of such highly expansive soils [3–5]. To control these adverse effects of expansive soils, conventional binders like cement and lime were used in the past [6, 7]. However, these binders have some disadvantages, including environmental concerns due to the emission of carbon dioxide during the manufacturing process. Alternative binders such as fly ash and slag are promoted as these by-products possess potential pozzolanic compounds. However, activators are generally used along with these binders to accelerate the reactivity. Alkali-activated fly ash is the cementitious material in which alumina and silica from fly ash dissolute in the alkali solution to form a geopolymeric gel network of sodium-alumino-silicate-hydrate (N–A–S–H) gel by reorientation [8]. However, the reactivity of alkali-activated fly ash with expansive soils is not fully understood.

The primary purpose of stabilizing the expansive soil is to control swell-shrink behavior. Accurate measurement of the volume of a shrunken soil specimen may be calculated using a mercury displacement method [9, 10]. However, it has disadvantages of handling and health issues. Due to these disadvantages, the ASTM standard D427-04 [11] has been withdrawn, and the wax method ASTM D4943-08 [12] was suggested as an alternative for the determination of volumetric shrinkage of soils. However, it was also withdrawn recently. Hence, there is a need to look for a reliable

V. N. K. Mypati · S. Saride (✉)
Civil Engineering, Indian Institute of Technology Hyderabad, Hyderabad, Telangana 502285, India
e-mail: sireesh@ce.iith.ac.in

 
C. N. V. Satyanarayana Reddy et al. (eds.), *Ground Characterization and Foundations*, Lecture Notes in Civil Engineering 167,
https://doi.org/10.1007/978-981-16-3383-6_28

method that excludes the volume of cracks in the volume of a shrunken soil specimen. The image processing methods are new promising techniques in determining the volume of soil specimens [13].

In the present study, alkali-activated fly ash was used as a stabilizer to control the shrinkage characteristics of expansive soil, and the image processing method was used to determine the volume of the shrunken soil specimen accurately by excluding the volume of developed cracks in the expansive soil.

## 2 Material Properties

The expansive soil was obtained from Amaravati city, Andhra Pradesh, India. The soil consists of 6% sand, 31% silt, and 63% clay. The properties of the expansive soil are listed in Table 1. The liquid limit and plasticity index of the expansive soil were 77% and 44%, respectively. According to the American Society for Testing and Materials, ASTM D2487 [14], the soil can be classified as clay with high plasticity (CH). The maximum dry unit weight (MDU) and optimum moisture content were found to be 14.2 kN/m$^3$ and 28.5%, respectively. The fly ash was obtained from the Narla Tatarao Thermal Power Station (NTTPS), Vijayawada, India. The chemical oxide composition of the fly ash was determined through X-ray fluorescence (XRF) analysis and is presented in Table 2. The fly ash can be classified as Class F, according to ASTM C618 [15]. Sodium hydroxide (NaOH) at 99% purity and $Na_2SiO_3$ (consisting of 10% $Na_2O$, 38% $Si_2O$, and 52% $H_2O$) is obtained from a local chemical supplier.

A liquid alkali activator ratio (LAA), defined as the ratio of $Na_2SiO_3$ and NaOH, of 50:50, is considered for preparing the liquid component, which consists of LAA and water, of the soil-fly ash mix [16]. Three different liquid components, 20, 25, and 30% by dry weight of the soil-fly ash mixture, were considered. A 20% fly ash by dry weight of the expansive soil is considered [17, 18]. Hence, the liquid to fly ash ratios are calculated to be 1.0, 1.25, and 1.5, respectively, as shown in Table 3.

**Table 1** Properties of expansive soil and fly ash

| Property | Soil | Fly ash |
|---|---|---|
| $p^H$ | 8.24 | 7.75 |
| Specific gravity ($G_s$) | 2.74 | 2.2 |
| Liquid limit (LL) | 77% | – |
| Plastic limit (PL) | 33% | – |
| Plasticity index (PI) | 44 | – |
| Shrinkage limit (SL) | 10% | – |
| Free swell index (FSI) | 110% | – |
| Silt | 31% | 67% |
| Clay | 63% | 30% |

**Table 2** Chemical oxide composition of expansive soil and fly ash

| Oxide | Expansive soil (ES) (%) | Fly ash (FA) (%) |
|---|---|---|
| $SiO_2$ | 59.02 | 52.09 |
| $Al_2O_3$ | 29.63 | 17.15 |
| CaO | 1.17 | 3.49 |
| $Fe_2O_3$ | 4.98 | 18.95 |
| $K_2O$ | 1.86 | 1.18 |
| MgO | 0.54 | 3.08 |
| $SO_3$ | 0.25 | 0.49 |
| $Na_2O$ | 0.23 | 0.34 |
| $P_2O_5$ | 0.68 | 0.42 |
| $TiO_2$ | 1.84 | 2.25 |
| $MnO_2$ | 0.03 | 0.17 |

**Table 3** Nomenclature of samples

| FA (%) | LAA (%) | LAA/FA |
|---|---|---|
| 20 | 20 | 1 |
| 20 | 25 | 1.25 |
| 20 | 30 | 1.5 |

# 3 Experimental Methods

A series of experiments, including volumetric shrinkage tests and scanning electron microscopy (SEM) analyses, were conducted at different curing periods. The following sections briefly describe the sample preparation for these experiments.

## *3.1 Sample Preparation Procedure*

The expansive soil was mixed with 20% fly ash by dry weight. The soil-fly ash mixture was mixed with LAA solution, which consists of 3M NaOH solution and liquid sodium silicate solution of ratio 50:50 by weight at three different LAA/FA ratios shown in Table 3. After mixing with respective LAA, samples were packed in a polythene cover to protect for moisture loss, and it was cured for 7-days, 14-days, and 28-days by keeping them in an environmental chamber maintained at 27 °C temperature and 95% relative humidity. Thereafter, samples were taken out from the environmental chamber on respective curing period, and further, specimens were prepared according to ASTM D4943 [19] to determine the volumetric shrinkage. The volumetric shrinkage strain was determined by using an image processing technique that involves capturing the image, converts the image to grayscale, and converts grayscale to a binary scale. The image capturing was carried out at constant altitude with uniform light intensity and high-resolution digital camera. The setup used in the present study was similar setup reported by Julina et al. [13]. Further, the image processing techniques were applied to the captured images with the free access Image J software. The *image segmentation threshold* image processing method was used to convert the gray image to a binary image, as shown in Fig. 1. The average surface areas of the top and bottom portion of the image processed samples excluding cracked areas were determined by Image J software. The height of the shrunken samples was measured by subtracting the average height from the original sample height, which measured from the top of the mold to the surface of the sample by using Vernier calipers. The volumetric shrinkage was determined by using Eq. 1. In the vernier method, the diameter and height of the sample were determined with Vernier calipers, and the shrinkage strains were determined by using Eq. 2.

$$\text{Volumetric shrinkage strain,}\quad (V_s) = \frac{(V - A_{sr}H)}{V} \times 100 \qquad (1)$$

where $V$ = initial volume of the soil specimen; $A_{sr}$ = average surface area of the top and bottom of shrunken soil specimen, and $H$ = height of shrunken soil specimen.

$$\text{Volumetric shrinkage strain,}\quad (V_s) = \frac{\left(V - \left(\frac{\Pi D_{avg}^2}{4} H\right)\right)}{V} \times 100 \qquad (2)$$

**Fig. 1** **a, b** are samples of control soil and control soil with 20% fly ash; **c–e** are oven-dried samples of treated soil with LAA/FA = 1, 1.25, and 1.5 at 28-days curing period

where $V$ = initial volume of the soil specimen; $D_{avg}$ = average diameter of shrunken soil specimen, and $H$ = height of shrunken soil specimen.

Further, scanning electron microscopy (SEM) images of stabilized samples of LAA/FA = 1, LAA/FA = 1.25, and LAA/FA = 1.5 were captured to investigate the microstructural changes at 28-days curing period. Before capturing the images, samples were coated with gold–palladium to increase the conductivity of the samples. After that, samples were mounted on the aluminum stub with a conductive tap on top. SEM images were captured on selected samples with Zeiss, EVO 60 field emission under backscatter mode at a voltage of 20 kV.

## 4 Results and Discussion

The volumetric shrinkage strains determined by the Vernier and Mercury methods and image processing method are shown in Fig. 2. Mercury method was performed only on treated samples as it was difficult to perform the test on untreated specimen due to severe cracking and splitting of the specimen. The percentage error in volumetric shrinkage between both the methods is relatively high for the untreated sample, which is of about 4.5% compared to the treated samples (2.5%). The error is due to the development of more shrinkage cracks in untreated specimen compared to the treated specimens and also due to the area of shrinkage cracks which were included in the Vernier method. The increase in volumetric shrinkage strains in the image processing method indicates the high accuracy over the Vernier/Mercury method. It

**Fig. 2** Comparison of volumetric shrinkages strains determined by image processing method and Vernier method of an untreated and treated sample at 28-days curing period

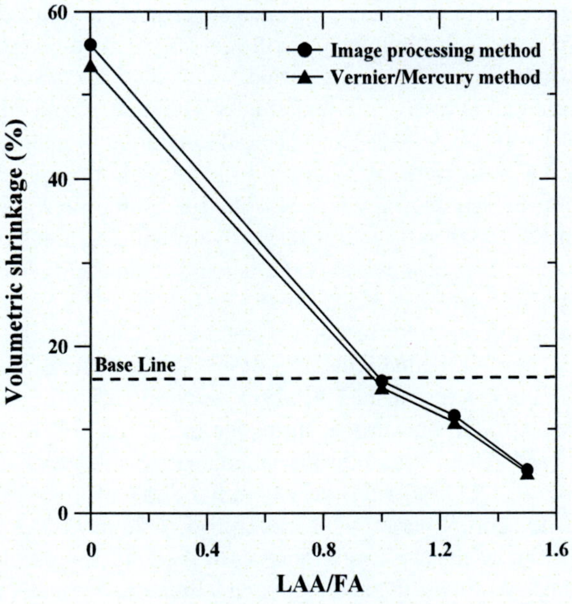

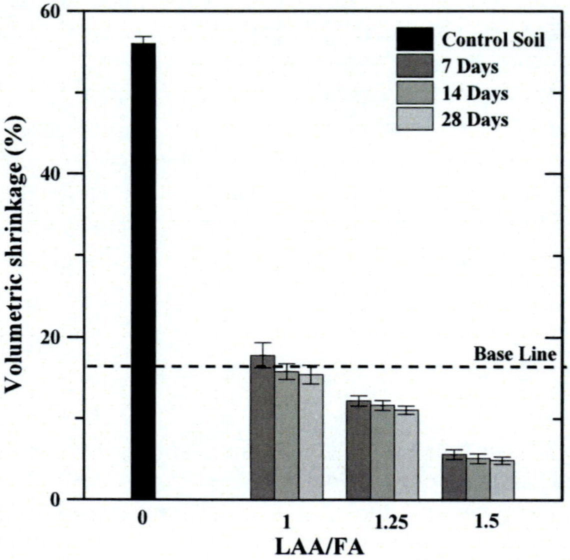


**Fig. 3** Volumetric shrinkage strains of control soil and expansive soil treated with different LAA/FA at 7-days, 14-days, and 28-days curing period

was reported that the degree of severity is less when the volumetric shrinkage strain is less than 17% [20]. Hence, this value is considered as a baseline in this analysis.

Figure 3 depicts the volumetric shrinkage of control and stabilized soil specimens at 7-days, 14-days, and 28-days curing period. There is a significant decrease in the volumetric shrinkage of control soil from 56 to 15%, 11%, and 5%, respectively, with the treatment of LAA/FA = 1, 1.25, and 1.5 for 28-days cured specimens. The decrease in volumetric shrinkage strains of stabilized soil is due to the activation of pozzolanic compounds in the geopolymerization process. The geopolymerization occurs during the curing period, which includes the dissolution of alumina and silica in alkali solution, reorientation of aluminates, and silicates to form a possible gel network N–A–S–H [21]. The gel network solidifies to form a geopolymer. As the LAA ratio increases, there is an increase in precipitation of the gel network in the soil-fly ash matrix, which was the reason for the significant decrease of volumetric shrinkage from 15 to 5%, respectively, with treatment from LAA/FA = 1 to LAA/FA = 1.5. Figure 4 presents the percent reduction in volumetric shrinkage strains of specimens treated with LAA/FA = 1, 1.25, and 1.5 and cured for 7-, 14- and 28-days with reference to the control specimen. The reduction in the volumetric shrinkage of samples treated with LAA/FA = 1.0 from control soil is about 73%. However, a further increase in the LAA/FA ratio from 1.25 to 1.5 has shown a minimal additional reduction in the vertical shrinkage from 7 to 17% for a curing of 28 days shown in Fig. 4. The decrease in the volumetric shrinkage of samples treated with LAA/FA = 1.0, 1.25, and 1.5 is very minimal of 17–5%, 12–11%, and 5.5–5%, respectively, as the curing period increases from 7 to 28 days. The minimal reduction in $V_s$ may be due to the slow solidification of the gel network during a curing period of 7–28 days. At times, the complete crystallization of the gel network may take years [22].

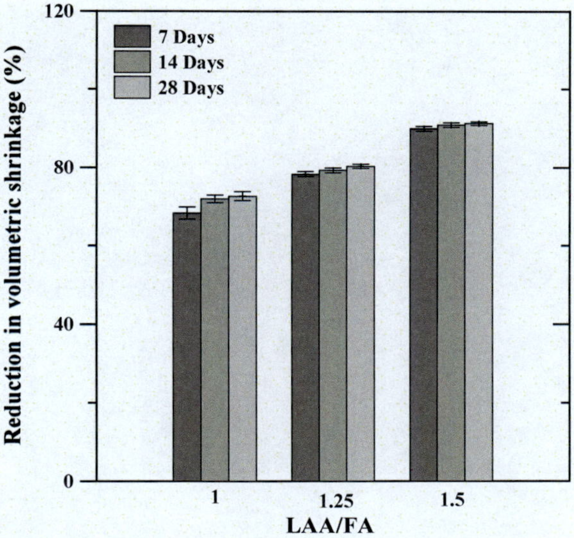


**Fig. 4** Percentage reduction in volumetric shrinkage of expansive soil treated with varying LAA/FA from 1 to 1.5 at different curing period

However, all the specimens treated with LAA/FA ratios have shown the volumetric shrinkage strains less than the baseline, indicating that the stabilization is effective for all LAA/FA ratios.

To understand this phenomenon, SEM images were captured to investigate the microstructural features of the stabilized soil. SEM images of stabilized soil samples cured at 28 days were chosen for the investigation, which was shown in Fig. 5. It was observed that from the sample treated with LAA/FA = 1.0 has partial dissolution of fly ash particles, which is due to the insufficient addition of LAA. At the same time, the other samples were found intact, indicating that the LAA/FA ratio is sufficient. Layers of control soil were disappeared in treated samples with LAA/FA.

## 5 Conclusions

The following conclusion can be drawn from the experimental studies conducted on alkali-activated fly ash-treated expansive soils.

1. Alkali-activated fly ash with a liquid alkali activator ratio of 50:50 was effectively controlled the volumetric shrinkage of expansive soil.
2. A three-molar concentration of NaOH is adequate to enable the geopolymer reactions in the fly ash stabilized expansive soil.
3. The image processing method is a promising method to determine the area of the shrinkage, excluding the cracked areas.

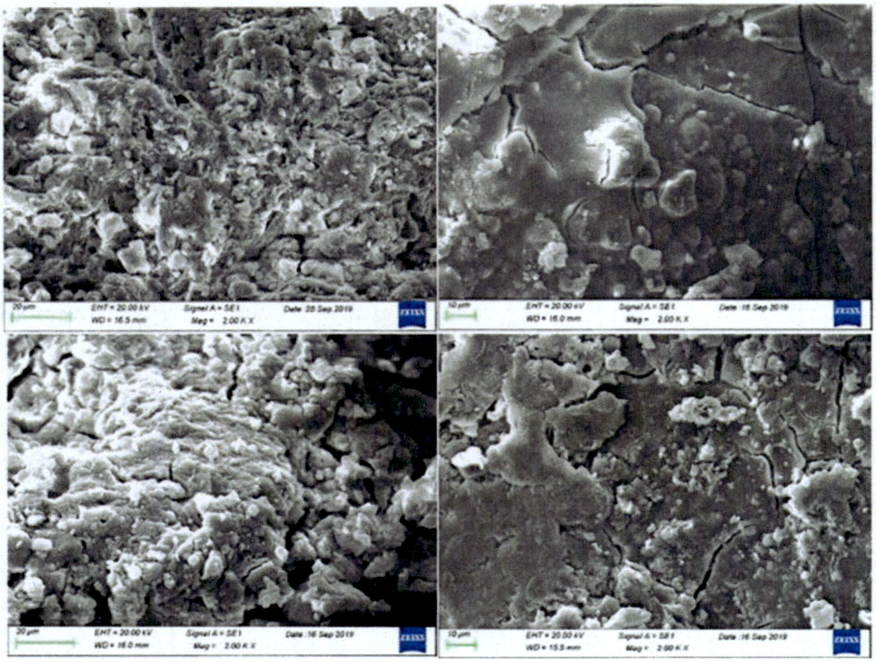

Fig. 5 Scanning electron microscopy images of **a** control soil, **b** soil with LAA/FA = 1.0, **c** soil with LAA/FA = 1.25, and **d** soil with LAA/FA = 1.5 at 28-days curing period

4. The volumetric shrinkage strains from image processing method were accurate compared to the veriner method. This method is advantageous especillay when the shrinkage cracks are severe, and the specimen is vulnerable to breaking.
5. The volumetric shrinkage was reduced significantly from 56 to 15%, 11%, and 5%, respectively, with a treatment of expansive soil with LAA/FA = 1.0, 1.25, and 1.5 at a curing period of 28 days.
6. The reduction in the volumetric shrinkage of samples treated with LAA/FA = 1.0 from control soil is about 73%. A further increase in the LAA/FA ratio from 1.25 to 1.5 has shown a minimal additional reduction in the vertical shrinkage from 7 to 17% for a 28-days curing period.
7. The LAA/FA = 1.25 may be considerd as the optimum dosage to control the volumetric shrinkage of expansive soil at a curing period of 28 days.
8. With a treatment of LAA/FA = 1.5, clay layers were disappeared in the control soil, and a dense crystalline product was seen in the SEM images.
9. SEM images also depict a right amount of dissolution of fly ash that occurred in the control soil treated with LAA/FA = 1.5.

## References


1. Haines, W.B.: The volume change associated with variations of water content in the soil. J. Agric. Sci. **13**(3), 296–310 (1923)
2. Bozozuk, M.: Soil shrinkage damages of shallow foundations at Ottawa, Canada. Eng. J. **45**(7), 33–37 (1962)
3. Jayatilaka, R., Lytton, R.L.: Prediction of expansive clay roughness in pavements with vertical moisture barriers. Rep. No. Federal Highway Administration (FHWA)/Texas (TX)-98/187-28 F, Texas Transportation Institute, Texas A&M Univ., College Station, TX (1997)
4. Sebesta, S.: Investigation of maintenance base repairs over expansive soils. Rep. No. Federal Highway Administration (FHWA)/Texas (TX)-03/0-4395-1, Texas Transportation Institute, Texas A&M Univ., College Station, TX (2002)
5. Zhang, Z., Tao, M., Morvant, M.: Cohesive slope surface failure and evaluation. J. Geotech. Geoenviron. Eng. **131**, 898–906 (2005)
6. Herrin, M., Mitchell, H.: Lime-soil mixtures. Highway Res. Board Bull. **304**, 99–138 (1961)
7. Sivapullaiah, P.V., Sridharan, A., Bhaskar Raju, K.V.: Role of amount and type of clay in the lime stabilization of soils. Ground Improv. **4**, 37–45 (2000)
8. Provis, L., Van Deventer, J.S.J.: Geopolymers: Structure, Processing, Properties, and Industrial Applications. Woodhead Publishing Limited, Cambridge, UK (2009)
9. Rao, K.S.S., Rao, S.M., Gangadhara, S.: Swelling behavior of a desiccated clay. ASTM Geotech. Test. J. **23**(2), 193–198 (2000)
10. Tripathy, S., Rao, K.S.S., Fredlund, D.G.: Water content—void ratio swell-shrink paths of compacted expansive soils. Can. Geotech. J. **39**(4), 938–959 (2002)
11. ASTM D427-04: Test method for shrinkage factors of soils by the mercury method. ASTM International, ASTM, West Conshohocken, PA (2004)
12. ASTM D4943-08: Standard test method for shrinkage factors of soils by the wax method. ASTM International, ASTM, West Conshohocken, PA (2008)
13. Julina, M., Thyagaraj, T.: Determination of volumetric shrinkage of an expansive soil using digital camera images. Int. J. Geotech. Eng. (2018). https://doi.org/10.1080/19386362.2018.1460961
14. ASTM D2487-17e1: Standard practice for classification of soils for engineering purposes (unified soil classification system). ASTM International, ASTM, West Conshohocken, PA (2017)
15. ASTM C618-19: Standard specification for coal fly ash and raw or calcined natural pozzolan for use in concrete. ASTM International, ASTM, West Conshohocken, PA (2019)
16. Saride, S., Jallu, M.: Effect of fly ash geopolymer on layer coefficients of reclaimed asphalt pavement bases. J. Transp. Eng. Part B: Pavem. **146**(3), 04020033 (2020)
17. Horpibulsuk, S., Rachan, R., Suddeepong, A.: Assessment of strength development in blended cement admixed Bangkok clay. Constr. Build. Mater. **25**(4), 1521–1531 (2011)
18. Yaghoubi, M.J., Arulrajah, A., Disfani, M.M., Horpibulsuk, S., Bo, M.W., Darmawan, S.: Effects of industrial by-product based geopolymers on the strength development of a soft soil. Soils Found. **58**, 716–728 (2018)
19. ASTM D4943-18: Standard test method for shrinkage factors of cohesive soils by the water submersion method. ASTM International, ASTM, West Conshohocken, PA (2018)
20. Punthataecha, K.: Volume change behavior of expansive soils modified with recycled materials. Ph.D. thesis, The University of Texas at Arlington, Arlington, TX (2002)
21. Singh, B.G.V.P., Kolluru, V.L.S.: Evaluation of sodium content and sodium hydroxide molarity on compressive strength of alkali activated low calcium fly ash. Cement Concr. Compos. **81**, 22–132 (2017)
22. Cristelo, N., Glendinning, S., Pinto, A.T.: Deep soft soil improvement by alkaline activation. Ground Improv. **164**(GI2) (2011)

# Strength and Microstructure Evolution of Soft Soils by Using Nano-silica

**Anuradha Patro and Rupashree Ragini Sahoo**

## 1 Introduction

In civil engineering soil is one of the most frequently used materials. Nearly all structure usually rest on soil. Presence of weak or soft soil in the construction sites is a major issue for any type of construction. To strengthen the weak soil properties, many methods like stabilization of soil, soil reinforcement, grouting, addition of admixtures, etc., are adopted. Addition of admixtures like lime, fly ash, cement, and bitumen based on type of soil improves the properties of soil to some extent. Use of industrial waste as additives is recently under study, but it arise a question of toxicity. So there is a need for finding a new innovative material. Improving the engineering properties soil by using additives, stability and strengthening of the soil increases.

One of the new innovative fields, nanotechnology, is recently been introduced to geotechnical engineering. Nanotechnology is the science that deals with the particles which are less than 100 nm. The sizes of nanoparticles have crucial part in behavior of soil exhibiting different properties. To enhance strength of soft soil use of industrial by-products, waste materials and innovative material are being. A series of laboratory test carried out by Pham and Nguyen [1] to study the fundamental geotechnical properties by adding nano-silica with the clayey soil and obtained that swelling of clay soil decreases by using of nano-silica. Experimental test carried out by Mohammadi and Niazian [2], use of nano-clay enhances the value of liquid limit and plastic limits of soil and also improves the shear strength parameters of additive mixed sample. Zhang et al.[10] (2004) indicated that the use of nano-silica as additive the Atterberg limits and strength capacity of clay increased and permeability decreases. Changizi et al. [3] presented the UCS and shear strength of clayey soil used polyester fiber of 20 mm long which has recycled, generated form polyethelene (PET) bottles, together with Nanosilica of particle size 11-13 nanometre as a different stabilizer. Moradi and

A. Patro (✉) · R. R. Sahoo
Civil Engineering Department, VSSUT, Burla 768017, India

 
C. N. V. Satyanarayana Reddy et al. (eds.), *Ground Characterization and Foundations*, Lecture Notes in Civil Engineering 167,
https://doi.org/10.1007/978-981-16-3383-6_29

Seyedi [4] studied on stability on liquefiable slity sands and nano silica under two distinct circumstances and impact of disruption on strength of stabilized samples was discussed. Mostafa et al. [6] had explorer the impact of Nano-Silica (NS), Silica Fume (SF) and Lime (L) on the stability of the sub grade pavement layer. With percent 2, 4, 6 and 8 for lime and 5, 10 and 15 for SF were used to prepare NS samples. Kumar and Manikandan [7] researched nanotechnology and investigated the efficacy of nano-sized silica and lime particles within the stabilization of soil discovered that addition of nano sized silica and lime additive to soil reduced the liquid limit and plastic index values and enhanced plastic limit value. Pashabavanpouri and Jahangiri [5] analysed the effect of clay added Nanosilica and lime with completely different percentage. The outcome shows that less addition of Nanosilica in to clay used lime results in a major improvement in properties of plasticity, strength, compaction as well as swelling of altered soil. Alireza et al. [8] had examined the stabilization of weak soil addition with lime and nano silica and conducted the CBR test, compaction test with totally different percentage of lime and nano silica mixture. Optimum quantity of lime within the mixture of soil lime was 5% during which the very best quantity of CBR strength was achieved.

## 2 Experimental Work and Methodology

The different type of material used, preparation of sample, and procedure of test have discussed. To analyze the engineering characteristics of soil, various test such as grain size analysis, Atterberg's limit test, specific gravity test, compaction test, UCS test, and CBR (soaked, unsoaked) test was conducted.

### *2.1 Material Used*

**Soil**: Soil sample used for research work was collected from a site at Krushna prasad block, Puri, Odisha, at a depth of 2ft below ground level. The soil was oven dried and lumps were crushed in to small segments and screen by 4.75 mm size sieve to separate pebbles, roots, gravel, etc. The soil was categorized according to the IS classification system (Table 1) [9].

**Nano-silica**

Nano-silica used in this experimental study was supplied by the Astrra Chemicals, Chennai. The tests were conducted at percentage of nano-silica contents of 0, 0.6, 0.8, 1.0%, and 1.5%. Physical composition and chemical characteristics of nano-silica are illustrated in Tables 2 and 3, respectively.

**Table 1** Geotechnical properties of soil

| | Properties | Value |
|---|---|---|
| 1 | Specific gravity | 2.40 |
| 2 | Liquid limit (%) | 58.01 |
| 3 | Plastic limit (%) | 24.74 |
| 4 | Plasticity index (%) | 33.27 |
| 5 | Classification of soil as per IS 1498 | CH |
| 6 | *Compaction properties* | |
| | MDD (g/cc) | 1.58 |
| | OMC (%) | 20.70 |
| 7 | UCS ($kN/m^2$) | 80.54 |
| 8 | *CBR (%)* | |
| | Unsoaked | 3.46 |
| | Soaked | 1.81 |

**Table 2** Physical composition of nano-silica

| Purity (%) | Avg. particle size (nm) | Specific surface area ($m^2/g$) | Specific gravity | Temped density (g/L) | Sieve residue | PH value |
|---|---|---|---|---|---|---|
| 99 | 17 | 202 | 2.4 | 44 | 0.02 | 4.12 |

**Table 3** Chemical properties of nano-silica

| Material | Content (%) |
|---|---|
| $SIO_2$ | 99.88 |
| C | 0.06 |
| Cl | 0.009 |
| $Al_2O_3$ | 0.006 |
| $TiO_2$ | 0.003 |
| $Fe_2O_3$ | 0.001 |

## 2.2 Test Method

### Preparation of sample

First the soil was air dried, and it was pulverized with the help of wooden hammer. Then, it was sieved with 4.75 mm I.S sieve. First the required content of nano-silica has been blended with clayey soil under dry condition. The different percentages of nano-silica are 0.6, 0.8, 1.0, and 1.5% of by the total soil weight taken for test, and variation of OMC, MDD, UCS, and CBR values was evaluated.

**Specimen preparation for compaction test**

The test was carried out by standard compaction method as per IS2720 (Part VII)-1980 to determine the OMC and MDD. The compaction tests were done on the soil-nano-silica. Different percentage of nano-silica was blended with weighted oven dry soil. The appropriate amount of water was mixed with soil-nano-silica mixture, and also the wet specimen was compacted in proctor mold in three layers using 2.6 kg of standard proctor rammer.

**Specimen preparation for unconfined compression test**

For UCS test, samples were prepared as per IS: 2720(Part 10)-1991 at OMC and MDD obtained from compaction test. The cylindrical specimen used for the test having diameter 3.7 mm and height 10.2 mm. Mixture filled in mold with three equal lays of having each layer carried 25 numbers of blows. After compaction, the sample is taken in extruder of cylindrical stainless tube from mold and placed in load frame machine. The force given should be produce rate of axial strain 0.5–2% per minute. Each test was carried out by at least three samples to minimize error, and average values were used. The tests were done with soil, and soil with nano-silica.

**Specimen preparation for California bearing ratio test**

Samples for CBR test were prepared as per IS 2720:(part 16)-1987.The mold used for the test have the diameter of 150 mm and height 175 mm. The samples were compacted in 5 layers, and each layer carried 55 numbers of blows by 4.5 kg weight rammer with free fall of 450 mm. For soaked CBR test, specimen kept submerged in water for 96 h before testing. Tests were conducted at a rate of penetration of 1.25 mm/min until 12.5 mm penetration. From the test results, the load-penetration curve were plotted and CBR value calculated. By using different percentages of nano-silica with soil, both soaked and unsoaked CBR tests were conducted.

## 3 Results and Discussion

A series of experimental work has been performed with combination of nano-silica in natural soil. There are very limited literature on studies of use of nano-silica on strength of soil. Results of experiments such as Atterberg's limits, compaction test, unconfined compressive strength test, CBR test (both soaked and unsoaked) of admixture mixed soil are compared with values of untreated soil. The consequences of nano-silica on various engineering properties were evaluated. The interaction at the interface between nano-silica and soil matrix is examined by using scanning electron microscopy (SEM). EDS test shows the chemical composition of materials.

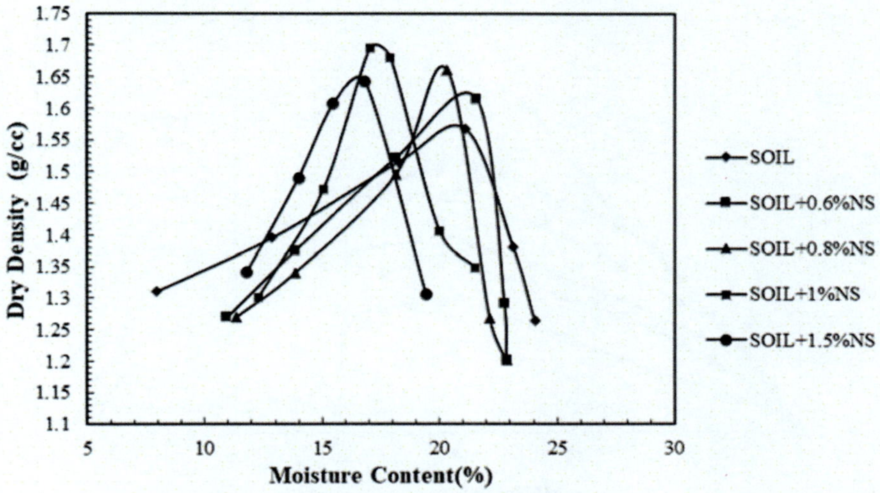


**Fig. 1** Variation of compaction curve of soil using different proportion of nano-silica

## *3.1 Compaction Test*

**Effect of nano-silica on compaction characteristics of soil**: The compaction test on soil treated with varying percentage of nano-silica is done by conducting standard proctor test result graphs that have been plotted and are shown in Fig. 1. It is observed that relative proportion of nano-silica has considerable effects on the OMC and MDD of compacted mix. Increase in nano-silica content from 0 to 1.5%, the MDD value increases and then decreases at a certain point. And the OMC value decreases with higher percentages of nano-silica. In addition of 1% of nano-silica with soil, the OMC value decreased to 16.1 and MDD value increased to 1.71 g/cc.

## *3.2 Unconfined Compressive Strength Test*

**Effect of nano-silica on UCS of soil**

The UCS test was conducted for nano-silica mixed soil, and the stress–strain results are shown in Fig. 2. Result from the study shows that, by increasing nano-silica percentage, the UCS value increases. So the optimum nano-silica content is found 1.5%, and corresponding UCS value is 208.17 KN/m$^2$. The UCS of optimum nano-silica of 1.5% stabilized soil increased by factor 2.58 when compared with that of natural soil.

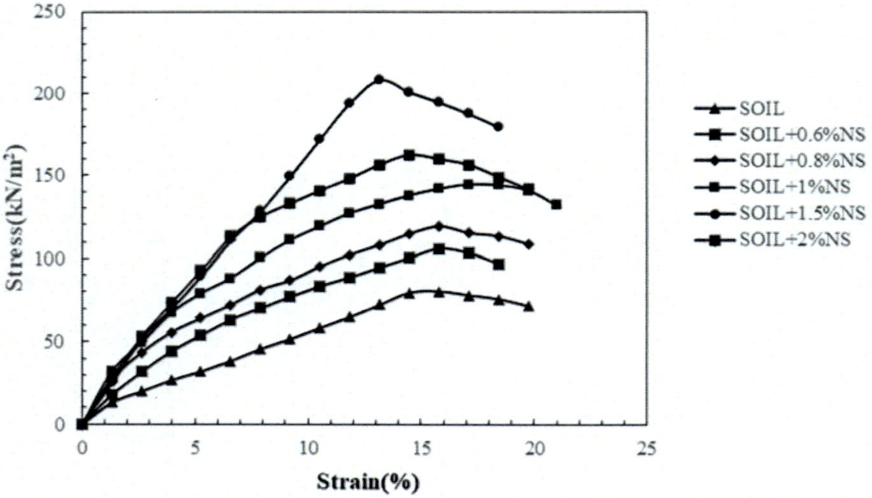


**Fig. 2** Stress–strain curve of soil using different content of nano-silica

## 3.3 *California Bearing Ratio Test (CBR)*

**Effect of Nano-silica on CBR of soil**

**Soaked CBR test**

CBR value is important engineering parameter to evaluate a sub-base and "sub-grade material for design of pavement." Results shows that addition of nano-silica increases the soaked CBR up to nano-silica content 1.0%, then the CBR value decreases. The soaked CBR value increases from 1.81 to 3.03%, when 1.0% nano-silica mixed with soil. Thus, maximum percentage of nano-silica for soaked CBR is 1.0%.The load-penetration curve for soaked soil sample treated with different percentages of nano-silica are shown in Figs. 3 and 4.

**Unsoaked CBR test**

The unsoaked CBR value test results of soft soil treated with different percentage of nano-silica are shown in Figs. 5 and 6. In addition of different percentages of nano-silica, increasing nano-silica content significantly increased the unsoaked CBR value when compared to virgin soil. The CBR value increases from 3.46 to 9.32%, when 1.0% nano-silica was added.

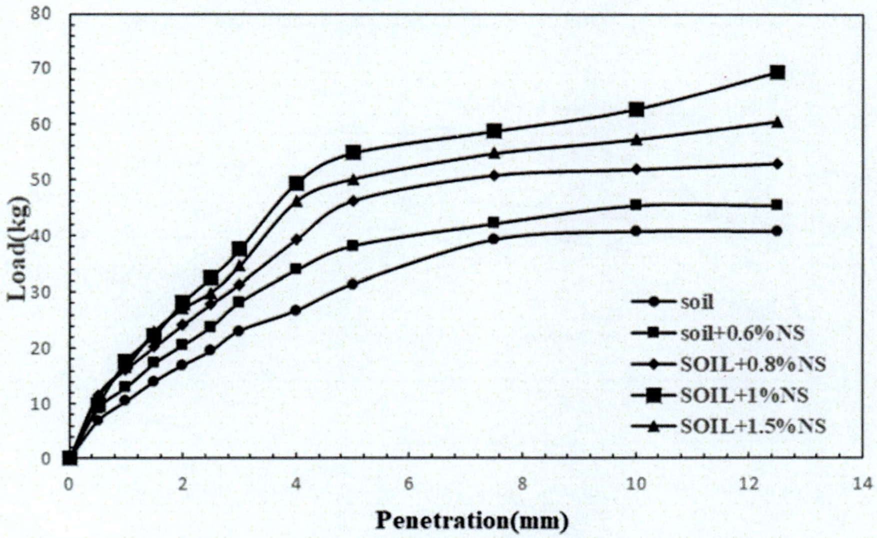


**Fig. 3** Load-penetration curve for nano-silica reinforced soil obtained from soaked CBR test

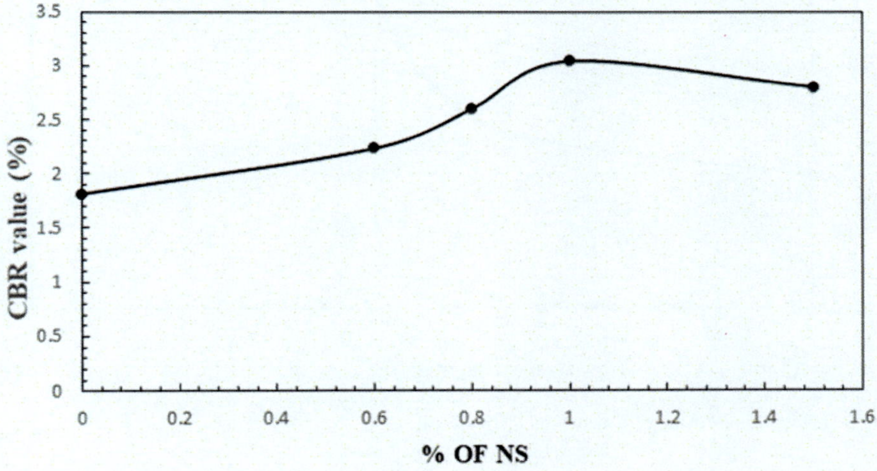


**Fig. 4** Effect of nano-silica on soaked CBR of soil

## *3.4 Scanning Electron Microscope (SEM) Test*

In this, samples "were carried out for" SEM test, i.e., untreated natural "soil, soil treated with" 1% nano-silica which is shown in Fig. 7a, b. Examining SEM image reveals the individual particle's surface texture and morphology. According to the results, adding nano-silica to soil may help to improve the shearing resistance. When water is added with nano-soil, composite viscous gel occurs because of mixing water

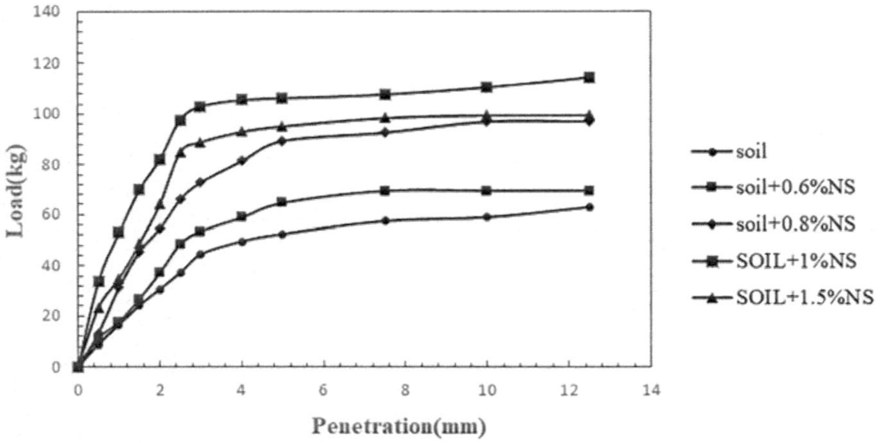


**Fig. 5** Load-penetration curve for nano-silica reinforced soil obtained from unsoaked CBR

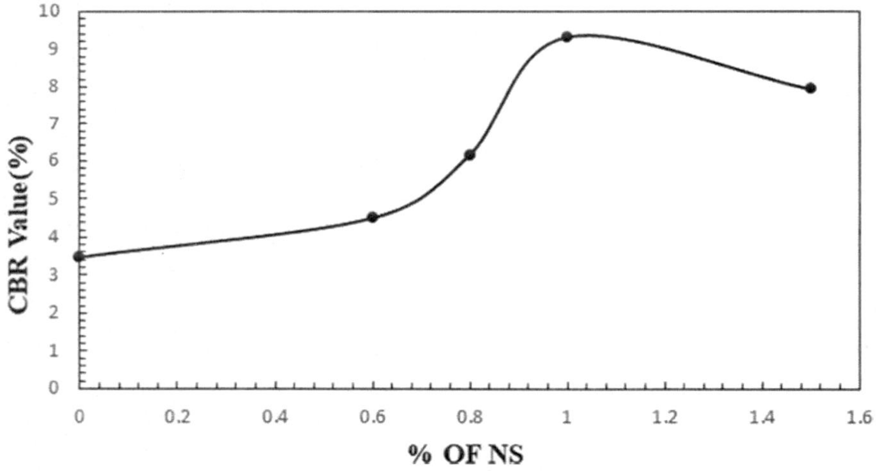


**Fig.6** Effect of nano silica on unsoaked CBR of soil

with nano-silica frictional strength between soil particles enhance due to presence of viscous gel. The nano-soil makes the soil particles distances lesser and produces denser soil matrix. It means increasing nano-soil results improve effective interfacial contact area between soil contents and enhance the interfacial friction and bond strength.

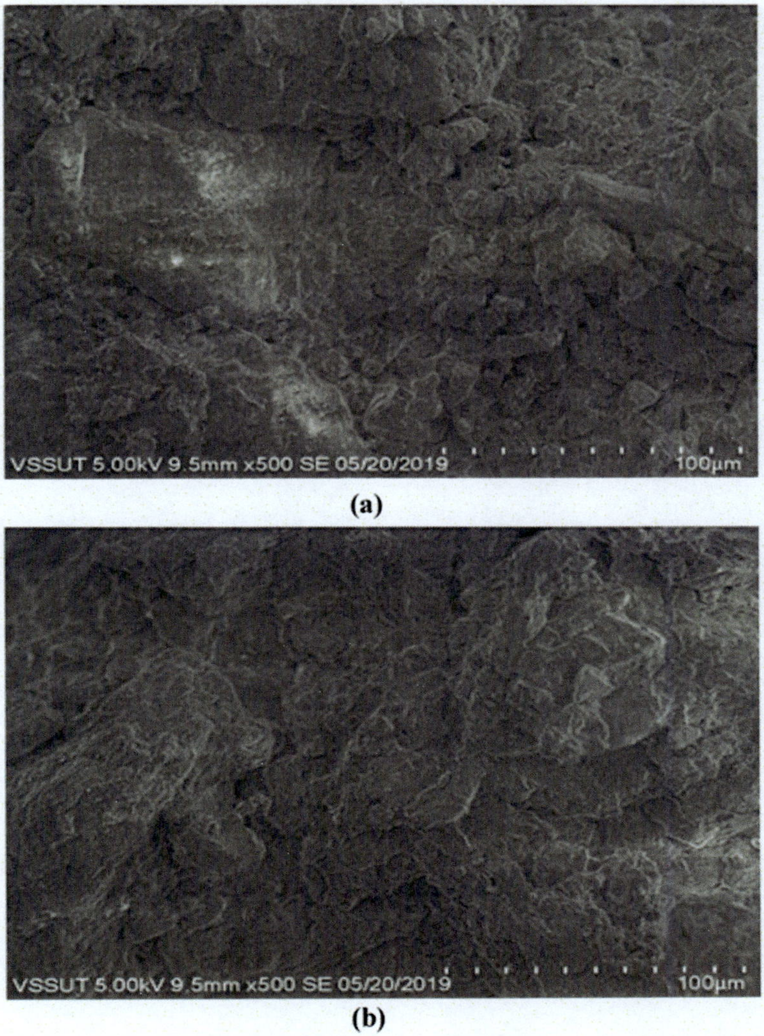


**Fig. 7** SEM images of soil specimens **a** untreated soil, **b** soil stabilized with nano-silica

## *3.5 Energy-Dispersive X-Ray Spectroscopy (EDS)*

Figure 8a, b shows the EDS of soil, soil + Nano-silica, samples. From EDS analysis, it is observed that the component of soil nearly equal to the components of soil with nano-silica mixture, it means nano-silica cannot do any chemical reaction with soil EDS test which is used for analysis of elements and chemical characterization of a sample. It depends on an interaction of few source of X-ray excitation and sample.

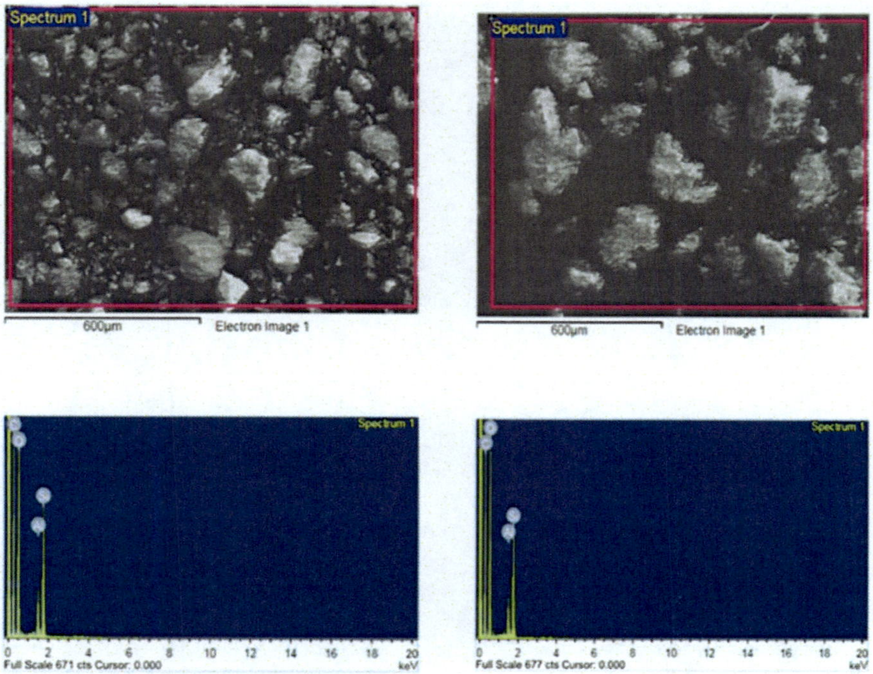


| Elem... | Weight% | Atomic% |
|---|---|---|
| C K | 43.43 | 51.81 |
| O K | 50.04 | 44.82 |
| Al K | 1.69 | 0.90 |
| Si K | 4.85 | 2.47 |
| Totals | 100.00 | |

| Elem... | Weight% | Atomic% |
|---|---|---|
| C K | 45.61 | 53.89 |
| O K | 48.71 | 43.21 |
| Al K | 1.45 | 0.76 |
| Si K | 4.24 | 2.14 |
| Totals100.00 | | |

**Fig. 8** **a** EDS of soil sample. **b** EDS of soil + Nano-silica

The basic study of this analysis is to know that each element has a unique atomic structure allowing unique set of peaks in its X-ray emission spectrum.

## 4 Conclusions

Nowadays, analysis on the application of nano-materials has become a large topic follow from economic advantage. To review the result of nano-material on engineering properties of soil, an experimental program was carried out by adding nano-silica on the soft soil. The following conclusions can be taken from the experimental outcomes.

1. "The MDD values of soft soil initially increased with addition of" nano-silica then decreased after optimum. "The MDD value" increases up to 1% of nano-silica; thereafter, its value consistently decreased. In the other hand, OMC has continuously decreased" with by using nano-silica in soil.
2. The UCS values of soft soil without curing increases up to 1.5% of nano-silica and after that decreases.
3. Both the CBR soaked and unsoaked values increase with use of up to 1% nano-silica. Soaked CBR value gives better result for both sub-grade and pavement construction.
4. Inclusion of nano-silica to the soil, it showed formation of a very dense matrix SEM image in which pores were filled to a large extent and interlock between the particles increases which helps to improve the strength of soft soil.

"In this paper, it is noticed that the method of soil stabilized with nano-silica and is a considerably applied method of ground improvement, which increases the shear strength, UCS, and both soaked and unsoaked CBR of soil. Because of this, it increases the stability of structures, i.e., foundation and roadbed. This improvement technique can be considered as a practical method for improvement of mechanical behaviors of soil in civil engineering project."

## References


1. Pham, H., Nguyen, Q.P.: Effect of silica nanoparticles on clay swelling and aqueous stability of nanoparticle dispersions. J. Nanopart. Res. **16**(1), 2137 (2014)
2. Mohammadi, M., Niazian, M.: Investigation of nano-clay effect on geotechnical properties of Rasht clay. J. Adv. Sci. Technol. **3**(3), 37–46 (2013)
3. Changizi, A.H.: Strength properties of soft clay treated with mixture of nano-SiO2 and recycled polyester fiber. J. Rock Mech. Geotech. Eng. (2015)
4. Moradi, G., Seyedi, S.: Effect of sampling method on strength of stabilized silty sands with colloidal nano silica. J. Civ. Eng. Res. **5**(6), 129–135 (2015)
5. Pashabavandpouri, S.J.: Effect of nano silica on swelling, compaction and strength properties of clayey soil stabilized with lime. J. Appl. Environ. Bio. Sci. **5**(7S), 538–548 (2015). ISSN: 2090-4274
6. Mostafa, A., Ouf, M.S., Elgendy, F.E.: Stabilization of sub grade pavement layer using silica fume and nano silica. Int. J. Sci. Eng. Res. **7**(3) (2016)
7. Kumar, S.A., Manikandan, R.: Influence of nanosized additives on the improvement of clay soil. Int. J. Adv. Sci. Eng. Res. **1**(1) (2016)
8. Alireza, S.G.S., Mohammad, M.S., Hasan, B.H.: Application of nanomaterial to stabilize a weak soil. In: International Conference on Case Histories in Geotechnical Engineering, Paper 5 (2013)
9. ASTM D2487: Standard practice for classification of soils for engineering purposes (unified soil classification system)
10. Rong, M.Z., Zhang, M.Q., et al.: Analysis of the interfacial interactions in polypropylene/silica nanocomposites. Polymer International (2003)

# Analysis of Desiccation Crack Patterns of Expansive Soil Treated with Lignosulphonate and Lime

**G. Landlin and S. Bhuvaneshwari**

## 1 Introduction

Cracking phenomenon of soil is a complex phenomenon and influenced by many parameters which include soil mineralogy, temperature variations, thickness of soil layer [1, 2]. The alternate wetting and drying of soils, physiochemical properties of the soil also affect the cracking pattern of the soil. Previous studies have indicated that the soil-crack morphology can significantly influence the inherent properties of the soil [3, 4]. The desiccation cracks formed on the upper surface of the soil can influence the properties of the soil related to the geotechnical, hydraulic, and geo-environmental applications [5]. Cracking patterns can develop through the alterations in moisture content, stiffness and tensile strength of the soil [6]. Cracks are developed on the soil surface as the result of the drying mechanism related to the evaporation of the water from the exposed soil surface, largely related to the volumetric shrinkage of the soil. When the threshold state is reached, desiccation cracks occur in the soil and manifest on the soil surface [7]. Crack propagation is also related to the tensile stresses developed during the water evaporation. When the generated tensile stresses exceed the inherent tensile strength of the soil, the cracks can propagate and weaken the soil [1, 8].

Cracking of soil initiates weaker soil zones and can lead to stability issues in dams, tailing ponds, earthen embankments, landfill, liners etc. [5, 9]. The higher hydraulic conductivity encountered due to the crack formation can affect the performance of the contaminant barriers very severely [10]. Stability of natural slopes and vertical cuttings and bearing capacity of the foundation are severely affected by moisture

G. Landlin (✉) · S. Bhuvaneshwari
SRM Institute of Science and Technology, Kattankulathur 603202, India
e-mail: landling@srmist.edu.in

S. Bhuvaneshwari
e-mail: bhuvanes1@srmist.edu.in

 
C. N. V. Satyanarayana Reddy et al. (eds.), *Ground Characterization and Foundations*, Lecture Notes in Civil Engineering 167,
https://doi.org/10.1007/978-981-16-3383-6_30

ingress due to the cracking behaviour. The crack propagation could be effectively reduced by increasing the soil strength and treating the soil with suitable additives like lime, cement and sand [11]. The morphological characteristics of the soil-crack pattern such as the crack area density, crack length and width are some of the parameters which help in understanding the connectivity of cracks [12–14]. The evaluation and analysis of these parameters can help in understanding the variation in the propagation of the crack pattern and characterization of the soil cracking behaviour and its influence on the geotechnical and geoenvironmental applications. The crack patterns and its propagation can be effectively studied by adopting imaging techniques [5, 15]. Further, numerical analysis has also been carried out for the study of crack pattern in soil [10, 16, 17].

In the present study, the crack pattern analysis has been carried out for the potentially swelling soil and additives amended soil. The conventional alkaline additive, lime and the biopolymer-based additive lignosulphonate had been amended to the soil and a comparative analysis is carried to understand the variation in the propagation of the crack pattern.

## 2 Materials and Methodology

### 2.1 Soil

The soil used for the present study is an expansive soil which was collected from depths of 2–3 m in Siruseri, Chennai. The classification and properties of the expansive soil are shown in Table 1. The index properties of the soil were evaluated based on Indian standards IS 2720 [18]. Based on the evaluated plasticity characteristics, the soil is classified as inorganic clay of high plasticity as per IS 1498 [19]. Based on the values of free swell index (FSI) and swell potential, the soil is classified as

**Table 1** Properties of soil

| S. No. | Property | Values |
|---|---|---|
| 1 | Specific gravity | 2.68 |
| 2 | Liquid limit | 64.08% |
| 3 | Plastic limit | 27.8% |
| 4 | Plasticity index | 36.28% |
| 5 | Shrinkage limit | 10.37% |
| 6 | Clay | 64% |
| | Silt | 27% |
| | Sand | 9% |
| | Gravel | 0% |
| 7 | Differential Free swell % | 111% |

highly expansive category [20]. The expansive soil is amended with lignosulphonate, a plant-based biopolymer and conventional additive lime. Calcium lignosulphonate (LS) is a byproduct of paper industry. Both additives were procured commercially.

## 2.2 Methodology

The crack patterns were observed based on the procedure given by Fang and Daniels [21]. Approximately 100 g of soil passing through 425 μm sieve is made into a paste with water content of 1.5 times the liquid limit. The wet soil paste is then spread on clean glass plate, such that the diameter is 110 mm and the maximum height of the wet mud pad does not exceed 1.27 mm as distinct crack patterns develop in thinner samples. The glass plate is left for air drying at room temperature of 28 °C. Similar procedure was carried out for both the treated and untreated soil samples. The crack patterns were observed for untreated soil (*S*), 0.5, 1.5, 3% LS treated sample (S0.5LS, S1.5LS, S3LS respectively) and 2, 4% lime treated sample (S2L, S4L).

In the present study, images were taken during 24 h (*t*1) and 72 h (*t*2) to analyze the crack patterns based on crack initiation to the complete dried state. The images were captured by DSLR camera fixed at 216 mm from the sample. The images of untreated soil taken during the first and third day of crack propagations are shown in Fig. 1. The images of LS treated and lime treated samples are shown in Figs. 2 and 3 respectively.

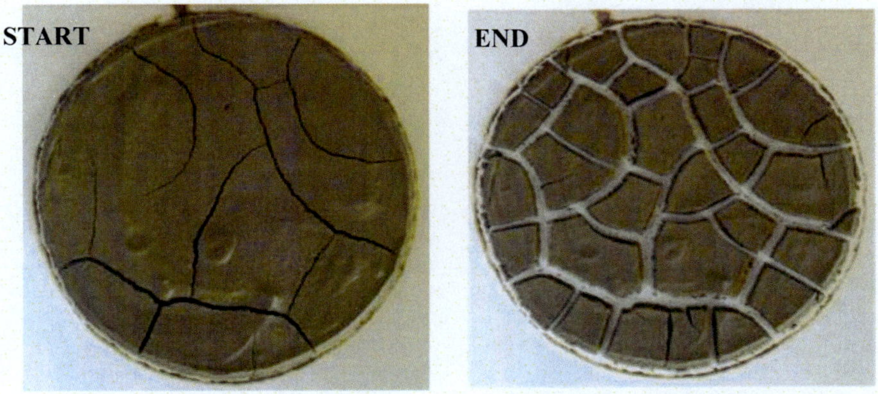


(a) S-t1 (b) S-t2

**Fig. 1** Start and end of crack propagation of untreated soil

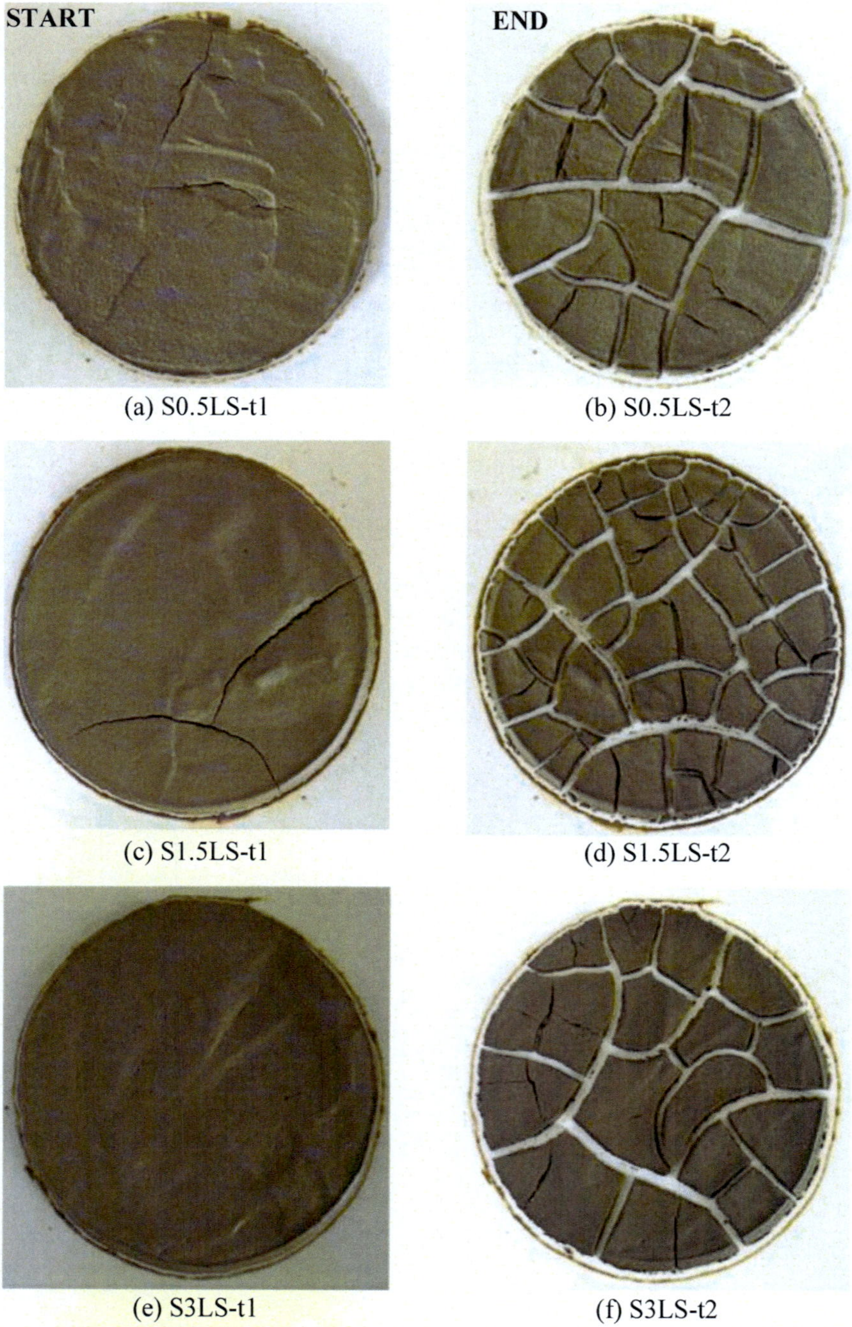


**Fig. 2** Start and end of crack propagation of lignosulphonate treated samples

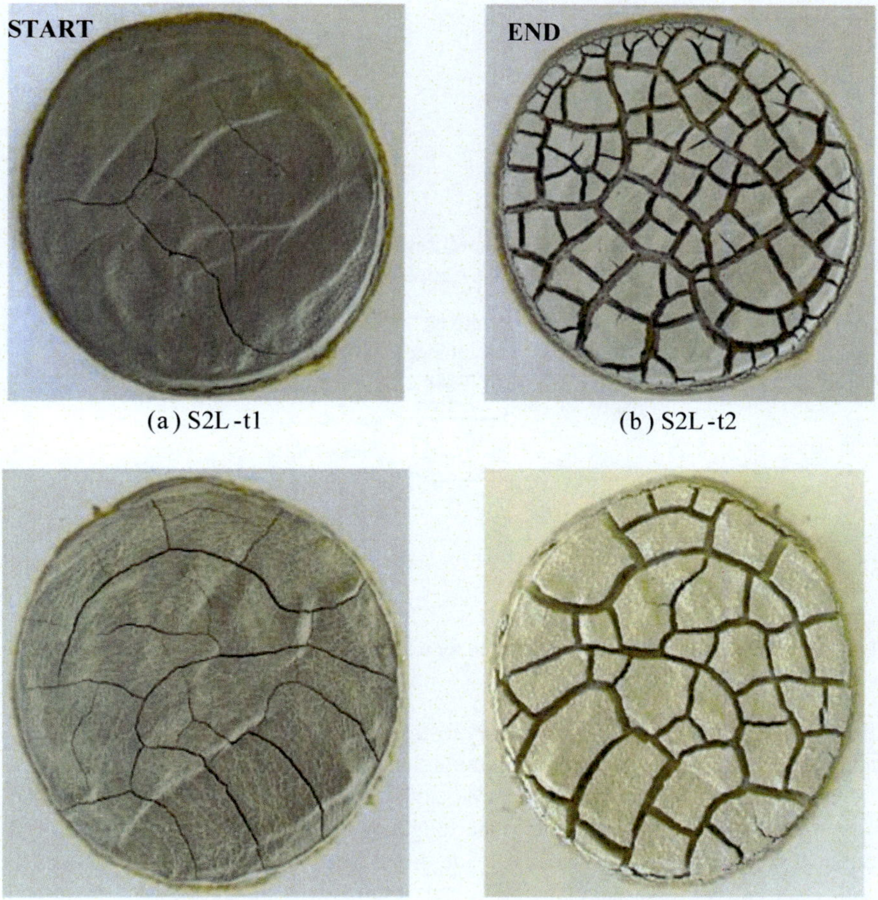


(a) S2L -t1

(b) S2L -t2

(c) S4L -t1

(d) S4L -t2

**Fig. 3** Start and end of crack propagation of lime treated samples

## 2.3 Image Analysis

The digital images were analyzed further for the crack morphology. Preprocessing the image was done in ImageJ software. To quantify the cracks in the image, a software called particle and crack analysis system (PCAS) was used. The crack patterns were analyzed for the crack morphology based on the methodology given in Fig. 4.

Using ImageJ, the noise in the image is reduced and scale (mm/pixel) of the image was calibrated. Then the image was converted into an 8-bit image (gray scale image). The gray scale image is turned into binary image using a threshold tool. The presence of white and black dots spread near the cracks due to the shadows and lighting. Were removed by using the erode and dilate tool. The binary image is cropped into a square

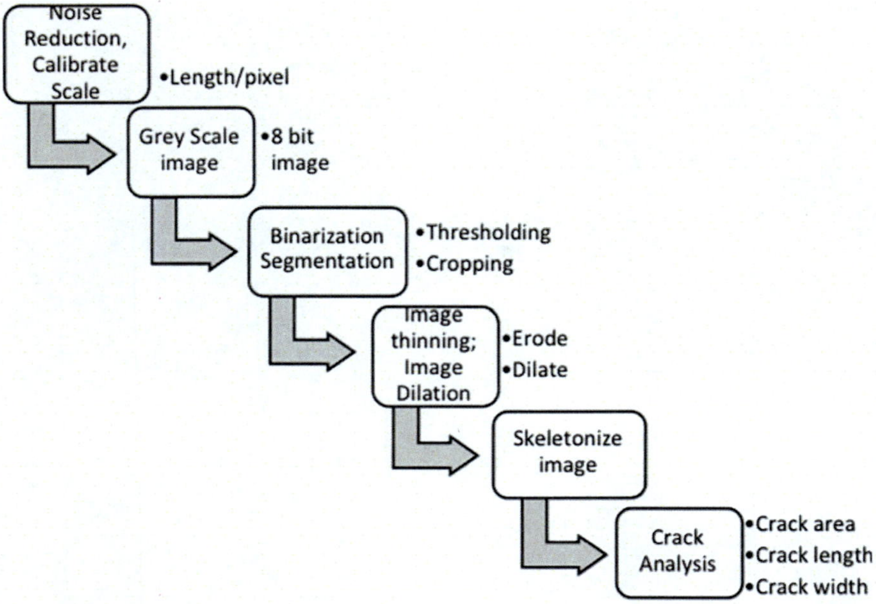


**Fig. 4** Schematic of the methodology adopted for the image processing of the crack patterns

image. This clearly defines the boundary of the cracks. The typical grayscale, binary and cropped image of the untreated sample St1 is shown in Fig. 5.

After the preprocessing, the image is skeletonized as shown in Fig. 5e. PCAS software is used to further quantify the length, width and area of the cracks. The boundary of every individual crack was identified (see Fig. 5f) and the length of each crack intersections was calculated. The maximum width at 100 random pixels on each crack was acquired and the area of the cracks was computed. All the values derived were in the unit of pixel. The derived values were divided by the calibrated scale to convert the values in mm.

## 3 Crack Pattern Analysis

The treated samples and the untreated samples were analyzed for the variation in the crack morphology. The crack parameters such as the length, width, number of clods ($N_{\mathrm{cl}}$), and crack intensity factor ($I_{\mathrm{cf}}$) were computed and compared for the various samples. The onset of cracks, progression and distribution of cracks were analyzed based on the analytic measurements. The various crack morphological parameters for both treated and untreated soil are depicted in Table 2.

At the end of 24 h ($t1$), the cracks for untreated soil are initiated, as could be observed from Fig. 1a and the morphological parameters listed in Table 2. The

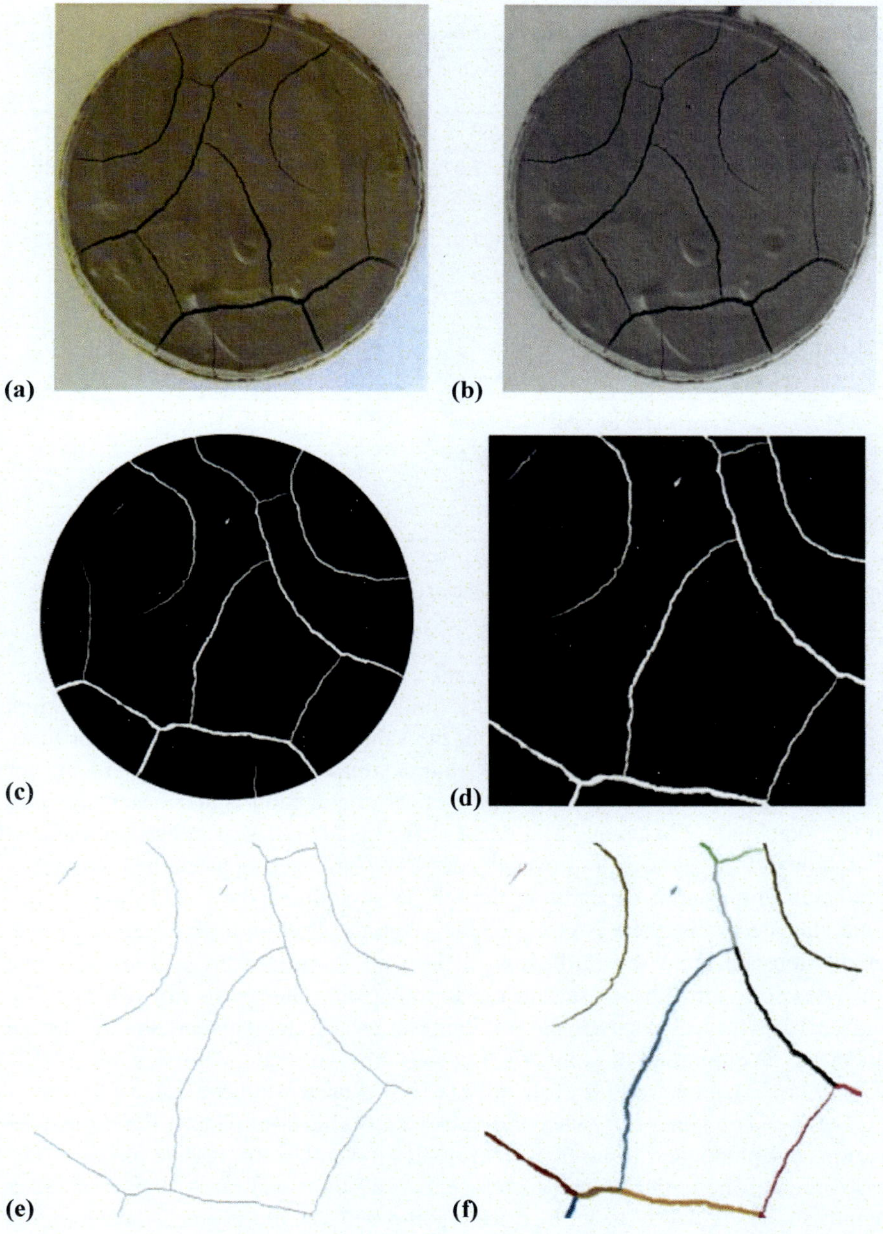

Fig. 5 Typical image analysis of the crack pattern of S-t1. **a** captured image, **b** gray scale image, **c** binary image, **d** cropped image, **e** skeletonized image, **f** crack identified image

**Table 2** Crack parameters of untreated and treated samples

| Sample designation | $N_{cl}$ | $N_c$ | $A_{cl}$ (mm$^2$) | $A_c$ (mm$^2$) | $L_c$ (mm) | $W_c$ (mm) | Avg. $W_c$ (mm) | $I_{cf}$ (%) |
|---|---|---|---|---|---|---|---|---|
| St1 | 8 | 15 | 5484.89 | 209.74 | 327.64 | 0.89 | 0.61 | 3.82 |
| St2 | 25 | 49 | 3463,94 | 2142.07 | 542.03 | 6.12 | 4.02 | 61.84 |
| S0.5LSt1 | 8 | 15 | 5077.82 | 192.16 | 243.14 | 1.14 | 0.73 | 3.78 |
| S0.5LSt2 | 20 | 38 | 3529.46 | 1427.01 | 503.24 | 4.6 | 2.89 | 40.43 |
| S1.5LSt1 | 1 | 3 | 6438.65 | 79.2 | 127.29 | 0.68 | 0.57 | 1.23 |
| S1.5LSt2 | 36 | 79 | 3606.60 | 1977.33 | 756.22 | 5.87 | 2.93 | 54.83 |
| S3LSt1 | 1 | 6 | 5551.02 | 4.12 | 16.57 | 0.57 | 0.28 | 0.07 |
| S3LSt2 | 19 | 42 | 4006.48 | 1370.82 | 526.43 | 5.06 | 2.78 | 34.22 |
| S2Lt1 | 2 | 7 | 6168.96 | 60.07 | 200.58 | 0.37 | 0.27 | 0.97 |
| S2Lt2 | 55 | 110 | 3878.99 | 2348.99 | 974.57 | 4.56 | 2.43 | 60.56 |
| S4Lt1 | 16 | 50 | 5927.39 | 247.86 | 572.82 | 0.75 | 0.38 | 4.18 |
| S4Lt2 | 26 | 50 | 4083.68 | 1262.75 | 526.42 | 3.21 | 2.08 | 30.92 |

$N_{cl}$ no. of clods, $N_c$ no. of cracks, $A_{cl}$ area of clods, $A_c$ area of cracks, $L_c$ length of cracks, $W_c$ maximum width of crack, $I_{cf}$ crack intensity factor

number of crack segments (8) and the crack area (209 mm$^2$) is comparatively higher than the lignosulphonate amended soils which depict a very slow crack initiation phenomenon as could be observed from the values of Table 1. However, for lime-treated soil, a lesser percentage (2%) behaves similar to LS amended soil. But for the higher percentage of lime, (4%), the crack propagation is at a faster rate with higher crack area (247 mm$^2$) and crack segments (16). These inferences indicate that, though lime addition improves the basic and engineering properties of the soil, the crack propagation depends on the tensile strength of the soil. In the case of lime amended soils, there is considerable alkalinity leading to flocculated particle orientation and increased brittleness of the soil. These properties make the lime amended soil vulnerable to surface crack propagation during the dry seasons [22]. Figure 6 depicts the changes in the crack area and crack segments for the soil samples under the time duration of $t1$ and $t2$. To cover the whole range of crack area (4.12–2348.99 mm$^2$), a logarithmic scale of base 10 was used to illustrate $A_c$.

At the end of 3 days ($t2$), when the crack patterns have sufficiently developed, the various morphological features of test samples indicate a progressive increase. The crack area of the untreated soil increases to 2147 mm$^2$ and crack segments, crack length and crack width depict a corresponding increase at the end 3 days duration ($t2$). However, for the 2% lime amended samples, the crack area and segments are higher compared to 4% lime. At lower contents of lime, there is a complete ion exchange reaction, which imparts a more flocculated structure and hence numerous small crack segments at the periphery. But in the case of 4% lime amendment, there is a more predominant cementation reaction due to hydration, indicating larger segments comparatively.

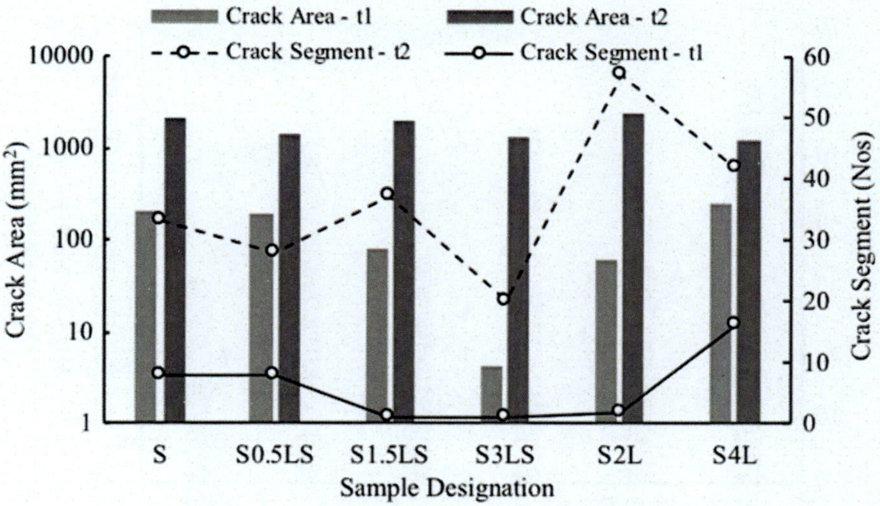


**Fig. 6** Variation of crack area and crack segments for treated and untreated samples

The entire changes in the crack morphology could be better indicated by a more collective parameter, crack intensity factor, $I_{cf}$ expressed as a percentage. The crack intensity factor is depicted in Figs. 7 and 8 for the time duration of $t1$ and $t2$. The untreated soil is prone to excessive cracks, as indicated by the $I_{cf}$ value of 3.82% at t1 time duration and 61.84% at $t2$-time duration. The soil is basically highly expansive

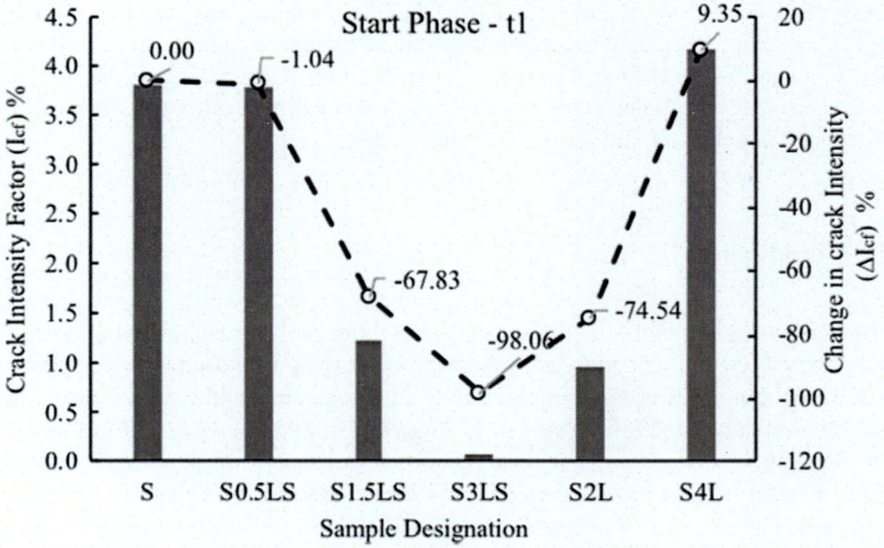


**Fig. 7** Variation of crack intensity factor and change in crack intensity for treated and untreated soil at $t1$-time duration

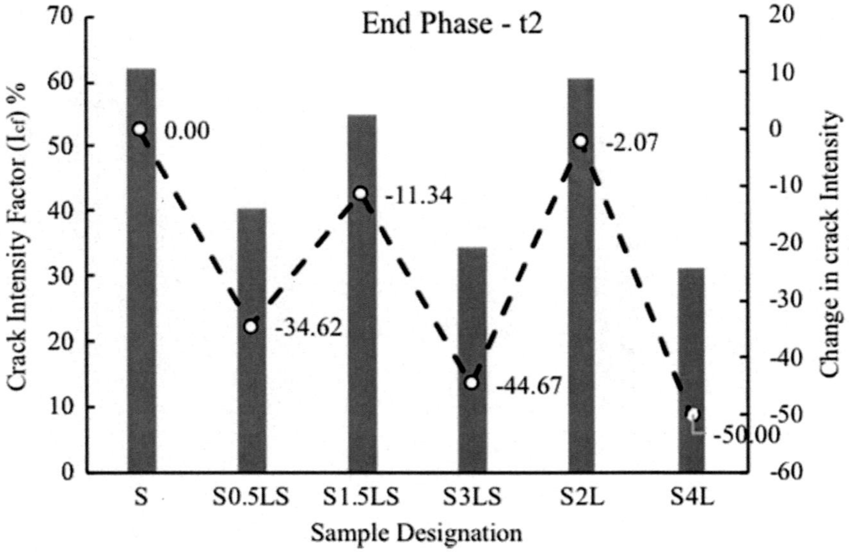


**Fig. 8** Variation of crack intensity factor and change in crack intensity for treated and untreated soil at *t*2-time duration

in nature as indicated by its index properties (Table 1) and the excessive crack propagation is reasonably expected. The increased percentage of LS amended soil depicts a slow crack propagation as indicated by the $I_{cf}$ values and $\Delta I_{cf}$ values. This could be attributed to a more aggregated particle orientation, lesser rate of water drying and higher inherent tensile strength, due to the dense bonding between LS additive and soil particles. The lime-treated soils are characterized by high brittleness which plays a major role in the crack propagation mechanism. The cracking phenomenon could be majorly attributed to the tensile strength development and shrinkage pattern of the soil, which is the further scope of this study.

## 4 Conclusions

The present study mainly focuses on the crack propagation phenomenon of potentially expansive soil and additives amended composites. Lignosulphonate a plant-based biopolymer and conventional additive, lime is adopted for the study. The crack patterns were analyzed for the various changes in the morphological features based on the image analysis tool. The results depict a reasonable variation between the crack patterns of the untreated soil, LS and lime amended soil. The lignosulphonate addition, imparts particle aggregation and higher moisture retention capacity for the treated soil which slows the rate of cracking and depicts a lesser cracked area compared to untreated soil. However, for the lime amended soil, the crack propagation is higher due to the increased flocculated orientation of the soil particles and increased brittleness of the composites. Nevertheless, these inferences have to be substantiated by the tensile strength variation and shrinkage values of the treated and untreated soil, which is the further scope of the study.

## References


1. Tang, C., Shi, B., Liu, C., Gao, L., Inyang, H.I.: Experimental investigation of the desiccation cracking behavior of soil layers during drying. J. Mater. Civ. Eng. 873–878 (2011)
2. Albrecht, B.A., Benson, C.H.: Effect of desiccation on compacted natural clays. J. Geotech. Geoenviron. Eng. 67–75 (2001)
3. Vogel, H.J., Hoffmann, H., Leopold, A., Roth, K.: Studies of crack dynamics in clay soil II. A physically based model for crack formation. Geoderma **125**, 213–223 (2005)
4. Uday, K.V., Singh, D.N.: Investigation on cracking characteristics of fine-grained soils under varied environmental conditions. Dry. Technol. **31**, 1255–1266 (2013)
5. Atique, A., Sanchez, M.: Analysis of cracking behavior of drying soil. In: 2nd International Conference on Environmental Science and Technology. IPCBEE, vol. 6, pp. 66–70 (2011)
6. Tang, C., Cui, Y., Tang, A., Shi, B.: Experiment evidence on the temperature dependence of desiccation cracking behavior of clayey soils. Eng. Geol. **141**(3–4), 261–266 (2010)
7. Tang, C., Shi, B., Liu, C., Suo, W.B., Gao, L.: Experimental characterization of shrinkage and desiccation cracking in thin clay layer. Appl. Clay Sci. 52(1–2), 69–77 (2011)
8. Peron, H., Huekel, T., Laloui, L., Hu, L.B.: Fundamentals of desiccation cracking of fine grained soils: experimental characterization and mechanics identification. Can. Geotech. J. 46, 1177–1201 (2009)
9. Yesiller, N., Miller, C.J., Inci, G., Yaldo, K.: Desiccation and cracking behavior of three compacted landfill liner soils. Eng. Geol. 57, 105–121 (2000)
10. Miler, C.J., Mi, H., Yesiler, N.: Experimental analysis of desication crack propagation in clay liners. J. Am. Water Res. Assoc. **34–3**, 677–686 (1998)
11. Leung, M., Vipulanandan, C.: Treating contaminated, cracked and permeable field clay with grouts. In; Proceedings of the Specialty Conference on Geotechnical Practice in Waste Disposal, pp. 829–843. Geotechnical Special Publication No. 13. ASCE, New York (1995)
12. Xiong, D., Lu, X., Xian, J., Zhou, H., Yang, Z.: Selection of judging indicators for surface morphology of soil crack under different development degrees in Yuanmou Arid-hot valley region. Wuhan Univ. J. Nat. Sci. **13**(3), 363–368 (2008)
13. Xiong, D., Long, Y., Yan, D., Lu, X., Ji, Z., Fang, H.: Surface morphology of soil cracks in Yuanmou Dry-hot Valley Region, Southwest China. J. Mt. Sci. **6**(4), 373–379 (2009)
14. Vogel, H., Hoffmann, H., Roth, K.: Studies of crack dynamics in clay soil experimental methods, results, and morphological quantification. Geoderma **125**(3–4), 203–211 (2005)
15. Tang, C., Shi, B., Cui, Y., Liu, C., Gu, K.: Desiccation cracking behavior of polypropylene fiber–reinforced clayey soil. Can. Geotech. J. **49**(9), 1088–1101 (2012)
16. Nahlawi, H., Kodikara, J.K.: Laboratory experiments on desiccation cracking of thin soil layers. Geotech. Geol. Eng. **24**, 1641–1664 (2006)
17. Tang, C.S., Shi, B., Gao, W., Chen, F., Cai, Y.: Strength and mechanical behavior of short polypropylene fiber reinforced and cement stabilised clayey soil. Geotext. Geomembr. **25**(3), 194–202 (2007)
18. IS 2720: Compendium of Indian standards on soil engineering. Bureau of Indian Standards, New Delhi (1989)
19. IS: 1498: Classification and identification of soils for general engineering purpose (1970)

20. Sridharan, A.: Classification procedures for Expansive soils. Proc. Inst. Civ. Eng. Geotech. Eng. **143**, 235–240 (2000)
21. Fang, H.Y., Daniel, J.N.: Introductory Geotechnical Engineering. Taylor and Francis, London (2006)
22. Kamruzzaman, A., Chew, S., Lee, F.: Structuration and destructuration behavior of cement-treated singapore marine clay. J. Geotech. Geoenviron. Eng. **135**(4), 573–589 (2009)

# Laboratory Assessment of Clogging Potential Using Soil Drilling Test

**Arya S. Babu and M. K. Sayida**

## 1 Introduction

The increasing demands of the growing population for buildings, transportation and other utilizes have triggered the need for tunnelling as well as other soil excavation activities. Clogging is one of the major troubles encountered by the various machines involved in the soil boring works [1]. It leads to economic losses, extension of span of the work and even standstills in the project [2]. Clogging is a phenomenon where the parts of the equipment get plugged by the soil. The equipment irrespective of their size, i.e. a small hammer used for compaction to massive tunnel boring machines, is subjected to clogging. Clogging potential is the property of a soil by virtue of which the soil adheres to a foreign surface and thereby clog an equipment.

The process of clogging involves a combination of the mechanisms of cohesion and adhesion of the soil. The clogging potential of a soil can be evaluated by analytical approach, semi-empirical approach or physical stimulation approach. In analytical approach, the adhesive and cohesive forces are determined to indirectly predict the clogging potential. The conventional as well as modified vane shear, direct shear apparatus [1, 3, 4], and other tests using pistons [5, 6], cones, etc., were performed to assess the adhesive and cohesive forces in soil. The semi-empirical approach makes use of water content and the Atterberg limits data of a soil to predict the clogging behaviour of the soil. A clogging potential diagram was developed by Thewes [2], and this diagram was further updated by Hollman and Thewes [4]. The diagram helps in clogging classification based on the liquid limit, plastic limit and water content of a soil [2, 7].

For physical simulation approach, the actual drilling process in the field is stimulated in the laboratory. Sass and Burbaum [8] developed an apparatus in the laboratory

A. S. Babu (✉) · M. K. Sayida
College of Engineering Trivandrum, Trivandrum, Kerala 695016, India



C. N. V. Satyanarayana Reddy et al. (eds.), *Ground Characterization and Foundations*, Lecture Notes in Civil Engineering 167,
https://doi.org/10.1007/978-981-16-3383-6_31

to replicate the conditions of the cutting wheel of a tunnel boring machine. Feinendegen et al. [9] devised an equipment and performed a cone pull-out test to evaluate the clogging potential. Zumsteg and Puzrin [10] used a Hobart mortar mixture and a novel shear plate device to determine the stickiness potential of clay pastes. Kang et al. [11] devised an equipment which facilitates soil drilling process and studied the effect of rotational velocity, penetration rate and size of drill bit on clogging potential. In the field, certain soil conditioners are applied to modify the soil properties and thereby reduce the clogging risks. The commonly used conditioners are water, foam, polymers, soap solution and salt solutions [1, 11, 12].

In this paper, the clogging potential of Thonnakkal soil is assessed by means of a laboratory soil drilling test. The research focuses on the analysis of influence of bentonite clay, water content and penetration rate of the drill bit on the clogging potential of Thonnakkal soil.

## 2 Experimental Studies

### *2.1 Material*

The soil used for the study was collected from Thonnakkal region of Thiruvananthapuram District, Kerala. The natural water content was determined by conducting the test according to IS 2720 (Part 2) 1973 [13]. The grain size analysis was performed as per IS 2720 (Part 4) 1985 [14]. The soil belongs to clayey sand with silt category. The specific gravity of the soil was tested as per IS 2720 (Part 3) 1980 [15]. The results of various soil tests conducted on Thonnakkal soil are listed out in Table 1. The bentonite clay used for the study was collected from a dealer based on Kutch region, Gujarat. The properties of the sodium bentonite used are presented in Table 2.

**Table 1** Properties of Thonnakkal soil

| Property | Value |
|---|---|
| Natural water content (%) | 20.4 |
| Clay (%) | 48 |
| Silt (%) | 14 |
| Sand (%) | 38 |
| Liquid limit (%) | 46 |
| Plastic limit (%) | 27 |
| Plasticity index (%) | 19 |
| Specific gravity | 2.5 |
| MDD (g/cc) | 1.8 |
| OMC (%) | 17 |

**Table 2** Properties of Bentonite

| Property | Value |
|---|---|
| Liquid limit (%) | 310 |
| Plastic limit (%) | 40 |
| Plasticity index (%) | 270 |
| Specific gravity | 2.86 |
| MDD (g/cc) | 1.22 |
| OMC (%) | 32.6 |

## 2.2 *Testing Apparatus*

Two different tests were performed which are the Atterberg limit tests and laboratory drilling test. The liquid limit was tested within a cone penetrometer test apparatus. In order to conduct the laboratory drilling test, a small-scale soil drilling equipment was fabricated. The photograph of the device is shown in Fig. 1.

The different parts of the equipment are electric power supply unit, drill bit, a rotational velocity controller and a motor. A high carbon steel drill bit with 50 mm diameter and 28 mm height was used. The drill bit is hollow cylindrical with its one end open. The open end of the drill bit has teethed edges. The whole setup is mounted on the loading frame of a tensile testing strength equipment. The penetration rate can be adjusted by changing the gears in the tensile strength testing equipment. The equipment facilitates the change in penetration rate, shape and size of drill bit as well as rotational velocity during the test.

## 2.3 *Methodology*

The dry Thonnakkal soil and bentonite were mixed at various proportions by weight to get soil mixtures M0, M20 and M40. The proportion of the two soils in these mixtures is presented in Table 3. The desired amount of water was added to these dry mixtures to prepare the samples. The Atterberg limit test was carried out as per IS 2720 (Part-5) 1985 [16].

The laboratory drilling test was carried out using a cylindrical mould of 10 cm diameter and 11.5 cm height. The soil sample was filled in the mould as per IS 2720 (Part-7) 1980 [17]. The tests were conducted at three different water content which lies between plastic limit and liquid limit for all the soil mixtures. The penetration rates used in the test are 20, 40 and 80 mm/min, and the rotational velocity was 30 rpm. The mould containing the soil was placed on the apparatus below the drill bit, and the drill bit was allowed to penetrate into the soil. The penetration as well as rotation happens during the working of the machine. The drill bit was pulled out of the soil once a particular depth of penetration was achieved. The drill bit was disconnected from the device, and the soil adhering to the inside of the bit was scrapped out and

**Fig. 1** Laboratory drilling test setup

**Table 3** Proportions of the soil mixture

| Mixture | Thonnakkal soil (%) | Bentonite (%) |
|---|---|---|
| M0 | 100 | 0 |
| M20 | 80 | 20 |
| M40 | 60 | 40 |

weighed. The clogging potential is expressed in terms of a parameter WSDB (weight of soil sticking to drill bit). WSDB is the ratio of the weight of soil sticking to the drill bit to the cross-sectional area of the drill bit.

**Table 4** LL, PL and PI values of soil mixtures

| Mixture | LL (%) | PL (%) | PI (%) |
|---|---|---|---|
| M0 | 46 | 27 | 19 |
| M20 | 105 | 31 | 74 |
| M40 | 154 | 34 | 120 |

# 3 Test Results and Discussion

## *3.1 Atterberg Limit Tests*

The effect of bentonite content on the LL, PL and PI of Thonnakkal soil was studied. The LL, PL and PI of the three soil mixtures are shown in Table 4. A rapid spike in LL happens when bentonite is introduced into Thonnakkal soil. The swelling nature of the bentonite leads to increased water absorption, which in turn favours LL increment. Researches in the past suggested that the soil with higher LL and PI has more chance of clogging. Hence, it can be concluded that bentonite has the tendency to increase the clogging potential of Thonnakkal soil.

## *3.2 Results of Laboratory Drilling Test*

The influence of parameters like water content, penetration rate of drill bit and bentonite content on the clogging potential was analysed.

**Effect of water content on WSDB**

The tests were carried out at three different water contents between PL and LL for all the three mixtures. The analysis was carried out using the parameter consistency index. The consistency index corresponding to a particular water content was calculated as $I_C = (LL - w)/(LL - PL)$, where LL is the liquid limit, $w$ is the water content, and PL is the plastic limit. The water content and corresponding consistency index of the various mixtures are presented in Table 5.

The variations of M0, M20 and M40 mixtures are shown in Figs. 2, 3 and 4. For all the mixtures, WSDB increases then reaches a peak value and then decreases, when consistency index is reduced. The highest consistency value is when water content is near PL while the lowest consistency value occurs when water content is near LL. The peak WSDB value for all the three soil mixtures lies in the consistency index range of 0.4–0.5. The peak value is nearer to the LL, but changes in WSDB values are rapid in between water content corresponding to peak WSDB value and PL. It can be inferred that the clogging can be controlled by maintaining a higher consistency.

**Table 5** Water content and consistency index of various mixtures

| Mixture | Water content (%) | Consistency index |
|---|---|---|
| M0 | 25 | 1.11 |
| | 35 | 0.58 |
| | 45 | 0.05 |
| M20 | 35 | 0.95 |
| | 65 | 0.54 |
| | 95 | 0.14 |
| M40 | 40 | 0.93 |
| | 90 | 0.54 |
| | 140 | 0.12 |

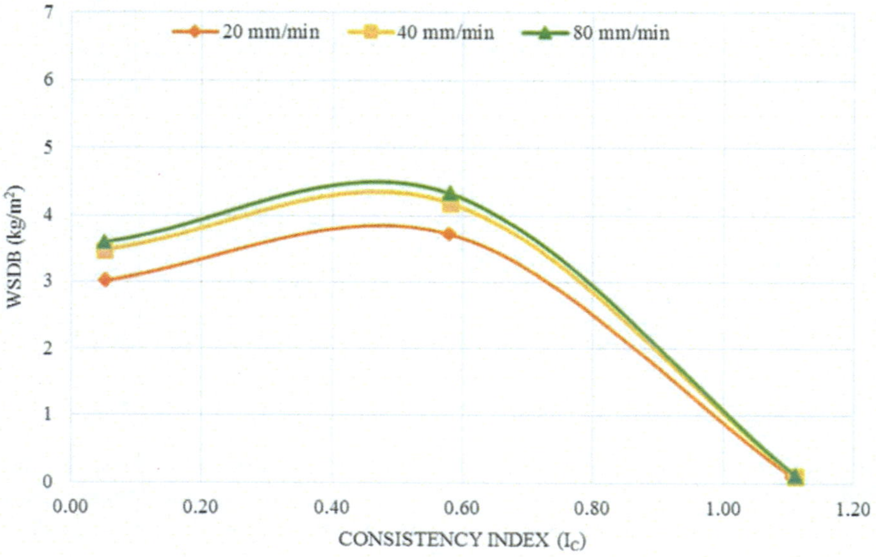


**Fig. 2** Variation in WSDB with consistency index of M0 mixture

## Effect of penetration rate on WSDB

The effect of penetration rate on WSDB of M0 mixture or Thonnakkal soil) is shown in Fig. 5. At $I_C = 0.58$ and 0.05, when the penetration rate is increased from 20 to 80 mm/min, an initial gradual increase followed by a constant WSDB value is observed. But when $I_C = 1.11$, WSDB values is constant (approx. zero) irrespective of the penetration rate.

The trend shown by M20 mixture (Fig. 6) is similar to that of M0 mixture, except that when the $I_C$ value is highest the WSDB value is not zero but it is between 1 and 1.4 kg/m$^2$. In M40 mixture (Fig. 7), initially an increase happens, and after a certain point, the rate at which the increase happens gets slowed down for $I_C = 0.53$ and

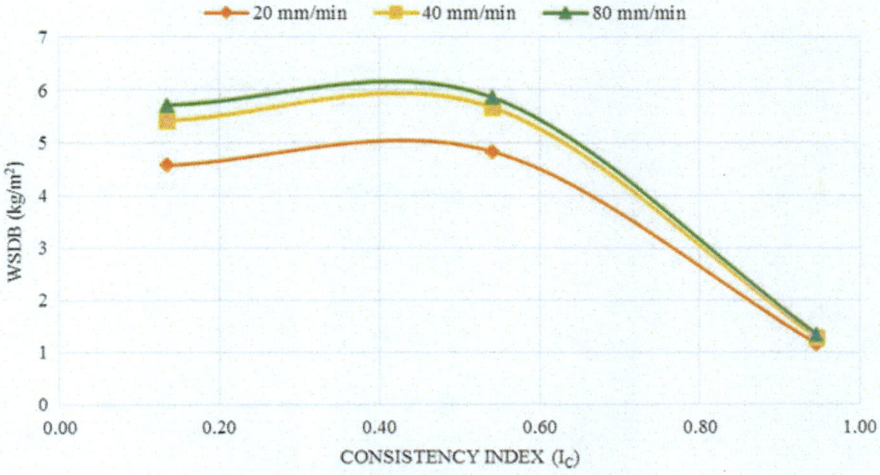


**Fig. 3** Variation in WSDB with consistency index of M20 mixture

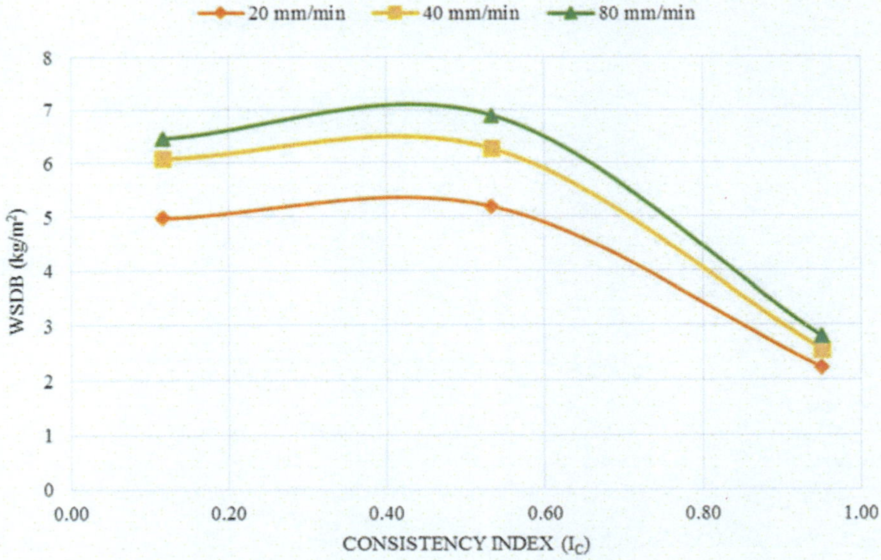


**Fig. 4** Variation in WSDB with consistency index of M40 mixture

0.12. When $I_C = 0.53$ and 0.12, the WSDB growth rate is 33% and 30%, respectively. For $I_C = 0.95$, a 26% increase is observed when penetration rate is increased from 20 to 80 mm/min.

In all the mixtures, the highest WSDB value occurs at 80 mm/min while the lowest is at 20 mm/min, but the changes occurring in WSDB values are more evident in between 20 and 40 mm/min range.

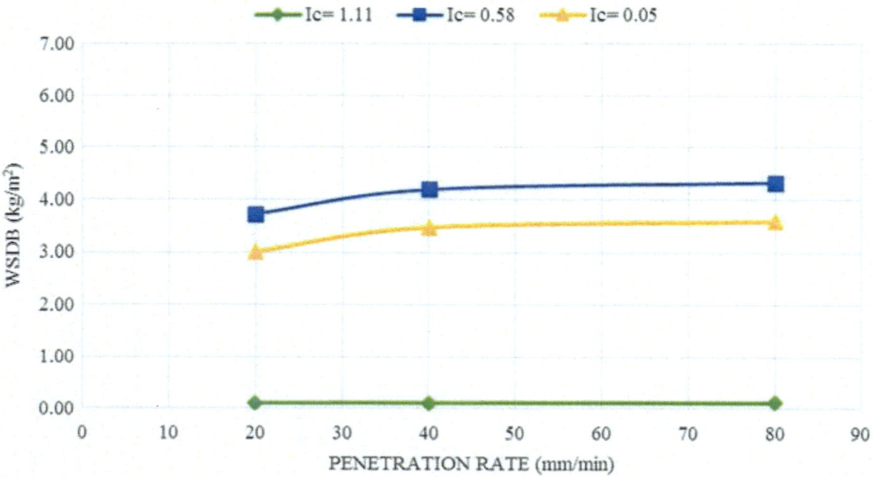


**Fig. 5** Variation in WSDB with penetration rate of M0 mixture

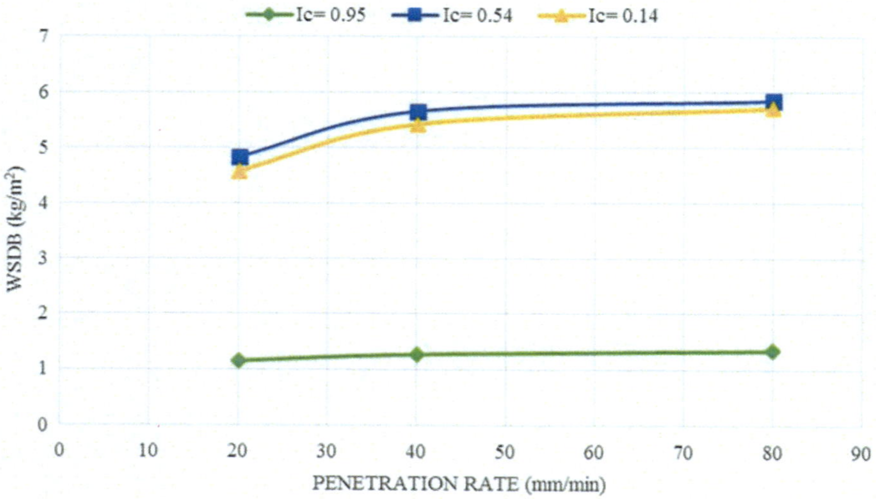


**Fig. 6** Variation in WSDB with penetration rate of M20 mixture

## Effect of bentonite content on WSDB

The effect of bentonite content on WSDB is analysed at different consistency indices and penetration rates. The variations happening at 20 mm/min penetration rate are shown in Fig. 8. The change in WSDB at 40 mm/min and 80 mm/min is shown in Figs. 9 and 10, respectively. It is clear from these figures that at a particular consistency, when the bentonite content is increased the WSDB value also increases. The WSDB value of Thonnakkal soil is low when compared to bentonite and Thonnakkal

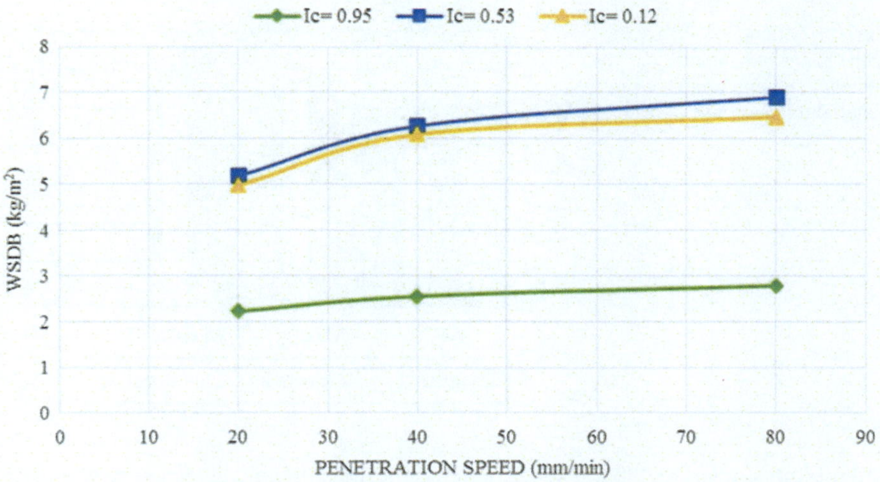


**Fig. 7** Variation in WSDB with penetration rate of M40 mixture

**Fig. 8** Variation in WSDB with change in bentonite content at 20 mm/min penetration rate

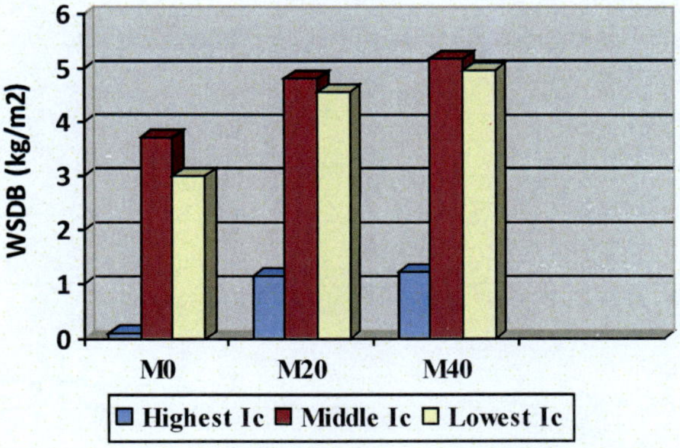


**Fig. 9** Variation in WSDB with change in bentonite content at 40 mm/min penetration rate

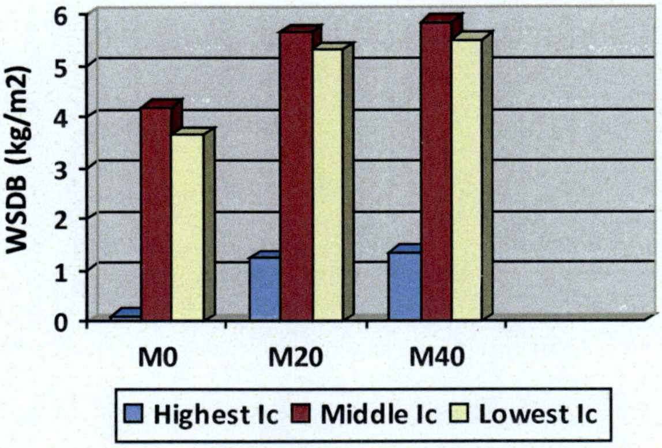

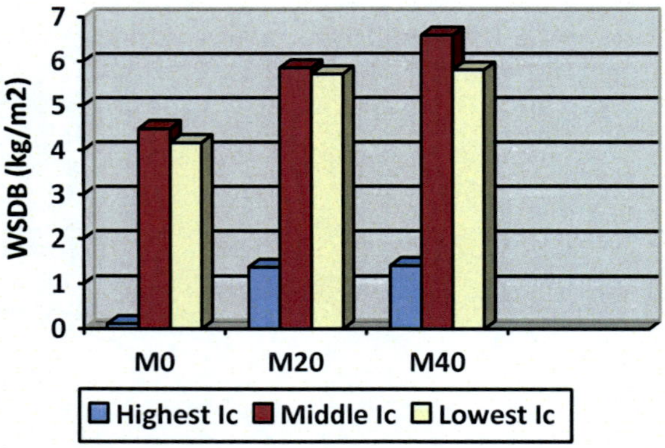


**Fig. 10** Variation in WSDB with change in bentonite content at 80 mm/min penetration rate

soil mixtures. It shows that presence of bentonite considerably changes the clogging nature of Thonnakkal soil.

The variation in peak WSDB value due to the introduction of bentonite into Thonnakkal soil are visible in Fig. 11. There is a rapid change in peak WSDB when the bentonite content is between 0 and 20%, and the increase becomes gradual in 20–40% range.

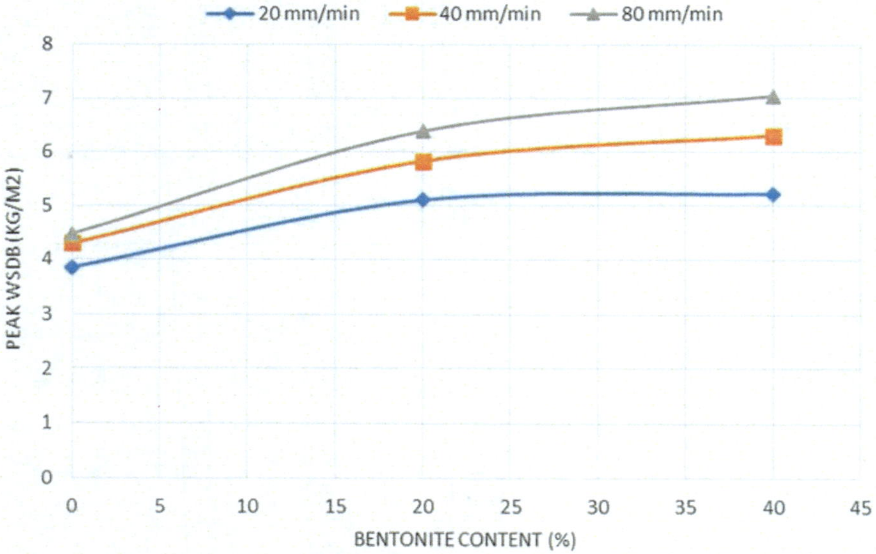


**Fig. 11** Variation in peak value of WSDB due to bentonite content

## 4 Conclusions

1. The Atterberg limits value can be considered as an indicator of clogging potential. Higher the LL and PI, higher is the tendency of the soil to clog.
2. The Atterberg limits of Thonnakkal soil is not much high, but this is not the case once bentonite is added into this. The LL and PI of the soil mixture other than M0 are very high.
3. The data obtained from the laboratory drilling test is more informative compared to the Atterberg limit test. A wide range of parameters which influence the clogging potential can be analysed.
4. Water content is a factor which influence clogging potential. The WSDB value increases from a water content near to PL and reaches a peak value, and then it starts reducing towards LL. The clogging can be limited by maintaining a higher soil consistency.
5. The increase in WSDB with the increase in bentonite points out that bentonite content can also elevate the tendency of Thonnakkal soil to clog.
6. The changes in peak WSDB value are more predominant in the bentonite range of 0–20%. There is increase in WSDB value in between 20 and 40% bentonite content also but this increase is very gradual.
7. The penetration rate of the drill bit is another factor affecting clogging potential. The increase in penetration rate increases the clogging potential of the soil.

In this paper, the maximum percentage of bentonite introduced into Thonnakkal soil was 40%, but the effect of higher percentages of bentonite should also be analysed in future studies. Similarly, the effect of size and shape of drill bit on clogging potential should also be studied.

## References

1. Alberto-Hernandez, Y., Kang, C., Yi, Y., Bayat, A.: Mechanical properties of clayey soil relevant for clogging potential. Int. J. Geotech. Eng. **12**(6), 529–536 (2017)
2. Hollmann, F., Thewes, M.: Assessment method for clay clogging and disintegration of fines in mechanised tunnelling. Tunn. Undergr. Space Technol. **37**, 96–106 (2013)
3. Littleton, I.: An experimental study of the adhesion between clay and steel. J. Terrramech. **13**(3), 141–152 (1976)
4. Yao, Y.S., Zeng, D.C.: Investigation of the relationship between soil-metal friction and sliding speed. J. Terramech. Sci. Direct **27**(4), 283–290 (1990)
5. Amir, K.B., Ali, M., Mohammad, G., Akbar, C.: Assessment of the adhesion potential of kaolinite and montmorillonite using a pull-out test device. Bull. Eng. Geol. Env. **76**(4), 1507–1519 (2017)
6. Thewes, M., Burger, W.: Clogging of TBM drives in clay–identification and mitigation of risks. In: International World Tunnel Congress and the 31st ITA General Assembly, Turkey, pp. 737–742 (2005)
7. Thewes, M.:"Adhäsion von Tonbödenbeim Tunnelvortriebmit Flüssigkeitsschilden (Adhesion of clay soil during tunneling with liquid shields)", PhD thesis, University of Wuppertal, Institute of Soil Mechanics and Foundation Engineering, Wuppertal, Germany (1999)

8. Sass, I., Burbaum, U.: A method for assessing adhesion of clays to tunneling machines. Bull. Eng. Geol. Env. **68**, 27–34 (2008)
9. Feinendegen, M., Ziegler, M., Spagnoli, G., Fernandez-Steeger, T., Stangek, H.: A new laboratory test to evaluate the problem of clogging in mechanical tunnel driving with EPB shields. In: ISRM International Symposium—EUROCK 2010, Switzerland, pp. 429–432 (2010)
10. Zumsteg, R., Puzrin, A.M.: Stickiness and adhesion of conditioned clay pastes. Tunn. Undergr. Space Technol. **31**, 86–96 (2012)
11. Kang, C., Bayat, A., Yi, Y.: Performance evaluation of TBM clogging potential for plain and conditioning soil using a newly developed laboratory apparatus. Int. J. Geotech. Eng. **14**(5), 463–472 (2018)
12. Kang, C., Bayat, A., Yi, Y., Wu, Y.: Assessment of the clogging potential of two clays. Appl. Clay Sci. **178**, 76–83 (2019)
13. IS 2720-Part 2: Methods of test for soils-determination of water content. Bureau of Indian Standards, New Delhi (1973)
14. IS 2720-Part 4: Methods of test for soils-grain size analysis. Bureau of Indian Standards, New Delhi (1985)
15. IS 2720-Part 3: Methods of test for soils-determination of specific gravity. Bureau of Indian Standards, New Delhi (1980)
16. IS 2720-Part 5: Methods of test for soils-determination of liquid and plastic limit. Bureau of Indian Standards, New Delhi (1985)
17. IS 2720-Part 7: Methods of test for soils-determination of water content-dry density relation using light compaction. Bureau of Indian Standards, New Delhi (1980)

# Analysis of Freezing Thawing Cycles on Unconfined Compressive Strength of Expansive Soil

**Mohmad Maaz M. Mansuri, Bhavita Dave, C. H. Solanki, and A. K. Desai**

## 1 Introduction

### *1.1 Unsaturated Soil*

Generally, soil mechanics can be divided into two main domains: One deals with the saturated portion of the earth, which means soil dealing with that portion of the earth whose voids filled with water or any other fluids. That means it contains two-phase system soil solids and water (or air or any other fluids). Other domain deals with the unsaturated or partly saturated soil which has some voids filled with water and some voids filled with air that means it contains a three-phase system, i.e. soil solids, water and air. Due to more than one fluid phase, it results in a change in material behaviour. According to Fredlund and Raharjdo [1], the unsaturated soil is generally neither in arid condition nor in a saturated state, but it has degree saturation of ranging from 0 to 100%. Unsaturated soils (i.e. water and air in the voids) form the largest category of soils which do not follow the behaviour of classical saturated soil mechanics [2]. Almost 60% of land area on earth is an arid or semi-arid region so that these soils strata are not fully saturated [2]. Moreover, conventional soil mechanics consider fully saturated soil condition for shear strength analysis as a worst-case condition, but there were some landslides happened, whilst soils are unsaturated also [3]. As artificial fills or earthen dam after construction does not remain in the same condition in terms of saturation during its life period. So that we need to analyse the

M. M. M. Mansuri (✉) · B. Dave · C. H. Solanki · A. K. Desai
Civil Engineering Department, S. V. National Institute of Technology, Surat 395007, India

C. H. Solanki
e-mail: chs@amd.svnit.ac.in

A. K. Desai
e-mail: akd@amd.svnit.ac.in



C. N. V. Satyanarayana Reddy et al. (eds.), *Ground Characterization and Foundations*, Lecture Notes in Civil Engineering 167,
https://doi.org/10.1007/978-981-16-3383-6_32

soil condition in an unsaturated state, it leads to the development of conventional soil mechanics. Theories developed for the saturated and dry condition cannot be applied directly for the unsaturated condition. Terzaghi came up with a concept contractile skin which is nothing, but a thin layer of the air–water interface and suggested that the contractile skin might be in the order of 10–6 mm in thickness.

### 1.2 Unsaturated Shear Strength

The shear strength of soil, whether saturated or unsaturated, may be defined as the maximum internal resistance per unit area the soil is capable of sustaining along the failure plane under external or internal stress loading. The shear strength of unsaturated soil appeals to more research in geotechnical engineering and practise from the last few decades. The shear strength of unsaturated soil governs main parameters of geotechnical engineering such as bearing capacity, earth pressure, slope stability etc. The shear strength of unsaturated clayey soil is strongly influenced by the physico-chemical interaction between water and clay minerals. Therefore, it is necessary to study the shear strength behaviour of unsaturated clayey soil [4].

Numerous experimental programmes were conducted by various researchers to understand the shear strength behaviour of unsaturated soils from different countries. The series of six undrained and unconfined compression tests were conducted on unsaturated compacted clayey soil comprising 52% sand, 18% silt and 30% clay [5]. Guadalix Red silty clay was investigated by performing the test under net normal stress of 120–600 kPa for suction lower than the air entry value and obtained elliptical failure envelope [6]. The effect of suction on shear strength was studied by testing the clayey soil from Ningxia Hui, China, with advanced triaxial test apparatus. The suction-controlled direct shear test was performed on unsaturated expansive clay and studied the stiffness of clay with suction [7]. The consolidated drained test with modified direct shear apparatus was performed for investigating the Regina clay and Glacial till (clayey till), respectively [8]. Most of the investigators studied the shear strength behaviour on various types of unsaturated soils and observed the non-linear behaviour of the strength envelope. But very few researchers studied the shear strength behaviour of unsaturated clayey soil in the Indian context.

### 1.3 Shear Strength Equation for Unsaturated Soils

For saturated soil, shear strength is commonly described by the M-C failure criterion.

$$\tau_f = c' + (\sigma - u_w) \times \tan \emptyset' \tag{1}$$

where

$\tau_f$ is the shear stress on the failure plane at failure,
$c'$ is the effective cohesion,
$(\sigma - u_w)$ is the effective normal stress on failure plane at failure, and
$\emptyset'$ is the effective angle of internal friction.

Unlike saturated soils, the mechanical behaviour of unsaturated soils depend on two independent stress-state variables. These variables are the stress tensor, $(\sigma - u)$, which is net normal stress, and matric suction $(u_a - u_w)$ [1]. Soil behaviour is independent of the individual values of $u_a$, $u_w$ or the total stress $(\sigma)$ so long as the stress-state variables, $(\sigma - u_a)$ and $(u_a - u_w)$ are invariant.

**Extended M-C failure envelope**

Fredlund et al. [9] proposed the equation which is shown below for determining the shear strength of partly saturated soils in terms of two independent stress-state variables, $(\sigma_s - u_a)$ and $(u_a - u_w)$

$$\tau_f = c' + (\sigma_s - u_a)\ \tan\emptyset' + (u_a - u_w)_f\ \tan\emptyset_b \tag{2}$$

where

$\tau_f$ is the shear stress on the failure plane at failure,
$c'$ is the effective cohesion, which is the intercept of the extended M-C failure envelope on the shear stress axis where the net normal stress and the matric suction $((u_a - u_w)_f)$ at failure are equal to zero,
$(\sigma_s - u_a)_f$ is the net normal stress state on the failure plane at failure,
$u_{af}$ is the pore-air pressure on the failure plane at failure,
$\emptyset'$ is the angle of internal friction associated with the net normal stress-state variable-$(\sigma_s - u_a)_f$,
$(u_a - u_w)_f$ is the matric suction on the failure plane at failure, and
$\emptyset_b$ is the angle indicating the rate of increase in shear strength relative to the matric suction-$(u_a - u_w)_f$.

This Eq. (2) is an extension of the M-C shear strength eq$^n$ for a saturated soil. For unsaturated soils, the shear strength envelope was initially given as a planar surface based on a limited set of data available at that time in the literature (Fredlund et al. 1978). Later experimental evidence by several investigators established that the shear strength for unsaturated soils is non-linear when tested over a broad range of suction [6, 10]. Fredlund et al. [9] equation is valid for interpreting data for both linear and non-linear shear stress.

## *1.4 Freezing–Thawing of Soil*

Jammu and Kashmir is the coldest region in India. Indian meteorological department reports the temperature varies from – 20 °C in winter to 35 °C in summer. *F–T* cycle causes frost action, which leads to frost heave and thaw weakening. The former is due to formation of ice crystal, ice lenses and later are due to melting of ice. Casagrande [11] observed that ice segregation did not appear in soils containing less than 1% of grains smaller than 0.02 mm. Freezing can be of two types open and closed system. De Groot [12] defined as open system as an exchange of matter, heat, work and energy with its surroundings. Jones defined open-system freezing for soil as the condition where pore water in excess of that available in the voids of the soil, is available to be moved to the surface of freezing to form segregated ice in frost susceptible soil. De Groot [12] defined as closed system as an exchange of heat, work and energy, but no matter. Jones defined closed system of soil as, beyond the originally present water in the voids of the soil, there is no source of water available, during the freezing process or near the zone of freezing, and the ice lenses may or may not form. In an open system, as the freezing front moves downward, water is drawn up from a free surface through the soil. This movement of water due to freezing level creates ice lenses in the soil. In closed system, there is a redistribution of moisture to the freezing front occurs, and this movement causes a decrease in dry density and an increase in water content near the freezing front.

## 2 Present Study—Unconfined Compressive Strength Test for Partially Saturated Soil

Geotechnical engineers are aware that the matric suction holds the soil together in UC tests [1]. In conventional analyses, in the undrained shear strength, the independent contribution of matric suction is not consciously considered. The matric suction is the function of the change in pore water pressure and the in-situ pore-water pressure, resulting from the stress relief during sampling in unsaturated fine-grained soil (UFG). Due to this, the measured undrained shear strength must be interpreted considering the influence of matric suction. Vanapalli et al. [8] proposed a technique to estimate $\varphi_b$ with respect to matric suction using unconfined compression test results for unsaturated fine-grained soils assuming a planar shear strength envelope.

$$\tan \emptyset_b = \left[\frac{\sigma_1}{2}\left(\cos \emptyset' + \sin \emptyset' \times \tan \emptyset'\right) - \left(\frac{\sigma_1}{2} \tan \emptyset' - c'(u_a - u_w)\right)\right] \quad (3)$$

The pore-air pressure can be assumed to be atmospheric, and the test results can be interpreted taking constant matric suction. Equation can be used to determine the contribution of matric suction in practical applications towards undrained shear strength, $\varphi_b$, knowing the UCS, $\sigma_1/2$ and matric suction, $(u_a - u_w)$ of the soil along with the effective shear strength parameters, $c'$ and $\varphi'$.

In this study, the main aim was to study the effect of freezing thawing cycles on UCS of partly saturated expansive soil as we know that partly saturated soil has three-phase system containing solid, liquid and gas phase. So, the effect of alternate freezing thawing and change in the degree of saturation greatly influences the shear

**Table 1** Engineering properties of soil

| Property of soil | Value |
|---|---|
| Specific gravity [16] | 2.69 |
| Grain size distribution [17] | |
| Gravel content (> 4.75 mm) (%) | 0 |
| Sand (0.075–4.75 mm) (%) | 9.82 |
| Coarse sand (2–4.75 mm) (%) | 0.54 |
| Medium sand (0.425–2 mm) (%) | 0.99 |
| Fine sand (0.075–0.425 mm) (%) | 8.29 |
| Silt (0.002–0.075 mm) (%) and Clay (< 0.002 mm) (%) | 90.18 |
| Liquid limit (wL) (%) [18] | 63.25 |
| Plastic limit (wp) (%) [18] | 25.92 |
| Plasticity index (*I*p) (%)[18] | 37.33 |
| Maximum dry density (kN/m$^3$) [19] | 16.79 |
| Optimum moisture content (%) [19] | 16.571 |
| Indian standards classification—IS 1498-1970 [20] | CH |

strength of the soil [13]. CH soil of South Gujarat region was used. UCS tests were performed on different combinations of dry density, degree of saturation and *F*–*T* cycles. The samples were prepared for the degree of saturation of 80, 60, 40 and 20%. Whilst the dry density was kept 100, 95 and 85% of MDD [14, 15]. Graphs were plotted to analyse the results.

## 3 Testing Material

In this study, the soil was collected from Jahangirpura, Surat region. The soil is black cotton (BC) soil in its physical appearance. The soil is classified as CH soil as per IS: 1498-1970. It is first oven-dried for 24 h. After that drying, crushing and sieving of the soil were done. Then, to identify the classification of soil type and various basic properties of soil, basic tests were performed. Table 1 shows the basic engineering properties of soil.

## 4 Experimental Setup

An UCS mould of size 38 mm as an internal diameter and height of 76 mm was used for the experimental study. Sample extractor was needed to extract the sample from the UCS mould safely. A chest freezer with the temperature control up to – 25 °C was used to run the freezing thawing cycles of soil specimens. UCS/Triaxial testing apparatus used for testing the sample.

## 5 Sample Preparation

From the Engineering properties of soil (OMC, MDD, G) amount of water required and amount of soil needed for each set of $S_r$ and DD (DD = 100, 95, 85% of MDD, $S_r$ = 60, 40, 20%) was calculated [14, 15]. Then, required amount of soil and water was thoroughly mixed in a bowl. Then, it was delivered into UCS mould having diameter 38 mm and height 76 mm.

Each sample was passed through required number of freezing–thawing cycles except the samples which were of zero cycle. Zero cycle samples were directly tested. Samples which were not of zero cycles were subjected to required numbers of freezing–thawing cycles in a closed system. After preparation of the sample, they were put in the airtight plastic container to obtain a closed system, so that no moisture can come in or go out to or from the system. Then for the freezing process, the container was kept in the refrigerator for 24 h. Freezing was done below − 23 °C.

After 24 h of freezing, the container was kept in the dark place for thawing at an average temperature varying between 25 and 30 °C according to IS code 4332-4 [21] for 23 h. Then, container was put in equilibrium period for 1 h. So, it took 48 h to complete one cycle of freezing and thawing. Similarly, various cycles (i.e. 0, 1, 3, and 5) were performed on a soil sample. For each combination (i.e. the same degree of saturation, same dry density and same cycle), three specimens were prepared and tested. Then, the average of those three readings was taken for the analysis. After completion of required no of cycles UCS Test were performed as per IS 2720 Part 10 [22], and the shear strength of each sample was calculated from the results of the UCS test. After completing the UCS test on a different degree of saturation and various dry density for each cycle (0, 1, 3, and 5) analysis of the results were done. Graphs were plotted to know the effect of freezing and thawing cycles on the unconfined compressive strength of the soil for the different degree of saturation and dry density.

As discussed earlier, in this study, for each density (i.e. 100, 95, 85% of MDD) taking degree of saturation from 100% $S_r$ keeping an interval of 20%, i.e. degree of saturation varying from 100, 80, 60, 40, 20% were taken. But, due to having a more water content in 100% degree of saturation and 80% degree of saturation sample was very soft and had a negligible shear strength. So, it was not easy to perform the test.

Freezing thawing cycles were started from zero cycles keeping an interval of one cycle for the ease of analysis up to where curve shows reverse result, i.e. in this study, it was up to five cycles.

# 6 Results and Discussion

## 6.1 *Comparison with the Degree of Saturation*

**UCS value versus No of Freezing–Thawing Cycles**

UCS value (kN/m$^2$) versus number of freezing–thawing cycles graphs were plotted to analyse the variation of UCS value with respect to number of *F–T* cycles. Comparison of UCS value with the degree of saturation was also plotted.

**Dry Density = 100%**

Reading were shown in Table 2 and Fig. 1 shows $q_u$ versus F–T Cycles for MDD-100% showing relation of Sr.

UCS value increased up to 3 cycles then decreased for 100% dry density for 60, 40 and 20% degree of saturation.

**Dry Density = 95%**

Experimental reading were shown in Table 3 while Fig. 2 shows $q_u$ vs F-T Cycle for Dry density of 95%.

**Table 2** Values of $q_u$ for MDD-100%

| $q_u$ (kN/m$^2$) | | | | |
|---|---|---|---|---|
| $S_r$\cycles (%) | 0 | 1 | 3 | 5 |
| 60 | 133.271694 | 201.3949 | 248.7607533 | 222.9693953 |
| 40 | 474.2798647 | 491.801 | 578.295202 | 530.04673 |
| 20 | 592.7762813 | 622.7191 | 708.1999633 | 307.044646 |

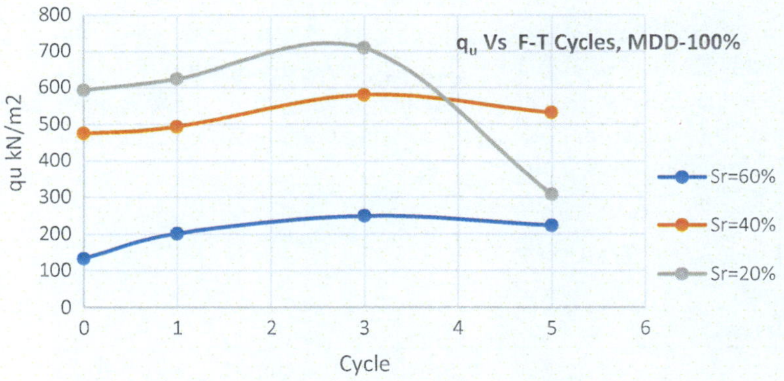


**Fig. 1** $q_u$ versus *F–T* Cycles for MDD-100%

**Table 3** Values of $q_u$ for MDD-95%

| $q_u$ (kN/m$^2$) | | | | |
|---|---|---|---|---|
| $S_r$\cycles (%) | 0 | 1 | 3 | 5 |
| 60 | 27.52386 | 38.80145 | 341.9235 | 0 |
| 40 | 365.1324 | 464.2771 | 460.5506 | 82.7677 |
| 20 | 1034.106 | 1182.349 | 807.2793 | 219.8967 |

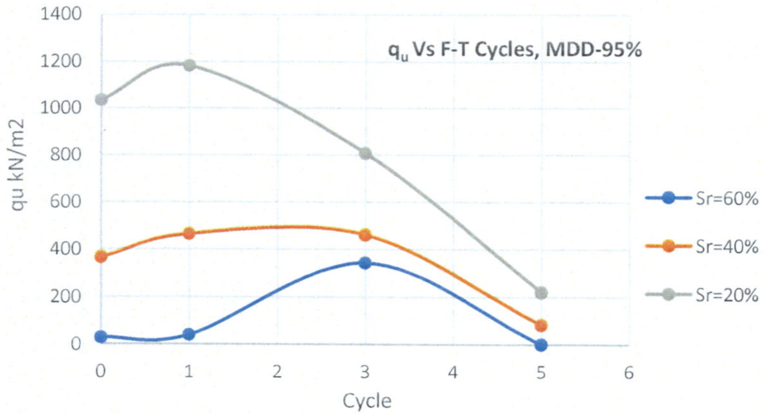


**Fig. 2** $q_u$ versus *F–T* cycles for MDD-95%

UCS value increased up to 1 cycle then decreased for 95% dry density for 40% and 20%, whilst for 60% degree of saturation, it increased up to 3 cycles (sudden increase for three cycles) and no value for five cycles.

**Dry Density = 85%**

Table 4 contain experimental readings of dry density 85%, while Fig. 3 shows relationship between $q_u$ vs F-T Cylce for the same dry density.

UCS value decreased as the number of cycle increased for 85% dry density and 40% and 20%, whilst for 60% degree of saturation it gives no value for all cycle as the sample becomes very soft.

**Table 4** Values of $q_u$ for MDD-85%

| $q_u$ (kN/m$^2$) | | | | |
|---|---|---|---|---|
| $S_r$\cycles (%) | 0 | 1 | 3 | 5 |
| 60 | 0 | 0 | 0 | 0 |
| 40 | 209.0113 | 171.2232 | 66.68488 | 41.77612 |
| 20 | 883.7054 | 567.5406 | 465.6501 | 424.3316 |

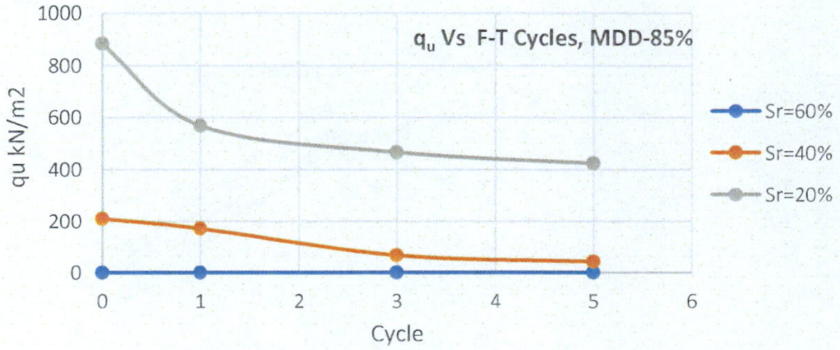


**Fig. 3** $q_u$ versus *F–T* cycles for MDD-85%

**Discussion from the above graphs (UCS vs. *F–T* Cycles)**

The freezing–thawing behaviour of an expansive soil at proctor density somewhat varies with each other. It showed variation in UCS for *F–T* cycle for the individual degree of saturation. For 60 and 40% saturation, it increased up to three cycles and retarded for five cycles, whereas for 20% saturation it decreases as the numbers of *F–T* cycle increases.

## 6.2 *Comparison with Number of* F–T *Cycles*

**Dry Density = 100% MDD**

Relation between $q_u$ vs $S_r$ with varying F-T cycle at Dry Density - 100% is shown in Fig. 4.

**Dry Density = 95% MDD**

Relation between $q_u$ vs $S_r$ with varying F-T cycle at Dry Density - 95% is shown in Fig. 5.

**Dry Density = 85% MDD**

Relation between $q_u$ vs $S_r$ with varying F-T cycle at Dry Density - 85% is shown in Fig. 6.

The graphs show the continous retarders in UCS value with incremental of the degree of saturation for each FT cycle.

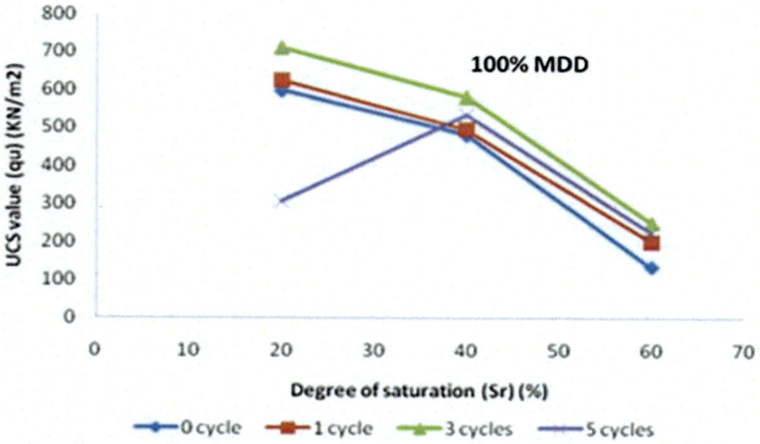


**Fig. 4** $q_u$ versus $S_r$, MDD-100%

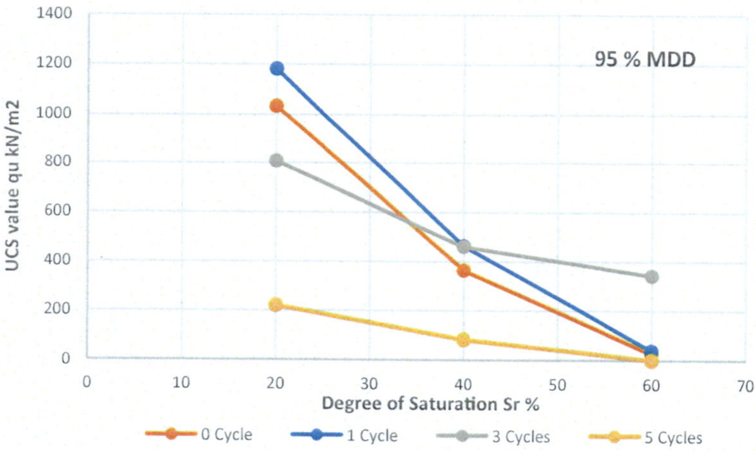


**Fig. 5** $q_u$ versus $S_r$, MDD-95%

## 7 Conclusions

Through the analysis data of the experimental results, the following conclusions were obtained:

1. The UCS parameter in the freezing–thawing process (temperature variation) has been effective such as, the volume of voids effective on the shape of voids. Pores, neck and direction of water flow either getting in or out of pores which acted in the volume.
2. The UCS value has been increased up to freezing–thawing cycle (3 nos.), and then it reduces for 100 and 95% of Proctor density. In contrast, for 85% of Proctor

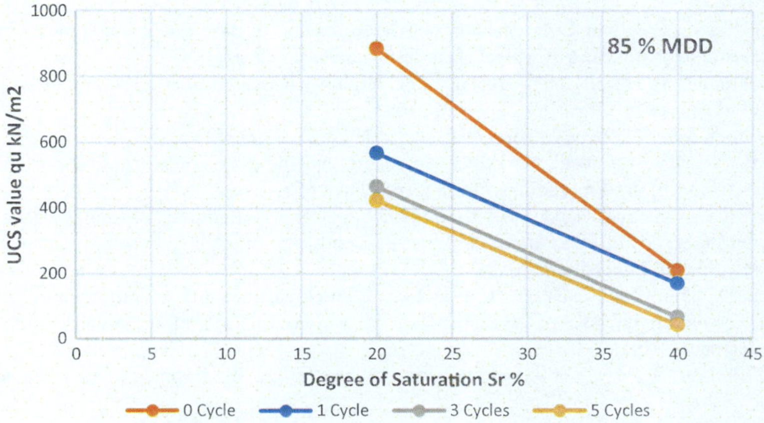


**Fig. 6** $q_u$ versus $S_r$, MDD-85%

density, the UCS values go on decreasing with increasing no of freezing–thawing cycles.

3. It is seen that the UCS value has been reduced as the degree of saturation increased. It is because at lower water content volume of voids has more portion of air instead of water, so the suction value is more at lower water content which gives additional strength to the soil.

## References

1. Fredlund, D.G., Rahardjo, H.: Soil Mechanics for Unsaturated Soil. A Wiley-Interscience Publication, Wiley Inc. (1993)
2. Fredlund, D.G., Rahardjo, H., Fredlund, M.D.: Unsaturated Soil Mechanics in Engineering Practice (2012)
3. Bharat, T.V.: NPTEL Lectures Fundamental aspects of Unsaturated Soil Mechanics and its Basic Principles (2017)
4. Pande, P.B., Khandeshwar, S.R., Bajad, S.P.: Shear strength behavior of an unsaturated clayey soil. In: Proceedings Indian Geotechnical Conference (2019)
5. Chantawarangul, K.: Comparative study of different procedures to evaluate effective stress strength parameters for partially saturated soils. M.Sc. Thesis. Asia Institute of Technology, Bangkok, Thailand (1983)
6. Escario, V., Juca, F.: Strength and deformation of partly saturated soils. In: Proceedings of 12th International Conference on Soil Mechanics and Foundation Engineering, vol. 1, pp. 43–46. Rio, Balkema, Rotterdam (1989)
7. Yongfu, X.U., Yongzhan, C., Songyu, L.: Triaxial test on unsaturated expansive soils. Chin. J. Geotech. Eng. **20**(3), 14–18 (1998)
8. Vanapalli, S.K.: Simple test procedures and their interpretation in evaluating the shear strength of unsaturated soils. PhD. Thesis. Department of Civil Engineering, University of Saskatchewan, Saskatoon, SK (1996)
9. Fredlund, D.G., Morgenstern, N.R., Widger, R.A.: The shear strength of unsaturated soils. Canadian Geotechnical Journal 15 (3):313–321 (1978)

10. Gan, J.K.M., Fredlund, D.G., Rahard, H.: Determination of the shear strength parameters of an unsaturated soil using the direct shear test. Can. Geotech. J. (1988)
11. Casagrande, A.: Discussion of frost heaving. Highway Research Board, Proceedings, vol. 11, p. 168–172 (1931)
12. De Groot, S.R.: Thermodynamics of Irreversible Processes (1951)
13. Vaidya, J., Dave, B., Joshi, N.: Effect of freezing-thawing cycles on suction measurement of unsaturated soil. In: Proceedings Indian Geotechnical Conference (2019)
14. Dave, B.S., Solanki, C.H., Desai, A.K.: Influence of Degree f Saturation on Strength and Consolidation Properties of Unsaturated Soil and its Centrifuge Modelling. Japanese Geotechnical Society Special Publication (2019)
15. Shah, D., Chinwala, N., Desai, H.: Effect of different densities and saturation on unconfined compressive strength of expansive and yellow plastic clay. In: Proceedings Indian Geotechnical Conference (2019)
16. IS:2720 (Part 3/Sec 1) 1980: Determination of specific gravity. Bureau of Indian Standards, New Delhi
17. IS:2720 (Part 4) 1985: Grain size analysis. Bureau of Indian Standards, New Delhi
18. IS:2720 (Part 5) 1985: Determination of liquid limit and plastic limit. Bureau of Indian Standards, New Delhi
19. IS: 2720 (Part 7) 1980: Determination of water content-dry density relation using light compaction. Bureau of Indian Standards, New Delhi
20. IS:1498-1970: Classification and identification of soils for general engineering purposes. Bureau of Indian Standards, New Delhi
21. IS:4332-part IV (1968): Wetting and drying, and freezing and thawing tests for compacted soil-cement mixtures, New Delhi (1980)
22. IS:2720 (Part 10) 1991: Determination of unconfined compressive strength. Bureau of Indian Standards, New Delhi

# Standard Penetration Test (SPT) Pitfalls and Improvements

**Selvam Sermalai, Manoj Mukundan, and Swathi Alagirisamy**

## 1 Introduction

Standard penetration test (SPT) is one of the in-situ tests conducted for onshore and near shore soil investigation projects. Correlations based on SPT N values are extensively used for the estimation of design soil parameters. In addition, the settlement calculation and liquefaction analysis of cohesionless soils are based on the SPT *N* values.

Review of current practices in India shows many drawbacks/ pitfalls in the measurement of SPT *N* values, which leads to errors in the measured SPT *N* values. A study is carried out to understand the effect of various SPT hammer operating systems. This paper presents the results of three case studies and summarizes the ways to improve quality and reliability of SPT test results.

## 2 Background

In early 1900, geotechnical soundings were used just for delineating soil and rock interface using cuttings from wash boring [1]. Around 1902, Colonel Charles R.

S. Sermalai (✉) · S. Alagirisamy
Geotechnical Engineer, Somehsa Geo-Coastal Pvt Ltd., Bangalore 560034, India
e-mail: selvam@somehsa.com

S. Alagirisamy
e-mail: swathi@somehsa.sg

M. Mukundan
Geotechnical Consultant, Somehsa Geosciences Pte. Ltd., Singapore 415979, Singapore
e-mail: manoj@somehsa.com



C. N. V. Satyanarayana Reddy et al. (eds.), *Ground Characterization and Foundations*, Lecture Notes in Civil Engineering 167,
https://doi.org/10.1007/978-981-16-3383-6_33

**Fig. 1** Colonel Charles R. Gow (1872–1949)

Gow [2] started using 1 in. diameter drive sampler (Figs. 1 and 2). Later, Mohr [3] developed 2 in. split-spoon sampler and standardized the testing procedure.

Terzaghi [4] liked Harry Mohr's split spoon sampler and named it as "Standard Penetration Test." Terzaghi and Peck [5] presented first published SPT correlations. The 2 in. diameter split spoon sampler became a nationwide standard in the USA in 1958 and the apparatus and procedure were officially adopted by ASTM as test method D1586 [6].

As per Indian practice, IS 2131–1981 [7] provides guidelines for conducting SPT and corrections to be applied to the measured SPT $N$ values. This code gives corrections for only overburden and dilatancy of soils.

IS 1893 (Part 1)–2016 [8] provides guidelines for correction factors for non-standard SPT equipment and procedures and a summary is given below.

$$N_{60} = N \times C_{60} \tag{1}$$

where

$N$ = measured (uncorrected) SPT blow count
$N_{60}$ = normalized SPT blow count
$C_{60} = C_{\mathrm{HT}} \times C_{\mathrm{HW}} \times C_{\mathrm{SS}} \times C_{\mathrm{RL}} \times C_{\mathrm{BD}}$

$C_{\mathrm{HT}}$ = correction factor for non-standard hammer weight

**Fig. 2** Gow's pipe sampler

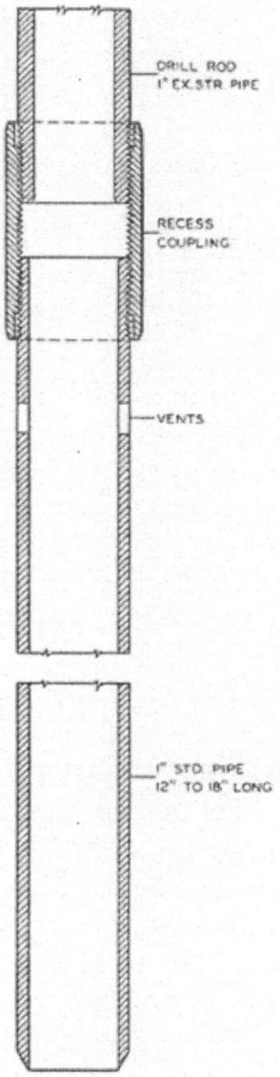


= 0.75; for Donut hammer with rope and pulley
= 1.33; for Donut hammers with trip/auto release mechanism

$C_{HW}$ = correction factor for non-standard hammer height of fall

= $(H * W)/48387$; $H$ = height of fall in mm; $W$ = hammer weight in kg

$C_{SS}$ = correction factor for non-standard sampler

= 1.1; sampler with room for liners, but used without liners (loose sand)
= 1.2; sampler with room for liners, but used without liners (dense sand)
= 0.9; sampler with room for liners, liners used (loose sand)
= 0.8; sampler with room for liners, liners used (dense sand)

$C_{RL}$ = correction factor for rod length

= 0.75; for rod length 0–3 m
= 0.80; for rod length 3–4 m

= 0.85; for rod length 4–6 m
= 0.95; for rod length 6–10 m
= 1.00; for rod length 10–30 m

$C_{BD}$ = correction factor for non-standard borehole diameter

= 1.00; for borehole diameter of 65–115 mm
= 1.05; for borehole diameter of 150 mm
= 1.15; for borehole diameter of 200 mm.

For cohesionless soils, correction for overburden is applied as below.

$$(N_1)_{60} = C_N \times N_{60} \tag{2}$$

where $C_N$ = correction factor for overburden pressure [7]

$$= 0.77 \times \log(2000/\sigma') \tag{3}$$

where $\sigma'$ = effective overburden pressure at the time of testing, in kN/m$^2$.

For fine sand and silt below water table (IS 2131–1981) [7], dilatancy correction is applied for SPT $N$ values corrected for overburden pressure, which is greater than 15, as below.

$$(N_1)''_{60} = 15 + {}^1\!/_2 * [(N_1)_{60} - 15] \tag{4}$$

## 3 Geotechnical Design and Importance of SPT *N* Value

SPT *N* values are extensively used in geotechnical design, using correlations developed by Terzaghi and Peck [5] and by others to arrive at different in-situ soil parameters. Also, SPT *N* values are used for estimating bearing capacity and settlement of foundations.

### *3.1 In-Situ Properties of Soils*

SPT N values are used to estimate various soil parameters such as, unit weight, relative density, angle of internal friction of cohesionless soils, unconfined compressive strength of cohesive soils, and stress–strain modulus of soils.

### 3.2 Allowable Bearing Capacity of Foundations

Allowable bearing capacity, for cohesionless soils, is given by Terzaghi and Peck [5], using SPT *N* value originally in graphical forms. Later many engineers, Meyerhof [9], Skempton [10], Teng [11], Parry [12], and Bowles [13] proposed their equations using SPT *N* value to estimate allowable bearing capacity of foundations, which are available in published literature.

### 3.3 Pile Design

IS 2911 (Part 1/Sect. 2): 2010 [14], Annex B-4 includes a method of using standard penetration test *N* value for arriving at ultimate pile capacity for a bored cast-in-situ piles.

### 3.4 Liquefaction Analysis

Liquefaction resistance of soil is determined using SPT *N* value, $(N_1)_{60.}$ IS 1893 (Part 1): 2016 [8], Annex-F presents simplified procedure for evaluation of liquefaction potential using SPT *N* values.

## 4 Pitfalls in SPT Measurements

SPT *N* values play a major role in the estimation of soil parameters, bearing capacity of foundations and liquefaction analysis. However, in practice, the equipment used and operational procedures considerably influence measured SPT *N* values. Following are some of the common pitfalls in the SPT measurements.

1. Using non-standard hammer weight (Standard weight of hammer is 63.5 kg).
2. Using variable drop heights of hammer due to manual operations (Standard drop height of hammer is 760 mm).
3. Restricted free fall of the hammer. Using more than 2 turns of rope around the drum and/or using wire cable restrict the free fall of the hammer weight. (Usually 1–1/2 to 2 wraps of rope around the drum help free fall of hammer).
4. Inclination of driving rods. Hammer weight not striking the drive cap (anvil) concentrically, resulting in increased SPT *N* values.
5. Not using a guide rod for hammer, resulting in incorrect SPT *N* values.
6. Incorrect drilling and sampling procedures.

(a) Drilling non-standard diameter boreholes. Larger diameter may result in decrease in blow count. (About 100–130 mm diameter borehole is a standard recommended practice).
(b) Inadequate cleaning of boreholes.
(c) Failure to maintain sufficient hydrostatic head in borehole. The water level in the borehole must be at least equal to the piezometric level especially in sandy strata. If not, soil at the bottom of the borehole may get disturbed to a loose state.
(d) Over washing the borehole ahead of casing, resulting in low blow count.
(e) Not seating sampler spoon on undisturbed soil, resulting in incorrect SPT $N$ value.
(f) Over driving the sampler, resulting in higher blow count.

7. Inadequate supervision. Accurate recording of drilling, sampling and depth shall be done by an experienced supervisor/Engineer. Sometimes, higher blow counts may be observed when gravel piece plugs (the layer is not gravel) sampler which may not be recognized by an inexperienced observer.

## 5 Improvements in SPT

Following are some of the measures taken, in order to minimize the errors in the measurements of SPT $N$ value.

1. Using Donut hammer with standard weight or measuring the weight of hammer to apply relevant correction.
2. Using Auto-trip hammer system. This ensures standard drop height and free fall of the hammer for all blows, and eliminates the errors due to variable drop height and restricted free fall of hammer.
3. Using "SPT Analyser" for the measurements of energy transferred to driving rods.
4. Maintaining the verticality of drill rods. This needs further improvements considering the safety of operations.
5. Ensuring adequate supervision by an experienced Engineer at project site.

## 6 Case Studies

A study is conducted to understand the effectiveness of different SPT hammer operating systems being used in India. The "donut hammer" with the following different operating system is studied at different project sites; Site-A, Site-B, and Site-C. SPT analyzer is used to measure energy transferred to driving rods in all the project sites. Three cases of hammer operation systems were considered.

Case-1: Donut hammer—manual operation using manila rope
Case-2: Donut hammer—winch operation using steel wire rope

Case-3: Donut hammer—Auto trip hammer, operated by winch.

## 6.1 SPT Analyzer

SPT analyzer measures energy transferred by SPT hammers using force and velocity measurements. It provides a means to measure transferred energy into drill string while performing a standard penetration test (SPT).

The force and velocity measurement are obtained from sensors comprising, two strain gauge bridges and two accelerometers, instrumented on to a SPT rod which is, in turn, is connected at the top of drill string, below the hammer anvil during SPT test.

SPT analyzer obtains, processes force and velocity time histories and display the normalized force units along with various other pertinent parameters such as transferred energy and driving system efficiency in real time during the test (Fig. 3). ASTM D4633-2010 [15] describes procedures for energy measurement during dynamic penetrometer testing. Clayton [16] explained the theory, measurement and significance of SPT energy measurement.

a) Instrumented Rod b) Pile Driving Analyser for Energy Measurement

**Fig. 3** **a** Instrumented rod. **b** Pile driving analyser for energy measurement

**Fig. 4** SPT Energy measurements at Site-A

## 6.2 Site-A

Site-A is for solar power plant project in Karnataka, India. Calyx Rotary drilling rig was used for drilling operations. Donut hammer was used and operated manually using manila rope (Case-1). The SPT hammer energy was measured using SPT analyzer, and the energy transferred to each blow was recorded (Fig. 4).

## 6.3 Site-B

Site-B is for construction of Coal Jetty for thermal power project in Tamil Nadu, India. Offshore soil investigation was conducted using jack-up barge and Calyx Rotary drilling rig placed over the jack-up barge. Donut hammer was used and operated by winch using steel wire rope (Case-2). The SPT hammer energy was measured using SPT analyzer, and the energy transferred to each blow was recorded (Fig. 5).

## 6.4 Site-C

Site-C is for high-grade steel plant in Andhra Pradesh, India. Soil investigation was done using Hydraulic Rig. Auto-trip hammer (Donut hammer with auto trip mechanism) was used for SPT tests (Case-3). The hammer was operated by winch using steel wire rope. The SPT hammer energy transferred was measured using SPT analyzer (Fig. 6).

**Fig. 5** SPT Energy measurements at Site-B

## 7 Results and Discussions

The results of SPT energy measurements for Site-A, Site-B, and Site-C are presented in Figs. 7, 8, and 9, respectively.

**Site-A**: Measured hammer energy (refer Fig. 7) varies from 30 to 75%, with an average energy ratio of 55%. The measured transmitted energy indicates high fluctuations of energy between hammer blows. The variations in energy are mainly due to manual operations. However, the average energy is about 55%, which is close to the standard energy level of 60%.

**Site-B**: Measured hammer energy (refer Fig. 8) varies from 35–42%, with an average energy ratio of 39%. The fluctuations of energy between hammer blows are less compared to Site-A. However, the measured energy level is low compared to the standard energy level of 60%. The low energy measured is mainly due to inclination of SPT rods and winch operation which limits free fall of SPT hammer.

**Fig. 6** SPT Energy measurements at Site-C

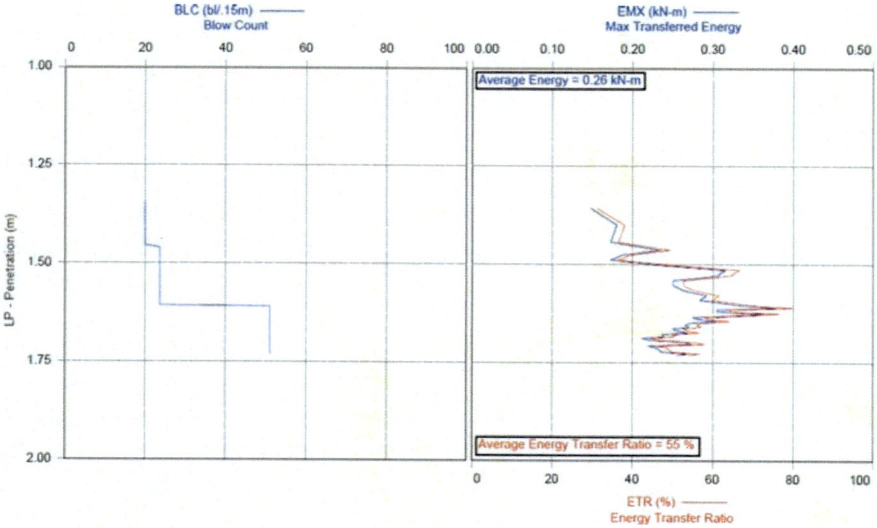


**Fig. 7** Results of SPT energy measurements at Site-A

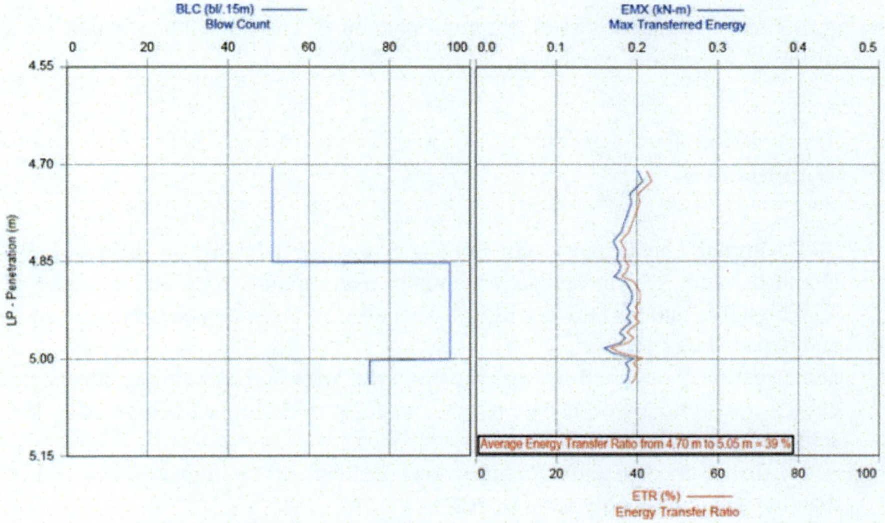


**Fig. 8** Results of SPT energy measurements at Site-B

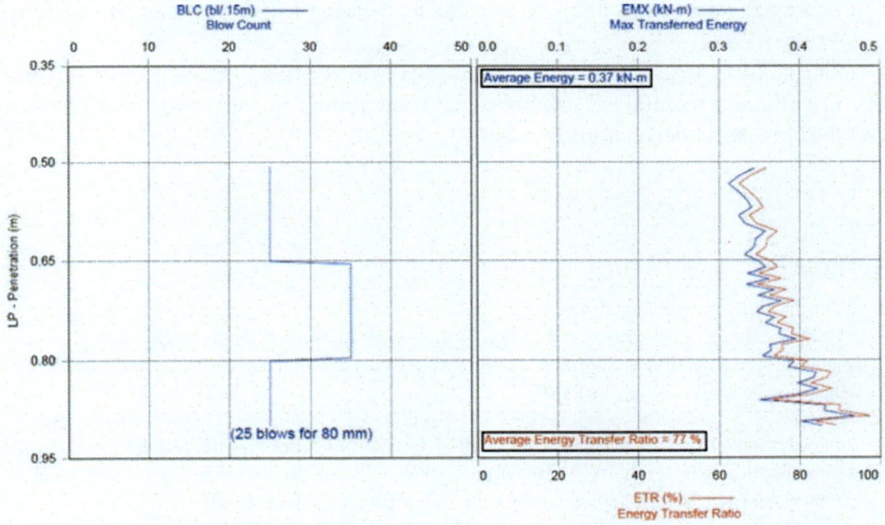


**Fig. 9** Results of SPT energy measurements at Site-C

**Site-C**: Measured hammer energy (refer Fig. 9) varies from 65 to 90%, with an average energy ratio of 77%. The measured energy is higher than the standard energy level of 60%. The higher measured energy is mainly due to the usage of auto-trip hammer system, which ensures standard drop height and free fall of hammer for all blows. The variation of energy between blows is due to inclination of driving rod during SPT test. We also observed that improper connection (loose) between driving rods results in lower energy transfer.

## 8 Conclusions

1. SPT hammer operating system influences energy delivered to driving rods. Donut hammer system operated manually using manila rope delivers energy with high fluctuations between blows. There is no consistency in energy transfer and depends on operators.
2. Donut hammer operated by winch using steel wire delivers almost consistent energy transfer. However, the energy transfer ratio is low as compared to the standard energy level of 60%.
3. Auto-trip hammer system (Donut hammer) delivers SPT hammer energy higher than the standard energy level of 60%.
4. The measurement of SPT hammer energy (using SPT Analyzer) helps to normalize the SPT $N$ value to $N_{60}$ which is used for geotechnical design. In-spite of variations in energy delivered by different SPT hammer operating systems, energy measurement helps to normalize measured SPT $N$ value to standard value of $N_{60}$.
5. The inclination of SPT driving rods results in lower energy transfer as compared to vertically maintained driving rods. Maintaining the verticality of SPT rods during testing needs improvement.

## References

1. David Rogers, J., Terzaghi, G.M.: The Origins of the Standard Penetration Test, Joint meeting, Association of Environmental and Engineering Geologists. American Society of Civil Engineers, Chicago, Illinois (2009)
2. Hvorslev, H.J.: Subsurface exploration and sampling of soils for civil engineering purposes. In: Waterways Experiment Station, Committee on Sampling and Testing, Soil Mechanics and Foundations Division, 465 p. American Society of Civil Engineers: U.S. Corps of Engineers Waterways Experiment Station, Vicksburg, MS. Vicksburg, Miss (1949)
3. Mohr, H.A.: Exploration of Soil Conditions and Sampling Operations: Bull 269. Harvard University, Graduate School of Engg. (1940)
4. Terzaghi, K.: Recent trends in subsoil exploration. In: 7th Conference on Soil Mechanics and Foundation Engineering, University of Texas, Austin (1947)
5. Terzaghi, K., Peck, R.B.: Soil Mechanics in Engineering Practice, 1st edn., 566 p. Wiley, New York (1948)
6. ASTM D 1586: Standard Test Method for Penetration Test and Split-Barrel Sampling of Soils, American Society for Testing and Materials, Philadelphia (1992)
7. IS 2131: Method for Standard Penetration Test for Soils, Bureau of Indian Standards, New Delhi (1981: Reaffirmed 2002)
8. IS 1893-Part 1: Criteria for Earthquake Resistant Design of Structures, Bureau of Indian Standards, New Delhi (2016)

9. Meyerhof, G.G.: Penetration tests and bearing capacity of cohesionless soils. JSMFD, ASCE **82**, SM 1, 1–19 (1956)
10. Skempton, A.W.: The bearing capacity of clays. Proc. Build. Res. Cong. **36**(3), 425–447 (1951)
11. Teng, W.C.: Foundation Design, Wiley, New York (1962)
12. Parry, R.H.: Estimating bearing capacity of sand from SPT values, J. Geotech. Eng. Div. ASCE **103**(GT9), 1014–1019 (1977)
13. Bowles, J.E.: Foundation Analysis and Design, Fifth Edition, The McGraw-Hill, New Delhi (1997)
14. IS 2911 (Part 1/Sec 2), Design and Construction of Pile Foundations - Code of Practice, Part 1 Concrete Piles, Section 2 Bored Cast In-situ Concrete Piles, Bureau of Indian Standards, New Delhi (2010)
15. ASTM D 4633: Standard Test Method for Energy Measurement for Dynamic Penetrometers, American Society for Testing and Materials, Philadelphia (2010)
16. Clayton, C.R.I.: SPT energy transmission: theory, measurement, and significance. Ground Eng. **23**(10), 35–43 (1990)

# A Comparison of Rock Mass Deformation Modulus from Empirical Correlations Versus Plate Load Test at Pare Hydro-electric Power Project, Papum Pare, Arunachal Pradesh

**Pawan Kumar Singh, Diganta Goswami, and Dibyajyoti Kalita**

## 1 Introduction

In rock mechanics and rock engineering, determination of deformability of rock masses becomes most important for the design and analysis of different structures constructed on rock as it describes the relationship between the applied load and the resulting strain. Since natural rock masses inherently come with weaknesses in the form of discontinuities such as joints, bedding planes, faults, it becomes important to characterize the rock mass in a precise manner. The International Society for Rock Mechanics [1] has defined the modulus of deformation of a rock mass ($E_m$) as the ratio of stress to the corresponding strain including both elastic and inelastic behaviour during the loading of a rock mass [1]. The static modulus of deformation is among the parameters that best represents the mechanical behaviour of a rock and a rock mass, in particular when it comes to underground excavations. The deformation modulus is, therefore, a cornerstone of many geomechanical analyses [2]. Deformation modulus of a rock mass can be estimated using various established empirical correlations and in-situ tests like plate loading test (PLT), plate jacking test (PJT) and Goodman Jack test, etc. However, most of the in-situ tests are difficult to operate, time consuming and expensive, due to which conducting large number of in-situ tests is not feasible, so in-situ tests are conducted only at a few locations of the whole project site. Lesser number of in-situ tests cannot properly characterize the rock mass as a whole. This factor drives an engineer to rely on the empirical correlations to estimate the deformation modulus in a simple and effortless manner. Several researchers have worked and still been working on the development of empirical correlations from which deformation modulus value can be obtained. In this paper, some of the correlations suggested by the researchers such as Hoek and Brown [3], Isik et al. [4], Sonmez et al. [5], etc., have been used.

P. K. Singh (✉) · D. Goswami · D. Kalita
Department of Civil Engineering, Assam Engineering College, Guwahati, Assam 781013, India



C. N. V. Satyanarayana Reddy et al. (eds.), *Ground Characterization and Foundations*, Lecture Notes in Civil Engineering 167,
https://doi.org/10.1007/978-981-16-3383-6_34

## 2 Description of the Site

The present study has been carried out at the dam site of river Pare situated in the state of Arunachal Pradesh, India. The detailed geological mapping encompasses 250 m upstream and 300 m downstream of the dam axis, which trend along N74°E–S74°W covering an area of 77,500 $m^2$. The Pare river flows towards SSE direction between the left and right abutment which are bounded by rocky slopes or high-level terraces trending in NNW–SSE direction. The study area lies in the Upper Siwalik formation (Tertiary Group) consisting of brownish and grey coloured, fine to medium-grained concretionary, soft, friable, pebble impregnated, salt-pepper textured sandstone, sand rock and pebble beds which are moderately jointed. The rock mass prevailing at the dam site has been found to be weak and was identified with bedding planes and four joint sets (Bedding Plane, J1, J2, J3, J4) having different orientations. The ground water table was at greater depth, but seepage was observed all throughout the drift.

## 3 Rock Mass Characterization at the Dam Site

The rock mass classification improves the quality of site investigation by quantifying the input data and enables better engineering judgement and due to this, the effectiveness of the project improves. For the design of slopes, foundation and underground excavation, almost any form of analysis is required for reliable estimates of strength and deformation characteristics of rock mass. The rock mass present at the site is characterized by using geological strength index (GSI).

### *3.1 Geological Strength Index (GSI)*

The geological strength index (GSI) was first developed by Hoek [6] to characterize the jointed rock mass based on the geological description of the rock without direct access. Out of all the rock mass classification systems, the GSI system considers both Hoek–Brown and Mohr–Coulomb strength parameters for design purpose. GSI incorporates the mass blockiness and the discontinuity conditions, which are the main parameters essential for the assessment of a rock mass.

The values of GSI were calculated for the location where in-situ tests were carried out. To estimate the value of GSI, the chart given by Marinos and Hoek [7] and equation given by Cai et al. [8] in terms of its quantification by block volume and joint condition factor are considered. The GSI values obtained in the studied area were in the range of 38–56 as observed from Table 1. The disturbance factor ($D$) for the rock mass present at the dam site is taken to be 0.7 for small scale blasting at the site.

**Table 1** Computation of GSI values at the dam site

| Joint | Planarity | $J_w$ | $J_s$ | $J_a$ | $J_c$ | $V_b$ | GSI |
|---|---|---|---|---|---|---|---|
| Bedding plane | Smooth planner | 1 | 1 | 1 | 1 | 10,000 | 45.3 |
| J1 | Smooth planner | 1 | 1 | 1 | 1 | 500 | 38 |
| J2 | Smooth planner | 1 | 1 | 1 | 1 | 500 | 38 |
| J3 | Rough undulating | 1.5 | 1.5 | 1 | 2.25 | 10,000 | 54 |
| J4 | Rough undulating | 1.5 | 1.5 | 1 | 2.25 | 20,000 | 56 |

$J_w$ discontinuity weaviness in m, $J_s$ small scale smoothness in cm
$J_a$ joint alteration factor, $J_c$ joint condition factor; $V_b$ block volume in cm$^3$

# 4 Estimation of Deformation Modulus ($E_d$) of Rock Mass

In spite of being sophisticated, the in-situ deformation modulus is considered to be one of the most important parameters required for geo-engineering investigation. Based on the statistical analysis, several empirical equations are developed for indirect estimation of deformation modulus. Accurate values can seldom be found because in-situ tests may be subjected to measurement errors due to the blasting process or from equipments or test site preparation. It is never recommended to directly go for an empirical correlation for obtaining the values of $E_m$ for a site, rather in-situ methods must be preferred as the results obtained from empirical correlations might sometimes deviate substantially from the actual values of the deformation modulus of the rock mass. There are two ways by which the deformation modulus of rock masses can be determined, viz. In-situ measurements and empirical correlations.

## *4.1 In-Situ Methods*

The deformation modulus values at the dam site were estimated using plate loading test (PLT). Plate load test was conducted as per IS 1888–1982 in which vertical load was applied to the test plate (plate dimension 0.45 m × 0.45 m, thickness 25 mm) by means of hydraulic jack against rolled steel joist capable of providing reaction and the settlement was recorded by two dial gauges of 0.01 mm sensitivity each positioned on either side of the plate and held by datum bars resting on immovable support on either side of the plate. The reaction for the jack was obtained from a kentledge placed on platform supported at a minimum distance of 3.5 times the size of test plate from its centre. The observations of the plate load test were obtained in the form of load-settlement data and subsequently, $E_m$ values are calculated using Eq. (1) as shown in Table 2.

As per IS Code 7317 1993 [9], $E_m$ values are calculated using the formula given below:

**Table 2** Deformation modulus of rock mass of Pare H.E Project, Pare, A.P.

| Applied pressure (MPa) | Deformation, $\delta$ (mm) | | Elastic rebound, $e$ (mm) | Deformability modulus (GPa) | |
|---|---|---|---|---|---|
| | Loading | Unloading | | $E_m$ | $E_i$ |
| 1.125 | 2.31 | 0.75 | 1.56 | 0.17 | 0.25 |
| 2.25 | 3.63 | 1.69 | 1.94 | 0.22 | 0.41 |
| 3.375 | 4.36 | 2.65 | 1.71 | 0.27 | 0.71 |
| 4.5 | 4.82 | 2.82 | 2 | 0.33 | 0.79 |
| 5.625 | 5.21 | 3.05 | 2.16 | 0.38 | 0.92 |

$E_i$ deformation modulus of intact rock

$$E_m = \frac{mP\left(1 - \nu^2\right)}{\delta\sqrt{A}} \tag{1}$$

where

$E_m$ = Deformation Modulus of the subgrade (Rock Mass)
$m$ = Displacement coefficient
$P$ = Total normal surface load
$\nu$ = Poisson's ratio
$\delta$ = average surface displacement of rock mass
$A$ = area of the loading plate.

## 4.2 Empirical Correlations

In this research paper, the empirical correlations which contain the parameter GSI are only considered. Deformation modulus of rock mass can be estimated in a much easier way with the help of empirical formulae suggested by various researchers depending on the quality of the rock mass. Though there are different empirical equations available for estimating the $E_m$ values of rock masses as mentioned earlier, not all the empirical equations are considered in this study.

**Deformation modulus ($E_m$) with GSI**. Nevertheless, there are different methods which define GSI but a general international standard has not been specified yet. Based on this index the value of $E_m$ can be determined indirectly as different correlations are suggested by various researchers (such as Beiki et al. [10], Hoek and Brown [3], Hoek and Diederichs [11], Isik et al. [4], Sonmez et al. [5]), which not only takes GSI into account but also $E_i$ and unconfined compressive strength of intact rock ($\sigma_c$). Table 3 shows the various correlations used in this study. Using these correlations, the values of $E_m$ were calculated corresponding to the GSI values obtained. After that, statistical analysis was performed for the development of the correlation between $E_m$ and GSI, respectively, as presented in Table 4.

**Table 3** Selected equations for estimating $E_m$ (GPa) based on GSI

| Empirical relation | Proposed by |
|---|---|
| $E_m = (s_c/100)^{0.5} 10^{(GSI-10)/40}$ | Hoek and Brown [3] |
| $\frac{E_m}{E_I} = (s^a)^{0.4}; s = e^{\frac{GSI-100}{9}}; a = \frac{1}{2} + \frac{e^{\frac{-GSI}{15}} - e^{\frac{-20}{3}}}{6}$ | Sonmez et al. [5] |
| $E_m = 100\left(\frac{1-\frac{D}{2}}{1+e^{\frac{75+25D-GSI}{11}}}\right)$ | Hoek and Diederichs [11] |
| $E_m = 0.0054GSI$ | Isik et al. [4] |
| $E_m = \tan\{1.56 + \ln(GSI)^2\}^{1/2} s^{1/3}$ | Beiki et al. [10] |

*D* disturbance factor; s, *a* rock mass constants

**Table 4** Statistical results of calculated theoretical values of $E_m$ based on GSI

| GSI value | Deformation modulus of rock mass ($E_m$, GPa) | | | |
|---|---|---|---|---|
| | Mean (Geometric) | Maximum | Minimum | Standard deviation |
| 45.3 | 0.12264 | 0.6984 | 0.0464 | 0.27 |
| 38 | 0.1309 | 0.4587 | 0.0633 | 0.16 |
| 38 | 0.1631 | 0.4587 | 0.0633 | 0.15 |
| 54 | 0.2637 | 1.1524 | 0.0688 | 0.43 |
| 56 | 0.2967 | 1.293 | 0.0694 | 0.48 |

## 5 New Correlations Developed

To indirectly estimate the deformation modulus for the rock masses present at the dam site, certain empirical relationships, applicable to the site conditions has been preferred from the literature available, using which the $E_m$ values were calculated corresponding to the GSI values obtained from the geomechanical classification, the quantitative chart and the equation suggested by Cai and Kaiser [8], respectively. A simple regression analysis has been performed to obtain the relationship between deformation modulus and GSI.

### 5.1 GSI Versus $E_m$

An attempt was made to develop a relationship between in-situ deformation modulus at the dam site and estimated GSI values by calculating the geometric mean as shown in Table 4. After that, a regression analysis was carried out in order to establish a relationship and found that the data fits best for the polynomial equation with a good coefficient of determination value of 0.731. The results obtained from the analysis have been enlisted in Table 5. The best fit curve, obtained from plotting the values

**Table 5** Relationships between deformation modulus ($E_m$, GPa) and GSI values

| Type of equation | Equation | Coefficient of Determination, $R^2$ |
|---|---|---|
| Exponential | $E_m = 0.05e^{0.031GSI}$ | 0.427 |
| Polynomial | $E_m = 0.001GSI^2 - 0.119GSI + 2.757$ | 0.731 |
| Linear | $E_m = 0.007GSI - 0.116$ | 0.486 |
| Logarithmic | $E_m = 0.338\ln(GSI) - 1.058$ | 0.456 |

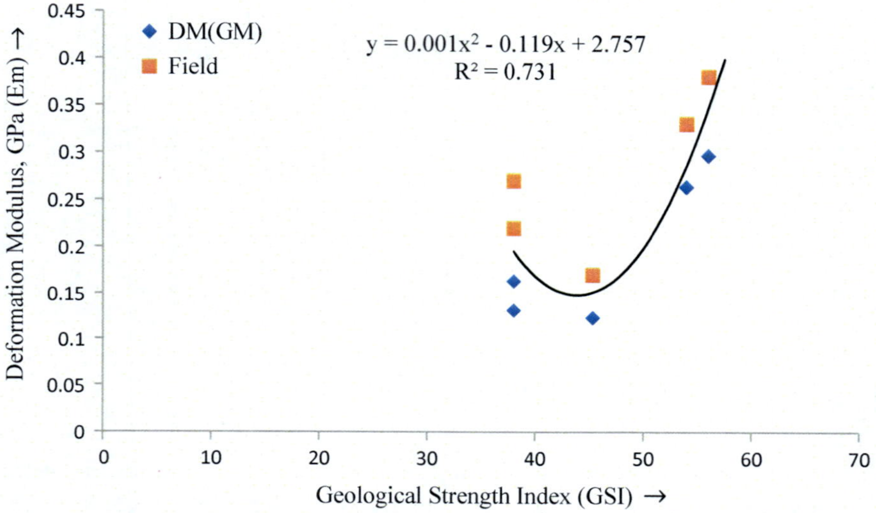


**Fig. 1** Relationship between GSI and $E_m$

of deformation modulus (Geometric mean and Field value) against GSI, has been shown in Fig. 1.

## 6 Validation of Equation Developed

The equation developed as shown in Table 5, having the highest coefficient of determination ($R^2$) is checked for accuracy so that the values of estimated $E_m$ go with good agreement with the values obtained at Pare Dam site.

The performance of the empirical equation obtained from regression analysis in the present study has been predicted using two parameters, viz. root means square errors (RMSE) and variance account for (VAF) as presented in Eqs. (2) and (3), respectively. If the values of RMSE and VAF were 0 and 100, respectively, then the model is accepted to be excellent [12].

$$\text{RMSE} = \sqrt{\frac{1}{N}\sum_{i=0}^{N}(y - y')^2} \tag{2}$$

$$\text{VAF} = \left[1 - \frac{\text{var}\left(y - y'\right)}{\text{var}\,(y)}\right] \times 100 \tag{3}$$

where

$y =$ The experimental $E_m$ values, and

$y' =$ The estimated $E_m$ values.

The RMSE and VAF values were obtained using Eqs. (2) and (3) for the developed correlation were found to be 0.02 and 92.67, respectively. From these values, it can be said that the equation developed from the regression analysis yields a good prediction of the deformation modulus of the rock mass for the present site.

## 7 Limitations of the Correlations Developed

From this research work, we have obtained the following limitations:

1. As a few numbers of plate load tests were conducted at the dam site, the field $E_m$ values obtained were very less. The equations developed would have been more precise if more field data would have been available.
2. As the joint spacing ranged widely, the $V_b$ values obtained from the chart given by Cai and Kaiser [13] may not be precise due to which the field GSI values obtained may not be accurate and as a result the estimated $E_m$ values based on the empirical equations of GSI may not tally exactly with the field values.
3. The GSI-based empirical correlations were developed for the GSI range of 32–56.
4. The developed correlation is valid only for weak rock masses.

## 8 Conclusions

In this study, new empirical correlations of deformation modulus of rock mass are developed based on GSI. From the plate loading test, the in-situ deformation modulus was calculated for the dam site and the values of in-situ GSI were calculated from the chart and the equation given by Cai and Kaiser [13]. Although it is considered that estimating the $E_m$ values empirically is far easier, consumes less time and is less expensive than the in-situ measurements but due to limited availability of data for a site, the heterogeneity of the rock mass, its anisotropic nature sums together and results in different outcomes corresponding to different equations. A database is created in order to calculate the $E_m$ values from several empirical correlations using the GSI values obtained from quantitative measures of block volume and joint condition factor. A simple regression analysis has been performed to develop a relationship between the estimated $E_m$ values with GSI and was found that the equation with the highest coefficient of determination is a polynomial function for GSI with a value of 0.731. Further, the RMSE and VAF values have been calculated and found out to be 92.673 and 0.02 for the developed correlation, which shows that the developed equation can be used with a good degree of accuracy for the given site.

## References


1. ISRM (International Society of Rock Mechanics): The complete ISRM suggested methods for rock characterization, testing and monitoring: 1974–2006, pp. 628. ISRM Turkish National Group, Ankara, Turkey (2007)
2. Palmstrom, A., Singh, R.: The deformation modulus of rock masses—comparisons between in situ tests and indirect estimates. Tunn. Undergr. Space Technol. **16**, 115–131 (2001)
3. Hoek, E., Brown, E.T.: Practical estimates of rock mass strength. Int. J. Rock Mech. Min. Sci. **34**(8), 1165–1186 (1997)
4. Isik, N.S., Ulusay, R., Doyuran, V.: Deformation modulus of heavily jointed-sheared and blocky greywackes by pressure meter tests: numerical, experimental and empirical assessments. Eng. Geol. **101**, 269–282 (2008)
5. Sonmez, H., Gokceoglu, C., Ulusay, R.: Indirect determination of the modulus of deformation of rock masses based on the GSI system. Int. J. Rock Mech. Min. Sci. **41**, 849–857 (2004)
6. Hoek, E.: Strength of rock and rock masses. ISRM News J. **2**(2), 4–16 (1994)
7. Marinos, P., Hoek, E.: GSI—a geologically friendly tool for rock mass strength estimation. In: Proceedings of the International Conference on Geotechnical and Geological Engineering (GeoEng), pp. 1422–1466, Melbourne, 19–24 Nov 2000
8. Cai, M., Kaiser, P.K., Uno, H., Tasaka, Y., Minami, M.: Estimation of rock mass deformation modulus and strength of jointed hard rock masses using the GSI system. Int. J. Rock Mech. Min. Sci. **41**(1), 3–19 (2004)
9. IS 7317: Code of practice for uniaxial jacking test for modulus of deformation of rock. Bureau of Indian Standards, New Delhi (1993)
10. Beiki, M., Bashari, A., Majdi, A.: Genetic programming approach for estimating the deformation modulus of rock mass using sensitivity analysis by neural network. Int. J. Rock Mech. Min. Sci. **47**, 1091–1103 (2010)
11. Hoek, E., Diederichs, M.: Empirical estimation of rock mass modulus. Int. J. Rock Mech. Min. Sci. **43**, 203–215 (2006)
12. Alemdag, S., Gurocak, Z., Gokceoglu, C.: A simple regression based approach to estimate deformation modulus of rock masses. J. Afr. Earth Sci. **110**, 75–80 (2015)
13. Cai, M., Kaiser, P.K.: Visualization of rock mass classification systems. Geotech. Geol. Eng. **24**(4), 1089–1102 (2006)

# Strength Behavior of Lime Stabilized Soil Reinforced with Waste Plastic Strips

**G. Gnana Prasanna and G. Venkata Krishna**

## 1 Introduction

Increasing population and developmental activities force people to convert agricultural fields to roadways and railways. As most of the cultivating fields comprise of highly compressible soils, there is need to improve the properties of soils. There are many methods to improve soil; one of the best methods is soil stabilization. The main problem with highly compressible soils is their high swelling and shrinkage property. Lime can be effectively used for reducing the swelling and shrinkage property of highly compressible/expansive soils.

Because of developmental activities, the problem is not only with the availability of soils but also with the waste plastics present in the landfills. Plastic products have wide range of applications. These are used as carriage bags, garbage bags, fluid containers, clothing, toys, films, wrapping materials, household and industrial products and building materials with its more advantageous characteristics like less weight, less cost and more load carrying capacity and are used by all types of age groups of people [1]. Consoli et al. [2] carried out an experimental study of the utilization of the polyethylene fibres derived from plastic wastes in the reinforcement of uncemented and artificially cemented sand and showed that the plastic waste improved the stress strain response of uncemented and cemented sands. This is perhaps one of the earliest attempts advocating the use of plastic waste. Consoli et al. [3] proposed a field application for such materials designed for increasing the bearing capacity of spread foundations when placed on a layer of fibre-reinforced cemented sand built over a weak residual soil stratum. Muntohar et al. [4] examined the Engineering behavior of the stabilized clayey/silty soil reinforced with randomly distributed discrete plastic waste fibers and proves that it is a very effective method for soil improvement. Kumar et al. [5] conducted Unconfined Compression tests on soft

G. Gnana Prasanna (✉) · G. Venkata Krishna
Department of Civil Engineering, Sree Vidyanikethan Engineering College, Tirupati 517102, India



C. N. V. Satyanarayana Reddy et al. (eds.), *Ground Characterization and Foundations*, Lecture Notes in Civil Engineering 167,
https://doi.org/10.1007/978-981-16-3383-6_35

clay soil and found that unconfined compressive strength of clay increases with the addition of fibers and it further increases when fibers are mixed in clay sand mixture. Sivakumar Babu and Chouksey [6] presented an approach for the recycling of plastic waste from water bottles as reinforcing material in geotechnical and civil engineering practice. Arya [7] carried out an experimental study to investigate the dry density and CBR behavior of waste plastic (PET) content on stabilized red mud, fly ash and red mud fly ash mix. Yetimoglu et al. [8] studied the load–penetration behavior of randomly distributed fiber-reinforced sand fills overlying soft clay subgrade by performing laboratory CBR tests and found that adding fiber inclusions in sand fill resulted in an appreciable increase in the peak piston load and it increased with an increase in fiber content. Dutta and Sarda [9] investigate the CBR behaviour of waste plastic strip reinforced stone dust/fly ash overlying saturated clay and reveals that addition of waste plastic strip in stone dust/fly ash overlying saturated clay resulted in an appreciable increase in the CBR and the secant modulus. Tang [10] investigate the effects of discrete short polypropylene fiber (PP-fiber) on the strength and mechanical behavior of uncemented and cemented clayey soil and found that an increase in the unconfined compressive strength (UCS), shear strength and axial strain at failure, decreased the stiffness and the loss of post-peak strength, and changed the cemented soil's brittle behavior to a more ductile one.

The usage of plastic in geotechnical engineering applications is one of the better methods to improve the engineering properties of soil and to control the percentage of plastic waste in environment. In the present study, an attempt is made to use this plastic waste for improving the engineering properties of lime stabilized soil by adding plastic waste strips as reinforcement.

# 2 Materials and Methodology

## *2.1 Materials Used*

### Soils

The soil used for the study is highly compressible soil and is abundantly available in and around Kallur, Tirupati, A.P. The soil properties are shown in Table 1.

### Lime

Quick lime is used in the present study. For obtaining optimum lime content, $P^{H}$ meter test had performed. The $P^{H}$ values of various soil and lime composition. The PH values of 12.4 are taken as optimum percentage of lime if 12.4 value is not attained. So, in successive, highest values of $P^{H}$ the lowest percentage are taken.

### Plastics

Plastic waste used in the study is thrown water bottles, collected from various places in and around Tirupati. These plastic bottles are made into small strips with aspect

**Table 1** Geotechnical properties of soil

| Parameters | | Values |
|---|---|---|
| Coarse-grained fraction (%) | Gravel | 4 |
| | sand | 21 |
| Fine-grained fraction (%) | Silt | 20 |
| | clay | 55 |
| Specific gravity | | 2.52 |
| Liquid limit (%) | | 78 |
| Plastic limit (%) | | 22 |
| Shrinkage limit (%) | | 17 |
| Plasticity index PI (%) | | 56 |
| Classification | | CH |
| Maximum dry density, MDD (g/cm$^3$) | | 1.56 |
| Optimum moisture content, OMC (%) | | 18 |
| Unconfined compressive strength (kg/cm$^2$) | | 3.09 |
| Cohesion (kg/cm$^2$) | | 1.545 |
| CBR % (unsoaked sample) | | 9.54 |
| CBR % (soaked sample) | | 2.18 |
| Free swell index (%) | | 80 |
| Cohesion (kg/cm$^2$) | | 9 |
| $\phi$ (°) | | 0 |

**Table 2** Basic properties of the plastic strips

| Parameters | Properties |
|---|---|
| Resin type | Polypropylene |
| Aspect ratio | 3 |
| Thickness | 200 μ |
| Specific gravity | 0.91 |
| Length × Breadth (UCS and Triaxial), mm | 5 × 15 |
| Length × Breadth (CBR), mm | 8 × 24 |

ratio 3 and mixed with lime stabilized soil. The properties of the plastic strips and the picture of the plastic strips are given in Table 2 and Fig. 1, respectively.

## 2.2 *Soil Composition Indications*

For easy identification of soil samples, soil composition indications are given in Table 3. For obtaining optimum lime content, $P^H$ meter test had performed. The $P^H$ values of various soil and lime composition. The $P^H$ values of 12.4 are taken as optimum

**Fig. 1** Water bottle and strips of plastic bottles

**Table 3** Soil composition indications

| Soil composition | Symbol |
|---|---|
| Soil | $S1$ |
| Soil + 12% lime | $S2$ |
| Soil + 12% lime + 0.25% plastic strips | $S3$ |
| Soil + 12% lime + 0.5% plastic strips | $S4$ |
| Soil + 12% lime + 0.75% plastic strips | $S5$ |
| Soil + 12% lime + 1% plastic strips | $S6$ |

percentage of lime if 12.4 value is not attained. Successive highest values of $P^{H}$ the lowest percentage are taken.

## 2.3 Methodology

Plastic waste strips are mixed at different percentages, i.e., 0.25, 0.5, 0.75 and 1.0% to the dry weight of soil. Unconfined compression test and California bearing ratio test are conducted to determine the strength and CBR values of soil and lime stabilized soil. Mixing of plastic strips in soil should be done carefully such that these strips are distributed uniformly in the soil. The mixing is done manually, and proper care is taken to prepare a homogeneous mixture.

**Unconfined Compression test**

The unconfined compression test gives the undrained shear strength of the soil in a simple and quick way and is determined at which the cylindrical specimen fails in compression or at 15% of strain.

$$q_u = P/A$$

where

$q_u$ = Unconfined compression strength
$P$ = Load at which the specimen fails in compression, kg
$A$ = Corrected area of the specimen, $cm^2$.

Figures 2a, b are showing soil mixed with plastic waste strips and the sample prepared. The load at which these samples fail under compression gives unconfined compression strength. The UCS tests are conducted in accordance with the ASTM D5102-09 [11].

**Triaxial Compression Test**

Triaxial compression test on soil measures the shear parameters of the soil. In this test, soil sample is subjected to stress all around the sample. The test is most widely used and is suitable for all types of soils. The unconsolidated undrained tests are conducted in accordance with the ASTM D2850-03 [12].

$$S = c + \sigma \tan \Phi$$

**California Bearing Ratio test**

The California bearing ratio test is conducted for evaluating the suitability of the subgrade and materials used in sub-base and base course of a flexible pavement. CBR is defined as the ratio of force per unit area required to penetrate a soil mass with a circular plunger of 50 mm diameter at the rate of 1.25 mm/min to that required for corresponding penetration of a standard material. The specimens are prepared in a cylindrical mold of 150-mm diameter and 175-mm height and compacted in three layers at its MDD and OMC based on the standard Proctor compaction. The tests

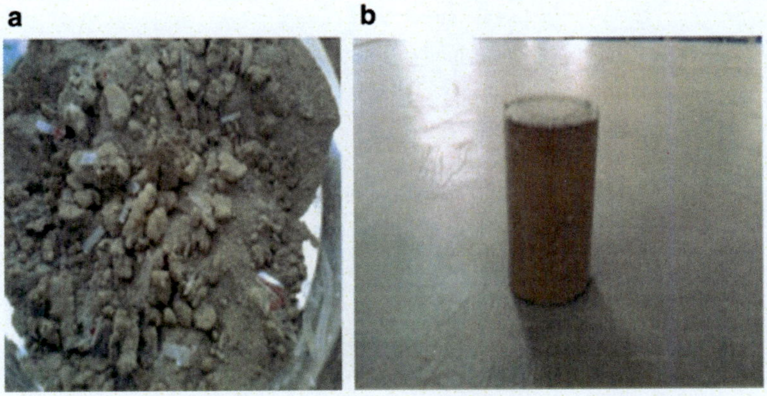

**Fig. 2** **a** Plastic waste strips mixed soil and **b** sample prepared with plastic waste mixed soil

were conducted in accordance with ASTM D1883-07 [13]. The mold is kept under CBR testing machine, and the load corresponding to the 2.5 mm and 5.0 mm is taken from load-penetration curve to determine the CBR values.

$$\text{CBR Value} = (\text{Test load/Standard load}) \times 100$$

Soaked CBR tests are also conducted for examining the performance of plastic waste mixed soil when it is in its worst condition. After compacting the plastic waste mixed soil in CBR mold, the setup is kept submerged in water for about 4 days. The specimen is covered with surcharge mass to simulate the effect of overlying material. After 96 h of submergence, it is taken out and tested to determine the soaked CBR value.

## 3 Results and Discussion

The results obtained from the unconfined compression tests, triaxial tests and California bearing ratio tests performed on plastic waste reinforced lime stabilized soils are discussed below.

### *3.1 Lime Stabilized Soil*

Plasticity of clay is changed on addition of lime, and the values are tabulated under Table 4.

**Table 4** Properties of the lime stabilized soil

| Parameters | Values |
|---|---|
| Liquid limit (%) | 45 |
| Plastic limit (%) | 18 |
| Shrinkage limit (%) | 10 |
| Plasticity index PI (%) | 27 |
| Classification | CL |
| Maximum dry density, MDD (g/cm$^3$) | 1.6 |
| Optimum moisture content, OMC (%) | 22 |
| Unconfined compressive strength (kg/cm$^2$) | 7.69 |
| Cohesion (kg/cm$^2$) | 3.84 |
| CBR % (unsoaked sample) | 15.33 |
| CBR % (soaked sample) | 6.32 |
| Cohesion (kg/cm$^2$) | 13 |
| $\phi$ (°) | 4 |

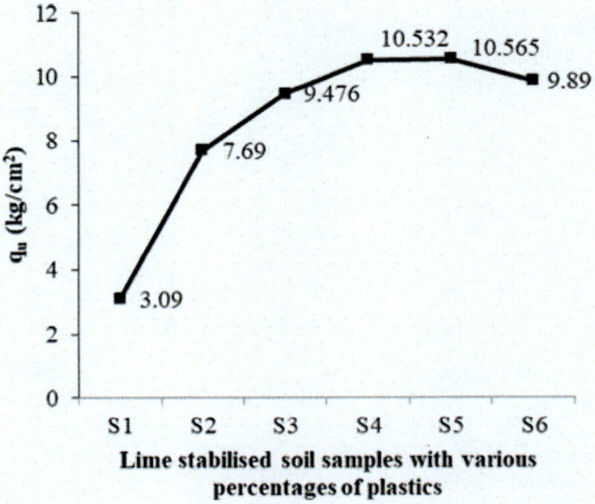


**Fig. 3** Variation of compressive strength with percentage of plastic strips

## 3.2 *Unconfined compression test*

Soil samples are tested for unconsolidated undrained compressive strength. Variation of compressive strength with percentage of plastic strips is shown in Fig. 3, and this indicates that unconfined compression strength of lime stabilized soil is increased due to addition of different percentages of plastic waste.

The increase in unconfined compressive strength of lime stabilized soil is due to developed friction between soil and plastics. And also the plastic strips not allow the soil to expand by arresting the movement of the particles. Lime decreases the swelling and shrinkage characteristics and increases the size of the particle of black cotton soils and thereby improves the strength, whereas plastic strips act as additional support to increase the stability, resist the soil movement and improve the strength.

## 3.3 *Triaxial Test*

The results of triaxial tests indicate that there is appreciable increase in $C$ and $\phi$ values when lime stabilized soil is reinforced with plastic strips. Variations of shear parameters with addition of plastic strips are shown in Figs. 4 and 5.

There is appreciable improvement in cohesion parameter of the soil, and little variation in the angle of internal friction is observed from the given results. This is due to improvement in the bond force that is developed between the plastic strips and soil.

In both UCS and triaxial compression tests, there is decrease in strength, and it is due to the loss of integrity between soil and plastic waste strips. This loss of integrity

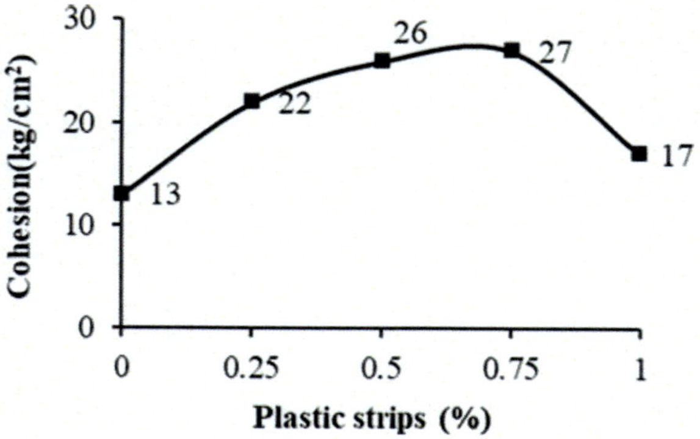


**Fig. 4** Increase of cohesion with increase in percentage of plastic strips

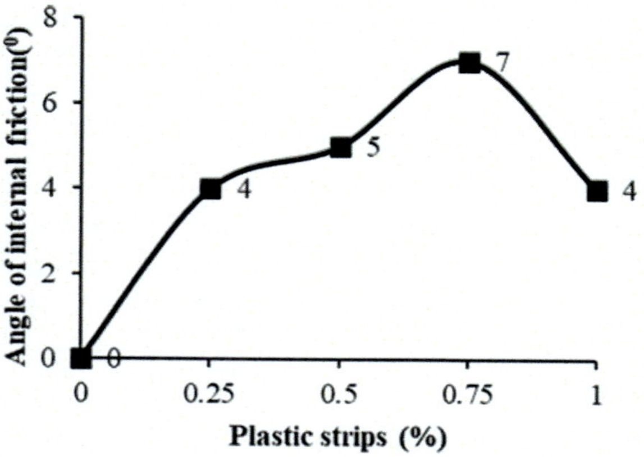


**Fig. 5** Variation of angle of internal friction with the increase in the plastic strips

is due to addition of a large amount of the fiber. Addition of large amount of fibers will cause a slippage and decrease the friction between soil and fiber.

## *3.4 California Bearing Ratio Tests*

**Unsoaked CBR Test**

In unsoaked CBR test, the specimens are prepared by mixing different percentages of plastic waste strips, and then, they are kept under CBR testing machine to determine the CBR values. CBR values of the tested specimens are represented in Fig. 6.

**Soaked CBR Test**

In soaked CBR test, after preparing the specimens, the specimens are immersed in water and soaked for 4 days to simulate the worst condition of soil. After soaking period is completed, the specimens are tested using CBR testing machine (Fig. 7).

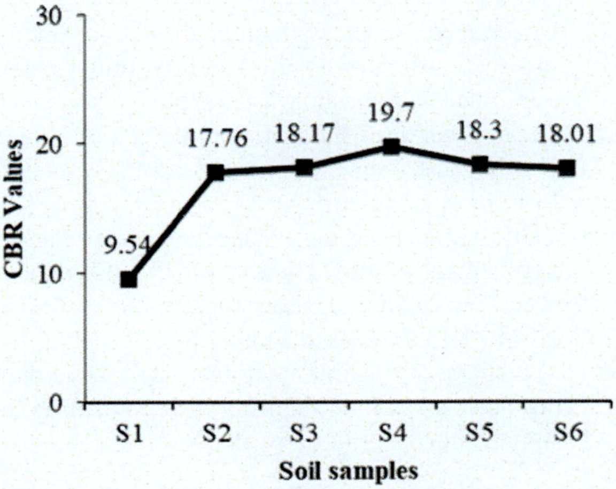


**Fig. 6** Variation of Unsoaked CBR values for different soil samples mixed with various percentage of plastic strips

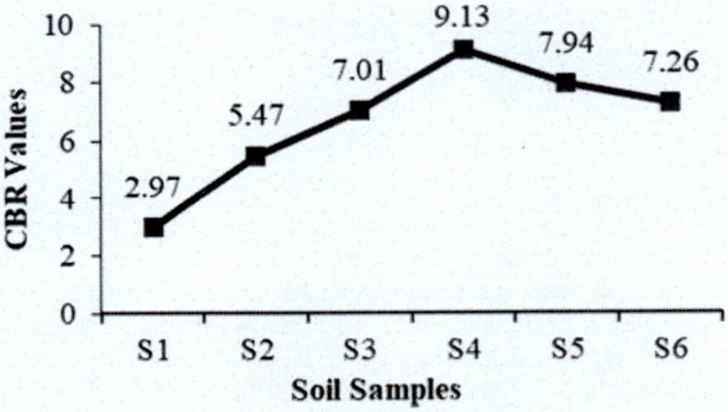


**Fig. 7** Variation of soaked CBR values with different soil samples mixed with various percentage of plastic strips

The CBR values of lime stabilized soils are increased with the increase in plastic waste content. This increase in CBR values is attributed to the resistance offered by the strips present in the soil. Generally, the plunger penetrates into the soil when the specimen is subjected to loading. But in case of plastic waste reinforced soil, the strips present in soil do not allow the plunger to penetrate into the soil so that the fiber-reinforced soil can take more load than plain soil. The penetration resistance of the soil is increased due to the inclusion of plastic waste strips and also due to confinement effect of the mold which is not there in unconfined compression test.

## 4 Conclusions

A series of UCS tests, triaxial tests and CBR tests are conducted on soil stabilized with lime and reinforced with different percentages of plastic strips. The conclusions made from the present study are given below:

a. Unconfined compressive strength of lime stabilized soil is increased to 2.48 times of unstabilized soil. And it is further increased to 1.34 times when it is reinforced with plastic waste strips.
b. Shear strength of the lime stabilized soil is increased by inclusion of waste plastic strips. Angle of internal friction is increased from 0° to 7° and cohesion from 19 to 27 kg/cm$^2$.
c. California bearing ratio values are increasing with increase in percentage of plastic waste strips for lime stabilized soils in both soaked and unsoaked conditions. The values increased from 9.54 to 19.70 for unsoaked condition and from 2.97 to 9.13 for soaked condition.
d. Thickness of flexible pavement will be reduced with inclusion of waste plastic strips in lime stabilized soil. Reduction in thickness of pavement saves significant cost of its construction.

This study has given a scope to reutilization of plastic waste in improving the geotechnical properties of soils and thereby safe disposal of plastic waste. Further studies have to be carried out using different additives to make expansive soils as suitable subgrade.

## References

1. An Overview of Plastic Waste Management: (2013) Central Pollution Control Board, Delhi
2. Consoli, N.C., Montardo, J.P., Prietto, P.D.M., Pasa, G.S.: Engineering behavior of sand reinforced with plastic waste. J. Geotech. Geoenviron. Eng., **128**(6), 462–472 (2002)
3. Consoli, N.C., Vendruscolo, M.A., Prietto, P.D.M.: Behavior of plate load tests on soil layers improved with cement and fiber. J. Geotech. Geoenviron. Eng. ASCE **129**(1), 96–101 (2003)
4. Muntohar, A.S.: Engineering properties of silty soil stabilized with lime and rice husk ash and reinforced with waste plastic fiber. J. Mater. Civ. Eng. **25**, 1260–1270 (2013)
5. Kumar, A., Walia, B.S., Mohan, J.: Compressive strength of fiber reinforced highly compressible clay. Constr. Build. Mater. **20**, 1063–1068 (2005)
6. Sivakumar Babu, G.L., Chouksey, S.K.: Stress–strain response of plastic waste mixed soil. Waste Manage. **31**, 481–488 (2011)
7. Arya, C.S.: A study on CBR behavior of waste plastic (pet) on stabilized red mud and fly ash. Int. J. Struct. Civil Eng. Res. **2**(3), 2319–6009 (2013)
8. Yetimoglu, T., Inanir, M., Inanir, O.E.: A study on bearing capacity of randomly distributed fiber-reinforced sand fills overlying soft clay. Geotext. Geomembr. **23**, 174–183 (2004)
9. Dutta, R.K., Sarda, V.K.: CBR behavior of waste plastic strip reinforced stone dust/fly ash overlying saturated clay. Turkish J. Eng. Environ. Sci. **31**, 171–182 (2006)
10. Tang, C.S., Shi, B., Gao, W., Chen, F.J., Cai, Y.: Strength and mechanical behavior of short polypropylene fiber reinforced and cement stabilized clayey soil. Geotext. Geomembr. **25**(3), 194–202 (2007)
11. ASTM D5102-09 (ASTM 2009) Standard test method for unconfined compressive strength of compacted soil-lime mixtures
12. ASTM D2850-03 (ASTM 2007) Standard test method for unconsolidated-undrained triaxial compression test on cohesive soils
13. ASTM D1883-07 (ASTM 2007) Standard test method for CBR (California Bearing Ratio) of laboratory-compacted soils

# Effect of Tyre Waste Addition on UCS of Bentonite–Sand and Bentonite-Rock Quarry Dust Mixes

**Nazrul Islam, Tinku Kalita, and Malaya Chetia**

## 1 Introduction

Disposal of hazardous waste is one of the greatest challenges to mankind for all times. With expanded natural contamination and its acknowledgement has prompted the requirement for designed engineered waste management facilities. The B:S mixes are mostly used as a liner/barrier material at the municipal solid waste disposal site and also in the nuclear waste repository. The compacted B:S mixes are mostly used as a liner in a landfill and in vertical cut-off walls because of their less vulnerability to desiccation cracking and frost action [1]. The bentonite has a high content of montmorillonite and is mostly used to achieve low permeability in B:S mixes. It has high swelling capacity which can fill up the voids between sand particles. Moreover, bentonite has high compressibility which is undesirable in most practical purposes. The B:S mixes provide low compressibility which is offered by sand in the matrix [2]. Apart from permeability, the strength is the important criteria of B:S mixes and is required to evaluate for the design of the liner in the landfill. The UCS test is one of the quickest, cost-effective and simplest tests for determining the compressive strength. The UCS is useful for determining the shear strength of undisturbed and remoulded samples as well as the bearing capacity of the soil [3]. There are many practical situations like in embankments, shallow footings and retaining walls where the UCS test is most appropriate to use. Kolawole et al. [4] reported that the UCS larger than 200 kPa is most suitable for liner material. It is commonly observed that locally available sand is mixed with bentonite to improve its specific designing qualities like maximum dry density (MDD) [1], shrinkage, shear strength and thermal conductivity. Sridharan et al. [5] proposed that 20% of bentonite is adequate to fill the voids made by the sand framework. Hoeks et al. [6] highlighted that an addition of

N. Islam · T. Kalita · M. Chetia (✉)
Assam Engineering College, Guwahati 781013, India
e-mail: mchetia.ce@aec.ac.in



C. N. V. Satyanarayana Reddy et al. (eds.), *Ground Characterization and Foundations*, Lecture Notes in Civil Engineering 167,
https://doi.org/10.1007/978-981-16-3383-6_36

12% bentonite can achieve the required hydraulic conductivity of $10^{-9}$ m/s. Mollins et al. [7] suggested that the bentonite content of 20% is sufficient to fill up the voids in S:B matrix. A decrease in the hydraulic conductivity was observed with increase in the bentonite content, but after 20% bentonite content, the decrease in the hydraulic conductivity was marginal [2]. Sand, the mostly used fine aggregate is commonly derived from the river banks. The consumption of natural sand is not only limited to geotechnical and geoenvironmental applications but also in construction industries. So, there is need of an alternative material in the near future to overcome the scarcity of sand. The rock quarry dust (Q) is a by-product from the rock crushing process. In quarrying process, the parent rock has been crushed into coarse aggregates; the fine residue generated is called quarry dust and is mostly found as waste. The disposal of such wastes has become a big topic amongst the researchers. Over the last two decades, there is a rapid rise in the number of motor vehicles across the globe. It is estimated that 0.6 million tonnes of tyre scraps are generated annually across India. Nearly 70% of these waste tyres are thrown in landfill or were illegally dumped [8]. So, the scientific utilization of such waste will be beneficial. Ghazavi and Sakhi [9] reported an increase in the angle of shearing resistance from 10 to 94% after addition of tyre chip of 15, 30 and 50% by volume to sand. The maximum improvement was obtained for 50% shreds content of size $4 \times 8$ cm. Gotteland et al. [10] reported an increase in the shear strength of soil-tyre crumb for tyre crumb content up to 34% by mass. Marto et al. [11] found that shear resistance of sand-tyre chips mixture is greater than the sand alone and for optimum tyre mix of 20% the increases in internal friction angle was from 32.8° to 34.2°. Reddy et al. [12] suggested an optimum of 30–40% tyre chips for a sand-tyre mixes is effective for the backfill material. However, there is not much research work available in the literature related to the effect of tyre dust addition on the UCS of B:S and B:Q mixes. The tyre dust consists of the tyre waste having a range of small particle sizes. As the bentonite shows high swelling and shrinkage behaviour, it is expected that the tyre dust can reduce the amount of desiccation cracks and increase the UCS of the mixes. It can also improve the ductility of the mixes. Therefore, this study investigates the effect of tyre dust addition on the UCS of B:S and B:Q mixes. In the first phase of experiments, the effect of full replacement of sand in B:S mix with rock quarry dust on the UCS and failure strain was studied. In the next phase, the UCS and failure strain of B:S:TD and B:Q:TD mixes were determined and compared to find out the optimum tyre dust content and mix. The study was also done to understand the effect of curing period on the UCS of B:S, B:Q, B:S:TD and B:Q:TD mixes.

## 2 Materials and Methods

Locally available sand (S), rock quarry dust (Q), tyre dust (TD) and commercially available bentonite clay (B) of high compressibility were selected for the study. The sand samples were collected from Kulshi riverbank and rock quarry dust was collected from Patharkuchi crusher unit in Kamrup, Assam, India. The basic tests

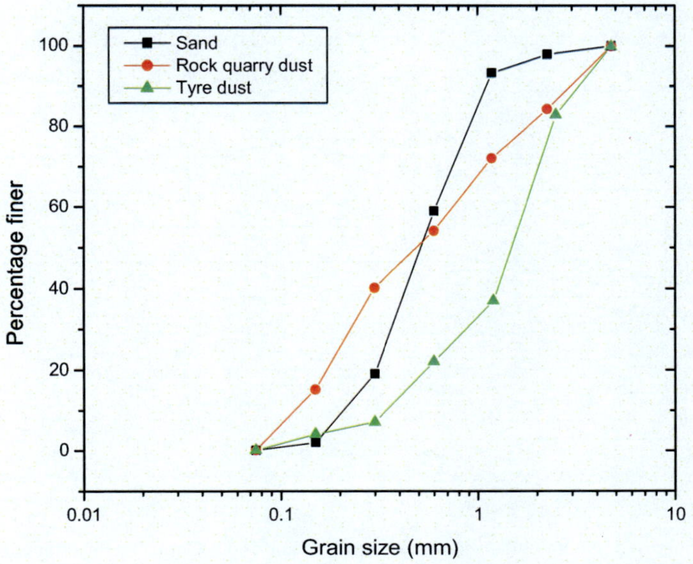


**Fig. 1** Grain size distribution curves of sand, rock quarry dust and tyre dust

were conducted as per Indian standard specifications. The Fig. 1 shows the grain size distribution curves of sand, rock quarry dust and tyre dust. It highlights that the percentage of finer particles is greater in rock quarry dust compared to sand. Both the sand and rock quarry dust are poorly graded. Based on the sizes and range, the tyre dust can be classified as tyre crumbs/granulated rubber [13].

The IS classification of the samples conforming to IS: 1498-1970 [14] was done using liquid limit and plasticity index for cohesive samples and based on particle size fractions for cohesion-less samples. Table 1 shows the physical properties and classification of sand, rock quarry dust and bentonite. All the properties were determined using the guidelines of the respective IS codes.

The mineralogical classification of sand and rock quarry dust was done using electron microscope. The shape and mineral characterization of sand and rock quarry dust are listed in Table 2.

The major mineral present in the sand is quartz with rounded to sub-rounded shape, whereas in rock quarry dust, the major mineral is quartz and substantial amount of feldspar is also present and the shape of minerals are angular to sub-angular.

The experimental investigation was conducted in three phases. In the first phase, the effect of replacement of sand completely by the rock quarry dust on the UCS of the B:S mix was studied. The bentonite content was kept as 30% and sand content as 70% in the mix of B:S and the mix was designated as B30:S70. In the second phase, the effect of tyre dust content on the UCS of B:S and B:Q mixes were investigated to find out the optimum tyre dust content for a particular curing period of the sample. In the third phase, the effect of the curing period on the UCS of the B:S and B:Q mixes was evaluated. The required samples were prepared by adding air-dried bentonite

**Table 1** Physical properties and classification of sand, rock quarry dust and bentonite

| Property | Sand | Rock quarry dust | Bentonite |
|---|---|---|---|
| Specific gravity, G | 2.67 | 2.72 | 2.65 |
| $D_{10}$ (mm) | 0.22 | 0.095 | – |
| $D_{30}$ (mm) | 0.38 | 0.23 | - |
| $D_{60}$ (mm) | 0.64 | 0.85 | – |
| Coefficient of uniformity, $C_u$ | 2.91 | 8.95 | – |
| Coefficient of curvature, $C_c$ | 1.02 | 0.655 | – |
| Liquid limit, $w_L$ (%) | – | – | 207 |
| Plastic limit, $w_P$ (%) | – | – | 62.5 |
| Plasticity index, PI (%) | NP | NP | 144.5 |
| Swelling index (%) | – | – | 883.3 |
| Classification | SP | SP | CH |

**Table 2** Shape and mineralogical characterization of sand and rock quarry dust minerals

| Minerals | Sand, S | | Rock quarry dust, Q | |
|---|---|---|---|---|
| | Amount (%) | Shape | Amount (%) | Shape |
| Quartz | > 95 | Rounded to sub-rounded | > 65 | Angular to sub-angular |
| Biotite | 2–3 | Flaky | 3–4 | Flaky |
| Muscovite | 1–2 | Flaky | 1–2 | Rounded, flakey |
| Feldsper | – | – | > 25 | Elongated, sub-angular to angular |
| Hornblende | – | – | > 1 | Elongated |
| Rock fragments | < 1 | Sub-angular | > 1 | Elongated, sub-angular |
| Others | < 1 | Sub-rounded | – | – |

to the air-dried sand or rock quarry dust in different proportions of the total weight of the mix. The TD was added as a proportion of total weight after the addition of sand or rock quarry dust. The TD proportion ranging from 2 to 16% was added. After thorough mixing, each sample was sealed in an airtight polythene bag and kept inside a desiccator for 24 h for uniform distribution of moisture before the UCS samples were prepared and test was performed. The UCS samples of the mixes were prepared at their corresponding MDD and OMC. The UCS of the samples was determined as per IS guidelines. The test samples prepared were sealed with an airtight polythene bag and kept in desiccator for 3, 7 and 14 days, and then, the tests were performed. The Fig. 2 shows the UCS test setup, and Fig. 3 shows the failure pattern of the B30:Q70 mix sample while performing the UCS test.

**Fig. 2** UCS test setup

**Fig. 3** Failure pattern of B30:Q70 sample

**Table 3** MDD and OMC values for different mixes

| Mix | OMC (%) | MDD (gm/cc) | Mix | OMC (%) | MDD (gm/cc) |
|---|---|---|---|---|---|
| B30:S70 | 15.10 | 1.82 | B30:Q70 | 12.60 | 2.02 |
| B30:S70 + 2%TD | 16.20 | 1.79 | B30:Q70 + 2%TD | 15.80 | 1.98 |
| B30:S70 + 4%TD | 17.00 | 1.75 | B30:Q70 + 4%TD | 16.01 | 1.87 |
| B30:S70 + 6%TD | 17.30 | 1.71 | B30:Q70 + 6%TD | 16.40 | 1.80 |
| B30:S70 + 8%TD | 17.50 | 1.68 | B30:Q70 + 8%TD | 16.82 | 1.76 |
| B30:S70 + 10%TD | 17.61 | 1.67 | B30:Q70 + 10%TD | 17.03 | 1.70 |
| B30:S70 + 12%TD | 17.72 | 1.65 | B30:Q70 + 12%TD | 17.40 | 1.68 |
| B30:S70 + 14%TD | 18.01 | 1.64 | B30:Q70 + 14%TD | 17.90 | 1.65 |
| B30:S70 + 15%TD | 18.05 | 1.63 | B30:Q70 + 15%TD | 18.00 | 1.64 |
| B30:S70 + 16%TD | 18.10 | 1.63 | B30:Q70 + 16%TD | 18.20 | 1.64 |

The compaction of the mixes was done by the Proctor's light compaction method as per IS: 2720-7 [15]. The OMC and MDD values are shown in the Table 3 for all the samples.

## 3 Results and Discussion

### 3.1 *Stress Versus Strain Characteristics of B:S and B:Q Mixes*

The B30:S70 and B30:Q70 mixes were prepared at their corresponding MDD and OMC. The UCS sample was prepared and the UCS test was performed just after the sample was made (curing period is equal to 0). The stress versus strain plots are compared in the Fig. 4.

It is evident from the Fig. 5 that the UCS of B30:Q70 mix is comparatively much higher than that of B30:S70 mix which is 228.13 kPa. The UCS of B30:S70 mix is 163.26 kPa. The UCS of B30:Q70 mix is 39.7% more than that of B30:S70 mix. The failure strain is found to be more in case of B:Q mix than that of the B:S mix.

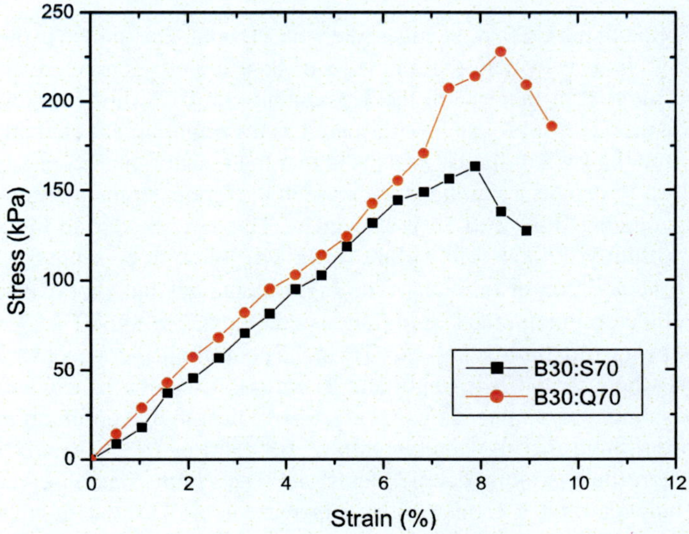


**Fig. 4** Stress versus strain plots of B30:S70 and B30:Q70 mixes for a curing period of 0 day

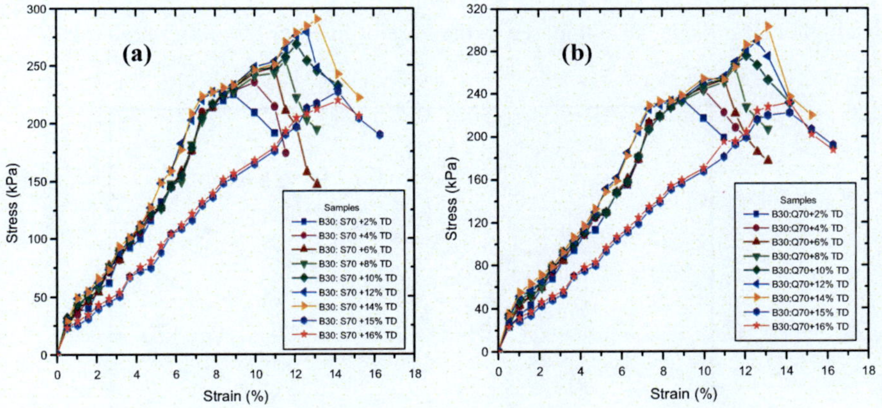


**Fig. 5** Stress–strain curve of **a** B30:S70 and **b** B30:Q70 with varying tyre dust content

## *3.2 Stress Versus Strain Characteristics of B:S:TD and B:Q:TD Mixes*

The tyre dust proportion ranging from 2 to 16% was added and the strength behaviour was studied. The variations of stress with strain were evaluated for B30:S70 and B30:Q70 mix with added tyre dust proportions. The stress versus strain variation of B30:S70 and B30:Q70 mixes for different TD content are shown in the Fig. 5a, b, respectively.

The variation of the UCS of the mixes with the TD content is shown in the Fig. 7a. From the Fig. 7a, it is evident that the UCS of both the mixes increases up to TD = 14% and the further increase in the TD content, the UCS of the mixes reduces abruptly. So, the TD = 14% can be considered as the optimum TD content for both the mixes. But, the UCS of the B:Q mix is higher than that of the B:S mix for TD = 14%. The Fig. 7b depicts a significant increase in the failure strain after the addition of tyre dust ranging from 2 to 16% by weight. This may be due to higher strain mobilized within tyre fibres. The failure strain was found to be same for both the mixes after the addition of tyre dust. The Fig. 6 compares the stress versus strain variations of the optimum mixes. It highlights that the mix B30:S70 + 14%TD can be replaced by the mix B30:Q70 + 14%TD in the geotechnical or geoenvironmental applications where the UCS is important. Therefore, it can be concluded that the sand can be completely replaced by the rock quarry dust in that optimum mix.

It has been found that the combination of B30:S70 or B30:Q70 mix with the addition of tyre dust exhibits ductile behaviour. The failure was observed by the increasing bulging of the specimen with the increase in the TD content in the mix as evident from the Fig. 8. The vertical failure pattern of the specimen can be clearly noticed. This may be due to the enhanced frictional resistance between the rough textures of the tyre dust and the sand or rock quarry dust particles. Also, the stress versus strain curve highlights that the reinforced mix has the potential to sustain up to a higher strain value as compared to the unreinforced or less reinforced mixes.

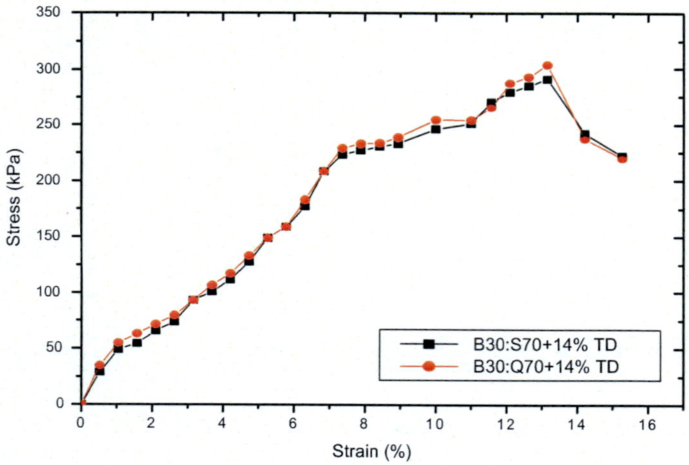


**Fig. 6** Stress versus strain plots of mixes with optimum tyre dust content

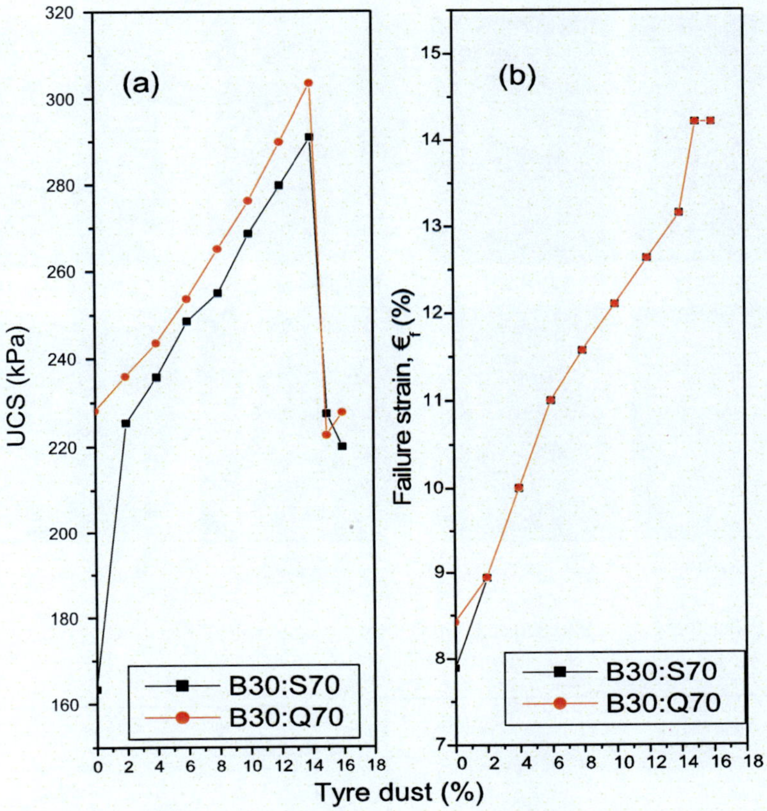


**Fig. 7** Effect of tyre dust content on **a** UCS and **b** failure strain of B30:S70 and B30:Q70 mixes

## 3.3 *Effect of Curing Period on UCS of B:S and B:Q Mixes*

The stress–strain variations of the B30:S70 and B30:Q70 mixes at a curing period of 0, 3, 7 and 14 days are presented in the Fig. 9a, b, respectively. The stress versus strain curve shows an increase in the peak stress value with increase in the curing period for all the test samples. This may be due to the thixotropy or time-dependent hardening. It is mainly due to the tendency of cohesive soil to regain their chemical equilibrium by reorienting the water molecules in the adsorbed water layer. Similar phenomenon was described as isothermal, reversible and time-dependent process of hardening which occurs at constant volume and at rest condition [16, 17]. Subba Rao [18] concluded that the increase in shear strength with curing period was due to the rearrangement of particles and formation of strong bonds during the aging process. The Fig. 10 shows a significant increase in the UCS value of the B30:S70 and B30:Q70 mixes from 0 to 14 days of curing period. The increase in the UCS

**Fig. 8** Failure patterns of B30:S70 and B30:Q70 mix samples with TD content

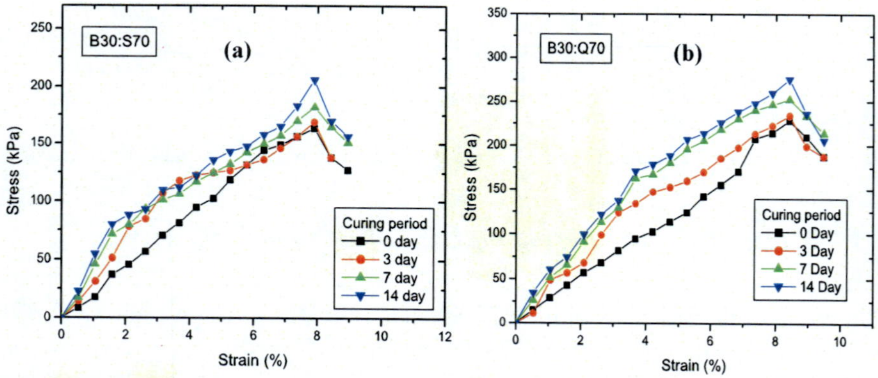


**Fig. 9** Stress versus strain plots of **a** B30:S70 and **b** B30:Q70 mixes at different curing period

for 14 days of curing period was 26.22% for B30:S70 and 20.93% for B30:Q70 mix from 0 day of curing period.

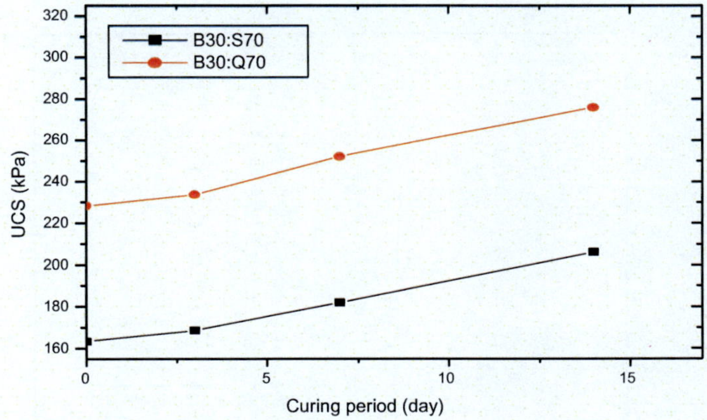


**Fig. 10** Variation of UCS of B30: S70 and B30: Q70 mixes with curing period

### 3.4 *Effect of Curing Period on UCS of B:S:TD and B:Q:TD Mixes*

The Fig. 11a, b shows the effect of the curing period on the stress versus strain characteristics of the optimum mixes B30:S70 + 14%TD and B30:Q70 + 14%TD. The Fig. 12 highlights that the UCS increases linearly for both B30:S70 + 14%TD and B30:Q70 + 14%TD mixes up to 14 days of curing period. The increase in the UCS for 14 days of curing period was 13.05% for the B30:S70 + 14%TD mix and 10.83% for the B30:Q70 + 14%TD mix which is approximately 50% of the increase in the UCS of their corresponding mixes without the addition of the tyre dust. The

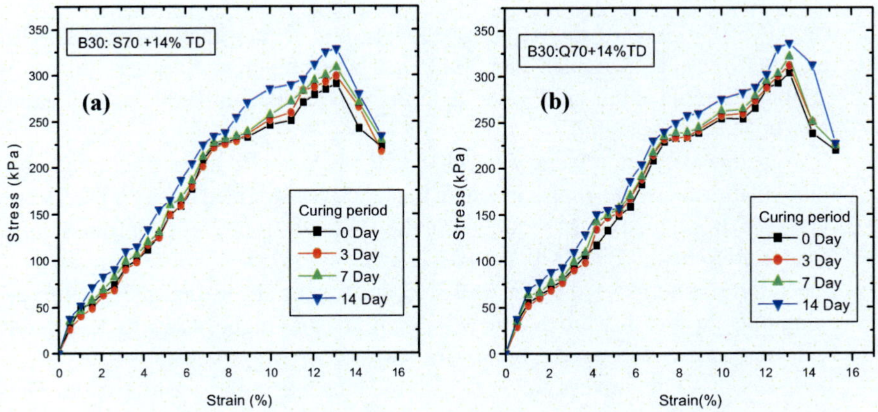


**Fig. 11** Stress versus strain curve of **a** B30:S70 + 14%TD and **b** B30:Q70 + 14%TD mixes of different curing period

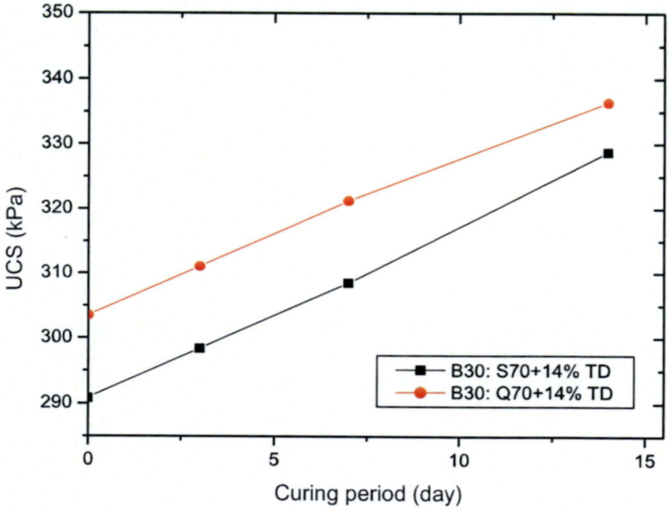


**Fig. 12** Variation of UCS with curing period of B30:S70 + 14%TD and B30:Q70 + 14%TD mixes

effect of the curing period on failure strain was found to be insignificant for these mixes.

## 4 Conclusions

The following conclusions can be made from this study:

1. The UCS of B30:Q70 mix is comparatively higher than that of the B30:S70 mix and the value is greater than 200 kPa. Therefore, the rock quarry dust can completely replace the sand in the B:S mix and the B:Q mix can be a suitable liner material in a landfill.
2. The tyre dust content influences the UCS of the B:S and B:Q mixes significantly. For both the mixes, the improvement in the UCS was obtained up to a TD content of 14%. After that, a sharp reduction in the UCS was found. The optimum mix was found to be the B30:Q70 + 14%TD mix.
3. The failure patterns of the B:S and B:Q specimen are brittle, and the mixes containing the tyre dust are ductile which can sustain a higher strain as compared to the specimens without the tyre dust.
4. The UCS of all the mixes increases with the increase in the curing period. The curing period does not have significant influence on the failure strain of the mixes.
5. All the B30:S70:TD and B30:Q70:TD mixes where, TD = 2–16%, have the UCS greater than 200 kPa. Moreover, for a particular TD content, the UCS of a B30:Q70:TD mix is greater than that of a B30:S70:TD mix, and hence, the mix

B30:Q70:TD can easily replace the mix B30:S70:TD. Therefore, these mixes can be used as liner material though other important properties like sorption capacity, permeability of those mixes etc. should be investigated.

## References


1. Dixon, D.A., Gray, M.N., Thomas, A.W.: A study of the compaction properties of potential clay-sand buffer mixtures for use in nuclear fuel waste disposal. Eng. Geol. **21**, 247–255 (1985)
2. Komine, H.: Simplified evaluation on hydraulic conductivities of sand–bentonite mixture backfill. Appl. Clay Sci. **26**(1–4), 13–19 (2004)
3. Awoleye, O.A., Bouazza, A., Rama-Rao, R.: Time effects on the unconfined compressive strength and sensitivity of clay. Eng. Geol. **31**, 345–351 (1991)
4. Kolawole, J., Osinubi, M., Nwaiwu, C.M.O.: Design of compacted lateritic soil liners and covers. J. Geotech. Geoenviron. Eng. **132**(2), 203–213 (2006)
5. Sridharan, A., Rao, A., Sivapullaiah, P.V.: Swelling pressure of clays. Geotech. Test. J. **9**(1), 24–33 (1986)
6. Hoeks, J., Glas, H., Hofkamp, J., Ryhiner, A.H.: Bentonite liners for isolation of waste disposal sites. Waste Manag. Res. **5**(1), 93–105 (1987)
7. Mollins, L.H., Stewart, D.I., Cousens, T.W.: Predicting the properties of bentonite sand mixtures. Clay Miner. **31**(2), 243–252 (1996)
8. Pasalkar, A., Bajaj, Y., Wagh, A., Dalvi, J.: Comprehensive literature review on use of waste tyres rubber in flexible road pavement. Int. J. Eng. Res. Technol. 685–689 (2015)
9. Ghazavi, M., Sakhi, M.A.: Influence of optimized tire shreds on shear strength parameters of sand. Int. J. Geomech. ASCE **5**, 58–65 (2005)
10. Gotteland, P., Lambert, S., Balachowski, L.: Strength characteristics of tyre chips-sand mixtures. Studia Geotechnicaet Mechanica **27**(1–2), 25–36 (2005)
11. Marto, A., Latifi, N., Moradi, R., Oghabi, M., Zolfeghari, S.Y.: Shear properties of sand-tire chips mixtures. Electron. J. Geotech. Eng. **18**, 325–334 (2013)
12. Reddy, S.B., Kumar, D.P., Krishna, A.M.: Evaluation of the optimum mixing ratio of a sand-tire chips mixture for geoengineering applications. J. Mater. Civ. Eng. 28(2) (2016)
13. ASTM D6270: Standard practice for use of scrap tires in civil engineering applications, American Society for Testing and Materials, West Conshohocken, United States (2008)
14. IS: 1498: Indian standard classification and identification of soils for general engineering purposes. Bureau of Indian Standards, New Delhi (1970)
15. IS: 2720-Part 7: Determination of water content-dry density relation using light compaction-Indian standard methods of tests for soils. Bureau of Indian Standards, New Delhi (1980)
16. Mitchell, J.K.: Fundamentals of soil behavior. Seconded, Wiley, New York (1993)
17. Błażejczak, D., Horn, R., Pytka, J.: Soil tensile strength as affected by time, water content and bulk density. Int. Agro Phys. **9**(3), 179–188 (1995)
18. Subba Rao, K.S., Tripathy, S.: Effect of aging on swelling and swell-shrink behavior of a compacted expansive soil. Geotech. Test. J. **26**(1), 36–46 (2003)

# Consolidation: Critical Appraisal of Settlement Versus Rate of Settlement (SRS) Approach with Fuzzy Logic

**Sudhir Kumar Tewatia, Malaya Chetia, Taslima Nasrin, and Kanishck Tewatia**

## 1 Introduction

Some simple and very short concepts are sometimes path breaking in the development of research in some subjects. One such concept was given by Terzaghi [1] as:

$$\sigma = \sigma' + u \tag{1}$$

where $\sigma$ is total pressure, $\sigma'$ is effective pressure, and $u$ is pore-water pressure. This was to pave the way for the development of modern soil mechanics, particularly in clays. Similarly, since the days of Newton, differential equations have been used for solving problems mostly as below

$$y = f(x);\ \ y' = \mathrm{d}y/\mathrm{d}x = f_1(x);\ \ y'' = \mathrm{d}^2y/\mathrm{d}x^2 = f_2(x) \tag{2}$$

That means differential equation was used as a function of $x$. Little practice was there to use it as a function of $y$. This, particularly, created the problems when $x$ is not known. Similarly happened in consolidation where all researchers were using $\delta$-$t$ or $\delta$-$\sigma$ plots. The solution failed when $t$ or $\sigma$ is not known at the time of load increment and that was the usual case in the field. Terzaghi [2] gave the average

S. K. Tewatia
Jimma University, Jimma, Ethiopia

M. Chetia · T. Nasrin (✉)
Assam Engineering College, Guwahati 781013, India

K. Tewatia
Mindsarovar Technologies Private Limited, Bangalore, India



C. N. V. Satyanarayana Reddy et al. (eds.), *Ground Characterization and Foundations*, Lecture Notes in Civil Engineering 167,
https://doi.org/10.1007/978-981-16-3383-6_37

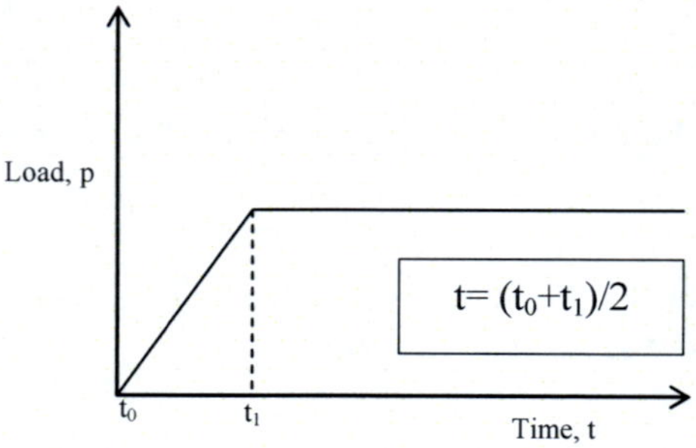


**Fig. 1** Uniform ramp loading [2]

time of loading concept in case of continuous uniform loading that, however, was rough/untrue assumption (Fig. 1).

Tewatia [3–19] suggested the $y - f(y')$ and $y - f(y'')$ plots. In consolidation, it is called the settlement versus rate of settlement, SRS approach. They use characteristics of degree of consolidation, $U$ versus theoretical velocity, $dU/dT$, and $U$ versus $dT/dU$ plots in linear and semi-log formats where $U$ is the degree of consolidation and $T$ is the time factor in Terzaghi's one-dimensional consolidation equation

$$U = 1 - \frac{8}{\pi^2} \sum_{N=0}^{N=\infty} \frac{1}{(2N+1)^2} \text{Exp}\left(-\frac{(2N+1)^2 \pi^2}{4} T\right) \quad (3)$$

Earlier, Eq. 3 was resolved in two parts; (1) parabolic 0–60% $U$ by Fox [20] as

$$T = \frac{\pi}{4} U^2 \quad (4)$$

and (2) exponential 60–100%$U$ as

$$U = 1 - \frac{8}{\pi^2} \text{Exp}\left(-\frac{\pi^2}{4} T\right) \quad (5)$$

Beginning of the secondary consolidation in the range of the primary consolidation was a hypothesis, but the SRS approach separated it quantitatively since its beginning. Creep and secondary consolidation (considered to be same) were defined separately due to high-resolution power of the SRS approach [14, 15]. Terzaghi's equation was resolved in two parts parabolic Eq. 4 and exponential Eq. 5. Nothing was known of the part where it changes its nature from parabolic to exponential. Only three types of consolidation settlements were known initial, primary, and secondary, but there may exist six phases. The $c_v$ was calculated using a substantial data and portion of the $\delta$-$t$ plot, and it was unnecessarily time consuming in the laboratory and field. All the methods failed when time, settlement, and pressure at the instant of load increment were not known. All the methods provided $c_v$ in the laboratory that was affected by secondary consolidation. To compare the calculated $c_v$ with true $c_v$ (i.e., $c_v$ that is not affected by secondary consolidation), one had to compare the calculated values of hydraulic conductivity with the measured values of hydraulic conductivity. All such problems are solved by the SRS approach very easily and quickly.

In spite of all such outstanding clear merits and its publications in various most reputed journals for about 24 years, the approach could not get momentum. Instead, the reputed professors of reputed universities started stealing this approach from ASTM Geotechnical Testing Journal [3], Springer Journal of Geotechnical and Geological Engineering [15], and ASCE International Journal of Geomechanics [17] and published in most reputed geotechnical journals like ASTM Geotechnical Testing Journal [21], Applied Clay Science [22] and Géotechnique [23], etc. There was a parliament question also in India in 1997 on research espionage of this technique. This paper is an attempt to find the merits and demerits of the SRS approach in consolidation with further possible improvements like application of fuzzy logic and parabolic fitting [24] for finding the $d\delta/dt$, etc. The fuzzy logic is an approach to computing, based on "degree of truth" in which the truth values of variables may be any real number between 0 and 1 both inclusive. It is used to represent the concept of partial truth, where the truth value may range between completely true and completely false.

## 1.1 Settlement Versus Rate of Settlement (SRS) Approach

By definition

$$U = \frac{\delta - \delta_0}{\delta_{100} - \delta_0} \tag{6}$$

$$T = \frac{c_v t}{H^2} \tag{7}$$

In Eq. 3, $U$ is differentiated with respect to $T$. The plots of $U$-$T$ and $U$-$dU/dT$ in various formats are given in the Fig. 2a, b [3]. In Fig. 2b, the symmetrical $S$ curve is divided into three parts: 1. Parabolic 0–40%$U$, 2. Exponential 60–100%$U$, and 3. Transition zone $U = 40$–$60\%$. From the Fig. 3a.

$$\delta_0 = \delta - \frac{s}{2.3026} \tag{8}$$

and in Fig. 4b.

$$\delta_{100} = \delta + \frac{s}{2.3026} \tag{9}$$

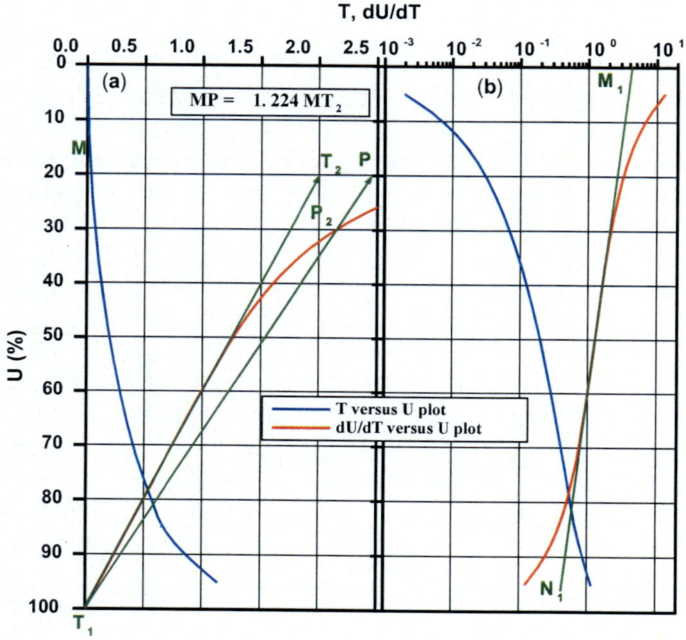


**Fig. 2** Theoretical $U$ versus ($T$, d$U$/d$T$) plots in **a** linear and **b** semi-log formats

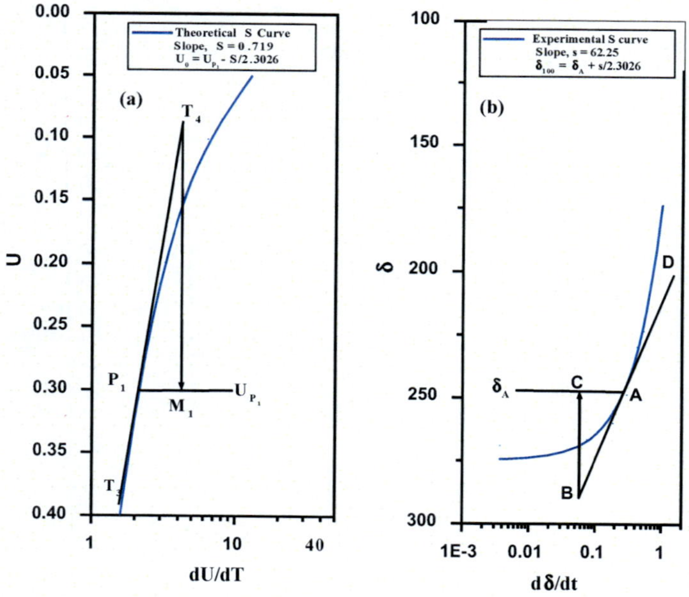


**Fig. 3** $S$ curve **a** theoretical plot for finding $\delta_0$ and **b** experimental plot for finding $\delta_{100}$

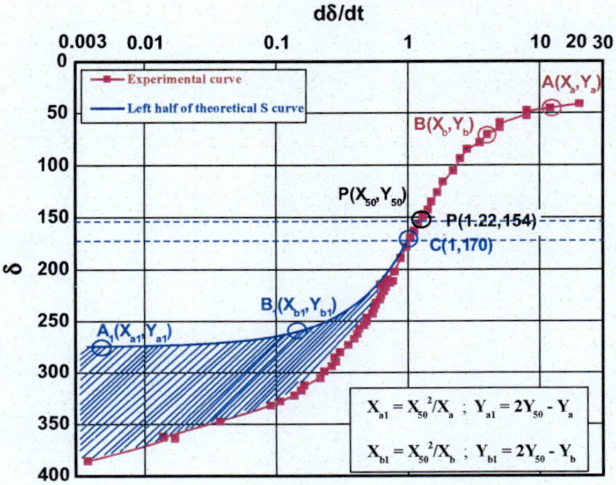


**Fig. 4** $\delta$ versus $\mathrm{d}\delta/\mathrm{d}t$ semi-log plot for Sawan Bhado (SB) dam soil for isolation of secondary consolidation [3, 4, 15]

where $s$ is the slope of experimental $\delta$-$\log_{10}(\mathrm{d}\delta/\mathrm{d}t)S$ curve. The slope at the point of inflection,

$$
\begin{aligned}
s_{50} &= 1.009(\delta_{100} - \delta_0) \approx (\delta_{100} - \delta_0) \\
&= \text{amount of primary consolidation}
\end{aligned}
\tag{10}
$$

The Fig. 3a shows the theoretical $U$ versus $\mathrm{d}U/\mathrm{d}T$ plot for finding $\delta_0$. The Fig. 3b shows the experimental $\delta$ versus $\mathrm{d}\delta/\mathrm{d}t$ plot for finding $\delta_{100}$. The Fig. 4 shows the quantitative isolation of secondary consolidation from experimental plot of *SB* dam soil [3, 4, 15].

$$Y_{a1} = Y_{50} + (Y_{50} - Y_a) \text{ and } \log X_{a1} = \log X_{50} + (\log X_{50} - \log X_a)$$

Hence,

$$Y_{a1} = 2Y_{50} - Y_a, \text{ and } X_{a1} = X_{50}^2/X_a \tag{11}$$

## 2 Determination of $c_\upsilon$, $\delta_0$ and $\delta_{100}$ Using $\delta$-($\mathrm{d}\delta/\mathrm{d}t$) Plot

Tewatia [3] derived

$$\delta = \frac{4H^2}{\pi^2 c_{\mathrm{v}}} v + \delta_{100} \tag{12}$$

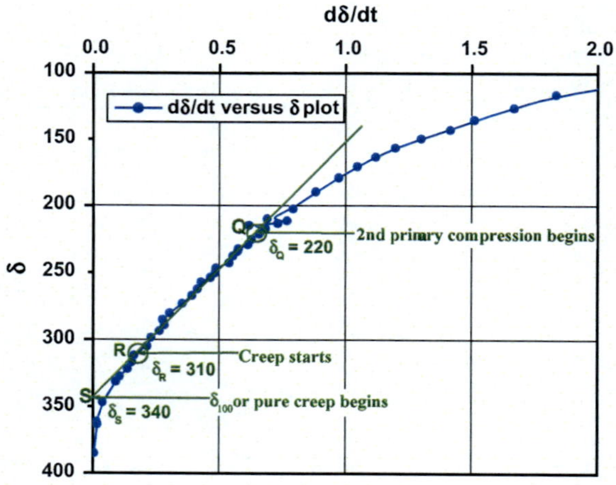


**Fig. 5** $\delta$ versus $\mathrm{d}\delta/\mathrm{d}t$ linear plot for SB dam soil for finding $c_v$ and $\delta_{100}$

where $H$ is the drainage path. The Eq. 12 is the equation of straight line in the form: $y = m\,x + c$, where $m$ is slope and $c$ is intercept on $y$-axis. The $\delta$-($\mathrm{d}\delta/\mathrm{d}t$) plot is a straight line having a slope, $m = (4H^2)/(\pi^2 c_\upsilon)$ and intercept on $\delta$ axis, $c = \delta_{100}$. Thus, $c_\upsilon$ can be evaluated as:

$$\delta = \frac{4H^2}{\pi^2 c_v} v + \delta_{100} \tag{13}$$

As per Terzaghi's assumptions if $c_\upsilon/H^2$ is constant, then $\delta$-($\mathrm{d}\delta/\mathrm{d}t$) curve should be a straight line. As secondary consolidation starts and runs superposed over primary consolidation, therefore, $c_\upsilon/H^2$ is not constant. Equation 12, therefore, gives a curve, considerable portion of which is straight line (Fig. 5). When this line is extrapolated to cut $\delta$ axis, it gives $\delta_{100}$ (Fig. 5), and $\delta_0$ can be determined as from the Fig. 2a [4]

$$\delta_0 = \delta_{100} - \left(\frac{\delta_{100} - \delta_{30}}{0.70}\right) \tag{14}$$

## 2.1 Determining True $c_v$ and Fuzzy Logic for $c_v$

The first step is to plot $\delta$-($\mathrm{d}\delta/\mathrm{d}t$) on semi-log scale as shown in the Fig. 4. Draw tangent in the middle straight-line portion and find its slope $s_{50}$ over one log cycle as in the Figs. 2b and 6 [3, 4, 15]. The $\delta_0$ is found using Eq. 8, and $\delta_{100}$ is determined using Eq. 10 as

$$\delta_{50} = 0.5(\delta_0 + \delta_{100}) = \delta_0 + 0.5(\delta_{100} - \delta_0) = \delta_0 + 0.5s_{50} \tag{15}$$

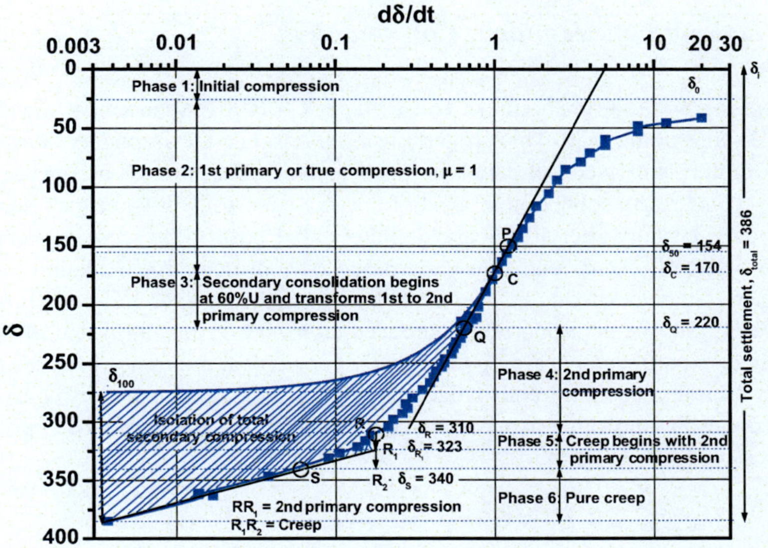


**Fig. 6** Six phases of consolidation for SB dam soil

$$c_v = \frac{0.8033\left(\frac{d\delta}{dt}\right)_{50}}{s_{50}} H^2 \quad (16)$$

$$\mu = \frac{t_{50}\left(\frac{d\delta}{dt}\right)_{50}}{(0.245)(s_{50})} \quad (17)$$

where $\mu$ is a fuzzy logic that measures the trueness of $c_v$ [3, 4, 15]. It's value varies between 0 and 1. Closer is $\mu$ to 1, closer is $c_v$ to the true $c_v$. The true $c_v$ is the one that is devoid of the effect of secondary consolidation. The $\mu$ and $c_v$ values determined by various methods for various soils are given in the Table 1.

**Table 1** $\mu$ and $c_v$ values determined by various methods for various soils.

| Method | $c_v$ ($\times 10^{-5}$ cm$^2$/s) | | | |
|---|---|---|---|---|
| | SB dam soil | BC soil | Bentonite–sand mix | Bentonite |
| | $w_L = 61\%$<br>$w_P = 29\%$ | $w_L = 69\%$<br>$w_P = 33\%$ | $w_L = 100\%$<br>$w_P = 30\%$ | $w_L = 495\%$<br>$w_P = 49\%$ |
| Casagrande [25] | 2.58 | 12.3 | 3.92 | – |
| Taylor [26] | 3.31 | 16.3 | 4.07 | 1.44 |
| SRS [3, 15]<br>Fuzzy value, $\mu$ | 4.82<br>1 | 24.5<br>0.97 | 5.57<br>0.94 | 2.81<br>0.92 |

### 2.2 Isolation of Secondary Consolidation

On semi-log plot in Fig. 2b, unlike *U*-*T* *S* curve, *U*-(d*U*/d*T*)*S* curve is symmetrical about the mid-point, *U*50. This property is used to isolate the secondary consolidation from the primary consolidation in Fig. 4. Up to 50%*U*, there is no (or insignificant) secondary consolidation in most of the inorganic soils. It is primary consolidation only. Usually after 50%*U* and definitely after 60%*U*, the secondary consolidation essentially starts and runs superposed over primary consolidation [3, 10, 15]. Therefore, the experimental curve deviates from theory after mid-point [usually 50–60%*U*]. So, the upper portion is retraced after 50%*U* (i.e., *δ50*) that gives the theoretical experimental curve up to 100%*U*. The vertical difference between these two curves is the secondary consolidation in Figs. 4 and 6. The secondary consolidation is defined as any other experimental compression that is not initial and primary compression [3, 4, 15, 27].

### 2.3 Six Phases of Consolidation Settlement

Literature shows three phases of consolidation only. They are (i) initial compression, (ii) primary compression, and (iii) secondary compression. The other phases could not be visualized because of the limitations of all available methods before the SRS approach. The SRS approach shows six phases of consolidation due to its high-resolution power. They are initial compression, first primary compression, transition from first primary compression to second primary compression, second primary compression, and transition from second primary compression to creep and lastly creep (Figs. 4, 5, 6 and 7).

### 2.4 Isolation of Creep

To isolate creep (Fig. 7), we draw a vertical line from *R* where creep starts. Extrapolate the straight line of later portion of the *S* curve. The point where vertical line cuts the extrapolated line is the point below which we get creep from the combination of 2nd primary compression and creep [15].

## 3 Limitations of SRS Approach

The plot of SRS has a weakness that it requires *δ*-*t* data to be recorded very precisely [28]. Unlike IS, ASTM or BS codes the settlement is recorded first at different intervals of time and later time in seconds is recorded when dial gauge needle coincides

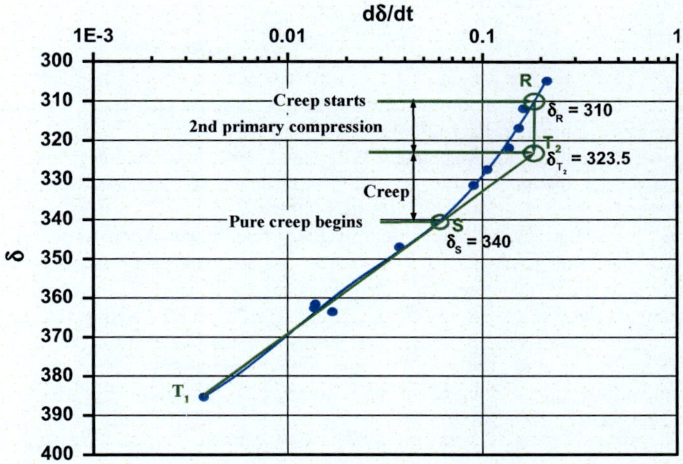


**Fig. 7** $\delta$ versus $\mathrm{d}\delta/\mathrm{d}t$ semi-log plot for SB dam soil

with the exact mark on the dial gauge [3, 4, 15]. The SB dam soil data were recorded like this, while other soils data were recorded as usual. Though this method works well in usual data recording system, but still the parabolic fitting of slope ($\mathrm{d}\delta/\mathrm{d}t$) can be used for better accuracy [28]. In this method, only three consecutive $\delta$-$t$ data points are fitted by a parabolic curve, and the slope at 2nd point is taken as the $\mathrm{d}\delta/\mathrm{d}t$. However, now very precise instruments for measuring even less than a micron settlement are available, so the precise data recording is no longer a problem.

# 4 Discussion and Conclusions

The $\delta$-($\mathrm{d}\delta/\mathrm{d}t$) method can determine $c_{\upsilon,}$ $\delta_{0,}\delta_{100}$ and (hence, using Eqs. 3, 6, and 7) entire unknown $\delta$-$t$ data, when the time and settlement at the instant of load increment are not known, by observing data just for a few minutes (theoretically zero time or a point) as $\delta$-($\mathrm{d}\delta/\mathrm{d}t$) plot (Fig. 5). It is not capable of giving some quantitative estimate to show how far the $c_v$ is from the true $c_v$ but a fuzzy logic. The method is capable of isolating not only creep, but the whole secondary consolidation that runs superposed over the primary consolidation. The $\delta$-log($\mathrm{d}\delta/\mathrm{d}t$) plot can give true $c_v$ as well as all the six phases of consolidation. No other method is available in literature that determines $c_v$ at less than 70%$U$. The $\delta$-log($\mathrm{d}\delta/\mathrm{d}t$)*or S* curve plot requires data upto 50%$U$ to determine the value of $c_v$. Thus, the *S* curve plot is the fastest rapid loading method for vertical consolidation that takes 1/4th time of Taylor method as $T_{50} \approx (T_{90})/4$.

## References

1. Terzaghi, K.: Mathematisch-naturewissenschaftliche Klasse. Akademie der Wissenschaften in Wien, Sitzungsberichte, Part Iia **132**(3/4), 125–138 (1923)
2. Terzaghi, K.: Theoretical Soil Mechanics. Wiley Inc., New York, London (1943)
3. Tewatia, S.K.: Evaluation of true $c_v$ and instantaneous $c_v$, and isolation of secondary consolidation. Geotech. Test. J. **21**(2), 102–108 (1998)
4. Tewatia, S., Tewatia, K.: Consolidation of Soils: Rate of Settlement Approach, 1st edn. Lambert Academic Publishing, LAP, Riga Latvia (2019). https://www.morebooks.de/store/gb/book/consolidation-of-soils:-rate-of-settlement-approach/isbn/978-620-0-31301-0; https://www.researchgate.net/publication/336739313_CONSOLIDATION_OF_SOILS_RATE_OF_SETTLEMENT_APPROACH_RSA_A_Scholarly_book_on_the_State-of-the-Art
5. Tewatia, K., Tewatia, S.K., Rath, S.: Discussion on 'Evaluation of consolidation results by the settlement rate approach' by Olek, B.S., Borecka, A., Wozniak, H.: (2016). Electron. J. Geotech. Eng. **21**(07), 1–9 (2019)
6. Tewatia, K., Tewatia, A., Rath, S.: Discussion of 'Fastest rapid loading methods of vertical and radial consolidations' by Tewatia, S.K., Bose, P.R., Sridharan, A.: (2013). Int. J. Geomech. **15**(1), 1 (2015)
7. Tewatia, S.K.: Time dependent behavior of clayey soils. Ph.D. Thesis, Department of Civil Engineering, Delhi College of Engineering, Delhi University, India (2010)
8. Tewatia, S.K.: Trend of settlement in primary and secondary consolidations. Geomech. Geo-Eng. **8**(2), 125–134 (2012)
9. Tewatia, S.K.: Equation of 3D consolidation in Cartesian co-ordinates. Int. J. Geotech. Eng. **7**(1), 105–108 (2013)
10. Tewatia, S.K.: Principle of super position of rate of settlement in 2D and 3D consolidations. Geotech. Geol. Eng. **33**(5), 1369–1378 (2015)
11. Tewatia, S.K., Venkatachalam, K.: Closure for discussion on 'improved $\sqrt{t}$ method to evaluate consolidation test results.' Geotech. Test. J. **21**(2), 155–156 (1998)
12. Tewatia, S.K., Venkatachalam, K., Sridharan, A.: T-chart to evaluate consolidation test results. Geotech. Test. J. **21**(3), 270–274 (1998)
13. Tewatia, S.K., Bhargav, D.N., Chouhan, G., Sridharan, A.: Velocity method for creep of clayey soils. Electron. J. Geotech. Eng. **7**(Bundle B) (2002)
14. Tewatia, S.K., Bose, P.R.: Discussion on 'a study on the beginning of secondary compression of soils' by Robinson, R.G. J. Test. Eval. **34**(5), 1–6 (2006)
15. Tewatia, S.K., Bose, P.R., Sridharan, A., Rath, S.: Stress induced time dependent behavior of clayey soils. Geotech. Geol. Eng. **25**(2), 239–255 (2007)
16. Tewatia, S.K., Sridharan, A., Singh, M., Rath, S.: Theoretical equations of vertical and radial consolidations by equating degrees of consolidation by settlement analysis and dissipation of pore pressure. Geotech. Geol. Eng. **30**(4), 1037–1043 (2012)
17. Tewatia, S.K., Bose, P.R., Sridharan, A.: Fastest rapid loading methods of vertical and radial consolidations. Int. J. Geomech. **13**(4), 332–339 (2013). https://doi.org/10.1061/(ASCE)GM.1943-5622.0000213
18. Tewatia, S.K., Huat, B., Rath, S., Tewatia, K.: U-chart for quick evaluation of radial and 3D consolidation test results. Ground Improv. **166**(4), 188–195 (2013)
19. Tewatia, S.K., Bose, P.R., Sridharan, A.: Closure to 'fastest rapid loading methods of vertical and radial consolidations'. Int. J. Geomech. **15**(1), 1–3 (2015)
20. Fox, E.N.: In: Proceedings of 2nd International Conference on Soil Mechanics and Foundation Engineering, vol. 1, pp. 41–42 (1948)
21. Al-Zoubi, M.S.: Settlement rate-settlement method for radial consolidation (SRSM-RC). Geotech. Test. J. **37**(5), 907–916 (2014)
22. Al-Zoubi, M.S.: Consolidation analysis using the settlement rate-settlement (SRS) method. Appl. Clay Sci. **50**, 34–40 (2010)
23. McKinley, J.D., Sivakumar, V.: Coefficient of consolidation by plotting velocity against displacement. Géotechnique **59**(6), 553–557 (2009)

24. Singh, S.K.: Confined aquifer parameters from temporal derivative of drawdowns. J. Hydraul. Eng. **127**(6), 466–470 (2001)
25. Casagrande, A.: Notes on Soil Testing for Engineering Purposes. Soil Mechanics Series No. 8, Publication No. 268, Howard University, Cambridge, Mass (1939)
26. Taylor, D.W.: Fundamentals of Soil Mechanics. Wiley Inc., New York (1948)
27. Olson, R.E.: State of the art: consolidation testing. In: Young, R.N., Townsend, F.C. (eds.) Consolidation of Soils: Testing and Evaluation, ASTM STP 892, pp. 7–70, West Conshohocken, PA (1985)
28. Tewatia, S.K.: Discussion and closure on 'comparison of the hyperbolic and Asaoka observational method of monitoring consolidation with vertical drains' by Tan and Chew, 1996'. Soils Found. Jap. Geotech. Soc. **38**(2), 224–227 (1998)

# Characterization of Dispersive Soils

**Sameer Vyas, Beena Anand, Rajeev Kumar, and S. L. Gupta**

## 1 Introduction

Dispersive soils are highly susceptible to erosion and piping phenomenon and undergo deflocculation in the presence of flowing water. It rapidly erode forming tunnels and deep gullies by a process in which the individual clay particles go into suspension in slow-moving water (colloidal erosion), damaging earth dams, canals, and other hydraulic structures [1, 2].

The failure initiated by piping makes the embankments constructed on dispersive soil susceptible. Dispersive piping in dams has occurred either on the first reservoir filling or, less frequently, after raising the reservoir to highest level.

Dispersive soils contain higher content of dissolved sodium ions in pore water than ordinary soils, The tendency for dispersive erosion in a given soil depends on variables such as mineralogy and chemistry of the clay, as well as dissolved salts in the water in soil pores and in the eroding water [3–5].

Dispersive soils cannot be differentiated from ordinary soils by conventional soil mechanics tests [6, 7]. An investigation in which four different laboratory tests for dispersion were performed on a considerable number of soils of diverse origins and properties has provided improved understanding of the properties of dispersive soil and strengthened identification criteria [8].

### 1.1 *Mechanism of Clay Dispersion*

Clay particles are negatively charged as the result of isomorphous substitution or broken edges, so they naturally repel each other (like charges repel). However,

S. Vyas (✉) · B. Anand · R. Kumar · S. L. Gupta
Central Soil and Materials Research Station, CSMRS, New Delhi 110016, India



C. N. V. Satyanarayana Reddy et al. (eds.), *Ground Characterization and Foundations*, Lecture Notes in Civil Engineering 167,
https://doi.org/10.1007/978-981-16-3383-6_38

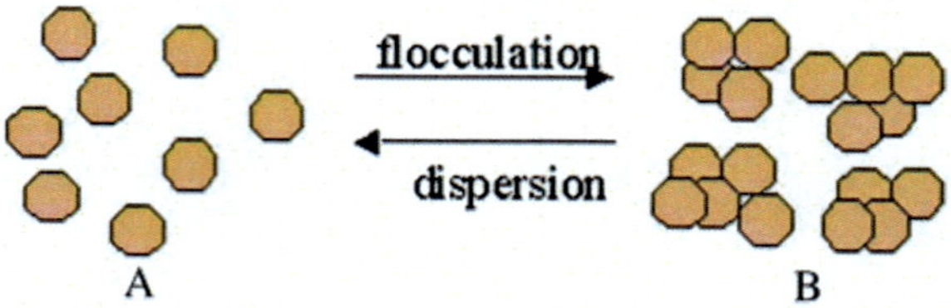


**Fig. 1** Soil particles in a dispersed and flocculated condition

the presence of adsorbed cations tends to mask this negative charge (repulsion) to varying degrees, depending on the type of cation. Small multi-charged cations (i.e., $Al^{+3}$, $Ca^{+2}$, $Mg^{+2}$) are strongly adsorbed by colloids (unlike charges attract), greatly reducing the negative charge. If the negative charge of the colloid is sufficiently reduced, flocculation will occur as represented and shown in Fig. 1a. If, on the other hand, the colloids are saturated with large weakly charged cations (i.e., $Na^{+}$, $K^{+}$), dispersion will occur as represented in Fig. 1b. Sodium, a weakly charged cation that encourages dispersion and calcium, a multi-charged cation that encourages flocculation, provides an example of the intricate relationship between soil chemistry and the physical condition of a soil. The importance of the physical condition and the influence of sodium creates the situation where knowing the concentration of this cation in a soil is essential.

The flocculation and dispersion reactions are the result of cation exchange. Cation exchange is the interaction between a cation in solution and another cation on the surface of any surface-active material, such as clay or organic matter. The cation exchange that occurs between sodium and calcium on clay mineral surfaces.

## *1.2 Diffuse Double Layers*

Clay particles some times behave as a colloidal particles in suspension. The diffuse double layer occurs at the interface between the clay surface and the soil solution. It is made up of the permanent negative charge of the clay and the cations or counter-ions in the soil solution that balance the negative charge. The counter-ions are influenced by two equal but opposing forces—the electrical force attracting the positive ion to the negative surface, and the diffusive or thermal forces (responsible for Brownian motion) which tend to move the cations away from the surface. The balance of these two forces gives rise to a distribution of cations in water adjacent to the clay surface. This distribution, described as a diffuse electrical double layer or simply diffuse double layer, is made up of the negative clay surface and the spread-out (diffuse) distribution of the counter-ions.

When the electrical diffused double layers of the two clay particles carrying similar outer charge come close to each other, they repel due to electrostatic repulsion and thus initiated colloidal dispersion phenomenon.

### 1.3 Dispersivity Due to Sodic Soils

When two clay particles with a high concentration of sodium counter-ions sit close to one another, their double layers overlap or interact as presented in Fig. 2.

As a consequence, the total concentration of the ions at the plane mid-way between the two particles is greater than that in the soil solution in which the particles are immersed. This creates a difference in osmotic pressure which will draw water between the particles, causing them to move further apart—this is the swelling associated with sodic soils. In the presence of free water (e.g., excess rainfall or low electrolyte irrigation water) at a soil surface, a sodic soil may move a stage further in disruption, so that the particles become dispersed in this water. Dispersion can be decreased to a considerable extent in the presence of high concentrations of electrolytes.

Most studies reported in the literature have shown that failures of structures built of dispersive clay soils occurred on first wetting. All failures were associated with the presence of water and cracking by shrinkage, differential settlement, or construction deficiencies. These failures emphasize the importance of early recognition and identification of dispersive clay soils; otherwise, the problems that they cause can result in sudden, irreversible, and catastrophic failures.

The common soil classification index tests do not distinguish between dispersive and non-dispersive clay soils. The recommended tests for the identification of dispersive clay soils are pinhole test, crumb test, double hydrometer test, and chemical analysis of pore water extract.

The objectives of this study is to established a confirmatory test for identification of soil dispersivity by chemical analysis of pore water extract and mineralogical identification. Conducting different test on soil samples and correlating the results, it was observed that the chemical test including determination of total dissolved salts (TDS), percent sodium, and mineralogical examinations are very important parameters which need to be conducted properly which is also going to help us in understanding their influence on other engineering tests.

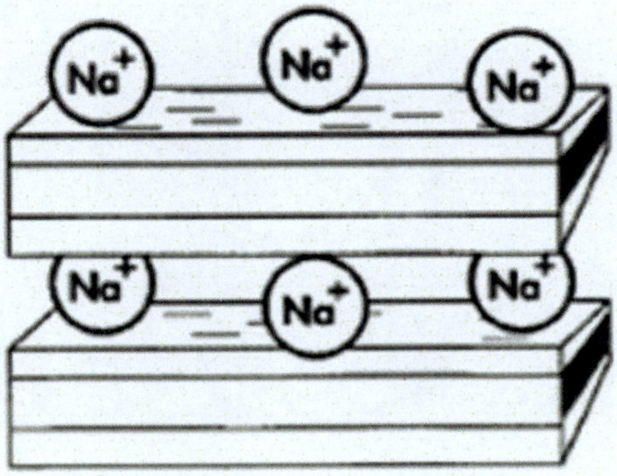


**Fig. 2** Negatively charged clay particle with adsorbed sodium counter-ions

# 2 Experimental

## 2.1 Material and Methodology

The four different types of soil samples were selected for this study based on engineering tests conducted on these soil samples.

## 2.2 Chemical Tests

The chemical analysis of soils pore water extract seem to the most reliable methods for characterizing soil dispersivity. The phenomenon of dispersivity is due to the presence of dissolved sodium content in pore water, and therefore, its percentage was calculated by estimating total dissolved cations (Na, K, Ca, and Mg) present in soils pore water extract. To obtain saturation extract, soil is mixed with distilled water until a saturated soil paste with water content near the liquid limit is obtained. The paste is allowed to set for a number of hours until equilibrium is attained between the salts in the pore water and those on the cation exchange complex. Subsequently, a small quantity of pore water is filtered from the soil paste using a vacuum. This extracted pore water is tested using EDTA titration method (For Ca and Mg) and using flame photometer (For Na and K) (Fig. 3).

The percent sodium and TDS (sum of the four metallic cations) are determined. The further interpretation were evaluated using graph as shown in Fig. 4

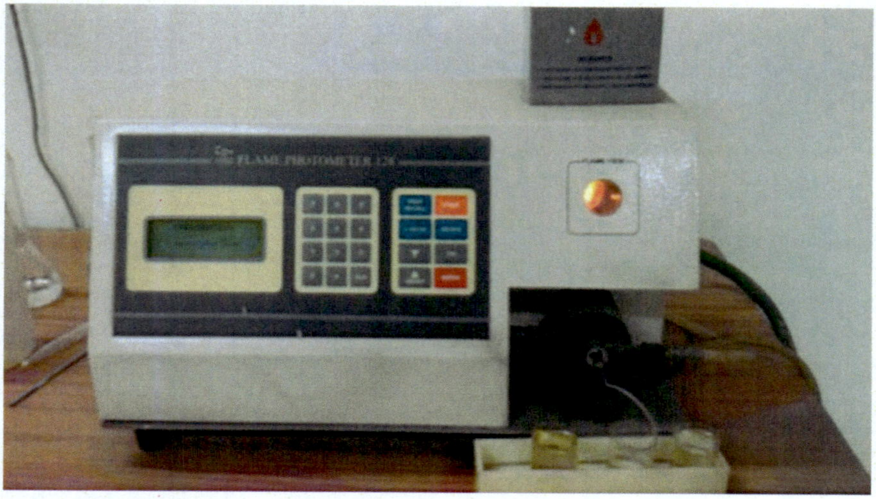

Fig. 3 Determination of sodium and potassium using flame photometer

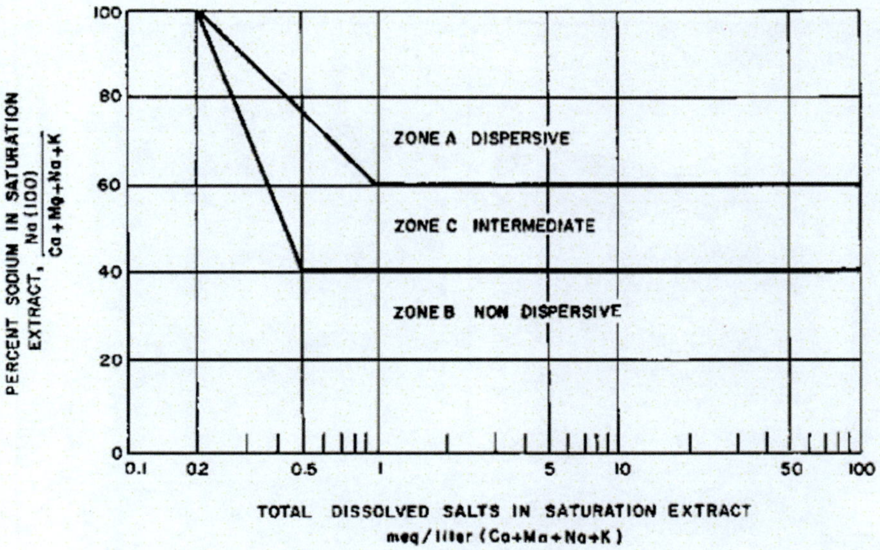


**Fig. 4** Plot of percent sodium verses total dissolved cations

After plotting the above graph for each samples, the results were categorized as follows:-

Zone A: Dispersive
Zone B: Non-dispersive
Zone C: Intermediate.

## 2.3 *Mineralogy Study Using X-Ray Diffraction (XRD Tests)*

XRD is a technique used extensively in the geological sciences, materials, environmental science, chemistry, physics, metallurgy, and other subjects. The instrumental setup is presented in Fig. 5

# 3 Results and Discussion

The results of chemical analysis of pore water extracts of all the four samples are presented in Table 1.

The first sample was fall in intermediate category (C) while all others samples were fall in category (A), which indicated their dispersive nature.

**The XRD pattern of soil samples is presented in** Figs. 6 and 7.

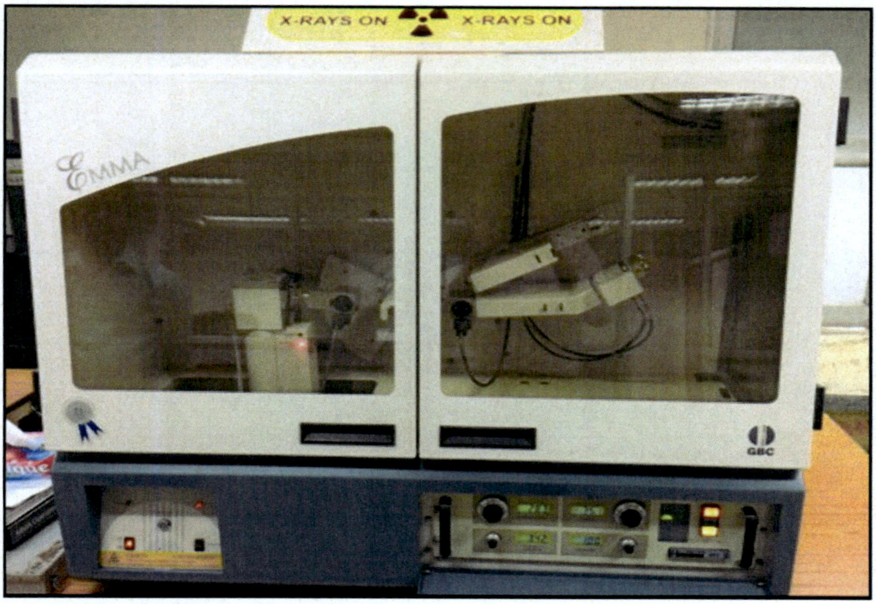


**Fig. 5** Mineralogy study with XRD

**Table 1** Test results of pore water analysis of soil samples

| Sample no | mEq/L | | | | Total cations | Sodium % | Remarks grade |
|---|---|---|---|---|---|---|---|
| | Na | K | Ca | Mg | | | |
| 1 | 2.19 | 0.47 | 0.8 | 0.4 | 3.86 | 56.73 | C |
| 2 | 3.20 | 0.69 | 0.6 | 0.2 | 4.69 | 68.23 | A |
| 3 | 4.12 | 1.01 | 1.0 | 0.6 | 6.73 | 61.21 | A |
| 4 | 3.44 | 0.80 | 0.6 | 0.2 | 5.04 | 68.25 | A |

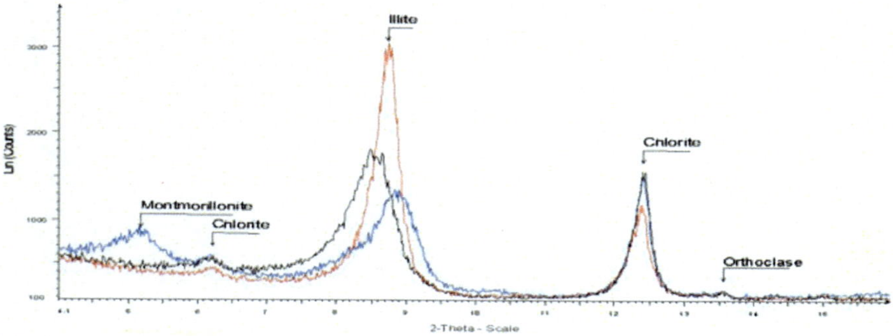


**Fig. 6** XRD pattern of dispersive and intermediate soil samples

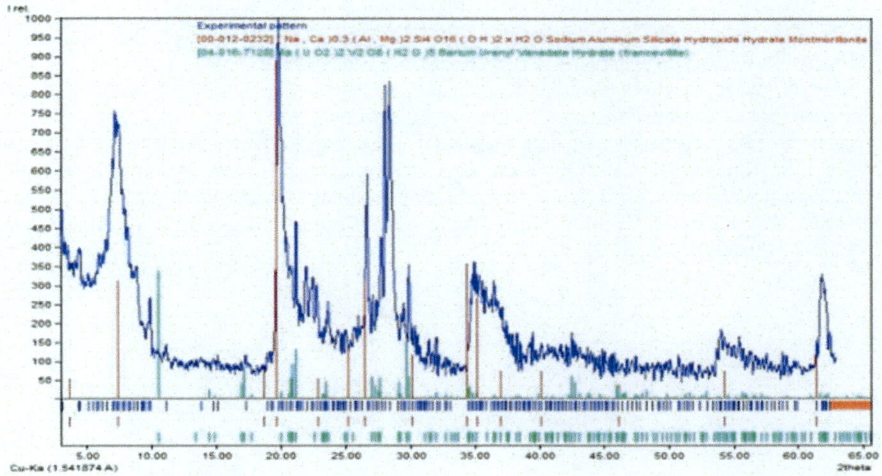

**Fig. 7** Typical XRD pattern of dispersive soil

The XRD pattern of soil samples shows that soil with dispersive and intermediate nature is rich in montmorillonite and illite clay minerals. Both of these clay minerals are having high colloidal activity and responsible for soil dispersion.

## 4 Conclusions

The mechanism of dispersivity of soil is, however, based on surface chemistry and depends on types of cations present in soil pore water in dissolved state. Therefore, the importance of chemical tests are of great concern. The results of percent sodium and TDS are need to be evaluated and correlated to clearly understand and designate a sample under the title of dispersive nature. The results of chemical analysis of pore water extract of soils and mineralogical examination with XRD clearly show that soils with high sodium content and contain montmorillonite minerals such as smectite and illite are dispersive and intermediate in nature.

## References


1. Sherard, J.L., Dunnigan, L.P., Decker, R.S.: Identification and nature of dispersive soils. J. Geotech. Eng. **102**(GT4), 298–312 (1976)
2. Sherard, J.L., Decker, R.S.: Introduction in dispersive clays, related piping and erosion in geotechnical projects. ASTM STP **623**, 362–369 (1977)
3. Ouhadi, V.R., Yong, R.N.: Experimental and theoretical evaluation of impact of clay microstructure on the quantitative mineral evaluation by XRD analysis. Appl. Clay Sci. J. **23**(1–4), 141–148 (2003)

4. Seta, A.K., Karathanasis, A.D.: Water dispersible colloids and factors influencing their dispersibility from soil aggregates. Geoderma **74**, 225–266 (1995)
5. Nagy, G., Nagy, L., Kopecskó, K.: Examination of the physico—chemical composition of dispersive soils. Period. Polytech. Civ. Eng. **60**(2), 269–279 (2016)
6. Sayehvand, S., Dehghani, M.: Identification and management of dispersive soils. EJGE **9** (2014)
7. Neetu, S., Sameer, V., et al.: Characterization of dispersive soils—a comparative evaluation between available tests. Int. J. Innov. Res. Sci., Eng. Technol. IJIRSET **4**(90) (2015)
8. Umsha, T.S., Dinesh, S.V., Sivapullaiah, P.V.: Characterization of dispersive soils. Mater. Sci. Appl. **2**, 629–633 (2011)

# Study of Maliya Marine Clay for a Highway Embankment

Anandji Kalyanji Shah

## 1 Introduction

An alignment of a highway embankment linking Ahmedabad to Kandla port passes through the Little Rann of Kachchh, where it encounters a 6.5 km wide belt of marine clay deposit of thickness up to 15 m+ . The Rann Little of Kachchh is an extremely flat, coastal alluvium plain—originating from Gulf of Kachchh; it runs WSW-ENE and then S–N for 96 km with varying widths of 8–50 km. The Rann of Kachchh was formerly a sea basin. Silting, upheaving, man-made, and natural bunds, and turning away and thinning of rivers created a Rann.

Along the highway embankment the Rann constitutes a tidal swamp, a soft marshy ground, and is extremely flat with ground level at almost 2.4 m R.L. It is within the reach of daily sea tides of 2.8 m R.L. The high tide level of the year is 3.3 m R.L., and the maximum flood level is at 5.1 m R.L. The maximum height of the highway embankment is kept at 6.9 m at an approach of the bridge over an intruding creek. It is 15 m + thick, normally consolidated, soft [0.25 kgf/cm$^2$ < $q_u$ < 0.5 kgf/cm$^2$], to very soft [$q_u$ < 0.25 kgf/cm$^2$, ~ 25 kPa]. A man can hardly walk over it. It is of low shear strength, low permeability, and high compressibility with $0.6 < C_c < 1.05$. This paper describes an engineering design for a typical embankment considering 7.5 m height and 12 m top width on this marine clay deposit incorporating sand drains and berms.

A. K. Shah (✉)
Applied Mechanics Department, Faculty of Technology and Engineering, The M. S. University of Baroda, Vadodara 390001, India

 
C. N. V. Satyanarayana Reddy et al. (eds.), *Ground Characterization and Foundations*, Lecture Notes in Civil Engineering 167,
https://doi.org/10.1007/978-981-16-3383-6_39

## 1.1 Site and Soil Profile

The nearest villages to this area are Maliya in the south, 40 km from Morbi, and village Surajbari in the north. The upper surface of the marine clay deposit is at ground level, and along the highway alignment, its thickness varies from 3 m at chainage 4800 m (from Maliya side) to 15.5 m+ at chainage 10,800 m. Then, onward it reduces to 3 m thickness at chainage 13,995 m. With facilities from Gujarat Engineering Research Institute, the author made one bore hole up to 12 m depth below G.L. at chainage 10,500 m. The upper portion of the deposit is relatively stiff and penetrates below G.L. for a depth up to 2.8 m.

## 1.2 Properties of Maliya Marine Clay Deposit

Presence of sufficient quantity of electrolyte (salt) in sea water gives rise to edge to face structure and flocculation during sedimentation process of fine-grained soil. Actually, marine clay deposits are arranged in a more random three-dimensional orientation [1]. This concept was substantiated by Rosenquist [2], through electron micrograph. In addition, slowly increasing geological loading simply pushes particle edges to other particle faces and bonds particles in a natural space frame [3]. This process allows slowly increasing load to be carried without substantial reduction of voids ratio or water content of sediments below.

**Properties of Maliya Clay in brief**

Marine clays are of high moisture content of 80% or more, near to or above liquid limit, except for upper relatively stiff crust portion with 40–50% (Fig. 1).

Mechanical analysis shows that silt size particles are predominant. Silt size particles: 60–75%, clay size particles: 22–35%, sand size particles: 2–13%.

Accordingly, Maliya marine clays are inorganic clay of high compressibility. They mostly fall in CH group on Casagrande's chart. A few fall in CI and MH group. As per PRA classification they fall on A-7 and A7-6 group. Standard proctor maximum dry density MDD is 16.35 kN/m$^3$ at optimum moisture content OMC of 18%, and compression index $C_c = 0.6$ to 1.05. Coefficient of consolidation is from (50–100) $\times$ 10$^{-5}$cm$^2$/s. From circular box shear strength test, cohesion intercept, $C_{cu} = 0$, and an angle of shearing resistance, $\phi_{cu} = 18^{\circ}$. Average shear strength, as determined by laboratory vane shear tests and unconfined compression tests, all on undisturbed samples procured from 0 to 12 m depth is 16.6–22.5 kPa.

**Stiff crust**. Hard surface layer overlying softer soils is generally the case with soft marshy ground [4]. According to them it is formed by the effect of desiccation at upper surface that is not permanently flooded. Moum and Rosenquist [5] state it to be the result of weathering effect. Maliya marine clay deposit is in a marshy swamp, under effect of daily tides in a semi-arid region. Accordingly, desiccation has caused an upper stiff crust of thickness from 1.2 to 3 m.

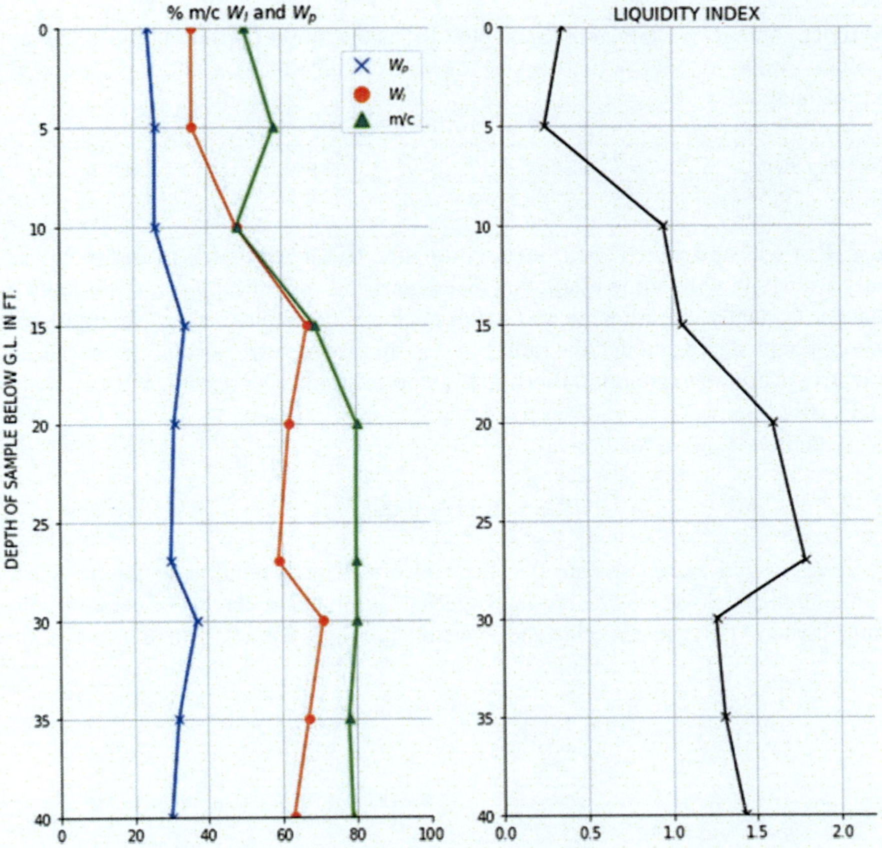


**Fig. 1** Consistency limits. w = moisture content, wL = liquid limit, wP = plastic limit

**Leaching**. Leaching reduces salt content from pore water. As per [6], the reduction of salt content largely reduces remolded strength. Bjerrum [7] states that it reduces shear strength of both the remolded samples, and undisturbed samples, and holds out the reduced strength due to leaching to be the cause of natural or otherwise undisturbed Norwegian clay slopes. Maliya marine clays do not seem to have been subjected to leaching.

**Sensitivity**. Sensitivity is the loss of strength on remolding at unaltered water content. It is more pronounced in marine clays. It is measured by the ratio of undisturbed strength to remolded strength at constant water content. In very soft samples, the strengths may be determined by vane shear tests. Current concepts attribute loss of strength partly to orientation of particles and partly due to reduction in inter-particle forces. Laboratory vane shear tests on undisturbed samples of Maliya marine clay were performed. Sensitivity values were seen to be in range 3–8, in one case 10.

**Activity**. Activity is defined by Skempton [8] as the ratio of plasticity index to clay fraction. Value for Maliya marine clay varies from 0.94 to 1.21, and it is termed as normal clay.

| | | | |
|---|---|---|---|
| Activity value | < 0.75 inactive | 0.75–1.4 normal | > 1.4 active |

**$S_{u/P}$, Ratio of undrained shear strength to effective overburden pressure**. It was only after 1945 with refined sampling techniques and vane shear tests; it was known that for normally consolidated soft deposits also $S_u$ (undrained shear strength) did increase with depth and that the ratio $S_{u/p}$ was approximately constant where p is an effective overburden pressure. Bjerrum [7] reported it to be increasing with plasticity index (Fig. 2).

Skempton [8] gave relation as

$$S_{u/p} = 0.11 + 0.0037 I_p$$

Terzaghi and Peck [9] state that the above statistical relation so far has been found applicable over a wide range of types of sedimented clays. For the normally consolidated Maliya marine clay, the value of $S_{u/P}$ ranges from 0.25 to 0.28, while the

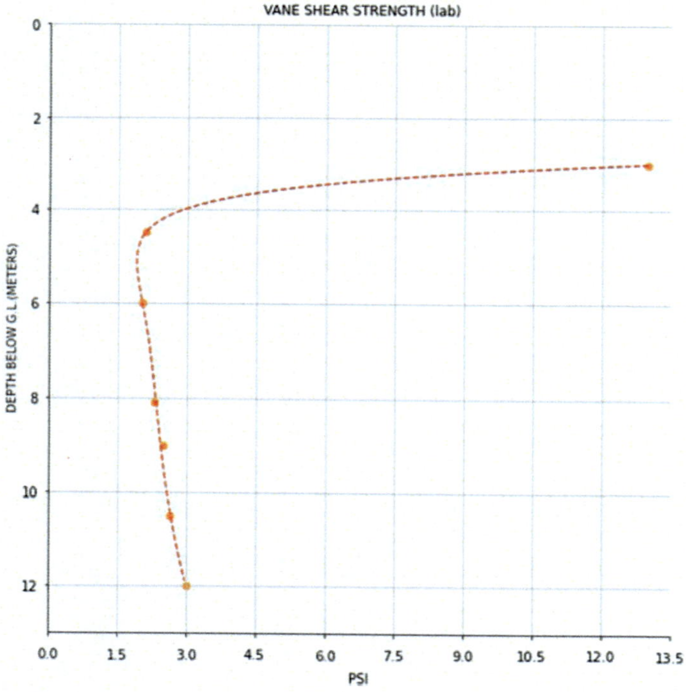


**Fig. 2** Vane shear strength

plasticity index is in range of 25–35%. For samples procured from depth 4.5–12 m below G.L., the average value of $S_{u/p}$ is found to be 0.286. Corresponding average value of $I_P$ is 32.5%. Skempton's relation gives a value of $S_{u/p}$ to be 0.23.

## 2 Shear Strength Tests

The undisturbed samples were collected in sampling tubes, one tube per 1.5 m depth up to 12 m below ground. The tubes were carefully sealed and brought to the laboratory.

### 2.1 *Circular Box Direct Shear CU Consolidated Test*

Consolidated undrained circular box direct shear tests were performed using undisturbed samples. In order to totally avoid the observed compression disturbance on pushing sample out of sampling tube, the tubes were cut into pieces of the size just sufficient to get required best samples of 2 cm height. The circular shape of the box mold ensured that the disturbance, if any, to the sample while transferring to the mold did not occur. Also the change in area of the contact area of the enforced plane of shear, up to failure would, at all strains, be less in case of circular mold as compared to rectangular mold.

**Procedure**. The diameter of the samples was 6.0325 cm. Load was applied in increments of 25 lb, varying from 0 to 125 lb. These correspond to increments of 5.65 psi, or 39 kPa, giving the normal stress, б, to be of (39, 78, 117, 156, 195) kPa.

The time required for 100% consolidation:

$$T = T_v\left[H^2/c_v\right] = 1.3\left[(1)^2/\left(50 \times 10^{-5}\right)\right] = 43.3 \text{ min}$$

The samples were consolidated under normal load for 45 min and then sheared at fast rate of 0.1275 cm/min, $dL/L_0 = 2.105$, say 2% per minute. The vertical dial gauge readings were recorded for obtaining the changes in the volume in the sample during the consolidation and also the shearing process. Bulk unit and dry unit weight at the beginning of the test for each sample and the moisture content at failure for a few samples were determined.

**Analysis of test results and discussion**. Shear stress versus horizontal strain curves are presented here for the undisturbed sample designated Maliya/1256 obtained from depth 10.5 m below G.L. The curve with 'yield point' is observed. Lambe [10] states that cohesion is generally mobilized and destroyed before other components of strength become active. Schmertmann and Osterberg [11] demonstrated that 'not only does the cohesion component develop very rapidly, but in some clays the cohesion reaches maximum and then decreases measurably at decreasing rate while friction

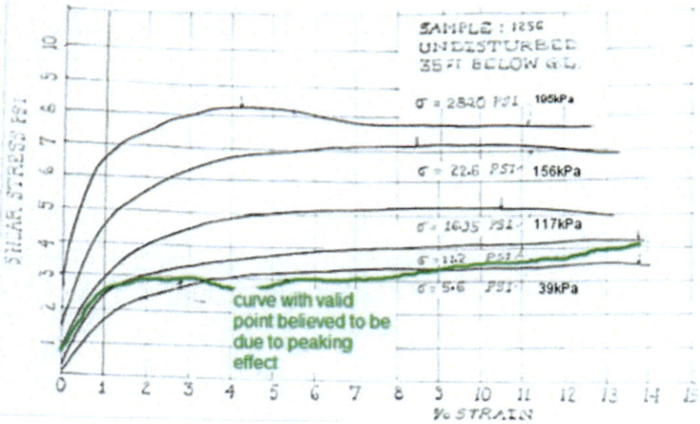


**Fig. 3** Peaking effect in stress–strain curve plot

tends to steadily increase'. Hvorslev [12] states that 'being destroyed' cohesion is not being fully compensated by friction, the cohesion passes its peak and decreases more rapidly. Pronounced peaking effect exhibits a 'noticeable yield point' in the stress–strain curve, as shown in Fig. 3.

Lines for shear strength τ versus normal stress б pass through origin for normally consolidated clays when samples are tested in the laboratory at the normal pressure б greater than the existing effective overburden pressure in the field. But in the case of the same samples when tested in the laboratory under normal stress which is less than the corresponding effective overburden, cohesion intercept appears on *y*-axis and it is found to be proportional to existing effective over burden pressure to which the sample in the field has been consolidated. In short, as per [12],

$$\tau = \sigma'_p \tan \phi'_c + \sigma'_f \tan \phi'_r$$

Now, for normally consolidated clays, $\sigma'_p = \sigma'_f$, hence

$$\tau = \sigma'_f (\tan \phi'_c + \tan \phi'_r)$$
$$= \sigma'_f \tan \phi'_s$$

And the strength line shear strength τ versus normal stress б passes through origin giving cohesion intercept to be zero for normally consolidated clays, as shown in Fig. 4. In this, $C_{cu} = 0$ and $\phi_{cu} = \text{avg}18.5°$. In other cases, the corresponding average values are: $C_{cu} = 21$ kPa and $\phi_c = 4°$.

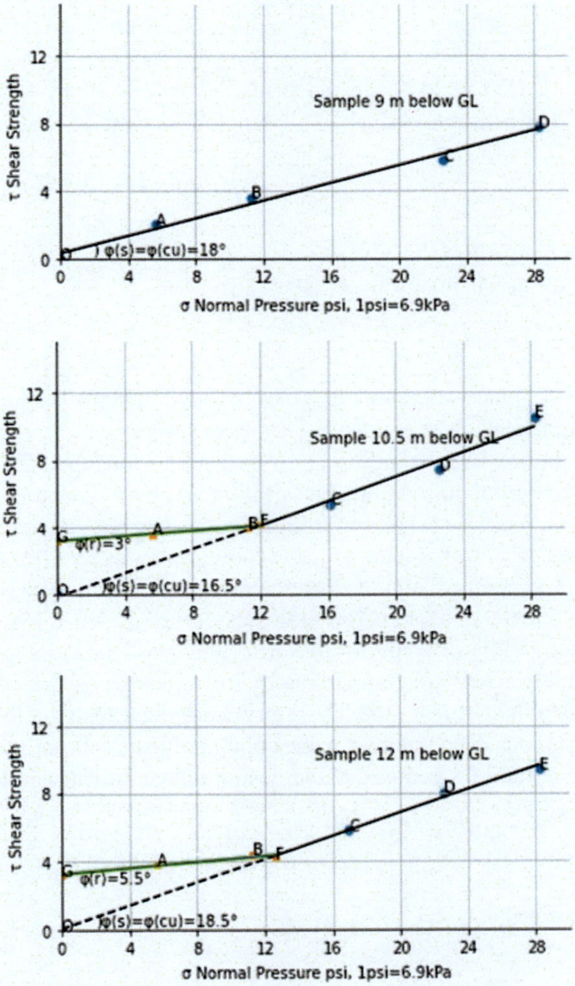


**Fig. 4** Shear strength versus normal pressure

## 2.2 *Vane Shear Test*

Skempton [8] concluded that 'the strength of normal consolidated clays does increase with depth.' This finding became possible due to refined sampling techniques and vane shear tests. According to Carlson [13] and ASCE STP [14], there is an increase in shear strength values obtained at higher rate of rotational speed. To find the shear strength of Maliya marine clay, the samples were subjected to rotational speed of (5–7)° per min, the minimum rotation rate that can be uniformly applied by hands in the laboratory. Shear strength develops at small strains and assuming constant shearing resistance over the plane with only one circular surface (bottom) together with cylindrical surface total torque

$$T = T_1 + T_2 = 2\pi r^2[L + r/3]\tau.$$

Also torque $T = \theta \times$ (spring factor)/180 where $\theta$ is in degrees.
Substituting values of the length and radius of the vanes and of the spring factor:

$$\tau = 0.0795\theta \text{ psi} = 0.548\theta \text{ kPa}.$$

| Depth in m below G.L | 0 | 1.5 | 3 | 4.5 | 6 | 8.1 | 9 | 10.5 | 12 |
|---|---|---|---|---|---|---|---|---|---|
| Shear strength ave in kPa | 41.4 | 68.3 | 90.4 | 14.5 | 14.1 | 15.8 | 17.3 | 17.9 | 20.7 |

In place, shearing resistance of foundation soil is found by an unconfined compression test on specimens. The shearing resistance is equal to half the unconfined compressive strength $q_u$. In case of unconsolidated undrained tri-axial test, the strength is found to be half of the maximum deviatoric stress. Skempton [15] states that it can be found out from a direct shear test in which no consolidation is allowed under a normal load and is sheared immediately after putting it in position. Unconfined compression tests were carried out on undisturbed specimens collected from 4.5 and 9 m depth below G.L. Results are shown in Fig. 6. The strength values corresponding to 3% strain value are (1.917 psi) 13.22 kPa and (2.2 psi) 15.18 kPa, respectively. They are comparable to the values (2.1psi) 14.5 kPa and (2.5psi) 17.25 kPa as obtained from the laboratory vane shear tests on the corresponding undisturbed samples. The strength values obtained from laboratory vane shear tests are utilized for stability analysis of the design embankment, and the strength value adopted is 12 kPa.

## 3 Consolidation Tests

Consolidation tests were performed on undisturbed samples from every 1.5 m depth up to 12 m below G.L. Diameter of the mold was 60.325 mm, and consolidation pressure ranged from 1/16 to 8 kgf/cm$^2$ (6.25–800 kPa). The load increment was 1, and the time duration for action of each of these was 24 h. Dial gauge reading were recorded to make a plot of these readings versus square root of time t. One such typical plot for pressure range from 107 to 214 kPa is shown in Fig. 5. Coefficient of consolidation $c_v$, compression index $C_c$, and consolidation ratio $r$ are calculated. Casagrande's graphical method is used to find pre-consolidation pressure for each sample. These graphical values are usually found to be smaller than actual ones. One reason for this discrepancy might be that in the field the pressure increment ratio is smaller and also that the duration increases the value of the pre-consolidation load.

The initial void ratio varies from 1 to 2.4, $c_v$ from (22–160) $\times$ 10$^{-5}$cm$^2$/s and compression index $C_c$ from 0.66 to 1.05. It is observed that when values of $c_v$ are plotted on log scale against the depth of the samples, a straight line is seen.

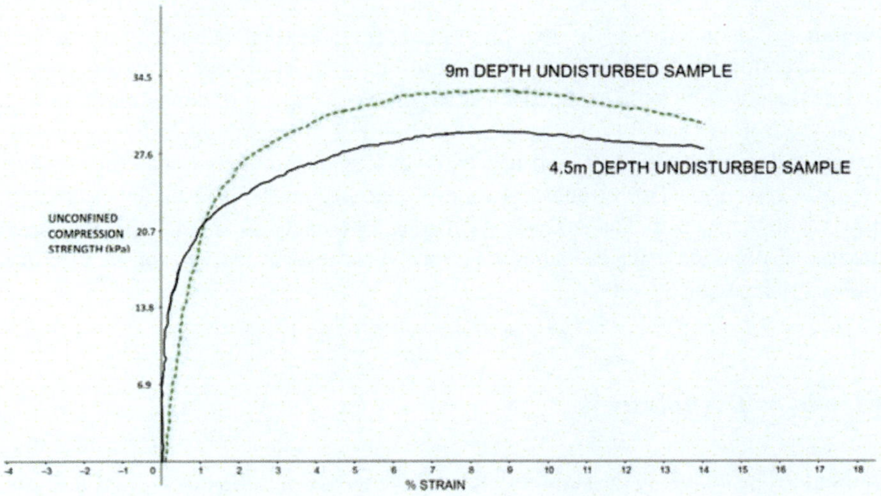


**Fig. 5** Unconfined compression test

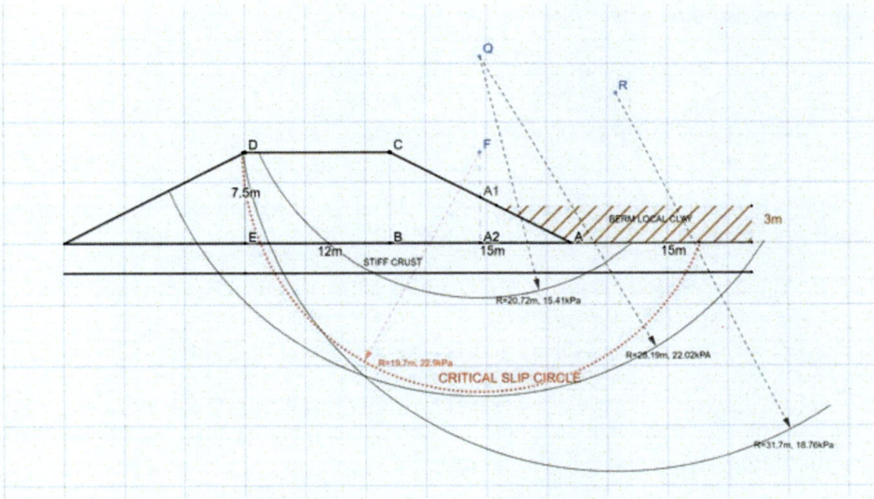


**Fig. 6** Stability analysis

## 4 Embankment and Stability Analysis

In the case under consideration, the embankment is to rest on soft to very soft normally consolidated thick marine clay deposit of low permeability, in marshy swamp at tidal alluvial flat Rann of Kachchh. The ground level is 2.4 R.L., and maximum flood level is 5.1 R.L. The embankment has to be built only in one season at least above this flood level for its safety. This means rapid construction over saturated clay sub-soil without any chance for the dissipation of pore pressures. For such situation, the end of construction condition is well accepted to be the most critical, as in due course of time there will be an increase in shear strength consequent of consolidation. As a result, the factor of safety will increase with time. The shear strength is determined by unconfined compression test, and it is equal to half of the unconfined compressive strength, $\tau = q_u/2$. This strength is used in stability analysis of natural saturated slopes with their bases. The analysis is called $\phi_u = o$ analysis. In this case, the shear strength as obtained using the laboratory vane shear tests is adopted for the stability analysis purpose.

## 4.1 $\phi_u = 0$ Analysis

Barber [16] in analyzing the stability of natural slopes concludes that $\phi_u = 0$ analysis applies only for end of construction type condition. Terzaghi and Peck [9] state that the results of an unconfined compression test on a perfectly undisturbed sample are approximately same as those of a consolidated undrained test performed under same confining pressure $p_3$ where $p_3 = (0.7–0.9)\ p_{v.}$

## 4.2 Stability of Slopes of Embankment

The upper portion, a relatively stiff crust, of 15 m+ soft marine clay deposit varies in thickness from 1.8 to 2.4 m. Its shear parameters in molded state are:

(1) At standard proctor density MDD 1.676 g/cm$^3$, 16.35 kN/m$^3$, OMC 18.7%: $C_{CU} = 3$psi (20.7 kPa), $\phi_{cu} = 23°$
(2) At proctor density and OMC + 3%: $C_{CU} = 1.8$psi (12.24 kPa), $\phi_{cu} = 24°$.

Shear strength develops at about 7–8%. Minimum shear strength $= q_u/2 = 7$psi (48.3 kPa) at 7% strain. This strength is considerably higher than the undisturbed strength of underlying soft foundation material.

## 4.3 Stability Analysis

The embankment with 7.5 m height and 12 m top width is examined for stability of slopes of 1.5:1 and 2:1. Twelve slip circles for former and eight slip circles for the latter are considered. Shear strength required for equilibrium is 22.7 kPa and 18.9 kPa, respectively. It is known that it is the base, in such cases, which governs the design of the embankment slopes.

## 4.4 Stability of the Base

**Mode of failure**. Base failure may occur in several different situations. The fill may bodily sink in the supporting soil—failure by sinking; failure by spreading, and failure by piping [4]. For the fills on soft homogeneous clay they state that "if the surface of the clay is very close to the base of the fill, the thickness of the clay stratum is at least half as great as the base width of the fill and the stratum is fairly homogeneous. The failure of a fill on such a base has a general character of a base failure along a midpoint circle. The average undrained shearing resistance for the base be determined." For the real fill having cohesion and friction or only friction, equivalent cohesion ($\phi = 0$) is used. Maliya marine clay base is homogeneous, close to the base of the fill, and it has the thickness nearly half of the base width of the embankment with 2:1 slope. 15–16 m thickness nearly meets these criteria. In Fig. 6, we present the midpoint circle ($\phi_u = 0$) analysis for case of the slope of 2H:1V in all five circles are examined.

**Shearing resistance by embankment and by base**. The material utilized for the examined embankment is considered to be a local material at MDD with omc + 3%. Its stress–strain curves are a bit flatter, and its peak strength is developed at 7% strain as compared to 2–4% strain in undisturbed soft clay. To avoid plastic state in foundation material and consequent heaving, and extreme subsidence, it will not be desirable to strain material in the foundation base, and hence, in the embankment material also beyond 3% strain. The contribution of the embankment toward shearing resistance at this strain is 7.1psi (49 kPa), say 48 kPa against average ultimate strength of 9psi (63 kPa). Average shearing resistance by the foundation is taken as 12 kPa.

**Critical midpoint circle analysis**. In this paper, the examination for a design embankment with 12 m top width, 7.5 m height, and side slopes of 2H:1V is presented, as shown in Fig. 6. The analysis for slope 2:1 is presented. Factor of safety is calculated, and it is 0.916 which is not ok. Terzaghi and Peck [4] state that "it is commonly required that the factor of safety with respect to base failure should be at least 1.5. Considering the unavoidable errors in estimating the average shearing resistance of the clay, this factor is very low. Nevertheless, in order to satisfy the requirement, high fills on soft clay must be provided with very gentle slopes. Hence, if the fill is very long, it may be economical to further reduce the factor of safety still further, to 1.2 or 1.1" and then monitor to detect any impending failure, if any, during construction.

In this case, it is noted that the shear resistance has, fairly well, been arrived at. The berms 3 m thick made from local clay and compacted 100 lb. per cft = 15.72 kN/m$^3$ are incorporated in design of an embankment. Factor of safety with 10.5 m is 1.28. It satisfies the safety requirement as suggested by Terzaghi and Peck [4], and hence, it is ok.

FOS CALCULATION IN CASE OF 10.5 BERM:

$$12.5\text{ m} \times 7.5\text{ m} \times 1\text{ m} \times 16.35\text{ kN per cu m} \times 13,\ 5\text{ m} = 19865.25\text{ kN - m}\ldots \tag{1}$$

$$(1/2) \times 7.5\text{ m} \times 3.75\text{ m} \times 1\text{ m} \times 16.35\text{ Kn per cu m} \times 5\text{ m} = 149.8\text{ kN - m}\ldots \tag{2}$$

$$7.5\text{ m} \times 3.75\text{ m} \times 1\text{ m} \times 16.35\text{ m kN per cu m} \times 3.75 = 172441\ \text{ kN - m}\ldots \tag{3}$$

$$\text{total of } [1] + [2] + [3] = 22739.4 = M\text{D} = \text{Driving moment}\ldots \tag{4}$$

Resisting moment

$$(41.576\text{ m} \times 1\text{ m} \times 12\text{ kPa} + 4.876\text{ m} \times 1 \times 48\text{ kPa} + 7.62\text{ m} \times 1\text{ m} \times 48\text{ kPa}) = 1100.7\text{ kN} \cdot 1100.7\text{ kN} \times 19.7\text{ m lever arm} = 21683.79\text{ kN - m}\ldots \tag{5}$$

$$(1/2)\ 7.25 \times 3.75\ 16.35\,\text{kN per cu m} \times 2.5\text{ m lever arm} = 574.8\text{ kN - m}\ldots \tag{6}$$

$$[5] + [6] = 22258.6\text{ kN - m}\ldots \tag{7}$$

$$\text{Deduct for 1.5 m cracking } 1.5\text{ m} \times 1\text{ m} \times 48\text{ kPa} \times 19.7\text{ m} = 1418.4\text{ kN - m}\ldots \tag{8}$$

$$[7]-[8] = 20840.2\text{ kN - m}\ldots \tag{9}$$

Now add for berm of 10.5 m × 3 m:

$$\text{Small triangle } (1/2) \times 6\text{ m} \times 7.5\text{ m} \times 15.72\text{ kN per cu m} \times 5.5\text{ m lever} = 1945.35\text{ kN - m}\ldots \tag{10}$$

$$[9] + [10] = 22785.5\text{ kN - m} = \text{total resisting moment}\ldots \tag{11}$$

$$10.5\text{ m} \times 3\text{ m} \times 1\text{ m} \times 15.72\text{ kN per cu m} \times 12.75\text{ lever arm } 6313.5\text{ kN - m} \tag{12}$$

Total resisting moment

$$M_{\text{R}} = [11] + [12] = 29099.0\text{ Kn - m}\ldots \tag{13}$$

$$\text{Fos} = [13]/[4] = 1.279\text{ say } 1.28$$

**Settlement Prediction**

Findings from consolidated tests on undisturbed samples are used. Charts are used to find vertical pressure at any depth. Vertical settlements are then computed using equation:

$$S = H(e_1 - e_2)/(1 + e_1)$$

in usual notation. The settlement is calculated for only center line of the embankment. The coefficient of consolidation $c_v$ is taken as $50 \times 10^{-5}\text{cm}^2/\text{s}$ and height as 15 m. With time in seconds given by:

$$t = \text{TH}^2/c_v$$

it works out to 147 T years, where $T$ is time factor whose numerical value corresponds to $U\%$, degree of consolidation $S/S_{\text{final}}$. Total settlement is 1.95 m (6.49 ft.). For 50% settlement (0.925 m), the time required is 29.4 years in case of no sand drains.

## 4.5 Sand Drains

Let us consider vertical sand drains of diameter $d = 0.3$ m and at triangular spacing 2.8 m as show in Fig. 7. The equivalent circular diameter $D$ is $2.85 \times 1.05 = 3$ m. Hence, the system can be thought of as a cylinder of diameter $D' = 3$ m with an internal drain of diameter $d = 0.3$ m. As per [16], time factor $T = c_v t/H^2 = c_v t/(D')^2$ which is 0.132 for 50% settlement. For comparison of time settlement of the embankment foundation, with parameters as given above and writing,

**Fig. 7** Sand drains pattern

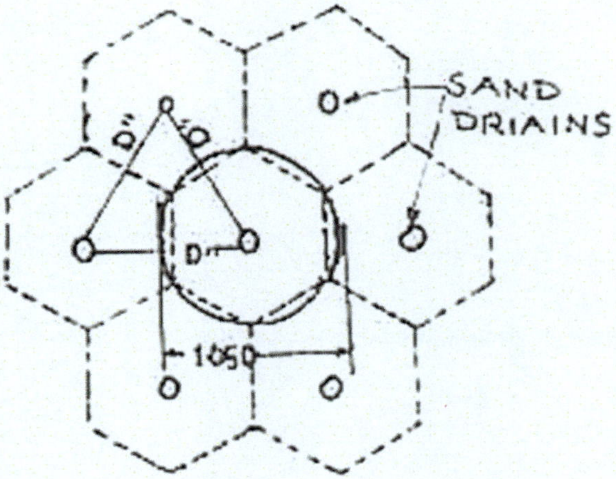

$$T = c_{vr}t/(D')^2$$

With $c_{vr} = 1.5c_v$, we arrive at $t = 47$ T. For 50% settlement, the time t required is 47 $\times$ 0.132 = 6.2 months as shown in Fig. 8. This is a considerable gain, 57X, improvement through the use of sand drains. Sand blanket of 0.6 m below the embankment is incorporated to discharge the drained water.

*Note* This work was submitted as a dissertation to The M.S. University of Baroda, by the author in partial fulfillment of the requirements for the degree of Masters of Engineering in 1967.

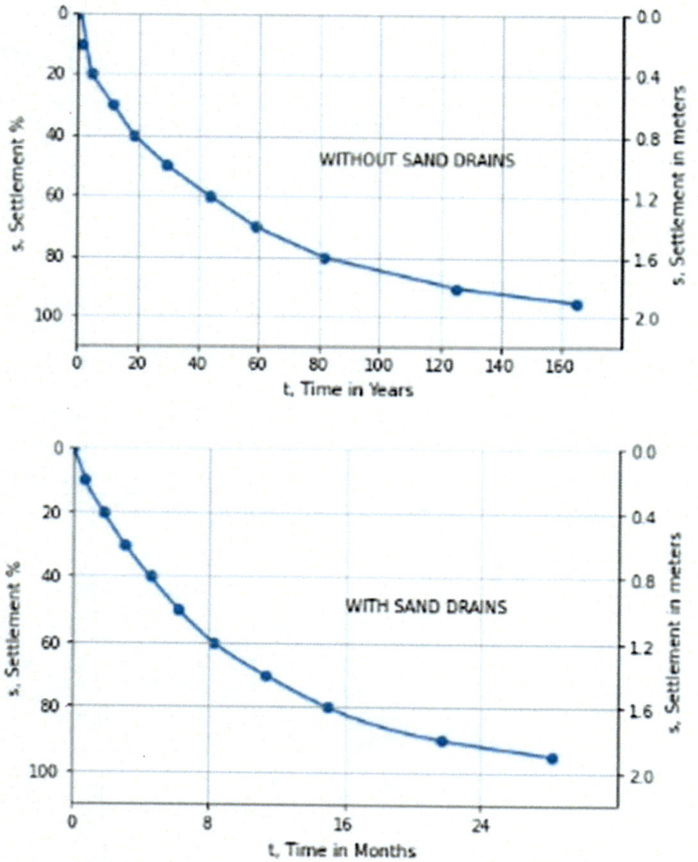


**Fig. 8** Settlement versus time

## 5 Conclusions

1. The Rann of Kachchh was formerly a sea basin. The huge amount of fine-grained soil brought by rivers filled it up. As a result of filling up due to silting, partially upheaving of the sea floor, man-made mounds, and moving away of rivers a sea basin got converted into the Rann.
2. The Maliya marine clays are of marine origin at the gulf end in a Little Rann of Kachchh. They mostly are of CH group. A few fall in CI and MH group. The thickness of deposit varies from 3 to 15 m plus. Its moisture content is at 80%, greater than liquid limit which is from 50 to 70%. The laboratory vane shear test gave the sensitivity value varying from 2 to 8.
3. From box shear test, it can be inferred that the clay is normally consolidated.
4. In the box shear test, shear strength line passes through origin when samples are tested under normal load intensities greater than in-situ effective burden. The average angle of shearing resistance $\phi_{\text{cu}}$ is18.2. But it gives cohesion intercept when tested under normal load intensities smaller than in-situ effective burden. In that case $C_{\text{cu}=}$ 21 kPa average and $\phi_{\text{r}} = 4.3°$.
5. "Yield point" is observed in stress–strain curve during box shear test. This is explained by the phenomena of pronounced peaking effect.
6. Undrained shear strength is seen to increase with depth of deposit. It increases from (2 psi) = 14 kPa to (3 psi) = 21 kPa for a depth range of 4.5 and 12 m.
7. The ratio of undrained shear strength $S_{\text{u},}$ to $p$, an effective over burden pressure is constant over the depth. Its value is found to be average 0.27 as determined by laboratory vane shear tests.
8. The upper portion of the deposit is relatively stiff as a result of desiccation.
9. It is seen that it is the base that governs the design in this case. Midpoint circle failure mode has been examined for the embankment with berms and 2H:1V slope.

The use of sand drains reduces the time from 29 years to 6.2 months for 50% primary settlement which is calculated to be (3.25 ft.), nearly 1 m.

**Acknowledgements** The author is highly indebted to late H. P. Oza, M.Sc. (Bom) S.M. (M.I.T), A. I.MAR. E. (Lon.), M.I.E. (India), Director, Engineering Research Institute, Baroda, now Gujarat Engineering Research Institute (GERI) Vadodara, the Guide for the dissertation, for his esteemed guidance, invaluable help, and continuous encouragement throughout the work. The author is much thankful to the staff of GERI for their help at the time of collecting undisturbed samples.

The author also owes to Late Prof. R. S. Dighe, Head of Applied Mechanics and Structural Engineering Department, for his very kind encouragement throughout the work. He was the basic incentive for the work. The author is also thankful to late Prof. N. Y. Hiryur. The author expresses his sincere thanks to Shri A. C. Shah, ARO (Assistant Research Officer), for his help and suggestions from time to time.

## References

1. Tan: as referred by Rosenquist (1957)
2. Rosenquist, I.Th.: Mechanical properties of soil-water systems. In: Proceedings of ASCE, vol. 85, J SM2, Paper 2000, pp. 31–53 (1959)
3. Crawford, C.B.: Cohesion in an undisturbed clay. Geotechnique **13**(2), 132–145 (1963)
4. Terzaghi, K., Peck, R.B.: Soil Mechanics in Engineering Practice, 1st edn. Wiley, New York (1948)
5. Moum, J., Rosenquist, I.Th.: On the weathering of young marine clay. In: Proceedings, 4th International Conference, SM & FE, vol. I, pp. 77–79 (1957)
6. Skempton, A.W., Northey, R.C.: The sensitivity of clays. Geotech. J. **3**(1), 30–52 (1952)
7. Bjerrum, L.: Geotechnical properties of Norwegian marine clays. Geotech. J. **4**(2), 46–59 (1954)
8. Skempton, A.W.: The colloidal activity of clays. In: Proceedings, 3rd International Conference, Soil Mechanics and Foundation Engineering, vol. I, pp. 57 (1953)
9. Terzaghi, K., Peck, R.B.: Soil Mechanics in Engineering Practice, 2nd edn. Wiley, New York (1967)
10. Lambe, T.W.: The structure of compacted clay. In: Proceedings of ASCE, vol. 184 SM2, paper1654 (1958)
11. Schmertmann, J.H., Osterberg, O.J.: An experimental study of the development of cohesion and friction with axial strain in saturated cohesive soils. In: Proceedings of ASCE, Research Conference on Shear Strength of Cohesive Soils, pp. 643 (1960)
12. Hvorslev M.J.: Uber die Pestigkeiseigenschfen gesterterbindinger Boden (on the properties of remolded cohesive soils). Thesis 159 pages, published by Denmark's Naturevigenska-beligeeSesfund, investieviden skabelige skrifter, Series A. Sr. 45, Copenhagen. Revaluated and restated in physical components of Shear Strength of the saturated clays byhim, the author, Proceedings, ASCE research conference on Shear Strength of cohesive soils (1960)
13. Carlson (Cadling), L.: Determination in-situ shear strength of undisturbed clay by means of a rotating auger. In: Proceedings, 2nd International Conference Soil Mechanics and Foundation Engineering, pp. 265–280 (1948)
14. ASCE STP: Special Technical Publication no.193; Use of vane tests in soft soils
15. Skempton, A.W.: Slip in the west bank of EAU brink cut. J. Inst. Civ. Eng. Lond. **24**, 267–287 (1953)
16. Barber, L.: Use of direct shear test in highway design. In: Symposium on Direct Shear Testing, ASTM STP NO. 131 (1952)

# Variability in Settlements of Foundations on Fine Grained Soils

**Bala Deepthi Madanu, Ch. Nageshwar Rao, and Madhav Madhira**

## 1 Introduction

The basic and essential component of foundation design are estimation of settlements. Permissible settlement is always the controlling criteria for the design of shallow foundation more than the safe bearing capacity. Settlement the vertical downward movement of the ground caused by foundation load is a combination of immediate settlement due to elastic deformation of soil, consolidation settlement resulting from dissipation of excess pore pressure with time and secondary settlement or creep because of plastic adjustment and movement of soil particles under constant effective stress. Elastic or immediate settlement depends upon foundation type, rigid or flexible and type of soil. Generally, the two methodologies adopted to estimate settlements are based on: (1) Laboratory tests, oedometer and triaxial and (2) In situ tests—SPT, CPT, Dilatometer, Pressuremeter, etc. and settlements need to be validated. Unfortunately, most of engineers do not validate their predictions.

This paper presents estimates of settlements using SPT results based on two different approaches. SPT is one of the most versatile and commonly used tests for geotechnical characterization of a site predominantly due to its ease, reliability and economy. SPT N values are far better than considering laboratory determined properties in estimating settlements of foundations. Schmertmann et al. [1] proposed a simple method based on variation of strain influence factor with depth and static cone tip resistance for estimating settlement. Burland and Burbidge's [2] procedure estimates settlement based on variation of SPT $N$ values within the depth of influence. In this paper, settlements are estimated using the above two methods and variability of settlements at the site are compared. Data required for this study is obtained from the site of Dunkani–Kharaghpur section at NH-06.

B. D. Madanu · Ch. Nageshwar Rao (✉) · M. Madhira
VNR Vignana Jyothi Institute of Engineering and Technology, Hyderabad 500090, India
e-mail: nageshwarrao_ch@vnrvjiet.in



C. N. V. Satyanarayana Reddy et al. (eds.), *Ground Characterization and Foundations*, Lecture Notes in Civil Engineering 167,
https://doi.org/10.1007/978-981-16-3383-6_40

## 2 Literature Review

Meyerhof [3–5], D'Appolonia et al. [6], Burland and Burbidge [2] derived settlements using SPT *N*. Schmertmann [7], Schmertmann et al. [1] derived settlements based on tip resistance from cone penetration test (CPT). Davis and Poulos [6] developed a three-dimensional approach of predicting settlement and compared them with model footing test results. Bowles [8] presents a relation for estimating settlement using elastic continuum theory considering both shape and depth factors. Burland and Burbidge [2] examined settlement variations with reference to depth to width ratio at few site locations. Mayne and Poulos [9] proposed displacement influence factors for homogenous (modulus constant with depth) and Gibson-type (modulus linearly increasing with depth) and a new solution for Gibson soil of finite thickness. Akpila [10] investigated bearing capacity and settlements of raft foundations on sandy soil using SPT.

## 3 Methodology

Borehole data from Six Laning of Dunkani–Kharaghpur section of NH-06 is considered in this paper. Boreholes are drilled at different chainages and corresponding standard penetration test (SPT) N values are given. Figure 1 shows the variations of SPT *N* with depth. From Fig. 1a–f, j SPT *N* can be observed to be nearly constant upto depths of 8.0–10 m and increases with depth beyond. Figure 1d, g–i shows SPT *N* values increasing and decreasing with depth. Settlements are estimated using these SPT *N* values for footing widths of 2.0, 4.0 and 10.0 m under uniform pressure of 100 kN/m$^2$ based on Schmertmann et al. [1] and Burland and Burbidge [2] methods. Generally, footing width greater than 6 m is considered as raft/mat foundations. Therefore, 10 m footing width is considered as raft foundation.

### 3.1 *Schmertmann et al. [1] Method*

Schmertmann [7] suggested a method to estimate the settlement of shallow foundations on soils considering a simplified variation of strain influence factor with depth. The compressible soils underneath the footing are divided into different sub-layers based on the variation of static cone resistance, $q_c$, with depth. The deformation modulus of soil is estimated from static cone resistance, $q_c$. This method, known as the "2*B*–0.6" method, is based on vertical strain distribution with depth underneath the footing to a depth of twice the footing width.

Studies showed that Schmertmann [7] method needs modification to the strain influence diagram that accounts for different shapes and load intensity. Schmertmann et al. [1] is an improvement over the previously given method.

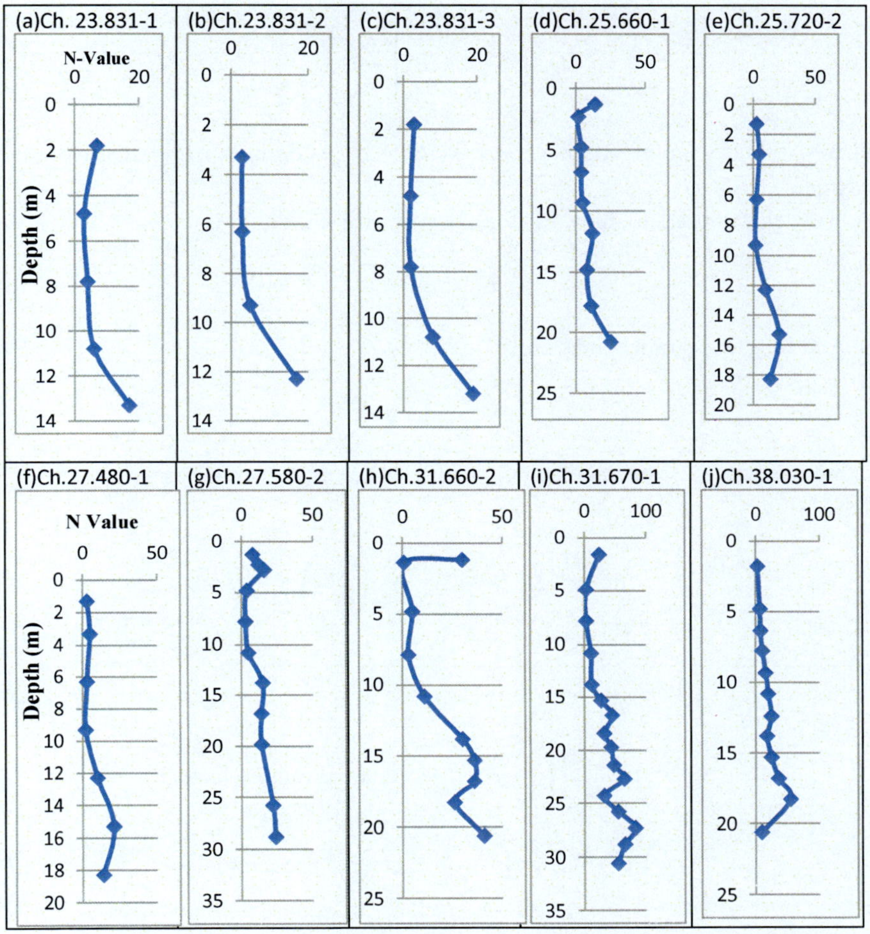


**Fig. 1** Depth versus SPT *N* at various chainages

The strain influence diagram for axisymmetric footings (square and round) is constructed as:

Strain influence factor $I_z$ is equal to 0.1 and 0 at depths $Z = 0$ and $2B$, respectively, from footing level. Maximum $I_z$ occurs at a depth of $B/2$ and has a value of:

$$I_{Zp} = 0.5 + 0.1\left[\frac{\Delta q}{\sigma'_{vp}}\right]^{0.5} \quad (1)$$

For strip footings ($L/B > 10$), i.e., plane strain conditions, $I_z = 0.2$ and 0 at depths 0 and $4B$, respectively. Maximum $I_z$ at a depth of $B$ has a value of

$$I_{\text{Zp}} = 0.5 + 0.1\left[\frac{\Delta q}{\sigma'_{\text{vp}}}\right]^{0.5} \quad (2)$$

where $\Delta q$ = net applied footing stress (kN/m$^2$), $\sigma'_{\text{vp}}$ = initial vertical effective stress (kN/m$^2$).

The deformation modulus, $E_s$, for axisymmetric condition is

$$E_s = 2.5\, q_c \quad (3)$$

And for plane strain conditions,

$$E_s = 3.5\, q_c \quad (4)$$

Settlement, $S$, is estimated as

$$S = C_1 C_2\left(q' - q\right)\sum_{i=1}^{n}\left(\frac{I_{\text{zi}}}{E_{\text{si}}}\right)\Delta Z_i \quad (5)$$

where

$$C_1 = 1 - 0.5\left(\frac{q}{q' - q}\right) = \text{depth correction factor} \quad (6)$$

$$C_2 = 1.0 + 0.2\log\left(\frac{t}{0.1}\right) = \text{creep correction factor} \quad (7)$$

$q'$ = stress at the level of foundation (kN/m$^2$); $q$ = initial effective overburden pressure at the foundation level (kN/m$^2$); $I_{\text{zi}}$ = strain influence factor of layer $i$; $E_{\text{si}}$ = deformation modulus of layer $i$ (kN/m$^2$); $\Delta z_{\text{i}}$ = thickness of layer $i$ (m); $\gamma$ = unit weight of soil (kg/m$^3$); $D_f$ = depth of foundation (m); $B$ = width of footing (m); $t$ = time in years.

Depending upon the variation in deformation modulus, $E_s$, soil below the footing is divided into a number of layers shown in Fig. 2 up to depth of influence 2$B$, since axisymmetric condition. Deformation modulus of each layer is taken as $E_{s1}, E_{s2}, E_{s3},$ etc. Thickness of each layer is taken as $\Delta_{z1}$, $\Delta_{z2,}$ $\Delta z_3$, etc. and corresponding strain influence factors are $I_{z1}, I_{z2}, I_{z3}$, etc. Average of the values of influence factors, $I_z$ at top and bottom of each layer is considered for settlement estimation. Schmertmann [7] suggested that if only SPT $N$ values are available, SPT $N$ is converted in to cone tip resistance, $q_c$ using $\frac{q_c}{N}$ ratio versus mean grain size, $d_{50}$, plot (Table 1).

If the soil mixture consists of sand and large traces of silt, then the ratio of 2 can be used and if large amount of sand is present then ratio of 3.5 is used. If combination

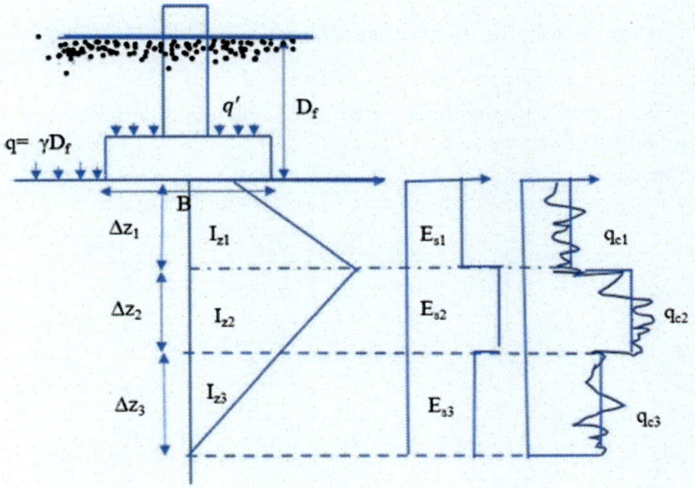


**Fig. 2** Strain influence diagram

**Table 1** Correlation between $q_c$ and SPT $N$ as per Schmertmann [7]

| Soil type | $\frac{q_c}{N}$ (kN/m$^2$) |
|---|---|
| Silts, sandy silts, slightly cohesive silt-sand mixtures | 2.0 |
| Clean, fine to medium sands and slightly silty sands | 3.5 |
| Coarse sands and sand with little gravel | 5.0 |
| Sandy gravels and gravel | 8 |

of coarse sand with little gravel is present then a ratio of 5.0 is used. If large amount of gravel is present then ratio of 8 is used.

## 3.2 *Burland and Burbidge [2] Method*

Burland and Burbidge [2] presented an empirical method based on average SPT blow count, foundation width and the zone of influence estimated based on the type of variation of SPT $N$, (i) constant or increasing and (ii) decreasing with depth.

**Case 1**: $N_{60}$ is approximately constant or increases with depth,

$$Z' = 1.4\left(\frac{B}{B_r}\right)^{0.75} B_r \tag{8}$$

where $B$ = footing width (m); $B_r$ = reference footing width = 0.3 m; $Z'$ = depth of influence (m).

**Case 2**: $N_{60}$ decreases with depth, $Z' = 2B$ or $Z''$ the depth to bottom of soft layer from footing level (whichever is less).

Settlement, $S_B$ is estimated as

$$S_B = \alpha_1\alpha_2\,\alpha_3\left(\frac{1.25\,\frac{L}{B}}{0.25+\frac{L}{B}}\right)^2\left(\frac{B}{B_r}\right)^{0.7}\left(\frac{q'}{p_a}\right)B_r \tag{9}$$

where

$$\alpha_1 = 0.14(\text{constant});\;\; \alpha_2 = \frac{1.71}{N_{60(a)}^{1.4}} = \text{compressibility index} \tag{10}$$

$$\alpha_3 = \frac{z''}{z'}\left(2-\frac{z'}{z'}\right) \leq 1 = \text{correction for depth of influence} \tag{11}$$

and

$$q' = q_0 \tag{12}$$

$L$ = length of footing (m), $B$ = width of footing (m), $B_r$ = reference width = 0.3 m, $N_{60(a)}$ = average of SPT $N$ within the depth of influence, $q_0$ = net applied stress at the level of the foundation (i.e., the stress at the level of the foundation minus the overburden pressure); $p_a$ = atmospheric pressure = 100 kN/m$^2$.

### 3.3 *Estimation of Settlement, $S_S$, by Schmertmann et al. [1]*

The soils present in the considered site are all silty clay. Hence

$$\frac{q_c}{N} = 2 \tag{13}$$

Cone tip resistance, $q_c$ is calculated from SPT $N$ using Eq. 13 from which deformation modulus, $E_s$, is determined using Eq. 3. The strata below the footing is divided into layers based on deformation modulus and values of $I_z$ found at the top and the bottom of each layer. $I_z$ values other than $I_{zp}$ and $I_{z(2B)}$ are obtained by interpolation. Average of $I_z$ of the values at the center of layer is calculated for each layer along with thickness of layer, $\Delta z$. Product of each layer thickness with the ratio of influence factor to deformation modulus is summed up for all the layers. The obtained sum is multiplied with correction factors for depth ($C_1$), and for creep, ($C_2$) for net stress of 100 kN/m$^2$ to estimate the total settlement. Strain influence factor diagrams

**Table 2** Estimation of settlement for chainage 23.831–1 using Schmertmann et al. [1] method

$B = 2$ m

| Layer No. | Thickness $\Delta z$ (m) | SPT $N$ | $q_c = 200$ N (kN/m$^2$) | Deformation modulus $E_s = 2.5q_c$ (kN/m$^2$) | Strain influence factor ($I_z$) | $(I_z/E_s) * \Delta z$ |
|---|---|---|---|---|---|---|
| 1 | 1 | 5 | 1000 | 2500 | 0.35 | 0.00015 |
| 2 | 0.5 | 5 | 1000 | 2500 | 0.56 | 0.00011 |
| 3 | 1.45 | 3 | 600 | 1500 | 0.35 | 0.00035 |
| 4 | 1.05 | 4 | 800 | 2000 | 0.105 | 0.00006 |
| | | | | | | 0.00065 |

Settlement, $S_S = 53.62$ mm

**Table 3** Settlements for different width of footings for chainage 23.831–3

| Width of footing (m) | Settlement (mm) |
|---|---|
| 2 | 53.62 |
| 4 | 101.3 |
| 10 | 164.0 |

are drawn for footing widths of 2.0, 4.0 and 10.0 m for all chainages and their corresponding settlements estimated. Table 2 presents typical settlement analysis for chainage 23.831–1 for footing width of 2 m.

The stratum is divided into 4 layers based on variation of deformation modulus, $E_s$ with depth (Table 2). Even though deformation modulus, $E_s$ is the same for the first two layers, they are considered as distinct layers because maximum influence factor, $I_z$ at depth of $\frac{B}{2}$ occurs in Layer 1. $\Delta z$, is thickness of each layer. Settlements are estimated on similar lines for 4 m footing width and raft foundation ($B = 10$ m) shown in Table 3.

Table 3 presents settlements obtained are 54.7 mm, 101.3 mm and 164 mm for 2.0 m, 4.0 m and 10.0 m footing widths, respectively.

### 3.4 Estimation of Settlement, $S_B$ by Burland and Burbidge [2]

The data from the same chainage 23.831–1(Fig. 1a) is analyzed. SPT N values in Fig. 1a are considered to be nearly constant up to 10 m depth and to increase with depth for depths beyond 10.0 m. The depth of influence, $Z'$, is determined accordingly using Eq. 8. Average of SPT $N$ is calculated within this depth of influence, $Z'$. The term $\alpha_2$ was obtained from Eq. 10. The parameters, net bearing stress of 100 kN/m$^2$ and atmospheric pressure of 100 kN/m$^2$, are not mentioned in Table 4. Settlements are estimated (Table 4) using Eq. 9.

**Table 4** Estimation of settlements at chainage 23.831–1 for 2 m, 4 m and 10 m footing widths

| Footing width $B$ (m) | 2 | 4 | 10 |
|---|---|---|---|
| Variation of SPT $N$ with depth | Constant | Constant | Increasing |
| Depth of influence, $Z$ (m) | 1.7 | 2.9 | 5.8 |
| Avg $N_{60}$ | 5 | 4.3 | 4.1 |
| Compressibility index, $\alpha_2$ | 0.1 | 0.21 | 0.23 |
| Settlement, $S_B$ (mm) | 28.3 | 56.3 | 112.9 |

Burland and Burbidge [2] method is difficult to apply for soil strata whose SPT $N$ values do not follow trends of case I (constant or increasing with depth) or case II (decreasing with depth) such as those depicted in Fig. 1d, g–i. For these cases, only the initial variation of SPT $N$ with depth is considered and settlements estimated.

In Table 4, depths of influence, $Z$ are 1.7 m, 2.9 m and 5.8 m for footing widths of 2 m, 4 m and 10 m, respectively. Depth of influence ($Z = 5.8$ m) is high for raft foundation (footing width, $B = 10$ m) and the corresponding average $N_{60}$ of 4.1 is less. But depth of influence ($Z = 1.7$ m) is less for 2 m footing width and average $N_{60}$ equal to 5 is high. Greater average $N_{60}$ implies soil is stiff and thus gives less settlement, $S_B = 28.3$ mm, whereas smaller average $N_{60}$ implies soil is loose and gives more settlement $S_B = 112.9$ mm.

## 4 Results and Discussion

Settlements were estimated for all chainages for footing widths of 2.0, 4.0 and 10.0 m using Schmertmann et al. [1] and Burland and Burbidge [2] methods and compared in Table 5.

Ratios, $S_s/S_B$, of settlements estimated using Schmertmann et al. [1] and Burland and Burbidge [2], are also listed in Table 5. Figure 3 presents the ratios for footing widths of 2.0, 4.0 and 10.0 m given in Table 5 in graphical form. Ratio, $S_s/S_B$ varies from 4.7 to 0.5, 5.3 to 0.6 and 4.9 to 0.8 for footing widths of 2.0 m, 4.0 m and 10.0 m, respectively. Results shown in Table 5 and Fig. 3 indicates the ratio, $S_S/S_B$ to be greater than 1.0 in most cases implying that Schmertmann et al. [1] method estimates settlements to be greater than those from Burland and Burbidge [2] methods for the three sizes of footing widths. In few cases wherein the SPT $N$ value is around 1 and 2 it is vice versa, i.e., Burland and Burbidge [2] method predicts larger settlements than those from Schmertmann et al. [1] method.

The ratio, $S_s/S_B$ is larger than 3.0 (Fig. 3) for chainages who's initial SPT $N$ are greater than 15. It implies that Schmertmann et al. [1] method accounts for the larger value, while Burland and Burbidge [2] method averages the high value and thus predicts relatively smaller value of settlement. The converse is true if the ratio, $S_s/S_B$ is less than 1 (as at some of the chainages) where in soil is very loose at shallow depths with SPT $N$ value less than 3. The effect of smaller values of $N$ is not felt

**Table 5** Settlements based on Burland and Burbidge [2] and Schmertmann et al. [1] methods

| Chainage | Settlements $S_s$ (mm) Schmertmann et al. [1] | | | Settlements $S_B$ (mm) Burland and Burbidge [2] | | | $S_s/S_B$ | | |
|---|---|---|---|---|---|---|---|---|---|
| Footing width (m) | 2 | 4 | 10 | 2 | 4 | 10 | 2 | 4 | 10 |
| 23.831–1 | 53.62 | 101.3 | 164.1 | 28.4 | 56.4 | 112.9 | 1.9 | 1.8 | 1.5 |
| 23.831–2 | 63.6 | 127.7 | 167.1 | 57.1 | 92.7 | 163.8 | 1.1 | 1.4 | 1.0 |
| 23.831–3 | 106.0 | 184.2 | 241.8 | 75.1 | 134.7 | 283.1 | 1.4 | 1.4 | 0.9 |
| 25.660–1 | 61.4 | 111.5 | 169.8 | 75.1 | 92.7 | 149.1 | 0.8 | 1.2 | 1.1 |
| 25.720–2 | 39.3 | 73.6 | 127.3 | 20.8 | 32.7 | 66.0 | 1.9 | 2.2 | 1.9 |
| 27.480–1 | 49.3 | 113.4 | 211.3 | 34.7 | 57.9 | 126.1 | 1.4 | 2.0 | 1.7 |
| 27.580–2 | 35.3 | 78.5 | 181.8 | 7.4 | 14.9 | 43.5 | 4.7 | 5.3 | 4.2 |
| 31.660–2 | 30.3 | 86.4 | 135.8 | 8.7 | 25.2 | 46.5 | 3.5 | 3.4 | 2.9 |
| 31.670–1 | 57.9 | 122.5 | 196.2 | 38.8 | 53.6 | 144.7 | 1.5 | 2.3 | 1.4 |
| 38.030–1 | 34.5 | 58.8 | 75.1 | 28.4 | 38.8 | 52.7 | 1.2 | 1.5 | 1.4 |
| 38.060–2 | 37.5 | 57.1 | 86.1 | 31.4 | 46.1 | 52.8 | 1.2 | 1.2 | 1.6 |
| 41.900–1 | 88.8 | 78.0 | 237.0 | 102.5 | 141.3 | 223.0 | 0.9 | 0.6 | 1.1 |
| 41.960–2 | 37.3 | 77.9 | 157.5 | 9.0 | 23.9 | 46.9 | 4.1 | 3.3 | 3.4 |
| 44.935–1 | 143.4 | 172.4 | 239.1 | 271.0 | 166.6 | 316.4 | 0.5 | 1.0 | 0.8 |
| 44.960–2 | 44.2 | 135.5 | 260.0 | 11.2 | 35.8 | 62.1 | 4.0 | 3.8 | 4.2 |
| 44.497–2 | 75.5 | 161.9 | 203.8 | 44.1 | 60.5 | 166.2 | 1.7 | 2.7 | 1.2 |
| 48.520–1 | 22.7 | 84.6 | 174.7 | 8.6 | 26.0 | 49.9 | 2.7 | 3.3 | 3.5 |
| 50.510–2 | 90.2 | 187.3 | 250.2 | 75.1 | 111.7 | 262.6 | 1.2 | 1.7 | 1.0 |
| 50.560–1 | 88.8 | 153.5 | 196.6 | 82.9 | 122.0 | 242.5 | 1.1 | 1.3 | 0.8 |
| 64.915–1 | 29.5 | 58.5 | 106.6 | 13.5 | 28.1 | 53.3 | 2.2 | 2.1 | 2.0 |

(continued)

**Table 5** (continued)

| Chainage | Settlements $S_s$ (mm) Schmertmann et al. [1] | | | Settlements $S_B$ (mm) Burland and Burbidge [2] | | | $S_s/S_B$ | | |
|---|---|---|---|---|---|---|---|---|---|
| Footing width (m) | 2 | 4 | 10 | 2 | 4 | 10 | 2 | 4 | 10 |
| 64.930–2 | 15.4 | 43.2 | 79.3 | 3.8 | 9.4 | 16.3 | 4.1 | 4.6 | 4.9 |
| 69.765–2 | 24.4 | 64.1 | 125.4 | 8.2 | 26.6 | 45.5 | 3.0 | 2.4 | 2.8 |
| 69.780–1 | 32.6 | 95.7 | 191.7 | 15.4 | 35.8 | 60.6 | 2.1 | 2.7 | 3.2 |

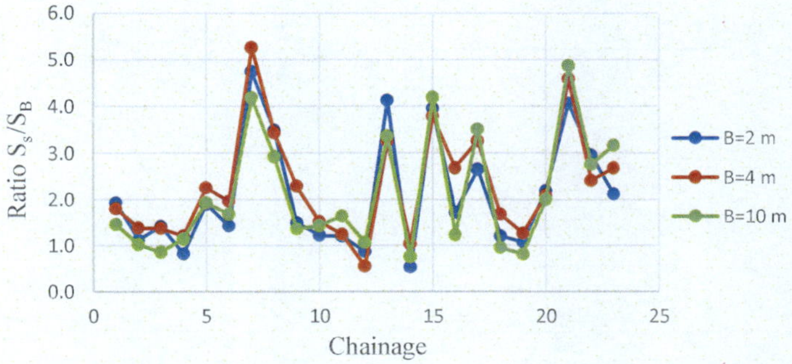


**Fig. 3** Ratio, $S_S/S_B$, between Schmertmann et al. [1] and Burland and Burbidge [2] methods for $B$ = 2, 4 and 10 m

in Schmertmann et al. [1] method possibly because of corresponding smaller strain influence coefficients, $I_z$, while the average value used in Burland and Burbidge [2] gets reflected. Estimates of settlements from Schmertmann et al. [1] and Burland and Burbidge [2] methods for all the cases, i.e., SPT $N$ is constant/increasing, decreasing or increasing and decreasing with depth are compared in Fig. 4.

Table 6 shows the angles of deviation for footing width of 2 m are 8.8° and 7.6° which are considered to be relatively close for both SPT $N$ increasing and decreasing with depth. But for 4 m footing width and raft foundation ($B$ = 10 m), angles of deviation from 45° are relatively large with 12.6 and 14.2 for SPT $N$ increasing with depth and 18.8 and 16.6 for SPT $N$ decreasing with depth.

## 5 Conclusions

1. Settlements based on Schmertmann et al. [1] method are in general larger than those based on Burland and Burbidge method [2], but if the soil at shallow depth is loose it is vice versa.
2. Burland and Burbidge method [2] can be used for footing widths less than 4 m for soil profile with SPT N variation with depth is not uniform, i.e., either increasing or decreasing since deviation angle is less.
3. Schmertmann et al. [1] method is preferable for footing widths greater than 2 m.
4. The settlement ratio, $S_S/S_B$, is greater than 3 for stiff strata at shallow depths. The settlements ratio, $S_S/S_B$, is less than 1 if the soil at shallow depths of footing is soft with SPT $N$ value less than 3.

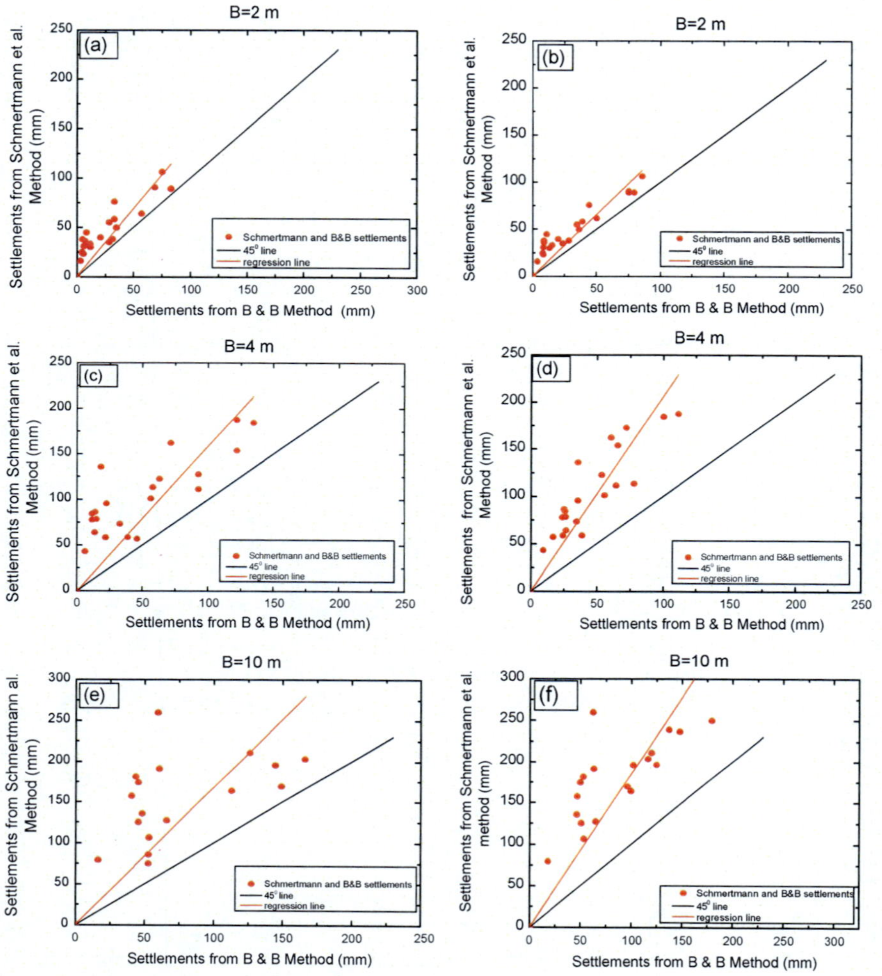


**Fig. 4** Settlements from Schmertmann et al. [1] versus settlements from Burland and Burbidge [2] methods: **a**, **c** and **e** for footing widths 2, 4 and 10 m and SPT *N* increasing with depth; **b**, **d** and **f** for footing widths 2, 4 and 10 m and for SPT *N* decreasing with depth

**Table 6** Angles of deviation of regression lines from 45° for the plots drawn between Schmertmann et al. [1] and Burland and Burbidge [2] methods

| Footing width (m) | Angles of deviation of regression lines from 45° Schmertmann et al. [1] and Burland and Burbidge [2] methods | |
|---|---|---|
| | Increasing SPT *N* with depth (°) | Decreasing SPT *N* with depth (°) |
| 2 | 8.8 | 7.64 |
| 4 | 12.6 | 18.8 |
| 10 | 14.2 | 16.6 |

## References

1. Schmertmann, J.H., Hartman, J.P., Brown, P.R.: Improved strain influence factor diagrams. J Geotech Eng. Div. **104**(8), 1131–1135 (1978)
2. Burland, J.B., Burbidge, M.C.: Settlement of foundations of sand and gravel. Proc., Inst. Civ. Eng. **78**, 1325–1381 (1985)
3. Meyerhof, G.G.: Penetration tests and bearing capacity of cohesionless soils. J. Soil Mech. Div. **82**(1), 1–12 (1956)
4. Meyerhof, G.G.: Shallow foundations. J. Soil Mech. Found. Div. **91**(2), 21–31 (1965)
5. Meyerhof, G.G.: General report: state-of-the-art of penetration testing in countries outside Europe. In: Proceedings of the 1st European symposium on penetration testing 2.1, pp. 40–48 (1974)
6. Davis, E.H., Poulos, H.G.: The use of elastic theory for settlement prediction under three dimensional conditions. Geotechnique **18**(1), 67–91 (1968)
7. Schmertmann, J.H.: Static cone to compute static settlement over sand. J. Soil Mech. Found. Div. **96**(3), 1011–1043 (1970)
8. Bowles, J.E.: Elastic foundation settlements on sand deposits. J. Geotech. Engrg. **113**(8), 846–860 (1987)
9. Mayne, P.W., Poulos, H.G.: Approximate displacement influence factors for elastic shallow foundations. J. Geotech. Geoenviron. Eng. **125**(6), 453–460 (1999)
10. Akpila: Settlement response of raft foundation on sand using SPT method. Can. J. Pure Appl. Sci. **8**(1), 2769–2774 (2014)
11. D'Appolonia, D.J., D'Appolonia, E., Brissette, R.F.: Settlement of spread footings on sand. J. Soil Mech. Found. Div. **96**(2), 754–761 (1970)

# Probabilistic Investigation on Seismic Bearing Capacity of Shallow Foundation on Unsaturated Fly Ash Deposit

Abhijit Anand and Rajib Sarkar

## 1 Introduction

Fly ash is a fine-grained, coal combustion residue by-product generated at coal-based thermal power plants, which if not disposed of properly, may become a hazardous waste [1]. The increase in urban population and consequently increased demand of power have put a lot of burden on the coal-based thermal power plants. In this pursuit, tremendous amount of fly ash is generated globally while the utilization of fly ash is limited. In this regard, one of the most promising techniques of bulk utilization of fly ash may be in its use as a geomaterial for reclamation of low-lying areas, land fill, construction of embankment and mine fill [2–6]. Bearing capacity and settlement are two key design parameters while evaluating the suitability of fly ash as a fill material. Various researchers [1, 7, 8] carried out model tests to study the load settlement characteristics of fly ash deposit/slopes.

Conventionally, shallow foundations are constructed in vadose zone which is partially saturated due to various environmental factors such as evapotranspiration and capillary action of soil. Classical bearing capacity theories neglect the suction stress while estimating the bearing capacity and therefore, yields a very conservative solution, which may lead to an uneconomical design of a footing. Recently, various researchers [9–12] investigated the problem of bearing capacity under unsaturated framework, and based on the results, it has been well established that the matric suction significantly affects the shear strength and hence the bearing capacity of a footing. The matric suction distribution profile depends on several factors, such as depth of water table and surface flux boundary conditions. Incorporation of influence of these parameters is vital while designing a foundation.

A. Anand · R. Sarkar (✉)
IIT(ISM) Dhanbad, Dhanbad, Jharkhand 826004, India
e-mail: rajib@iitism.ac.in



C. N. V. Satyanarayana Reddy et al. (eds.), *Ground Characterization and Foundations*, Lecture Notes in Civil Engineering 167,
https://doi.org/10.1007/978-981-16-3383-6_41

Often, foundations are located in seismically active zone where the bearing capacity of the foundation is affected significantly due to the seismic forces acting on the foundation. Various researchers have attempted to address the problem of seismic bearing capacity of a shallow footing by adopting a pseudo-static approach [13–15]. However, problem of seismic bearing capacity under unsaturated framework has not been given due consideration in the literature.

Therefore, in the present study, an attempt has been made to provide a solution to the seismic bearing capacity of a shallow strip footing under unsaturated framework, based on a finite element analysis. Variation of seismic bearing capacity with seismic horizontal coefficient ($k_h$) has been studied under a deterministic framework for different steady-state surface flux boundary condition. Effects of uncertainty in the random input parameters have been presented through a detailed probabilistic analysis.

# 2 Problem Definition and Material Properties

## 2.1 *Schematics of the Problem*

Figure 1 illustrates the graphical representation of the problem considered in the present study. A surface, strip footing of width ($B = 1$ m) has been considered resting on the surface of a fly ash deposit. Groundwater table (GWT) has been assumed to be at a depth of 10$B$ from the ground surface. Above the GWT, under a steady-state surface flux boundary condition, fly ash would not be uniformly saturated, and the effective degree of saturation would vary with depth. Due to the variable saturation,

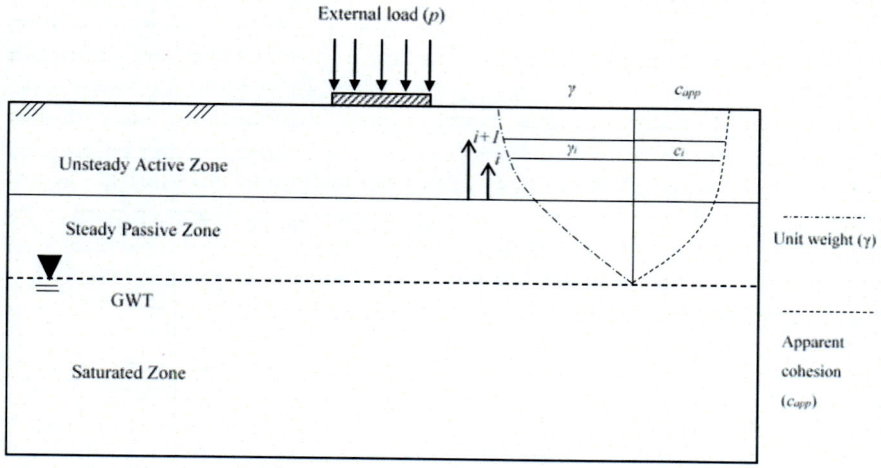


**Fig. 1** Schematics of the problem geometry considered in present study

effective unit weight ($\gamma_e$) and capillary cohesion ($c_{app}$) of fly ash deposit would also vary with depth as shown in Fig. 1. Partially saturated region above the GWT may be divided into $N$ number of layers, and capillary cohesion and effective unit weight may be assigned to each layer. In Fig. 1, $c_i$ and $\gamma_i$ represent the capillary cohesion and effective unit weight, respectively, at the $i$th layer above the groundwater table. The optimization of number of layers ($N$) and values of $c_i$ and $\gamma_i$ have been discussed in the subsequent sections.

### 2.2 *Fly Ash Properties Considered in the Study*

Based on an extensive literature survey on the geotechnical properties of several different Indian fly ash samples, Pandian [16] reported the range of parameters for Indian fly ash samples. For the present study, properties of fly ash samples are chosen based on the typical range of values reported in Pandian [16].

Water retention characteristic curve (WRCC) fitting parameters of the fly ash has been obtained from Prakash et al. [17] (Table 1).

## 3 Quantification of Suction Stress-Based Engineering Parameters

To incorporate matric suction into the original shear strength Mohr–Coulomb's (M-C) failure criterion, Fredlund et al. [18] extended the M-C criterion and proposed the following closed-form expression for the modified failure envelope

$$\tau_f = c' + (\sigma - u_a)_f \tan\phi' + (u_a - u_w)_f \tan\phi^b \tag{1}$$

where $c'$ = effective cohesion, $u_a$ = pore-air pressure, $u_w$ = pore-water pressure, $u_a - u_w$ = matric suction, $\phi^b$ = parameter representing internal friction angle due to presence of matric suction, $\phi'$ = angle of internal friction, $\tau_f$ = shear stress at failure, $\sigma$ = normal stress.

**Table 1** Properties of fly ash considered in the study

| Property | Values |
|---|---|
| Unit weight, $\gamma$ | 16.6 kN/m$^3$ |
| Friction angle, $\phi$ | 34° |
| True cohesion, $c$ | 0 kPa |
| $\alpha$ | 0.032 kPa$^{-1}$ |
| $n$ | 2.161 |
| $G$ | 2.0 |

It has further been well attested by various researchers, that the behaviour of unsaturated geomaterials can be modelled with sufficient accuracy using the water retention characteristic curve (WRCC) fitting parameters [10, 19]. In this regard, the effective stress state of a partially saturated soil can be expressed as [19]

$$\sigma' = \sigma - u_a - \sigma_s \tag{2}$$

where $\sigma'$ = effective stress, $\sigma$ = total stress and $\sigma_s$ = suction stress. The suction stress ($\sigma_s$) may be expressed as [20]

$$\sigma_s = -(u_a - u_w) \text{ for } u_a - u_w \le 0 \tag{3a}$$

$$\sigma_s = -S_e(u_a - u_w) \text{ for } u_a - u_w \ge 0 \tag{3b}$$

where $S_e$ = effective degree of saturation may be expressed as [20]

$$S_e = \frac{1}{\left[1 + \{\alpha(u_a - u_w)\}^n\right]^{\frac{n-1}{n}}} \tag{4}$$

where $\alpha$ = WRCC fitting parameter, closely related to the inverse of the air-entry value, $n$ = WRCC parameter representing the pore-size distribution.

Substituting the value of effective degree of saturation ($S_e$) into Eq. (3) yields the expression suction stress ($\sigma_s$). The contribution of matric suction on shear strength can be represented by considering additional resistance among the soil particles against shearing (apparent cohesion), arising predominantly due to the capillary effects, and may be expressed as

$$c_{app} = -\sigma_s \tan\phi' \tag{5}$$

Under a steady-state surface flux boundary condition (infiltration or evaporation), the degree of saturation and, therefore, matric suction does not remain constant with depth. Under a steady-state infiltration condition, the expression for matric suction may be expressed as [21]

$$\psi = (u_a - u_w) = -\frac{1}{\alpha}\ln\left[\left(1 + \frac{q}{k_s}\right)e^{-\alpha\gamma_w(D_w - z)} - \frac{q}{k_s}\right] \tag{6}$$

where $q$ = steady-state infiltration rate, $k_s$ = saturated hydraulic conductivity, $D_w$ = depth of water table from the ground surface, $z$ = distance of the point considered from the ground surface and $\gamma_w$ = unit weight of water.

### 3.1 Variation of Capillary Cohesion ($c_{app}$) with Depth

Variation of capillary cohesion with depth may be obtained by substituting the expression of matric suction ($\psi$) (Eq. 6) and effective degree of saturation ($S_e$) (Eq. 4) in Eqs. (3) and (5). After the substitution and simplification, the expression of capillary cohesion may be expressed as

$$c_{\text{app}} = \frac{1}{\alpha}\left[\frac{\ln\left\{\left(1+\frac{q}{k_s}\right)e^{-\alpha\gamma_w(D_w-z)} - \frac{q}{k_s}\right\}}{\left\{1+\left[-\ln\left\{\left(1+\frac{q}{k_s}\right)e^{-\alpha\gamma_w(D_w-z)} - \frac{q}{k_s}\right\}\right]^n\right\}^{1-1/n}}\right]\tan\phi' \quad (7)$$

### 3.2 Variation of Effective Unit Weight ($\gamma_e$) with Depth

As the effective saturation varies with depth, it is evident that the effective unit weight of the fly ash deposit would also vary with depth. Wang et al. [21] proposed the following closed-form formulation for effective unit weight due to variable saturation of geomaterials under a steady-state infiltration condition

$$\gamma_e = \frac{(\gamma_{\text{sat}} - \gamma_w)G_s + \{S_e - (1 - S_e)S_r\}(G_s\gamma_w - \gamma_{\text{sat}})}{G_s - 1} \quad (8)$$

where $S_r$ = residual saturation of fly ash, $G_s$ = specific gravity of fly ash particles and $\gamma_{sat}$ = saturated unit weight of fly ash. In the above formulation, substituting the expression for $S_e$ (Eq. 4) yields the variation of effective unit weight with depth.

## 4 Numerical Modelling

For numerical investigation on the seismic bearing capacity, commercially available finite element software package "*PLAXIS*$^{2D}$" has been considered in the present study.

### 4.1 Problem Definition and Boundary Conditions

Figure 2 depicts the developed finite element model and boundary conditions. As it has already been discussed that under steady-state surface flux boundary condition, the saturation of the soil does not remain constant, and therefore, fly ash deposit above

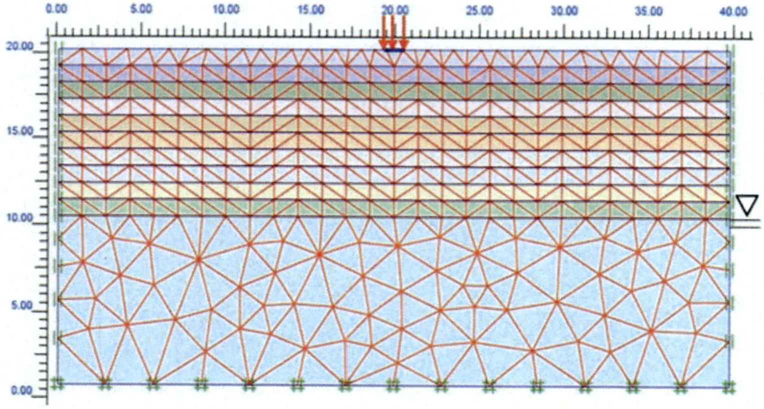


**Fig. 2** Typical layers considered in the finite element analysis and boundary conditions for the system considered

groundwater table (GWT) could not be modelled as a single homogeneous material. To overcome this, fly ash deposit above GWT has been divided into ($N$) number of layers and each layer has been assigned different unit weight and capillary cohesion obtained from Eqs. (7) and (8). In FE model, both horizontal and vertical fixities were applied to the bottom boundaries, while horizontal fixities were provided to the side boundaries. The domain size was chosen sufficiently large based on a sensitivity study to minimize the boundary effects. The number of discretized layers of unsaturated zones ($N$) was chosen to be 10 based on a sensitivity analysis. For the present study, a "*fine*" mesh refinement has been adopted based on the results obtained from a mesh optimization study.

## *4.2 Validation of the Developed Model*

In order to establish the efficacy of the developed model in predicting the seismic bearing capacity of strip footing, a set of pseudo-static analyses was carried out and the results are compared against the pioneering works reported in the literature [22–25]. For the validation purpose, the friction angle of the fly ash was assumed to be 40°, which is within the range of friction angle reported for fly ash samples [16]. Results are compared in terms of Terzaghi's [26] bearing capacity factor ($N_\gamma$) for different values of horizontal seismic acceleration coefficient. Figure 3 depicts the comparison between the $N_\gamma$ obtained from the present study and the values reported in the literature. Very close agreement establishes the efficacy of the present developed model in predicting the seismic bearing capacity under pseudo-static method.

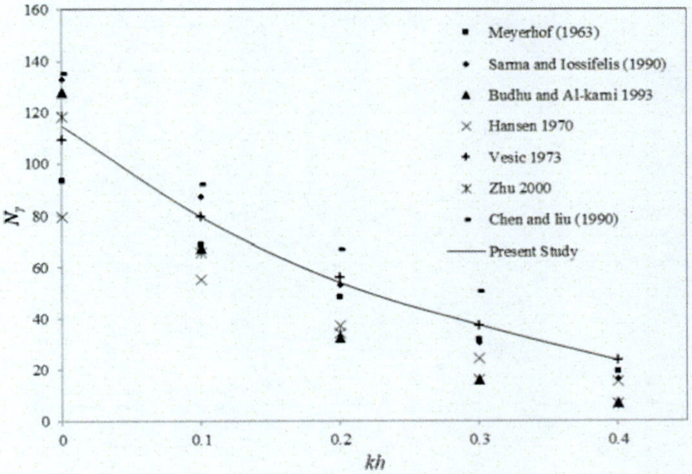


**Fig. 3** Comparison of the seismic bearing capacity factor ($N_{\gamma}$) obtained from the present investigation with results reported in literature

# 5 Results and Discussions

## 5.1 Deterministic Seismic Bearing Capacity

In order to study the influence of horizontal pseudo-static acceleration coefficient on the seismic bearing capacity, a set of numerical investigations was carried out considering different pseudo-static acceleration coefficient values ($k_{\mathrm{h}} = 0.0, 0.1, 0.2,$ 0.3, 0.4 and 0.5). Results are presented in dimensionless forms as Seismic Bearing Capacity Ratio (SBCR) which may be defined as

$$\mathrm{SBCR} = \frac{p}{\gamma B} \tag{9}$$

where $p =$ ultimate collapse load, $\gamma =$ unit weight of fly ash and $B =$ footing width.

Variation of SBCR with infiltration rate ratio ($q/k_{\mathrm{s}}$).

Variation of SBCR for different infiltration rate ratio ($q/k_s = 0.0, -0.2, -0.4, -0.6, -0.8$ and $-1.0$) has been shown in Fig. 4.

From Fig. 4, it is quite evident that the variation of steady-state infiltration rate ratio imparts substantial variation in the seismic bearing capacity of the footing on fly ash deposit. Furthermore, with increase in the magnitude of horizontal seismic acceleration coefficient ($k_{\mathrm{h}}$), an almost uniform reduction in SBCR was observed for all the values of $q/k_{\mathrm{s}}$ considered in the present study.

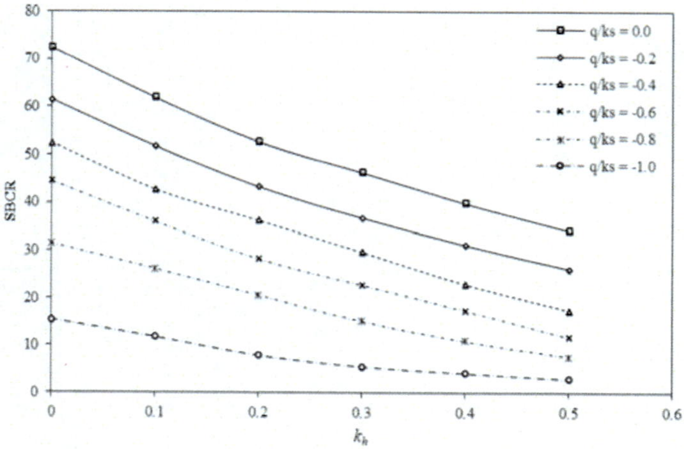


**Fig. 4** Variation of SBCR with $k_h$ for different infiltration rate ratio ($q/k_s$)

## 5.2 *Probabilistic Investigation*

SWCC fitting parameters are usually obtained by fitting a prescribed parametric model to a limited number of test data. This induces a substantial uncertainty to the measured WRCC parameter. In addition, inherent limitations of the measuring instrument to measure up to a limited range of matric suction also induced some uncertainty to the measurement of the parameters [17]. Occurrence of earthquake and rate of infiltration induced in soils due to environmental factors are also highly uncertain in nature. Therefore, a seismic bearing capacity analysis, based on deterministic set of parameters, may not yield reliable solution. To overcome above limitations, in the present study, the seismic bearing capacity analysis has been carried out under probabilistic framework. Uncertainty arising due to uncertainty in the input random variables has been quantified by development of prediction models based on analyses carried out using factorial design. The prediction models were constructed based on multiple regression analysis [27].

**Random parameters selection and their variability**

In the present study, four different significant random variables ($q/k_s$, $\alpha$, $n$ and $k_h$) have been considered for the probabilistic investigation. A factorial design approach has been adopted in which the variation of parameter is described by using a high point estimate and low point estimate [28]. The low point estimate and high point estimate and their distribution have been presented in Table 2.

For the low and high point estimates of WRCC fitting parameters, values and its distribution are chosen from the results presented in Prakash et al. [17]. For infiltration rate ratio, values are their distributions that are adopted from Anand and Sarkar [9].

**Table 2** Random parameters adopted in the present study

| Parameters | Distribution | High point estimate | Low point estimate |
|---|---|---|---|
| WRCC parameter ($\alpha$) | Log-normal | 0.017 | 0.040 |
| WRCC parameter ($n$) | Log-normal | 1.59 | 3.57 |
| Infiltration rate ratio ($q/k_s$) | Normal | – 1.0 | 0.0 |
| Horizontal seismic acceleration coefficient ($k_h$) | Normal | 0.0 | 0.5 |

For horizontal seismic acceleration coefficient ($k_h$), low point and high point estimates of 0 and 0.5 have been adopted which are conventionally adopted in pseudo-static investigation.

## Effect of random parameters on SBCR

Influence of random parameters on SBCR is presented in Table 3. The negative sign and positive sign represent the corresponding low point estimate and high point estimate of the random variable, respectively. Analysis of variance (ANOVA) results are presented in Table 4.

From the ANOVA results presented in Table 4, it is quite evident that the infiltration rate ratio ($q/k_s$) and horizontal seismic acceleration coefficient ($k_h$) are the most significant parameters affecting the seismic bearing capacity of the footing with their percentage contribution being 39.49% and 22.29%, respectively.

**Table 3** Model runs and corresponding SBCR

| Run | $q/k_s$ | $\alpha$ | $n$ | $k_h$ | SBCR |
|---|---|---|---|---|---|
| 1 | + | + | + | + | 4.76 |
| 2 | – | + | + | + | 2.34 |
| 3 | + | – | + | – | 51.92 |
| 4 | + | – | – | + | 48.76 |
| 5 | + | + | + | – | 21.38 |
| 6 | + | – | + | + | 17.53 |
| 7 | – | + | – | – | 16.98 |
| 8 | – | – | + | – | 16.98 |
| 9 | – | + | + | – | 16.98 |
| 10 | + | – | – | – | 16.98 |
| 11 | + | + | – | + | 17.10 |
| 12 | – | + | – | – | 33.37 |
| 13 | + | – | + | + | 2.34 |
| 14 | – | – | – | – | 45.18 |
| 15 | – | – | – | + | 2.34 |
| 16 | – | + | – | + | 2.34 |

**Table 4** ANOVA results for SBCR

| Terms | Factors | Percentage contribution (5) | $F$-value | $p$-value |
|---|---|---|---|---|
| Main | $q/k_s$ | 39.49 | 30.76 | 0.003 |
| | $\alpha$ | 11.23 | 8.75 | 0.032 |
| | $n$ | 3.55 | 2.77 | 0.157 |
| | $k_h$ | 22.29 | 17.37 | 0.009 |
| | $\alpha * q/k_s$ | 11.23 | 8.75 | 0.032 |
| Interaction | $n * q/k_s$ | 3.55 | 2.77 | 0.157 |
| | $k_h * q/k_s$ | 0.04 | 0.03 | 0.868 |
| | $n * \alpha$ | 0.00 | 0.00 | 0.996 |
| | $k_h * \alpha$ | 0.01 | 0.01 | 0.946 |
| | $n * k_h$ | 2.19 | 1.71 | 0.248 |

Effect of random variables on SBCR has been presented in Fig. 5. A thorough quantitative comparison of Fig. 5 suggests that infiltration rate ratio substantially affects the SBCR of a surface strip footing placed on an unsaturated fly ash deposit. Subsequently, the magnitude of seismic acceleration coefficient is the second most

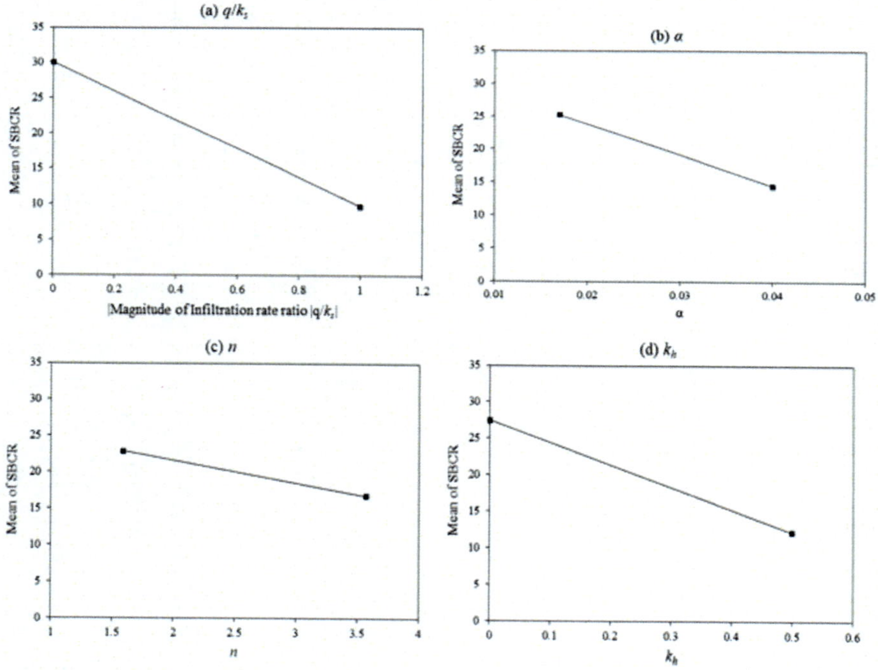


**Fig. 5** Effect of random variable **a** $q/k_s$, **b** $\alpha$, **c** $n$, and **d** $k_h$ on SBCR of unsaturated fly ash deposit

significant parameter. Among the SWCC fitting parameters, parameter $n$ has the least significance.

Development of prediction model

Prediction model has been developed for SBCR based on the results obtained from the factorial design analysis. Prediction model for SBCR can be expressed as

$$\begin{aligned}\text{SBCR} = 74.2 + 63.8\frac{q}{k_s} - 934\alpha - 3.77n - 5.6k_h - 943\frac{q}{k_s}\alpha \\ - 6.16\frac{q}{k_s}n - 2.6\frac{q}{k_s}k_h + \alpha n - 45\alpha k_h - 9.68nk_h \end{aligned} \quad (10)$$

Prediction model shown in Eq. (10) can be used by practicing engineers while solving the problem of seismic bearing capacity of shallow strip footing placed on a variably saturated fly ash deposit.

## 6 Summary and Conclusions

Present study investigates the seismic bearing capacity of a strip footing placed on an unsaturated fly ash deposit under a probabilistic framework. Influence of uncertainty in four different significant input variables on the overall seismic bearing capacity has been reported. Influence of steady-state infiltration rate ratio ($q/k_s$) and horizontal seismic acceleration coefficient ($k_h$) has been presented under a deterministic framework. Based on the results, the following major conclusions may be made

1. Seismic bearing capacity is substantially affected by variation of steady-state infiltration conditions, as well as magnitude of seismic acceleration coefficient ($k_h$).
2. Under deterministic framework, the reduction in SBCR with $k_h$ is uniform for all the infiltration rate ratio considered.
3. Prediction model developed in the present study may be useful for practicing geotechnical engineers while designing a shallow foundation on fly ash deposit in seismically active regions.
4. With increase in infiltration rate ratio, bearing capacity reduces substantially, and therefore, results of in situ field tests may be significantly affected due to the presence of matric suction or effective saturation of the geomaterials.

## References

1. Nadaf, M.B., Mandal, J.N.: Behaviour of reinforced fly ash slopes with cellular mattress and strips under strip loading. J. Hazard. Toxic Radioactive Waste **21**(4), 04017019 (2017)
2. DiGioia, A.M., Nuzzo, W.L.: Fly ash as structural fill. J. Power. Div. **98**(1), 77–92 (1972)

3. Ghosh, A., Subbarao, C.: Tensile strength bearing ratio and slake durability of class F fly ash stabilized with lime and gypsum. J. Mater. Civ. Eng. **18**(1), 106–115 (2006)
4. Kim, B., Prezzi, M., Salgado, R.: Geotechnical properties of fly ash and bottom ash mixtures for use in highway embankments. J. Geotech. Geoenviron. Eng. **131**(7), 914–924 (2005)
5. Kaniraj, S.R., Havanagi, V.G.: Behaviour of cement stabilized fiber reinforced fly ash soil mixtures. J. Geotech. Geoenviron. Eng. **127**(7), 574–584 (2001)
6. Leonards, G.A., Bailey, B.: Pulverized coal ash as structural fill. J. Geotech. Eng. Div. **108**(40), 517–531 (1982)
7. Gill, K.S., Choudhary, A.K., Jha, J.N., Shukla, S.K.: Experimental and numerical studies of loaded strip footing resting on reinforced fly ash slope. Geosynth. Int. **20**(1), 13–25 (2013)
8. Cui, W., Zheng, X., Zhang, Q.: Evaluation of bearing capacity of fly ash highway subgrade based on model test. J. Test. Eval. **46**(3), 943–955 (2018)
9. Anand, A., Sarkar, R.: Probabilistic investigation on bearing capacity of unsaturated fly ash. J. Hazard. Toxic Radioactive Waste **24**(4). https://doi.org/10.1061/(ASCE)HZ.2153-5515.0000547
10. Vahedifard, F., Robinson, J.D.: Unified method for estimating the ultimate bearing capacity of shallow foundations in variably saturated soils under steady flow. J. Geotech. Geoenviron. Eng. **142**(4), 04015095 (2016)
11. Oh, W.T., Vanapalli, S.K.: Interpretation of the bearing capacity of unsaturated fine-grained soil using the modified effective and the modified total stress approaches. Int. J. Geomech. **13**(6), 769–778 (2013)
12. Vanapalli, S.K., Mohamed, F.M.O.: Bearing capacity of model footings in unsaturated soils. In: Experimental Unsaturated Soil Mechanics, pp. 483–493 Springer, Berlin (2007)
13. Zhu, D.: The least upper bound solution for bearing capacity factor $N_\gamma$. Soils Found. **40**(1), 123–129 (2000)
14. Budhu, M., Al-Karni, A.: Seismic bearing capacity of soils. Geotechnique **43**(1), 181–187 (1993)
15. Sarma, S.K., Iossifelis, I.S.: Seismic bearing capacity factors of shallow strip footings. Geotechnique **40**(2), 265–273 (1990)
16. Pandian, N.S.: Fly ash characterization with reference to geotechnical applications. J. Indian Inst. Sci. **84**, 189–216 (2004)
17. Prakash, A., Hazra, B., Sreedeep, S.: Probabilistic analysis of unsaturated fly ash slope. J. Hazard. Toxic Radioactive Waste **21**(4), 06018002 (2019)
18. Fredlund, D.G., Morgenstern, N.R., Widger, R.A.: Shear strength of unsaturated soils. Can. Geotech. J. **15**(3), 313–321 (1978)
19. Lu, N., Likos, W.J.: Unsaturated Soil Mechanics. Wiley, New York (2004)
20. Lu, N., Godt, J.W., Wu, D.T.: A closed-form equation for effective stress in unsaturated soil. Water Resour. Res. **46**(5), W05515 (2010)
21. Wang, L., Sun, D., Chen, B., Li, J.: Three-dimensional seismic stability of unsaturated soil slopes using a semi-analytical method. Comput. Geotech. **110**, 296–307 (2019)
22. Chen, W.F., Liu, X.L.: Limit Analysis in soil Mechanics. Elsevier, Amsterdam (1990)
23. Meyerhof, G.G.: Some recent research on the bearing capacity of foundations. Can. Geotech. J. **1**(1), 16–26 (1963)
24. Hansen, J.B.: A Revised and Extended Formula for Bearing Capacity. Bulleting no. 28, Copenhagen: Danish Geotechnical Institute (1970)
25. Vesic, A.S.: Analysis of ultimate loads of shallow foundations. J. Soil. Mech. Found. Div. **99**(1), 45–73 (1973)
26. Terzaghi, K.: Theoretical Soil Mechanics. Wiley, New York (1943)
27. Myers, R.H., Montgomery, D.C., Anderson-Cook, C.M.: Response Surface Methodology: Process and Product Optimization Using Designed Experiments. Wiley, New York (2016)
28. Rosenblueth, E.: Point estimates for probability moments. In: Proceedings National Academy Science, pp. 3812–3814 (1975). https://doi.org/10.1073/pnas.72.10.3812

# Bearing Capacity of Reinforced CNS Soil Bed on Clay Soil with Inclined Reinforcement Considering Kinematics

**P. Rajashekar Reddy, G. V. N. Reddy, and E. Saibaba Reddy**

## 1 Introduction

Soil reinforcement for foundation beds is usually placed in horizontal layers to restrain tensile strains in the soil and increases the overall resistance of the composite medium through interfacial bond resistance. In many of the studies available related to the analysis and stability of reinforced soil structures, only bond resistance mobilized due to Jewell [1] axial pull is considered. However, in the reinforced soil structures, critical plane intersects the reinforcement layer obliquely; thus, the reinforcement deforms transversely under the action of oblique force/displacement. The soil beneath the reinforcement develops additional normal stress along with the soil reinforcement interface resulting in additional bond resistance. Rowe [2] considered a force in the reinforcement to act along the direction between the alignment of reinforcement and tangent to the slip surface. Reinforced soil structure has been studied by Uma Shankar and Madhav [3] considering kinematics and showed that reinforcement is subjected to transverse pull in addition to axial pull. Kumar and Madhav [4] presented an analysis for analyzing the reinforced earth wall with geosynthetic reinforcement inclined and established that factor of safety against pullout is enhanced due to mobilized normal stress acting on the bottom of reinforcement. It is noticed that geosynthetic reinforced structures accommodate large deformations before failure and at large deformation soil exhibit nonlinear behavior. In this paper, it is proposed to study the bearing capacity of nonlinear response of CNS soil bed reinforced with inclined reinforcement over in situ clay layer considering the transverse displacement of reinforcement.

P. R. Reddy (✉) · G. V. N. Reddy · E. Saibaba Reddy
JNTUH College of Engineering, Hyderabad, India



C. N. V. Satyanarayana Reddy et al. (eds.), *Ground Characterization and Foundations*, Lecture Notes in Civil Engineering 167,
https://doi.org/10.1007/978-981-16-3383-6_42

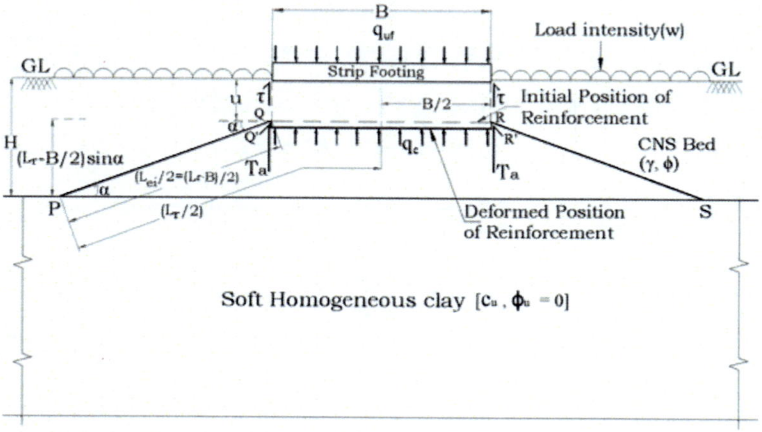


Fig. 1 Definition sketch

## 2 Methodology

### 2.1 General

A strip footing having width $B$ is resting on cohesive non-swelling soil stratum of thickness, $H$ with unit weight, $\gamma$ and angle of internal friction ($\phi$), relative fill stiffness factor ($\beta$) overlying soft clay with geotextile reinforcement with length $L_r$ is placed in the soil bed at a depth of $u$ from the bottom of footing with an inclination of $\alpha$. $\phi_r$ is the interface bond resistance between soil and reinforcement. Figure 1 shows deformation of the soil column and geotextile reinforcement due to punching shear failure of footing. The initial position of inclined reinforcement is represented by line PQRS, and it gets deformed to the new position defined by PQQ′R′RS. Full mobilization of shear resistance along with reinforcement-soil interface and nonlinear stress displacement response of soil bed is assumed.

### 2.2 Bearing Capacity of Cohesive Non-swelling Soil Bed on Clay Soil

The bearing capacity of footing resting on thin dense sand bed overlying soft clay considering punching mode of failure presented by Meyerhof [5] is

$$q_{\text{cns}} = cN_c + \frac{\gamma H^2}{B} k_s (1+2)\left(\frac{D}{H}\right) \tan\phi + \gamma D \leq 0.5\gamma \text{BN}_r \quad (1)$$

$N_c$, $N_r$ = Bearing capacity factors $\gamma$ = unit weight of soil. $\phi$ = angle of shearing resistance of soil, $K_s$ can be from the figure given by Meyerhof and Hanna [6].

## 2.3 Bond Resistance of Geotextile Reinforcement Placed Inclined in CNS Soil Bed

**Axial Pull**. Strip footing along with CNS soil column below the footing moves down due to punching effect resulting in mobilization of shear stresses on both sides of the soil column, and bond resistance will be mobilized at the soil reinforcement interface.

Tangential stress, $q_t$ offers direct resistance against pullout of reinforcement, and due to normal stress, $q_n$ resistance $q_n \tan\phi_r$ is mobilized as shown in Fig. 2.

Bearing capacity of soil bed with inclined reinforcement resting on soft clay considering axial pullout force mobilized in the reinforcement as

$$q_{\text{uir}} = cN_c + \frac{\gamma H^2}{B} K_s \tan\phi + 4\left\{\gamma\left(\frac{u}{B} + \frac{1}{2}\left[\left(\frac{Lr}{2B} - \frac{1}{2}\right)\sin\alpha\right]\right) + \frac{w}{B}\right\}$$
$$\left(\frac{Lr}{2B} - \frac{1}{2}\right)(\tan\phi_r \cos\alpha + \sin\alpha) \tag{2}$$

Normalizing the above equation with $c$

$$q^*_{\text{uir}} = N_c + \left(\frac{\gamma B}{c}\right)\left(\frac{H}{B}\right)^2 K_s \tan\phi + 4\left\{\frac{\gamma B}{c}\left(\frac{u}{B} + \frac{1}{2}\left[\left(\frac{Lr}{2B} - \frac{1}{2}\right)\sin\alpha\right]\right) + \frac{w}{c}\right\}$$
$$\left(\frac{L_r}{2B} - \frac{1}{2}\right)(\tan\phi_r \cos\alpha + \sin\alpha) \tag{3}$$

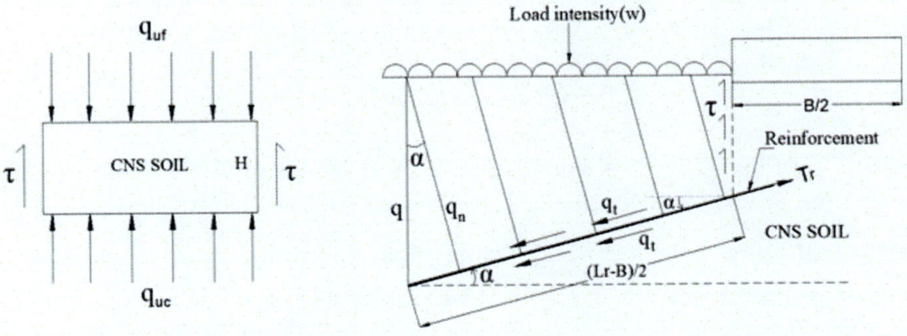


**Fig. 2** Stress on soil column and inclined reinforcement

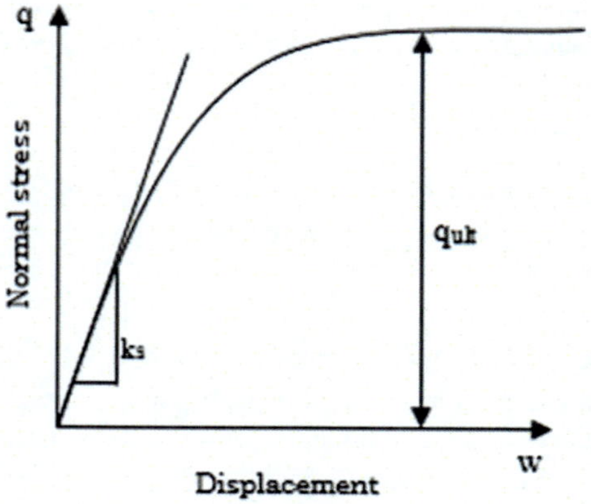


**Fig. 3** Normal stress displacement response of soil

**Transverse pull**. For the estimation of bearing capacity of the double layered soil considering the kinematics, punching shear failure is considered. The analysis is carried out assuming the response of the soil to the transverse displacement is nonlinear as shown in Fig. 3.

Uma Shankar and Madhav [3] developed a method to estimate the mobilized additional bond resistance for the analysis of sheet reinforcement subjected to transverse force/displacement. A transverse displacement, ($w_L$) of the reinforcement layer at the face of the footing is considered to estimate additional mobilized bond resistance. Because of transverse displacement, $w_L$ of the reinforcement, upward resisting force $P$ gets developed. Pullout force in the reinforcement increases due to this transverse displacement,

Tension mobilized in the reinforcement due to additional normal force, $P$ as

$$T = 2\gamma u_{avg} L_{ei} \tan \phi_r + P \sec \alpha \tan \phi_r \tag{4}$$

where $P$ is the transverse force in geotextile reinforcement layer mobilized due to transverse displacement ($w_L$) at the edge of footing, $P$ is calculated as follows.

$$P = \gamma u_{avg} L_{ei} P^* \tag{5}$$

where $P^*$ is the normalized transverse force in geotextile reinforcement obtained from Uma Shankar and Madhav [3]. In soil bed with relative stiffness, $\mu\left(\frac{k_s L}{\gamma H}\right)$, relative fill stiffness factor $\beta\left(\frac{k_s L}{q_{ult}}\right)$, the variation of normalized transverse force $P^*$ with normalized displacement $\left(\frac{w_L}{L}\right)$ is shown in Fig. 4 for $\phi = 30°$.

A parametric study has been carried out for $W_L = 0.001–0.01$, $\phi_r = 22.5°–30°$, $\gamma = 15–20$ kN/m$^3$, $\mu = 500–10{,}000$, $L_r = 2–10$ m, $\beta = 50–3000$ and $q_{ult} = 300–1000$ kN/m$^2$. The bearing capacity of nonlinear CNS bed with reinforcement inclined overlying soft clay soil is the sum of bearing capacity of clay layer, shear resistance developed in CNS bed, axial resistance of inclined reinforcement, and additional

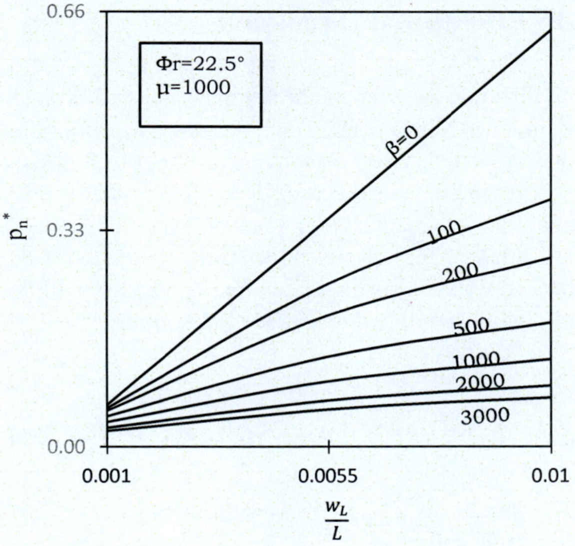


**Fig. 4** Variation of normalized transverse force, $P^*$ with normalized displacement, $(w_L/L)$-effect of relative fill stiffness factor

resistance mobilized due to kinematics.

$$q_{\text{uirkn}} = cN_c + \frac{\gamma H^2}{B} K_s \tan\phi + 4\left\{\gamma\left(\frac{u}{B} + \frac{1}{2}\left[\left(\frac{Lr}{2B} - \frac{1}{2}\right)\sin\alpha\right]\right) + \frac{w}{B}\right\}$$
$$\left(\frac{Lr}{2B} - \frac{1}{2}\right)(\tan\phi_r \cos\alpha + \sin\alpha)\left(1 + T^* + P^*\right) \tag{6}$$

Normalizing the above equation with $c$

$$q^*_{\text{uirkn}} = N_c + \left(\frac{\gamma B}{c}\right)\left(\frac{H}{B}\right)^2 K_s \tan\phi + 4\left\{\frac{\gamma B}{c}\left(\frac{u}{B} + \frac{1}{2}\left[\left(\frac{Lr}{2B} - \frac{1}{2}\right)\sin\alpha\right]\right) + \frac{w}{c}\right\}$$
$$\left(\frac{L_r}{2B} - \frac{1}{2}\right)(\tan\phi_r \cos\alpha + \sin\alpha)\left(1 + T^* + P^*\right) \tag{7}$$

$q^*_{\text{uir}}$ is the ratio of bearing capacity of geotextile reinforced CNS bed with reinforcement placed inclinedly considering axial tension in inclined reinforcement to that of undrained shear strength of underlying clay.

$q^*_{\text{uirkn}}$ is the ratio of bearing capacity CNS bed with nonlinear response and reinforcement placed horizontal considering the effect of the transverse force in addition to axial tension mobilized in horizontal reinforcement to that of undrained shear strength of in situ clay.

$q^*_{\text{uirkn}}$ is the ratio of bearing capacity CNS bed with nonlinear response and reinforcement placed inclined considering the effect of the transverse force in addition to axial tension mobilized in inclined reinforcement to that of undrained shear strength of in situ clay.

# 3 Results and Discussions

To illustrate the effect of the nonlinearity of CNS soil bed and transverse deformation of inclined reinforcement on normalized bearing capacity of strip footing supporting on CNS soil bed over clay for a wide range of following non-dimensional parameters are presented. $\gamma B/c = 1.8$, $L_r/B = 2.5$–$3.5$, $H/B = 0.5$, $w_L = 0.001$–$0.01$, $\beta = 0$–$3000$ computations are made for $\frac{\phi_r}{\phi} = 0.75$, $\phi = 30°$, $w/c = 0$ in addition to that $\alpha = 0°$–$20°$ are studied. For various values of $\alpha$, the effect of these parameters on normalized bearing capacity is quantified and compared with the normalized bearing capacity of nonlinear soil bed reinforced horizontally.

## 3.1 *Effect of Relative Stiffness of Soil Bed ($\mu$)*

The variation of $q^*_{uirkn}$ with $\alpha$ for different values of $\mu$, for $u/B = 0.15$, $H/B = 0.5$, $L_r/B = 3$, $\phi_r/\phi = 0.75$, $\phi = 30°$, $\beta = 500$, $w/c = 0$, $\gamma B/c = 1.8$, $w_L/L = 0.01$ is presented in Fig. 5. $q^*_{uirkn}$ increases from 6.98 to 10.03 an increase of 43.7% with increase in $\alpha$ from 0° to 20° due to inclination of reinforcement due to mobilization of additional normal force exerted on reinforcement and additional bond resistance mobilized due to transverse deformation of reinforcement for $\mu = 2000$. $q^*_{uirkn}$ increases from 7.98 to 9.5 an increase of 19%.

For the inclination of reinforcement, $\alpha = 10°$ with an increase in $\mu$ from 500 to 10,000 due to increasing in relative stiffness the transverse force required to mobilize transverse displacement increases, hence increase in bearing capacity. Effect of $\alpha$

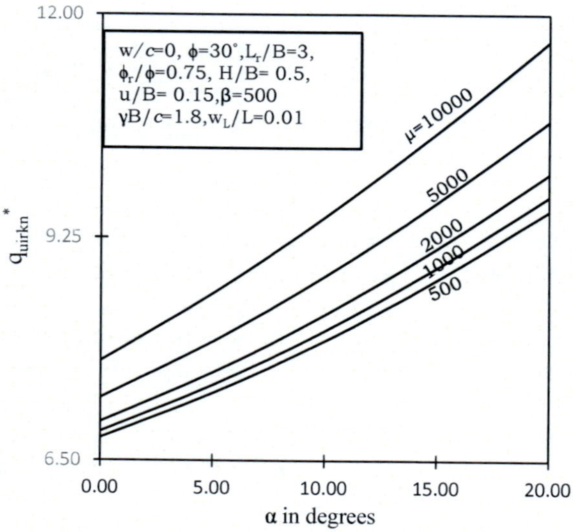


**Fig. 5** Variation of normalized bearing capacity with inclination of reinforcement($\alpha$)-effect of relative stiffness

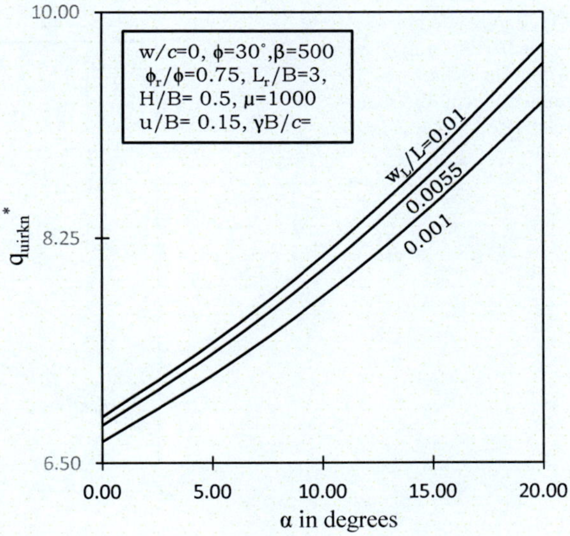


**Fig. 6** Variation of normalized bearing capacity with inclination of reinforcement($\alpha$)-effect of transverse displacement

on the increase of bearing capacity is significant, whereas effect of $\mu$ on normalized bearing capacity is considerable.

## 3.2 *Effect of Transverse Displacement* ($W_L$)

Variation of $q^*_{\text{uirkn}}$ with α for different values of $w_L/L$ for $L_r/B = 3$, $H/B = 0.5$, $u/B = 0.15$, $\phi_r/\phi = 0.75$, $\mu = 1000$, $\beta = 500$, $w/c = 0$, $\gamma B/c = 1.8$ is depicted in Fig. 6. $q^*_{\text{uirkn}}$ increases from 6.79 to 9.6 an increase of 41.4% with an increase in $\alpha$ from 0° to 20° for $w_{L/}L = 0.0055$ due to inclination of reinforcement resulting in mobilization of additional normal stress on reinforcement and also due to transverse deformation of reinforcement, additional bond resistance developed between the bottom of reinforcement and soil due to upward normal stress acting on bottom of reinforcement. $q^*_{\text{uirkn}}$ increases from 7.79 to 8.11 an increase of 4.1% for $\alpha =$ 10° with increase in transverse displacement from 0.001 to 0.01 due to upward normal stress acting on the bottom face of the reinforcement consequently additional bond resistance acting on the interface and thus additional tension mobilized in the reinforcement leading to increase in bearing capacity.

## 3.3 *Effect of Relative Fill Stiffness Factor* (β)

Variation of $q^*_{\text{uirkn}}$ with $\alpha$ for various values of $\beta$, for $H/B = 0.5$, $u/B = 0.15$, $L_{r/}B = 3$, $\phi = 30°$ $\phi_r/\phi = 0.75$, $\mu = 1000$, $\gamma B/c = 1.8$, $= 0.01$, $\alpha = 20°$, $w/c = 0$

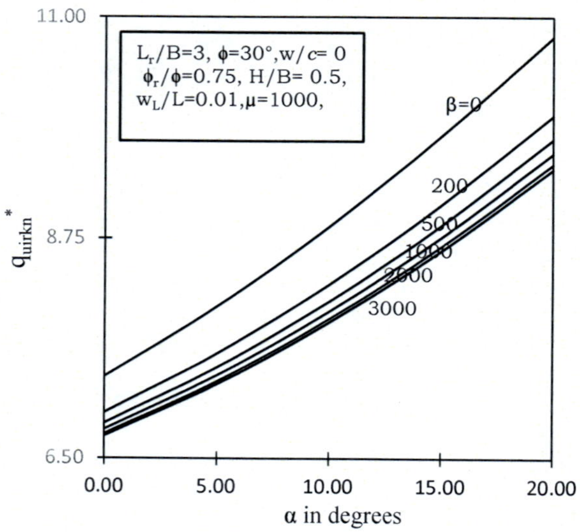


**Fig. 7** Variation of normalized bearing capacity with α inclination of reinforcement-effect of relative fill stiffness factor

is represented in Fig. 7. $q_{\mathrm{uirkn}}^*$ increases from 6.86 to 9.76 an increase of 42.3% with increase in $\alpha$ from 0° to 20° due to combined effect of additionally mobilized frictional resistance due to increase in normal stress component, $q_{\mathrm{n}}$ and additional bond resistance mobilized due to transverse deformation of reinforcement for $\beta =$ 200. $q_{\mathrm{uirkn}}^*$ increases from 7.89 to 8.87 an increase of 12.42% for a decrease in $\beta$ from 3000 to 0, due to stronger soil bed (more $q_{\mathrm{ult}}$) there by transverse displacement is localized near the failure surface.

### 3.4 *Effect of Various Improvement Techniques*

Variation of normalized bearing capacities $q_{\mathrm{uhr}}^*$ or $q_{\mathrm{uir}}^*$or $q_{\mathrm{uhrn}}^*$ or $q_{\mathrm{urikn}}^*$ with $\alpha$ for various improvement techniques, for $L_{\mathrm{r}}/B$, for $u/B = 0.15$, $\phi = 30°$ $\phi_{\mathrm{r}}/\phi = 0.75$, $H/B = 0.5$, $\mu = 1000$, $\gamma B/c = 1.8$, $w_{\mathrm{L}}/L = 0.01$, $w/c = 0$ are illustrated in Fig. 8. $q_{\mathrm{urikn}}^*$ increases 42.3% when compared with $q_{\mathrm{uhrn}}^*$ for 20° of inclination of reinforcement due to combined effect of increase in normal stress acting on reinforcement and additional bond resistance mobilized along bottom of reinforcement due to upward force acting normally on reinforcement caused due to transverse deformation of reinforcement. $q_{\mathrm{uri}}^*$ increases 20% when compared with $q_{\mathrm{uhr}}^*$ due to inclination of reinforcement resulting additional normal stress acting on reinforcement there by mobilization of additional bond stress along reinforcement.

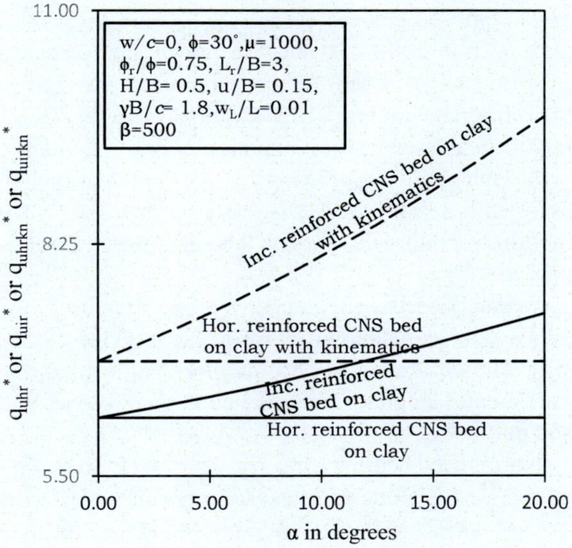


**Fig. 8** Variation of normalized bearing capacities versus $\alpha$-effect of various improvement techniques

**Table 1** Comparison of normalized bearing capacity of present study with available literature

| S. No | Author | Normalized bearing capacity | Results of present study for corresponding properties |
|---|---|---|---|
| 1 | Rajyalakshmi et al. [7] | 1.33 (Analy.) | 1.69 |

# 4 Validation of Results

As shown in Table 1, the predicted normalized bearing capacity value is more than the value of other author corroborating that the present approach is giving better results than the existing approach.

# 5 Conclusions

Bearing capacity analysis of strip footing resting on CNS soil stratum with geotextile reinforcement placed inclined overlying clay subjected to transverse displacement/pull is proposed by considering nonlinear shear stress displacement relation with horizontal reiforced CNS soil bed. Meyerhof's punching shear failure model for thin dense sand bed on clay is extended to include the axial tension in inclined reinforcement and mobilized additional bond resistance due to transverse deformation of inclined reinforcement theory proposed by Uma Shankar and Madhav [3] has been used.

Normalized bearing capacity, $q^*_{\text{urikn}}$ increases 19% with increase in stiffness of soil, $\mu$ from 500 to 10,000, for normalized displacement of 0.01 due to increase in stiffness of soil bed. As the reinforcement deforms transversely, the soil beneath the reinforcement mobilizes additional normal stresses resulting in mobilization of larger shear stresses for inclination of reinforcement, $\alpha = 10°$.

Normalized bearing capacity, $q^*_{\text{urikn}}$ increases 4.1% with increase in transverse deformation from 0.001 to 0.01, for $\alpha = 10°$. Soil beneath the reinforcement mobilizes additional normal, and therefore, shear stresses increases the bearing capacity.

Normalized bearing capacity, $q^*_{\text{urikn}}$ decreases from 8.87 to 7.89 a decrease of 12.4% with an increase in $\beta$ from 0 to 3000 for $\alpha = 10°$ due to weaker soil bed (less $q_{\text{ult}}$). The deformations become more uniform, and normal stresses are distributed over greater length of reinforcement, therefore less mobilization of bond resistance and pull out.

Normalized bearing capacity, $q^*_{\text{urikn}}$ is 31, 20.5, 18.2% more when compared with $q^*_{\text{uhr}}$, $q^*_{\text{uir}}$, $q^*_{\text{uhrkn}}$ due to increase in normal stress on reinforcement and additional upward normal stress acting on reinforcement due to transverse deformation of reinforcement, additional bond resistance developed along the bottom of reinforcement due to transverse deformation of reinforcement and increase in normal stress acting on reinforcement due to inclination of reinforcement.

## References

1. Jewel, R.A.: Keynote Lecture: links between the testing, modelling and design of reinforced soil. In: Proceedings of Ist Symposium on Earth Reinforcement Practice, vol. 2, pp. 755–772, Fukuoka, Japan (1992)
2. Kerry Rowe, R.: A review of the behavior of reinforced soil walls, keynote lecture. Proc. of Int. Symp. on Earth Reinforcement Practice (2), 801–830 (1992)
3. Uma Shankar, B., Madhav, M.R.: Analysis of inextensible sheet reinforcement subjected to transverse displacement/force Linear subgrade response. Geotext. Geomembr. **21**(2), 69–84 (2003)
4. Kumar, P.S.V.N., Madhav, M.R.: Pull-out of inclined reinforcement in reinforced earth wall. Indian Geotech. J. **41**(2), 95–99 (2011)
5. Meyerhof, C.G.: Ultimate bearing capacity of footings on sand layer overlying clay. Can. Geotech. J. **11**(2), 223–229 (1974)
6. Meyerhof, C.G., Hanna, A.M.: Ultimate bearing capacity of foundation on layered soils under inclined load. Can. Geotech. J. **15**(4), 565–572 (1978)
7. Rajyalakshmi, K., Madhav, M.R., Ramu, K.: Kinematics and bearing capacity of reinforced foundation bed on soft ground. Indian Geotech. J. **41**(3), 121–130 (2011)

# Mobilised Frictional Shear and Dead-Weight of Sand Wedge: Contributing to the Pull-Out Resistance of Belled Anchor Pile in Sand

**T. Deb and S. K. Pal**

## 1 Introduction

The scopes of anchor piles occupied a wide-space in foundation engineering applications in both off-shore and on-land structures. The anchor pile, whose enlarged base is making particular bell angle with the shaft, is known as belled anchor pile. A belled anchor pile may be specified by bell angle ($\alpha$), bell diameter ($D_b$), shaft diameter ($D_s$), embedment ratio ($L/D_b$) and diameter ratio ($D_s/D_b$). Depending upon the depth of embedment of the anchor in the soil and termination of failure surface either up to soil surface or within soil mass, the anchor is classified either as a (a) shallow anchor and (b) deep anchor. For shallow anchor, failure pattern is general type that reaches up to soil surface in collapsed stage, and failure occurs due to shear only; whereas in case of deep anchor, the effect of soil surface disappears.

In earlier days, pull-out resistance was predicted by Majer's soil cone theory [16] and Mors's earth pressure theory [17]. Turner [25] disagreed the soil cone method as predicted pull-out resistance of deeper foundation was highly overestimated. The studies on linear mathematical models of sand wedge were carried out by Downs and Chieurzzi [9], Clemence and Veesert [4], Sutherland et al. [23], Vermeer and Sutjiadi [28], Murray and Geddes [18] and Vanitha et al. [26], where pull-out resistance was the summation of frictional resistance and dead-weight of sand wedge. Balla [1] first introduced nonlinear slip surface to predict pull-out resistance of anchor. The nonlinear mathematical models of sand wedge were introduced by Matsuo [15], Rao and Kumar [20], Chottapadhyay and Pise [3], Saeedy [21] and Ghaly and Hanna [10], where the anchor failure mechanism was controlled by combination of dead-weight of sand wedge surrounding anchor contributing in passive resistance and

T. Deb (✉)
Khumulwng Polytechnical College, TTAADC Polytechnical Collage, Khumulwng, India

S. K. Pal
Department of Civil Engineering, NIT Agartala, Tripura 799046, India



C. N. V. Satyanarayana Reddy et al. (eds.), *Ground Characterization and Foundations*, Lecture Notes in Civil Engineering 167,
https://doi.org/10.1007/978-981-16-3383-6_43

mobilised frictional shear along slip surface, reverse to direction of wedge movement. Meyerhof and Adams (1968) suggested that pyramidal slip surface around the shallow anchor was initiated from the edge of anchor base and terminated in sand surface. Matsuo [15], Dickin [6], Tagaya et al. [24], Dickin and Leung [8], Ilamparuthi and Muthukrisnaiah [12], Ilamparuthi et al. [13] and Nazir et al. [19] experimentally investigated the nonlinear sand wedge around the shallow anchors from edge of anchor base to sand surface.

Though the present literature is providing a good number of mathematical models of sand wedge to predict pull-out resistance, but there is a scarcity in the availability of data on the variation of mobolised frictional shear and dead-weight of sand wedge based on pull-out resistance in reference of embedment ratios ($L/D_b$), diameter ratios ($D_s/D_b$) and bell angles ($\alpha$).

## 2 Objective of the Study

A 1-g 2D belled anchor panel possessing a bell angle ($\alpha$), base diameter ($D_b$) and shaft diameter ($D_s$) is embedded in dyed and non-dyed homogeneous dry sand deposits. Under vertical pull, as the applied stresses are acting along the vertical plane of the panel, a couple of symmetric nonlinear slip surfaces are formed in both the sides of each panel as shown in Fig. 1. These slip surfaces are generated from the anchor base and terminated in the sand surface. For the mathematical idealisation purpose, this slip surfaces can be interpreted as three-dimensional axisymmetric sand wedges

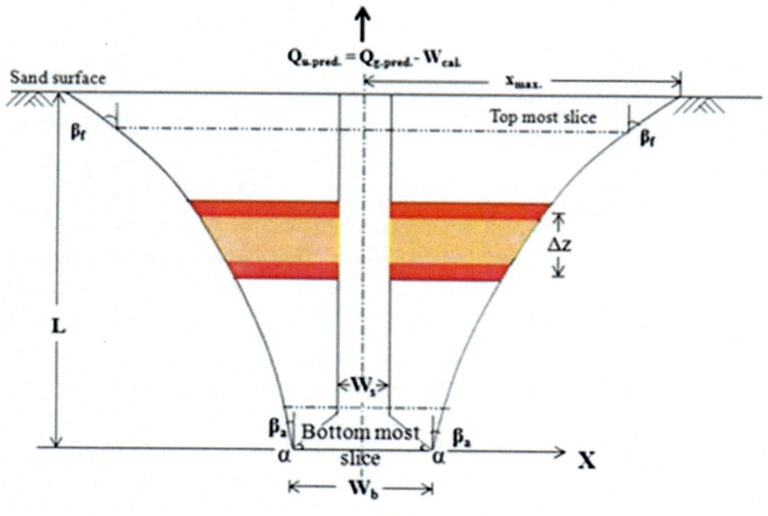


**Fig. 1** 1-g 2D belled anchor panel under vertical pull in dyed and non-dyed homogeneous dry sand deposits

of the same scale. Each model is subjected to static upward pull, i.e. $Q_{u.pred.}$ is the combination of as mobilised frictional shear and dead-weight of sand wedge.

The aim of the present study is to illustrate the (a) predicted values of pull-out resistance ($Q_{u.pred.}$) on the reference of horizontal extensions in slip surfaces, (b) ratio of mobilised frictional shear to pull-out resistance (× 100, %) and (c) ratio of dead-weight of sand wedge to pull-out resistance of belled anchors (× 100, %) on accounts of variation in embedment ratios, diameter ratios and bell angles in the dry sand deposit.

## 3 Materials, Models, Testing Tank and Sand Bed Preparation

The uniformly graded sand used in present study is procured from the local market. The preparation of foundation media with dry sand makes it easy to maintain the density of sand within the testing tank. Figure 2 shows grain size distribution curve of sand sample. The sand is having uniformity coefficient 1.1 and specific gravity 2.67. The maximum and minimum dry densities are found to be 14.20 and 16.50 kN/m$^3$, respectively. The placement density is fund to be 15.60 kN/m$^3$ at a certain calibrated height of free fall (700 mm) of sand by rainfall technique [5] and [7]. The angle of shearing resistance, i.e., $\phi = 33.5°$ at the placement density, according to UU triaxial test.

1 mm thick mild steel plate is used to fabricate the 2D panels. The length and height of each panel are 590 ± 5 mm and 650 mm, respectively. The thickness ($W_s$) of shaft part of the panel is 26 mm. The bell parts are having a range of thickness ($W_b$) as 92, 80, 68 and 56 mm, and hence, $W_s/W_b$ values are obtained as 0.28, 0.33, 0.38 and 0.46. All the panels are having bell angles ($\alpha$) of 45°, 54°, 63° and 72°. At the top of the panels, a threaded small cylinder is welded to connect it to the proving ring (1 kN capacity) and pulling shaft gently. In that small cylinder, two horizontally

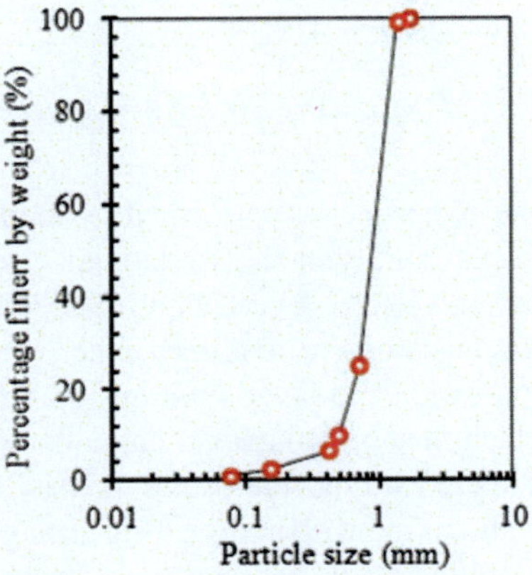


**Fig. 2** Grain size distribution of sand samples

projected steel strips are provided to hold dial gauges (LC = 0.01 mm) on them, and they are 180° apart from each other.

The testing tank is 700 mm (Length) × 600 mm (Width) × 700 mm (depth). The wall of model tank is made of plaxiglass in four sides. Vertical steel stiffeners are provided in three sides (except the front side) to prevent wall deflection outwards. To take uninterrupted measurements of failure points from outside, no stiffeners are provided in front wall. A 12 mm thick plaxiglass in front side is found to be stiffer enough to prevent outward wall deflection. The size of the tank is 6.5 times larger than the largest panel base as concerned in the present test, and so the tank size is large enough to avoid boundary effects.

To study the sand wedges, homogeneous sand media is prepared by placing successive layers of 3 mm thick red dyed and 18 mm thick non-dyed sand. The combination of dyed and non-dyed layer is chosen as well-suitable to prepare the foundation bed maintaining predetermined density of sand. The filling of testing tank is continued by dyed and non-dyed sand layers until each panel attains embedment ratio ($L/W_b$) 3, 4 and 5. When the thickness of upper most layer is found to be less than 3 mm, then it is adjusted along with the continuation of preceding layer (either dyed or non-dyed sand layer).

## 3.1 Experimental Program

For those anchors, slip surfaces terminate up to sand surface at collapsed stage are known as shallow anchors [14, 22] and [27]. Each panel is buried to attain embedment ratio ($L/W_b$) = 3, 4 and 5, and the panels are having thickness ratio ($W_s/W_b$) = 0.46, 0.38, 0.33 and 0.28, and these are possessing bell angle ($\alpha$) = 45°, 54°, 63° and 72°. To carry out the study, total 48 numbers of experiments are performed to investigate the variations in lateral extension of slip surfaces in both the sides of panels buried in dry sand deposits. The values of $L/W_b$ and $W_s/W_b$ as mentioned for 2D panel are exactly same used for the 3D anchor models as $L/D_b$ and $D_s/D_b$, respectively.

## 3.2 Experimental Set-Up, Test Procedure and Observations

The experiments performed as mentioned earlier are meant for the evaluation in the variations in the horizontal extension of failure points corresponding to the known vertical levels. Figure 3 shows the schematic diagram of experimental set-up consisting of loading frame, panel installed inside the testing-tank, dyed and non-dyed sand layers, pulling shaft and attached proving ring with it, position and attachment of dial gauges and other accessories. The steel channels are used to fabricate the loading frame, and the base is fixed with concrete floor by bolt connection. A nut (along with ball-bearing arrangement) is designed to rest on the reaction beam. A pulling shaft is mechanically working on nut and screw motion. The pulling shaft

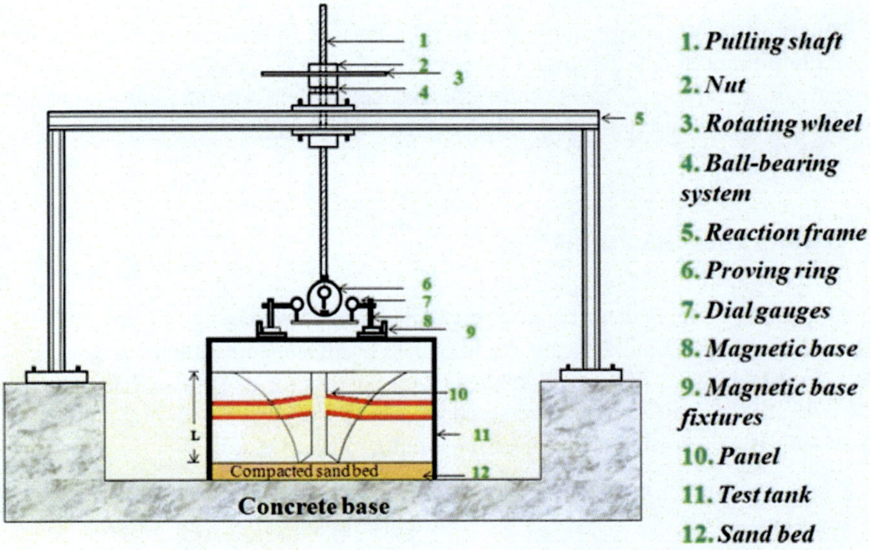


**Fig. 3** Schematic view of the experimental set-up

is attached with panel through proving ring. Thus, the panel is placed vertically at the centre of testing tank on compacted sand bed. Upward vertical movement of shaft is operated by manually controlled circular rotating wheel. The wheel is connected with nut arrangement. Due to the clock-wise motion of wheel, the panel shifts upward gradually. Initially, prior to each test, a compacted sand bed of 100 mm thick is prepared inside model tank over which the panel is placed. Each panel is subjected to vertical pull without causing any obliquity and tilt. The testing tank is filled up with dyed and non-dyed sand layers to attain desired embedment depth. The pictorial view of slip surfaces as obtained from laboratory experiments presented in Figs. 4a,

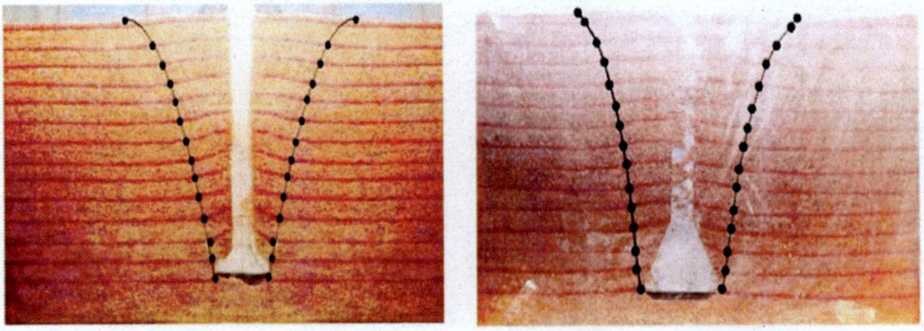

(a) (b)

**Fig. 4** Typical pictorial views of failure surface in both the sides of panels **a** 45-0.38-4 and **b** 63-0.28-3 b illustrates the formation of symmetric slip surfaces in both the sides of panels. The failure point at any horizontal level and vertical level can be distinguished by normal horizontal scaling from centre line of panel thickness to the failure point and by accounting the numbers dyed and non-dyed sand layers.

## 4 Model Identifications

Each model or panel is identified by a general coding form having four parts, which belong to bell angle, diameter ratio (or thickness ratio) and embedment ratio sequentially. For example, a 45° belled anchor panel possessing $W_s/W_b = 0.38$ at $L/W_b = 4$ is identified as 45-0.38-4. The symbol 72-0.28-3 represents a 72° panel and is possessing $W_s/W_b = 0.38$, and it is installed at $L/W_b = 3$.

## 5 Prediction of Pull-Out Resistance Based on Horizontal Slice Method

For the mathematical idealisation purpose and to calculate the pull-out resistance, the obtained slip surfaces are interpreted into axisymmetric three-dimensional figure of same scale. The whole wedge is the integration of number of horizontal slices of known thickness ($\Delta z = 21$ mm) as shown in Fig. 5. The pull-out resistance of each slice is the summation of mobilised frictional shear on slip surface of slices

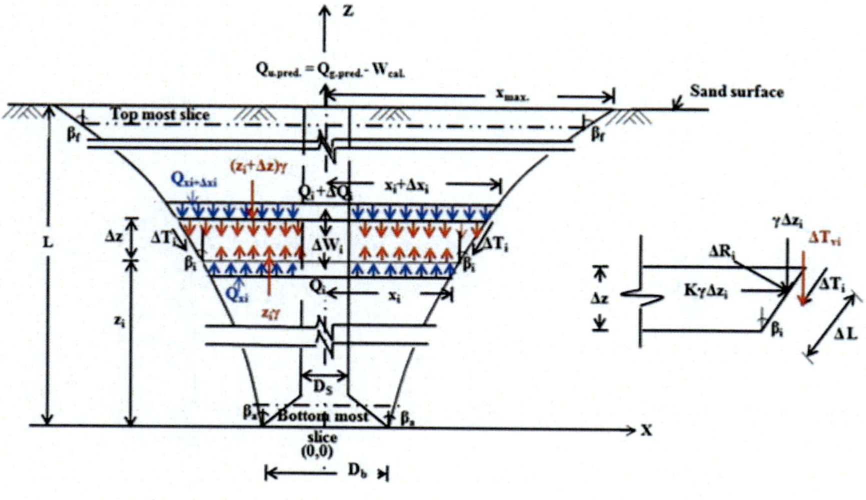


**Fig. 5** Free body diagram of three-dimensional *i*th slice wedge as per horizontal slice method

opposite to the direction of anchor movement and dead-weight of sand wedge. The mathematical analysis is presented elsewhere in details (Deb and Pal 2018).

Vertical component of shear resistance ($T_{\mathrm{vi}}$) for $i$th slice surface,

$$T_{\mathrm{vi}} = \left[ 2\gamma\pi\Delta Z\left( x_i + \frac{\Delta x_i}{2} \right)\{(Z_i + \Delta Z) - Z_i\}\{k_0 \cos\beta_i + \sin\beta_i\} \tan\phi \right] \quad (1)$$

Dead-weight of $i$th slice,

$$W_i = \left[ \pi\gamma\frac{\Delta Z}{3}\{x_i^2 + +(x_i + \Delta x_i)^2 + x_i(x_i + \Delta x_i)\} \right] \quad (2)$$

Considering vertical equilibrium for all elementary forces, gross pull-out resistance of $i$th slice as follows:

$$\begin{aligned} Q_{i.g} = &\left[ 2\gamma\pi\Delta Z\left( x_i + \frac{\Delta x_i}{2} \right)\{(Z_i + \Delta Z) - Z_i\}\{k_0 \cos\beta_i + \sin\beta_i\} \tan\phi \right] \\ &+ \left[ \pi\gamma\frac{\Delta Z}{3}\{x_i^2 + +(x_i + \Delta x_i)^2 + x_i(x_i + \Delta x_i)\} \right] \end{aligned} \quad (3)$$

Gross pull-out resistance for total wedge is found by summing up gross pull-out resistances for all the n number of slices,

$$Q_{\mathrm{g.pred.}} = \sum_{i=1}^{n} Q_{\mathrm{i.g.pred.}} \quad (4)$$

Net or predicted pull-out resistance,

$$Q_{\mathrm{u.pred.}} = Q_{\mathrm{g.pred.}} - W_{\mathrm{cal.}} \quad (5)$$

## 6 Results and Discussion

The effect of predicted pull-out resistances, ratio of mobilised frictional shear and dead-eight of sand wedge to the pull-out resistances is discussed based on the experimental observation on horizontal extension of failure points in slip surfaces of sand wedge and the application of horizontal slice method on the mathematical model. The range of the predicted pull-out resistance of all the models is from 22.88 to 383.70 N. The values of ratio of mobilised frictional shear to the pull-out resistance (×100,%) are from 13.50 to 30.81%, and ratios of dead-eight of sand wedge to the pull-out resistance (×100,%) are from 69.20 to 86.50%.

### 6.1 Pull-Out Resistance, Ratios of Mobilised Frictional Shear and Dead-Weight of Sand Wedge Based on Pull-Out Resistance of Belled Anchor Piles Influenced by Embedment Ratios ($L/D_b$)

In Fig. 6a, the typical plots present the pull-out resistance versus embedment ratio relationships for the models of $\alpha = 45°$, 54° and 63° and having $D_s/D_b = 0.33$, 0.28 and 0.38, respectively. These figure signifies that with the increase in the value of $L/D_b$ (i.e. 3, 4 and 5), the same belled anchor pile can achieve higher pull-out resistance, and this trend is true regardless the values of $D_s/D_b$ and $\beta$. For higher $L/D_b$ values, with higher anchor installation depth, gradually larger overburden pressure would act on the anchor base. As a result, gradually larger sand wedges are formed and offer higher pull-out resistances, as shown in Fig. 6b besides panel 45–0.33 at $L/T_b = 3$, 4 and 5. A similar pattern in the relationship of pull-out resistance with embedment ratio was also established by Dickin and Leung [7], Ghosh and Bera [11], Bera [2] and Nazir et al. (2014) in dry sand bed.

The typical Fig. 7a illustrates that the ratio of mobilised fictional shear to pull-out resistance (× 100, %) values gradually decrease with higher values of embedment ratio for all the 45° models. The typical Fig. 7b presents that for the same models, the ratio of dead-weight of sand wedge to pull-out resistance (× 100, %) values gradually increases with higher values of embedment ratio. This is due to the reason that for the sand wedges around the 45° model of $D_s/D_b = 0.33$, at $L/D_b = 3$, 4 and 5, the values of mobilised fictional shear is only 22.26, 18.84 and 16.76% of the pull-out resistance, respectively, whereas for this model and similar embedment depths, the dead-weight of sand wedge is 77.74, 81.16 and 83.24% of the pull-out resistance. The values of mobilised frictional shear, dead-weight of sand wedge

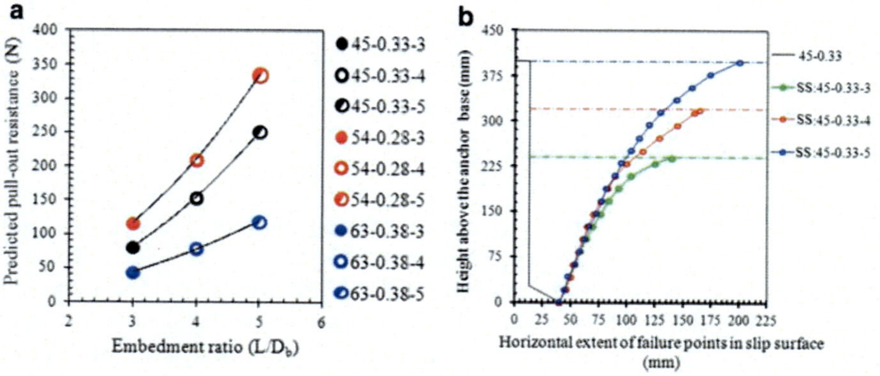


**Fig. 6** **a** Predicted pull-out resistance vs. embedment ratio relationships for 45°, 54° and 63° belled anchor pile model and having corresponding $D_s/D_b = 0.33$, 0.28 and 0.38. **b** Height above the anchor base vs. horizontal extent of failure points in slip surface relationship besides panel 45-0.33 at $L/T_b = 3$, 4 and 5

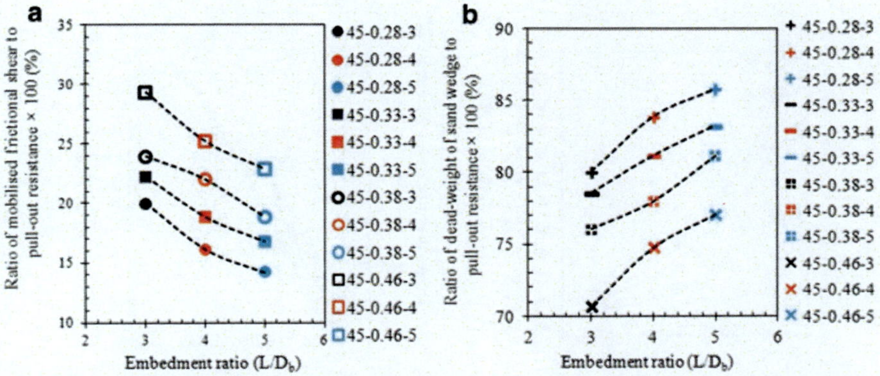


**Fig. 7** Relationships of **a** ratio of mobilised frictional shear to pull-out resistance (×100, %) versus embedment ratio and **b** ratio of dead-weight of sand wedge to pull-out resistance (×100, %) versus embedment ratio (×100, %), for all the models of 45°, having $D_s/D_b = 0.28$, 0.33, 0.38 and 0.46

and pull-out resistance increase with the formation of larger sand wedges at deeper embedment depths. For the primary and secondary increments in pull-out resistances 91.41 and 62.24%, the rise in dead-weight of sand wedge are 96.47 and 65.14%, and in mobilised fictional shear are 62.54 and 44.36%, respectively. This represents that for the rise in pull-out resistance, the contribution of dead-weight of sand wedge is more significant than the part of mobilised fictional shear.

### 6.2 Pull-Out Resistance, Ratios of Mobilised Frictional Shear and Dead-Weight of Sand Wedge Based on Pull-Out Resistance of Belled Anchor Piles Influenced by Diameter Ratios ($D_s/D_b$)

In Fig. 8a, the typical plots present the pull-out resistance versus diameter ratio relationships for the models of $\alpha = 63$, 72° and 45° at L/$D_b$ of 4, 5 and 3, respectively. From the figure, it can be noticed that as the anchors are possessing gradually higher values of $D_s/D_b$, i.e. from 0.28 to 0.33, from 0.33 to 0.38 and from 0.38 to 0.46 the pull-out resistance gradually decrease irrespective of $\alpha$ and $L/D_b$ values. In case of a particular $L/D_b$ value, gradually higher diameter ratio lower overburden pressure acts on anchor base; consequently, smaller sand wedges are generated. Figure 8b reveals that the pull-out resistance is gradually lesser for 72-0.28-5, 72-0.33-5, 72-0.38-5 and 72-0.46-5. With a gradual increase in diameter ratio, a decreasing pattern of pull-out resistance was also observed by [7].

The typical Fig. 9a illustrates that the ratio of mobilised frictional shear to pull-out resistance (× 100, %) values gradually increases with higher values of diameter ratios for all the 63° models. The typical Fig. 9b reveals that for the same models,

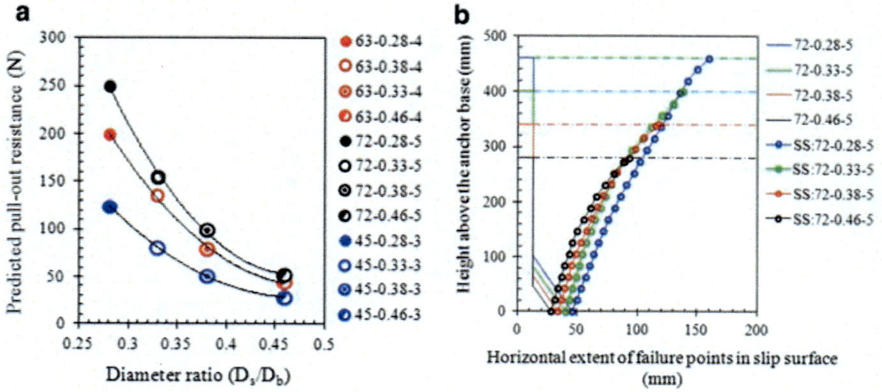


**Fig. 8** **a** Predicted pull-out resistance vs. diameter ratio relationships for 63°, 72° and 45° belled anchor models and at corresponding $L/D_b = 4$, 5 and 3. **b** Height above the anchor base versus horizontal extent of failure points in slip surface relationships besides 72° panel at $L/T_b = 5$, and these are possessing $T_s/T_b = 0.28$, 0.33, 0.38 and 0.46

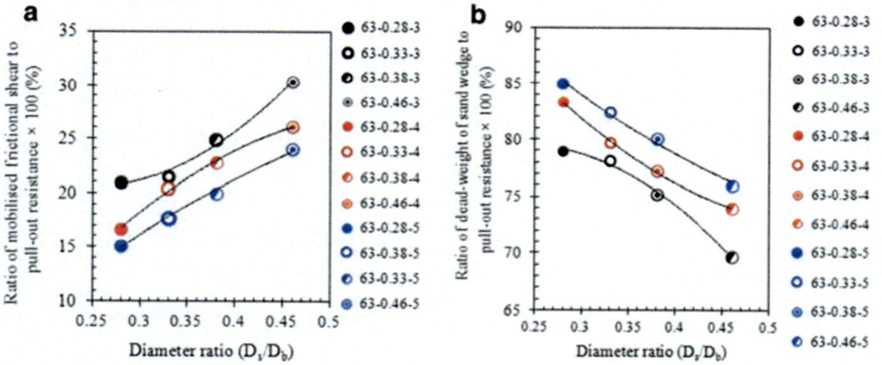


**Fig. 9** Relationship of **a** ratio of mobilised frictional shear to pull-out resistance versus diameter ratio (×100, %) and **b** ratio of dead-weight of sand wedge to pull-out resistance vs. diameter ratio (× 100, %), for all the models of 63°, installed at $L/D_b = 3$, 4 and 5

the ratio of dead-weight of sand wedge to pull-out resistance (× 100, %) values gradually decreases with higher values of diameter ratios. At the certain $L/D_b$ value, for the models having higher diameter ratios at shallow depth, gradually smaller sand wedges are formed on gradually smaller anchor bases. The values of mobilised frictional shear, dead-weight of sand wedge and pull-out resistance decrease with the formation of smaller sand wedges at shallow embedment depth (but at same $L/D_b$). In the Fig. 9b, for 63-0.28-4, 63-0.33-4, 63-0.38-4 and 63-0.46-4, the values of corresponding mobilised frictional shear are only 16.66, 20.36, 22.74 and 26.07%, and the respective values of dead-weight of sand wedge are 83.34, 79.64, 77.26 and 73.93% of pull-out resistance. The decrements in pull-out resistances from model

63-0.33-4 to 63-0.28-4, from 63-0.33-4 to 63-0.38-4 and from 63-0.38-4 to 63-0.46-4 are 31.76, 41.9 and 47.74%, whereas the corresponding rise in mobilised frictional shear are 19.91, 54.16 and 66.85%, and the decrease in respective dead-weight of sand wedge are 34.52, 42.99 and 48.91% based on pull-out resistance. This statistical analysis represents that the involvement of dead-weight of sand wedge to pull-out resistance value is significantly large than the mobilised frictional shear values, so these two factors (in %) show reverse trend to each other.

## 6.3 Pull-Out Resistance, Ratios of Mobilised Frictional Shear and Dead-Weight of Sand Wedge Based on Pull-Out Resistance of Belled Anchor Piles Influenced by Bell Angles (A)

In Fig. 10a, the typical plots present the pull-out resistance versus bell angle relationships for the models of $D_s/D_b = 0.33$, 0.38 and 0.46 and at $L/D_b = 4$, 3 and 5, respectively. From the figure, it can be noticed that as the anchors are possessing gradually higher values of α, i.e. from 45° to 54°, from 54° to 63° and from 63° to 72°, pull-out resistances attain gradually decreasing pattern irrespective of $L/D_b$ and $D_s/D_b$ values. At a particular embedment depth, for a certain $D_s/D_b$ value, in 72° anchors, the slant height of bell is significantly steeper, so under vertical tension, the influence zone above these anchor bases is very close to the anchor shaft; whereas, the slant height of bell for 45°, 54° and 63° anchors are much milder than that of 72° anchor, so the influence zone above these anchors base are horizontally extended reasonably far away surrounding the anchor shaft. So, the models having higher bell

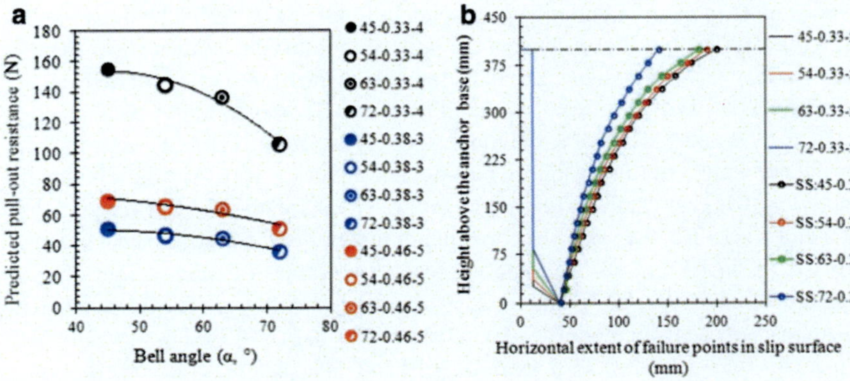


**Fig. 10** **a** Predicted pull-out resistance vs. bell angle relationships for belled anchor models having $D_s/D_b = 0.33$, 0.38 and 0.46 and at corresponding $L/D_b = 4$, 5 and 3. **b** Height above the anchor base versus horizontal extent of failure points in slip surface relationships besides panels possessing $T_s/T_b = 0.33$ at $L/T_b = 4$ and these having $\alpha = 45$, 54°, 63° and 72°

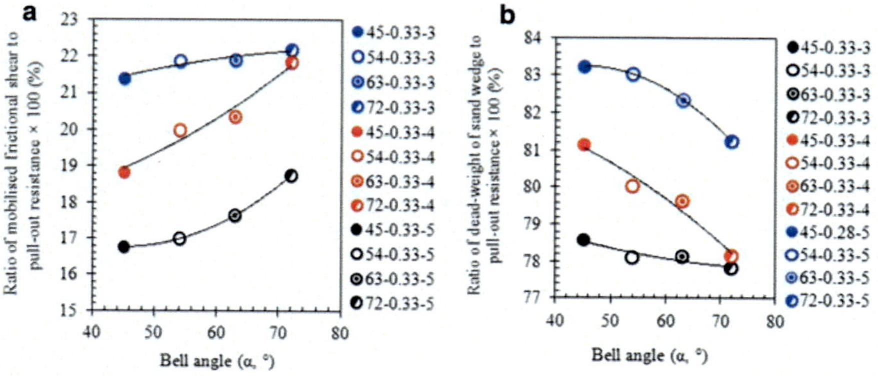


**Fig. 11** Relationships of **a** ratio of mobilised frictional shear to pull-out resistance versus bell angle (× 100, %) and **b** ratio of dead-weight of sand wedge to pull-out resistance versus bell angle (× 100, %), for all the models of having $D_s/D_b = 0.33$ and at $L/D_b = 3$, 4 and 5

angles form gradually smaller sand wedges; though the wedges for 45°, 54° and 63° are very close to each other but wedges for 72° are reasonably smaller than other wedges as shown in Fig. 10b, for panels possessing $T_s/T_b = 0.33$ at $L/T_b = 4$. So, the values of pull-out resistance decrease with the formation of smaller sand wedges. A similar failure trend was explained by Matsuo [15]. In this figure, for anchors having $D_s/D_b = 0.33$, $L/D_b$ of 4 and $\alpha$ values from 45° to 54°, 45° to 63° and 45° to 72°, pull-out resistance values are decreased 3.24, 8.92 and 27.70%. In general, all the 63° belled anchors are within 10% less than 45° anchors, whereas 72° belled anchors are within 30% less than 45° anchors. So, 45°, 54° and 63° anchors are found to be more efficient as a tension resistant structure than 72° anchors. A similar trend in pull-out resistance was also noticed by Nazir et al. (2014) for anchors of bell angles 30°–60° and Dickin and Leung [8] for anchors of bell angles 22°–72° in dry sand.

The typical Fig. 11a illustrates that the ratio of mobilised frictional shear to pull-out resistance (× 100, %) values gradually increases with higher values of bell angles, i.e. 45°, 54°, 63° and 72°, for all the models of $D_s/D_b = 0.33$ and a certain series of $L/D_b = 3$, 4 and 5. For the same models, the typical Fig. 11b illustrates that the ratio of dead-weight of sand wedge to pull-out resistance (× 100, %) values gradually decreases with higher values of bell angle. In Fig. 11a, b, for 45-0.33-5, 54-0.33-5, 63-0.33-5 and 72-0.33-5, the values of corresponding mobilised frictional shear are only 16.76, 16.98, 17.64 and 18.75%, and the respective values of dead-weight of sand wedge are 83.24, 83.02, 82.36 and 81.25% of pull-out resistance. The decrements in pull-out resistances from model 45-0.33-5 to 54-0.33-5, from 45-0.33-5 to 63-0.33-5 and from 45-0.33-5 to 72-0.33-5 are 6.48, 9.19 and 28%, whereas the corresponding rise in mobilised frictional shear are 1.31, 5.25 and 11.87% and the decrease in respective dead-weight of sand wedge are 0.26, 1.06 and 2.4% based on pull-out resistance. This statistical analysis represents that due to the major contribution of dead-weight of sand wedge than the mobilised frictional shear values to acquire pull-out resistance, the trend of two ratios (× 100, %) in reference to bell angles are differing each other.

## 7 Conclusions

The following significant conclusions may be drawn as listed below:

1. The variations in the horizontal extent of slip surfaces based on embedment ratios, thickness ratios and bell angles of 2D panels lead to the variations in predicted pull-out resistance in 3D models. The range of the predicted pull-out resistances of all the models is from 22.88 to 383.70 N.
2. The predicted pull-out resistances are increased with higher embedment ratios, lesser diameter ratios and bell angles.
3. The values of ratio of mobilised frictional shear to the pull-out resistance (× 100,%) are from 13.50 to 30.81%.
4. The values of ratio of dead-eight of sand wedge to the pull-out resistance (× 100,%) are from 69.20 to 86.50%.
5. The ratios of mobilised frictional shear to pull-out resistance (× 100, %) of anchors are decreased, and ratios of dead-weight of sand wedge to pull-out resistance of anchors (× 100, %) are increased due to higher embedment ratios, whereas the ratios of mobilised frictional shear to pull-out resistance (× 100, %) of anchors are increased and ratios of dead-weight of sand wedge to pull-out resistance of anchors (× 100, %) are decreased for higher values of diameter ratio and bell angle.

## References


1. Balla, A.: The resistance to breaking-out of mushroom foundations for pylons. In: Proceedings of 5th International Conference on Soil Mechanics and Foundation Engineering, Paris, vol. 1, pp. 569–576 (1961)
2. Bera, A.K.: Parametric study on uplift capacity of anchor with tie in sand. Kor. Soc. Civil Eng. **18**(4), 1028–1035 (2014)
3. Chottapadhyay, B.C., Pise, P.J.: Breakout resistance of horizontal anchors in sand. Jap. Soc. Soil Mech. Found. Eng. **26**, 126 (1986)
4. Clemence, S.P., Veesaert, C.J.: Dynamic uplift resistance of anchors in sand. In: Proceedings of the International Conference on Soil-Structure Interaction, Roorkee, India, pp. 389–397 (1977)
5. Dash, P.K., Pise, P.J.: Effect of compressive load on uplift capacity of model piles. J. Geotech. Geoenviron. Eng. **129**(11), 987–992 (2003)
6. Dickin, E.A.: Uplift behavior of horizontal anchor plates in sand. J. Geotech. Eng. **114**(11), 1300–1317 (1988)
7. Dickin, E.A., Leung, C.F.: Performance of piles with enlarged bases subjected to uplift forces. Can. Geotech. J. **27**, 546–556 (1990)
8. Dickin, E.A., Leung, C.F.: The influence of foundation geometry on uplift behaviour of piles with enlarged bases. Can. Geotech. J. **29**, 498–505 (1992)

9. Downs, D.I., Chieurzzr, I.: Transmission tower foundations. ASCE J. Power Div. **92**(P02), 91–114 (1966)
10. Ghaly, A., Hanna, A.: Ultimate uplift resistance of single vertical anchors. Can. Geotech. J. **31**, 666–672 (1994)
11. Ghosh, A., Bera, A.K.: Effect of geotextile ties on uplift capacity of anchors embedded in sand. Geotechnical Geology Engineering. **28**, 567–577 (2010)
12. Ilamparuthi, K., Muthukrishnaiah, K.: Anchors in sand bed: delineation of rupture surface. Ocean Eng. **26**, 1249–1273 (1999)
13. Ilamparuthi, K., Dickin, E.A., Muthukrisnaiah, K.: Experimental investigation of the uplift behavior of circular plate anchors embedded in sand. Can. Geotech. J. **39**, 648–664 (2002)
14. Krishnaswamy, N.R., Parashar, S.P.: Uplift behaviour of plate anchors with geosynthetics. Geotextiles Geomembranes **13**, 67–89 (1994)
15. Matsuo, M.: Study of uplift resistance of footing (I). Soil Foundation **7**(4), 1–37 (1967)
16. Majer, J.: Zur Berechnung Von Zugfundamenten. Osterreichische Bauzeitschrift 10, H.5 (1955)
17. Mors, H.: The behaviour of mast foundations subject to tensile forces. Bautechnik **10**, 367–378 (1959)
18. Murray, E.J., Geddes, J.D.: Uplift of anchor plates in sand. J. Geotech. Eng. **113**(3), 202–215 (1987)
19. Nazir. R., Moayedi, H., Pratikso, A., Mosallanezhad, M.: The uplift load capacity of an enlarged base pier embedded in dry sand. Arab. J. Geosci. **8**, 7285–7296 (2015). https://doi.org/10.1007/s12517-014-1721-3
20. Rao, K.S.S., Kumar, J.: Vertical uplift capacity of horizontal anchors. J. Geotech. Eng. **120**(7), 1134–1147 (1994)
21. Saeedy, H.S.: Stability of circular vertical earth anchors. Can. Geotech. J. **24**, 452–456 (1987)
22. Saran, S., Ranjan, G., Nene, A.S.: Soil anchors and constitutive laws. J. Geotech. Eng. **112**(12), 1084–1100 (1986)
23. Sutherland, H.B.: Uplift resistance of soils. Geotechnique **138**, 493–516 (1988)
24. Tagaya, K., Scott, R.F., Aboshi, H.: Pull-out resistance of buried anchors in sand. Soils Foundations **28**, 114–130 (1988)
25. Turner, E.A.: Uplift resistance of transmission tower footings. J. Power Div. **88**, 17–32 (1962)
26. Vanitha, L., Patra, N.R., Chandra, S.: Uplift capacity of pile group anchors. Geotech. Geol. Eng. **25**, 339–347 (2007)
27. Vesic, A.S.: Breakout resistance of objects embedded in ocean bottom. Soil Mechanics and Foundation. Report Number CR.69.031 prepared for U. S. Naval Civil Engineering Laboratory Port Hueneme, California under. Contract No. N6 2399–68-C-0043 (1969)
28. Vermeer, P.A., Sutjiadi, W.: The uplift resistance of shallow embedded anchors. In: Proceedings, 11th International Conference on Soil Mechanics and Foundation Engineering, San Francisco, CA, vol. 4, pp. 1635–1638 (1985)

# Influence of Moment on Load-Settlement Behaviour of Circular Footing Resting on Clayey Soil

**Sreedhu P. S. Potty, J. Jayamohan, and K. Kannan**

## 1 Introduction

Foundations for overhead catenary systems carrying electrical power in railway networks, transmission towers, and for large road and railway hoardings and other elevated commercial signs have to be designed mainly to resist large moments and relatively small vertical loads. Moments on the foundation base are mainly caused by horizontal forces acting on the structure. Horizontal forces act on a structure due to earth pressure, wind pressure, seismic force, hydrostatic pressure, etc. The non-uniformity of the soil pressure caused due to horizontal/eccentric loads tends to tilt the footing. This tilt increases with the increase of load eccentricity, and consequently, the bearing capacity is reduced. Short bored pile or pier foundations are widely used in situations where moment-carrying capacity is the dominant design requirement. But such foundations are very costly, and it cannot be adopted for small projects. Published literature on the moment-deformation behaviour of shallow foundation is very scarce.

Much research has been carried out in different foundations subjected to moments and eccentric loads [1, 2, 5, 7, 8]. Laboratory investigation and numerical analyses were carried out by [3, 4] to study the behaviour of one-sided skirted strip footing subjected to eccentric load. Reference [6] studied the behaviour of ring footing resting on loose sand and/or compacted randomly distributed fibre reinforced sand when subjected to eccentric, inclined and eccentric-inclined loadings by using finite element (FE) software PLAXIS 3D.

S. P. S. Potty · K. Kannan
Marian Engineering College, Thiruvananthapuram, Kerala, India

J. Jayamohan (✉)
LBS Institute of Technology for Women, Thiruvananthapuram, Kerala, India



C. N. V. Satyanarayana Reddy et al. (eds.), *Ground Characterization and Foundations*, Lecture Notes in Civil Engineering 167,
https://doi.org/10.1007/978-981-16-3383-6_44

This paper investigates the influence of moment acting on the structure, on the settlement and angular distortion of the footing by carrying out a series of laboratory scale load tests on model circular footing resting on clayey soil. Finite element analyses are carried out with the FE software *PLAXIS 2D,* and the results are compared with those obtained from laboratory studies for validation.

## 2 Laboratory Scale Load Test

The load tests are conducted in a combined test bed and loading frame assembly. The test beds are prepared in a tank of internal dimension 1000 mm length × 750 mm width × 750 mm depth. The test tank is constructed with 230 mm thick brick masonry walls. The model circular footing has a diameter of 100 mm, thickness 20 mm and is fabricated with mild steel. The clayey soil is filled in the test tank to the required level with compaction done in layers of 50 mm thickness. The water content of the clayey soil is maintained at 15.61%. To achieve the desired density of the soil, the layered filling technique is used. The clay was compacted by ramming. The compactive effort required to achieve the required density was determined by trial and error. The loading tests are carried out in a loading frame fabricated with ISMB 300. The vertical load is applied using a hand operated-mechanical jack of capacity 50 kN. The applied vertical load is measured using a proving ring of capacity 100 kN. Moment is applied by eccentric loading on the footing in addition to the vertical load. The eccentricity in all the tests is 245 mm. Eccentric load is measured using an additional proving ring of capacity 50 kN. The tilt of the model footing is measured using two dial gauges of 0.01 mm sensitivity kept diametrically opposite to each other 350 mm apart. The photograph of experimental setup is shown in Fig. 1. Locally available clay is used as foundation soil. The properties of clay are listed in Table 1. The improvement in load-settlement behaviour due to the addition of reinforced foundation bed is also investigated. Reinforced foundation bed is formed by placing a layer of sand reinforced with biaxial geogrid beneath the footing.

### 2.1 Parameters Used in the Study

The geometrical parameters in this study are presented in Fig. 2. The diameter and depth of circular footing are 'B' and 'D', respectively. Diameter of the model footing is 100 mm. Distance between two dial gauges is fixed as 350 mm. Eccentric load (P2) is increased at regular intervals. The eccentric distance ($e$) of eccentric load P2 is fixed as 245 mm. The experimental setup is shown in the Fig. 3, and parameters varied in Table 2. The thickness of reinforced foundation bed is represented as '$t$'. The number of geogrid layers (N) is varied as 1 and 2.

**Fig. 1** Experimental setup

**Table 1** Properties of clay

| S. No. | Properties | Values |
|---|---|---|
| 1 | Specific gravity | 2.68 |
| 2 | Optimum moisture content (%) | 18 |
| 3 | Maximum dry density (kN/m$^3$) | 15.61 |
| 4 | Liquid limit (%) | 58 |
| 5 | Plastic limit (%) | 22 |
| 6 | Shrinkage limit (%) | 16.2 |
| 7 | Permeability, k (m/s) | $3.03 \times 10^{-6}$ |
| 8 | Unconfined compressive strength, UCC (kN/m$^2$) | 140.08 |
| 9 | IS Classification | CH |
| 10 | Friction angle, $\Phi$ (°) | 5 |
| 11 | Cohesion, $c$ (kPa) | 25 |

## 3 Finite Element Analysis

In the present study, the experimental results obtained are validated by carrying out finite element analysis and comparing the results. PLAXIS 2D is a commercially available software for carrying out the finite element analyses. The geometric model in the finite element analysis is shown in Fig. 4.

There are different constitutive models available in the FE software for simulating the soil behaviour. Mohr–Coulomb model is adopted in the present study. The soil

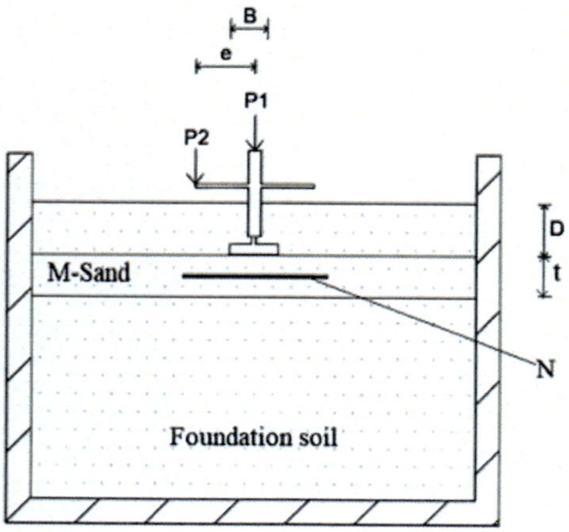


**Fig. 2** Geometric parameters

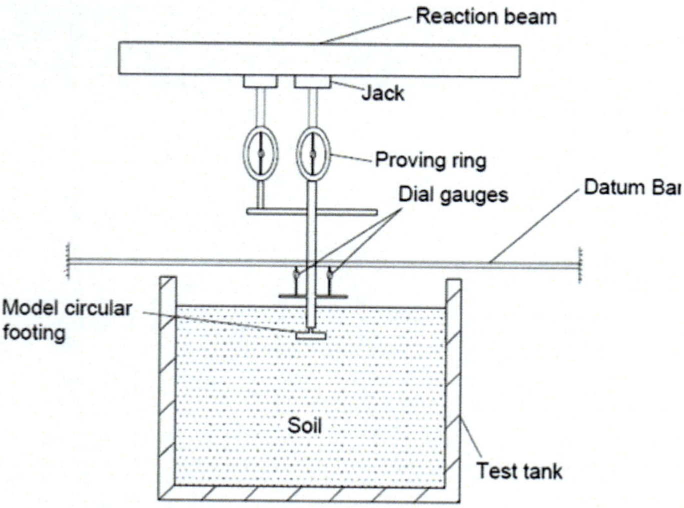


**Fig. 3** Schematic representation of test setup

**Table 2** Parameters varied

| Parameter | Vertical Load $P_1$ (N) | D/B | Moment M (N-m) | *t/B* |
|---|---|---|---|---|
| Values | 0, 90, 130, 170 | 0, 0.5, 1 | 0, 0.56, 1.13, 1.69 | 0.25, 0.5, 0.75 |

parameters obtained from direct shear tests; internal friction angle and cohesion intercept, are adopted in this nonlinear model. The axisymmetric model is used in the analysis, since circular footing is symmetric about its central axis. The nonzero prescribed displacements are used to simulate the settlement of the rigid footing. Figure 5 shows the typical deformed shape obtained after loading in the FE analysis.

**Fig. 4** Geometric model

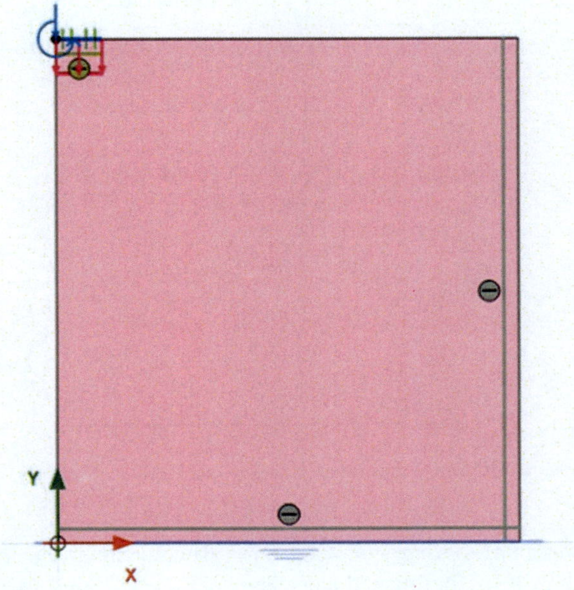


**Fig. 5** Typical deformed shaped

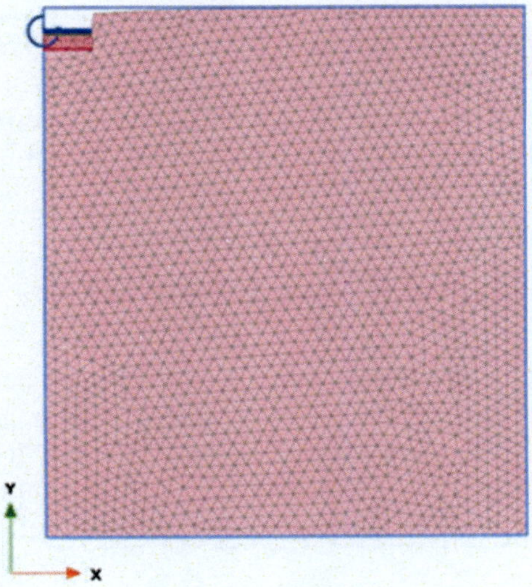


# 4 Results and Discussions

## *4.1 Influence of Depth of Footing on Rotation*

Moment versus tilt curves for constant vertical load $P_1 = 130$ N and different depths of footing are presented in Fig. 6. It is observed that depth of embedment influences

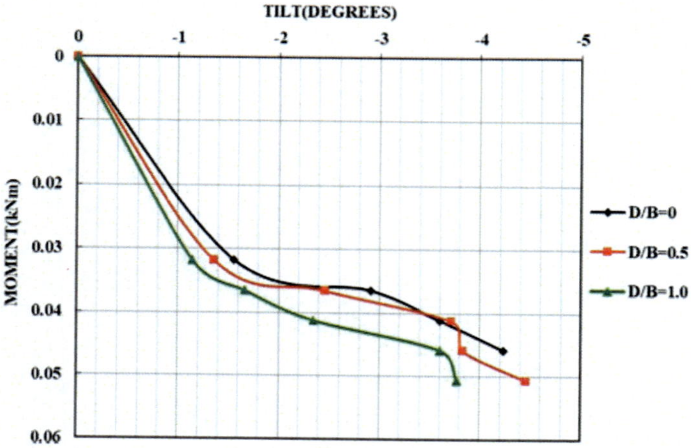


**Fig. 6** Moment versus rotation curves for P1 = 130 N and different depths of footing

the tilt of footing. For a constant vertical load, as the depth of embedment increases, the tilt of footing reduces. The rotation is found to be minimum for $D/B = 1.0$. Maximum rotation is observed when the footing is at the surface. When moment is applied, one side of footing moves down and other side moves up. The downward movement is restricted by contact pressure. The upward movement of footing is restricted by the weight of soil above the footing. As depth increases, the bearing resistance increases and the weight of soil above the footing increases. This increases the moment resisting capacity of footing.

### 4.2 *Influence of Vertical Loading on Rotation*

Moment versus rotation curve for constant depth $D/B = 0.5$ and varying vertical loads is given in Fig. 7. Vertical load influences the tilt of footing. For constant depth, when vertical load increases, the tilt increases upto $P_1 = 130$ N and further increase in vertical load tilt reduces. Maximum tilt is found at vertical load of 130 N.

### 4.3 *Influence of Moment Acting on the Structure on Settlement*

The load-settlement curve of circular footing resting on surface ($D/B = 0$) for varying the moments is given in Fig. 8. It is seen that the load-settlement behaviour is influenced by the moments acting on the structure. For a constant depth of footing, when the moment increases, the settlement also increases. The moment is applied

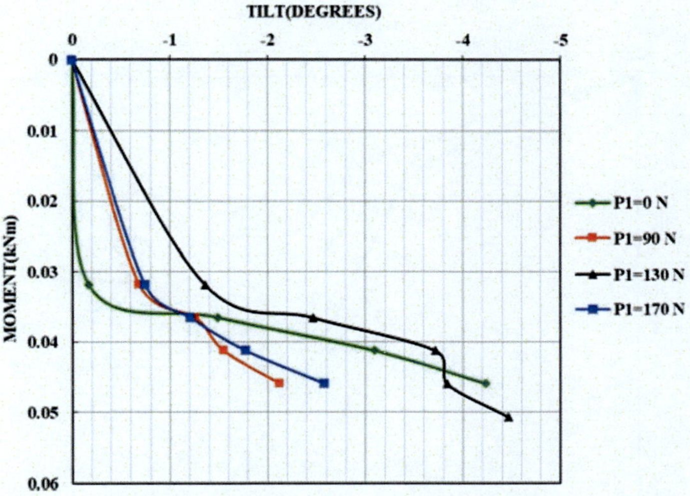


**Fig. 7** Moment versus rotation curves for $D/B = 0.5$ and varying vertical loads

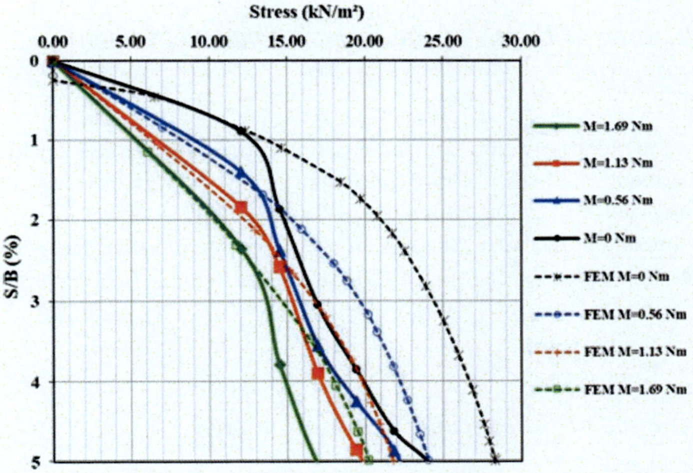


**Fig. 8** Load-settlement curves for $D/B = 0$ and varying the moments

by applying an additional eccentric load. Hence, as the moment increases, the total vertical load also increases.

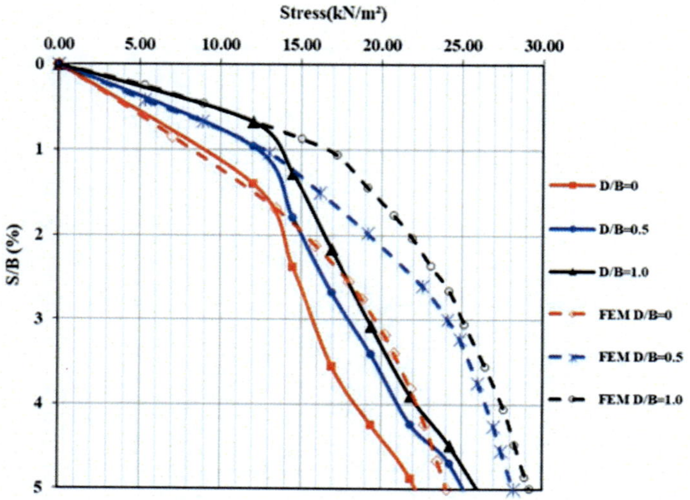


**Fig. 9** Load-settlement curves for constant moment $M = 0.56$ N-m and different depths of footing

## 4.4 *Influence of Depth of Footing on Load-Settlement Behaviour*

The load-settlement behaviour of circular footing resting at different depths for constant moment $M = 0.56$ N-m is presented in Fig. 9. It is observed that depth of embedment influences the settlement of footing. For a constant moment, as the depth of embedment increases, the settlement of footing reduces. The settlement is found to be minimum for $D/B = 1.0$. Maximum settlement is observed when the footing is at the surface.

## 4.5 *Influence of Reinforced Foundation Bed*

The improvement in load-settlement behaviour of model circular footing for $D/B = 0.5$ and $M = 1.69$ N-m due to the addition of reinforced foundation bed is presented in Fig. 10.

It is observed from the results that as the thickness of reinforced foundation bed increases the load-settlement behaviour improves. As the number of geogrid layers increases, the settlement reduces.

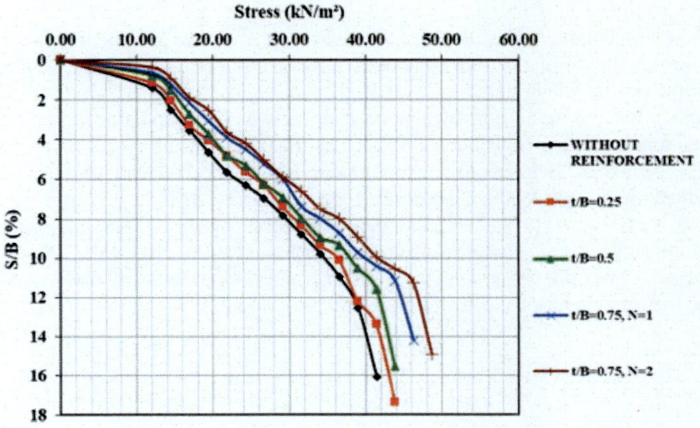


**Fig. 10** Load-settlement curves for $D/B = 0.5$ and $M = 0.56$ N-m in reinforced foundation bed

## 5 Conclusions

Moment-rotation behaviour and load-settlement behaviour of circular footing are investigated by carrying out a series of laboratory scale load tests and finite element analyses. The following conclusions are deduced from this study:

1. Depth of embedment reduces the tilt of footing and settlement of footing.
2. Moment acting on the footing adversely influences the load-settlement behaviour.
3. Vertical load acting on the footing significantly influences the tilt.
4. The rotation is found to be minimum for $D/B = 1.0$.
5. Maximum rotation is observed when footing is at surface.
6. The load-settlement behaviour under a combination of moment and vertical load improves due to the addition of reinforced foundation bed.

**Acknowledgements** The financial support received from TEQIP Four Funds of LBS Institute of Technology for Women, Thiruvananthapuram is gratefully acknowledged.

## References

1. Bransby, M.F., Randolph, M.F.: Combined loading of skirted foundations. Geotechnique **48**(5), 637–655 (1998)
2. Laman, M., King, G.J.W., Dickin, E.A.: Three-dimensional finite element studies of the moment-carrying capacity of short pier foundations in cohesionless soil. Comput Geotech. **25**, 141–155 (1999)
3. Saleh, N.M., Elsaied, A.E., Elleboudy, A.M.: Performance of skirted strip footing subjected to eccentric inclined load. Electron. J. Geotech. Eng. **13**, 1–13 (2008)

4. Saleh, N.M., Elleboudy, A.M., Elsaied, A.E.: Behaviour of skirted strip footing under eccentric load. In: Proceedings of the 17th International Conference on Soil Mechanics and Geotechnical Engineering, pp. 586–589 (2009)
5. Sawadan, K., Takemura, J.: Centrifuge model tests on piled raft foundation in sand subjected to lateral and moment loads. Soils Found. **54**(2), 126–140 (2014)
6. Sharma, V., Kumar, A.: Behavior of ring footing resting on reinforced sand subjected to eccentric-inclined loading. J. Rock Mech. Geotech. Eng. **10**, 347–357 (2018)
7. Taiebat, H.A., Carter, J.P.: Numerical studies of bearing capacity of shallow foundations on cohesive soil subjected to combined loading. Geotechnique **50**(4), 409–418 (2000)
8. Taiebat, H.A., Carter, J.P.: Bearing capacity of strip and circular foundations on undrained clay subjected to eccentric loads. Geotechnique **50**(1), 61–64 (2002)

# Horizontal Load—Deformation Behaviour of Shallow Circular Footing

T. S. Amritha Varsha, J. Jayamohan, and P. R. Anila Angel

## 1 Introduction

Foundations are the base from which all structures are constructed. The loads of a structure are transmitted to the ground through its foundation. The foundation design aims at providing a means of transmitting the loads from a structure to the underlying soil without causing any shear failure or excessive settlement of the soil under the imposed loads. Foundations are of different types such as shallow, deep. The bearing capacity of foundations has always been one of the subjects of major interest in soil mechanics and foundation engineering. Evaluations of bearing capacity of vertically loaded shallow foundations on various soils have been addressed by previous researchers. Vertical bearing capacity of shallow foundations can be found out by Terzaghi's analysis; Meyerhof's method, etc. and that of deep foundations can be found out by static pile load formula, pile load test, etc. Horizontal load bearing capacity of deep foundations can be found out by elastic theory of Reese and Matlock (1956), Plastic theory of Hansen (1951), Mayerhof (1973), etc. It is difficult to provide pile foundations for all structures due to its cost and difficulty in construction. Hence, it is essential to study horizontal load bearing capacity of shallow foundations.

Many researchers studied the vertical load bearing capacity of shallow foundations. But there are many structures in which horizontal load bearing capacity is of greater importance.

Taiebat and Carter [1] did numerical studies on shallow foundations on cohesive soil subjected to combined loading. They did 3D finite element analysis of circular foundations on cohesive soil under combined loading and proposed new equation for failure locus in terms of all three components. Horikoshi et al. [2] studied the

T. S. A. Varsha · P. R. A. Angel
Marian Engineering College, Thiruvananthapuram, Kerala, India

J. Jayamohan (✉)
LBS Institute of Technology for Women, Thiruvananthapuram, Kerala, India



C. N. V. Satyanarayana Reddy et al. (eds.), *Ground Characterization and Foundations*, Lecture Notes in Civil Engineering 167,
https://doi.org/10.1007/978-981-16-3383-6_45

performance of piled raft foundations subjected to static horizontal loads. Horizontal stiffness of foundation and proportion of load carried by raft and piles were studied. Phillip et al. [3] simplified lateral load analysis of fixed head piles and pile groups and developed characteristic load method to estimate deflection and bending moment and found that it is limited to cases where lateral load on pile is applied at ground surface. Sawadan and Takemura [4] conducted centrifugal model tests on piled raft, pile groups, rafts alone in Fig. 1 and showed vertical displacement due to horizontal load affects vertical resistance of piles. Zheng and Zhao [5] studied the effect of inclined loading on bearing capacity of strip footing on sand layer as in Fig. 2, and they did parametric study and proved that failure mechanism depends on geometric parameters and soil properties ($D/B$, $C/\gamma B$, $\varphi$). Abbasali et al. [6] studied influence of

**Fig. 1** Loading frame and test tank

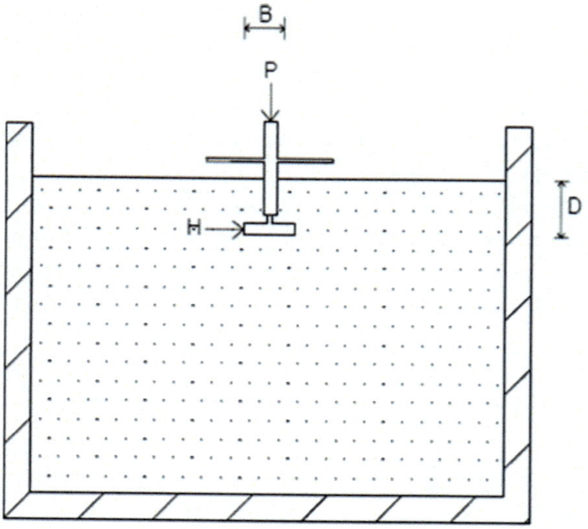


**Fig. 2** Geometric parameters

using composite soils under shallow foundations. 3D finite element analysis was done and studied bearing capacity improvement and shows that Sand clay mixture shows better performance than gravel clay mixture. In this paper, extensive investigations are carried out to study the load-settlement behaviour of circular footing subjected to horizontal loads. The reduction in lateral deformation of soil by providing micropiles beneath the footing is also studied. The study involved nonlinear finite element analyses using the FE software PLAXIS 2D, and the results were compared with those obtained from laboratory scale load tests. The parameters studied are influence of vertical load (*P*), embedment depth (*D*) on lateral displacement of footing and effect of combined lateral and vertical load on rotation of footing.

## 2 Laboratory Scale Load Test

The load tests are conducted in a combined test bed and loading frame assembly as shown in Fig. 1. The test beds are prepared in a tank which is designed keeping in mind the size of the model circular footing to be tested. The internal dimension of the test tank is 1000 mm length × 750 mm width × 750 mm depth, which has 23 cm thick brick masonry walls on the three sides. The front side of tank is formed using a framework of steel channels and angles. For test with clay alone, the weak soil is filled in the test tank to the required level with compaction done in layers of 5 cm thickness. The water content of the clayey soil is maintained at 25%. To achieve the desired density of the soil, the layered filling technique is used. The pre-determined density of clay is used to calculate the desired weight of soil required to fill the tank in layers of 50 mm height. A uniform density of 15.61 kN/m3 for clay was maintained in all the tests. The clay was compacted by ramming. The compactive effort required to achieve the required density was determined by trial and error. The loading tests are carried out in a loading frame fabricated with ISMB 300. The vertical load is applied using a hand operated—mechanical jack of capacity 50 kN. The applied vertical load is measured using a proving ring of capacity 100 kN. Arrangement for lateral loading is welded in the loading frame. Lateral load is measured using a proving ring of capacity 50 kN. The lateral displacement of the model footing is measured using two dial gauges of 0.01 mm sensitivity kept diametrically opposite to each other. The tilt due to combined vertical and horizontal loading is also measured using two dial gauges of 0.01 mm sensitivity kept diametrically opposite to each other. The model footing is placed exactly beneath the centre of loading jack to avoid eccentric loading (Table 1).

### *2.1 Geometric Parameters Used*

Diameter of circular footing (*B*) is kept constant and it is 10 cm. The vertical load (*P*) is varied as 0, 90, 130 and 160 N. Depth of embedment of footing (*D*) is varied as

**Table 1** Properties of clay

| S. No. | Properties | Values |
|---|---|---|
| 1 | Specific gravity | 2.68 |
| 2 | Optimum Moisture Content (%) | 18 |
| 3 | Maximum dry density (kN/m$^3$) | 15.61 |
| 4 | Liquid Limit (%) | 58 |
| 5 | Plastic Limit (%) | 22 |
| 6 | Shrinkage limit (%) | 16.2 |
| 7 | Permeability, $k$ (m/s) | $3.03 \times 10^{-6}$ |
| 8 | Unconfined compressive strength, UCC (kN/m$^2$) | 140.08 |
| 9 | IS classification | CH |
| 10 | Friction angle, $\Phi$ (°) | 5 |
| 11 | Cohesion, $c$ (kPa) | 25 |

$D/B = 0, 0.5$ and 1. Distance between two dial gauges is fixed as 30.5 cm. Horizontal load ($H$) is varying, and it is applied at regular intervals.

The arrangement for vertical and lateral loading is shown in Figs. 2, 3, and 4, respectively.

**Fig. 3** Arrangement for vertical loading

**Fig. 4** Arrangement for lateral loading

## 3 Finite Element Analysis

In the present study, the experimental results obtained are validated by carrying out finite element analysis. The software PLAXIS 2D is commercially available software for carrying out the finite element analyses. The geometric model in the finite element analyses is shown in Fig. 5.

There are different constitutive models available in the FE software for simulating the soil-behaviour. Mohr–Coulomb model is adopted in the present study. The soil parameters obtained from direct shear tests, internal friction angle and cohesion intercept, are adopted in this nonlinear model. The axisymmetric model is used in

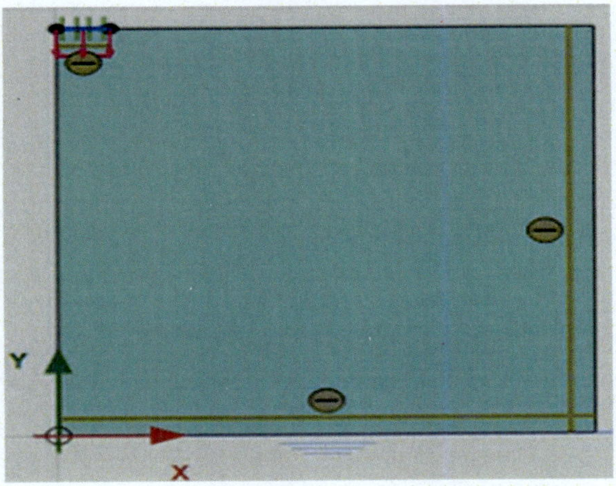


**Fig. 5** Geometric model

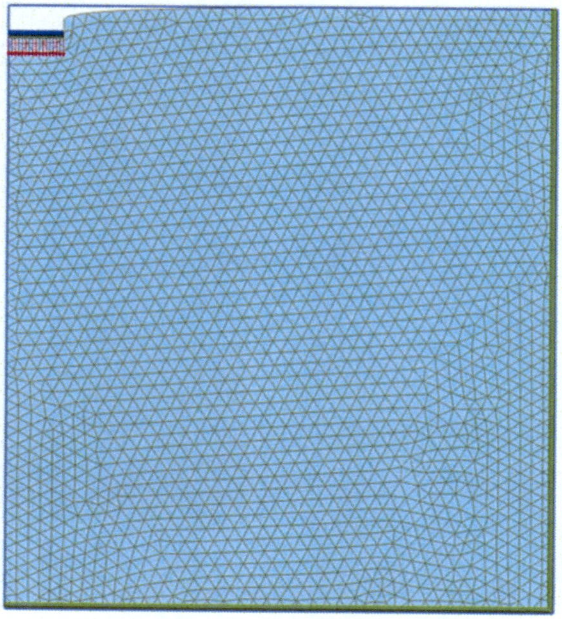

**Fig. 6** Typical deformed shape

the analysis since circular footing is symmetric about its central axis. The nonzero prescribed displacements are used to simulate the settlement of the rigid footing. Figure 6 shows the typical deformed shape obtained after loading in the FE software.

## 4 Results and Discussions

### *4.1 Influence of Vertical Load on Lateral Load-Deformation Behaviour*

The variation in lateral stress v/s settlement curve of circular footing resting in clayey soil of $D/B = 1.0$ where $D$ is embedment depth of footing and B is the width of footing by varying the vertical loads $P = 0$ N, $P = 90$ N, $P = 130$ N and $P = 160$ N is presented in Fig. 7. From the curves, it is observed that as lateral stress increases, lateral displacement increases. As vertical loading ($P$) increases, lateral displacement decreases. This is because as vertical load increases, normal stress beneath the footing increases and thereby increasing the force of friction which causes decrease in lateral displacement of footing. It is seen that there is slight variation between the results obtained from finite element analysis and experimental studies.

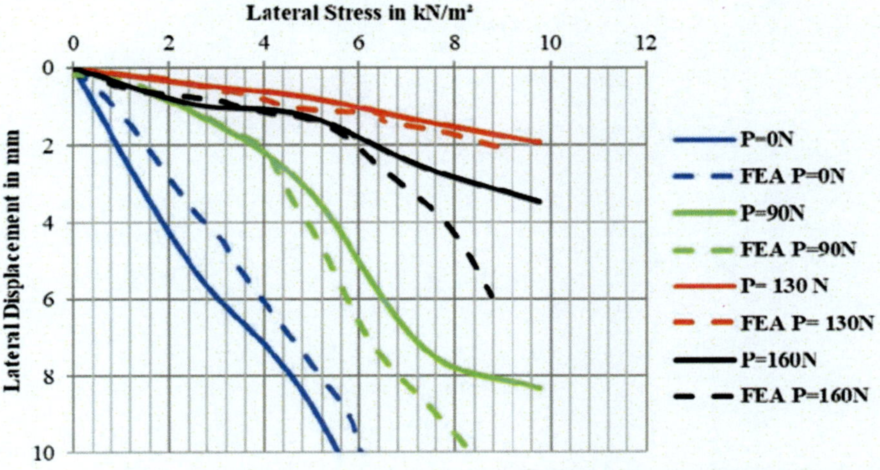


**Fig. 7** Influence of vertical load on lateral load—deformation behaviour when $D/B = 1$ for clay

## *4.2 Influence of Embedment Depth of Footing on Lateral Load—Deformation Behaviour*

From Fig. 8, it is observed that as lateral stress increases, lateral displacement increases. As depth of footing increases, lateral displacement decreases since the confinement of footing increases. Displacement is found minimum at $D/B = 1$. Displacement is observed maximum when the footing is at the surface. It is seen

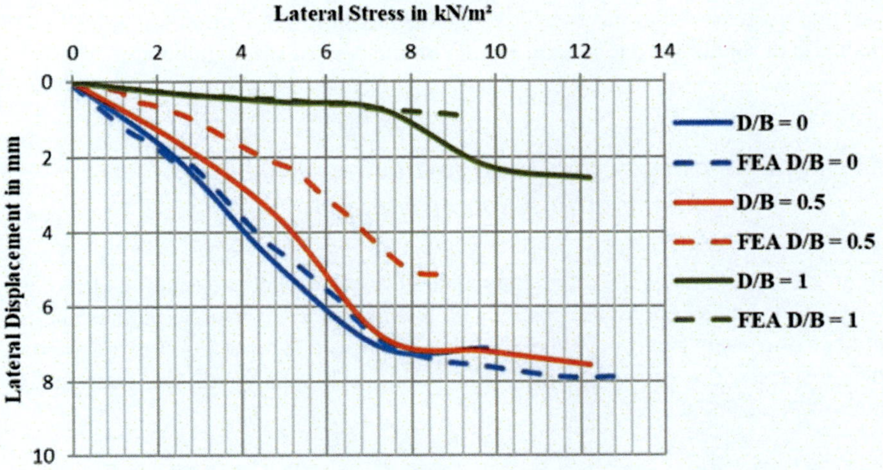


**Fig. 8** Influence of embedment depth of footing on lateral load-deformation behaviour when $P =$ 130 N for clay

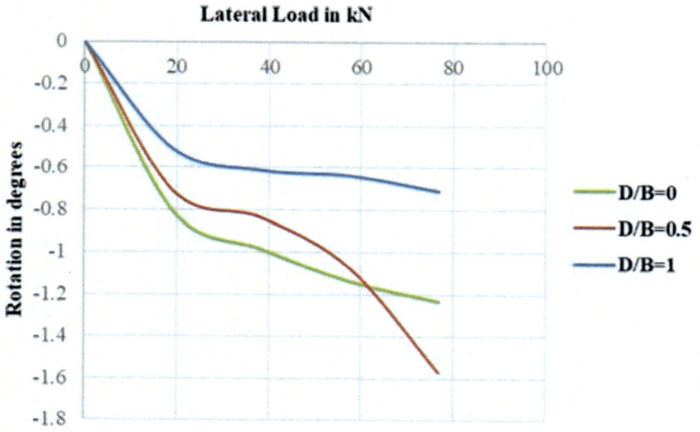


**Fig. 9** Influence of lateral load on rotation when $P = 130$ N for clay

that the results obtained from finite element analysis almost agree with experiment results.

## 4.3 Influence of Lateral Load on Rotation of Footing

Figure 9 shows lateral load versus rotation behaviour of circular footing at different embedment depth to width ratios (*D*/*B*) 0, 0.5 and 1 under vertical load of 130 N. From the curve, it is observed that as lateral loading increases, rotation increases. Rotation decreases with increase in embedment depth found minimum at $D/B = 1$. Decrease in rotation is due to increase in confinement of soil with depth.

## 4.4 Influence of Micropiles on Reduction in Lateral Displacement and Rotation of Footing

Figures 10 and 11 show the reduction in lateral displacement and rotation of footing by the provision of micropiles for $P = 90$ N and $D/B = 1$. With an increase in the number of rows of micropiles the behaviour improved. The addition of micropiles increases the stiffness of soil beneath the footing and thereby reducing rotation and lateral displacement.

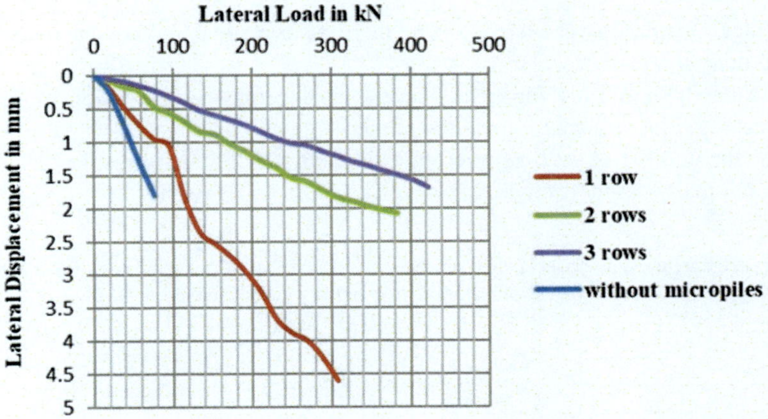


**Fig. 10** Influence of Micropiles on lateral load—lateral displacement behaviour

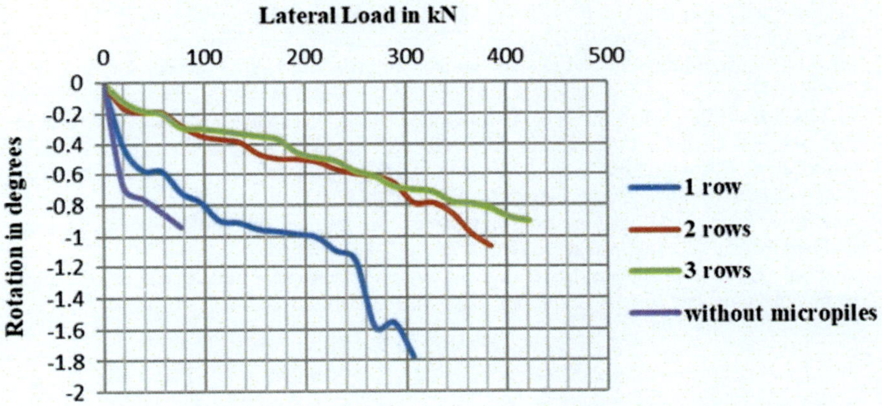


**Fig. 11** Influence of Micropiles on lateral load—rotation behaviour

## 5 Conclusions

The effects of lateral load on a model circular footing resting on clayey soil are investigated by carrying out a series of laboratory scale load tests and nonlinear finite element analysis using the FE software PLAXIS 2D. The influence of vertical load, embedment depth and addition of micropiles on the lateral load—deformation behaviour and lateral load—rotation behaviour is particularly studied. The results of laboratory scale load tests are validated by carrying out finite element analyses. Based on the results observed, the following conclusions are drawn.

1. Lateral displacement decreases with increase in vertical loading due to increase in normal stress.
2. As depth of footing increases, lateral displacement and rotation of footing decrease due to increase in confinement.
3. As the number of rows of micropiles increases, both lateral displacement and rotation of footing decrease.

**Acknowledgements** The financial support received from TEQIP Four Funds of LBS Institute of Technology for Women, Thiruvananthapuram, is gratefully acknowledged.

## References

1. Taiebat, H.A., Carter, J.P.: Numerical studies of the bearing capacity of shallow foundations on cohesive soil subjected to combined loading. J. Geotech. Eng. **122**(4), 409–418 (2000)
2. Horikoshi, K., Matsumoto, T., Hashizume, Y., Watanabe, T., Fukuyama, H.: Performance of piled raft foundations subjected to static horizontal loads. Int. J. Phys. Model. Geotech. **3**(2), 37–50 (2003)
3. Phillip, S.K., Brian, K.F., Chang, A.M.: Simplified lateral load analyses of fixed head piles and pile groups. J. Geotech. Eng. **123**(12), 711–723 (2007)
4. Sawadan, K., Takemura, J.: Centrifuge model tests on piled raft foundation in sand subjected to lateral and moment loads. Soils Found. **54**(2), 126–140 (2014)
5. Zheng, G., Zhao, J.: Ultimate bearing capacity of strip footings on sand overlying clay under inclined loading. Comput. Geotech. **106**(5), 266–273 (2019)
6. Abbasali, T. G., Khaleghnejad, M., et al.: Behaviour of eccentrically loaded shallow foundations resting on composite soils. J. Build.Eng. **102**(18), 303–315 (2019)

# Geotechnics of a Unique Irregular High-Rise Statue

**Ravi Sundaram, Abhay Gupta, and Sanjay Gupta**

## 1 Introduction

A gigantic 107-m-high statue of Lord Shiva in sitting posture has recently been constructed on a 500-m-high hillock at Ganesh Tekdi, Nathdwara in south western Rajasthan. It is the world's tallest statue of Lord Shiva and the fourth tallest statue in the world.

The base of the statue is rectangular in shape and covers an area of approximately 60 m × 45 m in plan area with a trident (trishul) standing by the side. It has an irregular configuration as per IS: 1893(Part-1)-2016 [1] and has irregularity in both horizontal and vertical planes. What makes it unique is the posture and expressions which sculptor detailing has given, and it imparts a divine look to the statue.

The paper presents results of the geotechnical and geological investigation for the statue and the foundation system adopted along with the structural details.

R. Sundaram (✉) · S. Gupta
Cengrs Geotechnica Pvt. Ltd., A-100 Sector 63, Noida 201309, India
e-mail: ravi@cengrs.com

S. Gupta
e-mail: sanjay@cengrs.com

A. Gupta
Skeleton Consultants Pvt Ltd., Noida 201301, India

 
C. N. V. Satyanarayana Reddy et al. (eds.), *Ground Characterization and Foundations*,
Lecture Notes in Civil Engineering 167,
https://doi.org/10.1007/978-981-16-3383-6_46

## 2 Project Details

### *2.1 Conceptual Planning*

The main elements for creating this monument are architecture and aesthetics of statue in the desired expressions. Structural engineering and design of main frame supporting the external shell and the design of shell are the key to a safe structure.

The site geology, geotechnical characteristics of the rock formation and slope stability are the basis of the foundation design aspects. A resourceful construction agency with rich experience and capabilities is an essential requisite to translate ideas into reality. A well-defined construction methodology for foundations, main gravity load and lateral load-resisting frame, external shell skin and the statue organ shapes ensures that the statue looks gracious and elegant.

Figure 1 presents a photograph of the statue [2] showing Lord Shiva in a pleasing appearance and the artist's impression.

**Fig. 1** Photograph (left) of the 107-m-(351 ft)-high statue shows Lord Shiva in sitting pose with trishul by the side. Photo on the right is the sculptor's imagination (model)

### 2.2 Implementation Steps

Key steps in implementing the design concepts include:

1. Statue mock-up finalisation
2. Site geology and geotechnical investigations to understand the stability of hillock slopes and deciding proper foundations with ground improvement using rock anchors.
3. Posture and expressions visualisation when viewed from ground to top and body proportions done by expert sculptor.
4. In-situ concrete casting for each and every curve and cuts using full scale FRP moulds, drawn from a full scale dummy made using high-density EPS and a six-arm robotic CNC machine.
5. Accurate 3D modelling of continuously varying geometry and structural framing to provide stability of large cantilevers.
6. Joint-less 20,000 $m^2$ of concrete skin casting using SCC and green cutting
7. Integration of structural steel frame and concrete skin to yield a pyramid like rigid structure to last for 250 years.
8. A 170-feet-high free standing trident by side of the statue made of concrete stem and steel top.
9. Wind tunnel testing to evaluate complex wind pressure from today to whole design life including future developments.

## 3 General Site Conditions

### 3.1 Regional Geology

Geologically, Udaipur district is comprised of basement rocks of Mangalwar Complex of Bhilwara Supergroup, followed by Aravalli Supergroup, post-Aravalli intrusives, Delhi Supergroup of rocks, post-Delhi intrusives and recent alluvium [3].

The rock deposits in Nathdawara area belong to the Nathdawara Group. The formations in the area belong to the middle Proterozoic Age. It comprises phyllite, calcareous phyllite and intercalatry bands of dolomite/calcitic marble. Table 1 presents the stratigraphic succession of the Aravallis in Udaipur District [4].

### 3.2 Geomorphology

Based on the general elevation, slope and landscape configuration in the terrain, the project area falls in the Aravalli landscape. The topography is hilly and undulating to semi-undulating. It experiences sub-humid to semi-arid climate and the average

**Table 1** Geological succession of the Aravalli Super-group in Udaipur District

| Group | Lithology |
|---|---|
| Jharol group | Phyllite, chlorite-schist, garnet-mica schist, calc-schist with marble, etc. |
| Nathdwara group | Phyllite, calcareous phyllite, intercalary bands of dolomite/calcitic marble |
| Bari Lake group | Meta-volcanics, conglomerate, arkose, quartzite, phyllite, schist, dolomite, quartzite etc. |
| Udaipur group | Phyllite, greywacke, mica-schist, quartzite, limestone and dolomite |
| Debari Group | Conglomerates, arkose, quartzite, phyllite, dolomite, limestone, chert and carbonaceous phyllites, Undifferentiated granites, basic sills/dykes |

**Fig. 2** Site overview

annual rainfall is 620–680 mm. It is mostly covered with very thin topsoil (< 0.5 m) and is underlain by rock.

The general trend of hill is N–S with inclined to steep dips. The area is traversed by numerous small to big nallahs (drains). Three small prominent main seasonal nallahs start from higher elevated area and merge into a downstream water pond. The general ground slope of the area is towards north east. Figure 2 is a photograph showing the site overview.

## 3.3 Site Conditions

The proposed project site covers an area of about 10,000 $m^2$. It is located in hilly region approximately 500 m away from national highway NH-8. The site is on sloping ground.

The ground rises from south to north with a level variation of about 20 m from the bottom of the hillock to the top. The general ground slope varies from 8° to 11°.

**Fig. 3** View of site: see the rock exposed at ground level

Some nallahs are seen in the surrounding area at lower elevation. Most of the nallahs have little or no flow except during the monsoon period.

Photographs of the site are illustrated on Fig. 3.

## 3.4 *Site Stratigraphy*

The geotechnical investigation of the site included five boreholes through the rock formation. A layout plan of the site showing the locations of the boreholes is illustrated on Fig. 4.

Drilling was done using NX size double tube core barrel with diamond bits. Water was circulated to lubricate and cool the bit as well as to flush out the cuttings. A photograph of the drilling in progress is illustrated on Fig. 5.

Quartzite (rock) is encountered from ground surface to the maximum explored depth of 15 m. The rock is moderately weak to strong and moderately to slightly weathered. Typical rock cores collected are illustrated on Fig. 6.

In general, the rock at shallow depth to about 1.5–4 m depth (RL 79–84.5 m) is weak and highly fractured and has iron stains. Core recoveries in this zone range from 0 to 28% and the RQD value ranges from 0 to 10%.

Below this, the fracture frequency reduces somewhat to about RL 78 to RL 70 m. Reddish brown iron stains/veins are observed in the rock mass. Core recoveries in this zone range from 40 to 90% and the RQD value ranges from 33 to 50%. Figure 7 presents typical boreholes drilled.

The rock below RL 78–70 m is strong and slightly weathered. The joints are widely spaced. Core recoveries in this zone range from 52 to 95% and the RQD value ranges from 56 to 87%.

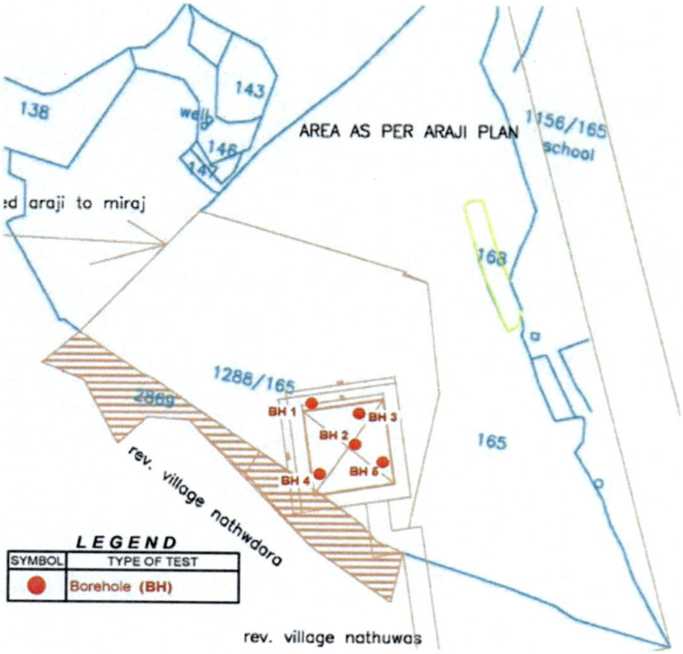


Fig. 4 Layout plan showing borehole locations

Fig. 5 Drilling in progress

**Fig. 6** Typical rock cores collected

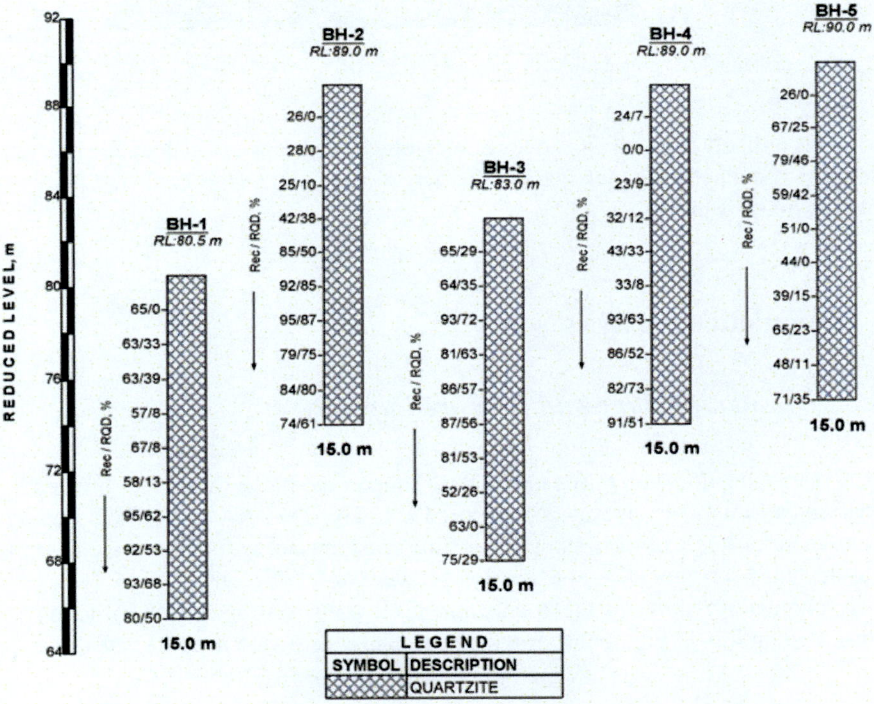


**Fig. 7** Typical boreholes profile

### *3.5 Assessment of Slope Stability*

In general, the hill slopes at the project site are fairly stable. The slopes within the plot and surrounding areas are not steeper than 8° to 11°. Further, rock is exposed at the ground surface all over the area. Hence, slope failure in slip-circle mode of

**Fig. 8** Foundation level preparation

failure is unlikely to occur. No zones of rock blocks that may get dislodged are seen. Below 3–4 m depth, the rock mass has widely spaced joints. Hence, block failure of the rock mass is unlikely.

# 4 Foundation Analysis

## *4.1 Foundation Type and Bearing Capacity*

RCC raft foundation with raised pedestals to support the statue was cast at 4 m depth. Bearing capacity analysis was done as per IS: 12070–1987 RA 2010 [5]. Based on assessment of RMR of the rock mass and the crushing strength of the rock, the raft was designed for a net allowable bearing pressure of 700 $KN/m^2$.

A photograph showing the foundation level prepared for raft construction is illustrated on Fig. 8. Figure 9 shows the pedestals constructed over the raft.

## *4.2 Rock Anchors*

For long-term stability of the statue and for stability against wind and earthquake loading, rock anchors were provided. The basic wind speed for the structure has been taken 47 m/s (IS: 875 Part 3–2015 [6]) and wind pressure is calculated for various heights for statue and trishul for a design wind speed of 70 m/s. As per IS: 1893 (Part1)-2016 [1], design was done considering the following parameters:

**Fig. 9** Foundation construction in progress

- Earthquake Seismic Zone: IV
- Importance factor of the structure: 2
- Response reduction factor: 3
- Damping ratio: 2%.

Rock anchors, 8 m long, embedded 7 m into the rock and 1 m into the raft were provided. The anchors were installed in 100 mm diameter drilled holes with a Fe500 32 mm TMT reinforcement bar grouted into the rock. Grouting was done using a GP-2 non-shrink grout. Anchors were designed to resist the entire lateral force due to wind and earthquake, ignoring the frictional resistance between rocks and concrete. The anchors were placed all along the periphery of concrete wall forming the hillock at the base of statue. A drawing showing anchor layout and other details is illustrated on Fig. 10.

### *4.3 Slope Protection Measures*

For long-term stability of slopes around the site, slope protection measures were adopted:

1. Proper drainage has been provided for the site roads.
2. Shoulder drains have been provided on either side of the roads and the water is drained away into carefully selected drainage ditches for disposal.
3. In areas where rock-slopes have been excavated, adequate precautions have been taken to ensure that the slope is stable.
4. A concrete paving has been provided all around the outer perimeter of the proposed structure. This pavement extends 1.5 m beyond the outer edge of the

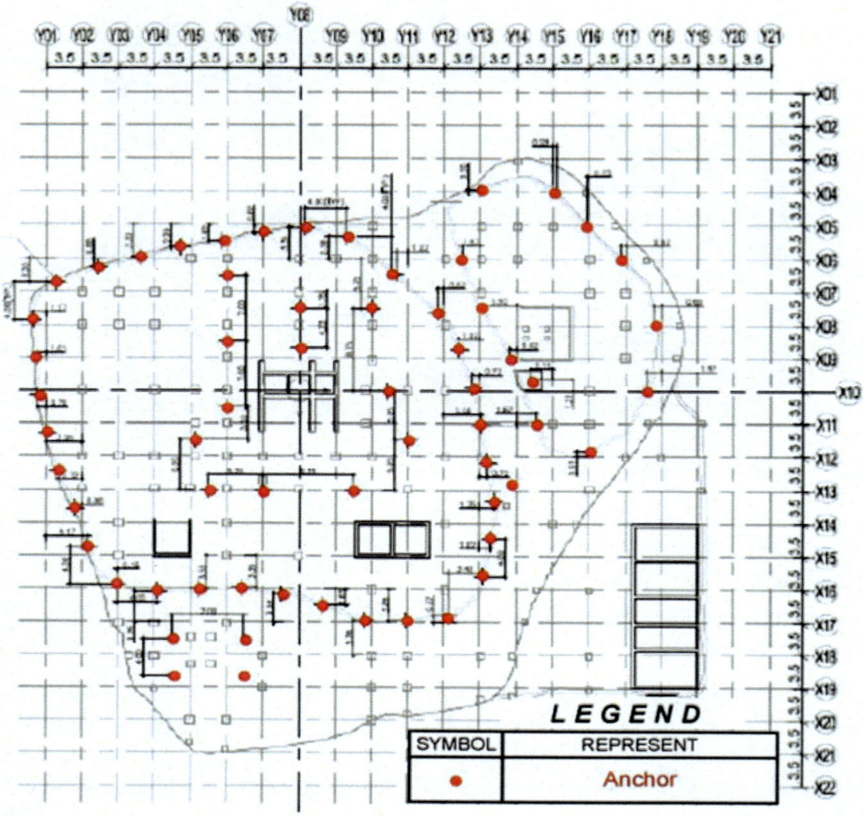


**Fig. 10** Anchors layout

exterior foundations. The paving helps in limiting water ingress into the soils beneath the foundations.

5. Entire base of statue has been provided with 8-m-deep HYSD bars anchoring rocks for 7 m penetration and 1 m inside the raft foundation. This will help binding the rocks as well as stability to entire structure.

## 4.4 Structure Details

The monument classifies as an irregular structure as per IS: 1893 (Part 1)-2016 [1]. The irregularity lies in both horizontal and vertical planes. So, steel members designed as per IS: 800-2007 [7] were used to provide flexibility of configuration and behaviour as well as ease of construction.

There is a huge platform representing seat for the Lord of about 33 m (110 ft) height over which Lord Shiva is sitting. The internal space has a lift and a staircase system

**Fig. 11** Construction in progress

for maintenance and monitoring purpose. Some internal floors are for common public use and a museum.

External skin of the statue was made through in-situ concrete pouring in specially made glass-fibre moulds. To properly support this skin and to transfer the wind forces/self-weight, special non-corrosive connections were designed. The thickness of skin shell varied from 200 to 400 mm. External geometry of statue at every 3 m interval has been used to prepare structural framing by superimposing all floor plans and arranging RCC walls designed as per IS: 456-2000 [8] and steel columns which can be taken through to maximum possible heights.

Figure 11 presents the statue under construction. Figure 12 presents the fully constructed statue and the development around it depicting Lord Shiva majestically overseeing and benevolently ruling over the surroundings.

## 5 Conclusions

The statue is a culmination of the efforts by geotechnical engineer, structural designer, sculptor, project managers and the construction agency. In addition to the geotechnical investigation, it involved installing about 70 rock anchors, 2600 tonnes of structural steel, casting about 13,000 $m^3$ of concrete and 20,000 $m^2$ of concrete skin.

**Fig. 12** View of the statue and surroundings

**Acknowledgements** The authors are extremely grateful to the Miraj Group of Industries, Nathdwara who developed this statue at Nathdwara for giving them an opportunity to be a part of this grand and mammoth project. M/s. Cengrs Geotechnica Pvt. Ltd. performed the geotechnical investigation and M/s. Skeleton Consultants have done the structural analysis and design as well as the wind tunnel testing. Prof Prem Krishna, former professor IIT Roorkee did the proof checking of design. M/s Shapoorjee Pallonjee Constructions did the execution. The project has received several awards so far for excellence in design and construction.

## References

1. IS: 1893-Part 1: Criteria for Earthquake Resistant Design of Structures, Bureau of Indian Standards, New Delhi (2016)
2. Miraj Group: Current Developments, http://www.mirajgroup.in/about/current-developments.html
3. Central Groundwater Board: Groundwater Scenario, Udaipur District, Rajasthan, Ministry of Water Resources CGWB Western Region, Jaipur (2013)
4. Department of Mines and Geology, Rajasthan: Geology and mineral resources of Udaipur District. www.dmg-raj.org/docs/Vol%20%2021-4-a.doc. Cached
5. IS: 12070: Code of Practice for Design and Construction of Shallow Foundations on Rocks, Bureau of Indian Standards, New Delhi (1987) Reaffirmed 2010.
6. IS: 875: Design Loads (other than Earthquake) for Buildings and Structures, Part 3—Wind Loads, Bureau of Indian Standards, New Delhi (2015)

7. IS: 800: General construction in steel—code of practice, Bureau of Indian Standards, New Delhi (2007)
8. IS: 456: Plain and reinforced concrete—Code of practice, Bureau of Indian Standards, New Delhi (2000). Reaffirmed (2005)

# Experimental Studies on Load-Settlement Behavior of Cohesionless Soil Using Bamboo Grid

**Bipasha Das and Nayanmoni Chetia**

## 1 Introduction

The advent of construction industry has led to various mega constructions; but due to lack of good quality of soil, it has become a headache for the engineers. There are many techniques to improvise all the types of soil and reduce the unfavorable ground condition. Many soil improvement processes are available. Various experiments are conducted on reinforced soil to improve bearing capacity of soil and reduce settlement. The high cost of synthetic polymer and durability of natural ones has made the scholars to explore more in this field. The cost of geosynthetic material is one of the deciding factors in its selection. Akinmusuru and Akinbolade (1981), Dixit (1985), Mandal and Manjunath (1994), Datye and Gore (1994), Sitharam and Hedge (2015), Dutta and Mandal (2016), Lal et al. (2017), Dong et al. (2010) are few scholars experimented with the natural geomaterials for bearing capacity of soil. Installing bamboo grid reinforced soil system which increases the frictional resistance hereby enhancing the bearing capacity of soil. In this study, the effects of various parameters such as the location of the top grid layer, number of grid layers, and aperture shape of the bamboo grid on bearing capacity and settlement are studied extensively. In this study, bearing capacity of soil with and without bamboo mat is compared theoretically, analytically and later validated the results with PLAXIS 2D®.

B. Das · N. Chetia (✉)
Department of Civil Engineering, Jorhat Engineering College, Jorhat, Assam 785007, India



C. N. V. Satyanarayana Reddy et al. (eds.), *Ground Characterization and Foundations*, Lecture Notes in Civil Engineering 167,
https://doi.org/10.1007/978-981-16-3383-6_47

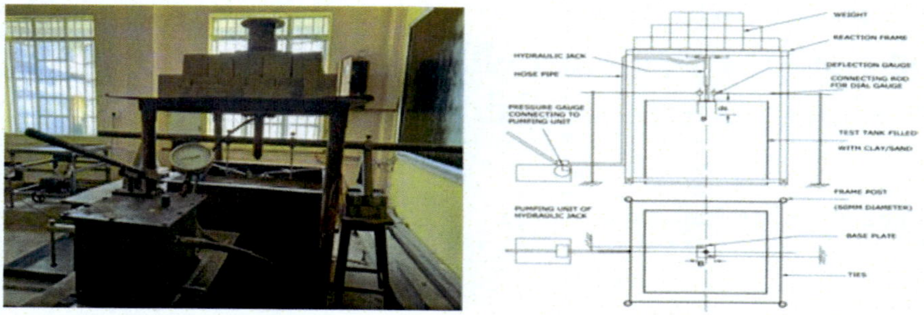


**Fig. 1** Test setup with its schematic diagram

# 2 Research Methodology

## 2.1 Test Setup

To conduct the plate load tests, a model steel tank of size 0.95 m × 0.95 m × 0.95 m is prepared at Jorhat Engineering College, Jorhat, Assam as per IS: 1888-1982. It consists of a loading frame, inverted hydraulic jack of capacity of 20 kN, pumping unit, and steel plates loaded with concrete cubes. Measurement of the magnitude of applied load pre-calibrated pressure is used. Deflection dial gauges of least count of 0.01 mm placed to measure the settlements of plates due to applied load. In this experiment, cohesion less sand is used. Here, the sand bed is prepared with 30 cm, and pre-calibrated compacting energy is applied to achieve the required loose, medium, dense relative densities. Figure 1 shows the test setup along with schematic used in this study.

## 2.2 Materials Used

**Sand**: The material used for this study is sand from Mariani (Bhogdoi River) of Jorhat district, Assam. The sand is cleaned to remove the unwanted materials from it. It is oven dried and sieved through IS sieve size of 4.75 mm, and portion passing is taken for the experiment. The properties of the sand, namely coefficient of curvature ($C_c$), uniformity coefficient ($C_u$), Zone of the sand (IS: 383-1970), and grading characteristics $D_{10}$, $D_{30}$, $D_{60}$ are determined according to the mentioned Indian Standard Codes. The sand is classified as poorly graded sand (SP) with specific gravity 2.65. The grain size distribution is shown in Fig. 2, and Table 1 shows the grading characteristics of the sand. Direct shear tests are done according to IS: 2720 (Part 13)-1986 to obtain the internal angle of friction (Φ). The unit weight of soil is 16.04 kN/m$^3$, and average relative density is 59.87%.

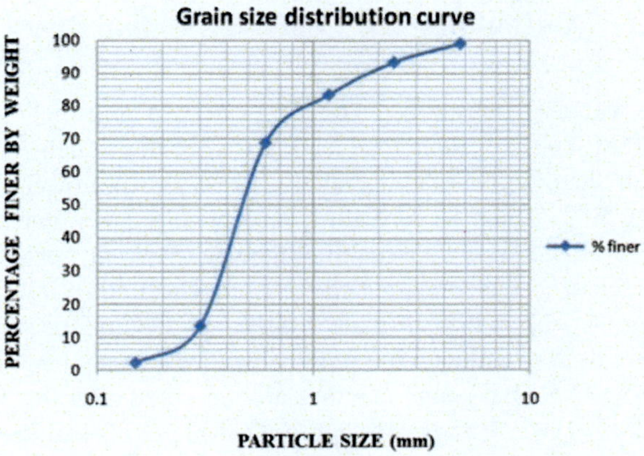


**Fig. 2** Particle size distribution curve

**Table 1** Properties of sand used

| Material | $D_{60}$ | $D_{30}$ | $D_{10}$ | $C_u$ | $C_c$ |
|---|---|---|---|---|---|
| Sand | 0.53 | 0.37 | 0.28 | 1.89 | 0.92 |

**Model Footing and Bamboo Grid**: Model circular footing (MF) for the tests used in this study is made up of mild steel plate having diameter 0.18 m and thickness 0.006 m. Model footing surfaces are smooth. Grids of *Bambusa tulda* (Jati Baah) treated with copper chrome boron (CCB) of length 0.9 m × 0.9 m, splints of width 3 mm, and thickness of 1 mm prepared in Rain Forest Research Institute. Jorhat, Assam is used. The bamboo strips used are configured in planar and orthogonal directional to get bidirectional and tridirectional patterns. The photographs of the model footing with its schematic and bamboo mats are given in Fig. 3a–c, respectively.

The durability of bamboo in civil engineering projects is a great concern. According to Gnanaharan (2000), the life cycle of bamboo in wet condition is 5 years. The insects and fungi play an important role in degrading the bamboo and its efficiency. A chemical compound copper chrome boron (CCB) is used to increase its durability by soaking the bamboo strips in it for 24 h. Moisture content hampers its efficiency as reinforcement as it needs to be coated with bitumen and sun dried for 24 h.

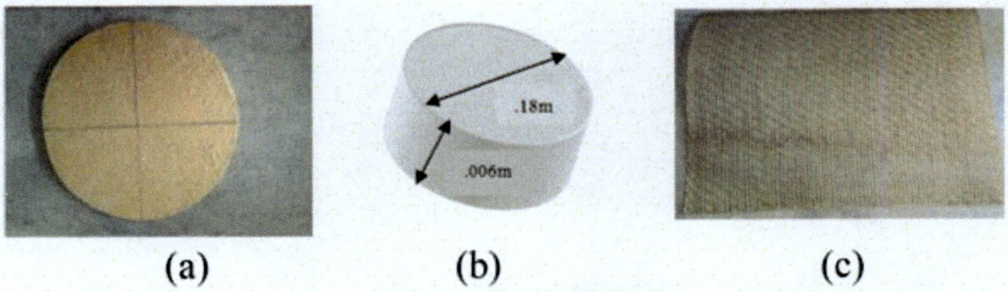


**Fig. 3** **a** Model footing, **b** schematic diagram, **c** bamboo mats

# 3 Methodology of Load-Settlement Test

Load tests are conducted on the model test tank, and test procedure is taken from IS: 1888-1982 for plate load test. Average relative density of 59.87% is maintained for all the tests. The average relative density and unit weight of sand beds are maintained as 59.87% and 16.04 kN/m$^3$, respectively. The number of drop is selected from the calibrated graph at a required density (59.87%) for all experiments. A sprit level is used to level the surface before placing the footing. It is placed centrally, under spindle of the jack so that the plate, reaction girder, and the spindle are coaxial. Dial gauges are used to estimate the settlements. Load is applied in cumulative equal increment of 10 kg. Dial gauge readings are measured after the time interval mentioned in IS: 1888-1982. Test is repeated until a high value of settlement is observed. For the further test, reinforcement is placed at the desired depth from the model footing.

## 3.1 *Calculation for Ultimate Bearing Capacity of Model Footing*

The ultimate bearing capacity of shallow circular foundation is determined by load-settlement curves. The comparative load displacement curve of model footing plate in loose, medium, dense sand with corresponding angle of internal friction as 27.5°, 31°, and 35° has been presented in Fig. 4. The ultimate bearing capacity for the model footing at different relative densities is calculated numerically by Terzaghi (1943), Meyerhof (1963), Hansen (1957, 1970), Vesic (1973), IS: 6403-1981 is listed in

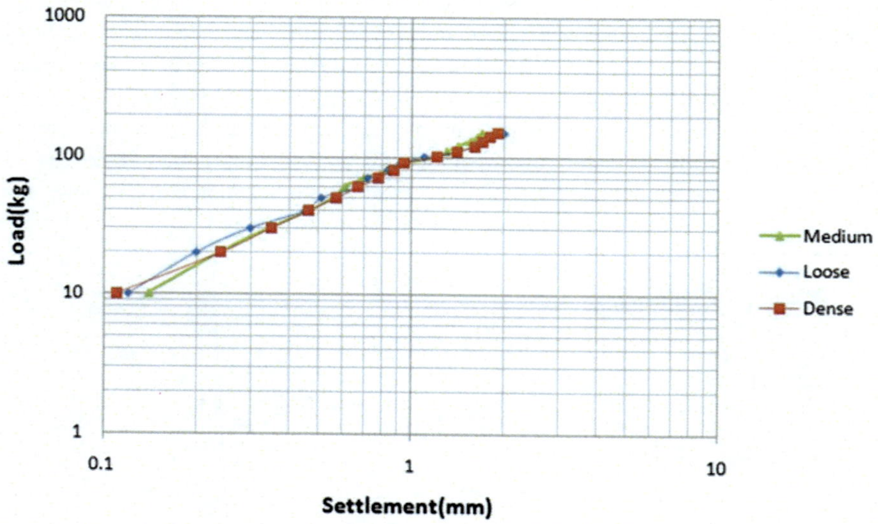


**Fig. 4** Comparison of load-settlement for circular footing

**Table 2** Comparison of ultimate bearing capacity

| Method | Ter | Mey | Hansen | Vesic | IS code | Exp |
|---|---|---|---|---|---|---|
| Loose | 14.75 | 18 | 10.08 | 15.37 | 14.84 | 12 |
| Medium | 25.82 | 38.26 | 18.75 | 26 | 26.1 | 23.8 |
| Dense | 46.13 | 81.38 | 36.26 | 49 | 46.6 | 41 |

Table 2. The relative densities are 34.8, 59.87, and 70.24%. Bearing capacity of model footing of diameter 0.18 m found out to be 23.80 kN/m$^2$ experimentally; this is extracted from load-settlement graph. Load intensity with respective settlement is plotted on a log–log scale. It matches with the results obtained from Terzaghi equation. Hence, it can be concluded that results obtained from the experimental set up is reproducible.

## *3.2 Effect of Bamboo Reinforcement at Different Depths*

The bamboo mat used as reinforcement having size 0.9 m $\times$ 0.9 m, strips of width 3 mm, and thickness 1 mm which are chemically treated with CCB are configured together for bidirectional (square aperture) and tridirectional (hexagonal) where the connection patterns are inter-woven, orthogonally and diagonally interlocked for maintaining the equal aperture size during the preparation of bamboo grids for the research. These bamboo mats are placed at a depth of 0.25B, 0.5B, B, 1.5B where B is the diameter of the model footing.

**Effect of Square aperture Bamboo at different depths**. The square aperture bamboo mats of size 0.9 m $\times$°0.9 m are placed at a depth of 0.25B, 0.5B, B, 1.5B where B is the diameter of the model footing. The improvement of bearing capacities due to installation of bamboo layer is represented by a non-dimensional factor called "*Improvement Factor*." The factor is defined as the ratio of ultimate bearing capacity of shallow foundation with bamboo layer to the ultimate bearing capacity of shallow foundation without it. It is clear that the ultimate load and ultimate bearing capacity decrease with the increase in the difference between the distance of bamboo layer and the model footing. Same is the case with the failure load. Figure 5 and 6 show square aperture bamboo grid placed at the required depth and load-settlement graph for model footing with bamboo reinforcement at different depths. Table 3 depicts the values obtained in (Fig. 7).

**Effect of Hexagonal aperture Bamboo at different depths**. The effect of hexagonal bamboo mats of size 0.9m $\times$°0.9 m when placed at 0.25B, 0.5B, B, 1.5B is discussed in Fig. 8 and Table 4. A graphically representation is given in Fig. 9 and (Fig. 10).

From the above discussion, it is clear that for the same size and aperture the TBG which is hexagonal aperture gives better performance compared to BBG of square

**Fig. 5** Square aperture bamboo grid at required depth

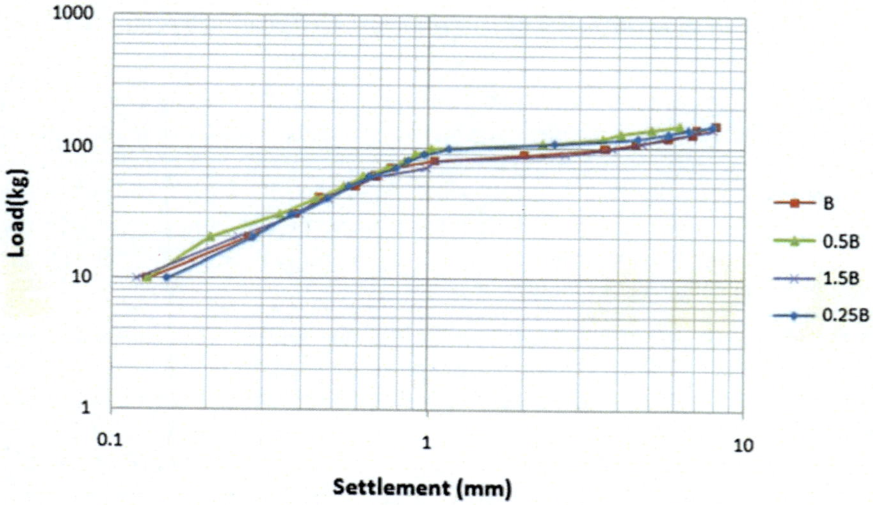


**Fig. 6** Load-settlement graph for model footing with bamboo layer at different depths

**Table 3** Ultimate bearing capacity of model footing with square aperture bamboo grid at various depths

| Depth | Failure load (kg) | Ultimate bearing capacity (kN/m$^2$) | IF |
|---|---|---|---|
| 0.25B | 120 | 47.15 | 1.98 |
| 0.5B | 110 | 43.22 | 1.81 |
| B | 80 | 31.4 | 1.32 |
| 1.5B | 70 | 27.5 | 1.15 |

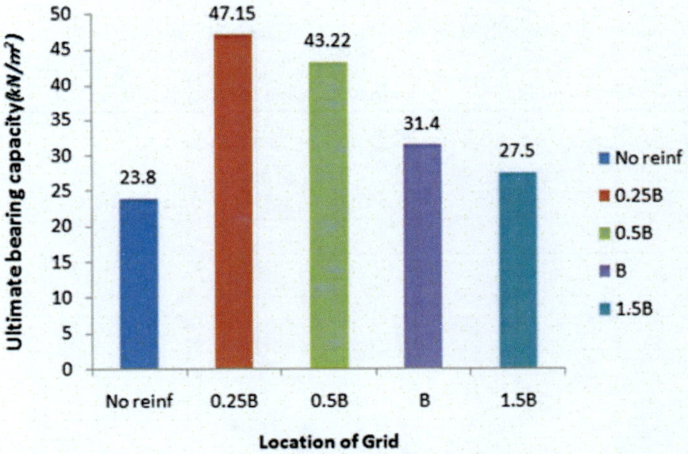


**Fig. 7** Comparison of bearing capacity with and without grid

**Fig. 8** Hexagonal aperture bamboo mat at required depth

aperture. The reason may be, due to its geometry, TBG is capable to transfer stresses in all direction, giving more bearing resistance compared to that BBG.

**Table 4** Ultimate bearing capacity of model footing with hexagonal aperture bamboo grid at various depths

| Depth | Failure load (kg) | Ultimate bearing capacity (kN/m$^2$) | IF |
|---|---|---|---|
| 0.25B | 170 | 66.8 | 2.81 |
| 0.5B | 150 | 58.9 | 2.47 |
| B | 130 | 51.08 | 2.13 |
| 1.5B | 80 | 31.4 | 1.32 |

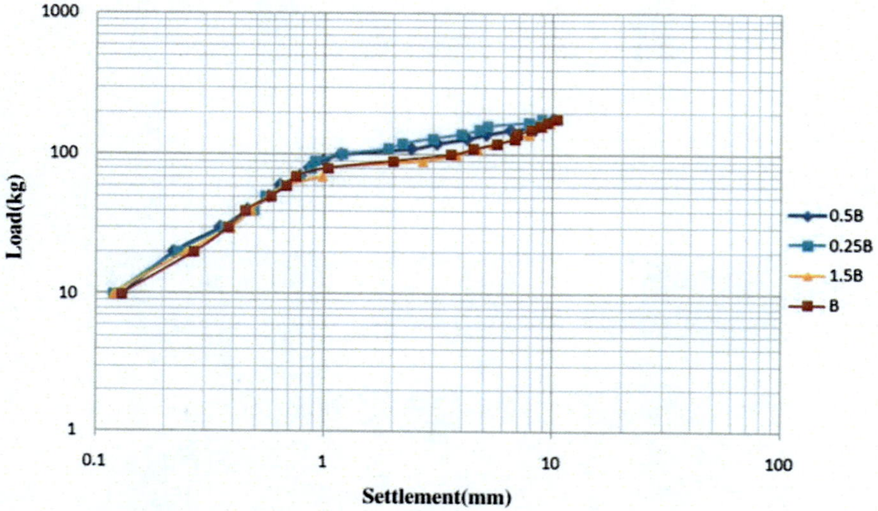


**Fig. 9** Load-settlement graph for model footing with bamboo layer at different depths

**Fig. 10** Comparison of bearing capacity with and without grid

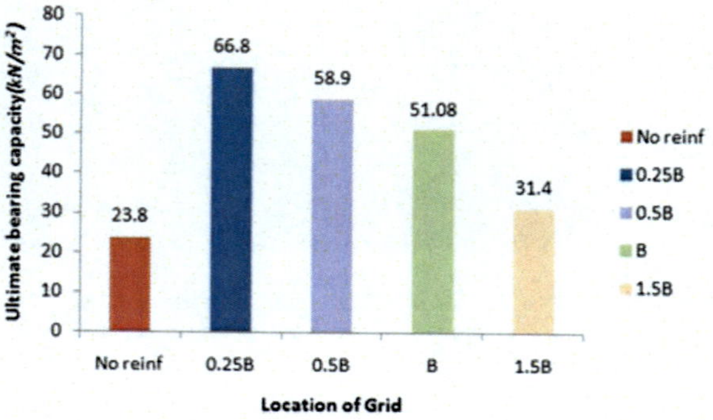

# 4 Numerical Analysis

The geometry of the finite element soil model adopted for the analysis is 0.95 m ×°0.95 m with the identical dimension of the model footing mentioned above. Comparison has been drawn for bearing capacity with and without bamboo mats. Later, bamboo mats placed at the required depth are shown with the results obtained. Here, ground water condition is neglected. Superstructure is replaced by an equivalent amount of loading and that equivalent amount of vertical pressure is applied. The various properties used in the study are given in Table 5. In this study, more emphasis is given to medium density sand. In this analysis, Mohr Coulomb analysis is used. Analysis is carried out in PLAXIS 2D® where the load–displacement curve is obtained.

## 4.1 Model Footing Without Bamboo Mat

Following geometrics are obtained when pressure is applied to the model footing during analysis. Figure 11a–c shows the model, displacement, and stress distribution of the soil bin, respectively. Figure 11d shows the load–displacement behavior of the foundation without bamboo mat under vertical loading. From this curve, ultimate bearing capacity is measured.

**Table 5** Properties used in PLAXIS 2D®

| Properties | Value |
|---|---|
| Type of soil | Sand |
| Soil model | Mohr–Coulomb |
| Void ratio | 0.5 |
| Foundation thickness (m) | 0.006 |
| Foundation diameter (m) | 0.18 |
| EA (kN/m) | 5,089,380.1 |
| EI (kNm$^2$) | 10,306 |
| w (kN/m/m) | 0.5 |
| Name of the geogrid | Bamboo |
| Pattern | Square hexagonal |
| EA (kNm$^2$) | 700–900 |

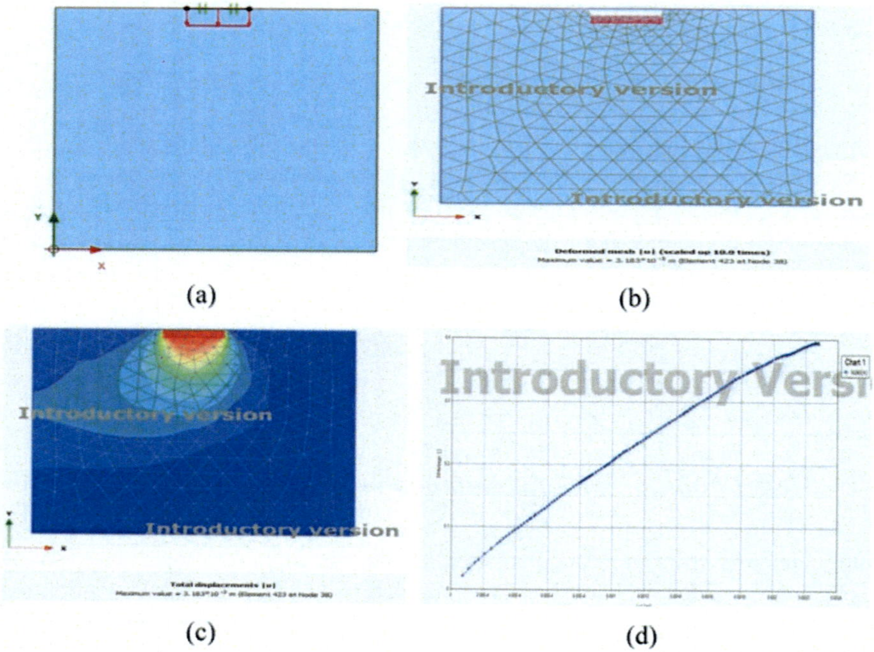


**Fig. 11** **a** Modeling in PLAXIS 2D, **b** displacement, **c** stress distribution, **d** load-settlement curve

## 4.2 *Model Footing with Square Aperture Bamboo Mat*

Following geometrics are obtained when maximum load is applied on the model footing with bamboo mat of size 0.9 m × 0.9 m during analysis. Figure 12a–c shows the model, displacement, and stress distribution of the soil bin with bamboo mat at a depth of B from the footing, respectively. Figure 12d depicts the load-settlement curve for the mentioned loading.

## 4.3 *Model Footing with Hexagonal Aperture Bamboo Mat*

Following geometrics are obtained when maximum load is applied on the model footing with bamboo mat of size 0.9 m × 0.9 m during analysis. Figure 13a–c shows the model, displacement, and stress distribution of the soil bin with bamboo grid at a depth of B from the footing, respectively. Figure 13d is with the load-settlement behavior for hexagonal aperture bamboo grid.

The model being axis symmetric load applied is given in kN/rad. Thus, the total load can be represented in the scale of kN by multiplying the value with $2\pi$. From the output results Table 6, it can be concluded that the TBG offers more bearing capacity when applied at any random depth (here B).

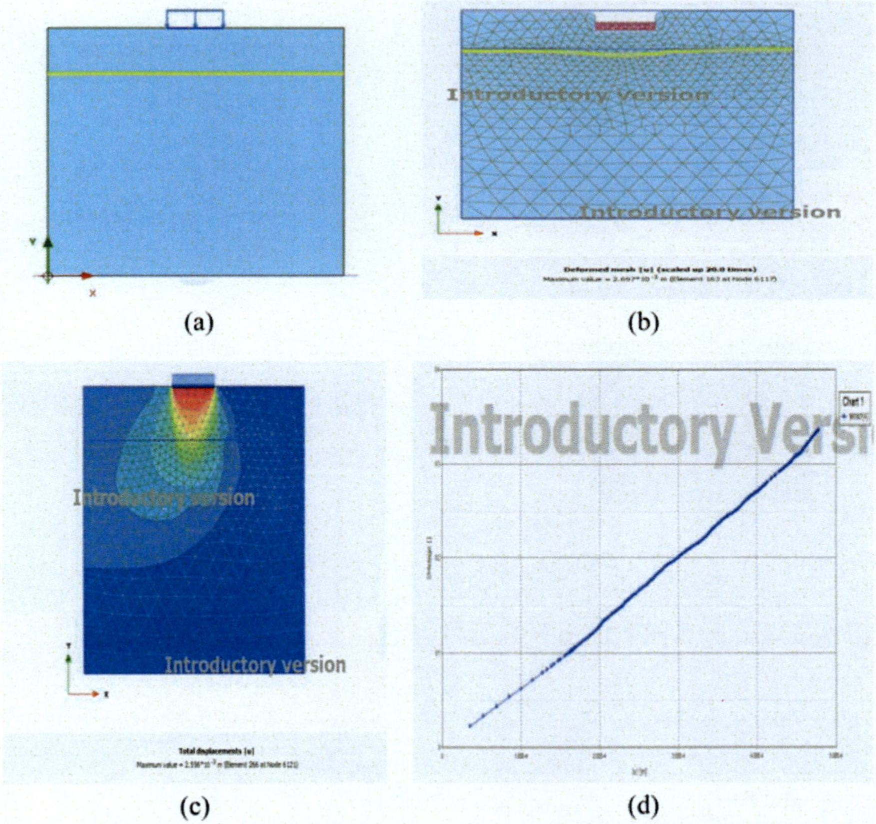


**Fig. 12** **a** Modeling in PLAXIS 2D, **b** displacement, **c** stress distribution, **d** load-settlement curve

## 4.4 *Effect of Number of Layers with PLAXIS 2D®*

The ultimate bearing capacity and settlement when different number of layers placed at required depths are studied with PLAXIS 2D®. With the above-mentioned properties, the analysis is observed for various variables depicted in (Fig. 14).

# 5 Conclusions

Based on the results from the study, the following conclusions can be drawn:

1. The bearing capacity increases with the increase in the relative density of soil and with insertion of bamboo mat.
2. The IF decreases significantly with the increase in the difference of the depth between footing and bamboo layer.

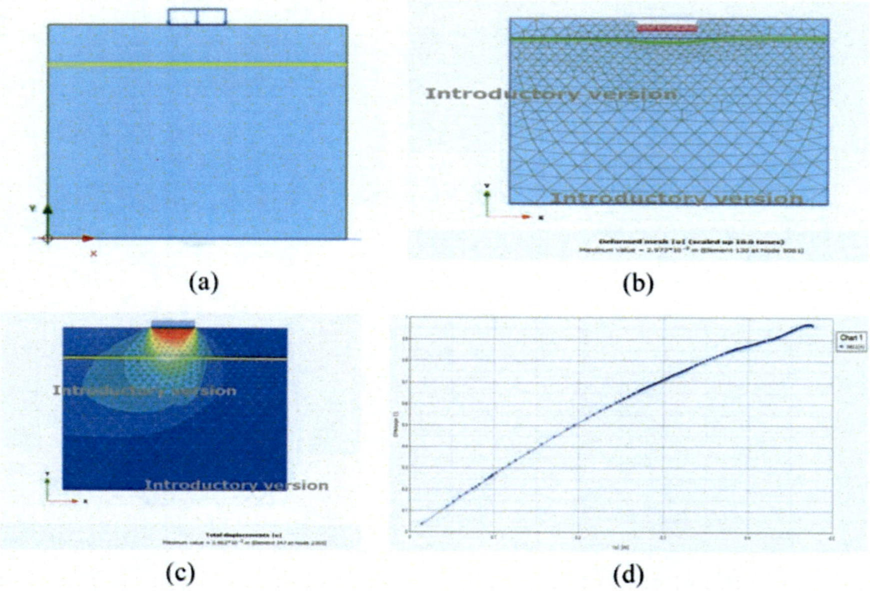


**Fig. 13** **a** Modeling in PLAXIS 2D, **b** displacement, **c** stress distribution, **d** load-settlement

**Table 6** Comparison of bearing capacity for medium density soil when bamboo grid placed at B

| Type of aperture | Ultimate bearing capacity (kN/m$^2$) |
|---|---|
| Bidirectional | 40 |
| Tridirectional | 62.4 |

3. Due to the aperture shape, the bearing resistance offered by TBG is more compared to that of the BBG for the same size of the reinforcement.
4. The settlement observed on the reinforced sand bed with three layers of TBG is less compared to the same number of layers of BBG. In case of bearing capacity, it is more in three layers of TBG. Thus, it can be concluded that ($N = 3$) of TBG is more effective than ($N = 4$) of BBG.

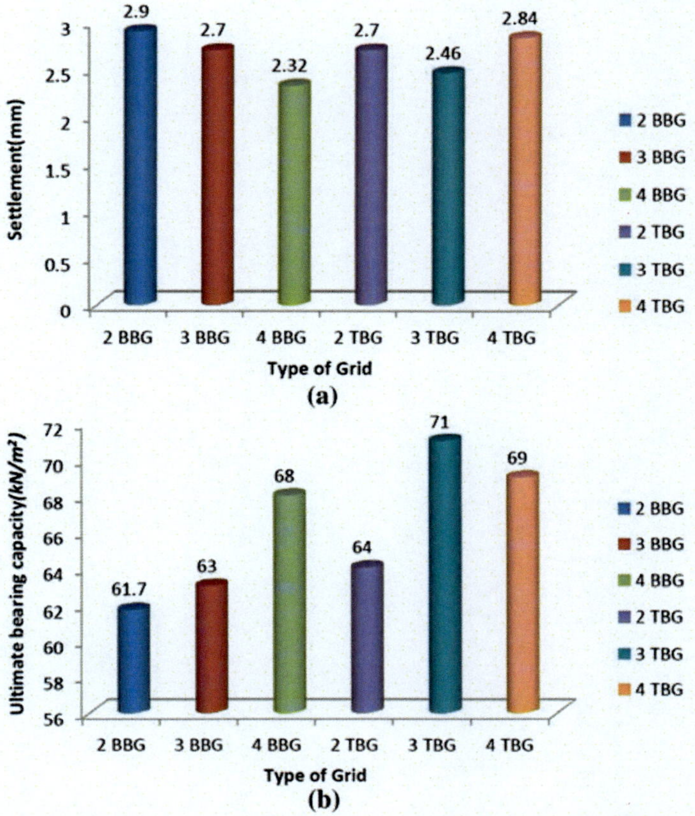


**Fig. 14** **a** Comparison between type of grid and settlement, **b** comparison between type of grid and settlement

# References


1. Panigrahi, B.: Improvement of bearing capacity of soil by using natural geotextile. Int. J. Geo-Eng. (2019)
2. Ahirwar, S.K., Mandal,J.N.: Behavior of bamboo grid-reinforced soil bed. Int. J. Geotech. Eng. (2018). ISSN: 1938–6362
3. PLAXIS 2D Tutorial Manual 2016
4. Shill, S.K.: Comparison of bearing capacity calculation methods in designing shallow foundations. Int. J. Eng. Technol. Manage. Appl. Sci. (2015). ISSN 2349-4476
5. Alhaji, M.: Bamboo as soil reinforcement—a laboratory trial. Leonardo J. Scie. (2008). ISSN 1583-0233
6. IS:1888-1982: Method of Load test on Soil, Bureau of Indian Standards, New Delhi

# Ultimate Bearing Capacity of Strip Footing on Reinforced Embankment Using Upper Bound Limit Analysis

Debashis Manna, G. Santhoshkumar, and Priyanka Ghosh

## 1 Introduction

In several occasions, foundations are placed on slopes such as highway or railway resting on embankments. It is understood from the literature [4, 5, 10, 11] that the bearing capacity of a footing resting on the sloping ground is generally found lower than that of a footing resting on the horizontal ground. Nowadays, the application of reinforced embankment has been a fascinating concept among geotechnical engineers. Hence, the bearing capacity of footings resting on such reinforced slopes can be worth exploring [3, 9, 12, 15]. The present investigation determines the ultimate bearing capacity of a surface strip footing resting on a single-layer reinforced embankment, as shown in Fig. 1a. The analysis was performed using the upper bound limit analysis, along with a multi-block failure mechanism. In this study, an influence factor ($R_f$) is introduced, which can be multiplied with the limit load of the footing resting on the horizontal soil bed to determine the effect of the embankment slope on the bearing capacity. Similarly, an increment factor ($E_f$) is introduced to determine the effect of reinforcement on the bearing capacity. The influence of setback distance ($S_L$), slope angle ($\beta$), cohesion ($c$), angle of internal friction ($\phi$) of soil, and embedment depth ($S_v$) on the magnitude of $E_f$ and $R_f$ is explored (Fig. 1a). The results are provided based on a parametric study so that it can be used to calculate the limit load of a strip footing resting on a reinforced embankment. While determining the influence of the reinforcement on the bearing capacity, the reinforcement is assumed to be a strong one, i.e., the tensile strength is much higher than the force induced in the reinforcement. Hence, the reinforcement is considered to fail due to slippage. Soil is assumed to follow the Mohr–Coulomb failure criterion along with the associated

D. Manna · G. Santhoshkumar · P. Ghosh (✉)
Department of Civil Engineering, IIT Kanpur, Kanpur 208016, India
e-mail: priyog@iitk.ac.in



C. N. V. Satyanarayana Reddy et al. (eds.), *Ground Characterization and Foundations*, Lecture Notes in Civil Engineering 167,
https://doi.org/10.1007/978-981-16-3383-6_48

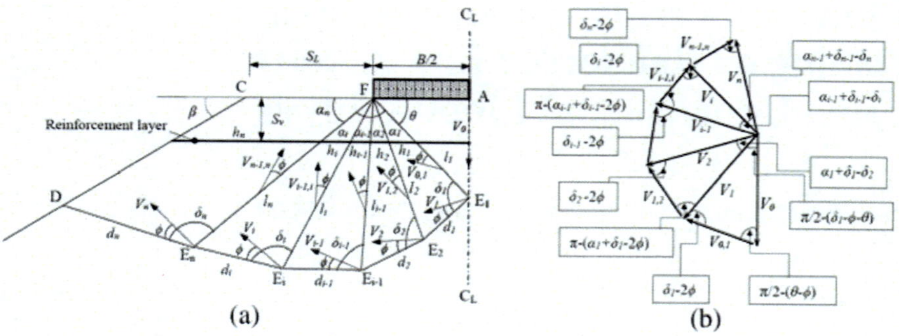


**Fig. 1** **a** Failure mechanism and velocity vectors. **b** Velocity hodograph

flow rule. It can be conceived that the magnitude of $R_f$ greatly depends on $c$, $\phi$, $S_L$, and $S_v$.

## 2 Problem Definition

A perfectly rough surface strip foundation of width $B$ rests on the top of a single-layer reinforced soil embankment with a setback distance $S_L$ on either side of the foundation, as shown in Fig. 1a. The reinforcement is placed horizontally with an embedment depth $S_v$. The objective is to determine the bearing capacity of the foundation using the classical upper bound limit analysis based on a kinematically admissible collapse mechanism, as shown in Fig. 1a. The $c$-$\phi$ soil in the embankment is assumed to follow the Mohr–Coulomb failure criterion along with the associated flow rule. It is also assumed that the collapse of the footing occurs prior to the failure of the embankment.

## 3 Analysis

### *3.1 Failure Mechanism*

Following the work of Biswas and Ghosh [1, 2], a kinematically admissible multi-block failure mechanism is assumed in the present study, as shown in Fig. 1a. Taking the advantage of the symmetricity, the analysis was carried out considering half-domain, as shown in Fig. 1a. In this analysis, $n$ number of rigid blocks are considered on either side of the plane of symmetry ($C_L$-$C_L$). The collapse mechanism can be defined by the geometric variables $\alpha_i$, $\delta_i$, and $\theta$, as shown in Fig. 1a. The triangular trapped wedge $AE_1F$ below the footing base is assumed to move along with the footing at the same velocity ($V_0$). The vertical movement of this trapped wedge causes a lateral movement of the remaining rigid blocks on the left side of the footing. However, the outermost rigid block ($FE_nDC$) turns out to be quadrilateral due to the presence of the sloping face on the left side. The absolute velocity of the *i*th block ($FE_iE_n$) can be presented as $V_i$, whereas the relative velocity between the *i*th block and the $(i-1)$th block can be considered as $V_{i-1,i}$ and so on. The movement of the rigid blocks with different velocities can be confirmed from the velocity hodograph shown in Fig. 1b. The interfaces among the blocks are considered as the velocity discontinuity lines.

It can be seen from Fig. 1a that the reinforcement cuts the rigid blocks with different lengths. Hence, the rate of internal energy dissipation ($D$) in case of slippage of the reinforcement can be expressed as [7]

$$D = 2l_e(\sigma_n f_b \tan\phi + f_c c)V_l \tag{1}$$

where $l_e$ is the effective length of the reinforcement, $\sigma_n$ is the normal stress acting on the reinforcement, $V_l$ is the relative velocity between the reinforcement and the soil mass, and $f_b$ and $f_c$ are the bond coefficients as recommended by Michalowski [8].

## 3.2 Ultimate Bearing Capacity

As per the upper bound limit analysis, the ultimate failure load of the strip footing can be determined by equating the rate of the external work done with the rate of the internal energy dissipation. Hence, the limit load ($P_{urf}$) on the footing can be expressed as a function of different geometrical variables of the failure mechanism such as $\alpha_i$, $\delta_i$, and $\theta$. The least upper bound solution can be obtained by conducting a rigorous optimization study. Hence, the ultimate bearing capacity of the footing ($q_{urf}$) can be expressed as

$$q_{urf} = \frac{P_{urf}}{B} = \left[\frac{c\cdot\left(f_3+f_4+f_5+f_c\cdot M_{fc}\right)+0.5\cdot\gamma\cdot B\left(-f_1-f_2+2\cdot\mu\cdot\frac{S_v}{B}\cdot M_{f\gamma}\right)}{\left(1-\mu\cdot\frac{S_v}{B}\cdot M_{fp}\right)}\right] \tag{2}$$

where

$$f_1 = 0.5\tan\theta$$

$$f_2 = 4\sum_{i=1}^{n}\frac{A_i}{B^2}\cdot\frac{V_i}{V_0}\cdot\sin\left(\delta_i-\theta-\phi-\sum_{j=1}^{i-1}\alpha_j\right)$$

$A_i$ is the area of the $i$th rigid block.

$$f_3 = \frac{\cos\phi \cdot \cos(\delta_1 - \theta - \phi)}{\cos\theta \cdot \sin(\delta_1 - 2\phi)}$$

$$f_4 = 2\cos\phi \cdot \sum_{i=1}^{n} \left(\frac{d_i}{B} \cdot \frac{V_i}{V_0}\right)$$

$$f_5 = 2\cos\phi \cdot \sum_{i=2}^{n} \left(\frac{l_i}{B} \cdot \frac{V_{i-1,i}}{V_0}\right)$$

$$M_{fc} = 2 \cdot \frac{S_v}{B} \cdot \sum_{i=1}^{n} \left[\frac{h_i}{S_v} \cdot \frac{V_i}{V_0} \cdot \cos\left(\delta_i - \theta - \phi - \sum_{j=1}^{i-1} \alpha_j\right)\right]$$

$$M_{f\gamma} = 2\sum_{i=1}^{n} \left[f_{d_i} \cdot \frac{h_i}{S_v} \cdot \frac{V_i}{V_0} \cdot \cos\left(\delta_i - \theta - \phi - \sum_{j=1}^{i-1} \alpha_j\right)\right],$$

$f_{di}$ is the ratio of the height of the soil mass above the reinforcement level at the midpoint of the $i$th reinforcement to the width of the footing

$$M_{fp} = 2\sum_{i=1}^{k} \left[\frac{h_i}{S_v} \cdot \frac{V_i}{V_0} \cdot \cos\left(\delta_i - \theta - \phi - \sum_{j=1}^{i-1} \alpha_j\right)\right]$$

$$\text{for } 0 \leq \left(\theta + \sum_{i=1}^{k} \alpha_i\right) \leq 90^\circ$$

$$\mu = f_b \tan\phi$$

## 4 Results and Discussion

The analysis was performed by writing an in-house code in MATLAB, and the results are presented in terms of increment factor ($E_f$) and influence factor ($R_f$), which can be defined as

$$E_f = \frac{q_{urf}}{q_u} \quad (3)$$

where $q_u$ refers to the bearing capacity of a strip footing placed on an embankment without any reinforcement.

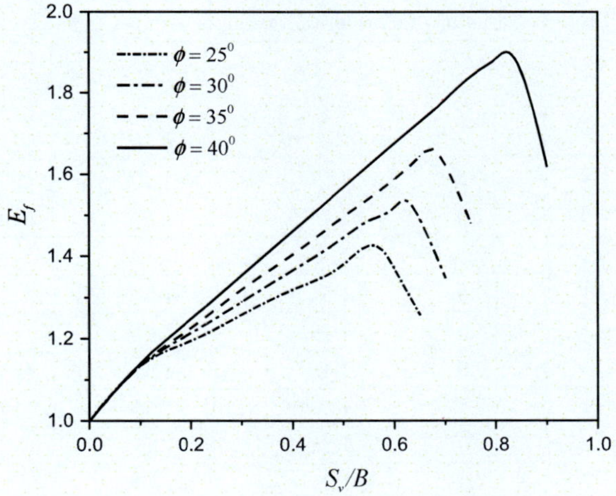


**Fig. 2** Optimum depth of reinforcement for different values of $\phi$ with $S_L/B = 1$, $\beta = 20°$, $c/\gamma B = 0.75$ for $\phi = 25°$ and $c/\gamma B = 0.5$ for $\phi = 30°, 35°, 40°$

$$R_f = \frac{|q_{urf}|_{\beta>0}}{|q_{urf}|_{\beta=0}} \tag{4}$$

## 4.1 Optimum Depth of Reinforcement

The range of the optimum depth of the reinforcement for different values of $\phi$ was suggested by Michalowski [8] for the horizontal semi-infinite ground. On the contrary, the present study involves a soil embankment with sloping ground surfaces. Hence, the optimal depth for different values of $\phi$ is calculated by performing a parametric study, and the results are presented in Fig. 2. Accordingly, the depth of the reinforcement layer ($S_v$) is kept as $0.5B$ (for $\phi < 40°$) and $0.75B$ (for $\phi = 40°$).

## 4.2 Increment Factor ($E_f$)

The variation of increment factor ($E_f$) with various input parameters is given in Table 1. From Table 1, it can be noted that the increment factor for the footing decreases with an increase in the slope angle of the embankment ($\beta$). It can also be seen that the increment factor is greatly affected by the variation of $c$ and $\phi$. The magnitude of $E_f$ is found to increase with an increase in the value of $\phi$, whereas $E_f$ decreases with an increase in $c$.

**Table 1** Variation of $E_f$ with various input parameters

| $\Phi$ | $c/\gamma B$ | $E_f$ | | | |
|---|---|---|---|---|---|
| | | $S_L/B = 1$ | | $S_L/B = 2$ | |
| | | $\beta = 10°$ | $\beta = 20°$ | $\beta = 10°$ | $\beta = 20°$ |
| 25° | 0.5 | 1.47 | 1.40 | 1.44 | 1.41 |
| | 1.0 | 1.46 | 1.37 | 1.41 | 1.38 |
| | 2.0 | 1.43 | 1.35 | 1.39 | 1.37 |
| 30° | 0.5 | 1.54 | 1.45 | 1.53 | 1.45 |
| | 1.0 | 1.49 | 1.41 | 1.51 | 1.41 |
| | 2.0 | 1.46 | 1.38 | 1.47 | 1.39 |
| 35° | 0.5 | 1.60 | 1.50 | 1.60 | 1.47 |
| | 1.0 | 1.55 | 1.45 | 1.55 | 1.44 |
| | 2.0 | 1.50 | 1.42 | 1.51 | 1.42 |
| 40° | 0.5 | 1.98 | 1.84 | 1.96 | 1.79 |
| | 1.0 | 1.88 | 1.75 | 1.87 | 1.73 |
| | 2.0 | 1.79 | 1.67 | 1.78 | 1.66 |

**Table 2** Variation of $R_f$ for different values of $\phi$ with $\beta = 20°$, $S_L/B = 2$

| $c/\gamma B$ | $R_f$ | | |
|---|---|---|---|
| | $\phi = 25°$ | $\phi = 30°$ | $\phi = 35°$ |
| 0.5 | 0.88 | 0.77 | 0.66 |
| 1.0 | 0.87 | 0.77 | 0.66 |
| 2.0 | 0.85 | 0.77 | 0.67 |

## 4.3 Influence Factor ($R_f$)

The variation of influence factor ($R_f$) for different values of $c$ and $\phi$ with $\beta = 20°$, $S_L/B = 2$, is shown in Table 2. It can be observed from Table 2 that the magnitude of $R_f$ decreases as $\phi$ increases, i.e., the reduction in the bearing capacity becomes higher with an increase in the value of $\phi$. However, in case of cohesion, the influence factor does not get much affected by the variation of $c/\gamma B$. From Table 2, it can be noticed that the maximum decrease in $R_f$ is about 25% as $\phi$ varies from 25° to 35°.

## 4.4 Collapse Mechanism

The multi-block critical collapse mechanisms generated for different values of $\phi$ with $c/\gamma B = 1$, $S_L/B = 1$, $\beta = 20°$, and $S_v/B = 0.5$ are shown in Fig. 3. It can be seen that the extent of the failure zone increases with an increase in the value of $\phi$.

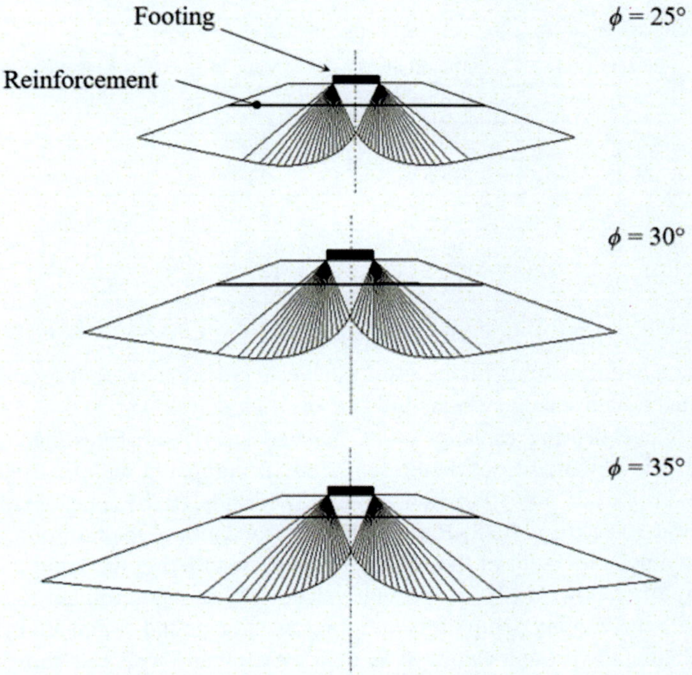


**Fig. 3** Collapse mechanisms for different values of $\phi$ with $c/\gamma B = 1$, $S_L/B = 1$, $\beta = 20°$ and $S_v/B = 0.5$

## 5 Comparison

Several classical theories [6, 14] are available for the determination of the bearing capacity factors ($N_c$ and $N_\gamma$) for a strip footing resting on soil without any reinforcement and sloping ground surface. In Table 3, the present values of $N_c$ and $N_\gamma$ are compared with the upper bound results of Soubra [13]. In Table 4, the present results are compared with the available classical theories [6, 8, 14]. The values provided by Meyerhof [6] are found to be lower than the present values, whereas the current

**Table 3** Comparison of $N_c$ and $N_\gamma$ values with Soubra [13]

| $\phi$ (°) | $N_c$ | | $N_\gamma$ | |
|---|---|---|---|---|
| | Present study | Soubra [13] | Present study | Soubra [13] |
| 20 | 14.84 | 14.86 | 4.48 | 4.49 |
| 30 | 30.15 | 30.24 | 21.45 | 21.51 |
| 40 | 75.36 | 75.77 | 119.31 | 119.84 |
| 50 | 267.20 | 270.09 | 1033.04 | 1042.48 |

**Table 4** Comparison of $N_\gamma$ values with available literature

| $\phi$ (°) | Present study | Meyerhof [6] | Vesic [14] | Michalowski [8] |
|---|---|---|---|---|
| 20 | 4.48 | 2.87 | 5.39 | 4.52 |
| 25 | 9.78 | 6.77 | 10.88 | 9.77 |
| 30 | 21.45 | 15.67 | 22.40 | 21.34 |
| 35 | 48.83 | 37.15 | 48.03 | 48.50 |
| 40 | 119.31 | 93.69 | 109.40 | 118.19 |
| 45 | 324.02 | 262.74 | 271.80 | 320.53 |

results find a better match with the results of Vesic [14]. The present results provide a closer match with the upper bound results of Michalowski [8].

The ultimate bearing capacity of an isolated strip footing resting on a reinforced soil bed obtained from the present study is compared with that reported by Michalowski [8]. For $\phi \geq 30°$, Michalowski [8] considered the cohesion of soil and the surcharge as zero and $0.25\gamma B$, respectively. The width of the reinforcement was taken four times the width of the footing. Hence, by utilizing the input parameters adopted by Michalowski [8], the present values of $E_f$ are determined for $\phi = 30°$ and 40° with the varying depth of the reinforcement layer and compared in Fig. 4. It can be seen that the present values of $E_f$ match reasonably well with those reported by Michalowski [8] for different values of $S_v/B$.

**Fig. 4** Comparison of $E_f$ with Michalowski [8] for cohesionless soil

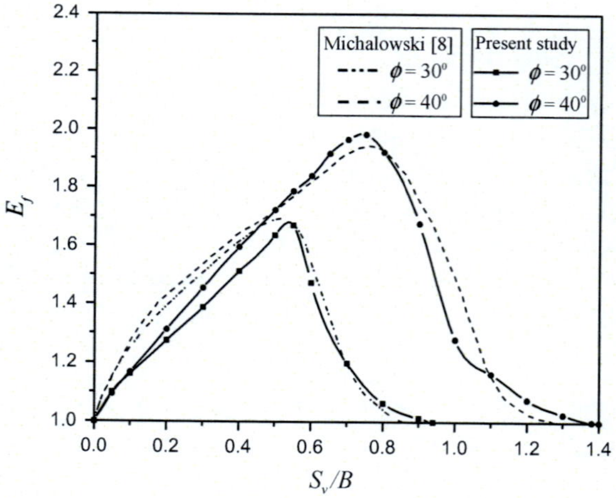

## 6 Conclusions

The ultimate bearing capacity of an isolated strip foundation resting on a reinforced embankment is investigated using the upper bound limit analysis along with a kinematically admissible multi-block failure mechanism. The results are presented in terms of influence factor ($R_f$) to represent the effect of the slope on either side of the footing and increment factor ($E_f$) to capture the effect of the reinforcement. The magnitude of $R_f$ is found to increase with an increase in $S_L$, but decrease with an increase in $\phi$. The value of $E_f$ is found to increase with an increase in $\phi$ and $S_v$. In contrast, the cohesion causes a decrement in the value of $E_f$. The value of $E_f$ is also found to decrease with an increase in $\beta$, whereas the setback distance ($S_L$) does not show any significant effect on $E_f$. The present results are found to match reasonably well with the results reported in the literature.

## References


1. Biswas, N., Ghosh, P.: Interaction of adjacent strip footing on reinforced soil using upper bound limit analysis. Geosynth. Int. **25**(6), 599–611 (2018)
2. Biswas, N., Ghosh, P.: Bearing capacity for isolated surface strip footing resting on multilayered soil bed. Indian Geotechnical Journal **49**(1), 37–49 (2019)
3. Javdanian, H.: On the behaviour of shallow foundations constructed on reinforced soil slope–a numerical analysis. Int. J. Geotech. Eng. **14**(2), 188–195 (2017)
4. Leshchinsky, B., Xie, Y.: Bearing capacity for spread footings placed near $c'$-$\phi'$ slopes. J. Geotech. Geoenviron. Eng., ASCE **143**(1), 06016020 (2017)
5. Meyerhof, G.G.: The ultimate bearing capacity of foundations on slopes. In: Proceedings, 4th International Conference on Soil Mechanics and Foundation Engineering, Vol. 3, pp. 384–386. Butterworths, London (1957)
6. Meyerhof, G.G.: Some recent research on bearing capacity of foundations. Can. Geotech. J. **1**(1), 16–26 (1963)
7. Michalowski, R.L.: Limit analysis in stability calculations of reinforced soil structures. Geotext. Geomembr. **16**, 311–331 (1998)
8. Michalowski, R.L.: Limit loads on reinforced foundation soils. J. Geotech. Geoenviron. Eng. **130**(4), 381–390 (2004)
9. Naeini, S.A., Rabe, B.K., Mahmoodi, E.: Bearing capacity and settlement of strip footing on geosynthetic reinforced clayey slopes. J. Central South Univ. **19**, 1116–1124 (2012)
10. Saran, S., Reddy, B.S.: Bearing capacity of eccentrically loaded footings adjacent to cohesionless slopes. Indian Geotech. J. **20**(2), 119–142 (1990)
11. Saran, S., Sud, V.K., Handa, S.C.: Bearing capacity of footings adjacent to slopes. J. Geotech. Eng. **115**(4), 553–573 (1989)
12. El Sawwaf, M.: Behaviour of strip footing on geogrid reinforced sand over a soft clay slope. Geotext. Geomemb. **25**, 50–60 (2007)
13. Soubra, A.H.: Upper bound solution for bearing capacity of foundations. J. Geotech. Geoenviron. Eng. **125**(1), 59–68 (1999)
14. Vesic, A.S.: Analysis of ultimate loads of shallow foundations. J. Soil Mech. Found. Div., ASCE **99**(1), 45–73 (1973)
15. Yoo, C.: Laboratory investigation of bearing capacity behaviour of strip footing on geogrid-reinforced sand slope. Geotext. Geomembr. **19**(5), 279–298 (2001)

# 3D Numerical Study of the Behavior of Piled Raft Foundation on Soft Clay with Uniform and Varying Pile Lengths

Rajib Modak, Debjit Bhowmik, and Baleshwar Singh

## 1 Introduction

In the past few decades, piled raft foundation has widely been used as a foundation option as it has proven to be an economic and efficient foundation type [8, 17]. In this type of foundation, the load is shared among the structural elements, i.e., piles and the raft. Unlike the traditional foundation design where the loads are assumed to be either carried by the raft or by the piles, the design of the piled raft foundation involves significant contribution of both the raft and the piles in load bearing [3, 12]. Taking into account of the relative proportion of load carried by each structural element, different design philosophies were suggested by many researchers [14–17, 20]. The piled raft foundation consists of three elements; the piles, raft and the soil. The design of such foundation system largely depends on the pile–soil–raft interactions. From the literature, many reports can be found where the various soil–structure interactions involved in the load sharing mechanism of piled raft foundation were evaluated to understand the behavior of the foundation system [5, 6, 17, 19, 23]. The use of piled raft foundation in soft soils is very much limited mainly due to the low bearing capacities and foundations undergoing settlements beyond the permissible limits [17, 18]. In spite of these concerns, successful attempts of piled raft foundations on soft clays have been reported in the literature [18, 27].

R. Modak (✉) · B. Singh
Department of Civil Engineering, Indian Institute of Technology Guwahati, Guwahati 781039, India
e-mail: rmodak@iitg.ac.in

B. Singh
e-mail: baleshwar@iitg.ac.in

D. Bhowmik
Department of Civil Engineering, NIT Silchar, Silchar 788010, India



C. N. V. Satyanarayana Reddy et al. (eds.), *Ground Characterization and Foundations*, Lecture Notes in Civil Engineering 167,
https://doi.org/10.1007/978-981-16-3383-6_49

In the conventional design of a piled raft foundation, pile group with uniform pile lengths are provided underneath the raft. However, in such cases, it is found that the peripheral piles carry majority of the design loads [11, 25], resulting in non-uniform distribution of loads among the piles for a rigid raft, and in case of a flexible raft, central pile undergoes more settlements compared to peripheral piles [9, 20] resulting in higher differential settlements.

From the literature, it can be observed that most of the reports on piled raft foundations focused on providing piles with similar length underneath the raft [5, 6, 9, 19, 20, 22, 24]. Few literatures were devoted to investigate the behavior of piled raft foundation with varying pile length [14, 21, 26]. Therefore, the effect of varying the lengths of piles distributed underneath the flexible raft with different pile spacing on the ultimate load capacity, load sharing, average and differential settlement of the foundation system needs to be investigated. The results obtained from this study can be used for an effective piled raft foundation design on soft clays.

## 2 Numerical Model

Three dimensional (3D) numerical model of piled raft foundation was developed using a general purpose FEM-based software *ABAQUS*. The model consisted of soil continuum, foundation elements geometry, interface element and load 150 kPa uniformly distributed on raft. Figure 1 shows the numerical model used for the parametric study. To model the soil and the foundation elements, eight-node hexahedral brick elements were used. The soil mass was considered to be isotropic, homogenous and elasto-plastic in nature. Properties of the soil were simulated in the soil block through the "Mohr–Coulomb" model where failure or yielding is considered to take place when the shear stress mobilized in any plane is equal to the soil shear strength. For modeling the soil, parameters like modulus of elasticity, cohesion,

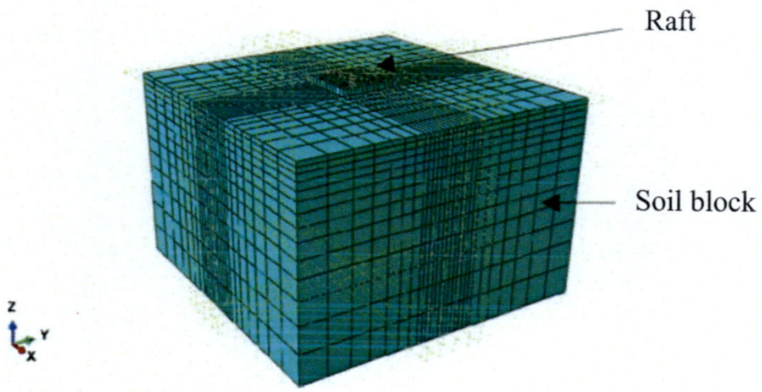


**Fig. 1** Finite element mesh of piled raft model used in the parametric study

angle of internal friction and Poisson's ratio were used. The raft and the piles material were considered to be linear elastic as their Young's modulus is higher than the soil Young's modulus. Rigid connection between the piles and raft was considered. To stimulate the interactions between the foundation elements and the soil, the master–slave surface option was used. The soil was considered as the slave surface, whereas the piles and the raft were treated as master surface [24]. The soil-structure contact was considered perfectly rough with no relative displacement between structural elements nodes and soil nodes in contact with it [21]. The size of the soil block is 32.5 x 32.5 x 29m (Length x Breadth x Height) *ABAQUS* was evaluated by taking into consideration of the soil at the boundary is not likely to be affected by appreciable amount of stresses and strains generated due to the load on the foundation. Laterally, the soil mass was considered five times the foundation width ($B_r$) and in the vertical direction it was taken as $2B_r$ from the tip of the pile (longest pile, in case of varying pile lengths). For soil in the proximity of the foundation, finer mesh was generated that gradually transformed to a coarser mesh at the boundary. The boundary conditions were given such that the movement of bottom portion of the model was restrained in all directions and no rotation. The side faces were restrained in horizontal directions and allowed movement in z-direction [14].

## 3 Parametric Study

The ultimate load-carrying capacity, load sharing, average and differential settlements of piled raft foundation founded on soft clay under undrained condition were evaluated for both uniform and varying length piles underneath the raft at different pile spacing. For modeling the piled raft foundation on soft clay, the soil, raft and pile properties used are shown in Table 1. The basic parameters used for the design of piled raft are given in Table 2. Raft and pile modulus of elasticity values were taken from [29]. The raft thickness was back calculated corresponding to raft soil stiffness ratio ($K_r$) value 17.8, where $K_r = 5.57E_r(1 - {\mu_s}^2){B_r}^{0.5}{t_r}^3/E_s(1-{\mu_r}^2){L_r}^{0.5}{L_r}^3$ for a relatively flexible raft lies in the range of 0.001 (fully flexible) to 1000 (rigid raft)

**Table 1** Material properties used in the parametric study

| Materials | Properties | Values |
|---|---|---|
| Soil | Young's modulus, $E_s$ (MPa) | 10 |
| | Poisson's Ratio, $\mu_s$ | 0.45 |
| | Bulk unit weight, $\gamma_b$ (kN/m$^3$) | 16 |
| | Undrained cohesion, $c_u$ (kPa) | 20 (soft clay) |
| | Angle of internal friction, ø (°) | 0 |
| Raft | Young's modulus, $E_r$ (GPa) | 25 |
| | Poisson's Ratio, $\mu_r$ | 0.2 |
| Pile | Young's modulus, $E_p$ (GPa) | 25 |
| | Poisson's Ratio, $\mu_p$ | 0.2 |

**Table 2** Piled raft geometrical configurations

| Parameters | Value |
|---|---|
| Raft width ($B_r$) | 6.5 m |
| Raft width ($L_r$) | 6.5 m |
| Raft thickness ($t_r$) | 0.75 m |
| Pile length ($L_P$) | 8 m[a] |
| Pile diameter ($d_p$) | 0.45 m |
| Pile spacing ($s_P$) | 4, 6 |
| Number of piles | 9 |

[a]Standard value for uniform length piled raft foundation

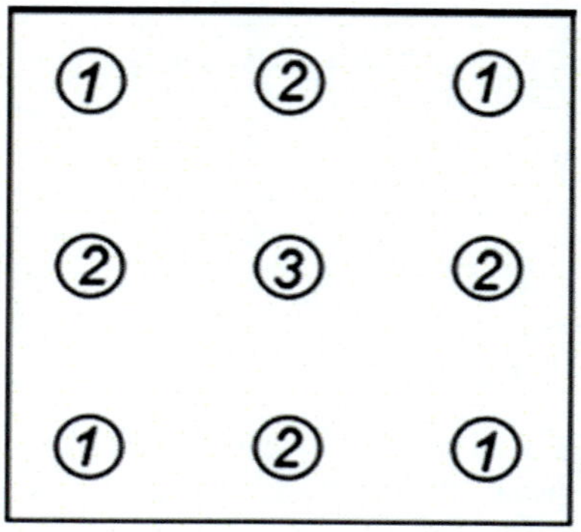


**Fig. 2** Plan of piled raft foundation

[10]. All the piled raft foundation models were subjected to a uniformly distributed load of 150 kPa. Total nine piles were arranged in a square pattern (3 rows × 3 columns) with a clear distance of 0.5 m from the raft edge to the outermost pile [13]. The piles in the group were termed based on their locations as corner piles labeled '1,' edge piles '2' and an interior pile '3' as shown in Fig. 2. Based on those locations, the pile lengths were varied while designing piled raft foundations having varying length of piles. The total pile length for all the piled raft foundation models with varying pile length was kept equal to the total pile length as in case of the piled raft foundation model with uniform length piles, i.e., $9L_P$ ($L_P = 8$ m). Based on the length of pile varied across the pile group, six different piled raft configurations were adopted as shown in Table 3, Type A, B, C, D, E and F along with the piled raft foundation with uniform pile length, Type U. For each piled raft configuration, four pile length arrangements ($A_1$, $A_2$, $A_3$ and $A_4$) were taken into consideration. The lengths of piles used for different piled raft configurations are shown in Table 4.

These 24 piled raft foundation models with varying lengths of piles along with the piled raft foundation model with uniform length piles (Table 4) were analyzed with pile spacing ($s_P$) to pile diameter ($d_P$) ratio equal to 4 and 6, i.e., $s_P/d_P$ ratio of 4 and 6 [29]. The ultimate load-carrying capacity values for all the piled raft foundation models were evaluated by using double tangent method [28] for the respective load

**Table 3** Different Piled raft configurations adopted based on pile length varied across the pile group

| Type | Center pile | Edge pile | Corner pile |
|---|---|---|---|
| A | Shortest | Intermediate | Longest |
| B | Shortest | Longest | Intermediate |
| C | Longest | Intermediate | Shortest |
| D | Longest | Shortest | Intermediate |
| E | Intermediate | Longest | Shortest |
| F | Intermediate | Shortest | Longest |
| U | Uniform | Uniform | Uniform |

**Table 4** Pile lengths based on the location of piles used for different piled raft configurations

| Type | $A_1$ | $A_2$ | $A_3$ | $A_4$ |
|---|---|---|---|---|
| A | 1: 1.25 $L_P$<br>2: 0.875 $L_P$<br>3: 0.85 $L_P$ | 1: 1.3125 $L_P$<br>2: 0.8125 $L_P$<br>3: 0.5 $L_P$ | 1: 1.25 $L_P$<br>2: 0.8125 $L_P$<br>3: 0.75 $L_P$ | 1: 1.125 $L_P$<br>2: 0.9375 $L_P$<br>3: 0.75 $L_P$ |
| B | 1: 1.25 $L_P$<br>2: 0.85 $L_P$<br>3: 0.875 $L_P$ | 1: 1.3125 $L_P$<br>2: 0.5 $L_P$<br>3: 0.8125 $L_P$ | 1: 1.25 $L_P$<br>2: 0.75 $L_P$<br>3: 0.8125 $L_P$ | 1: 1.125 $L_P$<br>2: 0.75 $L_P$<br>3: 0.9375 $L_P$ |
| C | 1: 0.8125 $L_P$<br>2: 1.125 $L_P$<br>3: 1.25 $L_P$ | 1: 0.875 $L_P$<br>2: $L_P$<br>3: 1.5 $L_P$ | 1: 0.8125 $L_P$<br>2: $L_P$<br>3: 1.75 $L_P$ | 1: 0.75 $L_P$<br>2: $L_P$<br>3: 2 $L_P$ |
| D | 1: 1.125 $L_P$<br>2: 0.8125 $L_P$<br>3: 1.25 $L_P$ | 1: $L_P$<br>2: 0.875 $L_P$<br>3: 1.5 $L_P$ | 1: $L_P$<br>2: 0.8125 $L_P$<br>3: 1.75 $L_P$ | 1: $L_P$<br>2: 0.75 $L_P$<br>3: 2 $L_P$ |
| E | 1: 0.5 $L_P$<br>2: 1.5 $L_P$<br>3: $L_P$ | 1: 0.75 $L_P$<br>2: 1.25 $L_P$<br>3: $L_P$ | 1: 0.875 $L_P$<br>2: 1.125 $L_P$<br>3: $L_P$ | 1: 0.8125 $L_P$<br>2: 1.15625 $L_P$<br>3: 1.125 $L_P$ |
| F | 1: 1.5 $L_P$<br>2: 0.5 $L_P$<br>3: $L_P$ | 1:1.25 $L_P$<br>2: 0.75 $L_P$<br>3: $L_P$ | 1: 1.125 $L_P$<br>2: 0.875 $L_P$<br>3: $L_P$ | 1: 1.15625 $L_P$<br>2: 0.8125 $L_P$<br>3: 1.125 $L_P$ |
| U | 1: $L_P$<br>2: $L_P$<br>3: $L_P$ | | | |

settlements curves obtained after the analysis. The average settlement ($s_{avg}$) of the piled raft foundation was evaluated using Eq. 1 [21], and the differential settlement ($s_{diff}$) was calculated using Eq. 2 [2].

$$s_{avg} = (2s_{centre} + s_{corner})/3 \tag{1}$$

$$s_{diff} = (s_{centre} - s_{corner}) \tag{2}$$

where $s_{\text{center}}$ is the settlement at the raft center and $s_{\text{corner}}$ is the settlement at the raft corner. The load sharing between the structural elements can be represented in term of a coefficient $\alpha_{pr}$ known as pile raft coefficient [7]. $\alpha_{pr}$ can be defined as the ratio of total design load carried by the piles ($Q_g$) to the load applied on the foundation ($Q$).

$$\alpha_{pr} = Q_g / Q \tag{3}$$

$\alpha_{pr} = 0$ represents a foundation with no piles, i.e., a shallow foundation; $\alpha_{pr} = 1$ represents a fully piled foundation, with the raft having no contact with the ground; and in the range of $0 < \alpha_{pr} < 1$ represents the case of piled raft foundation.

# 4 Results and Discussion

## *4.1 Ultimate Load-Carrying Capacity of Piled Raft Foundation*

This section discusses the effect of pile spacing and varying the pile lengths across the pile group on the ultimate load-carrying capacities of all the piled raft foundation models. The ultimate load-carrying capacities of all the models with $s_P/d_P$ ratio of 4 and 6 are in Figs. 3 and 4, respectively. It can be noted that the load-carrying capacity decreased as the pile spacing increased for all the piled raft foundation models. Similar results were observed previously by [4, 24]. This decrease in capacity of the piled raft system can be attributed to the reduction in the pile group effect with increased pile spacing. However, it should be noted that in order to make up for the reduction in the capacity of the piled raft system, the raft contribution increases with increased pile spacing.

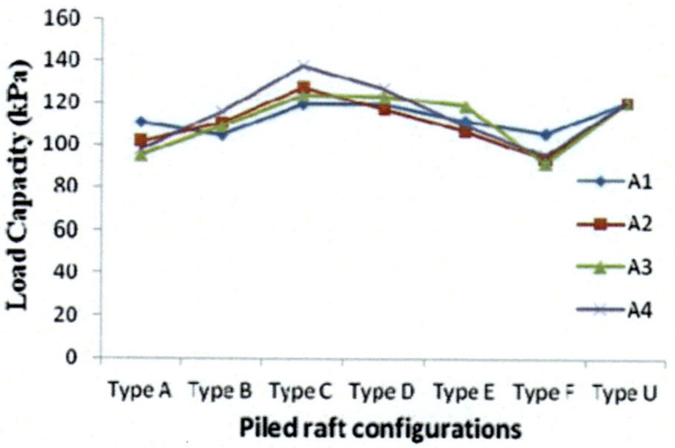


**Fig. 3** Ultimate load-carrying capacities of piled raft foundation models ($s_P/d_P$ ratio 4)

**Fig. 4** Ultimate load-carrying capacities of piled raft foundation models ($s_P/d_P$ ratio 6)

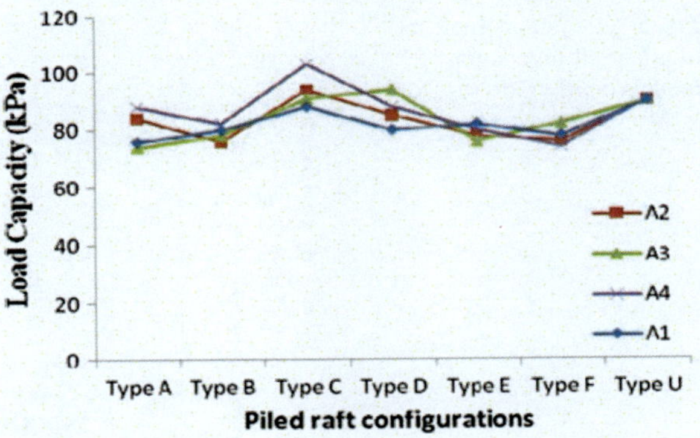


It can also be seen that for both $s_P/d_P$ ratio of 4 and 6, in comparison to the piled raft foundation with uniform length piles (Type U), the piled raft configuration with longest pile at the center, intermediate length piles at the edges and shortest pile at the corners (Type C) with A4 arrangement showed the highest improvement in ultimate load carrying. This is in accordance with the practical implication of the piled raft foundation design of Messeturm Tower in Frankfurt [17], where longer length piles were provided at the central region of the raft and pile lengths gradually decreased toward the raft edges. For both the pile spacing considered, Type A and Type F showed relatively lesser values of ultimate load-carrying capacities in comparison to other piled raft configuration models.

## 4.2 *Load Sharing of Piled Raft Foundation*

For the safe and economic design of piled raft foundation, it is very essential to determine the relative proportion of the design load carried by each structural component, i.e., the raft and the pile group [17]. Pile raft coefficient ($\alpha_{pr}$) indicates the portion of load carried by the piles in piled raft foundation system. The remaining load is considered to be carried by the raft in contact with the soil underneath. The total load carried by the piles is evaluated by determining the summation of the loads on different pile heads. From Figs. 5 and 6, it can be seen that with the increase in pile spacing, $\alpha_{pr}$ decreases, i.e., the proportion of total load carried by the raft increases. This is due to the fact that with the increase in pile spacing, the raft–soil contact stresses increases, thus, rafts contribution in load sharing increases [24]. The results are similar to those observed by Bandyopadhyay et al. [1]. For both $s_P/d_P$ ratio of 4 and 6, $\alpha_{pr}$ was observed to be higher in case of Type A and Type F configuration piled raft models in comparison to Type U configuration model. On the other hand, Type C configuration piled raft models showed the lowest $\alpha_{pr}$ values compared to other piled raft foundation models, indicating more contribution of raft in load sharing.

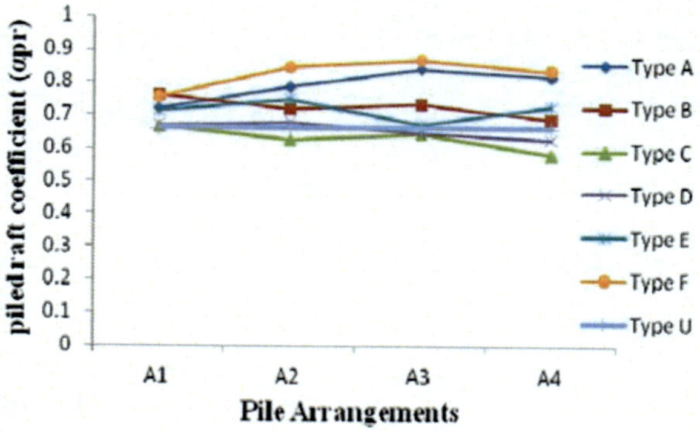


**Fig. 5** Pile raft coefficients of piled raft foundation models ($s_P/d_P$ ratio 4)

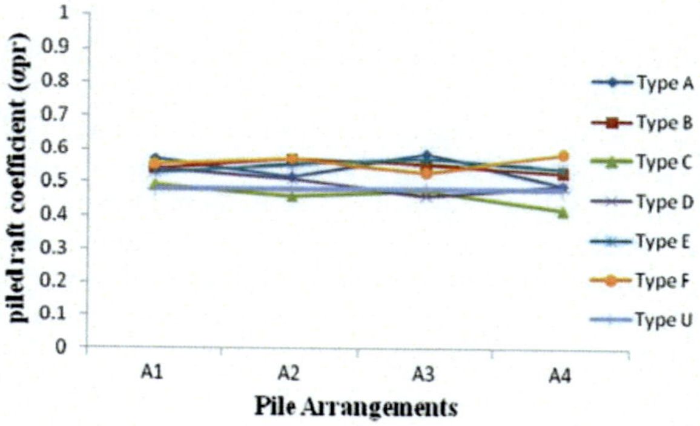


**Fig. 6** Pile raft coefficients of piled raft foundation models ($s_P/d_P$ ratio 6)

## 4.3 Average Settlements

The average settlement of all the piled raft foundation models with different pile spacing is shown in Figs. 7 and 8. It is evident from the figures that the average settlement reduced with increased pile spacing. Similar results were encountered by [2, 22]. This is due to the fact that with the increase in pile spacing, the area covered

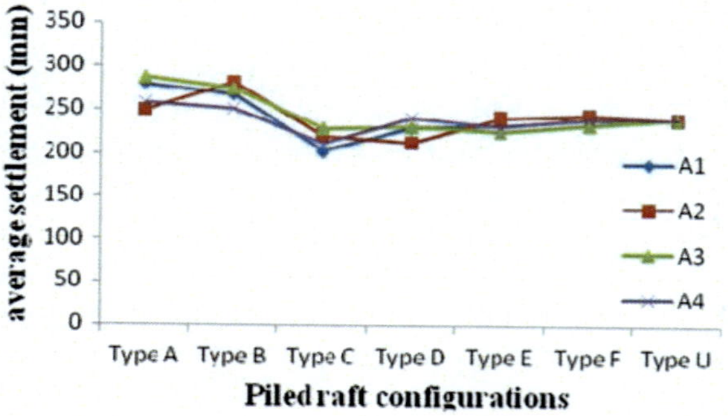


**Fig. 7** Average settlements of piled raft foundation models ($s_P/d_P$ ratio 4)

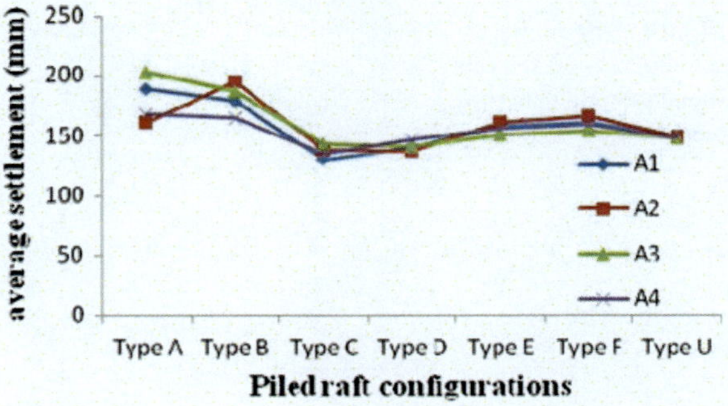


**Fig. 8** Average settlements of piled raft foundation models ($s_P/d_P$ ratio 6)

by the piles underneath the raft increases and the load gets uniformly distributed, resulting in decrease in average settlements. Type A and Type B configuration piled raft models shows higher values of average settlements, whereas Type C piled raft foundation models show lower average settlement values compared to Type U piled raft foundations.

## 4.4 *Differential Settlements*

Figures 9 and 10 show the effect of pile spacing and varying the pile lengths across the pile group on the differential settlements of all the piled raft foundation models. It can be seen that with the decrease in pile spacing, the differential settlement decreases. Similar results were observed by [9, 14, 20]. This may be due to increase in central raft sagging, having lesser pile support with increased pile spacing [20]. It can also be seen that the differential settlements were lowest in Type C, A4 arrangement piled raft model in comparison to Type U piled raft model.

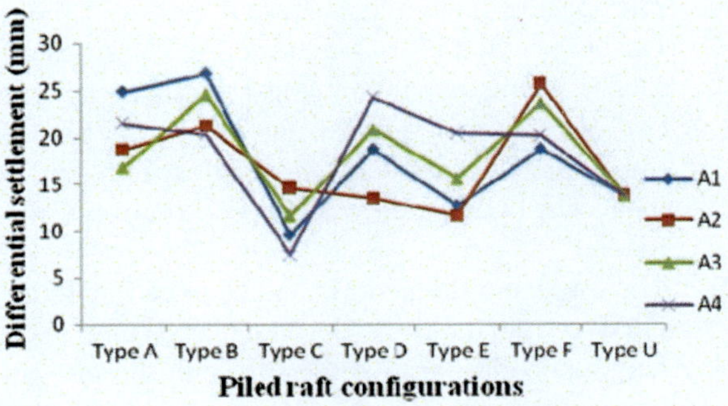


**Fig. 9** Differential settlements of piled raft foundation models ($s_P/d_P$ ratio 4)

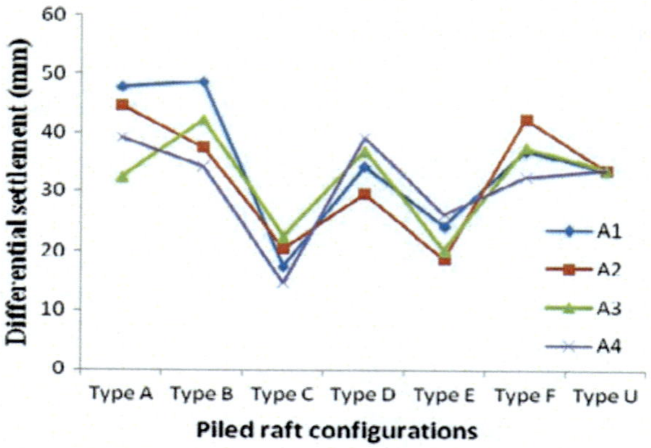


**Fig. 10** Differential settlements of piled raft foundation models ($s_P/d_P$ ratio 6)

## 5 Conclusions

The conclusions that can be drawn from the 3D finite element analysis carried out on piled raft foundations founded on soft clay are as follows:

1. The ultimate load-carrying capacity of piled raft foundation system for any piled raft configuration with uniform and varying length of piles decreases with the increase in pile spacing. This is due to the decrease in pile–pile interaction effects in the pile group with the increased pile spacing.
2. Reduction in average settlement and increment in differential settlement with increase in pile spacing were observed for all piled raft configurations. The decrease in average settlement is due to the uniform distribution of piles covering a larger area underneath the raft, whereas the increase in the differential settlement can be attributed to lesser pile support than required to resist sagging at the central portion of the raft with increased pile spacing.
3. Proportion of the design load carried by the piles decreases with the increase in pile spacing due to the increase in raft–soil contact pressure.
4. For both $s_P/d_P$ ratios of 4 and 6, Type C, A4 arrangement piled raft foundation model, having maximum length of pile at the center ($2L_P$), intermediate length ($L_P$) at edge and shortest pile length ($0.75L_P$) at corner showed the highest improvement in the load-carrying capacity and decrease in differential settlement when compared with Type U, uniform pile length piled raft foundation model.
5. The average settlement was noted to be minimum for Type C, A1 arrangement piled raft foundation model with longest pile ($1.25L_P$) at the center, intermediate length pile ($1.125L_P$) at the edge and shortest pile ($0.8125L_P$) at the corner in comparison with Type U, uniform pile length piled raft foundation.
6. Hence, in soft clay instead of providing a conventional piled raft foundation with uniform length piles underneath the raft, a piled raft foundation with Type C pile raft configuration can be suggested to get better performance with the same volume of pile material.

## References


1. Bandyopadhyay, S., Sengupta, A., Parulekar, Y.M.: Behavior of a combined piled raft foundation in a multi-layered soil subjected to vertical loading. Geomech. Eng. **21**(4), 379–390 (2020)
2. Cho, J., Lee, J.H., Jeong, S., Lee, J.: The settlement behavior of piled raft in clay soils. Ocean Eng. **53**, 153–163 (2012)
3. Choudhury, D., Katzenbach, R.: ISSMGE Combined Pile-Raft Foundation Guideline. Technische Universität Darmstadt, Germany, Institute and Laboratoty of Geotechnics (2013)
4. Chow, Y.K., Yong, K.Y., Shen, W.Y.: Analysis of piled raft foundations using a variational approach. Int. J. Geomech. **1**(2), 129–147 (2001)
5. Clancy, P., Randolph, M.F.: An approximate analysis procedure for piled raft foundations. Int. J. Numer. Anal. Meth. Geomech. **17**(12), 849–869 (1993)
6. Clancy, P., Randolph, M.F.: Simple design tools for piled raft foundations. Geotechnique **46**(2), 313–328 (1996)
7. de Sanctis, L., Mandolini, A.: Bearing capacity of piled rafts on soft clay soils. J. Geotech. Geoenviron. Eng. **132**(12), 1600–1610 (2006)
8. El-Mossallamy, Y.: Innovative application of piled raft foundation in stiff and soft subsoil. In: Deep Foundations 2002: An International Perspective on Theory, Design, Construction, and Performance, pp. 426–440.
9. Horikoshi, K., Randolph, M.F.: A contribution to optimum design of piled rafts. Geotechnique **48**(3), 301–317 (1998)
10. Horikoshi, K., & Randolph, M.F.: On the definition of raft—soil stiffness ratio for rectangular rafts. Géotechnique **47**(5), 1055–1061 (1997)
11. Katzenbach, R., Leppla, S., Choudhury, D.: Foundation Systems for High-Rise Structures. CRC Press (2016)
12. Katzenbach, R., Moormann, C.: Recommendations for the design and construction of piled rafts. In: International Conference on Soil Mechanics and Geotechnical Engineering, pp. 927–930 (2001)
13. Magade, S.B., Ingle, R.K.: Influence of clear edge distance and spacing of piles on failure of pile cap. Iran. J. Sci. Technol. Trans. Civil Eng. **44**(4), 1265–1281 (2020)
14. Mali, S., Singh, B.: 3D numerical modeling of large piled-raft foundation on clayey soils for different loadings and pile-raft configurations. Studia Geotechnica et Mechanica **42**(1), 1–17 (2020)
15. Mandolini, A., Di Laora, R., Mascarucci, Y.: Rational design of piled raft. Procedia Eng. **57**, 45–52 (2013)
16. O'Brien, A.S., Burland, J.B., Chapman, T.: Chapter 56 Rafts and piled rafts. In *ICE manual of geotechnical engineering* (pp. 853–886). Thomas Telford Ltd.
17. Poulos, H.G.: Piled raft foundations: design and applications. Geotechnique **51**(2), 95–113 (2001)
18. Poulos, H.G.: Piled raft and compensated piled raft foundations for soft soil sites. In: Advances in Designing and Testing Deep Foundations: In Memory of Michael W. O'Neill, pp. 214–235 (2005)
19. Randolph, M.F.: Design of piled raft foundations. In: Proceedings of the International Symposium on Recent Developments in Laboratory and Field Tests and Analysis of Geotechnical Problems, Bangkok, pp. 525–537 (1983)
20. Randolph, M.F.: Design methods for pile groups and piled rafts. In: International Conference on Soil Mechanics and Foundation Engineering, pp. 61–82 (1994)
21. Reul, O., Randolph, M.F.: Design strategies for piled rafts subjected to nonuniform vertical loading. J. Geotech. Geoenviron. Eng. **130**(1), 1–13 (2004)
22. Rodríguez, E., Cunha, R.P., Caicedo, B.: Behaviour of piled raft foundation systems in soft soil with consolidation process. In: Proceedings 9th International Conference on Physical Modelling in Geotechnics, London (2018)

23. Russo, G., Viggiani, C.: Factors controlling soil-structure interaction for piled rafts. Darmstadt Geotechnics. Darmstadt Univ. Technol. **4**, 297–322 (1998)
24. Sinha, A., Hanna, A.M.: 3D numerical model for piled raft foundation. Int. J. Geomech. **17**(2), 04016055 (2017)
25. Ta, L.D., Small, J.C.: An approximation for analysis of raft and piled raft foundations. Comput. Geotech. **20**(2), 105–123 (1997)
26. Tan, Y.C., Chow, C.M., Gue, S.S.: Piled raft with different pile length for medium-rise buildings on very soft clay. In: Proceedings of the International Conference on Soil Mechanics and Geotechnical Engineering, vol. 16, no. 4, p. 2045. AA Balkema Publishers (2005)
27. Tan, Y.C., Cheah, S.W., Taha, M.R.: Methodology for design of piled raft for 5-story buildings on very soft clay. In: Foundation Analysis and Design: Innovative Methods, pp. 226–233) (2006)
28. Vesic, A.S.: Analysis of ultimate loads of shallow foundations. J. Soil Mech. Foun. Div. **99**(sm1) (1973)
29. Viggiani, C.: Analysis and design of piled foundations. In: Proceedings, pp. 47–75

# Analysis of Decomposed Components of Raft and Piles of Piled-Raft Foundation in Sandy Soil

**Tusshar Sharma and Baleshwar Singh**

## 1 Introduction

Pile-raft foundation is a combination of shallow foundation (raft or roof) and deep foundation (pile group). In this type of foundation, the role of the raft is to provide the required bearing capacity, and the pile is mainly used as a settlement reducing agent, but it can also increase the bearing capacity. Over the past three decades, the use of piled-raft foundations to support different types of structures has greatly increased. Compared with other alternatives, this trend of using piled-raft foundations can be attributed to the potential economic advantages of such foundations. In addition, the piled-raft foundation can meet the most important design requirements at a lower cost, which makes the piled-raft foundation more preferred than other types of foundations. Moyes et al. [1] reported that piled-raft foundations satisfy the require serviceability performance while providing cost savings estimated to about 30% compared with conventional piled foundations systems. The pile with both skin friction and base resistance offers a high resistance to settlement. Hence, a group of piles underneath the raft can significantly reduce the settlements in piled-raft foundation. In general, the raft alone can provide the required bearing capacity but it cannot control the settlement. Therefore, the piles are crucial to reduce the settlement of the raft. Due to combining raft and piles in one system, piled-raft foundations are regarded as very complex systems. The complexity of this type of foundations is caused by the presence of many interaction factors involved in the system such as pile-to-pile, pile-to-raft, raft-to-raft, and pile-to-soil interactions. Earlier, due to the lack of analytical solutions for determining the load distribution between the pile and the

T. Sharma (✉) · B. Singh
Department of Civil Engineering, Indian Institute of Technology Guwahati, Guwahati 781039, India

B. Singh
e-mail: baleshwar@iitg.ac.in

 
C. N. V. Satyanarayana Reddy et al. (eds.), *Ground Characterization and Foundations*, Lecture Notes in Civil Engineering 167,
https://doi.org/10.1007/978-981-16-3383-6_50

pile cap (connecting the pile as a group), the pile cap was not considered to bear any load. The pile is considered to be able to withstand all loads, thus providing additional safety design. But in fact, the raft/pile cap also bears some loads. This leads to an uneconomical design, resulting in the need for additional piles. This type of foundation considers both piles and rafts as load-bearing members. This method greatly reduces the need for piles and makes the design very economical.

## 2 Numerical Modeling

In this section, the main characteristics of the finite element model (FEM) used here are summarized. All numerical analyses were performed using the commercial code PLAXIS 3D (2012 version).

### 2.1 Geometric Configuration

The piled-raft modeled in the finite element (FE) analysis consists of a square raft with a width ($B_r$) of 15 m and piles of different structures. The diameter ($B_p$) and length ($L_p$) of all piles are 0.6 m and 20 m, respectively. For the configuration of the piles, three different pile spacings ($S_p$) were considered: $3B_p$, $5B_p$, and $7B_p$. For unpiled rafts, the size of single piles and piles is the same as the size used for piled-rafts. Assuming that the pile, unpiled raft, and piled-raft are all linear elastic materials, their elastic modulus and Poisson's ratio are equal to 30 GPa and 0.15, respectively. Please note that the stiffness conditions of rafts and piles are very close to the stiffness conditions compared to soil. The areas selected for the geometry are all 120 × 120 m in the horizontal direction, and consider the soil depth of 80 m (vertical) underground (Fig. 1). The grid model of the piled raft is shown in Fig. 2.

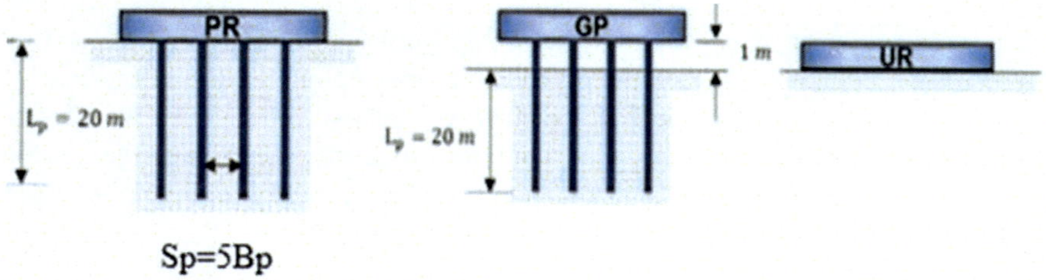


**Fig. 1** Types of foundations considered in FE analyses: piled-raft, pile group, and raft foundation

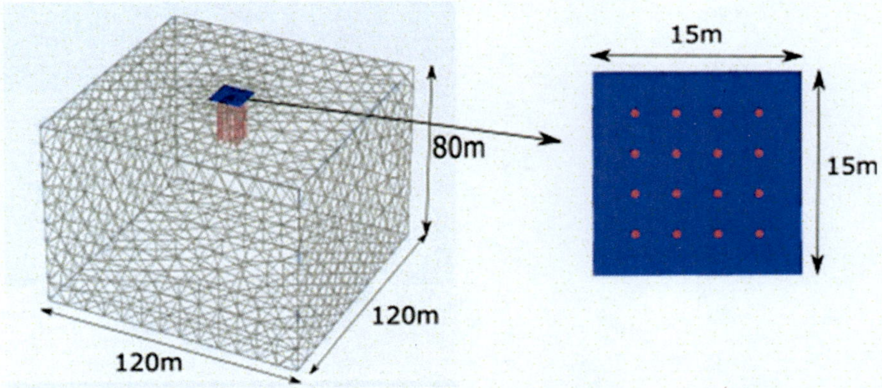


**Fig. 2** Finite element model for a piled-raft

**Table 1** Material properties used in the validation

| Material | Properties | Unit | Value |
|---|---|---|---|
| Soil | Young's modulus, $E_s$ | kPa | 37,108 |
| | Poisson's ratio, $\nu_s$ | – | 0.25 |
| | Unit weight, $\gamma$ | kN/m | 15.5 |
| | Friction angle, $\phi$ | ° | 35 |
| Raft | Young's modulus, $E_r$ | GPa | 30 |
| | Poisson's ratio, $\nu_p$ | – | 0.15 |
| Pile | Young's modulus, $E_p$ | GPa | 30 |
| | Poisson's ratio, $\nu_p$ | – | 0.15 |

## 3 Model Validation

The current numerical model was tested using a 2 m thick raft problem (size 15 m 15 m) with 16 piles with a diameter of 0.6 m and a pile length of 20 m [2]. The purpose of this validation is to ensure accurate finite element modeling for this study. The material properties of soil, raft, and piles are shown in Table 1. The results of this study are consistent with the results presented (Fig. 3).

## 4 Parametric Study

The settlement and load-sharing behavior of the piled-raft foundation have been studied by varying the raft thickness ($t$), raft embedment depth ($m$), and pile spacing ($S_p$). The different piled-raft configurations used in the study are listed in Table 2.

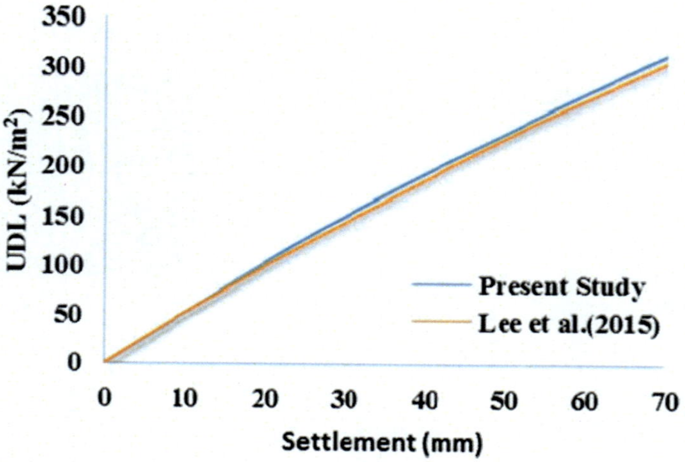


**Fig. 3** Comparison of load–settlement curves of piled-raft

**Table 2** Foundation parameters considered in the study

| Type | Size (m) | Pile configuration | Pile spacing(Sp) |
|---|---|---|---|
| Unpiled raft (UR) | $B_r = 15$, $t = 0.5, 1, 2$ | – | – |
| Group pile (GP) | $B_p = 0.6$<br>$L_p = 20{,}15{,}10$<br>$B_r = 15$, $t = 0.5, 1, 2$ | 2 × 2 (4 piles)<br>3 × 3 (9 piles)<br>4 × 4 (16 piles) | $3B_p$, $5B_p$, $7B_p$<br>$3B_p$, $5B_p$, $7B_p$<br>$3B_p$, $5B_p$, $7B_p$ |
| Piled-raft (PR) | $B_p = 0.6$<br>$L_p = 20{,}15{,}10$<br>$B_r = 15$, $t = 0.5, 1, 2$ | 2 × 2 (4 piles)<br>3 × 3 (9 piles)<br>4 × 4 (16 piles) | $3B_p$, $5B_p$, $7B_p$<br>$3B_p$, $5B_p$, $7B_p$<br>$3B_p$, $5B_p$, $7\ B_p$ |

**Table 3** Soil properties for different relative densities

| Dr (%) | $e$ | $\gamma$ (kN/m$^3$) | $E$ (kPa) | Φ° |
|---|---|---|---|---|
| 30 | 0.69 | 15.5 | 32,508 | 32 |
| 50 | 0.63 | 16.1 | 37,108 | 35 |
| 70 | 0.57 | 16.7 | 42,261 | 38 |

The properties of the soil for different relative densities and study are summarized in Table 3.

## 5 Results and Discussion

In this chapter, various interactions and pile behavior will be discussed. The vertical load pile-raft with various pile configurations and spacings is analyzed, and the results are discussed here. The load is converted into a uniformly distributed strip load, and the load is gradually applied to determine the level of settlement.

## 5.1 Load–Settlements Curve

The pile-raft load–settlement curves of all pile configurations are studied. In the load-deformation curve, the raft and pile are decomposed to closely analyze their behavior, and how much each member increases as the load increases. According to finite element analysis, the load–settlement curves are of unpiled raft (UR), group pile (GP), and piling raft (PR). Figure 4 shows that the bearing capacity of piled-rafts is higher than that of unpiled rafts and group piles, which is reasonable and can be expected to a certain extent. In addition, under a certain settlement level, the carrying capacity of GP is higher than that of UR. Above this upper limit, the carrying capacity of UR will be higher than that of GP. This result is also reasonable, considering that the load-carrying capacity of the piles is mobilized earlier than that of the raft due to smaller settlement level. The load–settlement curve of piled-rafts represents a combination of the load responses of both raft and piles and thus can be decomposed into raft (Rpr) and pile (Ppr) components, as plotted in Fig. 5. The load–settlement curve of piled-raft represents the load response combination of piled-raft and pile body, so it can be decomposed into raft (Rpr) and pile (Ppr) components, as shown in Fig. 5. As shown in Fig. 5, the bearing capacity of the pile is earlier than that of the raft, showing a higher bearing rate. With the further increase of the settlement, the load carried by the raft continues to increase, and the ratio of the bearing capacity is higher than that of the pile. For other cases with different DR and Sp values, similar results and trends were observed. Due to the interaction between rafts and piles,

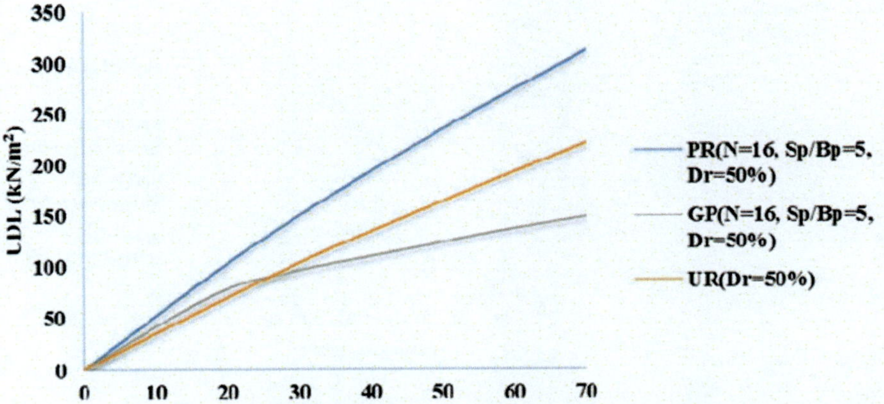


**Fig. 4** Comparison of load–settlement curves of piled-raft, unpiled raft, and group piles

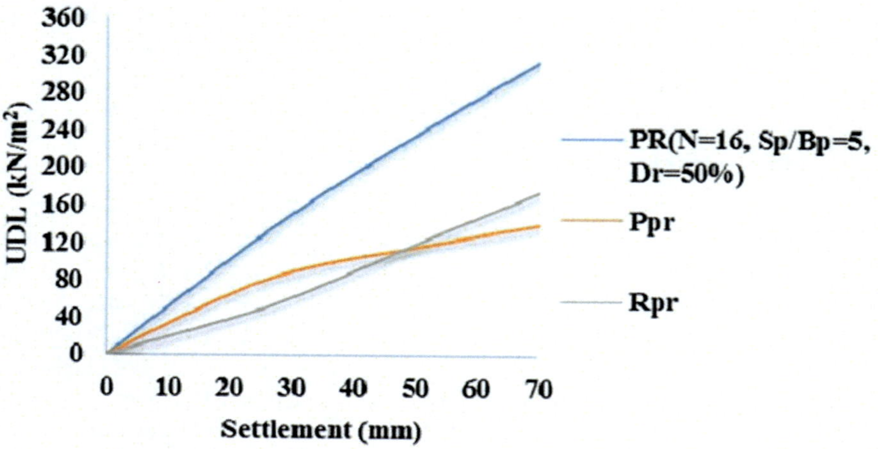


**Fig. 5** Comparison of load–settlement curves of decomposed foundation components of raft and pile

the bearing capacity of piles and piles in piled-rafts is different from unpiled rafts and group piles. Figure 6 compares the decomposed load–settlement curve of the raft (Rpr) and pile components (Ppr) with the load–settlement curve of the unrafted raft (UR) and group pile (GP). For different DR and Sp, Fig. 7 shows that under all soil and foundation conditions, Rpr's carrying capacity is less than UR. As the pile spacing increases, the load-bearing capacity of Rpr decreases, which indicates that the pile bears a higher proportion of the load.

**Fig. 6** Decomposed load–settlement curves of Rpr compared with UR

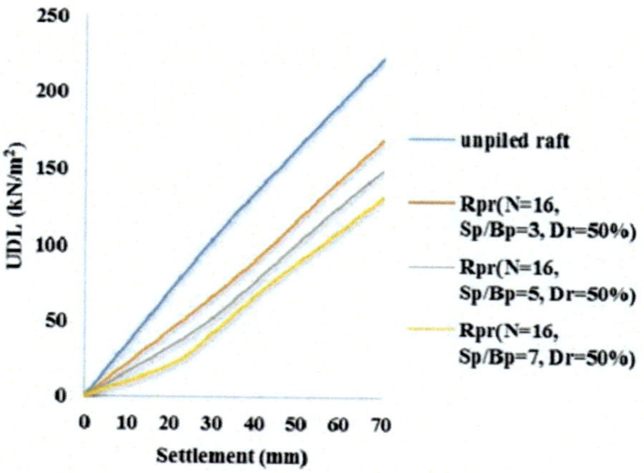

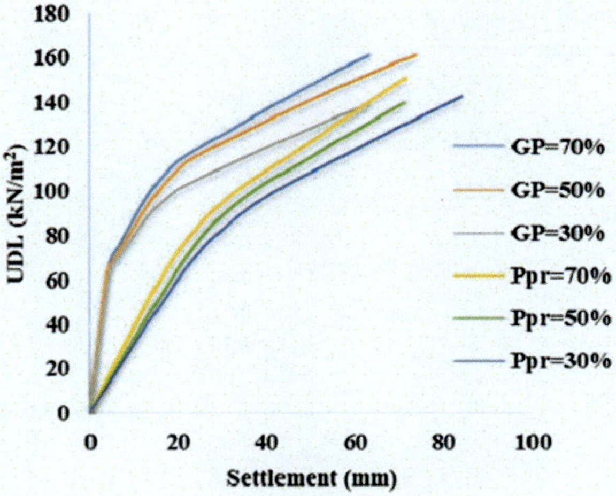


**Fig. 7** Decomposed load–settlement curves of Ppr compared with GP

## 5.2 Load Sharing Ratio

The load sharing ratio is determined by the load–settlement curve of the raft and pile decomposed by the finite element analysis method, and the load-sharing rate αp and its variation with settlement under different soil and foundation conditions are obtained. For different DR and Sp conditions, these results are shown in Figs. 8 and 9, respectively, and plotted as a function of normalized sedimentation $s/B_r$. Since the bearing capacity of small-diameter piles is earlier than that of large rafts, the αp value is initially high and then decreases nonlinearly with the increase in settlement. In particular, a significant decrease in αp was observed in the initial settlement range until $s/B_r = 0.001$. After this initial settlement range, the value of αp decreases continuously, while the observed decrease in αp is small and converges to certain values. Under different DR conditions, no significant changes in αp were observed, because

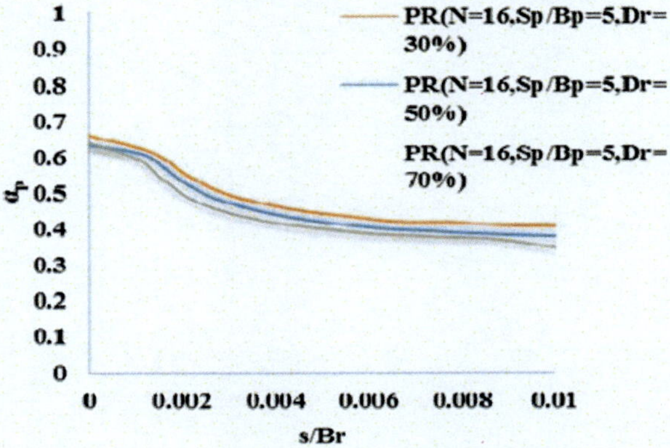


**Fig. 8** Variation of $\alpha_P$ with changes in relative density

**Fig. 9** Variation of $\alpha_P$ with changes in pile spacing

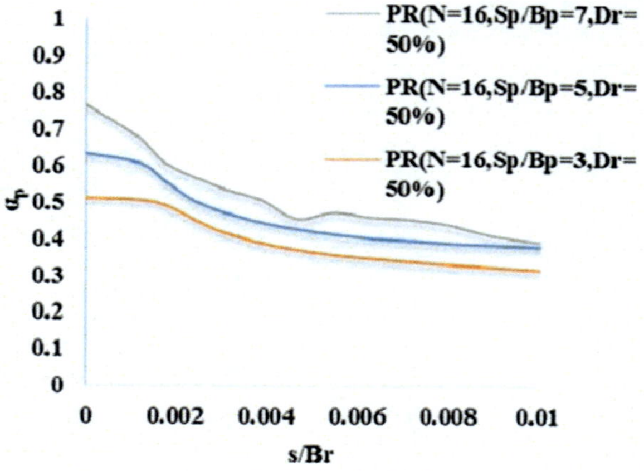


changing the DR will affect the load response of the raft and pile, thereby compensating for the change in the load-bearing capacity of the raft and pile. However, the influence of the pile spacing (Sp) shown in Fig. 9 is somewhat obvious.

As pile spacing increases, the values of $\alpha_p$ increase, meaning that the pile load capacity tends to decrease as pile spacing decreases, due to higher pile group interactions between neighboring piles.

## 5.3 *Effect of Pile Length*

The effect of pile length on the load–settlement relationship of piled-raft foundations supported by 2 × 2, 3 × 3, and 4 × 4 pile groups is shown in Fig. 10. It can be seen

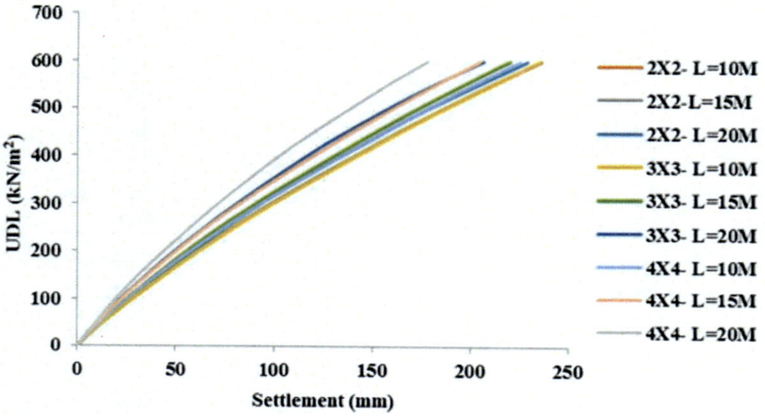


**Fig. 10** Effect of pile length on the load–settlement relationship of piled-raft supported by 2 × 2, 3 × 3, and 4 × 4 pile group

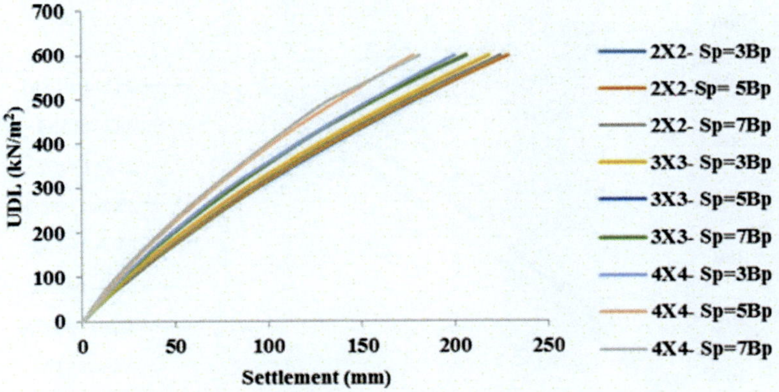


**Fig. 11** Effect of pile spacing on the load–settlement relationship of piled-raft supported by 2 × 2, 3 × 3, and 4 × 4 pile group

that before the yield point of the system, the stiffness of the pile-raft system increases as the pile length increases and becomes larger as the number of piles supporting the raft increases. After the yield point, the stiffness of the pile-raft foundation is not affected by the change in pile length because it is equal to the stiffness of the individual raft. On the other hand, as the pile length increases before and after the yield point, the bearing capacity of the piled-raft foundation increases significantly.

## 5.4 *Effect of Pile Spacing*

There is no effect of pile spacing on the load–settlement relationship of a piled-raft supported by fewer piles. The results show that the pile spacing within the range of $S = 3B_p$, $5B_p$, and $S = 7B_p$has no effect on the load–settlement relationship of the piled-raft supported by the 2 × 2 pile group. On the other hand, as can be seen from Fig. 11, the spacing between the piles affects the load–settlement curve of the piled-raft foundation supported by the 3 × 3 and 4 × 4 pile groups, as the number of piles supporting the raft increases. When the number of piles supporting the raft increases and the pile spacing increases, better distribution of the piles under the raft will provide important enhancements to the performance of the pile-raft foundation.

## 5.5 *Effect of Raft Thickness*

Within the range of raft thicknesses (0.5 m, 1 m, and 2 m) considered in this study, it was found that the raft thickness has no effect on the load–settlement relationship

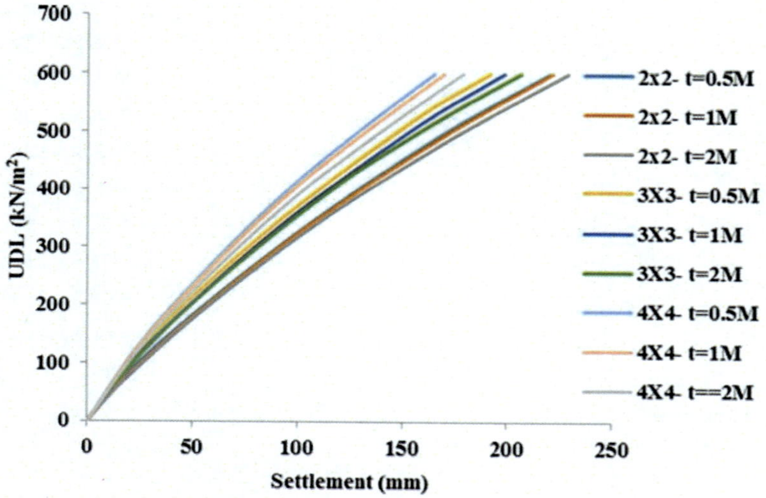


**Fig. 12** Effect of raft thickness on the load–settlement relationship of piled-raft supported by 2 × 2, 3 × 3, and 4 × 4 pile group

of piled-raft foundations either at small settlement or at large settlement levels as shown in Fig. 12.

## 6 Settlement Contours for Different Pile Configuration

In Figs. 13, 14, and 15, we can see how the soil settled under three different conditions. We can see that the soil underneath and the soil around the pile settled more than the rest of the soil in the area. Also, the part settled almost evenly, which indicates that the pile dominates the soil settlement in the pile-raft box because the pile is harder than the pile-raft and the soil.

## 7 Conclusions

In this study, a finite element analysis was conducted to study the load-sharing behavior of piled-rafts buried in sand. The numerical model is used for parameter research to study the influence of some important parameters on the performance of piled-raft foundations in small-scale and large-scale settlements. The effect of pile-raft interaction on load-sharing behavior is the focus of research. Various foundation and soil conditions were considered in the analysis, including pile configuration, pile spacing, and relative density. From the results reported in this article, the following conclusions are drawn:

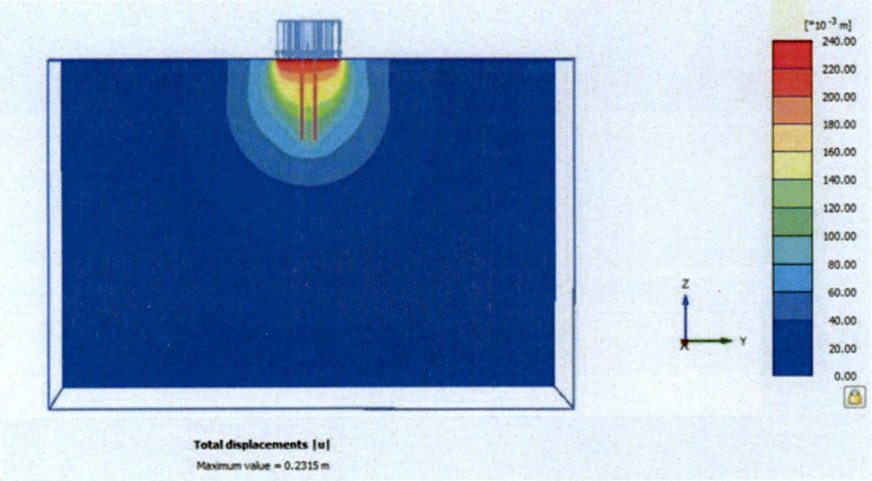


**Fig. 13** Settlement contours for 2 × 2 pile configuration

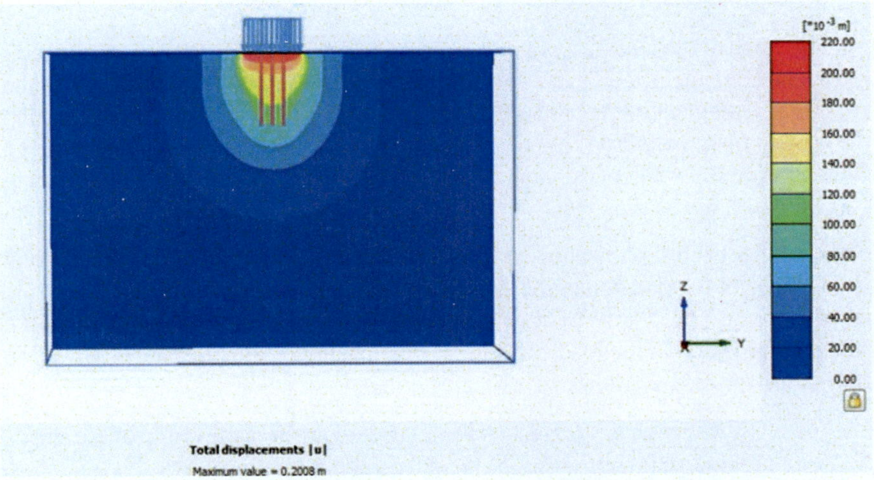


**Fig. 14** Settlement contours for 3 × 3 pile configuration

1. From the decomposed load–settlement curve, the bearing capacity of the pile is earlier than the bearing capacity of the raft, which shows a higher bearing rate. As the settlement increases further, the raft tends to carry a greater load than a pile compared to a pile. Due to the interaction between rafts and piles, in piles of rafts, the bearing capacity of each pile is smaller than that of unpiled rafts and group piles. As the pile spacing increases, the pile load ratio becomes higher.

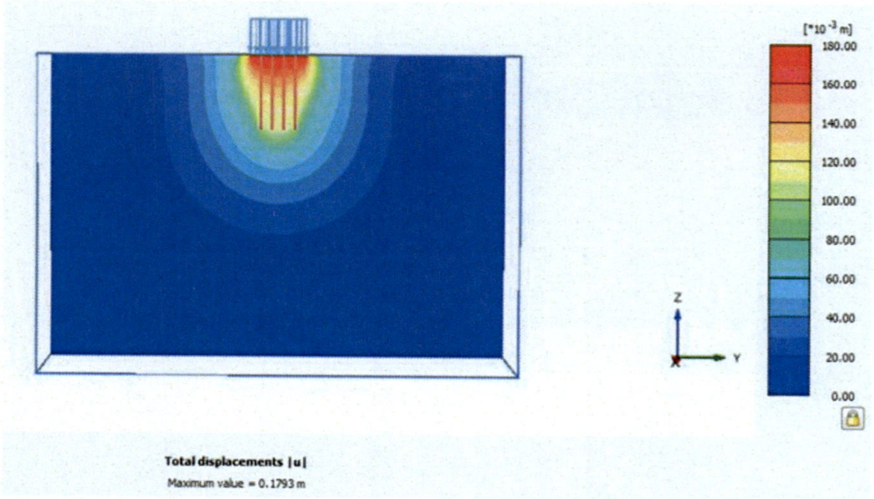


**Fig. 15** Settlement contours for 4 × 4 pile configuration

2. The value of the load-sharing ratio αp is initially high and decreases nonlinearly as the settlement increases. A significant decrease in αp was observed in the initial settlement range until s/Br was equal to 0.001. The effect of pile spacing is very obvious, while the effect of DR is not obvious. As the pile spacing increases, the value of αp increases, which indicates that when the pile spacing decreases, the bearing capacity of the pile tends to decrease.

The results of this study showed that the influence of certain parameters on the load–settlement relationship of the small settlement area is different from that of the large settlement area. The most important observations regarding the influence of the studied parameters on the load–settlement relationship of the piled-raft foundation can be summarized as follows:

1. The raft thickness has no effect on the load–settlement relationship of the small-scale or large-scale settlement of the piled-raft foundation.
2. The effect of pile spacing on the load–settlement relationship of the pile-raft foundation in small-scale settlements is negligible, and it has a significant effect on large settlements.
3. Pile length has an important influence on the load–settlement relationship of pile-raft foundation in large and small pile foundations.

## References

1. Moyes, P., Poulos, H.G., Small, J.C., and Badelow, F.: Piled raft design process for a high-rise building on the gold coast, pp. 1–9. Australia (2005). www.coffey.com
2. Lee, et al.: Estimation of load-sharing ratio for piled rafts in sands that include interaction effects. Can Geotech J (2015)

# Evaluation of Initial Stiffnesses and Ultimate Resistances of Shaft and Base of a Pile from Initial Load Test

**Vedhasri Sadula, CH. NageshwarRao, and Madhav Madhira**

## 1 Introduction

Due to rapid development in construction industry and speedy urbanization, high-rise buildings and heavy bridges that carry heavy loads require pile foundations that transfer these loads to deeper competent layers for stability and serviceability. Pile derives resistance through its shaft and base in case of axial loading. Piles are classified as end-bearing piles, friction piles or combination of end-bearing-friction piles based on the relative magnitude of shaft and base resistances.

Piling system is chosen considering ground conditions, characteristics of load, total and differential settlements and additional requirements of project. The ultimate capacity of the pile is estimated considering the strengths and unit weights of soil layers with depth, overburden pressure and other relevant parameters. The estimated capacities always need to be validated by conducting initial maintained load test. The estimated capacity may differ with the actual at site since the values of strength, stiffness, interface resistance between pile and soil, lateral earth pressure coefficient, etc. vary with depth and soil stratification and can be at variance from the design parameters considered for estimating the ultimate capacity of pile. The estimation of axial capacity of piles involves considerable uncertainties in selection of appropriate design parameters, and the design rules are not always consistent with the installation procedures/processes involved.

As per IS 2911–Part 4 [4], the ultimate and safe load on a pile are to be estimated by performing pile load test (PLT). Initial pile load test is conducted on a "test pile" up to its failure or at least two and half times the design load to determine the ultimate and safe load with applied factor of safety.

V. Sadula (✉) · CH. NageshwarRao · M. Madhira
VNR Vignana Jyothi Institute of Engineering and Technology, Hyderabad 500090, India

CH. NageshwarRao
e-mail: nageshwarrao_ch@vnrvjiet.in

 
C. N. V. Satyanarayana Reddy et al. (eds.), *Ground Characterization and Foundations*, Lecture Notes in Civil Engineering 167,
https://doi.org/10.1007/978-981-16-3383-6_51

Chin [3] developed a simple procedure to estimate the ultimate load capacity and initial stiffness of a pile from the load–displacement curve. Based on linear relationships between base load and settlement, Poulos [7] presented an approach for the prediction of load–displacement behavior of single pile and piers until failure. Meyerhof [6] examined the ultimate load of pile groups and the settlement of pile foundations in cohesive and non-cohesive soils. Armaleh and Desai [1] developed a method for estimating pile response using one dimensional finite element model for simulating nonlinear point resistance response. Based on fifty instrumented loading tests, new load-transfer curves were developed by Bohn [2], for different types of piles and ground types without the need of conducting pressure meter tests. Madhav and Vijay [5] proposed a new analytical solution to estimate the nonlinear responses of shaft and base of a pile using hyperbolic relationships.

## 2 Statement of the Problem

Estimating ultimate resistances and initial stiffnesses of shaft and base of pile using load-displacement curve from initial pile load test is important, as it enables verification of the a-priori predictions based on geometry of pile (i.e., diameter, length and shape), method and mode of construction and other uncertainties involved during pile installation at the site. The present study determines first the ultimate, $P_{\mathrm{ult}}$, and initial stiffness modulus, $k_p$, of pile from initial pile load test data. The ultimate resistances of shaft, $\tau_{\max}$, and base, $q_u$, and the initial stiffnesses of shaft, $k_\tau$, and base, $k_b$, of a pile under suitable combination of loads are then estimated. The load ($L$)–displacement ($\delta$) data using estimated pile responses ($k_\tau$, $\tau_{\max}$, $k_b$ and $q_u$) is predicted and compared with the real test data for validation.

## 3 Methodology

A single pile of diameter, $d$, length, $L$ and with an axial load, $P$, considered is shown in Fig. 1. The pile-soil system is modeled as Winkler-type model with distinct nonlinear responses for the base and shaft soil resistances.

A single rigid pile subjected to vertical loading derives its resistance by base and shaft resistances. The idealized representation of pile load test results of a single rigid pile, shaft and base resistances are shown in Figs. 2, 3 and 4, respectively. The initial slopes of the curves in Figs. 2, 3 and 4 represent the initial stiffness moduli of pile, $k_p$, shaft stiffness, $k_\tau$, and base stiffness, $k_b$, respectively. $P_{\mathrm{u}}$ is the ultimate load on the pile, while $q_u$ and $\tau_{\max}$ are the ultimate base and shaft resistances, respectively.

The ultimate capacity, $P_u$, and initial stiffness, $k_p$, of the pile are the two key known parameters required to estimate the four parameters, $k_\tau$, $\tau_{\max}$, $k_b$ and $q_u$ of the base and shaft responses. Chin's [3] method based on hyperbolic response between load

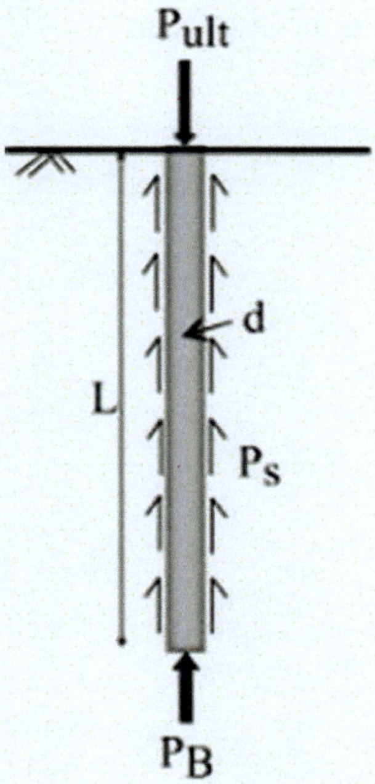


**Fig. 1** Definition sketch

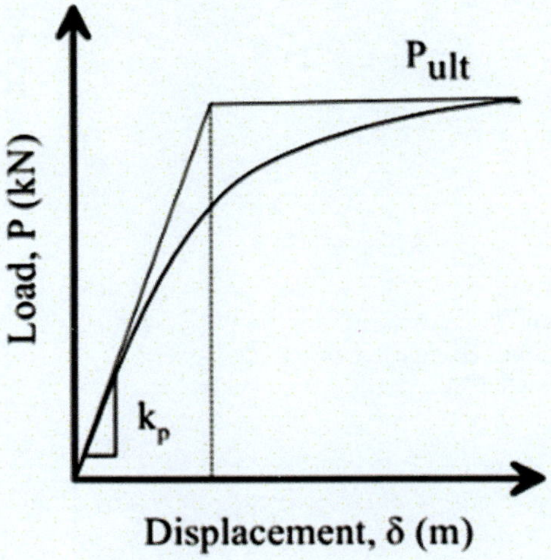


**Fig. 2** Typical pile load versus displacement response

and displacement is adopted to estimate $P_u$ and $k_p$. The load–displacement curve from initial pile load test is shown in Fig. 5.

The ratio $\delta/P$ is plotted with $\delta$ is shown in Fig. 6, and an equation

$$\delta/P = C_1\delta + C_2. \tag{1}$$

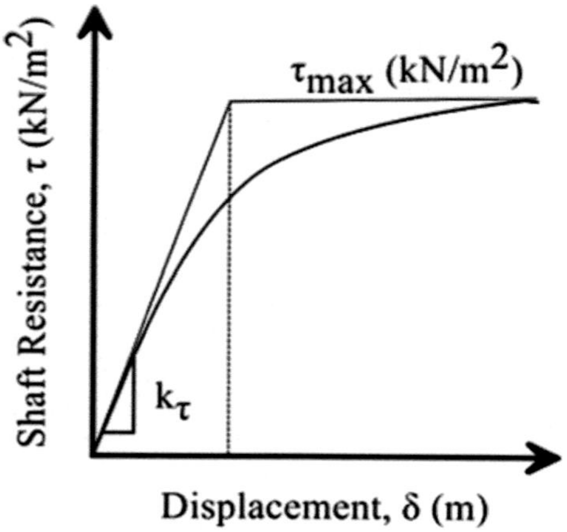


**Fig. 3** Assumed shaft resistance, $\tau$ versus displacement, $\delta$

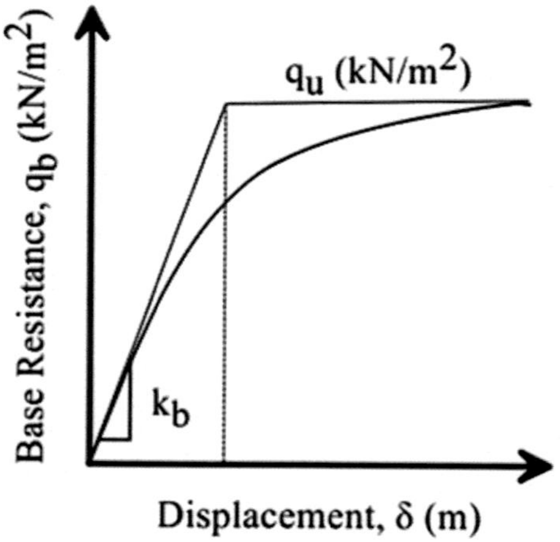


**Fig. 4** Assumed base resistance, $q_b$ versus displacement, $\delta$

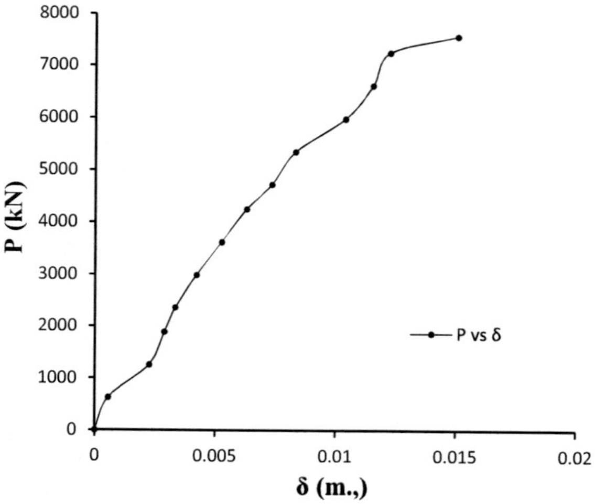


**Fig. 5** Load versus displacement curve from initial PLT

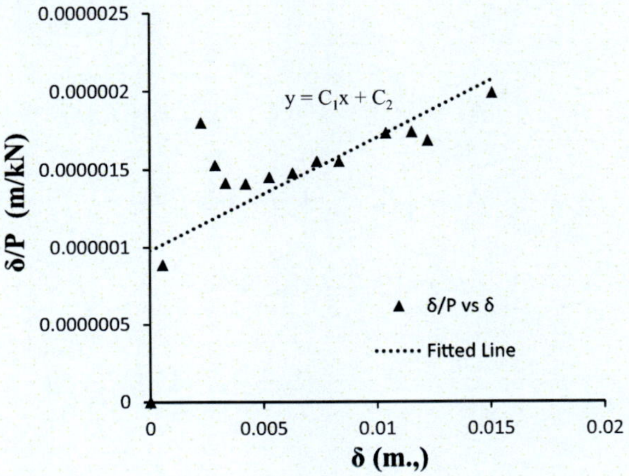


**Fig. 6** Chin's method for estimation of $P_u$ and $k_p$

is fitted where intercept $C_1 = 1/P_u$ and the slope $C_2 = 1/k_p$.

A set of nonlinear coupled equations derived by Madhav and Vijay [5] are given below:

(1) Equation for $k_b$:

$$k_b = k_p^* - (4L/d)\,k_\tau \tag{2}$$

(2) Equation for $k_\tau$:

$$k_\tau = (C_4 d_1 - C_2 d_2)\big/(C_1 C_4 - C_2 C_3) \tag{3}$$

(3) Equation for $q_u$:

$$q_u = P_u^* - (4L/d)\,\tau_{\max} \tag{4}$$

(4) Equation for $\tau_{\max}$:

$$\tau_{\max} = \frac{\left\{P_2^* \cdot A_2 - \left[\left(k_p^*/\,k_\tau\right) - (4 \cdot L/d)\right] \cdot\ P_u^* \cdot\ k_\tau \cdot \delta_2\right\}}{\left\{(4L/d).\left[A_2 -\ k_p^* \cdot \delta_2 + (4 \cdot L \cdot\ k_\tau \cdot \delta_2/d)\ \right]\right\}} \tag{5}$$

where $k_p{}^* = k_p/(\pi d^2/4)$—normalized stiffness of the pile; $k_\tau$ and $k_b$ are shaft and base stiffnesses, respectively. $P_u{}^* = P_u/(\pi d^2/4)$—normalized $P_u$ of the pile; $\tau_{\max}$ and $q_u$ are maximum shaft and base resistances, respectively.

$d_1, d_2, A_1, A_2, C_1, C_2, C_3$ and $C_4$ are parameters used in equations as

$$d_1 = (P_1^*/\delta_1) - \left[\left(k_p^* \cdot P_u^*\right)/A_1\right] \tag{6}$$

$$d_2 = (P_2^*/\delta_2) - \left[\left(k_p^* \cdot P_u^*\right)/A_2\right] \tag{7}$$

$$A_1 = \left[P_u^* - (4 \cdot L \cdot \tau_{max}/\, d)\right] + \left[k_p^* - (4 \cdot L \cdot k_\tau)/d\right] \cdot \delta_1 \tag{8}$$

$$A_2 = \left[P_u^* - (4 \cdot L \cdot \tau_{max}/\, d)\right] + \left[k_p^* - (4 \cdot L \cdot k_\tau)/d\right] \cdot \delta_2 \tag{9}$$

$$C_1 = (4 \cdot L/d) \cdot \left\{(\tau_{max}/\, \tau_{max} + k_\tau \cdot \delta_1) - \left(P_u^*/\, A_1\right) + (4 \cdot L \cdot \tau_{max}/\, d \cdot\, A_1)\right\} \tag{10}$$

$$C_2 = -\left(4 \cdot L \cdot k_p^*/d \cdot A_1\right) \tag{11}$$

$$C_3 = (4 \cdot L/d) \cdot \left\{(\tau_{max}/\, \tau_{max} + k_\tau \cdot \delta_2) - \left(P_u^*/\, A_2\right) + (4 \cdot L \cdot \tau_{max}/\, d \cdot\, A_2)\right\} \tag{12}$$

$$C_4 = -\left(4 \cdot L \cdot k_p^*/\, d \cdot A_2\right) \tag{13}$$

The above equations were derived for 1/3rd and 2/3rd of $P_u$. However, some of the load tests are not carried out to a significant load but are limited to a small fraction of the ultimate load. Hence, new sets of equations are derived for lesser load combinations, viz.,

Case 1: $P_1 = 0.3 * P_u$ and $P_2 = 0.6 * P_u$
Case 2: $P_1 = 0.3 * P_u$ and $P_2 = 0.5 * P_u$
Case 3: $P_1 = 0.25 * P_u$ and $P_2 = 0.5 * P_u$
Case 4: $P_1 = 0.25 * P_u$ and $P_2 = 0.3 * P_u$.

Pile parameters are estimated by substituting appropriate load combinations of $P_1$ and $P_2$ from Eqs. 2 to 5.

## 4 MATLAB Analysis

MATLAB is used to simplify the process of solving coupled second degree polynomial equations, having two unknown parameters.

The four unknown parameters, $k_\tau$, $\tau_{max}$, $k_b$ and $q_u$ denoted as 1, 2, 3 and 4, respectively, are estimated using four nonlinear coupled equations. The above flowchart depicts the procedure followed in MATLAB software.

The procedure must be repeated for different assumed values of unknown parameter-1 to get corresponding unknown parameter-2. A set of values of $\tau_{max}$ are estimated using Eq. 5 assuming $k_\tau$. Similarly, another set of $k_\tau$ are estimated using Eq. 3 assuming $\tau_{max}$. $\tau_{max}$ was plotted against $k_\tau$, obtained from the above two procedures, and the intersection of the two curves gives the $\tau_{max}$ and $k_\tau$ values for

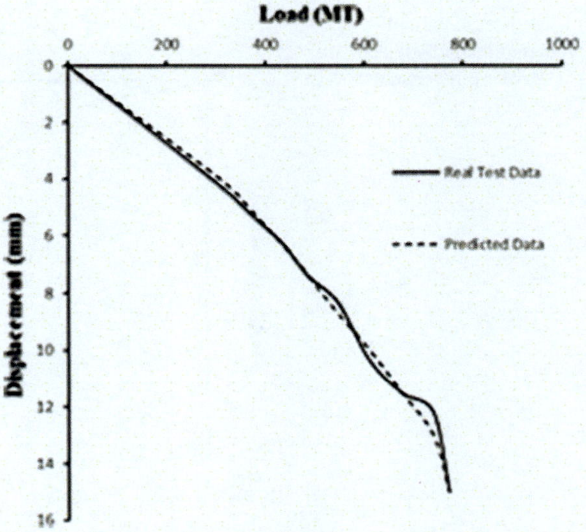


**Fig. 7** Measured and predicted load–displacement curves of a pile located at Bur balasan

the pile under consideration. $\tau_{\max}$ and $k_\tau$ are substituted in Eqs. 2 and 4 to obtain the values of $k_b$ and $q_u$, respectively.

## 5 Application of the Proposed Method

The present method is applied to 6 test piles for which the load–displacement curves from initial loading tests which are available. The ultimate resistance and stiffness modulus of each test pile are determined using Chin's [3] method. The four unknown parameters of each pile ($k_\tau$, $\tau_{\max}$, $k_b$, and $q_u$) are estimated using appropriate load combination (Case 1, Case 2, Case 3 or Case 4). Load–displacement curves of 6 test piles considered in this study are shown in Figs. 7, 8, 9, 10, 11 and 12. The diameter and length of all 6 piles are 1.0 m, and 24.0 m, respectively.

The ultimate load, $P_u$, and the initial stiffness, $k_p$, are listed in Table 1. The load combinations used for these test piles are given in Table 2.

The estimated parameters of each pile obtained from the analysis are summarized in Table 3.

## 6 Results and Discussion

The main purpose of this study is to estimate the shaft and base resistance parameters of pile, i.e., $k_\tau$, $\tau_{\max}$, $k_b$ and $q_u$ using load–displacement data from initial pile load

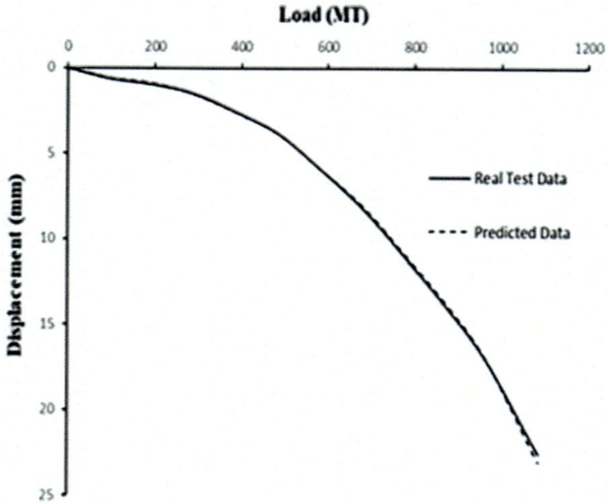


**Fig. 8** Measured and predicted load–displacement curves of a pile located at Dharaikuri

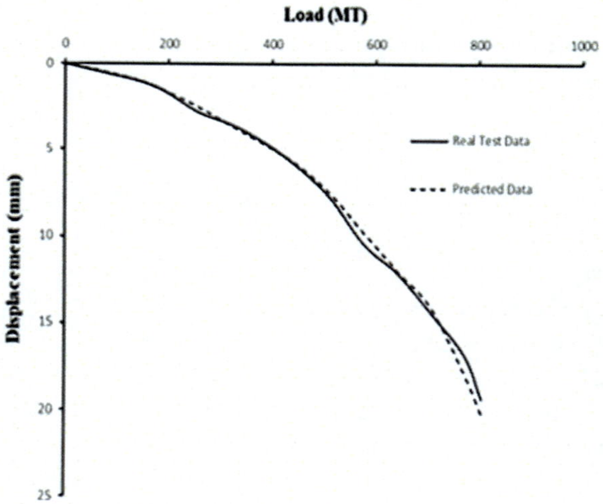


**Fig. 9** Measured and predicted load–displacement curves of a pile located at Karala

test. The estimated shaft and base resistances of piles obtained for the first time using this method are presented in Table 3.

## 7 Conclusions

This paper presents a simple method to estimate the ultimate resistances and initial stiffnesses of shaft and base, based on the analysis of initial pile load test data. A set of equations for different load combinations are developed for the purpose of determining initial shaft and base stiffnesses and ultimate shaft and base resistances. The

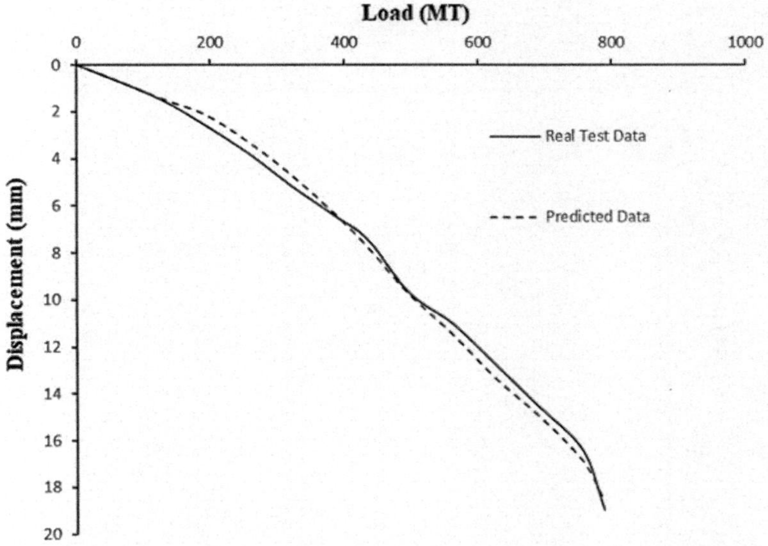


**Fig. 10** Measured and predicted load–displacement curves of a pile located at Kartowa

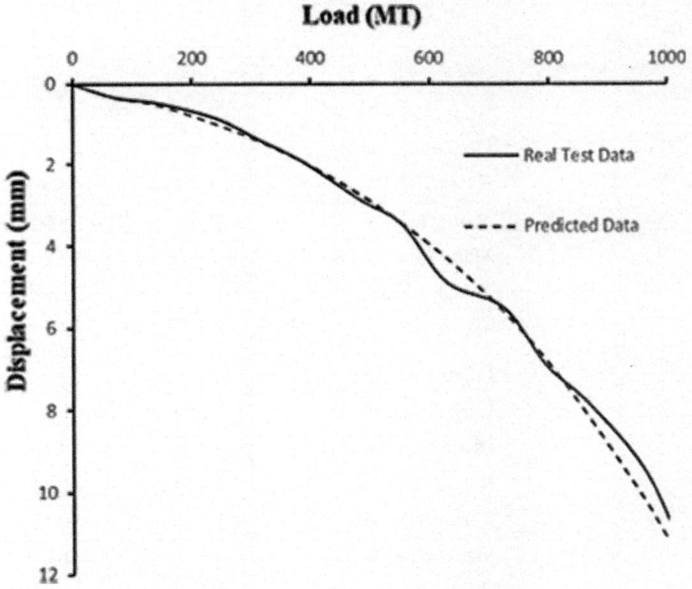


**Fig. 11** Measured and predicted load–displacement curves of a pile located at Mahananda

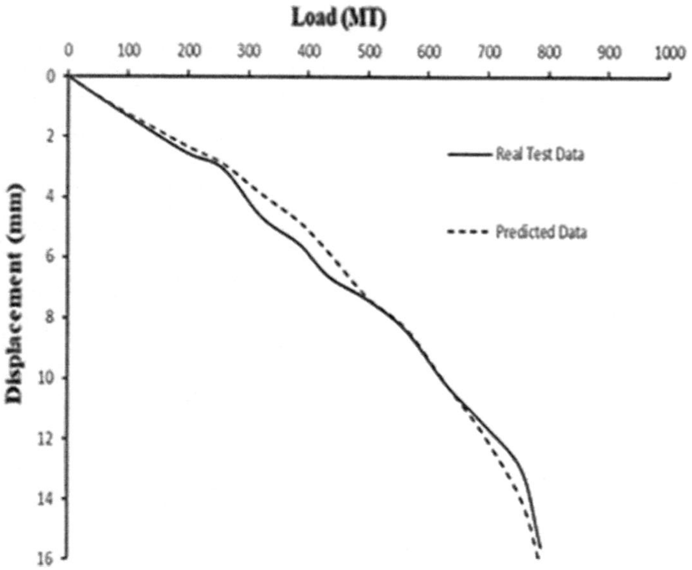


**Fig. 12** Measured and predicted load–displacement curves of a pile located at Talma

**Table 1** Ultimate load and initial stiffness

| Pile reference | $P_u$ (MN) | $k_p$ (MN/m) |
|---|---|---|
| Buribalasan, W. B | 20.0 | 1,666.7 |
| Dharaikuri, W. B | 16.7 | 3,333.3 |
| Karala, W. B | 12.5 | 2,500.0 |
| Kartowa, W. B | 20.0 | 1,428.6 |
| Mahananda, W. B | 16.7 | 3,333.3 |
| Talma, W. B | 20.0 | 1,666.7 |

**Table 2** Load combinations for piles

| Pile reference | $P_1$ | $P_2$ |
|---|---|---|
| Buribalasan, W. B | 0.25 * $P_u$ | 0.3 * $P_u$ |
| Dharaikuri, W. B | 0.3 * $P_u$ | 0.6 * $P_u$ |
| Karala, W. B | 0.25 * $P_u$ | 0.3 * $P_u$ |
| Kartowa, W. B | 0.3 * $P_u$ | 0.6 * $P_u$ |
| Mahananda, W. B | 0.3 * $P_u$ | 0.6 * $P_u$ |
| Talma, W. B | 0.25 * $P_u$ | 0.3 * $P_u$ |

*Note* Load combination is selected based on available initial load test data

**Table 3** Estimated initial stiffnesses and ultimate resistances of shaft and base

| Pile reference | $k_\tau$ (MN/m$^2$/m) | $\tau_{\max}$ (kPa) | $k_b$ (MN/m$^2$/m) | $q_u$ (MPa) |
|---|---|---|---|---|
| Buribalasan, W. B | 14 | 20 | 778 | 23.5 |
| Dharaikuri, W. B | 34 | 38 | 980 | 17.6 |
| Karala, W. B | 26 | 40 | 687 | 12.1 |
| Kartowa, W. B | 14 | 30 | 474 | 22.6 |
| Mahananda, W. B | 26 | 35 | 1748 | 17.9 |
| Talma, W. B | 15 | 28 | 682 | 22.8 |

proposed method is applied to six pile load test data using suitable load combination. The estimated responses of pile are used to predict the load–displacement responses which compare closely with the measured ones.

# References


1. Armaleh, S., Desai, C.S.: Load-deformation response of axially loaded piles. J. Geotech. Eng., 1483–1500 (1987). https://doi.org/10.1061/113:12(1483)
2. Bohn, C., Lopes dos Santos, A., Frank, R.: Development of axial pile load transfer curves based on instrumented load tests. J. Geotech. Geoenviron. Eng., ASCE **143**(1), 1–15 (2017)
3. Chin, F.K.: Estimation of the ultimate load of piles not carried to failure. In: Proceedings of the South East Asian Conference on Soil Engineering, pp. 81–90 (1971)
4. IS 2911-Part 4: Design and construction of pile foundations—Code of practice—Load test on piles. Bureau of Indian Standards, New Delhi (2013)
5. Madhav, M.R., Vijay, K.: A Method to estimate shaft and base responses of a pile from pile load test. Geotech. Eng. J. SEAGS & AGSSEA **50**(3), 118–123 (2019)
6. Meyerhof, G.G.: Bearing capacity and settlement of pile foundations. J. Geotech. Eng. ASCE **102**(3), 1–19 (1976)
7. Poulos, H.G.: Load-settlement prediction for piles and piers. J. Soil Mech. Found. Div. ASCE **98**(SM8), 843–848 (1972)

# Behavior of Single Pile Subjected to Eccentric Loading in Cohesionless Soils

**N. Dhana Sree, E. Saibaba Reddy, and V. Padmavathi**

## 1 Introduction

A single pile carries large loads from superstructure if its capacity is sufficient. Engineering structures such as bridges, retaining walls, and offshore structures which are supported by pile foundations frequently subjected to eccentric loads. These pile foundations experience excessive settlements if it is subjected to eccentric loads and the structure may tilt or rotate due to this eccentricity. Behavior of rigid piles and pile groups under eccentric vertical loads in sand and in clay has been studied by Kishida and Meyerhof (1965) and Saffery and Tate (1961), respectively. Response of Four Pile Group Subjected to Eccentric Loading is studied by Padmavathi et al. [5]. Response of 3 × 3 Pile Groups in Silt Subjected to Eccentric Lateral loading has been studied by Kong et al. [4]. The effect of the eccentricity on the ultimate capacity by model tests on pile groups is investigated by Komatsu et al. (2004). This paper investigates the response of single pile subjected to vertical eccentric loading in sand beds at 60% and 70% relative densities. The load–displacement responses of a single pile under eccentric loading are investigated by model tests. Load–displacement responses of the single pile are analyzed to quantify the effects of diameter, the effect of relative density, and the effect of eccentricities. Piles used in this investigation are shown in Fig. 1.

N. Dhana Sree (✉) · E. Saibaba Reddy · V. Padmavathi
JNTUH College of Engineering, 2 CBIT, Hyderabad, Telangana, India



C. N. V. Satyanarayana Reddy et al. (eds.), *Ground Characterization and Foundations*, Lecture Notes in Civil Engineering 167,
https://doi.org/10.1007/978-981-16-3383-6_52

Fig. 1 Steel pipe piles

## 2 Methodology

Model tests were performed on single pile installed in sand bed placed in a circular tank of diameter 60 cm and height 71 cm. The test setup (Fig. 2) consists of the following components: (a) Test tank, (b) Model piles, (c) Foundation medium, i.e., sand, (d) Arrangement for application of loads and (e) Measuring devices (four dial gauges). The pile cap is made of square in shape of size 12 cm, 15 cm, 17 cm, 20 cm for 42 mm, 49 mm, 75 mm, 100 mm diameter piles, respectively. The sand was compacted in five layers with relative densities of 60% and 70%, respectively. The properties of the sand used in this are shown in Table 1, and it is classified as poorly graded sand (SP) as per IS classification. Piles of diameter 42 mm, 49 mm,

(a) (b)

**Fig. 2** Schematic of test setup: **a** cylindrical tank, **b** measuring devices

**Table 1** Properties of sand

| Soil properties of sand | Value |
|---|---|
| Coefficient of Uniformity ($C_u$) | 3.63 |
| Coefficient of Curvature ($C_c$) | 0.85 |
| Specific Gravity ($G$) | 2.65 |
| Cohesion (c) | 0 |
| Maximum dry unit weight ($\gamma_d$) | 1.85 g/cc |
| Minimum dry unit weight ($\gamma_d$) | 1.57 g/cc |
| IS classification | SP |
| Angle of shearing resistance | $D_r$ 60% = 36° |
| | $D_r$ 70% = 38° |

75 mm, 100 mm for constant length of 500 mm have been studied in the model test tank. A total of 32 tests were performed on the single pile which were subjected to normalized eccentricities (*e*/*s*) and axial load. Tests data was obtained using a proving ring and four number of dial gauges (Fig. 2).

## 3 Experimental Procedure

The sand was filled in 6 layers, i.e., 14 cm, 6 cm, 8 cm, 14 cm, 14 cm, 11 cm from the bottom of the tank, respectively. The total weight of sand filled in the tank for the required relative densities of 60% and 70%, respectively, is shown in Table 2. Total number of 32 tests were done in the tank on single pile models with different diameters, axially and at different eccentricities [4, 7]. The single pile model is installed on the sand bed such that the load is applied at required position after the sand is compacted up to depth of 20 cm. The soil is compacted to required density by rammer weighing 8.9 kg [5]. This process is continued for the rest of the layers with required weight of sand till the height of 67 cm and compacted cautiously by maintaining the pile in the vertical position, and leveling is done to maintain it horizontally, at required eccentricity and relative density. The pile verticality and eccentricity were checked for each before the test. The relative density during the

**Table 2** Total weight of soil required for each pile

| Pile diameter (mm) | Total weight of soil (kg) | |
|---|---|---|
| | $D_r = 60\%$ | $D_r = 70\%$ |
| 42 | 326.33 | 331.13 |
| 49 | 325.9 | 330.9 |
| 75 | 323.94 | 329.52 |
| 100 | 321.1 | 326.7 |

test is cross checked by small bins which are already placed in the tank, and it is almost same. A load is applied through the hydraulic jack which is supported to the loading frame. Hydraulic jack is used for applying loads vertically on the pile. The proving ring which is placed between the hydraulic jack and the pile cap is used to record the values at small increments when a load is applied through the hydraulic jack, which is supported to the loading frame. Four number of dial gauges are placed above the pile cap.

### *3.1 Calculation of Load Carrying Capacity of Pile*

The method for estimating the load carrying capacity of a pile foundation can be static method, dynamic method, in-situ penetration tests, pile load tests.

The ultimate bearing capacity of a single pile in sand is obtained by,

$$Q_u = Q_p + Q_s$$

$$Q_u = A_p q N_q + f_{av} Pl$$

$$Q_u = A_p q N_q + (k\,\sigma_o \tan\delta)\,Pl$$

## 4 Results and Discussions

### *4.1 Comparison of Effect of Relative Density*

**Case—1: Pile Diameter 42 mm**

Figure 3 shows the effect of relative density with settlement for different loads is plotted for eccentricity, $e = 0$ mm. Ultimate loads for relative densities 60% and 70% increase from 2.96 to 4.79 kN [3]. The ultimate load is increased by 61.82% from relative density 60% to relative density 70%. The ultimate load and the settlement of the pile increase with increase in relative density. Similarly, for $e = 2$ mm (Fig. 4), $e = 4$ mm (Fig. 5), $e = 7$ mm (Fig. 6), the ultimate load of the pile increased with increase in relative density and it is shown in Table 3.

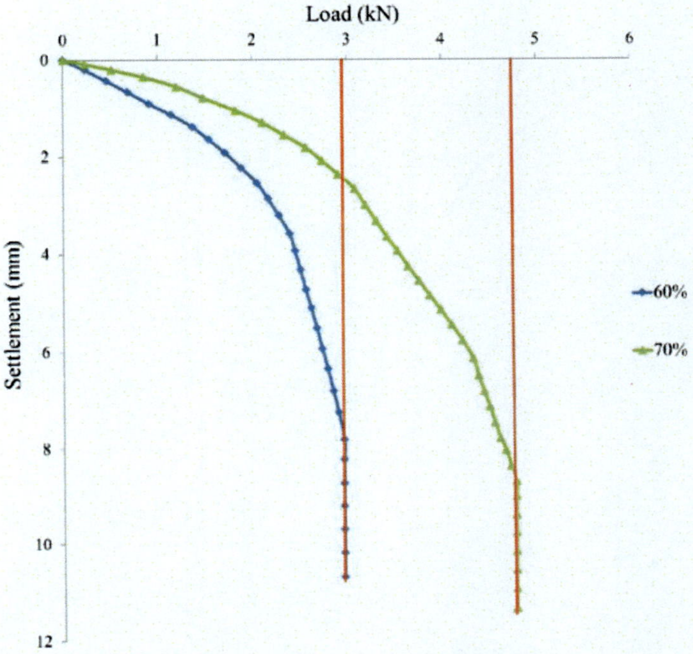


**Fig. 3** Effect of $D_r$ on pile capacity (42 mm diameter pile loading at center)

## 4.2 *Comparison of Effect of Eccentricity*

### Case—1: Pile Diameter 42 mm

Figure 7 depicts load versus settlement responses of a single pile for $e = 0$ mm, 2 mm, 4 mm, and 7 mm at relative density $D_r = 60\%$, for diameter 42 mm. The ultimate load for axially loaded pile of diameter 42 mm is 2.96 kN and for eccentricities of 2 mm, 4 mm, 7 mm are 2.96 kN, 2.45 kN, 2.05 kN, respectively. The ultimate loads for eccentricities 2 mm, 4 mm, 7 mm are decreased by 0%, 17.22%, 30.74%, respectively, when compared to axial load. The ultimate load of the pile decreases, and the settlement decreases with increase in eccentricity of the applied load. The pile capacity for $D_r = 70\%$ at e = 0 mm, 2 mm, 4 mm, and 7 mm is 4.79 kN, 4.79 kN, 4.10 kN, 3.53 kN, respectively, is shown in (Fig. 8). The ultimate load carrying capacity and settlement is observed same for both axial load and 2 mm in Fig. 8. The ultimate loads for eccentricities 2 mm, 4 mm, 7 mm are decreased by 0%, 14.40%, 26.30%, respectively, compared to axial load and are shown in Table 4.

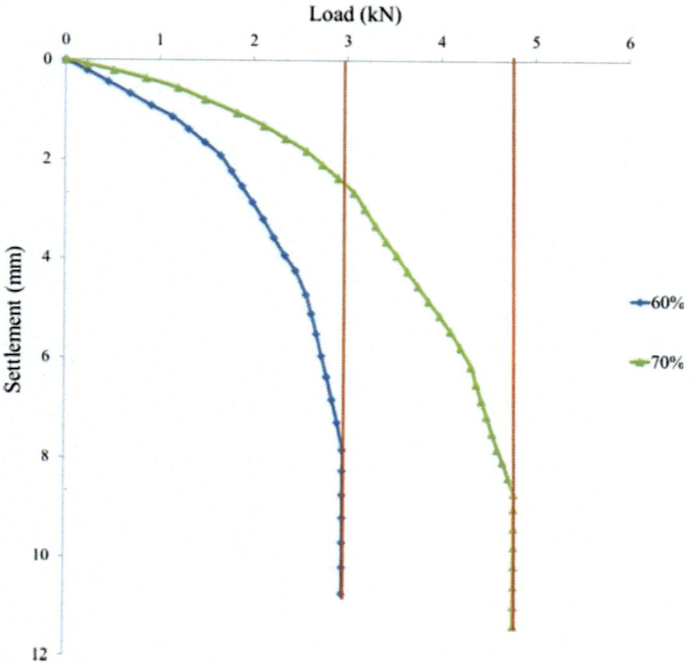


**Fig. 4** Effect of $D_r$ on pile capacity (42 mm diameter pile loading at e = 2 mm)

## 4.3 Comparison of Effect of Diameter

**Case—1:** $D_r = 60\%$

Figure 9 depicts load versus settlements response of a single pile at relative density 60% for different pile diameters, i.e., 42 mm, 49 mm, 75 mm, 100 mm. The ultimate load for axially loaded piles of diameter 42 mm, 49 mm, 75 mm, 100 mm is 296 kN, 404 kN, 9 kN, 1505 kN, respectively. The ultimate load of the pile increases with increase in pile diameter. This is due to the contact area of sand at the end of the pile increased with increase in the pile diameter, so the end bearing resistance is increased. The surface area increases with increase in the pile diameter as a result the frictional resistance of the pile is increased. So, the load carrying capacity is increased with increase in end bearing resistance and frictional resistance by increase in pile diameter. The effect of diameter on pile capacity for axial capacity at $D_r = 70\%$ for different piles is depicted in (Fig. 10). The ultimate load for axially loaded piles of diameter 42 mm, 49 mm, 75 mm, 100 mm is 479 kN, 644 kN, 1237 kN, 2134 kN, respectively.

**Case—2:** $D_r = 70\%$

See Fig. 10.

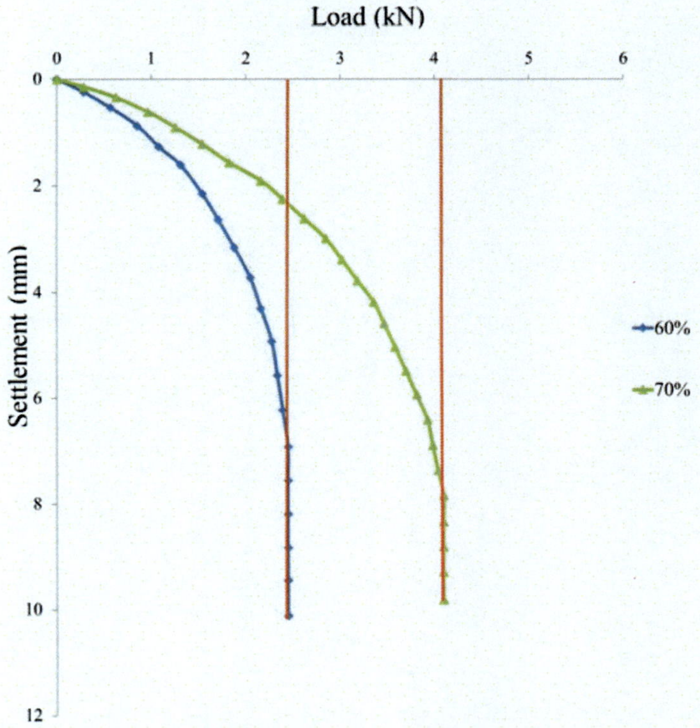


**Fig. 5** Effect of $D_r$ on pile capacity (42 mm diameter pile loading at e = 4 mm)

## 5 Conclusions

The test results were compared and conclusions were made based on the practical observation and are presented below.

1. The ultimate load carrying capacity of pile decreases upto 31% with increase in eccentricity from 0 to 7 mm when compared to its axial capacity for a pile of $\phi = 42$ mm.
2. Similarly, it is observed that there is decrease upto 24% with increase in eccentricity from 0 to 8 mm when compared to its axial capacity for pile of $\phi =$ 49 mm.
3. The ultimate load carrying capacity of pile decreases upto 23% with increase in eccentricity from 0 to 12 mm when compared to its axial capacity for pile of $\phi = 75$ mm and for $\phi = 100$ mm pile at eccentricity from 0 to 16 mm there is decrease upto 24% when compared with axial capacity of the pile.
4. For all the loading conditions, the load carrying capacity of pile increases from 62 to 72% for $\phi = 42$ mm pile with increase in relative density from $D_r = 60\%$ to 70% and for $\phi = 49$ mm pile the load carrying capacity of the pile increases from 59 to 62% with increase in relative density from $D_r = 60$ to 70%.

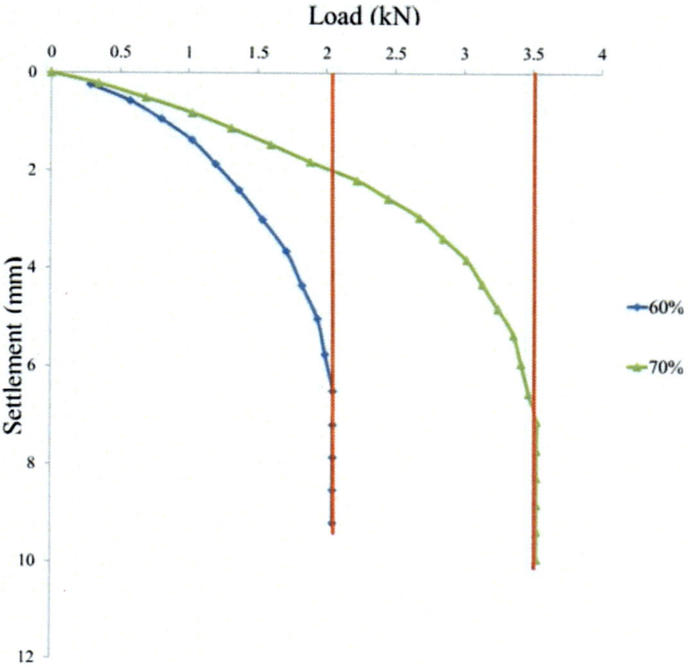


**Fig. 6** Effect of $D_r$ on pile capacity (42 mm diameter pile loading at e = 7 mm)

5. Similarly, it is observed that the load carrying capacity of pile increases from 33 to 37% for $\phi = 75$ mm pile and from 39 to 41% it increases for $\phi = 100$ mm pile with increase in relative density from $D_r = 60\%$ to 70%.
6. Load carrying capacity of pile increases with increase in pile diameter.
7. The ultimate load carrying capacity of the pile observed experimentally is 15% more than theoretical values.

**Table 3** % Increase in pile capacity for different pile diameters at different eccentricity, *e*

| Pile diameter (mm) | Load applied at *e* (mm) | Ultimate load (kN) @ $D_r$ = 60% | Ultimate load (kN) @ $D_r$ = 70% | Increase in pile capacity (kN) | % increase in pile capacity |
|---|---|---|---|---|---|
| 42 | 0 | 2.96 | 4.79 | 1.83 | 61.82 |
| | 2 | 2.96 | 4.79 | 1.83 | 61.82 |
| | 4 | 2.45 | 4.10 | 1.65 | 67.34 |
| | 7 | 2.05 | 3.53 | 1.48 | 72.19 |
| 49 | 0 | 4.04 | 6.44 | 2.40 | 59.40 |
| | 3 | 3.63 | 5.87 | 2.24 | 61.70 |

(continued)

**Table 3** (continued)

| Pile diameter (mm) | Load applied at *e* (mm) | Ultimate load (kN) @ $D_r$ = 60% | Ultimate load (kN) @ $D_r$ = 70% | Increase in pile capacity (kN) | % increase in pile capacity |
|---|---|---|---|---|---|
| | 5 | 3.40 | 5.41 | 2.01 | 59.11 |
| | 8 | 3.06 | 4.96 | 1.90 | 62.09 |
| 75 | 0 | 8.99 | 12.37 | 3.37 | 37.44 |
| | 4 | 8.36 | 11.36 | 3.00 | 35.88 |
| | 8 | 7.61 | 10.32 | 2.71 | 35.61 |
| | 12 | 6.92 | 9.23 | 2.31 | 33.38 |
| 100 | 0 | 15.05 | 21.34 | 6.29 | 41.79 |
| | 6 | 13.67 | 19.26 | 5.59 | 40.89 |
| | 11 | 12.51 | 17.53 | 5.02 | 40.12 |
| | 16 | 11.36 | 15.80 | 4.44 | 39.08 |

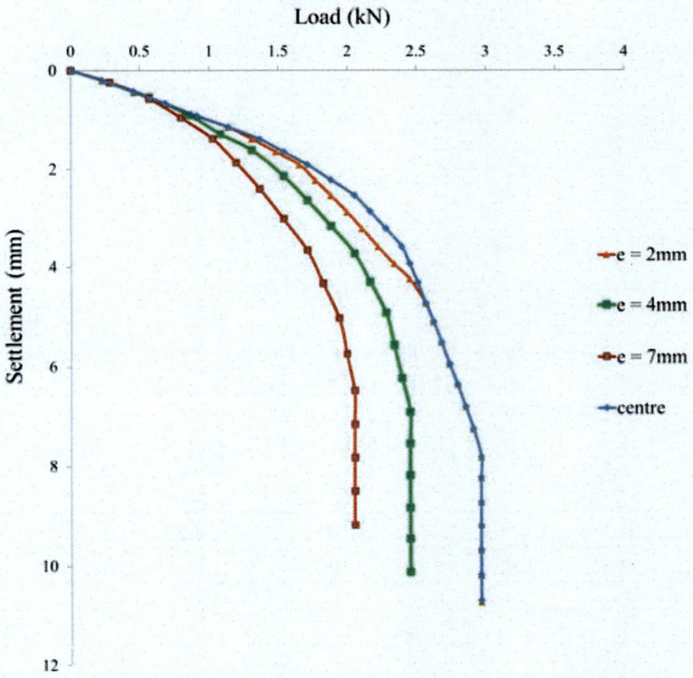


**Fig. 7** Effect of e on pile capacity (42 mm diameter pile $D_r$ = 60%)

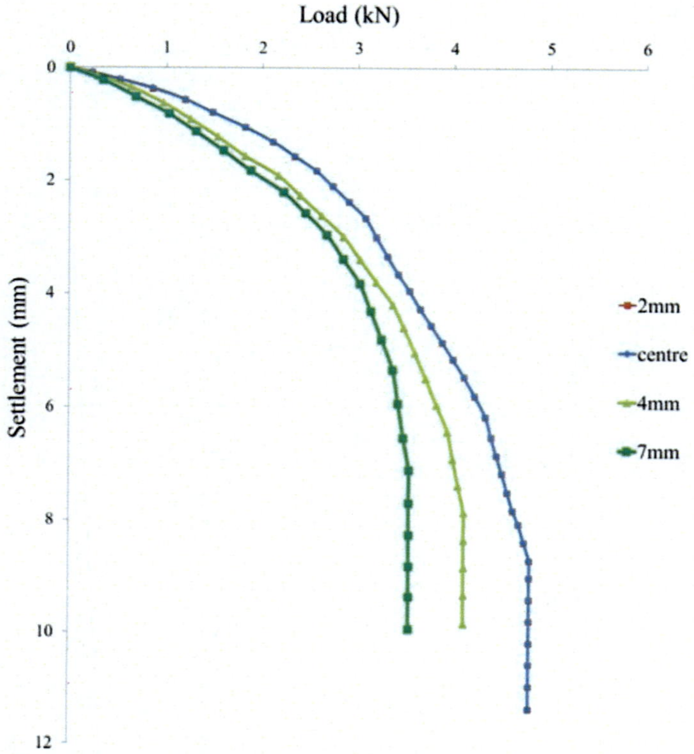


**Fig. 8** Effect of e on pile capacity (42 mm diameter pile $D_r = 70\%$)

**Table 4** % Decrease in pile capacity for different pile diameters at different eccentricity, *e*

| Pile diameter (mm) | Load applied at *e* (mm) | Ultimate load $D_r =$ 60% (kN) | Decrease in pile capacity (kN) | %Decrease in pile capacity | Ultimate load $D_r =$ 70% (kN) | Decrease in pile capacity (kN) | % Decrease in pile capacity |
|---|---|---|---|---|---|---|---|
| 42 | 0 | 2.96 | 0 | 0 | 4.79 | 0 | 0 |
| | 2 | 2.96 | 0 | 0 | 4.79 | 0 | 0 |
| | 4 | 2.45 | 0.51 | 17.22 | 4.10 | 0.69 | 14.40 |
| | 7 | 2.05 | 0.91 | 30.74 | 3.53 | 1.26 | 26.30 |
| 49 | 0 | 4.04 | 0 | 0 | 6.44 | 0 | 0 |
| | 3 | 3.63 | 0.41 | 10.14 | 5.87 | 0.57 | 8.85 |

(continued)

**Table 4** (continued)

| Pile diameter (mm) | Load applied at *e* (mm) | Ultimate load $D_r$ = 60% (kN) | Decrease in pile capacity (kN) | %Decrease in pile capacity | Ultimate load $D_r$ = 70% (kN) | Decrease in pile capacity (kN) | % Decrease in pile capacity |
|---|---|---|---|---|---|---|---|
| | 5 | 3.40 | 0.64 | 15.84 | 5.41 | 0.46 | 7.14 |
| | 8 | 3.06 | 0.98 | 24.25 | 4.96 | 0.91 | 14.13 |
| 75 | 0 | 8.99 | 0 | 0 | 12.37 | 0 | 0 |
| | 4 | 8.36 | 0.64 | 7.11 | 11.36 | 1.01 | 8.16 |
| | 8 | 7.61 | 1.39 | 15.44 | 10.32 | 2.05 | 16.57 |
| | 12 | 6.92 | 2.08 | 23.11 | 9.23 | 3.14 | 25.38 |
| 100 | 0 | 15.05 | 0 | 0 | 21.34 | 0 | 0 |
| | 6 | 13.67 | 1.38 | 9.16 | 19.26 | 2.08 | 9.74 |
| | 11 | 12.51 | 2.54 | 16.87 | 17.53 | 3.81 | 17.85 |
| | 16 | 11.36 | 3.69 | 24.51 | 15.80 | 5.54 | 25.96 |

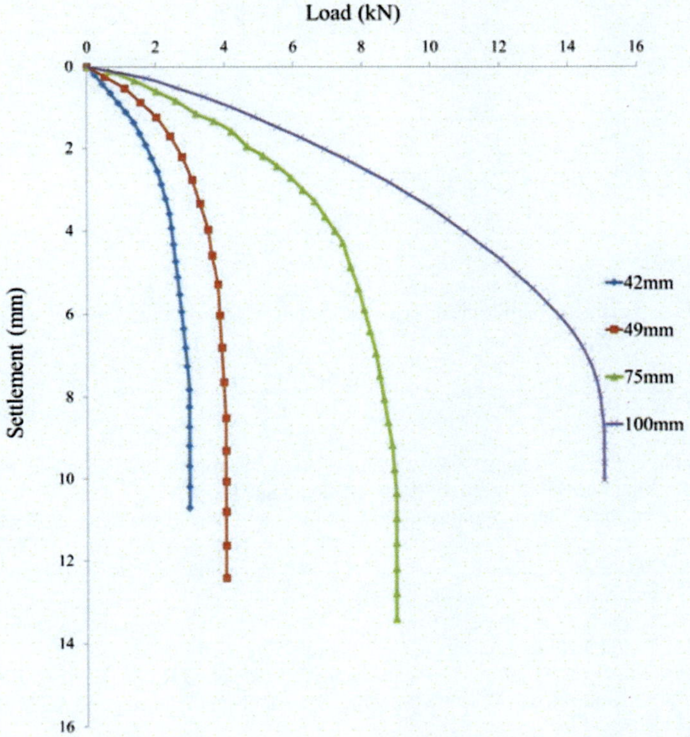


**Fig. 9** Effect of diameter on pile capacity (axial loading at $D_r = 60\%$)

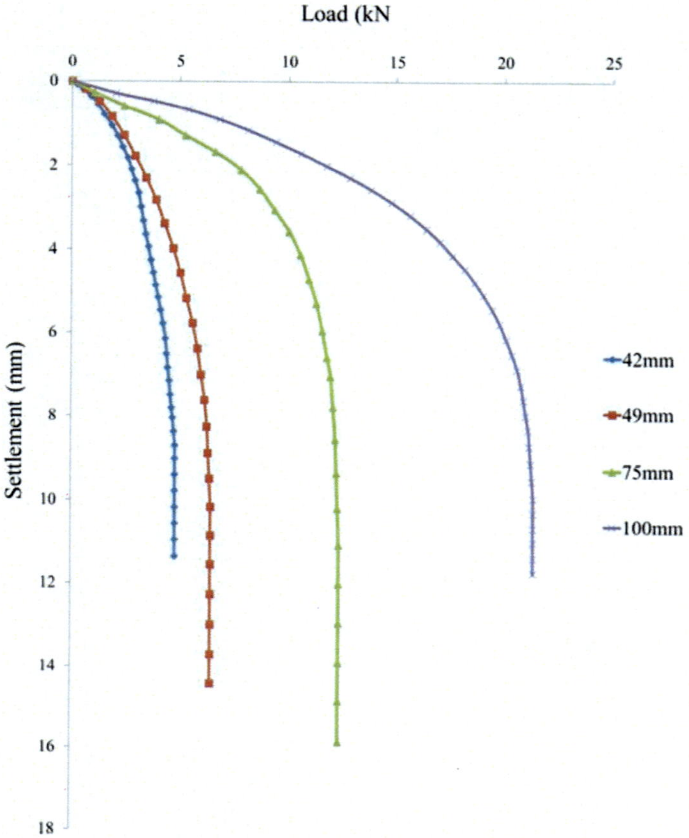


**Fig. 10** Effect of diameter on pile capacity (axial loading at $D_r = 70\%$)

# References


1. Al-Mhaidib, A.I.: Experimental investigation of the behaviour of pile groups in sand under different loading rates. J. Geotech. Eng. **24**, article no.889 (2006)
2. Amir, A., Ahmad, A.: Design and reanalysis of pile cap under eccentricity. J. Geotech. Eng. (2016)
3. Meyerhof, G.G., Sahap Yalcin, F.A., Mathur, S.K.: Ultimate pile capacity for eccentric inclined load. J. Geotech. Eng. **109**(3), 408–423 (1983)
4. Kong, L.G., Chen, R.P., Wang, S.H., Chen, Y.M,: Response of 3 × 3 pile groups in silt subjected to eccentric lateral loading. J. Geotech. Geo Environ. Eng. **141** (2015)
5. Padmavathi, M., Koteswara Rao, L., Padmavathi, V., Madhav, M.R., Sanjeeva, P.: Response of Four Pile Group Subjected to Eccentric Loading, Springer Nature Switzerland AG (2019) SUCI, pp. 26–35 (2019)
6. Zhang, Q.-q., Zhang, Z.-m.: Simplified calculation approach for settlement of single pile and pile groups. J. Comput. Civil Eng. ASCE **26**(6), 750–758

7. Di Laora, R., de Sanctis, L.: Bearing capacity of pile groups under vertical eccentric load. Acta Geotechnical **102** (2018)
8. Nogami, T., Chen, H.L.: Simplified approach of axial pile group analysis. J. Geotech. Eng. **110**(9) (1984)

# Evaluation of Bearing Capacity for Cast In-Situ Bored Piles

P. S. Raisa and S. Chandrakaran

## 1 Introduction

The use of bored cast in-situ piles has multiplied around the world. Bored cast in-situ piles have a moderate bearing capacity, low cost, reduced vibration during installation and allow easy length adjustments. The complex nature of the embedment ground of piles and lack of suitable analytical models for predicting the pile bearing capacity are the main reasons for the geotechnical engineer's tendency to pursue further research on this subject [1].

The prediction of bearing capacity of a bored cast in-situ pile is a complex problem. It is necessary to consider factors such as the boring method, installation process, quality of concrete, ground conditions, and experienced expertise while designing piles. The method of installation has a great impact on pile foundation, i.e., drilling can cause vibration and disturb the surrounding soil. Even after the installation of the pile, changes may occur in the soil naturally with time. The appropriate pile capacity can be obtained only by conducting a pile load test. The conduction of the pile load test for small projects is not economical. In such cases, other methods can be adopted for the prediction of pile bearing capacity. Various methods have been developed for predicting the pile bearing capacity by considering soil-pile interaction, soil stratigraphy, and soil resistance. Interpretation of bearing capacity can be done by graphical methods, load test, dynamic analysis, dynamic testing methods, static analysis, and SPT methods.

The static method in the IS 2911-Part 1/Sec 2 to determine the bearing capacity of bored cast in-situ pile contains many parameters that need to be evaluated using

P. S. Raisa (✉)
Department of Civil Engineering, Karunya Institute of Technology and Sciences, Coimbatore 641114, India

S. Chandrakaran
Department of Civil Engineering, NIT Calicut, Calicut 673601, India



C. N. V. Satyanarayana Reddy et al. (eds.), *Ground Characterization and Foundations*, Lecture Notes in Civil Engineering 167,
https://doi.org/10.1007/978-981-16-3383-6_53

trigonometric functions or graphs and tables. The static method uses the concept of critical depth for cohesive and cohesionless soil to find the angle of internal friction.

Recently, using the results of penetrometer tests like standard penetration test and cone penetration test to estimate the bearing capacity of piles had been the subject of a considerable number of researchers, and several approaches have been proposed [2]. Pile capacity by SPT is one of the easiest and earliest applications used. Due to SPT's simplicity of execution, a field engineer finds the method to be one of the most amiable and reliable ones [3]. There are two types of SPT methods to find capacity, direct and indirect methods. Direct methods apply $N$ values directly, whereas in indirect methods, friction angle and undrained shear strength values are required to find capacity. By using SPT data, the simplest form of the equation can be formed. With such an equation, pile load test can be avoided, i.e., only SPT value and area of the pile are necessary. Also, the SPT method can be applied to find the capacity of any kind of soil.

Benali et al. [4] developed a new method for pile capacity estimation based on the SPT test in different stratigraphy. The Eslami and Fellinius rule was applied and calibrated with 24 pile cases [4]. To obtain the unit base resistance of piles from SPT results, the failure zone and failure mechanism should be specified around the base of the pile [4].

Shah et al. [5] performed regression analysis on the parameters to obtain a specified link between bearing capacity and other parameters like cohesion, angle of internal friction, diameter, and depth of pile. The results show that there is no need to refer to a myriad of factors that need to be evaluated using trigonometric functions or by the use of graphs and tables unlike the equations recommended by the Indian Standards [5].

## 2 Database Records

Pile load test data and corresponding soil investigation report of fifteen bored cast in-situ piles installed along the coastal areas of Kerala were collected. The sites in these areas are covered by laterite, silty sand, and clayey sand. The soil report indicates a weak bearing strata at shallow depth, leading to the construction of a pile foundation. The diameter of piles varies from 0.5 to 1 m, and embedment length varies from 6 to 35 m. Out of 15 sites, five sites have cohesive soil, and other sites are covered by cohesionless soil. The summary of pile data is given in Table 1.

**Table 1** Summary of pile data collected

| Pile no. | Location | Soil | Pile diameter (mm) | Pile length (m) | Test load (T) | Total settlement (mm) |
|---|---|---|---|---|---|---|
| 1 | Chala bypass, Kannur | Cohesionless soil | 700 | 18.87 | 380 | 25.32 |
| 2 | Pottamal, Calicut | Cohesionless soil | 700 | 12 | 225 | 1.36 |
| 3 | Kannur | Cohesionless soil | 750 | 11.8 | 270 | 9.04 |
| 4 | Cochin- coast guard | Cohesionless soil | 600 | 15.15 | 368 | 3.24 |
| 5 | Cochin- coast guard | Cohesionless soil | 500 | 12.85 | 260 | 4.38 |
| 6 | Kunnamangalam, Calicut | Cohesive soil | 500 | 6.65 | 80 | 1.61 |
| 7 | Kunnamangalam, Calicut | Cohesive soil | 700 | 8.48 | 157 | 5.03 |
| 8 | Pallipuram, Cochin | Cohesionless soil | 600 | 34.6 | 210 | 4.27 |
| 9 | Azheekkal, Kannur | Cohesionless soil | 500 | 27 | 63 | 0.146 |
| 10 | Annakara, Thrissur | Cohesionless soil | 500 | 12 | 45 | 1.91 |
| 11 | Elamkunnapuzha | Cohesive soil | 500 | 8.2 | 33 | 12 |
| 12 | Elamkunnapuzha | Cohesive soil | 500 | 11.5 | 33 | 12 |
| 13 | Calicut | Cohesive soil | 500 | 7.5 | 42 | 3.627 |
| 14 | Kannur | Cohesionless soil | 600 | 24.37 | 181 | 0.73 |
| 15 | Trivandrum | Cohesionless soil | 1000 | 34.96 | 450 | 5.8 |

# 3 Determination of Bearing Capacity

## *3.1 Pile Capacity by Interpretation*

In some cases, the piles not loaded to failure, so the interpretation methods are used to interpret the failure load. Interpretation methods are graphical methods plotted using load and settlement data obtained from the pile load test. The chosen interpretation methods are Van der Veen's [6], Hansen's [7], Chin's [8], Decourt's [9], and Tangent intersection method. From the results of the failure load, the failure criterion is chosen as Ahmad and Pise's method. The results show that the capacity of piles is over-estimating in Decourt's and Van der Veen's, while Chin's and Mazurkiewicz's give comparatively reliable results. The bearing capacity obtained by the tangent intersection method is under-estimating.

### 3.2 Pile Capacity by Empirical and Analytical Methods

In this paper, we have chosen the static method, $\alpha$ method, $\beta$ method, and four SPT methods to compare and validate the results. The summary of these methods is given in Table 2.

The results of empirical and analytical methods are as shown in Table 3.

The following points are noted in the bearing capacity obtained by empirical and analytical methods,

1. The capacity of only five piles can be estimated by $\alpha$ method as this method applies to only cohesive soils. The $\beta$ method can be applied to both cohesive and cohesionless soils.
2. The SPT methods estimate bearing capacity without considering cohesion and excessive pore water pressure occurring around the pile. Thus, this method of SPT may not be reliable for low-permeable soils like clay and silt [5].
3. Out of four SPT method, Bazaraa and Kurkur does not contain any failure criteria. It is purely based on empirical analysis.
4. The end bearing SPT value varies for all four methods. This may cause uncertainty in bearing capacity values.
5. The bearing capacity estimated by the $\beta$ method and Aoki and D'alencar is over-estimating.
6. Meyerhof method gives reliable values of capacity compared to all other chosen methods.

### 3.3 Proposed Method

A new SPT method has been proposed by trial and error to find the bearing capacity of bored cast in-situ pile. In this paper, the most commonly used seven methods to interpret the failure load are selected. The failure load obtained by interpretation methods was analyzed and compared to decide the failure criterion.

The selection of the failure zone and failure criterion is necessary to find the unit base resistance. The ratio of *Qp*/*Qm* (Predicted bearing capacity divided by measured bearing capacity) was calculated, and the failure criterion is selected as Ahmad and Pise's method. Ahmad and Pise's method is also known as Modified Chin's method.

The SPT value of soil varies along the depth due to the heterogeneity of soil. This variation of N value can cause uncertainty in bearing capacity. The application of end bearing N value varies for each method. The end bearing and shaft SPT values are chosen by calculating the average of the N value along the depth. There are two methods to find the average, namely arithmetic mean and geometric mean.

**Table 2** Summary of analytical and empirical methods

| S. No. | Method | Unit base resistance | Unit shaft resistance | Remarks |
|---|---|---|---|---|
| 1 | IS code method (IS 2911-2005) | For cohesive soil $Q_b = C_p N_c A_p$ in (KN)<br>For cohesionless soil $Q_b = P_D N_q A_p$ in (KN) | For cohesive soil $Q_s = \sum_{i=1}^{n} \alpha_i C_i A_{si}$ in (KN)<br>For cohesionless soil $Q_s = \sum_{i=1}^{n} K_i\ P_{Di}\ \tan \delta_i A_{si}$ in (KN) | $\alpha$-adhesion factor in IS-2911<br>$\delta = 3\phi/4$ |
| 2 | $\alpha$ method (only for cohesive soil) | $Q_b = C_u N_c A_p$ in (KN) | $Q_s = \sum_{i=1}^{n} \alpha_i C_i A_{si}$ in (KN) | $\alpha$-adhesion factor from Dennis and Olsen curve |
| 3 | $\beta$ method (cohesive and cohesionless soil) | $Q_b = \sigma_v N_q A_p$ in (KN) | $Q_s = f_s A_s = A_s\,\beta\,\sigma_z\ f_s = K_o \tan \delta\,\sigma_z = (1 - \sin\ \phi) \tan \delta\,\sigma_z$ in (KN) | $\delta = 3\phi/4$ |
| 4 | Aoki and De'Alencar [10] | $r_t = (k/1.75)\ N_b$ in (MPa)<br>$N_b$ = average of 3 values of $N$ around pile base | $r_s = (ak/3.5)\ N_s$ in (KPa)<br>$N_s$ = average value of $N$ along pile embedment depth | Failure criteria: Van der Veen method<br>For sand: $a = 14$, $k = 1$<br>For clay: $a = 60$, $k = 0.2$ |
| 5 | Meyerhof [11] | $r_t = n_b\ N_b$ in (MPa)<br>$N_b$ = average of $N$ between 8B above and 4B below pile base, $N_b \le 50$ | $r_s = n_s\ N_s$ in (KPa)<br>$N_s$ = average value of $N$ around pile embedment depth | Failure criteria: min slope of load-settlement curve<br>$n_b = 0.12–0.40$<br>$n_s = 1 - 2$ |
| 6 | Bazaraa and Kurkur [12] | $r_t = n_b\ N_b$ in (MPa)<br>$N_b$ = average of $N$ from 1B to 3.75B around pile base | $r_s = n_s\ N_s$ in (KPa)<br>$N_s$ = average value of $N$ around pile embedment depth | $n_b = 0.06–0.2$<br>$n_s = 2 - 4$ |

(continued)

The arithmetic mean is calculated by the formula,

$$N_a = \left(\frac{(N_1 + N_2 + \cdots + N_n)}{n}\right) \tag{1}$$

The geometric mean is calculated by the formula,

$$N_g = (N_1 \times N_2 \times \cdots \times N_n)^{1/n} \tag{2}$$

**Table 2** (continued)

| S. No. | Method | Unit base resistance | Unit shaft resistance | Remarks |
|---|---|---|---|---|
| 7 | Decourt [9] | $r_t = k_b\, N_b$ in (MPa)<br>$N_b$ = average value of $N$ around pile toe | $r_s = \alpha(2.8N_s + 10))$ in (KPa)<br>$N_s$ = average value of $N$ around pile embedment depth | Failure criteria: Van der Veen method<br>For sand:$k_b = 0.325$, $\alpha = 0.5 - 0.6$<br>For clay: $k_b = 0.08$, $\alpha = 0.1$ |

*Note* $A_p$—C/S area of pile tip in m$^2$; $A_{si}$—surface area of pile shaft in $i$th layer in m$^2$; $N_c$ and $N_q$—bearing capacity factors; $C_p$—average cohesion at pile tip KPa; $C_u$—cohesion at pile tip in KPa; $C_i$—cohesion at $i$th layer in KPa; $\sigma_v$—average effective overburden pressure; $\sigma_z$—effective overburden pressure; $\delta$—effective angle of internal friction; $K$—coefficient of earth pressure; $P_D$—effective overburden pressure at pile tip; $P_{Di}$—effective overburden pressure at $i$th layer

**Table 3** Pile bearing capacity calculated by empirical and analytical methods

| Pile no. | Test load (Ton) | Is code method (Ton) | $\alpha$ method (Ton) | $\beta$ method (Ton) | Aoki and D'alencar method (Ton) | Bazaraa and Kurkur's method (Ton) | Decourt's method (Ton) | Meyerhof's method (Ton) |
|---|---|---|---|---|---|---|---|---|
| 1 | 380 | 558 | | 2517 | 1745 | 651 | 834 | 360 |
| 2 | 225 | 394 | | 2336 | 1183 | 529 | 870 | 234 |
| 3 | 270 | 282 | | 2310 | 1488 | 493 | 876 | 288 |
| 4 | 368 | 415 | | 3167 | 1366 | 542 | 820 | 295 |
| 5 | 260 | 303 | | 2551 | 1017 | 421 | 586 | 214 |
| 6 | 80 | 126 | 85 | 845 | 245 | 220 | 174 | 122 |
| 7 | 157 | 254 | 164 | 1939 | 641 | 318 | 319 | 279 |
| 8 | 210 | 306 | | 1659 | 964 | 435 | 650 | 179 |
| 9 | 63 | 85 | | 683 | 288 | 160 | 267 | 47 |
| 10 | 45 | 52 | | 397 | 214 | 78 | 142 | 69 |
| 11 | 33 | 45 | 43 | 361 | 167 | 105 | 153 | 61 |
| 12 | 33 | 48 | 48 | 277 | 126 | 103 | 114 | 66 |
| 13 | 42 | 67 | 57 | 449 | 185 | 140 | 161 | 67 |
| 14 | 165 | 389 | | 1649 | 1009 | 454 | 718 | 280 |
| 15 | 450 | 915 | | 3339 | 2472 | 699 | 1788 | 572 |

The average calculated by geometric mean shows pertinent results compared to the arithmetic mean. Thus, geometric mean was used to find SPT value at the end bearing and shaft.

After analyzing the different ways of selecting the failure zone of influence, it is found practical to use the Eslami and Fellenius rule [1]. The end bearing $N$ value is taken as the average of N between 4B below and 8B above the pile tip. The use of Eslami and Fellenius rule gave consistent results of capacity.

From the overall capacity obtained by Ahmad and Pise method, the end bearing and shaft resistance were determined using IS 2911. Thus, by attaining the percentage of shaft and end resistance, a trial and error method was followed to form the new formula. By following the geometric averaging and Eslami and Fellenius rule, the new formula is proposed as below.

For cohesionless soil:

$$Q_u = Q_b + Q_s = 60 * N_b * A_b + 1.35 * N_s * A_s \tag{3}$$

For cohesive soil:

$$Q_u = Q_b + Q_s = 48 * N_b * A_b + 1.20 * N_s * A_s \tag{4}$$

where $Q_u$ is the overall capacity in KN, $Q_b$ is the end bearing capacity in KN, $Q_s$ is the shaft resistance capacity in KN, $N_b$ is the average of $N$ between 4B below and 8B above the pile tip, $N_s$ is the average value of N along pile embedment depth, $A_b$ is the cross-sectional area of pile tip in m$^2$, and $A_s$ is the surface area of pile shaft in m$^2$.

## 4 Validation

The capacity obtained by the proposed method is compared with the empirical and analytical methods by the Log-Normal distribution method. This method can estimate the performance prediction of all the methods. In the Log-Normal distribution method, a graph is plotted between $Q_p/Q_m$ and probability density function.

At first, $Q_p/Q_m$ is calculated, and the natural logarithm for $Q_p/Q_m$ is measured for each pile. To find the probability density function, the mean ($\mu_{\ln}$) and standard deviation ($\sigma_{\ln}$) for the natural logarithm of $Q_p/Q_m$ are calculated as follows:

$$\mu_{\ln}\left(\frac{Q_p}{Q_m}\right) = \frac{1}{n}\sum_{i=1}^{n} \ln \frac{Q_p}{Q_m} \tag{5}$$

$$\sigma_{\ln}\left(\frac{Q_p}{Q_m}\right) = \left[\frac{1}{n-1}\sum_{i=1}^{n}\left[\ln\left(\frac{Q_p}{Q_m}\right) - \mu_{\ln}\right]\right] \tag{6}$$

The probability density function is calculated as follows,

$$F(x) = \frac{1}{\sqrt{2pi}\sigma_{\ln}x} \exp\left[-\frac{1}{2}\left[\frac{\ln(x) - \mu_{\ln}}{\sigma_{\ln}}\right]^2\right] \quad (7)$$

where $Q_p$ is the predicted value by various methods chosen, and $Q_m$ is the measured value of pile capacity by pile load test, $x = \left(\frac{Q_p}{Q_m}\right)$, $\mu_{\ln}$ is the mean of $\ln\left(\frac{Q_p}{Q_m}\right)$, and $\sigma_{\ln}$ is the standard deviation on $\ln\left(\frac{Q_p}{Q_m}\right)$.

The Log-Normal distribution graph shows a wide difference in the capacity obtained by all the methods and proposed method. The under and over-estimation of capacity can be clearly understood. The scattering of the proposed method was found to be low, and it showed better precision. Figure 5 shows the Log-Normal distribution of all the methods chosen for analysis.

Based on the Log-Normal analysis, the probability that the prediction falls within ±25% of all the methods is estimated by the below formula.

$$P(\%) = 100 \int_{0.75}^{1.25} f(x)dx \quad (8)$$

The Log-Normal approach also predicts the degree of scattering of uncertainty. The more the standard deviation, the higher is the degree of scattering and uncertainty. Here, β method has the higher degree of uncertainty. The results of performance prediction for the methods are presented in Table 4.

**Table 4** Probability that performance lies within ± 25% error

| S. No. | Method | Probability of estimating within ±25 error |
|---|---|---|
| 1 | IS code method | 62 |
| 2 | $\beta$ method | 12.65 |
| 3 | Aoki and De'alencar method | 24 |
| 4 | Bazaraa and Kurkur Method | 33.85 |
| 5 | Decourt method | 24.35 |
| 6 | Meyerhof method | 51.2 |
| 7 | Proposed method | 70.5 |

## 5 Conclusions

The bearing capacity determination of pile is always a complex problem faced by engineers and researchers. Among all the methods for capacity calculation, the SPT method is found to be flexible in terms of estimation, cost, and time. In some cases, unpredictable values are obtained. The geological changes that may occur in soil and the surrounding resources with period of time is one the reasons causing failure of the pile. This uncertainty of capacity may be due to errors in calculation, instrumental error, and unskilled workforce.

A wide range of peaks and troughs was observed in pile capacity obtained by different methods. This can predict capacity as a tedious problem.

Though $\alpha$ method gives reliable results of capacity, its application is limited to cohesive soils. Hence, this method was not considered for comparison with the proposed formula.

The scattering of the curve indicates the over and under-estimation of the methods. In Fig. 1, it can be noted that the scattering is low for the proposed method. The $\beta$ method has the highest degree of scattering, and hence, the capacity is over-estimating.

The calculation of the probability of error shows that the proposed method has better precision performance compared to other chosen methods. Meyerhof method and IS code method also give a reliable value of bearing capacity.

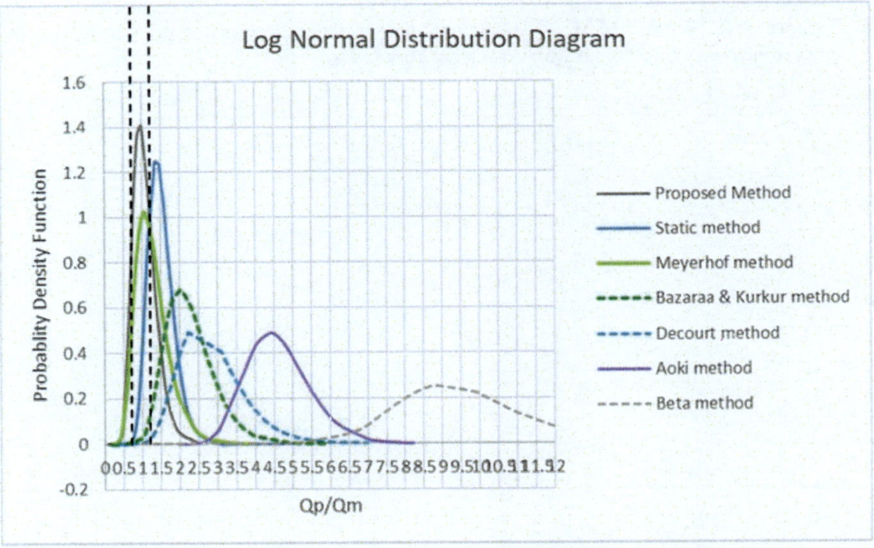


**Fig. 1** Log-Normal distribution for different methods of pile bearing capacity

## References

1. Shariatmadari, N., Eslami, A., Karimpour-Fard, M.: Bearing capacity of driven piles in sands from SPT–applied to 60 case histories. Iran. J. Sci. Tech. **32**, 125–140 (2008)
2. Karimpour-Fard, M., Eslami, A.: Estimation of vertical bearing capacity of piles using the results CPT and SPT tests. In: 4th International Conference on Site Characterization, pp. 1055–1062. Taylor & Francis Group, Brazil (2012)
3. Thounaojam, S., Sultana, P.: Prediction of bearing capacity of bored cast-in-situ piles. IOSR J. Mech. Civ. Eng. 01–06 (2015). e-ISSN: 2278-1684
4. Benali, A., Nechnech, A., Bouafia, A.: Bored pile capacity by direct SPT methods applied to 40 case histories. Civ. Environ. Res. **5**, 118–122 (2013)
5. Shah, A., Advani, P., Patel, J., Soni, H.: Estimation of load carrying capacity of bored cast in-situ piles. Int. J. Eng. Technol. Sci. Res. **4**(9), 814–819 (2017)
6. Van der Veen, C.: The bearing capacity of a pile. In: Proceedings of the Third International Conference Soil Mechanics Foundation Engineering, vol. 02, pp. 84–90. Zurich (1953)
7. Hansen, J.B.: Discussion on hyperbolic stress-strain response—cohesive soils. ASCE J. Soil Mech. Found. Eng. **89**(SM4), 241–242 (1963)
8. Chin, F.K.: Estimation of the ultimate load of piles not carried to failure. In: Proceedings of the 2nd Southeast Asian Conference on Soil Engineering, pp. 81–92. Singapore (1970)
9. Decourt, L.: Behavior of foundations under working load conditions. In: Proceedings of the 11th Pan-American Conference on Soil Mechanics and Geotechnical Engineering, vol. 04, pp. 453–488. Foz Dulguassu, Brazil (1999)
10. Aoki, N., De'Alencar, D.: An approximate method to estimate the bearing capacity of piles. In: Proceeding of the Fifth Pan-American Conference on Soil Mechanics and Foundation Engineering, pp. 367–376. Buenos Aires, Argentina (1975)
11. Meyerhof, G.G.: Bearing capacity of settlement of pile foundations. The eleventh Terzaghi lecture. ASCE J. Geotech. Eng. **102**(3), 195–228 (1976)
12. Bazaraa, A.R., Kurkur, M.M.: *N*-values used to predict settlements of piles in Egypt. In: Proceedings of In Situ'86, pp. 462–474. New York (1986)

# Effect of Bentonite Support Fluid on Pile Capacity

**Keerthi Sabu and Benny Mathews Abraham**

## 1 Introduction

Stabilization of soil in pile borehole is of prime importance for installation of cast-in-situ concrete piles. Various support fluids like bentonite (calcium- or sodium-based) and polymer are commercially available for stabilizing boreholes. Among these, bentonite support fluid is most commonly used in almost all piling sites due to its ease in availability, ease in handling and economy. So a study on the effect of these bentonite support fluids at pile–soil interface will be highly beneficial for engineers. Effect of bentonite layer on frictional resistance at pile surface is to be studied, and results could possibly affect the pile capacity calculations recommended by Indian Standards. Behaviour of bentonite fluid in various types of soil is also important.

The history and formulation of excavation fluids and their uses for the construction process were studied by Jefferis et al. [4]. The effect of water content, normal stress and rough surface on the relationship between shear stress and shear displacement of clay–concrete interface was analysed by Shakir and Zhu [8]. An interface simple shear apparatus was used for their studies. They have compared the results obtained from the simple shear tests and that obtained from the direct shear tests. It also involved using bentonite and polymer slurries as an interface layer between soil and concrete [9].

K. Sabu (✉)
Research Scholar, Division of Civil Engineering, Cochin University of Science and Technology, Cochin 682 022, India

Asst. Professor, Federal Institute of Science and Technology, Hormis Nagar, Angamaly, Ernakulam 683 577, India

B. M. Abraham
Retd. Professor, Division of Civil Engineering, School of Engineering, Cochin University of Science and Technology, Cochin 682 022, India

 
C. N. V. Satyanarayana Reddy et al. (eds.), *Ground Characterization and Foundations*, Lecture Notes in Civil Engineering 167,
https://doi.org/10.1007/978-981-16-3383-6_54

Lam et al. [5] presented a field trial in which the construction and testing of three piles were carried out at Stratford in east London. Maintained load tests done on the three test piles showed that the polymer fluids had better load-settlement behaviour and longer excavation open time without compromising the performance of the foundation. Lam et al. [6] conducted shear tests on concrete–sand interface using both polymer and bentonite support fluids and have compared their performance. From their findings, it was found that polymer support fluids do not form filter cake layer at the interface and hence provide an agreeable alternative to the bentonite slurry, which is commonly used in piling.

Rugang et al. [7] have presented the effect of water-based drilling fluid components on filter cake structure. Physical modelling of pile was done by Shrivastava et al. [10] to find out the uplift capacity of bored pile. Firstly, a modelled pile was bored without the use of drilling fluid and by using different percentage of bentonite and also using different percentage of polymer. The pull-out test was then performed on the fabricated test set-up. The test results were compared with the theoretical uplift capacity suggested by IS 2911 (Part 1/sec 2) [2]. Yoo and Han [11] have conducted shear tests to examine the performance of drilling fluids at concrete–soil and concrete–rock interface, particularly in the presence of seawater. From the results, it was found that polymer drilling fluid outperformed bentonite.

### *1.1 Need for the Study*

For the excavation to remain stable, the formation of a layer of filter cake with bentonite slurry is essential. But in the case of friction piles, the presence of filter cake adversely affects the load-carrying capacity of piles as it can reduce the interfacial resistance. Many investigations have been carried out to study the effect of bentonite filter cake on interface resistance, but no studies came out with clear conclusions that can be practically applied in the field.

Bentonite reduces the side resistance even with thin mud layer produced by exposing the hole for slurry for short time. Studies to examine the effect of thin layer slurries between soil and concrete are limited and need more investigation. Direct shear and simple shear tests are more practical than large-scale tests and most favourable for detailed studies. Studies on concrete–soil interface resistance will be beneficial for estimating quantitatively the effect of the presence of bentonite filter cake on the actual load-carrying capacity of pile foundations.

## 2 Materials and Methods

Bentonite was the main material used for the study. For representing the pile surface, cement mortar blocks were precast, for which ordinary Portland cement was used. For studying the effect of bentonite on frictional resistance at pile–soil interface, different soil types were used for conducting direct shear tests. Sand (fine, medium, coarse), clay and red earth were used to represent soil surrounding the pile.

Sand used for the study was river sand, procured from Kalady, which is a branch of the Periyar River—was dried and sieved into different fractions. River sand of three grades—fine (75–425 μm), medium (425–2 mm) and coarse (2–4.75 mm) as per BIS 1498 classifications—was used in the present study [1]. The grain size distribution curves of different fractions of sand are shown in Fig. 1.

43 grade ordinary Portland cement confirming to IS 8112 was used for making cement mortar blocks of size 60 × 60 × 12.5 mm [3]. The cement was kept in airtight container to avoid any change in properties with the time of storage. The physical properties of cement are presented in Table 1.

The bentonite used in this study is a commercially available, highly expansive sodium bentonite. Bentonite shows a great affinity towards moisture. The percentage of water present in the sample of bentonite varies depending on the climatic conditions. So bentonite was preserved in highly airtight polythene bags. Marine clay was collected from a site at Cheranalloor, Ernakulam, Kerala. Samples

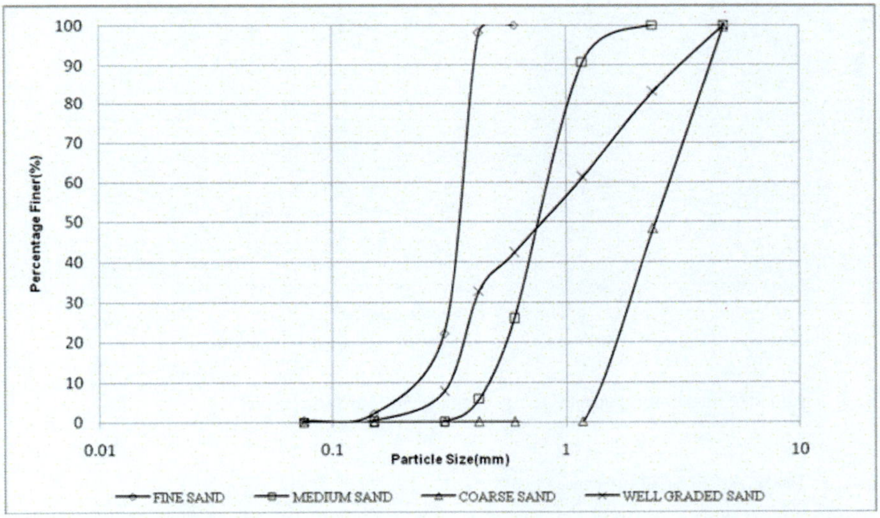


**Fig. 1** Grain size distribution curves of different fractions of sand

**Table 1** Properties of the cement used

| S. No. | Property | Characteristic value |
|---|---|---|
| 1 | Standard consistency | 37% |
| 2 | Initial setting time | 118 min |
| 3 | Final setting time | 216 min |
| 4 | Blaine's specific surface | 298,524 $mm^2/g$ |
| 5 | Specific gravity | 3.14 |

were collected from boreholes advanced by auger method. Proper attention was given while collecting clay samples, as it should not get mixed up with bentonite and was properly collected in polythene bags, so that its natural moisture content is preserved while conducting the laboratory tests. Red earth used for the study was obtained from a site at Ernakulam, Kerala. It was collected in polythene bags so that its properties will not change at the time of laboratory tests. Properties of bentonite, marine clay and red earth used for the study and its grain size distribution are shown in Fig. 2 and Table 2, respectively.

The cement mortar block was precast using a mild steel mould (60 × 60 × 25 mm) and plate (60 × 60 × 12.5 mm) assembly. Concrete mix was prepared by standard procedure which was followed in casting mortar cubes for compressive strength

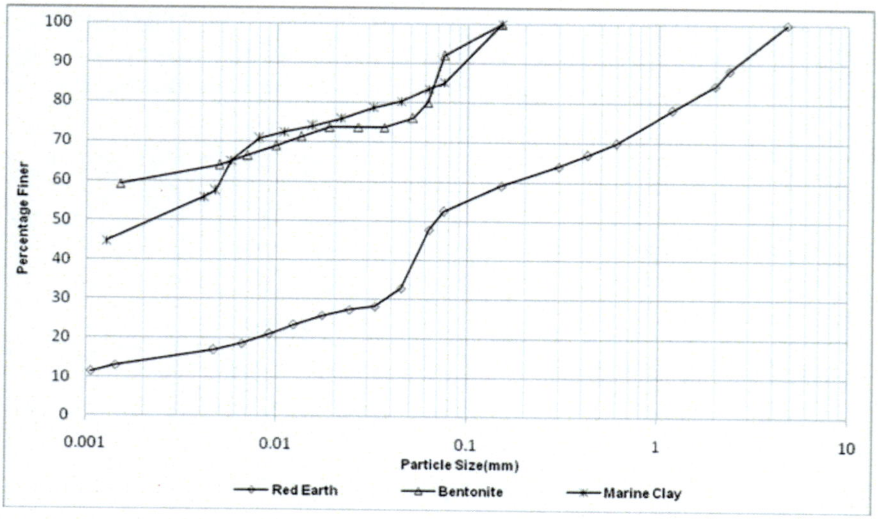


**Fig. 2** Grain size distribution curves of red earth, bentonite and marine clay

**Table 2** Properties of bentonite, marine clay and red earth used for study

| S. No. | Property | Bentonite | Marine clay | Red earth |
|---|---|---|---|---|
| 1 | Natural moisture content (%) | – | 76 | 27 |
| 2 | Liquid limit (%) | 386 | 115 | 65 |
| 3 | Plastic limit (%) | 56 | 46 | 41 |
| 4 | Plasticity index (%) | 330 | 59 | 24 |
| 5 | Shrinkage limit (%) | 7 | 23 | 36 |
| 6 | Free swell index | 1300% | 3.6 cc/g | – |
| 7 | Clay (%) | 61 | 48 | 16 |
| 8 | Silt (%) | 31 | 37 | 37 |
| 9 | Sand (%) | 8 | 15 | 47 |

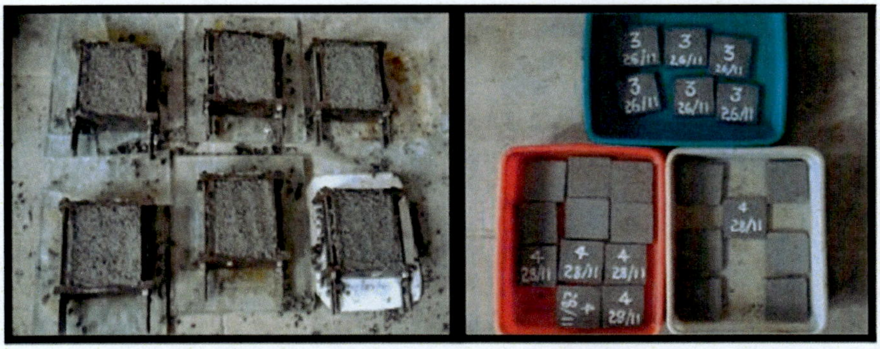

(a) (b)

**Fig. 3** Cement mortar blocks **a** cast-in moulds **b** kept for curing

test. Immediately after mixing, the mortar was filled into the mild steel mould and plate assembly. Filling was done by giving compaction manually. After 24 h, cement mortar blocks were removed from the mould and kept for 28 days of curing as shown in Fig. 3.

## 3 Methodology

For finding the interface frictional resistance, several laboratory tests like model pile tested in a triaxial cell, direct shear tests (concrete on sand or sand on concrete), model pile in centrifuge, etc., can be adopted [8, 9]. But direct shear and simple shear tests are more practical than large-scale tests and considered optimal for detailed studies.

Among these tests, direct shear test using a direct shear apparatus with shear box size 60 mm × 60 mm × 50 mm was selected for the study. A cement mortar block of 60 mm × 60 mm × 12.5 mm was placed in the lower half of shear box, and a layer of bentonite was spread over this. After that, sand (fine/medium/coarse) was filled into the shear box and made to a height of 25 mm and the shear test was conducted. In practical cases, pile may be surrounded by different types of soil with different properties, different moisture content, different densities, etc. In order to comply with these varying site conditions, direct shear tests were conducted on the following types of soil at different conditions.

Fine, medium and coarse sand at a unit weight of 1.5 g/cc, in dry condition.
Fine, medium and coarse sand at a unit weight of 1.55 g/cc, in dry condition.
Fine, medium and coarse sand at a unit weight of 1.5 g/cc, in saturated condition.
Fine, medium and coarse sand at a unit weight of 1.55 g/cc, in saturated condition.
Well-graded sand at a unit weight of 1.5 g/cc, in dry condition.
Marine clay only.
Red earth only.

Fine, medium and coarse sand at a unit weight of 1.5 g/cc (dry) on cement mortar surface.
Well-graded sand at a unit weight of 1.5 g/cc (dry) on cement mortar surface.
Marine clay on cement mortar surface.
Red earth on cement mortar surface.
Bentonite layer b/w fine/medium/coarse sand at a unit weight of 1.5 g/cc (dry) and cement mortar surface.
Bentonite layer b/w well-graded sand and cement mortar surface.
Bentonite layer b/w marine clay and cement mortar surface.
Bentonite layer b/w red earth and cement mortar surface.

After conducting shear tests for all these combinations, plots of normal stress vs shear stress were made for each case. Shear strength parameters $c$ and $\phi$ were found out from these plots. Values of these shear strength parameters were analysed for studying the effect of bentonite layer at pile–soil interface.

# 4 Results and Discussions

## 4.1 Direct Shear Tests on Sand

Bearing capacity of piles can be determined by approximating the angle of friction between pile and surrounding soil, δ to the angle of shearing resistance, $\phi$. Direct shear tests were conducted with fine, medium and coarse sand fractions (confirming to BIS 1498 classifications) to determine the value of $\phi$ [1]. Tests were conducted at a density of 1.5 g/cc in both dry and saturated conditions. Results obtained are presented in Table 3. The test results showed that $\phi$ values were either same or tend to decrease when sample is saturated. In all the cases, the variation in magnitudes of

**Table 3** Direct shear test results on fine, medium and coarse sand

| Type and soil condition | Angle of shearing resistance, $\phi$ (°) for a density of 1.5 g/cc |
|---|---|
| *Fine sand* | |
| (a) Dry state | 29 |
| (b) Saturated state | 26 |
| *Medium sand* | |
| (a) Dry state | 31 |
| (b) Saturated state | 28 |
| *Coarse sand* | |
| (a) Dry state | 34 |
| (b) Saturated state | 34 |

**Table 4** Values of angle of shearing resistance for the bentonite at different water contents sandwiched b/w medium sand and cement mortar surface

| Water content (%) | Angle of shearing resistance, $\phi$ (°) |
|---|---|
| 260 | 19 |
| 353 | 18 |
| **386**[a] | **17** |
| 484 | 23 |
| 543 | 24 |
| 736 | 24 |

[a]Liquid Limit of the bentonite

$\phi$ was found to be only up to three degrees. Hence, it was decided to proceed with this study by testing the soil samples in dry state, keeping a dry density of 1.5 g/cc.

### 4.2 Determination of Critical Conditions for Bentonite Layer

Bentonite support fluid forms a layer of filter cake on the surface of piles. For the present study, laboratory direct shear tests were conducted by manually applying a layer of bentonite over the precast cement mortar blocks. Bentonite layer could be applied at various consistencies. Hence, before proceeding to the actual tests, a critical condition for bentonite filter cake was found out.

Direct shear tests were conducted using bentonite at liquid limit water content, at water contents less than the liquid limit water content and also at water contents greater than the liquid limit water content, sandwiched between medium sand and cement mortar blocks. Results showed that critical value for angle of shearing resistance was at the liquid limit water content of bentonite as shown in Table 4. Hence, it was decided to proceed with the study by choosing bentonite at liquid limit water content at cement mortar–soil interface.

### 4.3 Direct Shear Tests with Bentonite Layer Between Sand and Concrete

Different combinations of sand–cement mortar surface with and without bentonite layer were selected, and direct shear tests were conducted to study the variation in the value of angle of shearing resistance, $\phi$. Direct shear tests were conducted for fine, medium and coarse sand, with bentonite layer manually applied at liquid limit water content on cement mortar surface, and sand was filled at a density of 1.5 g/cc over bentonite layer. The values of angle of shearing resistance obtained in each case are given in Table 5. To get a generalized idea of the percentage decrease in $\phi$

**Table 5** Variation of $\phi$ with gradation of sand and type of interface

| Sample description | Angle of shearing resistance, $\phi$ (°) for | | | |
|---|---|---|---|---|
| | Fine sand | Medium sand | Coarse sand | Well graded sand |
| Sand–sand interface | 29 | 31 | 34 | 32 |
| Sand–cement mortar block interface | 28 | 29 | 31 | 30 |
| Bentonite @LL b/w sand–cement mortar surface | 13 | 17 | 25 | 20 |

value, the direct shear tests were also done on well-graded sand and its results are also tabulated (Table 5). Results showed that the angle of shearing resistance was considerably reduced due to the presence of bentonite layer at cement mortar–sand interface.

## 4.4 *Direct Shear Tests with Bentonite Layer Between Cement Mortar Block and Marine Clay/red Earth*

In piling sites, engineers come across different varieties of soils. Hence, in this study, shear tests were also conducted on commonly available soil types—marine clay and red earth. For conducting direct shear tests with red earth/marine clay, the soil samples had to be placed over a layer of bentonite. But it was not practically possible to compact them after placing over the bentonite layer. In order to avoid such a problem, samples of red earth/marine clay of size 60 × 60 × 12.5 mm were prepared in advance by using the same mild steel mould (60 × 60 × 25 mm) and plate (60 × 60 × 12.5 mm) assembly which were used to cast cement mortar blocks. Hand compaction was applied to prepare specimens of marine clay and red earth at their natural moisture contents. It was possible to obtain samples with an average dry density of 1.6 g/cc and 1.2 g/cc for red earth and marine clay, respectively.

It can be seen from Fig. 4 that, for both sand and red earth, the angle of shearing resistance values obtained in the presence of bentonite support fluid was found to be much less than those obtained for pure sand/red earth case or the case of sand/red earth–cement mortar surface combination. In the case of marine clay, variations in the values of cohesion instead of the $\phi$ values were analysed. Figure 5 shows the variation of this shear strength parameter in the presence of bentonite layer, which also showed a decreasing trend, confirming the adverse effect of bentonite support fluid at pile–soil interface.

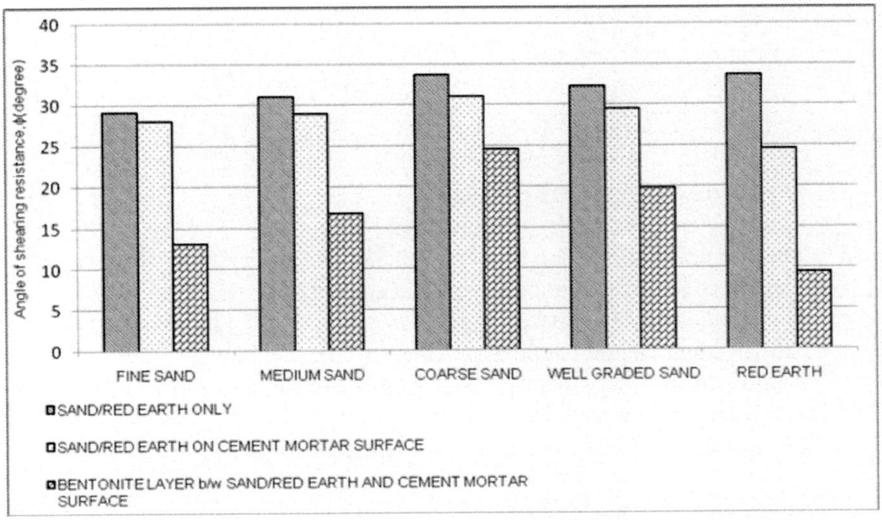


**Fig. 4** Effect of type of interface on angle of shearing resistance for different soil types

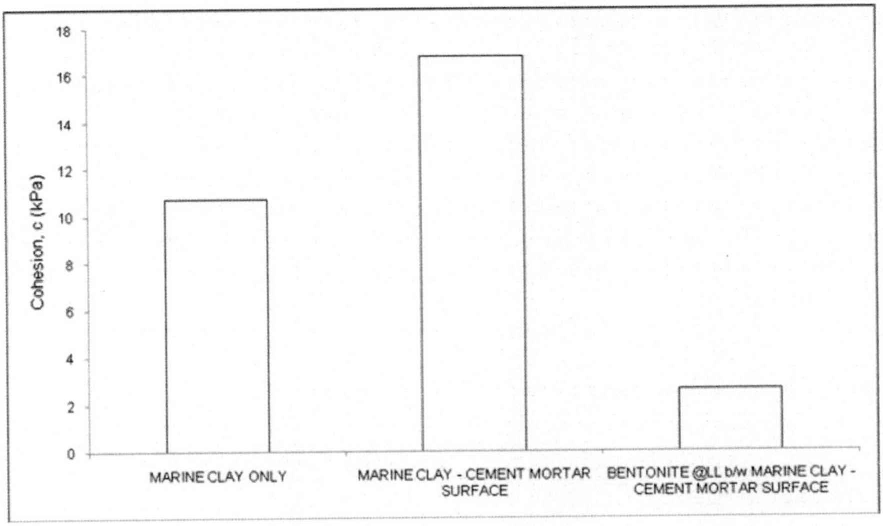


**Fig. 5** Effect of type of interface on cohesion

## 4.5 *Effect of Support Fluids on Frictional Resistance*

The main objective of the present study was to estimate the effect of bentonite support fluid on the interface frictional resistance in the case of concrete piles. The frictional resistance for piles installed in silty clayey sand is given by,

$$\text{Frictional Resistance} = \left[\sum_{i=1}^{n} K_i\ P_{Di} \tan\delta\right] A_{si} + \sum_{i=1}^{n} \alpha_i C_i\ A_{si}$$

where

| | |
|---|---|
| $K$ | The lateral earth pressure coefficient. |
| $P_D$ | The effective overburden pressure at pile toe. |
| $\tan\delta$ | The coefficient of friction between soil and pile surface. |
| $A_s$ | The effective surface area of the pile in contact. |
| $D$ | Stem diameter of the pile. |
| $\gamma$ | Effective unit weight of soil at pile toe. |
| $\delta$ | Angle of wall friction between pile and the surrounding soil. |
| $\alpha$ | The adhesion or reduction factor. |
| $C_i$ | The average undrained cohesion for a particular layer. |

Values of coefficient of friction depend on the type of soil and pile material. Indian Standards recommend angle of wall friction, $\delta = \phi$, where $\phi$ is the angle of shearing resistance [2].

For pilings assisted by support fluids, while calculating the skin friction, approximation of $\delta = \phi$ is to be relooked, considering the results of the laboratory investigations, discussed in the previous sections. Since, at the sand–concrete interface, support fluids form a thin layer, the effect of these support fluids on interface frictional resistance has to be considered. For this purpose, a pile of 60 cm diameter and 50 m length (very common in Cochin area) was assumed with surrounding soil having a bulk unit weight of 18 kN/m$^3$. Values of $K$ were taken as 1, $\alpha = 0.3$, and effective overburden pressure was taken for a depth of 20 times diameter of the pile. Variation in frictional resistance for piles in sand with bentonite slurry is presented in Table 6 and that of red earth and marine clay in the presence of bentonite which is presented in Table 7.

**Table 6** Variation of frictional resistance for pile in sand

| Sample description | Frictional resistance (kN) | | | |
|---|---|---|---|---|
| | Fine sand | Medium sand | Coarse sand | Well-graded sand |
| Without support fluid | 1159 | 1203 | 1304 | 1228 |
| With support fluid | 505 | 651 | 994 | 781 |

**Table 7** Variation of frictional resistance for pile in red earth or marine clay

| Sample description | Frictional resistance (kN) | |
|---|---|---|
| | Red earth | Marine clay |
| Without support fluid | 992 | 113 |
| With support fluid | 361 | 17 |

## 5 Conclusions

While calculating the ultimate capacity of bored/driven piles in cohesionless soils, value of angle of wall friction, $\delta$ is the requisite. But as per Indian Standard recommendations, this value is approximated to angle of shearing resistance, $\phi$ of the soil surrounding the pile. A set of direct shear tests has been carried out in order to study the effects of bentonite support fluid on the pile–soil interface frictional resistance. The results indicated that bentonite layer will adversely affect the interface frictional resistance. From the study, it is observed that

1. The value of $\delta$ at pile–sand interface (when bentonite layer is not considered) reduces up to 28% from the actual $\phi$ value of sand surrounding the pile.
2. The value of $\delta$ at pile–sand interface (when bentonite layer is considered) reduces up to 72% from the actual $\phi$ value of sand surrounding the pile.

On the other hand, in case of bored/driven piles in cohesive soils, value of shear strength parameter, $c$ is the requisite. From the study, it was found that

1. The value of cohesion, $c$ for the case of bentonite layer at pile–clay interface is found to be about 75% less, compared to the case when the effect of bentonite layer was not considered.

Similarly, it was observed that with the presence of bentonite slurry, frictional resistance decreased by 24–56% in case of sand–concrete interface, 64% in case of red earth–concrete interface and 84% in case of marine clay–concrete interface. This reduction in values will ultimately affect the load-carrying capacity of piles and hence the foundation performance. It is hoped that these findings will be very much beneficial for engineers while estimating the pile capacity.

## References

1. IS 1498-1970: Indian Standard Classification and Identification of Soil for General Engineering Purposes. Bureau of Indian Standards, New Delhi (2004)
2. IS 2911-Part I-Sec 2: Indian Standard Code of Practice for Design and Construction of Pile Foundations—Concrete Piles—Bored Cast in-situ Piles. Bureau of Indian Standards, New Delhi (2010)
3. IS 8112-2013: Indian Standard Ordinary Portland Cement, 43 grade-specification. Bureau of Indian Standards, New Delhi (2013)
4. Jefferis, S., Lam, C., Troughton, V.: Polymer systems for fluid supported excavations. In: Second Conference 2009, Geotechnical Issues in Construction: Short Paper Series, vol. X513, pp. 7–12. CIRIA, UK (2009)
5. Lam, C., Troughton, V., Jefferis, S., Suckling, T.: Effect of support fluids on pile performance—a field trial in east London. J. Ground Eng. **43**(10), 28–31 (2010)
6. Lam, C., Jefferis, S.A., Martin, C.M.: Effects of polymer and bentonite support fluids on concrete–sand interface shear strength. Geotechnique **64**(1), 28–39 (2014)
7. Rugang, Y., Guancheng, J., Wei, L., Tianqing, D., Hongxia, Z.: Effect of water-based drilling fluid components on filter cake structure. J. Powder Technol. **262**, 51–61 (2014)

8. Shakir, R.R., Zhu, J.G.: Behavior of compacted clay-concrete interface. J. Front. Architect. Civ. Eng. China **3**(1), 85–92 (2009)
9. Shakir, R.R., Zhu, J.G.: An examination of the mechanical interaction of drilling slurries at the soil-concrete contact. J. Zhejiang Univ. Sci. A Appl. Phys. Eng. **11**(4), 294–304 (2010)
10. Shrivastava, A.K., Jain, D., Vishwakarma, S.: Frictional resistance of drilling fluids as borehole stabilizers. Int. J. Geo-Eng. **7**(1), 7–12 (2016)
11. Yoo, C., Han, Y.-S.: Effect of drilling fluid on concrete-soil/rock interface shear strength in seawater drilling environment. J. Mar. Georesour. Geotechnol. **37**(8), 936–944 (2018)

# Predicting Residual Stress State Around Bored Cast-In-Situ Piles Utilizing Cavity Contraction and Expansion Solutions

**Alpha Lukose and Sudheesh Thiyyakkandi**

## 1 Introduction

Bored piles and drilled shafts are widely used as foundation systems for buildings, bridges, transmission towers, signages, retaining structures, etc., because of their minimally invasive nature and constructability through hard strata. The skin resistance offered by bored pile primarily depends on the residual stress state in close vicinity to the pile during the loading, which is influenced by the installation process and the behaviour of the thin layer of soil around the pile (i.e. shear band zone) upon the loading [5, 11, 13]. However, the existing popular methods for predicting unit skin friction such as depth-dependent $\beta$ method [17], FHWA rational method [7] and IS code method [9] are rather empirical or semi-empirical in nature and do not consider the fundamental mechanics of the stress state evolution processes. In the recent years, advancements have been made in this direction by using numerical approaches coupled with constitutive models [11], computational algorithms [12, 13], and using load transfer functions to describe pile-soil interaction [25]. Still, all these methods ignore the change in stress state caused by the method of construction (dry or wet/slurry methods).

Since the soil state in the immediate locale of the pile will be significantly altered during each stage of construction process and loading, tracking the evolution of soil state during each stage will enable more realistic prediction of residual confining stress and hence the unit skin friction of shaft. The excavation of hole and subsequent concrete placement can be considered analogous to cylindrical cavity contraction and expansion problems; and further concrete hydration and axial loading are similar to

A. Lukose (✉) · S. Thiyyakkandi
Department of Civil Engineering, Indian Institute of Technology, Palakkad, Kerala 678623, India
e-mail: 101704002@smail.iitpkd.ac.in

S. Thiyyakkandi
e-mail: sudheesh@iitpkd.ac.in



C. N. V. Satyanarayana Reddy et al. (eds.), *Ground Characterization and Foundations*, Lecture Notes in Civil Engineering 167,
https://doi.org/10.1007/978-981-16-3383-6_55

cavity contraction and expansion, respectively. A semi-analytical solution procedure, based on cylindrical cavity contraction and expansion phenomena, for predicting the residual stress state around a shaft in sand and thus its unit skin friction is presented in this paper. This solution procedure enables the full incorporation of non-linear soil behaviour during the construction and loading phases. The numerical algorithm involves discretization of the plastic zone of expanding and contracting cavities to thin cylindrical shell elements, similar to that adopted by Salgado and Randolph [19] for steady-state cavity expansion case, and an iterative approach is adopted from elastic–plastic boundary towards the cavity wall to determine the stress state and properties of each element. Unlike self-similarity technique which allows properties to be determined only at the limit/steady-state state, this solution procedure enables prediction of entire soil state at any stage of expansion or contraction. The validity of this solution approach was verified by predicting the response of two well-documented field test data of bored piles in sand.

## 2 Evolution of Lateral Stress Around Pile

The ultimate unit shaft friction, $f_s$, along a bored pile in sand at a given depth, $z$, can be expressed as:

$$f_s = \sigma'_{hf} \tan \delta' = K\,\sigma'_{vo} \tan \delta' = \beta \sigma'_{v0} \tag{1}$$

where $\sigma'_{hf}$—effective horizontal stress at failure; $\delta'$—soil–shaft interface friction angle, generally assumed to be equal to the angle of internal friction of soil ($\phi'$); $K$—coefficient of lateral earth pressure; and $\sigma'_{v0}$—initial effective vertical stress. The term $K \tan \delta'$ is collectively expressed as $\beta$ [16] in all the "$\beta$-methods" available in literature. The effective horizontal stress at failure ($\sigma'_{hf}$) can be expressed as: $\sigma'_{hf} = \sigma'_{h0} + \Delta\sigma'_h$; where $\sigma'_{h0}$ is the initial effective horizontal stress and $\Delta\sigma'_h$ is the stress change due to the installation processes (drilling, concreting and hardening) and axial loading of pile. The initial effective horizontal stress can be obtained as $\sigma'_{ho} = \left[(1 - \sin \phi')\mathrm{OCR}^{\sin \phi'}\right]\sigma'_{vo}$ [14]. The stress history of surrounding soil (i.e. variation of $\sigma'_h$) can be captured by modelling the entire process of bored pile construction and loading with the aid of cylindrical cavity contraction and expansion theories as illustrated in Fig. 1.

Typical cavity pressure–contraction/expansion response during the different stages of construction and axial loading is schematically depicted in Fig. 1a. During the first stage of construction, i.e. excavation of hole of radius $a_0$, the surrounding soil deforms inward slightly (contraction from $a_0$ to $a_e$), thus allowing the transfer of radial stress to hoop stress around the hole through arching (Fig. 1a, b). The effective radial stress ($p_e'$) at the hole wall is given by,

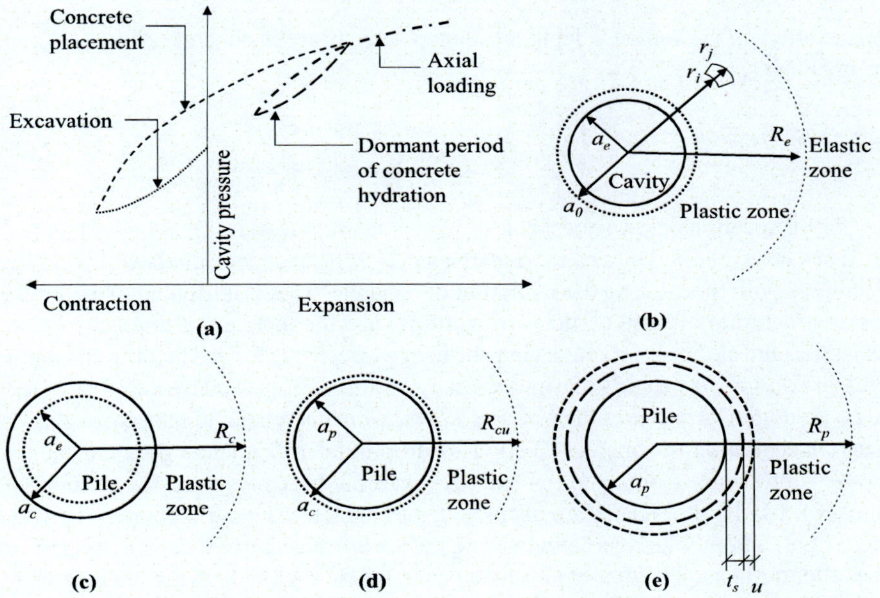


**Fig. 1** Pile construction and loading—cavity contraction/expansion analogy. **a** Cavity contraction/expansion curves. **b** Excavation (contraction). **c** Concrete placement (expansion). **d** Dormant period of hydration (contraction). **e** Axial loading

$$p'_e = \begin{cases} 0 & \text{for dry method} \\ \gamma_b h_b & \text{for wet method, above water table} \\ \gamma_b h_b - \gamma_w h_w & \text{for wet method, below water table} \end{cases} \quad (2)$$

where $\gamma_b$ and $\gamma_w$ are the unit weight of slurry (mineral or polymer) and water, respectively; $h_b$ is the height of slurry column above the point under consideration; and $h_w$ is the height of water table above the point under consideration. The cavity contraction may be purely elastic or elastic–plastic with the formation of a reverse plastic zone (radius $= R_e$) based on the magnitude of unloading (Fig. 1b), which depends upon the depth under consideration, location of water table, method of construction (wet or dry), etc.

The placement of fresh concrete in the shaft hole increases the lateral pressure on the cavity wall and displaces the cavity wall outward slightly (i.e. expansion from $a_e$ to $a_c$; Fig. 1c) with the development of plastic zone ($R_c$). Though different approaches are available, predicting the lateral pressure exerted by fresh concrete is very complex as it depends on several factors such as intrinsic properties of the materials, mix proportions, consistency, slump, placement rate, length and diameter of shaft, temperature, among others [2, 3, 10, 21]. Lings et al. [10] found that variation of lateral concrete pressure with depth is in conformity with the fluid concrete hydrostatic line down to certain depth, known as critical depth $\left(h_{\text{cr}} \approx \frac{1}{3} \times \text{depth of excavation}\right)$, below which it follows the slope of slurry pressure line. Accordingly, lateral pressure

($p_c$) exerted on the wall at a depth ($z$) immediately after the concrete placement can be obtained as:

$$p_c = \begin{cases} \gamma_c z & \text{for } z \leq h_{\text{cr}} \\ (\gamma_c - \gamma_b)h_{cr} + \gamma_b z & \text{for } z > h_{\text{cr}} \end{cases} \tag{3}$$

For the location below water table, effective lateral pressure, $p_c' = p_c - \gamma_w h_w$.

It has been reported by several researchers [1–5, 10] that concrete lateral pressure decreases with time during the hydration of concrete. This is attributed to the physical restructuring of solid particles (thixotropy) and the increase of internal friction as well as the autogenous and drying shrinkage effects [2, 3, 5]. The drop in lateral pressure during the plastic stage of cement hydration (i.e. dormant period) is reported to be predominantly due to the physical effect, which is related to the coarse aggregate concentration in concrete [2, 3]. This drop in lateral concrete pressure causes cavity contraction (from $a_c$ to $a_p$) and decreases the horizontal stresses around the pile (Fig. 1d). By the end of dormant period, the concrete skeleton becomes relatively rigid [2, 3] and further contraction during the accelerated stage of cement hydration (i.e. after dormant period) is expected to be negligible even though the internal pressure within concrete drop to zero. Assaad and Khayat [2] have observed a concrete pressure drop of 40–50% during the dormant period in their experimental study.

Finally, during the axial loading phase, shear strains develop in soil surrounding the shaft owing to the vertical displacement of the shaft [7, 11]. This shear strain tends to be prevalent within a thin cylindrical shear band of thickness, $t_s$, formed around the pile, which depends upon the soil gradation, typically in the range of 5–20 times the mean grain size, $D_{50}$ [21, 22]. During the loading, the pile and shear band is considered together as an indefinite cylinder with initial radius ($r_{pf} = a_p + t_s$), which undergoes a radial expansion/displacement of $u$ that depends on the dilation potential of the soil (Fig. 1e). The magnitude of this radial displacement ($u$) is given by [13]:

$$u = t_s \tan \psi' \frac{\gamma_{\text{cs}}}{2} \tag{4}$$

where $\psi'$ is the dilation angle of soil and $\gamma_{\text{cs}}$ is the shear strain to attain critical state.

During the axial loading, as the shear band expands, the surrounding soil deforms elastically until the onset of plastic yielding at the cavity wall ($r_{pf}$) and beyond that, a plastic zone is formed around the cavity, the radius of which ($R_p$) increases with further loading (Fig. 1e). The maximum radial stress generated at the outer boundary of shear band during the loading stage, which is influenced by the residual soil state subsequent to the pile installation, governs the ultimate unit skin resistance of pile at that depth. The proposed cavity contraction and expansion solutions procedure for the same is described below.

# 3 Cylindrical Cavity Contraction and Expansion Solutions

## 3.1 Equilibrium Equation, Yield Criterion and Flow Rule

The stress equilibrium around a cylindrical cavity can be expressed as:

$$\frac{d\sigma_r'}{dr} + \frac{\sigma_r' - \sigma_\theta'}{r} = 0 \tag{5}$$

where $\sigma_r{}'$ and $\sigma_\theta{}'$ are the effective radial and hoop stresses. It is assumed that pore water pressure ($p_w$) remains constant throughout the installation and loading, i.e. $\frac{dp_w}{dr} = 0$. The conventional geomechanics sign convention of compressive stresses and strains as positive is adopted here. The Mohr–Coulomb yield criteria for loading and unloading are given by Eqs. (6) and (7), respectively.

$$\sigma_r' = \sigma_\theta' N \tag{6}$$

$$\sigma_\theta' = \sigma_r' N \tag{7}$$

where $N$ is the flow number; $N = \frac{1+\sin\phi'}{1-\sin\phi'}$. The friction angle ($\phi'$) of soil within the plastic zone of an expanding and contracting cavity vary with the volumetric plastic strain, which is assumed to vary according to the Bolton (1986) model [6] for sand:

$$\phi_{ij}' = \phi_c' + D_\psi \left\{ I_{D_{ij}} \left[ Q + \ln\left( \frac{p_a}{100 p_{ij}'} \right) \right] - R_Q \right\} \tag{8}$$

where $\phi_{ij}'$ is the friction angle within a thin shell element, *ij* (Fig. 1b); $\phi_c'$ is the critical state friction angle; $I_{D_{ij}}$ is the relative density within element, *ij*; $p_a$ is the reference stress = 100 kPa; $Q$ and $R_Q$ are fitting parameters ($Q \approx 10$ and $R \approx 1$); $p_{ij}'$ is the mean effective stress within element, *ij*; and $D_\psi = 3$ for triaxial test and 5 for plane strain test. The dilatancy ($\psi_{ij}{}'$) and friction angles are assumed to hold the relationship given by Bolton [6]:

$$\psi_{ij}' = \frac{\phi_{ij}' - \phi_c'}{0.8} \tag{9}$$

A non-associated flow rule proposed by Zhang and Salgado [24] for Mohr–Coulomb soils is assumed to hold true:

$$N_{ij} = N_c D_{ij} \tag{10}$$

where $N_{ij}$ is the flow number within element, *ij*; $N_c$ is critical state flow number; and $D_{ij}$ is the strain rate ratio within *ij* given by:

$$D_{ij} = 1 - \left(\frac{\dot{\varepsilon}_v}{\dot{\varepsilon}_1}\right) \tag{11}$$

where $\dot{\varepsilon}_v$ is the volumetric strain rate $= \varepsilon_v^{(i)} - \varepsilon_v^{(j)}$ and $\dot{\varepsilon}_1$ is the major principal strain rate $= \varepsilon_1^{(i)} - \varepsilon_1^{(j)}$; for expansion, $\dot{\varepsilon}_1 = \dot{\varepsilon}_r$ and for contraction, $\dot{\varepsilon}_1 = \dot{\varepsilon}_\theta$. The strain components are defined as logarithmic strains to account for large strain in the plastic zone:

$$\left.\begin{aligned} \varepsilon_r^{(i)} &= -\ln\left(\frac{r_j - r_i}{(r_j - u_j) - (r_i - u_i)}\right); \quad \varepsilon_\theta^{(i)} = -\ln\left(\frac{r_i}{r_i - u_i}\right); \\ \varepsilon_v^{(i)} &= -\ln\left(\frac{r_j^2 - r_i^2}{(r_j - u_j)^2 - (r_i - u_i)^2}\right) \end{aligned}\right\} \tag{12}$$

## 3.2 Numerical Solution Procedure

For obtaining the evolution of soil state during the various stages of bored pile construction and loading, which will aid to estimate the residual confining stress around pile, an elaborate numerical procedure is adopted and is briefed below:

(i) Select required input parameters: $a_0$, $p_0' = \sigma_{h0}'$, $I_{D0}$, $\phi_c'$, $t_s$, $\gamma_{cs}$, and shear modulus, $G$.

**Excavation of Shaft Hole**. When the internal pressure decreases from $p_0'$, unloading happens, which will be purely elastic until reverse yielding occurs. The cavity pressure at the beginning of reverse yielding ($p_r'$) is:

$$p_r' = \frac{p_0'(N+1)}{N(N+1)} \tag{13}$$

(ii) If the effective radial stress after excavation ($p_e'$) is greater than $p_r'$, the corresponding contracted cavity radius ($a_e$) and radial and hoop stress fields around the cavity can be calculated as:

$$a_e = a_0\left(1 + \frac{(p_e' - p_0')}{2G}\right) \tag{14}$$

$$\sigma_r' = p_0' + (p_e' - p_0')\left(\frac{a_e}{r}\right)^2; \quad \sigma_\theta' = p_0' - (p_e' - p_0')\left(\frac{a_e}{r}\right)^2 \tag{15}$$

If $p_e' < p_r'$, the plastic unloading happens and a reverse plastic zone forms around the cavity. The accurate determination of stress and strain fields around the cavity considering the effect of volumetric plastic strain requires cavity radius ($a_e$) or the plastic radius ($R_e$) be known. However, since both $a_e$ and $R_e$ are initially unknown, a recursive approach with the assumed plastic radius is adopted to obtain the actual $a_e$ and $R_e$ and thus to estimate the variation of soil properties and stress state within the plastic zone concurrently. For each assumed plastic radius, plastic zone is discretized into thin elements, and an iterative approach beginning from the elastic–plastic boundary is employed as summarized below.

(iii) Assume a reverse plastic radius, $R_{e1}$. Divide this plastic zone into thin shell elements (*ij*), each of thickness d*r*, where *j* marks the outer boundary of each element. The first element is considered close to boundary with inner radius, $r_i = r_j - dr$, and outer radius, $r_j = R_{e1}$, where the entire soil state is known:

$$\left.\begin{array}{l} \sigma'_{r_j} = p'_r = \sigma'_R; \quad \sigma'_{\theta_j} = \sigma'_R \cdot N_p \\ \varepsilon_r^{(j)} = -\left(p'_0 - \sigma'_R\right)/2G; \quad \varepsilon_\theta^{(j)} = -\varepsilon_r^{(j)} \\ \varepsilon_v^{(j)} = \varepsilon_r^{(j)} + \varepsilon_\theta^{(j)} \quad \text{and} \quad u_j = \varepsilon_\theta^{(j)} \cdot R_{e1} \end{array}\right\} \tag{16}$$

The friction angle of this first element ($\phi_{ij}'$) is considered same as that at $r_j$ for initial iteration and then:

- Determine the inner radial stress ($\sigma'_{r_i}$) using Eq. (17), which is obtained by combining Eqs. (5) and (7). Subsequently, determine inner hoop stress ($\sigma'_{\theta_i}$) using Eq. (7) and mean stress ($p'_{ij}$) for the element by Eq. (18), obtained by modifying Davis (1968) equation [8] to apply for the unloading case.

$$\sigma'_{r_i} = \sigma'_{r_j} \cdot \left(\frac{r_j}{r_i}\right)^{(1-N_{ij})} \tag{17}$$

$$p'_{ij} = \frac{1}{3} \cdot \left[\bar{\sigma}_r\left(1 + N_{ij}\right)\left(1 + \mu_{ij}\right)\right] \tag{18}$$

where $\bar{\sigma}_r$-average radial stress for the element and $\mu_{ij} = 0.5\left(1 + \sin\phi'_{ij} \cdot \sin\psi'_{ij}\right)$.

- By substituting Eq. (12) in Eq. (11), an expression with inner radial displacement, $u_i$, as the only unknown is obtained (Eq. 19), where $D_{ij}$ is obtained from Eq. (10). Solve Eq. (19) for $u_i$ using numerical method (e.g. Newton–Raphson method).

$$\begin{aligned} &\varepsilon_v^{(j)} + \left(D_{ij} - 1\right)\varepsilon_\theta^{(j)} \\ &= \ln\left\{\left[\frac{\left(r_j - u_j\right)^2 - \left(r_i - u_i\right)^2}{r_j^2 - r_i^2}\right] \cdot \left[\frac{r_i - u_i}{r_i}\right]^{(D_{ij}-1)}\right\} \end{aligned} \tag{19}$$

- Estimate the radial, hoop, and volumetric strains (${\varepsilon_r}^{(i)}$, ${\varepsilon_\theta}^{(i)}$ and ${\varepsilon_v}^{(i)}$) using Eq. (12). Subsequently, using ${\varepsilon_v}^{(i)}$, calculate a new void ratio ($e_{ij}$)

and hence relative density ($I_{D_{ij}}$) of the element from which the new $\phi_{ij}'$ can be obtained using Eq. (8).

- The above steps are iterated to obtain sufficient convergence for $\phi_{ij}'$. All the parameters at inner boundary '*i*' will be thus obtained, which will be the outer boundary (*j*) for the next element inwards.

(iv) Repeat Step (iii) for the next element and proceed element by element inwards. After achieving convergence for each element, $r_i - u_i$ is compared with the initial cavity radius ($a_0$). If $r_i - u_i > a_0$, carry forward with the next element, whereas if $r_i - u_i \approx a_0$, cavity wall has been reached.

(v) Check if the difference between estimated cavity stress, $\sigma'_{r_i}$ and $p_e'$ (Eq. 2), is satisfactorily small. If not, repeat the above Steps (iii and iv) by gradually incrementing the plastic radius as $R_{e2}, R_{e3}, \ldots, R_{en}$ until $\sigma'_{r_i} \approx p'_e$. This iterative procedure will simultaneously give the stresses, strains and the varied soil properties within the plastic zone corresponding to the effective cavity pressure of $p_e'$.

**Concrete placement**. Cavity expansion resulting from concrete placement is considered here. The effective lateral pressure immediately after the concrete placement ($p_c'$) will certainly exceed the minimum pressure to cause plastic yielding $\left(p'_y = \frac{p'_e(N+1)}{N(N+1)}\right)$. The numerical procedure followed is similar to that adopted for plastic contraction and is summarized below:

(vi) Consider a plastic radius, $R_{c1}$, greater than $a_e$. Divide this plastic zone into thin shell elements (*ij*), each with thickness d*r*.

(vii) The stresses, strains and radial displacement values ($\sigma_R'$, $\sigma_\theta'$, $\varepsilon_R$, $\varepsilon_\theta$ and $u_R$) at $R_{c1}$ can be determined using Eq. (20) as follows:

$$\left.\begin{array}{c} \sigma'_R = \frac{2N\sigma'_{re}}{1+N}; \quad \sigma'_\theta = \frac{\sigma'_R}{N}; \quad \varepsilon_R = \frac{\sigma'_R - \sigma'_{re}}{2G} \\ \varepsilon_\theta = -\varepsilon_R; \quad u_R = -\varepsilon_\theta . R_{c1} \end{array}\right\} \tag{20}$$

where $\sigma_{re}'$ and $N$ are the radial stress and flow number at $R_{c1}$ following excavation.

(viii) The first element (*ij*) is considered with the outer boundary $r_j = R_{c1}$ and $\phi_{ij}'$ of the element is initially assumed to be the same as that at $R_{c1}$ at the end of excavation phase.

- Determine $\sigma'_{r_i}$ utilizing Eq. (21) (obtained by combining Eqs. 5 and 6), $\sigma'_{\theta_i}$ using Eq. (6) and $p_{ij}'$ by Eq. (22); Davis (1968) equation [8, 19]:

$$\sigma'_{r_i} = \sigma'_{r_j}\left(\frac{r_j}{r_i}\right)^{\frac{N_{ij}-1}{N_{ij}}} \tag{21}$$

$$p'_{ij} = \frac{1}{3}\left[\bar{\sigma}_r\left(1 + \frac{1}{N_{ij}}\right)(1 + \mu_{ij})\right] \tag{22}$$

- Obtain $u_i$ by solving Eq. (23), derived by plugging Eq. (12) in Eq. (11).

$$\varepsilon_v^{(j)} + (D_{ij} - 1)\varepsilon_r^{(j)} = \ln\left\{\left[\frac{(r_j - u_j)^2 - (r_i - u_i)^2}{r_j^2 - r_i^2}\right] \cdot \left[1 + \frac{u_i - u_j}{r_j - r_i}\right]^{(D_{ij}-1)}\right\} \tag{23}$$

- Estimate the radial, hoop and volumetric strains (${\varepsilon_r}^{(i)}$, ${\varepsilon_\theta}^{(i)}$ and ${\varepsilon_v}^{(i)}$) using Eq. (12). Subsequently, using ${\varepsilon_v}^{(i)}$, calculate a new void ratio ($e_{ij}$) and hence relative density ($I_{D_{ij}}$) of the element from which the new $\phi_{ij}'$ can be obtained using Eq. (8).
- The above steps are iterated to obtain sufficient convergence for $\phi_{ij}'$. This iterative process yields all the parameters at the inner face of the element ($i$).

(ix) Go to the next element and follow Step (viii). Once convergence is achieved for each element, $r_i - u_i$ is compared with $a_e$. If $r_i - u_i > a_e$, proceed to the next element. If $r_i - u_i \approx a_e$, cavity wall has been reached and corresponding new cavity radius, $a_c = r_i$.

(x) Compare the computed cavity pressure (${\sigma_{ri}}'$) with ${p_c}'$ (Eq. 3); if the difference is not adequately small, repeat the above steps by increasing the plastic radius in small increments ($R_{c2}$, $R_{c3}$, …, $R_{cn}$) until the close agreement is achieved. Corresponding to ${p_c}'$, the distribution of stresses, strain and soil properties within the plastic region will be known.

**Plastic phase of Concrete hydration**. Drop of pressure during the dormant period of concrete hydration will cause inward deformation of the surrounding soil (i.e. cavity radius reduces from $a_c$ to $a_p$). The variation of stresses and strains with the radial distance at the end of pressure drop can be obtained using the procedure adopted for excavation.

**Axial loading**. During axial loading, the pile and shear band is considered to form a cylindrical cavity with initial radius ($r_{pf} = a_p + t_s$), which will be displaced through a radial distance, $u$ (Eq. 4). The procedure adopted for modelling the concrete placement can be utilized here to obtain the soil state (stresses, strains and soil properties) corresponding to the full expansion of shear band.

## 4 Results and Discussions

### *4.1 Comparison with the Experimental Results*

Two well-documented field load test results of bored pile/drilled shaft were utilized for the validation of the proposed approach. A shear band thickness ($t_s$) of $5D_{50}$

was adopted for the prediction. Unit weight for bentonite slurry and concrete was assumed as 10.1 and 24 kN/m$^3$, respectively. A pressure drop of 50% during the dormant period of concrete hydration is assumed here [2]. The first case is an axial load test data on drilled shaft (TS2; length, $L = 5.5$ m and diameter, $D = 1.22$ m) in typical Florida sand reported by Thiyyakkandi et al. [20]. The soil in test site was predominantly sand with silt (SP-SM) with a thin layer of clay (0.75 m) on top. The water table at the time of load testing was 1.8 m below the ground surface. The $\beta$ profile for the shaft was produced by the proposed approach using the following input parameters [15, 20]: $D_{50} = 0.16$–$0.2$ mm; $e_{max} = 0.80$; $e_{min} = 0.43$; $G$ (kPa) $= 8000 + 650.\ z$ ($z$-depth; estimated from pressuremeter test); $\gamma_{cs} = 160\%$; top sand layer (0.75–3.80 m): $\phi_c' = 31°$, $I_{D0} = 40\%$ and bottom sand layer (3.80–5.50 m): $\phi_c' = 34°$, $I_{D0} = 50\%$. Figure 2a displays the experimental and predicted $\beta$ values (for $D_{50}$ of 0.16 and 0.2 mm).

The second case is a bored pile load test data reported by Viana da Fonseca and Santos [23]. The pile ($L = 6$ m and $D = 0.6$ m) was bored in saprolitic granite soil, which extended up to 20 m below the ground surface. The water table was located 10 m below the ground surface. The soil parameters considered for the prediction are as follows [13, 23]: $D_{50} = 0.10$–$0.25$ mm; $G_0$ (MPa) $= 120 + 5.z$; $G = 0.5\ G_0$; $\gamma_{cs} = 60\%$; $\phi_c' = 32°$, $I_{D0} = 40\%$; $e_{max} \approx 0.90$ and $e_{min} \approx 0.60$ (estimated from $D_{50}$ using the correlations proposed by Patra et al. [18]). The comparison between the predicted $\beta$ profiles (for $D_{50} = 0.10$ and 0.25 mm) and experimental results is shown in Fig. 2b. It can be seen from Fig. 2 that the predicted range of $\beta$ values is consistent with the experimental data, except some deviation in second case and this signifies good predictive capability of the presented approach.

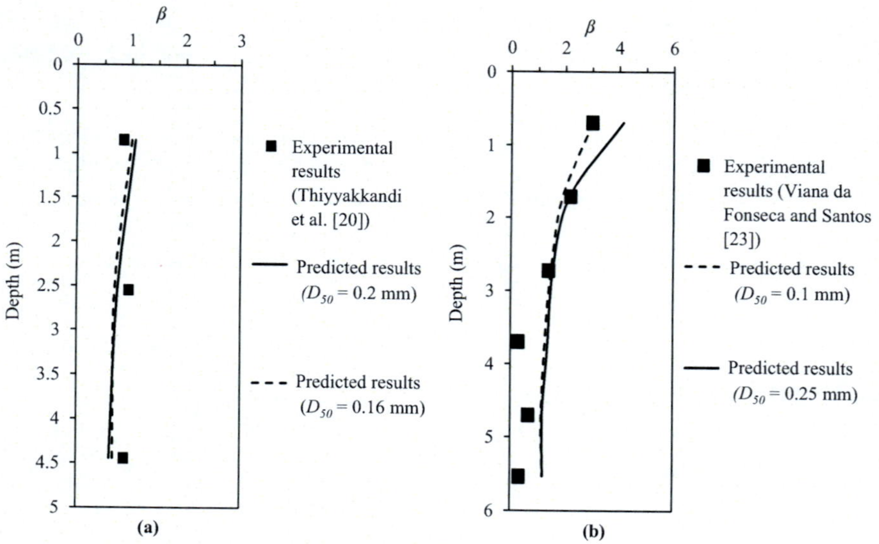


**Fig. 2** Comparison of predicted versus experimental $\beta$ values. **a** Thiyyakkandi et al. [20]; **b** Viana da Fonseca and Santos [23]

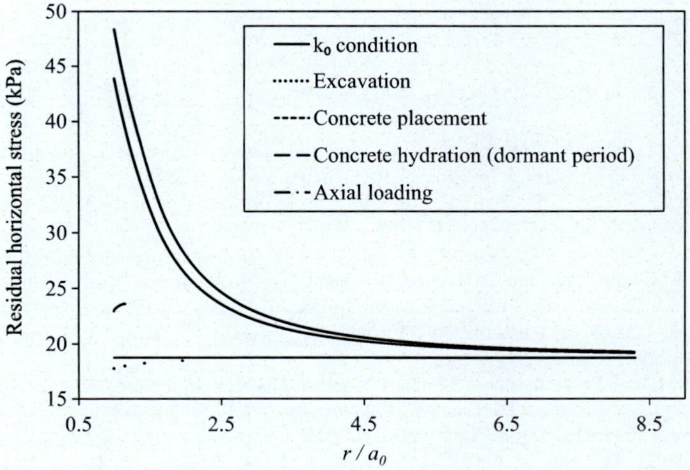


**Fig. 3** Residual horizontal stress around a shaft at different stages of pile installation and loading (for Shaft TS2 at 2.55 m depth; Thiyyakkandi et al. [20])

### *4.2 Stress Distribution Around Shaft*

Figure 3 shows the predicted stress state around Shaft TS2 [20], at a depth of 2.55 m. The stress state around the pile at different stages of installation and testing can be effectively predicted as shown. The different stages considered are: (a) in-situ $k_0$ condition; (b) excavation of shaft hole stabilized using bentonite slurry; (c) concrete placement; (d) plastic stage of concrete hydration; and (e) axial load test.

## 5 Conclusion

This paper has presented a semi-analytical approach in the framework of cavity contraction and expansion theories for tracking the evolution of stress state around a drilled shaft/bored pile during different stages of construction and axial loading. Unlike the existing self-similarity-based techniques, the presented solution enables prediction of soil state at any stage of expansion/contraction, incorporating the effect of volumetric plastic strain on the soil properties and stresses, throughout the plastic zone. The state of soil around a pile, which is significantly altered by the installation processes (excavation, concrete placement, and hydration) as well as the loading, could be captured using the presented method. The predictive capability of this solution method was validated by reproducing the two well-documented experimental results on drilled shafts/bored piles.

## References


1. Alexandridis, A., Gardner, N.: Mechanical behaviour of fresh concrete. Cem. Concr. Res. **11**(3), 323–339 (1981)
2. Assaad, J.J., Khayat, K.H.: Effect of mixture consistency on formwork pressure exerted by highly flowable concrete. J. Mater. Civ. Eng. **18**(6), 786–791 (2006)
3. Assaad, J., Khayat, K.H.: Variations of lateral and pore water pressure of self-consolidating concrete at early age. ACI Mater. J. **101**(4), 310–317 (2004)
4. Assaad, J., Khayat, K.H., Mesbah, H.: Variation of formwork pressure with thixotropy of self-consolidating concrete. ACI Mater. J. **100**(1), 29–37 (2003)
5. Bernal, J.B., Reese, L.C.: Study of the lateral pressure of fresh concrete as related to the design of drilled shafts. No. FHWA/TX-84/45+ 308-lF. University of Texas at Austin, Center for Transportation Research (1983)
6. Bolton, M.D.: The strength and dilatancy of sands. Geotechnique **36**(1), 65–78 (1986)
7. Brown, D.A., Turner, J.P., Castelli, R.J., Americas, P.B.: Drilled Shafts: Construction Procedures and LRFD Design Methods. Federal Highway Administration, United States (2010)
8. Davis, E.H.: Theories of plasticity and failure of soil masses. In: Lee, I.K. (ed.) Soil-Mechanics—Selected Topics, pp. 341–380. Butterworths, London (1968) (Chapter 6)
9. IS 2911(Part 1/Sec 2): Design and Construction of Pile Foundations. Bureau of Indian Standards, New Delhi (2010)
10. Lings, M.L., Ng, C.W., Nash, D.F.: The lateral pressure of wet concrete in diaphragm wall panels cast under bentonite. Proc. Inst. Civ. Eng. Geotech. Eng. **107**(3), 163–172 (1994)
11. Loukidis, D., Salgado, R.: Analysis of the shaft resistance of non-displacement piles in sand. Geotechnique **58**(4), 283–296 (2008)
12. Mascarucci, Y., Miliziano, S., Mandolini, A.: A numerical approach to estimate shaft friction of bored piles in sands. Acta Geotech. **9**(3), 547–560 (2014)
13. Mascarucci, Y., Miliziano, S., Mandolini, A.: 3M analytical method: evaluation of shaft friction of bored piles in sands. J. Geotech. Geoenviron. Eng. **142**(3), 04015086 (2016)
14. Mayne, P.W., Kulhawy, F.H.: Ko-OCR relationships in soil. J. Soil Mech. Found. Div. **108**(6), 851–872 (1982)
15. McVay, M., Bloomquist, D., Thiyyakkandi, S.: Field Testing of Jet-Grouted Piles and Drilled Shafts (BDK75-977-41). Final report submitted Florida Department of Transportation, Tallahassee, FL (2014)
16. Meyerhof, G.G.: Bearing capacity and settlement of pile foundations. J. Geotech. Eng. Div. **102**(GT3), 197–228 (1976)
17. O'Neill, M., Hassan, K.: Drilled shafts: effects of construction on performance and design criteria. In: Proceedings: International Conference on Design and Construction of Deep Foundations, p 1. FHWA, Orlando (1994)
18. Patra, C., Sivakugan, N., Das, B.: Relative density and median grain-size correlation from laboratory compaction tests on granular soil. Int. J. Geotech. Eng. **4**(1), 55–62 (2010)
19. Salgado, R., Randolph, M.: Analysis of cavity expansion in sand. Int. J. Geomech. **1**(2), 175–192 (2001)
20. Thiyyakkandi, S., McVay, M., Lai, P., Herrera, R.: Full-scale coupled torsion and lateral response of mast arm drilled shaft foundations. Can. Geotech. J. **53**(12), 1928–1938 (2016)
21. Uesugi, M., Kishida, H., Tsubakihara, Y.: Behavior of sand particles in sand-steel friction. Soils Found. **28**(1), 107–118 (1988)
22. Vardoulakis, I., Sulem, J.: Bifurcation Analysis in Geomechanics. Blackie Academic and Professional, London (1995)
23. Viana da Fonseca, A., Santos, J.: Behaviour of CFA, Driven and Bored Piles in Residual Soil. International Prediction Event-Experimental Site-ISC'2. FEUP/IST, Portugal (2008)
24. Zhang, J., Salgado, R.: Stress-dilatancy relation for Mohr-Coulomb soils following a non-associated flow rule. Geotechnique **60**(3), 223–226 (2010)
25. Zhang, Q.-Q., Feng, R.-F., Yu, Y.-L., Liu, S.-W., Qian, J.-G.: Simplified approach for prediction of nonlinear response of bored pile embedded in sand. Soils Found. **59**(5), 1562–1578 (2019)

# Estimation of Mobilized Shaft Resistance of Bored Piles from Pile Load Test

**Gouthami Manthena, Srinivas Kadali, and Madhav Madhira**

## 1 Introduction

Deep foundations such as piles are generally used to transfer heavy loads to a greater depth. Bored pile is one type of pile foundations. It is the most preferred type of pile in India because of its ease of construction, low vibration and flexibility of size to suit for different soil and loading conditions. Design of these bored piles involves consideration of two aspects, viz. ultimate capacity and settlements under applied loads. Coyle and Reese [4] explain the behaviour of bored piles in clayey soils and Reese et al. [9] describes the pile behaviour for drilled shafts in clayey soil to understand in detail. Many approaches are available to predict the bearing capacity of piles. But in situ pile load tests such as static and dynamic pile load tests with proper instrumentation are necessary to validate the estimate and to understand the load transfer behaviour of pile [8].

Many studies have been carried to study and analyse the load transfer mechanism and settlements using in situ pile load tests. Yousif [11] has performed pile load tests on bored piles to study the behaviour of piles in stratified soils, and some correlations were developed with respect to CPT and SPT values. Seo and Prezzi [10] analyse load transfer mechanism and settlement of various piles [5] for different soil conditions. Comodromos et al. [3] examined the bearing capacity and stiffness of single and pile group. Han and Prezzi [6] performed similar analysis in multi-layered soils. Analysis based on Plaxis 3D was also adopted by Mert and Ozkan [7] to predict pile capacity and settlement of bored piles.

Present study is carried out on bored piles in Sudan where the strata consist of clayey sand overlying highly plastic clay. Variation of axial load with displacement

G. Manthena · S. Kadali (✉) · M. Madhira
VNR Vignana Jyothi Institute of Engineering and Technology, Hyderabad 500090, India
e-mail: srinivas_kadali@vnrvjiet.in



C. N. V. Satyanarayana Reddy et al. (eds.), *Ground Characterization and Foundations*, Lecture Notes in Civil Engineering 167,
https://doi.org/10.1007/978-981-16-3383-6_56

and distribution of axial load with depth have been reported [11] based on which the shear stress mobilization with displacement has been evaluated.

## 2 Methodology

The test site is located on the bank of the Blue Nile as shown in Fig. 1 [11]. In situ and laboratory tests were performed to determine the index properties of soil. The results obtained from the laboratory tests; the ultimate bearing capacity of pile was estimated. This will be used to know the application of load during pile load test.

Four boreholes were drilled, and four CPT soundings were performed up to 15 m depth. Five bored piles (TPA1, TPA2, TPA3, TPB1 and TPC1) were considered for the analysis. Each of these test piles installed at various soil conditions with different geometrical details. The piles (TPA1, TPA2 and TPA3) are of diameter 0.2 m, with embedded lengths of 3.5, 5 and 6 m. The piles (TPB1 and TPC1) are of diameter 0.3 and 0.4 m with embedded length of 3.5 m. These details are presented in Table 1.

**Fig. 1** Test location

**Table 1** Details of test piles

| S. No. | Pile reference | Diameter of pile (m) | Length of pile (m) |
|---|---|---|---|
| 1 | TPA1 | 0.2 | 3.5 |
| 2 | TPA2 | 0.2 | 5 |
| 3 | TPA3 | 0.2 | 6 |
| 4 | TPB1 | 0.3 | 3.5 |
| 5 | TPC1 | 0.4 | 3.5 |

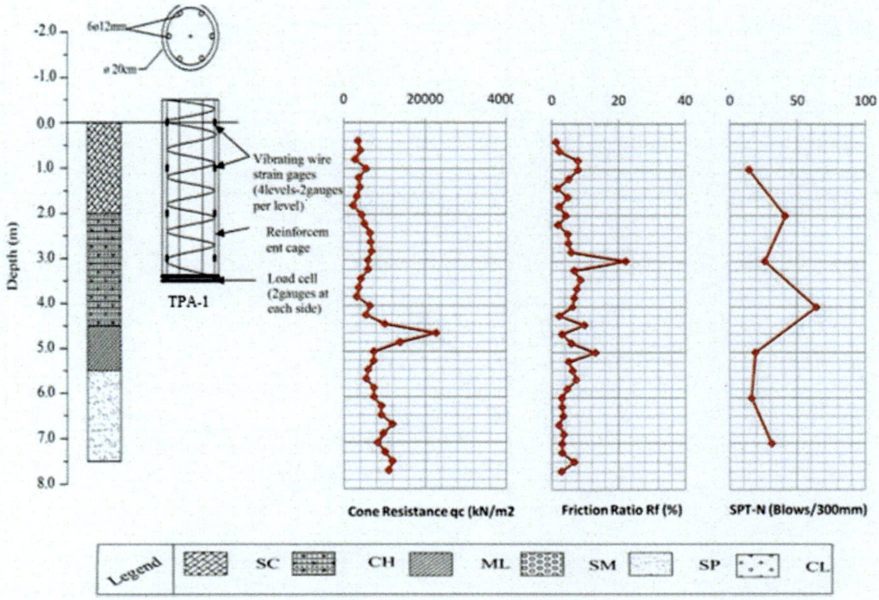


**Fig. 2** Instrumented details of pile TPA1, subsoil condition (SPT N values, CPT) [11]

For the sake of brevity, the subsoil conditions along the embedded length of the pile TPA1 only are presented in Fig. 2. This figure details the subsoil conditions in addition to SPT and CPT results [1, 2] of test piles. The top layer is medium dense light brown clayey sand. Beneath this layer is very stiff to hard light to dark brown silty clay of high plasticity in which the pile is resting.

Test pilesTPA1, TPB1 and TPC1 encounter the same soil conditions along the shaft length. The remaining test piles, TPA2 and TPA3, experienced additional soil strata such as low plastic silt followed by silty sand. The summary of all the test pile (TPA2, TPA3, TPB1 and TPC1) subsoil conditions is presented in Table 2.

**Table 2** Soil profile for test piles (TPA1, TPA2, TPA3, TPB1 and TPC1)

| Pile reference | Embedded length (m) | | | | |
|---|---|---|---|---|---|
| | SC | CH | ML | SM | CL |
| TPA1 | 0–2 | 2–3.5 | – | – | – |
| TPA2 | 0–2 | 2–4.5 | 4.5–5 | – | – |
| TPA3 | 0–2 | 2–4.5 | 4.5–5.5 | 5.5–6 | – |
| TPB1 | 0–2 | 2–3.5 | – | – | – |
| TPC1 | 0–2 | 2–3.5 | – | – | – |

*Note SC*—clayey sand, *CH*—high plastic clay silt, *ML*—low plastic silt, *SM*—silty sand, *SP*—poorly graded sand and *CL*—low plastic clay

**Fig. 3** Load versus strain gauge reading of all the test piles [11]

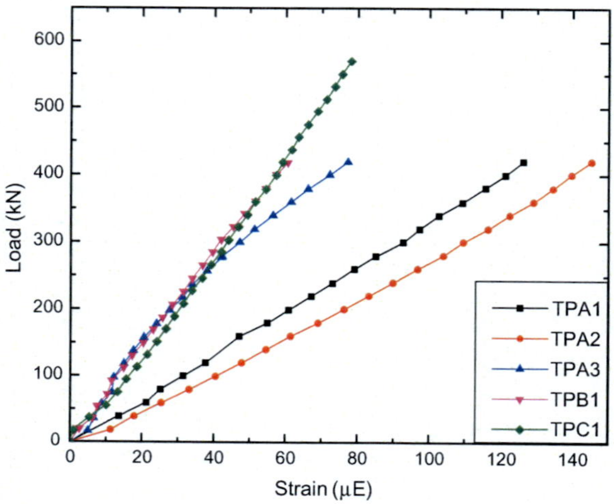


Hydraulic jack was used to apply the load at pile head with several load increments. Each load was maintained for a minimum of one hour or until the rate of settlement of the pile top decreased to 0.25 mm/h. The rate of settlements at pile head was measured using dial gauges. All the test piles were instrumented with strain gauges at four levels, each with two gauges (ref Fig. 2) to get the strains at each level for determining loads. The load versus strains readings of all the test piles are shown in Fig. 3. From the strain gauge readings, loads are estimated, and their variations along the shaft length with several load increments are plotted in Fig. 4.

## 2.1 *Load Versus Depth Curves*

The loads transferred along the shaft length of TPA1, TPA2, TPA3, TPB1 and TPC1 is presented in Fig. 4. In case of TPA1, TPA2 and TPB1, more than 50, 85 and 70% of load has been transferred in the first segment, i.e. from 0 to 1 m. For TPA3, 45–70% of the applied load has been transferred in the first segment. However, for TPC1, maximum load distribution has been observed in the second segment, i.e. up to 85% of load transmitted in the 1.0–2.0 m segment. Remaining loads were transferred to the successive segments.

The shaft resistance mobilized over each segment defined as the partial length of the pile between the corresponding strain gauge locations was calculated based on the loads estimated. The mobilized shaft resistance ($\tau_i$) of each segment was calculated by dividing the pile into number of segments corresponding to strain gauge levels (Fig. 5). The load reduction over the segment is $P_i - P_{i+1}$, where $P_i$ and $P_{i+1}$ are the loads on top and the bottom of the segment. The mobilized shaft resistance, $\tau_i$, is

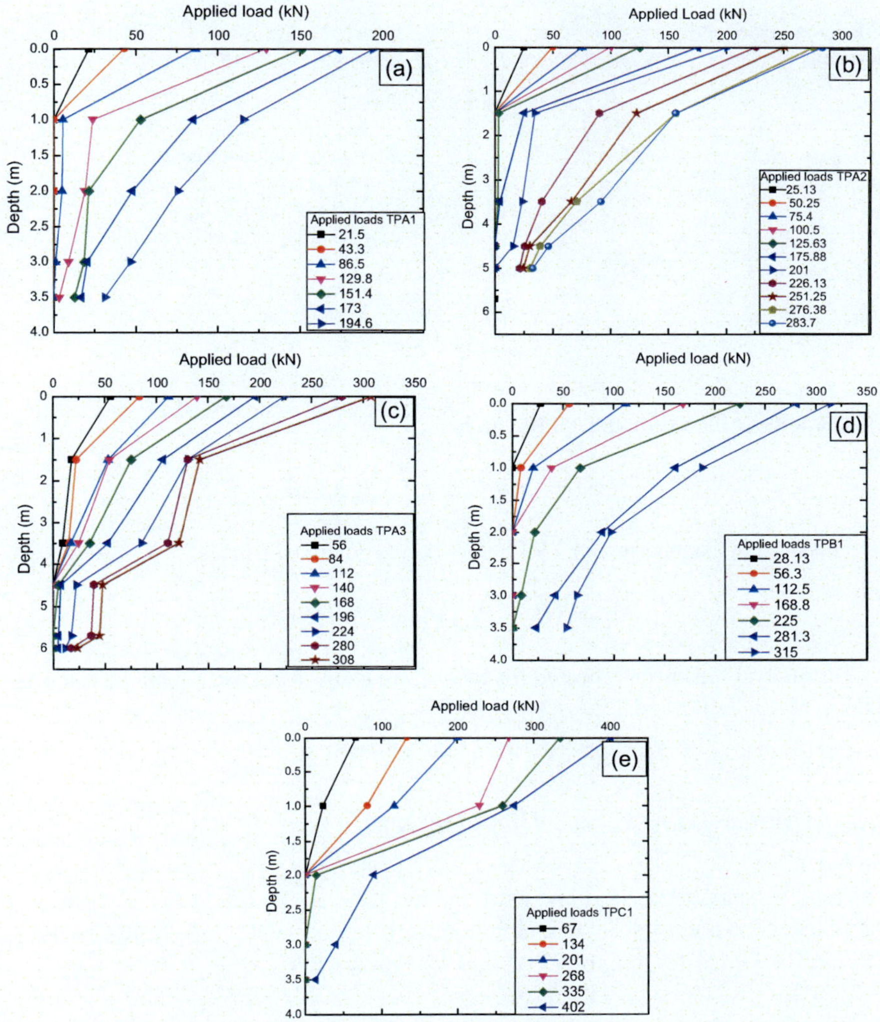


**Fig. 4** Load distribution with depth of piles. **a** TPA1, **b** TPA2, **c** TPA3, **d** TPB1 and **e** TPC1

$$\tau_i = \frac{P_i - P_{i+1}}{A_s} \tag{1}$$

where $A_s = \pi D L_i$.

Test pile $A1$ is divided into four segments corresponding to strain gauge locations as 0–1 m, 1–2 m, 2–3 m and 3–3.5 m. The stress evaluated for each segment is indicated at the centre of the segment $x_1, x_2, x_3, \ldots, x_n$. The applied load ($P_i$) was 21.6 kN at pile top. The load at depth 1.0 m, $P_{i+1}$ was 0 kN. The load transferred ($P_i - P_{i+1}$) over the segment 0–1 m is 21.6 kN.

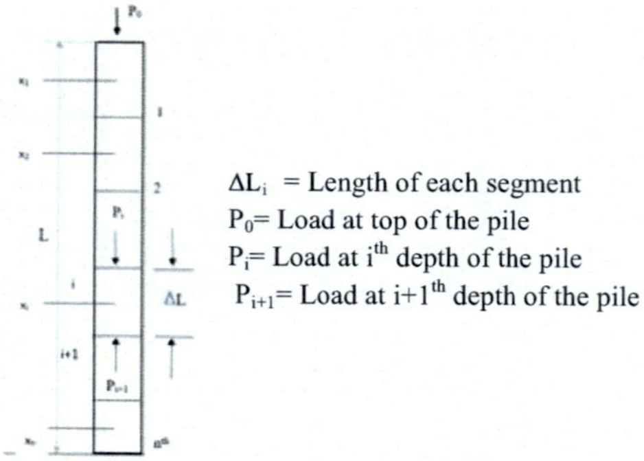


**Fig. 5** Sketch of pile segments

The area of the segment ($A_s$) 0–1 m is

$$A_s = \pi \times 0.2 \times 3.5 = 2.2\,\text{m} \tag{2}$$

The shaft resistance, $\tau_1$, of TPA1 at $x_1$ (i.e. segment 0–1 m) is

$$\tau_1 = (P_i - P_{i+1})/A_s = 9.83\,\text{kPa} \tag{3}$$

The analysis has been repeated for each strain gauge level and for all the test piles TPA2, TPA3, TPB1 and TPC1.

### 2.2 Applied Load Versus Displacement

The load versus settlement plots for all the test piles are shown in Fig. 6. Observed maximum settlements close to or at failure were 3 mm for TPA1, 5.3 mm for TPA2, 3.9 mm for TPA3, 4.3 mm for TPB1 and 2.7 mm for TPC1.

## 3 Results and Discussion

### 3.1 Shear Stress Distribution Along the Length of Pile

The estimated shear stresses in the segments of all piles (TPA1: 0–1 m, 1–2 m, 2–3 m, 3–3.5 m; TPA2: 0–1.5 m, 1.5–3.5 m, 3.5–4.5 m, 4.5–5 m; TPA3: 0–1.5 m, 1.5–3.5 m, 3.5–4.5 m, 4.5–5.7 m, 5.7–6 m; TPB1: 0–1 m, 1–2 m, 2–3 m, 3–3.5 m; TPC1; 0–1 m, 1–2 m, 2–3 m, 3–3.5 m) are shown in Fig. 7 from the strains measured for each applied load. Figure 7a shows variations of shear stress distributions with depth of test pile TPA1 for applied loads, 21.6, 43.3, 86.5, 129.8, 151.4 and 173 kN. For test pile TPA2, for an applied of 21.6 kN, the shear stress was 34.4 kPa over

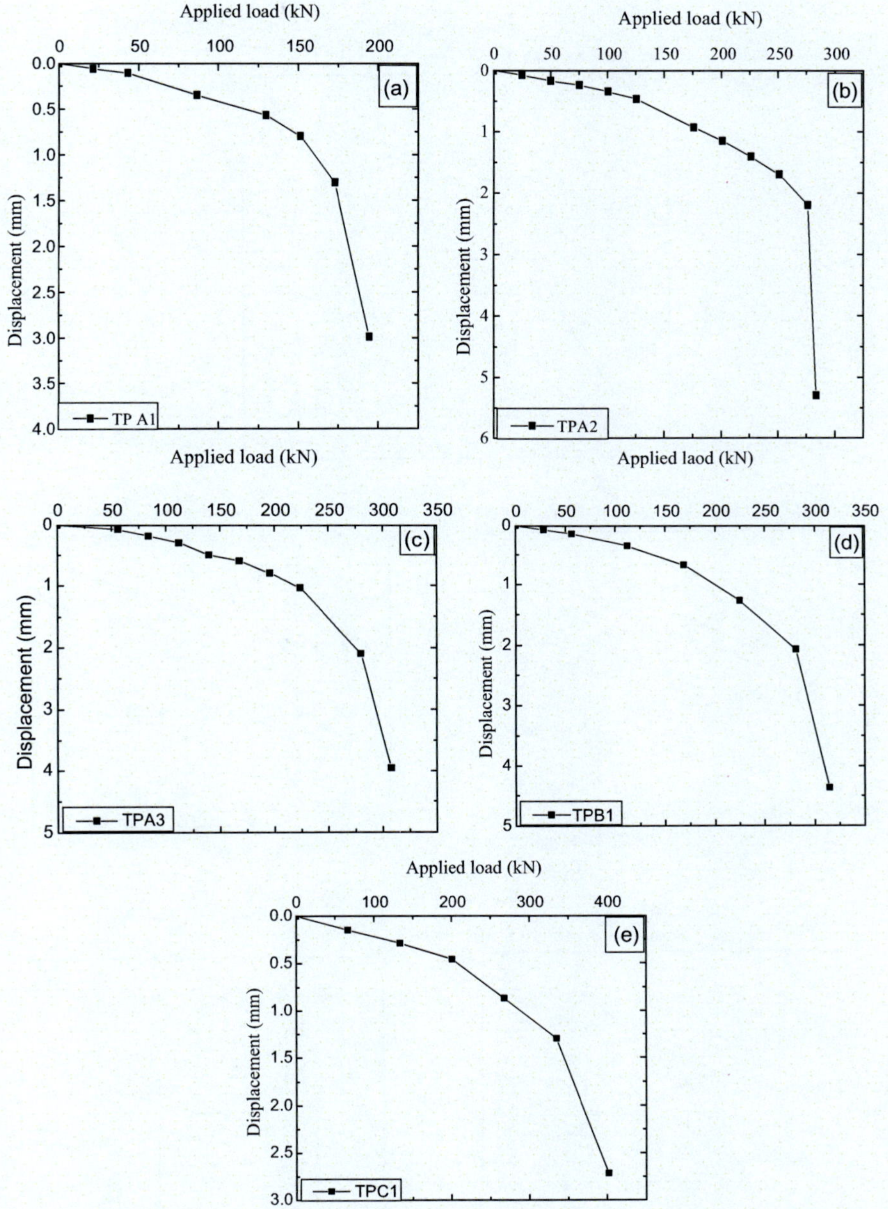


**Fig. 6** Load–displacement plots for piles **a** TPA1, **b** TPA2, **c** TPA3, **d** TPB1 and **e** TPC1 [11]

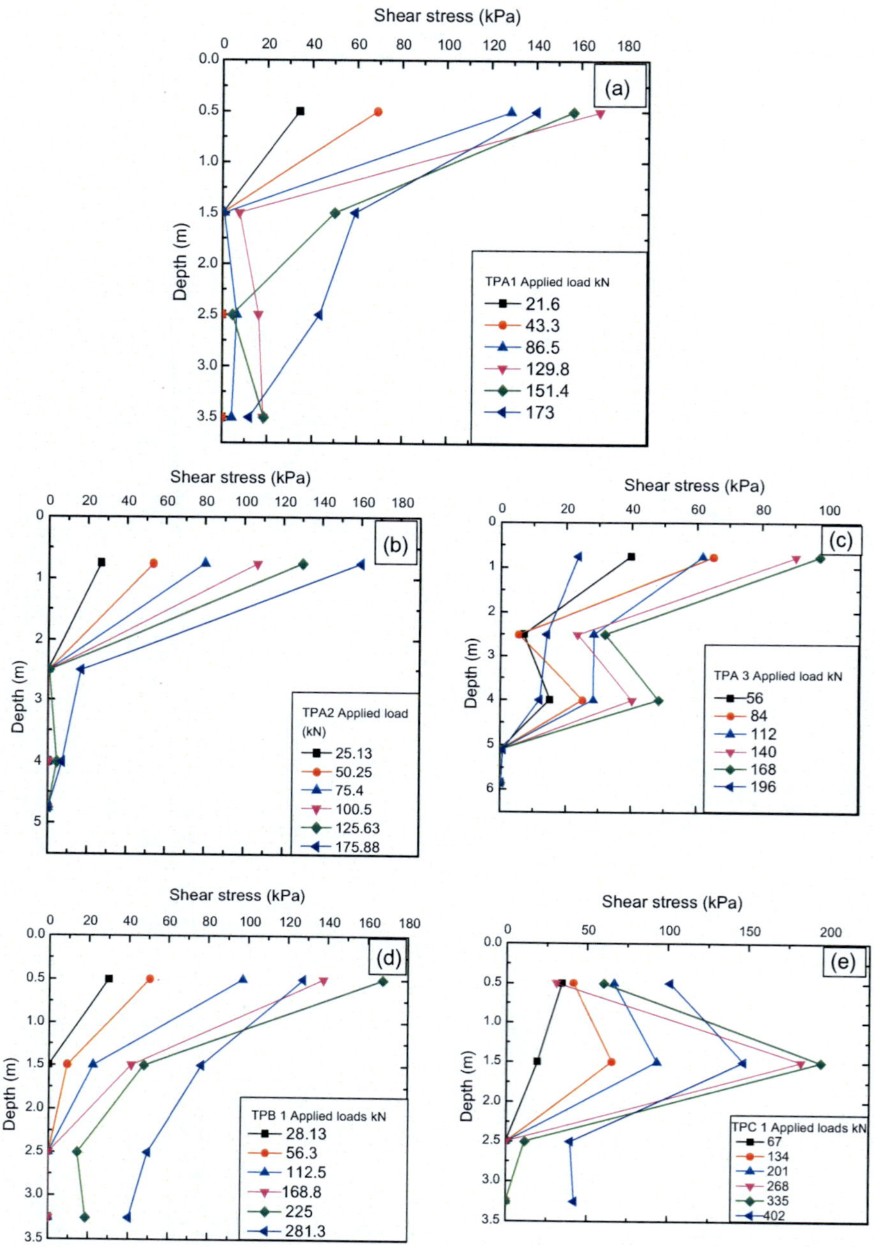


**Fig. 7** Shear stress variation along the pile length in segments **a** TPA1, **b** TPA2, **c** TPA3, **d** TPB1 and **e** TPC1

**Table 3** Shaft resistances at maximum applied load

| Time | Segments (m) | Shaft resistance (kPa) |
|---|---|---|
| TPA1 | 0–1 | 125 |
| | 1–2 | 64 |
| | 2–3 | 46 |
| | 3–3.5 | 50 |
| TPA2 | 0–1.5 | 135 |
| | 1.5–3.5 | 52 |
| | 3.5–4.5 | 72 |
| | 4.5–5 | 42 |
| TPA3 | 0–1.5 | 176 |
| | 1.5–3.5 | 16 |
| | 3.5–4.5 | 117 |
| | 4.5–5.7 | 4 |
| | 5.7–6 | 113 |
| TPB1 | 0–1 | 135 |
| | 1–2 | 95 |
| | 2–3 | 36 |
| | 3–3.5 | 23 |
| TPC1 | 0–1 | 101 |
| | 1–2 | 147 |
| | 2–3 | 40 |
| | 3–3.5 | 42 |

0–1 m segment (Fig. 7b). TPA3 shows similar behaviour (Fig. 7c) up to 2.5 m depth for applied loads of 56, 84, 112, 140, 168 and 196 kN.

For test pile TPB1, stress variation can be observed from Fig. 7d for loads of 28.13, 56.3, 112.5, 168.8, 225 and 281.3 kN. Stress was 29.9 kPa at 0.5 m (i.e. 0–1 m) for 28.13 kN load. The stress becomes 0 kPa at 1.5 m, and shear stress was fully mobilized over 2.5 m long segment. For the test pile TPC1, the applied loads were 67, 134, 201, 268 and 335 kN. In this case, maximum stress was mobilized beyond 2.5 m. All the test pile performances were similar. Shear stress has been fully mobilized over the top 2.5 m depth in most cases. Table 3 presents the mobilized shaft resistances of all the test piles at maximum applied load.

### *3.2 Mobilization of Shear Stress with Displacement*

The displacements are interpolated from the load–displacement plots for all the test piles. The shear stress–displacement plot is derived and shown in Fig. 8. The initial slopes of the shear stress–displacement curves for all the five test piles (TPA1, TPA2,

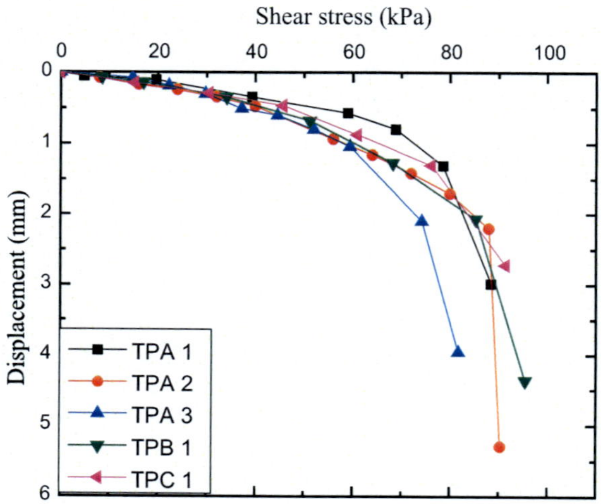


**Fig. 8** Shear stress versus displacement responses of all test piles

TPA3, TPB1 and TPC1) are linear up to shear stress of 40 kPa and close to each other indicating a unique response. The initial shaft–ground interface stiffness is of the order of 80 MN/m$^3$. The shear stress–displacement responses become nonlinear for higher shear stresses. The ultimate shear stress gets mobilized within a displacement of hardly 3 mm. The ultimate or maximum shear stress ranges between 85 and 90 kPa except for pile TPA-3 for which the ultimate shear stress is less than 80 kPa.

## 4 Conclusions

In this study, axial load distribution with depth arrived at by measuring strains at depth has been analysed to arrive at mobilization of shaft response with displacement. The shear stress at low to smaller stresses increases linearly with displacement and nonlinearly with higher displacement reaching the ultimate value at about 3 mm displacement.

## References


1. ASTM D1586/D1586M-18: Standard Test Method for Standard Penetration Testing of Soils
2. ASTM D3441-16: Standard Test Method for Mechanical Cone Penetration Testing of Soils (1998)
3. Comodromos, E.M., Anagnostopoulos, C.T., Michael, K.G.: Numerical assessment of axial pile group response based on load test. Comput. Geotech. **30**, 505–515 (2003)
4. Coyle, H.M., Reese, L.C.: Load transfer for axially loaded piles in clay. J. Soil Mech. Found. Div. **92**(2), 1–26 (1966)

5. Elkasabgy, M., Naggar, M.H.E.I.: Axial compressive response of large-capacity helical and driven steel piles in cohesive soil. Can. Geotech. J. **56**(2), 187–197 (2015)
6. Han, F., Prezzi, M.: Axial resistance of closed-ended steel-pipe piles driven in multilayered soil. J. Geotech. Geoenviron. Eng. ASCE **143**(3) (2017)
7. Mert, M., Ozkan, M.T.: Settlement analysis of axially loaded bored piles: a case history. Int. J. Geotech. Geol. Eng. **13**(5) (2019)
8. Reese, L.C., Stoke, K.H.: Instrumentation for tests of piles subjected to axial loading. Transp. Res. 1169 (1984)
9. Reese, L.C., Vijayvergiya, V.N., Hudson, W.: Load distribution for a drilled shaft in clay shale. U.S. Department of Transportation Federal Highway Administration Bureau of Public Roads, vol. 89(5) (1969)
10. Seo, H., Prezzi, M.: Analytical solutions for a vertically loaded pile in multilayered soil. Geomech. Geoeng. **2**(1), 51–60 (2007)
11. Yousif, E.: Use of penetration and pile load tests for evaluation of the capacity of bored piles in some Sudanese soil. Thesis. Building and Road Research Institute University of Khartoum (2012)

# Numerical Analysis of Jointed Piles

**B. Swathi, V. Balakumar, and S. S. Chandrasekaran**

## 1 Introduction

Presently, lot of infrastructure developments take place in the coastal areas where the strata is dominated by soft compressible clay layers, perhaps marine clays, organic in nature in certain cases. The geological history of the site can vary depending upon the location of site. Such compressible deposits might have formed naturally or the site could have been formed by recent hydraulic fills. In general, large diameter-bored piles become an automatic choice. The length of the piles becomes more than 50–60 m which poses quality problems. Moreover, bentonite solution has to be used to stabilize the boreholes which can form sediments with slush and cleaning the bottom of the pile becomes very difficult. Also, the installation process by drilling can disturb the consistency of the soil, which can change the properties, affecting the capacity of the pile. Hence, the option becomes precast pile which is devoid of such quality issues. Further, such piles are subjected to negative skin friction and accounting for it can make the design very uneconomical. The length of precast pile becomes high which leads to handling problems. The precast piles can be coated with a slip layer (hot blown bitumen which will reduce the negative skin friction to considerable extent. The practical issue had given rises to precast jointed piles. The behaviors of piles installed through such layers are influenced by the installation methods, Debeer [1], Viggiani (1989), [2] which is seldom considered in the design.

Segmental driven precast piles often recommend solutions, Causer et al. [3] and Shin [4]. Precast jointed piles have number of segments in which are jointed by

B. Swathi (✉)
Structural Engineering, Vellore Institute of Technology, Vellore, India

V. Balakumar
Simplex Infrastructures Limited, Chennai, India

S. S. Chandrasekaran
School of Civil Engineering, Vellore Institute of Technology, Vellore, India

 
C. N. V. Satyanarayana Reddy et al. (eds.), *Ground Characterization and Foundations*, Lecture Notes in Civil Engineering 167,
https://doi.org/10.1007/978-981-16-3383-6_57

patented joints to achieve the desired length so that the pile can be terminated in a hard stratum governed by the set to generate the desired capacity. There are several types of splices joints available which are patented. The overall effectiveness of splices was examined and investigated by many researchers Bruce et al., Gran, Doherty and Gavin [5] and Abou-Jaoude and Zhang [6]. In the cases of hollow spun piles, Irawan et al. [7], which belong to the jointed pile family is extensively used wherever possible depending upon the soil condition a feasibility.

In the case of jointed segmental piles, the negative skin friction can get Morsy et al. [8] and Gandhi et al. considerably reduced by applying the slip layer. Therefore, in order to overcome the difficulties, the institutions in Sweden developed pile joints so that the length need not be a constraint, Bredenberg and Broms and Brown and Shie [9]. Another important issue is axial pile deformation due to lateral soil movement as studied by Poulos and Chen [10], Doherty and Gavin [5] and Kroff et al. [11] soil displacement resulting from deep excavation. An important issue is the location of the joint below the cut-off level namely the bottom of the pile cap. Further pile section being smaller, unplanned excavations particularly in soft soils can cause large displacements in the pile and the joint can get overstressed. This paper discusses the main features of design, the significance of joint and its location below the raft and pile cap bottom. Studies on such issues appear to be very scarce.

### *1.1 Need for Study*

In the case of the marine deposits, there can be variation in the consistency at different locations which is unavoidable. Isolated very stiff to hard pockets are very common in the marine deposits and driving may become impossible. In such cases, the last joint can come closer to the raft bottom. Therefore, it becomes necessary to study the stress around the joint when the pile is subjected to peak loading such as test load and the load settlement of the jointed portion. In this, present study undergoes an excavation is assumed to be sufficient as two-dimensional analysis applicable and to determine the response of pile due to lateral soil movement.

## 2 Parametric Study

Parametric studies were conducted by varying the joint locations as $L/2$, $L/4$, $L/6$ and $L/8$ was carried out, where $L$ is the length of the pile. Figure 1 represents the layout of jointed pile. A series of 3D finite element analyses were performed on a single free head jointed pile in both homogeneous stiff clay and medium dense sand separately. The response of the jointed piles under pure axial and lateral load was analyzed initially. Further, the study was carried out on combined effect of both axial and lateral load. The ultimate load capacity was estimated, considering an appropriate

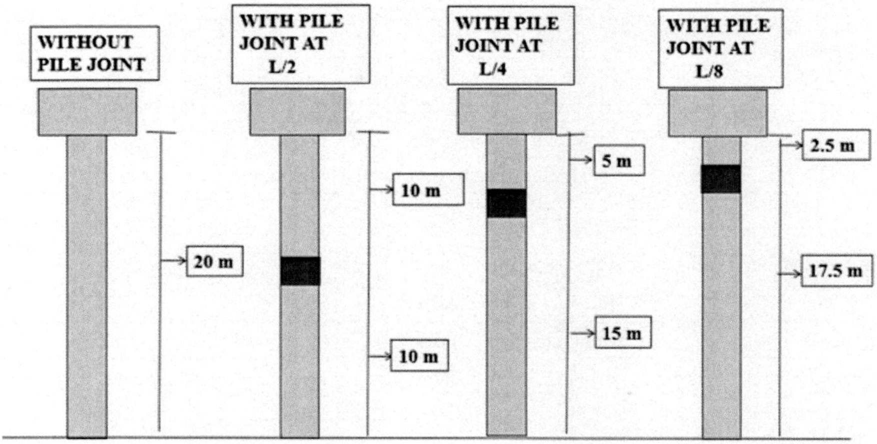


**Fig. 1** Pile joint layout

*N* adopting Meyerhof's expression, the capacity was calculated as 1000 kN and the corresponding lateral load was considered as 5% of axial load.

# 3 Finite Element Analysis

## *3.1 Introduction*

All numerical analyses in this investigation were performed by using 3D finite element program ABAQUS 6.14. The program is supported by a preprocessor to develop three dimensional meshes consisting of bar and beam type prismatic elements, 8-node or 20-node continuum brick elements with zero thickness type interface elements as well as a post-processing tool that is capable of plotting the original mesh, deformed mesh, displacement vectors, extracting nodal displacements, element stresses etc.

## *3.2 Finite Element Modeling*

Three-dimensional modeling of the jointed pile–soil system has been carried out in finite element analysis by using "ABAQUS 6.14". Figure 1 presents the finite element model of soil mass and jointed piles. The soil and pile were modeled using eight-node brick element soil strata having three degree of freedom of translations in the respective coordinate direction at each node. Out of these, the linear elastic and the Mohr–Coulomb model are used to simulate the behaviors of the soil, whereas piles

**Table 1** Numerical details of the FEM model

| Property | Pile | Steel | Soil |
|---|---|---|---|
| Material | M40 concrete (square piles) | Mild steel (high durability) | Medium sand/stiff clay |
| Width | 400 mm | 400 mm | 12 m |
| Depth | 20 m | 600 mm | 40 m |

**Table 2** Material properties of the FEM model

| Property | Pile | Steel | Sand | Clay |
|---|---|---|---|---|
| Material | Concrete | Mild steel | Medium dense | Stiff clay |
| Elastic modulus (MPa) | 3.1e4 | 210e3 | 100 | 80 |
| Density (kg/m$^3$) | 2400 | 7850 | 1500 | 1330 |
| Poisson ratio | 0.2 | 0.3 | 0.3 | 0.25 |
| Friction angle | – | – | 37° | – |
| Cohesion (kPa) | – | – | – | 200 |

have been idealized to behave as linearly elastic. Tables 1 and 2 present the physical and material properties of jointed pile and soil, respectively. The depth of soil block is 2 times of the length of the pile was chosen since it is the optimum limit; beyond that, additional bearing capacity does not appreciable.

The interaction between the sand and pile was modeled by defining tangential and normal contact behavior in the FE model. The normal force is transferred only when pile and soil are contacts tightly; otherwise, it becomes zero thickness. This type of normal contact behavior can be modeled adopting hardening model with contact option being provided. Perfect contact is provided between pile–steel–pile and pile–soil.

The needed mesh refinement was carried out keeping in mind the economics of computational time, efforts and the needed level of accuracy in the results. The fixed boundary condition is provided at the bottom surface and along the four edges of the soil block. All nodes on bottom surface were restrained in all three directions provided rigidly, and then, the loads are imposed on the jointed pile and the load settlement and shaft stress around the pile values are obtained on each node of piles near edges to the soil surface are used in plotting the graphs (Fig. 2).

### *3.3 2D Analysis of Excavation*

A two-dimensional finite element program was used to simulate the excavation using PLAXIS 2D. In the program, 15–node plane strain model of jointed portion is assumed to be linearly elastic and the soil is treated as an elasto-plastic material, obeying a Mohr–Coulomb failure criterion. The interface thickness between the soil

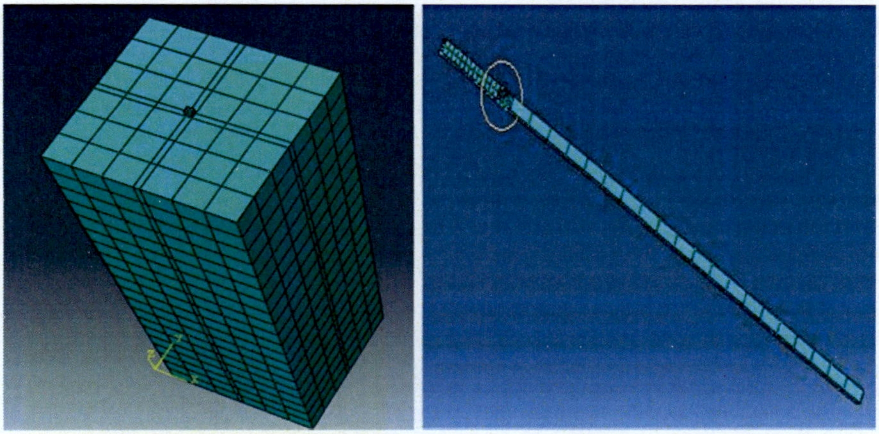

**Fig. 2** Finite element mesh for soil mass and jointed piles

and pile is kept as rigid, i.e., no-slip occurred. In mesh, vertical and bottom boundaries are fixed in all three directions. In the second phase, while the soil layers were excavated from top to bottom in various steps with each step involving the removal of consecutive 2 m until the desired depth of excavation was achieved in both cases (Fig. 3).

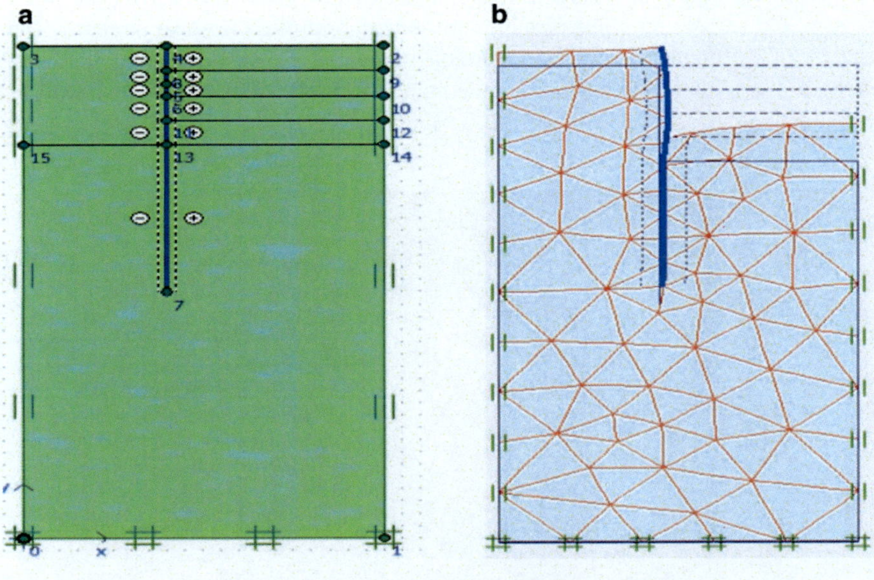


**Fig. 3** **a** Model excavation of jointed pile. **b** Deformed mesh PLAXIS 2D

## 4 Result and Discussion

The present work is intended to study the effect of joint position. The results of the numerical analyses were found to be a quick and reliable mode to get the desired results compared to model tests or observational studies under the present conditions of requirement. Figure 4 shows displacement contour for jointed pile. The shaft stress around pile, settlement criteria, deflection pattern of the jointed pile system was analyzed.

### *4.1 Load Settlement Response Due to Vertical Load*

Figures 5 and 6 present comparison on the load settlement response of jointed pile with joint at various locations. The performance of jointed pile is compared with a full-length pile without considering the joint. In the present study, the length of the pile is taken as 20 m (two segments of 10 m) long elements with one joint mainly because the performance is very critical only in the top portion, most specifically the first joint from the ground level.

These graphs indicate that for joints at middle portion of the pile, the load settlement response shows a very small variation from that of the pile without joint in case of medium dense sand, but in case of stiff clay, variation was higher. This may be perhaps that in case of stiff clay, the installation method generates excess pore water pressure and remolding effects leading to the temporary loss in shear strength. The capacity of the single pile obtained from the numerical analysis was compared with results of the jointed pile analysis in medium sand and this comparison shows the percentage variation of 25 mm of *L*/2, *L*/4 and *L*/8 is 25, 15.7 and 5.2%, whereas

**Fig. 4** Displacement contour for jointed pile

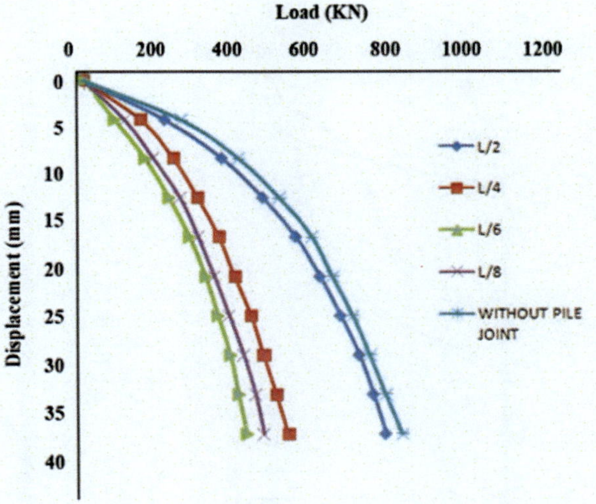


**Fig. 5** Load settlement response for various joint position in medium sand

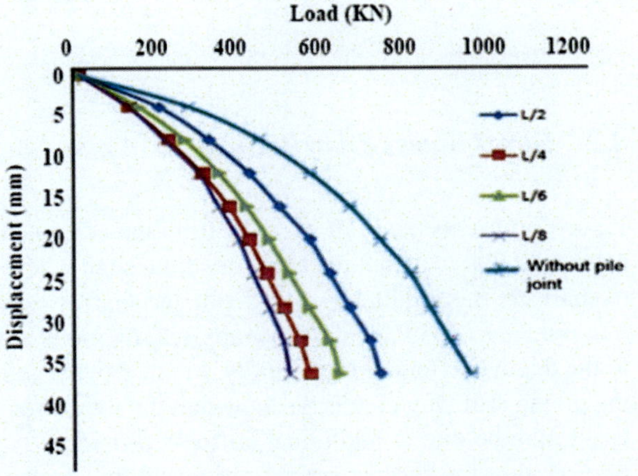


**Fig. 6** Load settlement response for various joint position in stiff clay

the single pile value of 23%. As the joint position moves toward ground level, the load settlement response become more or less identical and indicate a much lesser stiffness. Under the identical settlement, the reduction of load taken in order of 40%, the joint moves to *L*/4. This behavior indicates that as the joints move close to the ground, the capacity of the pile reduces that at 25 mm settlement the percentage variation is much higher than 12 mm. This may be because of loss of friction at higher settlement.

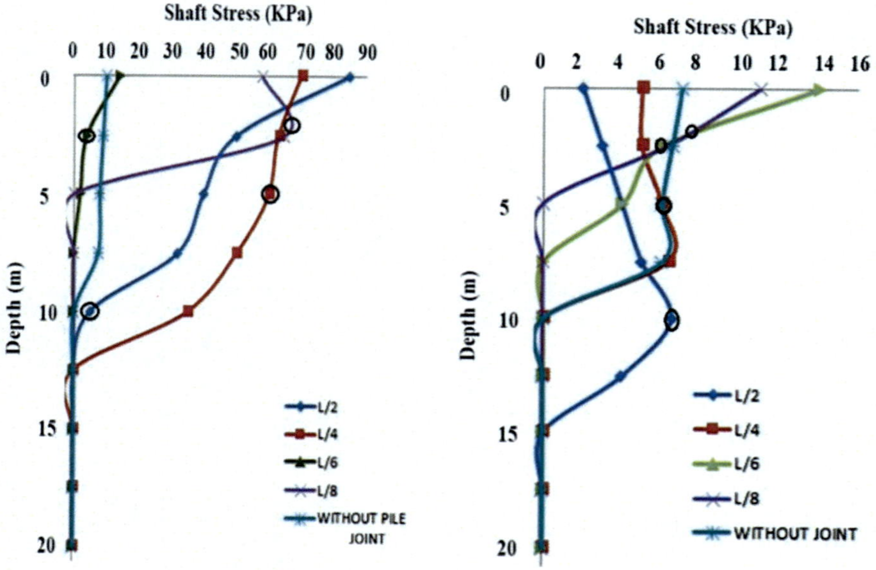


**Fig. 7** Shaft stress distribution curve in medium sand and stiff clay

## 4.2 Shaft Stress Distribution for Axial Load

Figure 7 presents the shaft stress distribution of jointed pile obtained from axial load approach. It is seen that in case of medium sand when the joint at *L*/2, the rate of fall of shaft stress is gradual. At 5 m depth, the shaft stress reduces from 85 to 50 kPa and 8 m reduce to 30 kPa and forms rapidly. The joints at *L*/4 the shaft stress remain up to the depth 6 m and reduce rapidly. In other two cases, the analysis of stress around the pile in stiff clay condition show that the trends remain more or less similar. This trend may be due to additional stiffness provided by joints. At the end of the pile portion, the stress value attains zero for all locations of the joint pile section.

## 4.3 Lateral Load Deflection Response

The load deflection response of jointed pile under the lateral load needs a separate consideration mainly because the joint position as lots of influences as it comes closes to pile cap. The reaction forces that developed at the nodes used to calculate the lateral load corresponding to the applied lateral deflection.

Figures 8 and 9 present the load deflection response of jointed pile medium sand and stiff clay. As discussed in this section, the percentage of reduction in settlement reduces with a decrease in the loading condition, and the initial loading condition in the settlement achieved shallow values. It appears that the joint position at *L*/8 gets

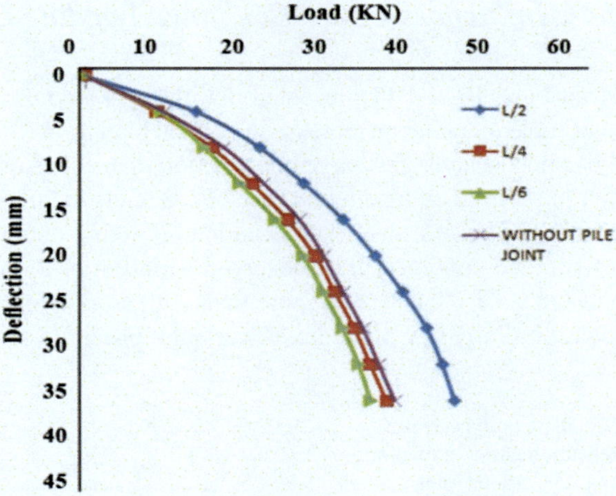


**Fig. 8** Lateral load deflection curve in medium sand for various joint locations

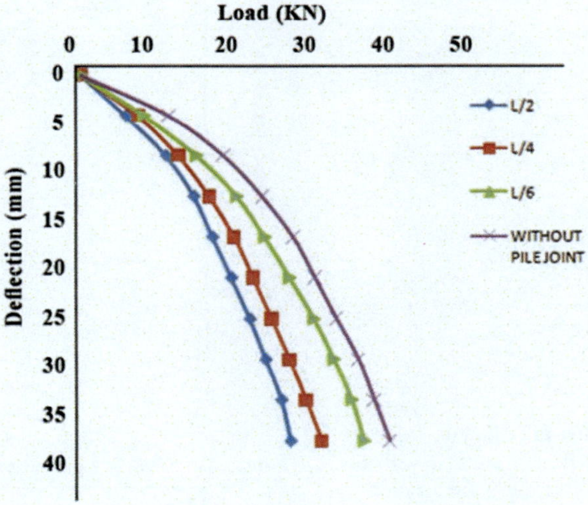


**Fig. 9** Lateral load deflection curve in stiff clay for various joint locations

failure and reduces stiffness of joint soil system. At any load, the deflection when the joint comes close is far higher than the deflection with joint at the center position; similar behavior was observed in the case of stiff clay condition. At 5 mm deflection, very little change will occur in the load deflection, while the deflection of medium sand *L*/2, *L*/4, *L*/6 and without joint has increased 42, 39, 16 and 20% than stiff clay, respectively.

### 4.4 Influences Under Combined Loading Condition

From Figs. 10 and 11 presented that there is a considerable increase in the lateral load capacity under an increase in vertical load level. Under the lesser load condition, the cohesive layer gets large deformation than cohesionless layer. It can be observed that the combined load in the case of *L*/8 tends to fail due to the location of the joint near to the cut-off level. The axial load is acting on top of the pile; hence, it acts as a fixed end condition. The observed variation of 25 mm displacement of *L*/2, *L*/4 and *L*/6 is 41, 34 and 20% more than without pile joint due to application of lateral load. Similarly, for jointed pile, variation observed in 50, 42, 26% above the valve

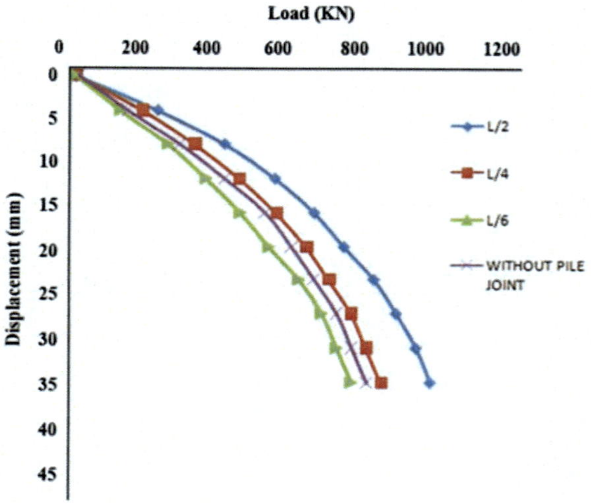


**Fig. 10** Combined load deflection curve in medium sand for various joint locations

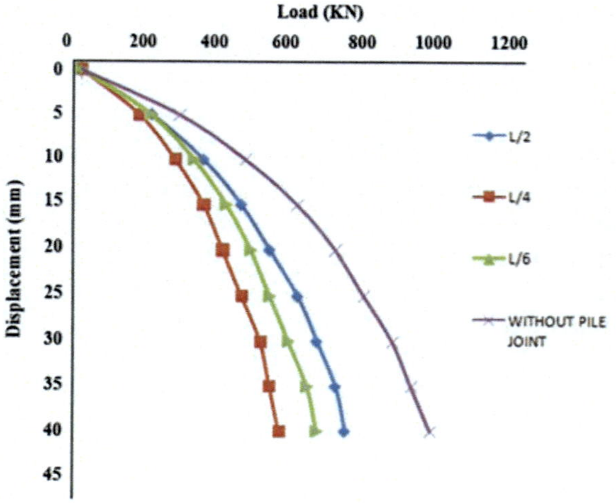


**Fig. 11** Combined load deflection curve in stiff clay for various joint locations

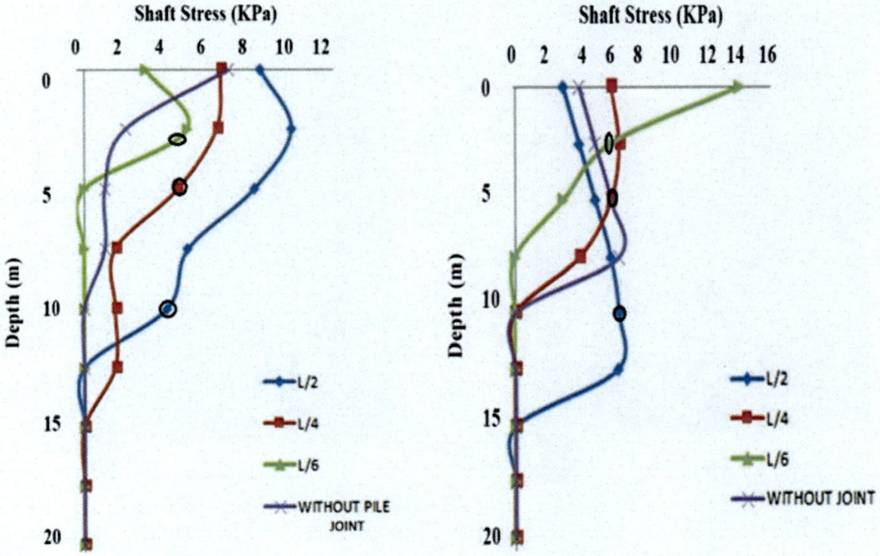


**Fig. 12** Shaft stress distribution curve in medium sand and stiff clay-various joint

for 12 mm of deflection. Increase of lateral load capacity of the pile with increased earth pressure on verge of the pile, due to confinement of soil particles.

## *4.5 Shaft Stress Distribution for Combined Loading*

Figure 12 show that design shaft stress in piles is higher under combined loading. Due to the influence of combined effect, the maximum shaft stress occurs in the case of *L*/6 condition, i.e., 14 kPa in stiff clay, whereas the induced stress will be increased, particularly in the jointed section. Alternatively, the joint portion at *L*/4 shows nearly equal in both cohesive and cohesion less soil and starts reducing rapidly at depth of 10 m. As increase depth of the pile, the induced stress in the pile will decrease.

## *4.6 Influence in Pile Deflection Due to Excavation*

The effect of excavation close to the existing pile is projected in Fig. 13. The joint is located at *L*/6. An excavation of 2 m induces a deflection 19.2 mm in the top segment, whereas in the case of pile without any joint, the maximum deflection occurs after an excavation of 8 m. Hence, the position of joints play an important role during excavation for the pile cap.

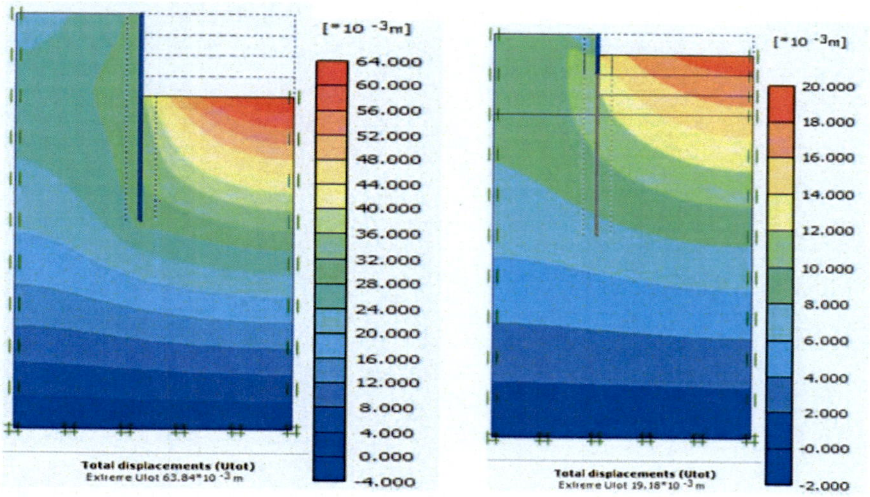


**Fig. 13** Displacement contour for unjointed pile and jointed pile

## 5 Conclusions

1. While jointed piles provide a safe solution whenever pile length larger, but the joint position plays a very important role. So, the length of the segments has to be carefully optimized. Geotechnical report has to be reviewed carefully for deciding the length of each segment.
2. In case of jointed pile the displacement is higher than single pile, when the joint is at mid location. As the displacement decrease and bending moment reaches maximum value, that particular location tends to brittle failure.
3. The numerical data shows that the shaft stress resistance increases under lateral load is applied. The pile body tends to rotate around the inflection point and produces a negative deflection closed to the jointed portion. Hence, the joint at *L*/8 portion fails down due to applied lateral load.
4. It can be concluded that under static loading condition, the location of joint portion should not be above *L*/6 from the cut off level. The percentage variation around nearly 41 and 39% in combined effects. So that the ideal condition for the location of the joint is *L*/2 to *L*/4, where *L* is the length of the pile.
5. It also clear that the maximum pile deflection follows soil movement closely at a distance from the excavation face. The pile deflection is very close to the jointed section, reflecting the fact that the pile is relatively flexible.

## References


1. DeBeer, E.E.: Different behaviour of bored and driven piles. In: Proceedings, International, Seminar, Ghent, on Deep foundation on Bored Piles and Auger Piles. Balkema, Rotterdam (1988)
2. Viggiani, C.: Further experiences with Auger piles in Naples Area. In: Deep Foundation on Bored and Auger Piles Ghent, pp. 77–94 (1993)
3. Causer, E., Rune, K.: The behaviour of connections in segmental precast concrete pile systems. J. Geotech. Eng. **11**(2019), 53e548. (2012)
4. Shin, Y., Kim, M., Ko, J., Jeong, S.: Proposed design chart of mechanical joints on steel PHC composite piles. Mater. Struct. **47**, 1221–1238 (2014)
5. Doherty, P., Gavin, K.: The shaft capacity of displacement piles in clay. A state-of-the-art review. Geotechn. Geolog. Eng. **29**(4), 389–410 (2014)
6. Abou-Jaoude, G., Zhang, Y.: The role of soil stiffness nonlinearity in 1D simulations. Geotechnique **65**(3) (2015). ISSN 0016-8505
7. Irawan, C., Suprobo, P., Raka, I.G.P., Djamaluddin, R.: A review of prestressed concrete pile with circular hollow section (spun pile). J. Teknol. **72**(5) (2015)
8. EI-Mossallamy, Y.M., Hefny, A.M., Demerdash, M.A., Morsy, M.S.: Numerical analysis of negative skin friction on piles in soft clay. HBRC J. **9**(1), 68–76 (2013)
9. Brown, D.A., Shie, C.F.: Some numerical experiments with a three-dimensional finite element model of a laterally loaded pile. Comp. Geotech. **12**, 149–162 (1991)
10. Poulos, H.G., Chen, L.T.: Pile response due to excavation induced lateral soil movement. J. Geotech. Geoenviron. Eng. **123**(9) (1997)
11. Kroff, M., Mair, R.J., Van Tol, F.A.F.: J. Geotech. Geoenviron. Eng. **142**(8) (2016)

# Genetic Algorithm Based Optimization and Design of Pile Foundation

**Bhargav Jyoti Borah and Sasanka Borah**

## 1 Introduction

Piles are deep foundations of relatively smaller diameter shaft which are driven into the ground or introduced into the soil by suitable means so as to support the load coming on it from the superstructure when a good bearing stratum is not available near the ground surface or at shallow depths. In such situations, load has to be transmitted to a firm strata capable of supporting such loads even though such strata may be at an appreciable depth below the ground surface. Piles are used in a wide variety of important structures and are not only subjected to vertical loads coming from the superstructure, but also to lateral loads coming from inclined loads, wind, waves, earthquakes, uplift forces, etc. [1].

Indian Standards [2–5] classifies concrete piles as driven cast-in situ piles, bored cast in situ piles, driven precast piles and precast piles in pre-bored holes. The capacity of the pile foundation is dependent on the material and geometry of each pile, their spacing in pile group, the load bearing strength and type of the surrounding soil supporting the pile, the method of pile installation, and the direction of applied loading [6]. The soil bearing capacity is affected by many factors such as type and strength of soil, foundation dimensions, soil weight, surcharge, type of loading etc. [7]. The variability of the properties of the supporting soil, soil profiles and multi-layered soil profiles, which generally exists in nature, affects the load bearing capacity of the pile drastically. The choice of a particular pile is governed by certain situations, namely, site conditions, economy, time considerations etc. Hence, in order to encounter such problems with large variability, the aid of modern high-speed computers and software has become popular without which this job would be tedious.

B. J. Borah (✉) · S. Borah
Department of Civil Engineering, Assam Engineering College, Guwahati 781013, India

S. Borah
e-mail: sasankaborah.ce@aec.ac.in



C. N. V. Satyanarayana Reddy et al. (eds.), *Ground Characterization and Foundations*, Lecture Notes in Civil Engineering 167,
https://doi.org/10.1007/978-981-16-3383-6_58

Conventional design approach for piles requires determining the load carrying capacity of piles which can be obtained from dynamic formulae, static formulae, pile load test or penetration test depending upon a particular situation giving due consideration to soil exploration data and chemical properties of soil. Depending upon the load coming from the superstructure, the pile or group of piles are designed in such a way so as to transfer the load to the supporting soil safely. In geotechnical design part, the ultimate load carrying capacity and the allowable load carrying capacity of the pile must be checked, whereas in the structural design part the designed foundation must be checked for bearing capacity, driving and handling stresses. Foundation designs are usually a trial-and-error procedure, in which a trial design (foundation dimensions and reinforcements) is chosen and is checked against the geotechnical and structural requirements, which is, followed by revision of the trial design, if necessary [8]. But due to various assumptions and approximations associated with such method of estimation of load bearing capacity and design methodology, the pile design often comes out to be conservative. A conservative design may lead to an uneconomical design. An optimization approach in the foundation design process may confirm the economic design. Hence, an optimized design methodology is the need of the hour. Optimized design approach for single piles evaluate the ultimate bearing capacity of the pile within stipulated limits and changing the pile dimensions iteratively until a reasonable pile geometry is found to support the total load coming on it. The pile is then designed structurally for bearing the loads acting on the pile satisfying all the structural requirements.

The research done in foundation engineering for the application of optimization methods using computer aided design methods is scarce. In particular, researches dealing with pile optimization assume different ideas mainly with groups of piles and simplification of the structural capabilities of reinforcement which often have limited practical considerations. Chow and Thevendran [9] used pile length as the main design variable to minimize differences in bearing loads between the piles in the pile group. Hoback and Truman [10] introduced a weightless optimality rule into the original optimality criteria approach to treat design variables, (e.g., the spacing and battering of the piles) that have no measurable effect on the objective function. Huang and Hinduja [11] adopted a quasi-Newton method to optimize the shape of a pile foundation with the assumption of a linear force–deflection relationship for the pile-soil system. Valliappan et al. [12] applied the generalized reduced gradient method to optimize pile foundation design with the lowest cost objective. Their design variables included pile length, diameter, number and pile cap. The allowable total and differential settlements were the only constraints. Kim et al. [13, 14] used GA and recursive quadratic programming, respectively, to optimize the layout of a pile foundation, with minimum differential settlement being the objective and with the assumption of linear pile-soil interaction. Chan et al. [15] presented an automatic optimal design method using a hybrid GA for pile group foundation design with the concrete volume of the piles and the cap as the objective function. Letsios et al. [16] proposed the formulation of an optimization problem using German Foundation Code DIN 4014 and Euro code 7 (EC7) design procedure. Both standards are based mainly, on four pile design criteria: (i) Axial bearing Capacity, (ii) Acceptable settlements,

(iii) Strength of pile as a structural element and (iv) Lateral bearing capacity and acceptable horizontal displacements. Pérez et al. [1] dealt with the optimization of three different materials for single piles: Steel pile, concrete pile and steel fiber reinforced concrete pile by developing a program using interior point algorithm implemented in MATLAB and an objective function based on the cost for each of the pile materials. Darius et al. [17] presented optimization of grillage foundation using GAs for integrating MATLAB environments with the goal of optimization so as to obtain the optimal layout of pile placement in the grillages without taking cost into account. Islam and Rokonuzzaman [8] used CAD to demonstrate design of foundation which is done in two different stages: Geotechnical design and the structural design. Most of the design concentrated on bearing capacity and settlement criteria ignoring the economy of the design. Nikolaou and Pitilakis [18] developed a stand-alone program based on MATLAB for the calculation of bearing capacity and settlements of shallow foundations using several well-known formulas from the literature and design codes that are preferred in engineering practice.

All the above research has contributed greatly to the problems of optimizing pile foundation design. However, most of these methods might be difficult to apply in practice owing to the fact that the adopted design method and constraints are not always code-based, Also the objective function fails to consider all the variables involved in pile foundation design and construction. In many cases, practical considerations or limits are not considered in design which causes problems in practical implementations. These shortcomings and the wide gap between conceptual optimization and design against the actual field applicability, ultimately, fail to cater to the needs of a designer.

This study focuses on bridging such gaps in conceptual design and practical applications by optimizing pile foundation design using a GA and Indian Standards code [2–5, 19, 20] specifications. The GA is used to optimize the pile dimensions based on Indian Standards [2–5] methodology within recommended limits and user defined practical specifications. A standalone software code has been developed using MATLAB programming for automating the entire optimization and design process. An estimate [21] for construction is also presented which may further ease the choice of selection of a particular design.

This study primarily makes an attempt to develop a GA for geotechnical design optimization of pile foundation for a given set of soil and load parameters.

## 2 Methodology

### *2.1 General*

This study makes an attempt to develop a GA for optimization and design of a pile foundation. The optimization methodology used in optimization in the said software is GA. GA is a search-based optimization technique that works on the principles of genetics and natural selection which is used to find optimal solutions to difficult problems which otherwise would take a considerable amount of time to solve [22]. It is frequently used to solve optimization problems particularly in research. The pile design, both geotechnical and structural, is as per Indian standards code [2–5, 19, 20] based design method. A preliminary estimate of conceptual design is also presented which is as per Assam Public Works Department Schedule of Rates [21]. Popular programming software MATLAB has been used to design the interface of the said software and link the relevant codes for optimization and design to the interface.

### *2.2 Genetic Algorithm (GA)*

The GA is a stochastic search-based optimization algorithm inspired by the process of evolution by means of natural selection, primarily fit for nonlinear optimization problems, where gradient based methods cannot be applied due to lack of smoothness of the objective function [22].The algorithm makes use of processes such as selection, crossover and mutation to improve upon a set of solutions and converge toward an optimal solution. Each solution consists of a set of properties, and it is through manipulation of these properties that the GA can converge toward a good solution. In GA, a population is a group of solutions, also referred to as individuals. For a given optimization problem, a group of potential solutions is initially generated, i.e., the first generation. Within a population, each solution or individual has a corresponding fitness value calculated with a fitness function. The fitness value indicates how good, or fit, a solution is. Within each population, some numbers of the fit solutions are chosen to carry over to the next generation of solutions as clones. These solutions guarantee that the best fitness value of each generation will either be maintained from one generation to another or improved upon, while providing properties of higher quality within a population.

Crossover is a reproduction function. Within each generation, a smaller group of solutions is selected to combine their properties in order to create new solutions, and these new solutions are considered a new generation. The crossover group is called parents, and the "genes" of two (or more) such parents are combined in a randomized way to produce one (or more) new solution, their child. It is these children that make up a new generation, and since parent solutions mostly have good fitness values, the new generation is expected to be better in terms of fitness [22].

Each solution may be subject to mutations; random changes in one or more properties (genes) of a solution according to some probability, referred to as the mutation rate [22]. Mutations exist in order to maintain and introduce diversity into a population. With more diversity, the algorithm has a lower risk of ending up in suboptimal local minima. A flowchart of the simple, general version of GA is shown in Fig. 1.

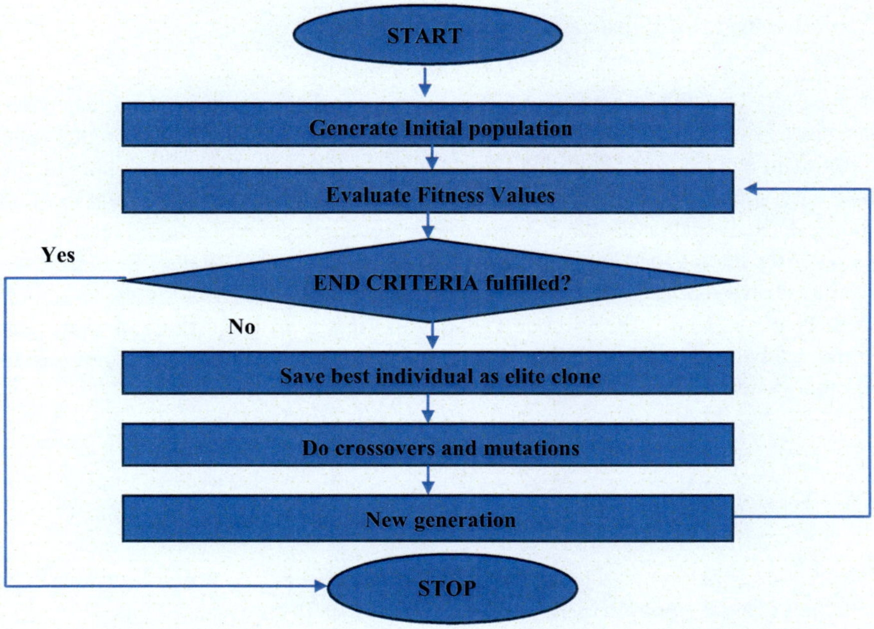


**Fig. 1** Flowchart describing the most basic GA

## 2.3 *Deign Considerations*

IS 2911(Part 1/Sec 2): 2010 [3] recommends that pile foundation is to be designed in such a way that the load from the structure can be transmitted to the sub-surface with adequate factor of safety against shear failure of sub-surface and without causing such settlement (differential or total), which may result in structural damage and/or functional distress under permanent/transient loading. The pile shaft should have adequate structural capacity to withstand all loads (vertical, axial or otherwise) and moments which are to be transmitted to the subsoil and should be designed according to IS 456: 2000.

## 2.4 *Pile Capacity*

In this study, the ultimate load capacity of a single pile is obtained by using static analysis, which depends on the reliability of the soil properties for various strata in which the pile rests. The minimum factor of safety on static formula shall be 2.5 as recommended by the code [3].

### 2.5 Analysis of Laterally Loaded Piles

A pile may be subjected to lateral force for a number of causes, such as, wind, earthquake, water current, earth pressure, effect of moving vehicles or ships, plant and equipment, etc. [3]. The lateral load capacity of a single pile depends not only on the horizontal subgrade modulus of the surrounding soil but also on the structural strength of the pile shaft against bending, consequent upon application of a lateral load. While considering lateral load on piles, effect of other co-existent loads, including the axial load on the pile, should be taken into consideration for checking the structural capacity of the shaft. The IS code [3] suggests that a group of three or more pile connected by a rigid pile cap should be considered to have fixed head condition. In all other conditions, the pile should be taken as free headed.

### 2.6 Structural Capacity

The IS code [3] suggests that the piles should have necessary structural strength to transmit the loads imposed on it, ultimately to the soil. In case of uplift, the structural capacity of the pile, that is, under tension should also be considered.

### 2.7 Estimation

An estimate of the anticipated cost of work to be carried out is generated by the software. In this study, the software generates a design output which includes the structural design of the pile and various specifications involved in its construction. The structural design serves as the necessary drawing for estimation, the other concrete and steel parameters along with workmanship requirements specified by Indian Standards [3] serves as the specifications and the Assam Public Works Department Schedule of Rates [21], as the Rate of Items, while preparing the estimate. All guidelines for estimation provided by Assam Public Works Department has been followed for estimation.

## 3 Results and Discussions

### 3.1 Software

The software developed has been designed to optimize pile dimensions based on the column load and the supporting soil parameters. The optimized results can then be used to structurally design the pile foundation or use user specified data for design of the pile. All design procedure has been adopted as per guidelines provided by Indian Standards code [2–5, 19, 20]. A detailed step-by-step procedure for using this software for optimization and design of pile foundation has been given below:

1. Under the 'OPTIMIZATION' section the optimization technique is selected as 'GA' from the drop down list provided.
2. Under the 'OPTIMIZATION PARAMETERS FOR PILE' section the Minimum diameter of the pile in mm, Maximum diameter of the pile in mm, Minimum Length of the pile in m, Maximum Length of the pile in m, Biaxial Moment acting in $x$-direction and $y$-direction, respectively, in kNm and the Column Load in kN are entered.
3. Next, the number of soil layers (viz. 1, 2, 3) of stratified soil are selected in the radio buttons provided under 'SOIL PARAMETERS' section.
4. Based on the number of soil layers in stratified soil selected the software highlights only those particular layers for which soil parameter input are needed for computation. In this particular case, three numbers of soil layer have been selected. Hence the software opts for soil parameters of three soil layers.
5. Under the layer section, the soil type, i.e., Cohesive soil or Granular soil for Layers 1 through 3 is selected. Depending upon the type of soil layer selected the software highlights the soil parameters required for that particular soil layer which are required for computation. In this particular case, Cohesive soil type has been selected for Layers 1 through 3.
6. The 'Layer length' in $m$ is entered. '$y$' = effective unit weight of the soil at pile tip, in kN/m$^3$; '$Ki$' = coefficient of earth pressure applicable for the $i$th layer; 'phi' = angle of wall friction between pile and soil for the $i$th layer in degrees for granular soil layer are entered. '$c_p$' = average cohesion at pile tip, in kN/m$^2$; 'alpha' = adhesion factor for the $i$th layer depending on the consistency of soil, '$c_i$' = average cohesion for the $i$th layer, in kN/m$^2$ for cohesive soils are entered. In this problem, for layer 3, which is a cohesive soil layer, the software opts the user for $c_p$ = average cohesion at pile tip, in kN/m$^2$; alpha = adhesion factor for the 1st, 2nd and 3rd layers of the soil, respectively, $c_i$ = average cohesion for the 1st, 2nd and 3rd layers, in kN/m$^2$ for cohesive soil, respectively.
7. The software takes all the input parameters and using the GA perform a number of iterations to obtain the optimized pile dimensions within the limits. The optimized results are displayed under the 'OPTIMIZED RESULTS OF PILE' section as Length of pile in m and Diameter of pile in mm. In this problem, the optimized results are: Length of pile is 14 m and Diameter of pile is 450 mm.
8. User may opt to use the optimized pile dimension for structural design by checking the 'Use Optimized Parameters for Design' check box under 'STRUCTURAL DESIGN PARAMETERS FOR REINFORCED CONCRETE PILE' section or provide the Length and diameter of the pile under the same section. In this case, optimized parameters have been used for design.

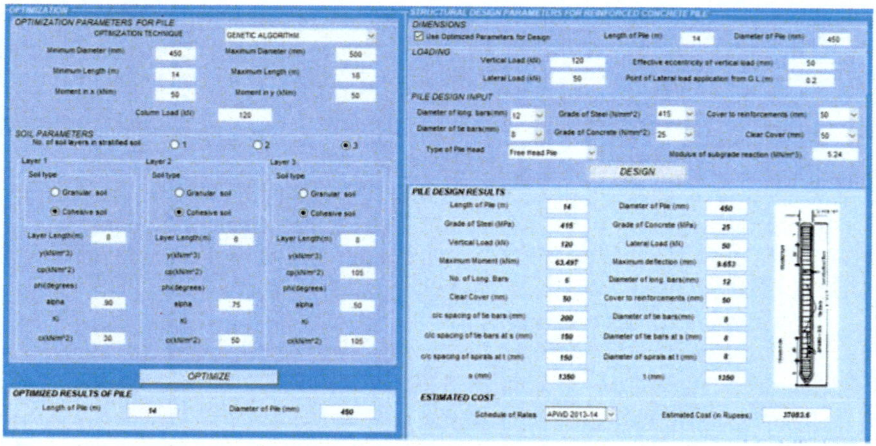


**Fig. 2** Pile foundation optimization and design software interface

9. The Vertical Load in kN (if optimized parameters not used), Effective eccentricity of vertical load in mm, Lateral Load in kN, Point of Lateral Load application from Ground Level in m and Modulus of subgrade reaction in MN/m$^3$ are entered. The Grade of Steel in N/mm$^2$ (viz. Fe 415, Fe 500), Grade of concrete in N/mm$^2$ (viz. M25, M30), Clear Cover in mm (viz. 50, 75 mm), Cover to reinforcement in mm (viz. 50, 75 mm), Diameter of Longitudinal bars in mm (viz. 12, 16, 20 mm) and Diameter of tie bars in mm (viz. 8, 10 mm) from the respective drop down lists provided under ' STRUCTURAL DESIGN PARAMETERS FOR REINFORCED CONCRETE PILE' section are selected. The structural parameters used in this problem are represented in Fig. 2.
10. The software generates the Design results under 'PILE DESIGN RESULT' section with a labeled Diagram of the pile with reinforcement detailing.
11. Under the 'ESTIMATED COST' section, the Schedule of Rates from the drop down list provided (Here, APWD 2013–14 [21]) is selected. The software generates an estimated cost of the pile in Rupees.

A stand-alone executable file is created with the use of Matlab Compiler in order to allow for users without a Matlab license to install and run the software on different operating systems.

The entire process can be summarized in the flow diagram (see Fig. 3).

## 3.2 *Calibration of Software*

A few numerical examples' data has been adopted from a few renowned books [23–25] in the field of geotechnical engineering for calibrating the developed software.

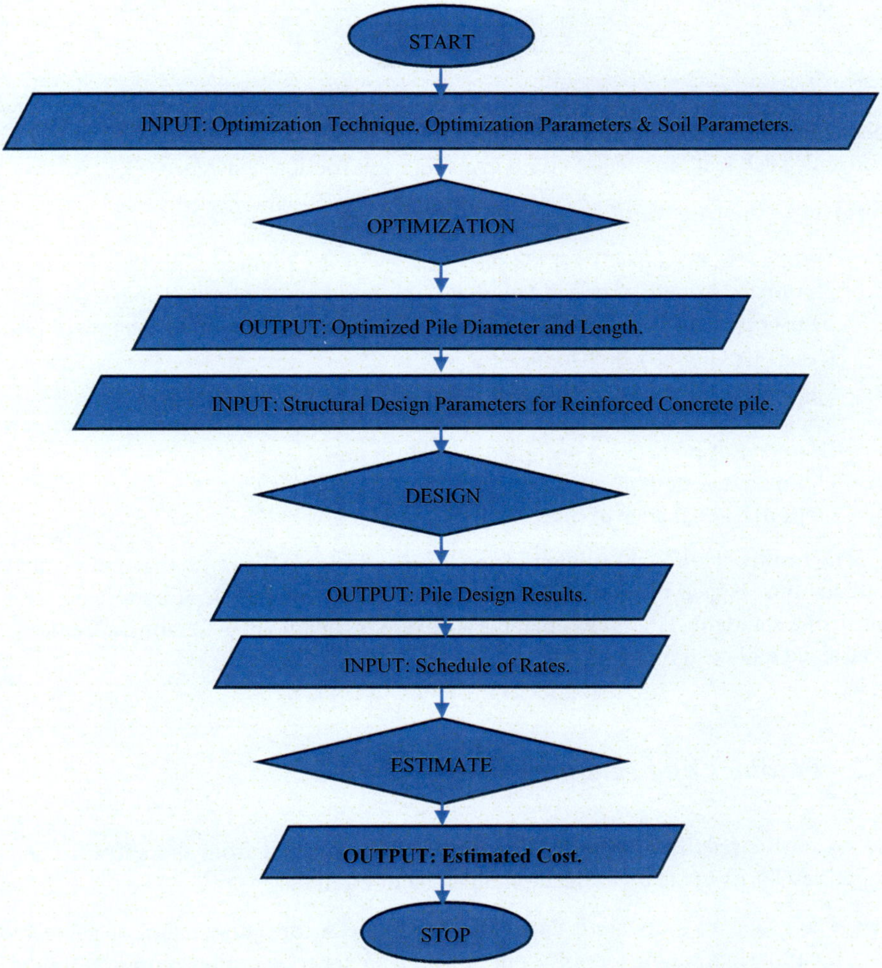


**Fig. 3** Flow diagram of pile foundation optimization and design

Methodology discussed in the previous section has been adopted for optimizing and designing the pile. This also serves as a manual verification of the results generated by the software.

The basic form of the GA used is as follows:

**Optimize**:

$$Q_u = A_p \left( \frac{1}{2} D_\gamma N_\gamma + P_D N_q \right) + \sum_{i=0}^{n} K_i P_{Di} \tan \delta_i A_{si} \quad (1)$$

Or

$$Q_u = A_p N_C c_p + \sum_{i=0}^{n} \alpha_i c_i A_{si} \tag{2}$$

depending upon the nature of the soil layers. The first term gives the end-bearing resistance ($Q_p$) and the second term gives the skin friction resistance ($Q_s$).

**Subject to the constraints**:

1. Ultimate Load Capacity of the pile ($Q_u$) ≥ 2.5 × Column Load ($Q$).
2. Optimized Length of pile ($L$) in m ≥ Minimum Length of pile (Lmin) specified by user in m and Optimized Length of pile ($L$) in m ≤ Maximum Length of pile ($L_{max}$) specified by user in m.
3. Optimized Diameter of pile ($D$) in mm ≥ Minimum Diameter of pile ($D_{min}$) specified by user in mm and Optimized Diameter of pile ($D$) in mm ≤ Maximum Diameter of pile ($D_{max}$) specified by user in mm.
4. Optimized Length of pile ($L$) in m > 0.
5. Optimized Diameter of pile ($D$) in mm > 0.

The results of software generated and manual design tallies with those of manual calculations. Hence the working process of the software has been correlated with those of a standard. This shows that the software is capable of generating acceptable optimized and design results

### 3.3 *Findings and Interpretations*

Based on the results of optimization and design obtained from the software and supported by its manual verification, it can be inferred that:

1. The proposed software is capable of using a GA for optimization of geotechnical design parameters of a pile for a given set of variables representing the supporting soil parameters.
2. The proposed software is capable of automating Optimization process by incorporating code based and user defined input parameters and limits in Geotechnical Design.
3. Acceptable structural design is achieved by incorporating optimized parameters and/or user defined parameters for design satisfying all Indian Standards [3, 20] codal provisions.
4. The software is capable of generating Structural Design based on the optimized results and/or user defined parameters following all Indian Standards [3, 19, 20, 26, 27] specifications.
5. The software is capable of providing a user friendly interface for providing a real time optimization and design environment for Engineers, providing users partial or full control over the entire Optimization and Design process.

6. The software is also capable of generating preliminary estimates of conceptual design, based on Schedule of Rates [21] and recommended guidelines, for better selection of a particular design.

### *3.4 Implications*

The software developed through this study has adopted the design methodology and constraints considering all the variables involved in pile foundation design and construction recommended by Indian Standards code [2–5, 19], which is widely used in practice in India. The software also provides provision for a real time optimization and design environment along with provisions for user-defined practical considerations or limits in design which may provide ease in practical implementations of the conceptual design. This may cater to shorten the gap between conceptual optimization and design outputs and the actual field applicability. The software also provides an estimate of the design produced so as to allow the designer to better judge the applicability of a particular design.

## 4 Conclusion and Further Scope

The software codes developed has its own limitations. However, based on the assumptions in this study, the code developed using a GA has suitably demonstrated its usefulness in the optimization of the pile design process. This code may later on be developed for being applied in the optimization of piles in group action. This will required further attention to the application of GA in Geotechnical Engineering.

## References

1. Pérez, L., Eduardo, J.: Structural Design Optimisation of Single Piles. http://hdl.handle.net/10589/138889. Last accessed: 25 July 2020 (2017)
2. IS 2911 (Part 1/Sec 1): Design and Construction of Pile Foundations—Code of Practice Part 1 Concrete Piles Section 1 Driven Cast In-situ Concrete Piles. Bureau of Indian Standards, New Delhi (2010)
3. IS 2911 (Part 1/Sec 2): Design and Construction of Pile Foundations—Code of Practice Part 1 Concrete Piles Section 2 Bored Cast In-situ Concrete Piles. Bureau of Indian Standards, New Delhi (2010)
4. IS 2911 (Part 1/Sec 3): Design and Construction of Pile Foundations—Code of Practice Part 1 Concrete Piles Section 3 Driven Precast Concrete Piles. Bureau of Indian Standards, New Delhi (2010)
5. IS 2911 (Part 1/Sec 4): Design and Construction of Pile Foundations—Code of Practice Part 1 Concrete Piles Section 4 Precast Concrete Piles in Pre-bored Holes. Bureau of Indian Standards, New Delhi (2010)

6. Abdurrahman, S.: Mathematical models and solution algorithms for computational design of RC piles under structural effects. Appl. Math. Model. **35**(1), 3611–3638 (2011)
7. Patel, V.: Software for design of shallow foundation using Matlab. Int. J. Eng. Sci. Res. Technol. **3**(4), 1912–1919 (2014)
8. Islam, S.M., Rokonuzzaman, M.: Optimized design of foundations: An application of genetic algorithms. Aust. J. Civ. Eng. **16**(1), 46–52 (2018)
9. Chow, Y.K., Thevendran, V.: Optimization of pile groups. Comput. Geotech. **4**(1), 43–58 (1987)
10. Hoback, A.S., Truman, K.Z.: Least weight design of steel pile foundations. Eng. Struct. **15**(5), 379–385 (1993)
11. Huang, Z., Hinduja, S.: Shape optimization of a foundation for a large machine tool. Int. J. Mach. Tool Des. Res. **26**(2), 85–97 (1986)
12. Valliappan, S., Tandjiria, V., Khalili, N.: Design of raft-pile foundation using combined optimization and finite element approach. Int. J. Numer. Anal. Meth. Geomech. **23**(10), 1043–1065 (1999)
13. Kim, H.T., Koo, H.K., Kang, I.K.: GA-based optimum design of piled raft foundations with model tests. J. SE Asian Geotech. Soc. **33**(1), 1–11 (2002)
14. Kim, K.N., Lee, S.H., Kim, K.S., Chung, C.K., Kim, M.M., Lee, H.S.: Optimal pile arrangement for minimizing differential settlements in piled raft foundations. Comput. Geotech. **28**(4), 235–253 (2001)
15. Chan, C.M., Zhang, L.M., Ng, J.T.M.: Optimization of pile groups using hybrid genetic algorithms. J. Geotech. Geo-environ. Eng. **135**(4), 497–505 (2009)
16. Letsios, C., Lagaros, N.D., Papadrakakis, M.: Optimum Design of Pile Foundations. In: Papadrakakis, M., Topping, B.H.V. (eds.) Sixth International Conference on Engineering Computational Technology 2008, Paper 180, pp. 35–40. Civil-Comp Press, Stirlingshire, Scotland (2008)
17. Darius, M., Juozas, K., Dmitrij, Š.: Optimization of grillages using GAs for integrating MATLAB and FORTRAN environments. Mokslas—Liet. Ateitis Sci—Future Lith. **4** (6), 564–568 (2012)
18. Nikolaou, K., Pitilakis, D.: SoFA: A matlab-based educational software for the shallow foundation analysis and design. Comput. Appl. Eng. Educ. **25**(2), 214–221 (2017)
19. IS 6403: Indian Standard Code of Practice for Determination of Breaking Capacity of Shallow Foundations. Bureau of Indian Standards, New Delhi (1981)
20. IS 456: Plain and Reinforced Concrete Code of Practice. Bureau of Indian Standards, New Delhi (2000)
21. Assam Public Works Department Schedule of Rates 2013–14 (Building), Commissioner & Special Secretary. Public Works Department, Assam (2013)
22. Abedin, A., Ligai, W.: Automating and optimizing pile group design using a GA, https://www.semanticscholar.org/paper/Automating-and-optimizing-pile-group-design-using-a-Abedin-Ligai/e28a0b906783c0b4f75019a4fb208931efe5b366#references. Last accessed: 25 July 2020 (2018)
23. Murthy, V.N.S.: Textbook of Soil Mechanics and Foundation Engineering, 2nd edn. CBS Publishers & Distributors, New Delhi (2009)
24. Punmia, B.C., Jain, A.K., Jain, A. K.: Soil Mechanics and Foundations, 16th edn. Laxmi Publications (P) Ltd., Guwahati (2014)
25. Ranjan, G., Rao, A.S.R.: Basic and Applied Soil Mechanics, 2nd edn. New Age International Publishers (P) Ltd., Guwahati (2014)
26. SP:34(S&T): Handbook on Concrete Reinforcement and Detailing. Bureau of Indian Standards, New Delhi (1987)
27. SP:16: Design Aids for Reinforced Concrete to IS: 456-l978. Bureau of Indian Standards, New Delhi (1987)

# A Numerical Study About the Development of Stressed Zone Around Single Pile When Moved Away from the Crest of the Slope Under Static Lateral Load

**S. V. Sivapriya and S. R. Gandhi**

## 1 Introduction

The behaviour of laterally loaded pile mainly deals with the soil-structure interaction, which is much more complex in a non-horizontal ground. The need for laterally loaded pile in a sloping ground exists in the area of a man-made embankment of railway track, hoarding board or off-shore structures. This slope can be stable or unstable; this study mainly focuses on a stable slope.

The soil-structure interaction of laterally loaded pile in horizontally was initially studied in detail by Reese and Matlock [1] extensively and they proposed equations for deflection, slope, load and moment. Their analytical study remains the basic along with Brom's [2] study. When a pile shows the behaviour of 'flexible' it undergoes large deflection before the soil yields, on corollary for 'rigid' piles less importance is given for soil yielding [3]. Due to the development, the structures are constructed in the sloping ground; where the foundation has to be taken to a deeper depth to transfer the load to the soil. The foundation of that particular structures suffers lateral load from waves, wind and also from the soil movement especially when they are located in an off-shore [4]. The governing factors of the laterally loaded pile in the sloping ground are its consistency, shear strength, slope, location of pile etc. With an increase in relative density from 30 to 35% in a slope of 1V:2H, the increase in lateral capacity is 17% and for 1 V:1.5H slope it increases by 15% [5].

The effect of slope, when pile moved away from the slope crest is widely studied and inferred. According to Jiang et al. [6], when a pile is placed beyond 7D (D-diameter of the pile) from the slope crest the effect of the slope is neglected. Similar

S. V. Sivapriya (✉)
Department of Civil Engineering, SSN College of Engineering, Chennai 603110, India

S. R. Gandhi
NIT, Surat, Gujarat 395007, India



C. N. V. Satyanarayana Reddy et al. (eds.), *Ground Characterization and Foundations*, Lecture Notes in Civil Engineering 167,
https://doi.org/10.1007/978-981-16-3383-6_59

to Muthukummaran's [7] study indicates the effect of slope becomes void when it is moved beyond 12 D from slope crest.

This study mainly deals with the formation of the passive wedge in front of the pile and the slope effect. The study carried out by the authors indicate beyond 2.5R [8], the effect of slope is not significant; from their study, the formation of passive wedge is inferred.

## 2 Analysis

A finite element tool was used in doing the analysis, and Table 1 shows the input parameters for the tool. A liner-elastic perfectly plastic soil model proposed by Mohr–Coulomb was used, with 10 node tetrahedral. The embedded pile *in-built* model with outer diameter 16 mm, an inner diameter of 14 mm and length of 450 mm was used and the soil-structure interaction is studied by introducing an interface element such that it did not affect the stiffness properties. The generated model is validated with the experimental work by placing the pile in horizontal ground condition (Fig. 1). The pile was placed in terms of 'R'—stiffness factor from the crest of the slope

**Table 1** Input parameters

| Element | Parameter | Unit | Variable |
|---|---|---|---|
| Soil | Unit weight | $kN/m^3$ | 17.842 |
| | Young's modulus | kPa | 8025 |
| | Poison's ratio | – | 0.495 |
| | Cohesive strength | kPa | 30 |
| Pile | Unit weight | $kN/m^3$ | 27 |
| | Young's modulus | kPa | 70,000 |
| | Poison's ratio | – | 0.2 |

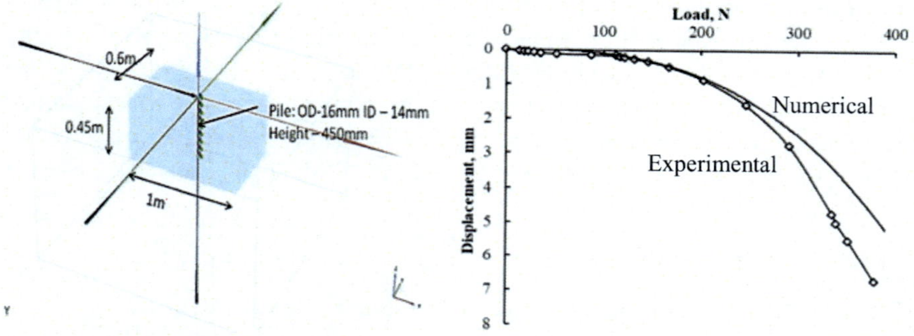


a. Model Generated

b. Validated with experiment

**Fig. 1** Generated model with validation

towards embankment; the value of 'R' was found using the data from the laboratory experiment by having horizontal ground as 94 mm. There are two ways to find the lateral load-bearing capacity of the pile: (i) giving the lateral load and finding the displacement or (ii) giving prescribed displacement and finding the load. In the current analysis, prescribed displacement of 5 mm (as per IS 2911 part IV, [9]) was given, and the stress zone formed around the pile was studied [8, 10].

## 3 Results and Discussion

The pile located in a stable slope to transfer the lateral load from the structure to the soil—called as an active pile, which is considered in the current study. The lateral capacity of pile increases with an increase in passive resistance offered by the soil to the pile when loaded laterally. This study mainly focuses on a passive wedge—denoted as stress zone formed in front of the pile (Fig. 2). A slope of 1 V:3H is formed which starts from the centre of the tank (0.5 m): initially, the pile is kept at the crest of the slope.

At a point, the length of the passive wedge in lateral direction will not fall in the slope area, the entire formation will happen in the embankment: until that point, the influence of slope exists, beyond which it does not have any significance. The angle of a passive wedge formation and depth at which the deflection of the pile vanishes termed as 'critical depth' is measured (Table 2). The pile was moved away from the crest of the slope at the rate of 1, 2, 2.5, 3, 4 and 5 times 'R'. Beyond 2.5 times the stiffness of the soil (2.5R), the effect of slope becomes annulled when it moved away from the slope [1], as the wedge formation started falling in the embankment. It is observed by noticing the compressed zone which is the stress zone, formed around

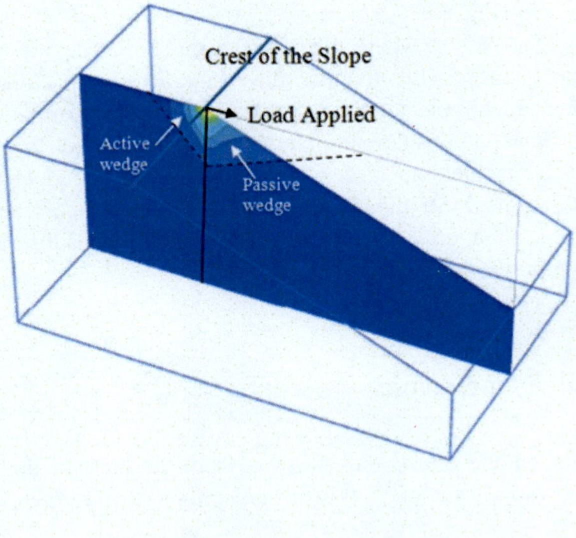


**Fig. 2** Formation of wedges in front of the piles

**Table 2** Measured values

| Location of the pile | Angle, deg | Critical depth in terms of R |
|---|---|---|
| Horizontal ground | 28 | 1.22 |
| Crest of the slope | 26 | 1.31 |
| 1R | 25 | 1.28 |
| 2R | 16 | 1.22 |
| 2.5R | 14 | 1.22 |
| 3R | 19 | 1.22 |
| 4R | 27 | 1.22 |
| 5R | 28 | 1.22 |

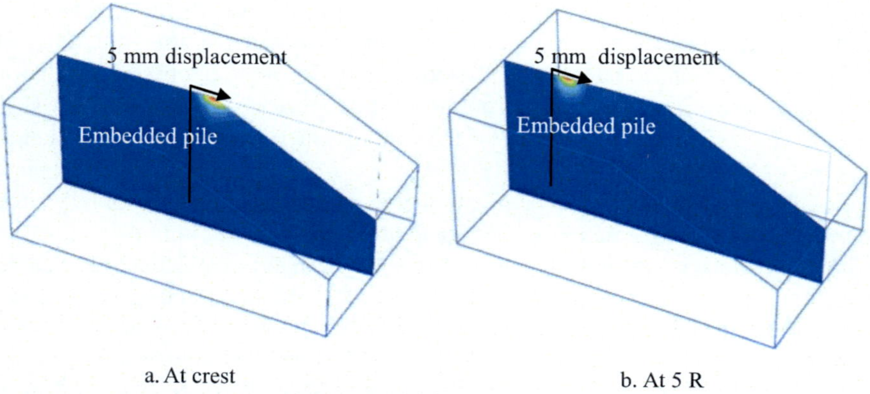


**Fig. 3** Stress zones formed around the single pile

the pile. To understand the stress, the soil is bisected at the centre of the pile in the $x$-axis (Fig. 3).

The passive wedge formed at an angle of 28 degrees for a horizontal ground which got reduced when the pile moved away from the slope till 2.5R and then increases and reaches the horizontal value passive wedge formation almost at 5R. The depth of critical depth is termed in terms of stiffness factor 'R'. For a horizontal ground, it is 1.22R, which increases when the pile is placed in a crest of the slope to 1.31R. This is mainly due to the reduction in passive resistance, which reduces the capacity of the pile. Then, after it starts decreasing and reaches the value similar to the horizontal ground at 2.5R pile position.

## 4 Conclusions

The single pile is moved away from the crest of the slope and placed at various position in the embankment; prescribed displacement of 5 mm was applied at pile head towards the slope through finite element tool. The following observations are made,

1. The maximum steeper angle of stress zone formed around the pile is observed when the ground condition is horizontal and it became flattered when moved away from the crest of the slope. When the pile is placed at 2.5R location, the stress zone starts to decrease and its length got reduced due to the increase in passive resistance.
2. The critical depth is maximum for pile at the crest and the depth of insertion reduces in that location. The critical depth decreases beyond "2R" location and remains constant.

## References


1. Meyer, B.J., Reese, L.C.: Analysis of Single Piles Under Lateral Loading. State Department of Highways and Public Transportation, Texas (1979)
2. Broms, B.B.: Lateral resistance of piles in cohesionless soils. J. Soil Mech. Found. **90**(3), 123–156 (1964)
3. Poulos, H.G., Aust, M.I.E.: Load-deflection prediction for laterally loaded piles. Aust. Geomech. J. **1**(1), 1–8 (1973)
4. Sivapriya, S.V., Balamurukan, R., Jai Vigneshwar, A., Prathibha Devi, N., Shrinidhi, A.: Eccentricity effect in sandy slope of laterally loaded single pile. Civ. Environ. Eng. **15**(2), 92–100 (2019)
5. Almas, B.N., Muthukkumaran, K.: Experimental investigation on single model pile in sloping ground under lateral load. Int. J. Geotech. Eng. **3**(1), 133–146 (2009)
6. Jiang, C., He, J.L., Lin, L., Sun, B.-W.: Effect of slope and loading direction on laterally loaded piles in cohesionless soil. Adv. Civ. Eng. **14** (1), 1–7 (2014)
7. Muthukkumaran, K.: Effect of slope and loading direction on laterally loaded piles in cohesionless soil. Int. J. Geomech. **14**(1), 1–7 (2014)
8. Sivapriya, S.V., Gandhi, S.R.: Soil-structure interaction of pile in a sloping ground under different loading conditions. Geotech. Geol. Eng. **37**(6), 1–10 (2019)
9. Bureau of Indian Standards Code.: Indian standards : 2911 (Part 4)—Lateral Load Pile Test, pp. 1–21 (1985)
10. Sivapriya, S.V., Gandhi, S.R.: Experimental and numerical study on pile behaviour under lateral load in clayey slope. Indian Geotech. J. **43**(1), 105–114 (2013)

# A Critical Review of Some Important Aspects of the Indian Practice of Geotechnical Design of Bored Piles

**Jimmy Thomas, Gitty Rose Eugine, and Joyis Thomas**

## 1 Introduction

The most commonly used type of deep foundation in India is bored and cast in situ concrete piles. The geotechnical design of bored piles in India is carried out in accordance with various codes like IS 2911 (Part 1/sec2) [1], IS 14593 [2], IRC 78 [3] and Indian Railways Bridge Substructure and Foundation Code [4]. The objective of this paper is to critically review some important aspects of the geotechnical design of bored piles as stipulated by the Indian codes and to compare with international practices. The following aspects are examined in this paper:

- Critical depth beyond which unit skin friction and unit end bearing resistance remains more or less constant with depth.
- The coefficient of earth pressure for calculation of skin friction resistance.
- Minimum depth of embedment for piles passing through a weak stratum and terminated in a competent bearing stratum.
- Effect of weak stratum below pile tip on end bearing resistance.

J. Thomas (✉)
Kochi 682301, India

G. R. Eugine
College of Engineering Kidangur, Kottayam 686583, India

J. Thomas
Advisian, Perth, WA 6000, Australia



C. N. V. Satyanarayana Reddy et al. (eds.), *Ground Characterization and Foundations*, Lecture Notes in Civil Engineering 167,
https://doi.org/10.1007/978-981-16-3383-6_60

# 2 Critical Depth

## 2.1 Background

The unit skin friction resistance and the unit end bearing resistance at any depth depend on the vertical effective stress. Since vertical effective stress increases linearly with depth, one would expect that unit skin friction and unit end bearing resistance should also increase linearly with depth. However, experimental research by Kerisel [5] and Vesic [6, 7] suggested that the unit skin friction and unit end bearing resistance increased linearly up to a critical depth, and beyond this, critical depth skin friction and end bearing resistance remained more or less constant. This idea was accepted by most engineers, and several leading textbooks and codes and design guidelines endorsed the concept of critical depth. Several researchers like Kulhawy [8], Randolph [9] and Fellenius [10] have challenged the concept of critical depth, saying this is a myth or fallacy arising from problems in interpretation of experimental data. A review of the relevant literature suggests that design practice is moving away from the concept of critical depth and limiting skin and end bearing resistance.

## 2.2 Indian Codes of Practice and Critical Depth

IS 2911 (Part 1/sec 2) Annex B clause B-1 for piles in granular soils [1] stipulates that *in working out pile capacity by static formula, the maximum effective overburden at the pile tip should correspond to the critical depth, which may be taken as 15 times the diameter of the pile shaft for* $\phi \leq 30°$ *and increasing to 20 times for* $\phi \geq 40°$. The IRS bridge substructure and foundations code [4] recommend that design is to be carried out in accordance with IS 2911, and hence, the above recommendations will be applicable for pile foundations for railway bridges.

The design of foundations for road bridges is carried out in accordance with IRC 78 [3]. In clause 1.1 of Annex 5 of IRC:78, which gives the procedure for calculation of end bearing resistance, it is stated that *effective overburden pressure at pile tip limited to 20 times diameter of pile for piles having length equal to more than 20 times diameter*. However, in clause 2 of Annex 5, which describes the method for calculating skin friction, no mention is made about critical depth.

## 2.3 Arguments for Critical Depth

The concept of a limiting value of unit end bearing resistance and unit skin friction for piles in sand was first proposed by Kerisel [5]. Experimental studies carried out by Vesic [6, 7] played a major role in getting wide acceptance for the concept of critical depth and limiting values of end bearing and skin friction. Vesic [6, 7] carried out a series of laboratory and field tests on instrumented piles with diameter ranging from 50 to 450 mm and with a range of length–diameter ratios installed in dry, moist and saturated using different installation techniques—buried, jacked-in and driven. End bearing resistance and skin friction were measured separately. He found that for shallow depth of penetration, there was a linear increase in both unit skin friction and unit end bearing resistance. However, at greater depths, both unit skin friction and unit end bearing resistance became more or less constant with depth. The maximum value of resistance appeared to depend on relative density of sand and the method of installation of piles. The depth up to which a linear increase observed in unit resistance was in the range of 10–20 times the pile diameter and appeared to depend on the relative density of sand and the method of installation of pile.

Vesic [6, 7] suggested that the explanation for the observed phenomena is that the vertical stress in the vicinity of the pile is different from the initial overburden pressure. He postulated that both unit skin friction and unit end bearing resistance are linear functions of the vertical stress in the vicinity of the pile at failure, which is not necessarily equal to nor proportional to the overburden pressure. Vesic [6, 7] puts forward the notion that if the unit skin friction and unit end bearing resistance attained a constant value at greater depth, it is because the vertical stress at failure which also becomes constant at greater depth. The departure of the vertical stress around and below the pile from the initial overburden stress was attributed by Vesic [6, 7] to arching. When the pile is loaded, the sand below the tip and around the pile is compressed downward. If the depth of embedment of the pile is sufficiently large, and if the base displacement is sufficiently small, there is a distance above the pile tip above which the effect of the downward movement of the soil is not significant. Above this level, the vertical stress at failure in the vicinity of the pile will be equal to the initial overburden stress, and the unit skin friction and unit end bearing resistance will increase linearly with depth up to this level.

Tavenas [11] conducted load tests on instrumented steel H-piles and precast concrete piles driven into a thick homogenous layer of medium dense sand. The piles were driven and tested incrementally to depths of 6, 9, 12, 15, 18 and 21 m. The results showed that for both H-piles and the precast concrete piles, the capacity was not a linear function of the embedded depth in sand. In particular, for the precast concrete pile, a critical depth was observed to be 23 times the width of the pile below which both unit skin friction and end bearing resistance remained perfectly constant.

Hanna and Tan [12] presented results of tests on small-scale model piles which showed that the unit skin friction and end bearing resistance increased linearly with depth up to length–diameter ratios of 30–40, and thereafter, the resistance remained virtually constant until length–diameter ratio of 112 corresponding to the maximum length of pile tested. Meyerhof [13] in his Terzaghi lecture stated that large-scale experiments and field observations showed that point resistance and average skin friction of a pile in homogenous sand will increase with depth up to a certain critical depth only, and below this critical depth, both end bearing resistance and average skin friction remain practically constant owing to the effects of soil compressibility, arching, crushing and other factors.

Textbooks like Pile Foundation Analysis and Design by Poulos and Davis [14], Pile Foundations in Engineering Practice by Prakash and Sharma [15], Geotechnical Engineering Handbook edited by Smoltczyk [16] and design manuals by Naval Facilities Engineering Command [17] and US Corps of Engineers [18] recommended the concept of critical depth and limiting unit end bearing and skin resistance for the design of pile foundations. Canadian Foundation Engineering Manual [19] states that both unit skin friction and unit end bearing resistance may continue to increase with depth, but at a decreasing rate and recommended that for practical design purposes, it is advisable to adopt limiting values for long piles in cohesionless soils.

Vesic [6], Poulos and Davis [14], Prakash and Sharma [15] suggest that where a sand layer is overlain by a clay layer, the critical depth is to be measured from the top of the sand layer. This is not explicitly mentioned in any of the Indian codes of practice. Some coastal regions in India like Kochi have thick deposits of marine clays underlain by sands, and it is a common practice to support structures on end bearing piles embedded into the sand layer. In such cases, if the critical depth is measured from the ground surface, instead of from the top of the sand layer, the calculated end bearing and skin friction in the sand layer would be too low.

## 2.4 Arguments Against Critical Depth

Kulhawy [8] argued that the concept of critical depth below which the end bearing and skin resistance are constant is not correct. He attributed the trend of unit skin friction increasing linearly up to a certain depth and then remaining more or less constant with depth observed in studies to the effects of over consolidation. If the soil is normally consolidated, the unit skin friction will increase linearly with depth. However, if the sand is over-consolidated, the behavior changes appreciably. Kulhawy [8] suggested that some over consolidation is the rule, rather than exception in most soil deposits. Using a hypothetical example of a sand stratum whose upper part is over-consolidated and lower part is normally consolidated, Kulhawy [8] demonstrated how the variation of the coefficient of earth pressure at rest ($K_0$) will result in unit skin friction which will remain more or less constant for certain range of depth. Based on this demonstration, Kulhawy [8] argued that the apparent limiting value of unit skin friction is only a coincidence due to the product of $\beta$ ($\beta = K \tan \delta$, where $K$ is the coefficient of earth pressure, and $\delta$ is the pile-soil interface friction angle) and $z$ (depth) nullifying the effects of each other over a limited range of depth.

Kulhawy [8] also showed based on Vesic's theory of the bearing capacity of deep foundations [20] that with increasing depth, the unit end bearing resistance increases at a decreasing rate. This was attributed primarily to the effect of decreasing rigidity with depth and partly to the reduction in the angle of shearing resistance with increasing confining stress. However, Kulhawy [8] suggested that there is no critical depth beyond which the unit end bearing resistance is constant.

Kraft [21] presented a detailed critical review of the concept of critical depth and limiting skin and end bearing resistance of piles. He commented that an explanation to dismiss the existence of limiting values as proposed by Kulhawy [8] is easy to develop, but such an argument does not eliminate the experimental evidence that gave rise to the creation of limiting values in the first place. Kraft [21] suggested that while arching can explain a trend of resistance increasing with depth at a decreasing rate, it does not necessarily follow that a limiting unit resistance is reached. He listed factors which could result in a trend of unit resistance increasing with depth at a decreasing rate—angle of shearing resistance of soil and pile-soil interface friction angle may decrease with depth due to an increase in effective overburden stress; soils may exhibit more contractive behavior with depth as a result of increasing confining stresses; because of longer duration of pile driving, there may be larger reduction in lateral stresses. Kraft [21] discussed several *artificial reasons*, which could influence the interpretation of pile load test data leading to an apparent limiting value of unit skin friction and end bearing resistance. These data interpretation problems include residual stresses, toe effects, failure criterion adopted for pile capacity, variability in soil stratigraphy and density of pile instrumentation, boundary effects in model tests and scale selected for plotting of test data. Kraft [21] concluded that no data from full-scale field load tests could be found which provided conclusive evidence for the existence limiting values. Although model tests show evidence of limiting shaft and tip resistance, it is likely that the data are being influenced by the boundary effects of the test chambers and limiting values do not exist in general. However, the rate of increase of resistance, especially end bearing resistance, decreases with an increase in overburden stress in homogeneous sands.

Randolph [9] observed that the basis for the concept of critical depth is empirical, and quantitative analytical justification for limiting values of skin friction and end bearing resistance was lacking. He suggested that two factors may contribute to end bearing resistance increasing at a decreasing rate with depth—reduction in angle of shearing resistance with increasing stress level and a reduction in rigidity index, since shear stiffness of soil increases with stress level at a slower rate than frictional strength. He also suggested that the so-called limiting value of skin friction below a certain depth, is not a true limit on skin friction, but due to the degradation of shaft friction at shallower depths as the pile was advanced.

Fellenius and Altaee [10] stated that *the critical depth is a fallacy which originates in the failure to interpret the results of full and model scale pile tests properly*. In the case of load tests on instrumented piles, the instruments would probably register only the loads applied to the pile during the test and disregard any loads present in the pile before the test. However, residual loads are induced in both driven and bored piles during and after installation through several phenomena—wave action during driving, soil quakes along the pile and reconsolidation of the soil after the installation disturbance. Residual loads are present in all piles even before measurements are taken and analysis performed, and their effect is commonly overlooked. In the interpretation of the results of load tests on full-scale instrumented piles, if the residual loads are not accounted for in the analysis, the measured load distribution may show a trend of resistance linearly increasing up to a depth and remaining constant below that depth. In the case of small-scale model tests, the residual loads may not be significant. However, because of the very low-stress levels, the sand will exhibit a dilatant response resulting in significantly higher values of lateral pressure at shallow depths, and the results may show a distribution of skin friction which first increases and then remains constant, which could be mistakenly interpreted as the existence of a critical depth.

Poulos et al. [22] expressed the view that limiting values of unit shaft friction and end bearing resistance probably do not exist although the rate of increase with depth is not linear. However, it remains convenient from a design view point into impose an upper limit for shaft and base resistance, and it is unlikely that this common practice would be discarded easily. Fleming et al. [23] stated that *modern approaches to pile design have generally moved away from limiting values of end bearing pressure, but accept that in a uniform sand deposit there will be a gradually decreasing gradient of design end bearing pressure with depth.* Tomlinson and Woodward [24] state that for long piles the assumption of a constant unit end bearing resistance below a depth of 10 to 20 diameters has been shown to be over-conservative. Many widely used methods for pile design like FHWA [25], API [26] and ICP [27] methods do not recommend use of a critical depth. However, it is accepted that unit skin friction and end bearing resistance does not necessarily increase linearly with depth and limiting values of skin and end bearing resistance may be specified.

### *2.5 Discussion*

From the review of literature, it is evident that the concept of critical depth is no longer considered to be credible and the practice of limiting the value of effective vertical overburden stress for calculation of unit skin friction and end bearing resistance to its value at a depth of 10–20 times the pile diameter is unduly conservative. However, unit skin friction and end bearing resistance may not continue to increase linearly with depth for large depths, and it is prudent to restrict the maximum design values to appropriate limits based on relevant experience.

## 3 Coefficient of Earth Pressure

### *3.1 Theoretical Background*

A widely used method for the calculation of skin friction resistance of piles in sand is using an expression of the form Eq. (1).

$$Q_{\mathrm{s}} = \sum_{i=1}^{n} K_i \sigma_{vi} \tan \delta_i \; A_{\mathrm{si}} \tag{1}$$

where $Q_s$ is the total skin friction resistance, $K_i$ is the coefficient of earth pressure for the $i$th layer, $\sigma_v$ is the vertical stress at the center of the $i$th layer, $\delta_i$ is the pile-soil interface angle for $i$th layer, and $A_{si}$ is the surface area of the pile in $i$th layer. The coefficient of earth pressure depends on number of factors including the relative density and stress history of the soil, geometrical and material characteristics of the pile and method of installation [24].

Prior to the installation of the pile, $K$ will be equal to $K_0$, the coefficient of earth pressure at rest, which is largely a function of the relative density and stress history of the soil. The original in situ state of stress will be altered during the installation of a bored pile depending on a number of factors—use of temporary casing or permanent liners, use of mineral or polymer slurries, method of concreting and quality of construction practices. When installation of bored piles is carried out using casings or liners which are advanced ahead of boring, the initial at rest earth pressure conditions may not be altered significantly. However, in case a borehole stabilized by drilling fluids, the $K$ value will decrease. But, when wet concrete is placed, because of the fluid pressure exerted by the fresh concrete, the $K$ value may become equal to or even exceed $K_0$. When the pile is loaded, shear stresses are induced in soils adjacent to the pile shaft. If relative density of the soil and the stress conditions are such that the soil tends to dilate, which is being restrained by the pile and surrounding soil mass, the lateral stresses on the pile will increase. Hence, the lateral stresses acting on the pile shaft at failure depend on a large number of factors, and hence, it is very difficult to determine $K$ accurately.

### 3.2 *K Values from Literature*

The $K$ values for bored piles recommended by various references are summarized in Table 1.

From Table 1, it may be seen that the values recommended by IS 2911 and IRC: 78 are appreciably higher than the values recommended by other references.

## 4 Minimum Embedment into Bearing Stratum and Effect of Weak Layer Below Pile Tip

### 4.1 *Theoretical Background*

When a pile passes through relatively weak strata and terminated in a competent bearing stratum, it is generally considered that some embedment into the bearing stratum is required to mobilize the full end bearing resistance which can be offered by the bearing stratum. Similarly, if there is a weaker stratum below and close to the pile toe, the end bearing resistance will be reduced. Both effects are related to the

**Table 1** Coefficient of earth pressure for bored and cast in situ concrete piles

| References | $K$ |
|---|---|
| IS 2911 (Part 1/sec 2) [1] | 1.0–1.5 |
| IRC:78 [3] | 1.5–1.8 |
| BS 8004 [28] | 0.7 |
| AASHTO [29]/FHWA [25] | $K_0$ |
| Canadian foundation engineering manual [19] | $K_0$ |
| Bangladesh national building code [30] | |
| Dry construction—good workmanship | $K_0$ |
| Slurry construction—good workmanship | $K_0$ |
| Slurry construction—poor workmanship | (2/3) $K_0$ |
| Casing under water | (5/6) $K_0$ |
| Tomlinson and Woodward [24] | 0.70 $K_0$ to $K_0$ |
| Fleming et al. [23] | 0.7 |
| Prakash and Sharma [15] | 0.5 |
| Viggiani [31] | |
| Loose sand | 0.5 |
| Dense sand | 0.4 |
| Kulhawy [8] | (2/3) $K_0$ to $K_0$ |

shape of the failure surface for the pile toe. The shape of the failure surface assumed in different theories of ultimate bearing capacity of piles is shown in Fig. 1 [31].

Here, (a) is considered by Prandtl and Caquot, (b) by De Beer and Meyerhof, (c) by Berezantsev and (d) by Bishop, Skempton and Vesic. It may be seen that for

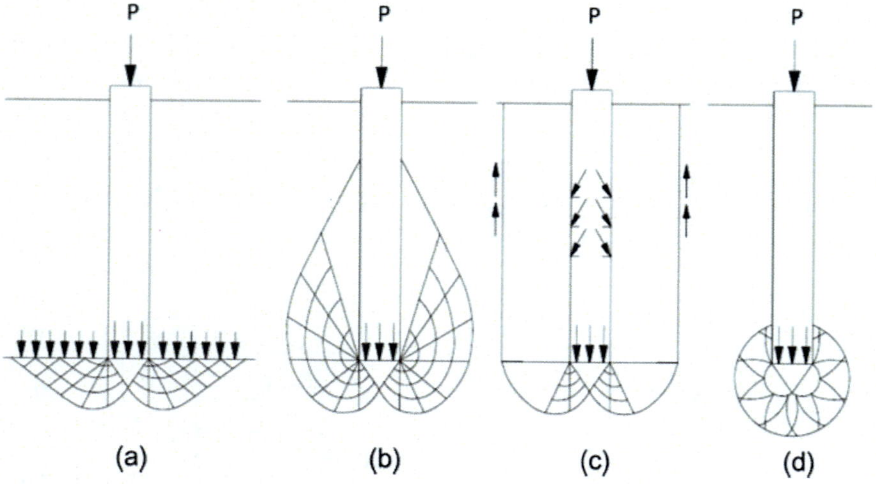


**Fig. 1** Failure surfaces under deep foundations, after [31]

most theories, the influence zone is in the vicinity of the pile toe, except in the case of theories of De Beer and Meyerhof, wherein the failure surface extends to a large distance above the pile tip. In this context, it is also interesting to note the comment by Fellenius [32] that the concept of bearing capacity does not appear to a pile toe, and instead, the load-settlement relationship is a function of the compressibility of the soil below pile toe and effective overburden stress.

### 4.2 Recommendations on Minimum Embedment into Bearing Stratum and Effect of Weak Layer Below Pile Toe

The recommendations on minimum embedment of pile into a bearing stratum and effect of weak layer below pile toe extracted from various sources are summarized in Table 2.

### 4.3 Discussion

From Table 2, it may be seen that most references suggest that the influence zone for end bearing resistance extends to a distance of one to three pile diameters above the pile toe and three to four diameters below the pile toe. An exception is the recommendation of Meyerhof [13], where the influence zone is considered to extend to a distance of ten times the diameter above and below the pile toe. This could possibly due to the shape of the failure surface considered by Meyerhof. The recommendation of IS 2911 for a minimum embedment of two times the pile diameter into a bearing stratum is generally satisfactory. However, Indian codes do not have a provision to account for possible reduction in end bearing capacity due to the presence of weak layer below pile toe. It is advisable to include a recommendation for appropriate reduction in end bearing capacity if there is a weak layer within three diameters below the pile toe.

## 5 Conclusions

A review of the literature concerning some important aspects of the design of bored piles was carried out, and the following conclusions may be drawn:

1. The concept of critical depth is no longer considered to be credible and the practice of limiting the value of effective vertical overburden for calculation of unit skin friction and end bearing resistance to its value at a depth of 10–20 times the pile diameter is unduly conservative. However, unit skin friction and

**Table 2** Recommendations on minimum embedment into a bearing stratum and effect of weak layer below pile toe on the end bearing resistance of piles

| References | Recommendation |
|---|---|
| IS 2911 (Part 1/sec 2) [1] | For piles passing through cohesive strata and terminating in a granular stratum, a penetration of at least twice the diameter of the pile shaft should be given into the granular stratum |
| BS 8004 [28] | The design of the pile foundation should consider the influence of different soil/rock layers below the pile toe |
| AASHTO [29] | When a shaft is tipped in a strong soil layer overlying a weak layer, the base resistance shall be reduced if the shaft base is within a distance of 1.5 times the shaft diameter of the top of the weak layer |
| Eurocode 7 [33] | The strength of a zone of ground above and below the pile base should be taken into account when calculating the pile base resistance. This zone may extend several diameters above and below the pile base. Any weak ground in this zone has a relatively large influence on the base resistance. Punching failure should be considered if weak ground is present at a depth of less than four times the base diameter below the base of the pile |
| AS 2159 [37] | Where a pile is funded on a stratum that overlies a softer or weaker stratum, allowance shall be made for the possible reduction of end bearing resistance due to the presence of weaker or softer stratum |
| Canadian foundation engineering manual [19] | Where a weak soil layer overlies a dense sand layer, the full toe capacity is not mobilized until the pile penetrates six diameters into the dense sand. Where a weak layer underlies a dense sand layer, toe capacity would be affected if the pile toe is less than three times the pile toe diameter above the weak layer |
| API [26] | In the case of piles embedded in a layer with adjacent weak layers, reduction in end bearing resistance need not be considered if the pile achieves penetration of two to three times into the layer, and the tip is approximately three times the diameter above the bottom of the layer to preclude punch through. Where these distances are not achieved, some modification in the end bearing resistance may be necessary |

(continued)

**Table 2** (continued)

| References | Recommendation |
|---|---|
| German geotechnical society [34] | The zone governing base resistance is from one diameter above and four diameters below the pile base for pile diameters up to 600 mm and from one diameter above and three diameters below the pile base for pile diameters greater than 600 mm |
| Salgado [35] | Pile has to penetrate a bearing stratum by at least two diameters if that layer's bearing resistance is to be fully developed |
| Coduto [36] | Soil between the depth of about one diameter above the toe and about three diameters below the toe has the most influence on the toe bearing capacity |
| Meyerhof [13] | In the case of a bearing stratum overlain by a weaker stratum, full toe resistance is mobilized if the penetration into the bearing stratum is at least ten times the pile diameter. Similarly, in the case of a bearing stratum underlain by a weak stratum, toe resistance will be reduced if the pile tip is within a distance of ten times the pile diameter from the top of the weak stratum |

end bearing resistance may not continue to increase linearly with depth for large depths, and it is prudent to restrict the maximum design values to appropriate limits based on relevant experience.

2. The coefficient of earth pressure recommended by Indian codes seems to be appreciably higher than the values used internationally. A suitable revision of the Indian codes appears to be desirable.
3. The recommendation of IS 2911 with respect to the minimum embedment of pile into a bearing stratum seems reasonable.
4. Indian codes do not have a provision to account for the influence of a weak layer below the pile toe on the end bearing resistance. It is advisable to include a recommendation to suitably reduce the end bearing resistance when there is a weak layer within a distance of three times diameter below the pile toe.

## References

1. IS 2911 (Part 1/sec 2): Design and Construction of Pile Foundations—Code of Practice, Part 1 Concrete Piles, Section 2 Bored Cast In-Situ Concrete Piless. Bureau of Indian Standards, New Delhi (2010)
2. IS 14593: Design and Construction of Bored Cast in-Situ Piles Founded on Rocks—Guidelines. Bureau of Indian Standards, New Delhi (1998)

3. IRC: 78: Standard Specifications and Code of Practice for Road Bridges, Section: VII, Foundations and substructure. Indian Roads Congress (2014)
4. Indian Railway Standard: Code of Practice for the Design of Substructures and Foundations of Bridges. Research Designs and Standards Organization, Lucknow (2013)
5. Kerisel, J.: Deep foundations in sands: variation of ultimate bearing capacity with soil density, depth, diameter and speed. In: Fifth International Conference on Soil Mechanics and Foundation Engineering, Paris, vol. 2, pp. 73–83 (1961)
6. Vesic, A.S.: Bearing capacity of deep foundations in sand. Highway Res. Rec. **39**, 112–153 (1963)
7. Vesic, A.S.: A study of bearing capacity of deep foundations. Georgia Institute of Technology, Georgia (1967)
8. Kulhawy, F.H.: Limiting tip and side resistance: Fact or fallacy? In: Proceedings of Symposium on Analysis and Design of Pile Foundations, ASCE, San Francisco (1984)
9. Randolph, M.: Pile capacity in sand—the critical depth myth. Australian Geomechanics **24**, 30–34 (1993)
10. Fellenius, B.H., Altaee, A.A.: Critical depth: how it came to being and why it does not exist. Proc. Instn Civ. Engrs. Geotech. Engng. **113**, 107–111 (1995)
11. Tavenas, F.A.: Load tests results on friction piles in sands. Can. Geotech. J. **8**(7), 7–22 (1971)
12. Hanna, T.H., Tan, R.H.S.: The behavior of long piles under compressive loads in sand. Can. Geotech. J. **10**(3), 311–340 (1973)
13. Meyerhof, G.G.: Bearing capacity and settlement of pile foundations. J. Geotech. Eng. Div. ASCE **102**(GT3), 197–228 (1976)
14. Poulos, H.G., Davis, E.H.: Pile Foundation Analysis and Design. Wiley (1980)
15. Prakash, S., Sharma, H.D.: Pile Foundations in Engineering Practice. Wiley (1990)
16. Kempfert, H.G., Eigenbrod, K.D., Smoltczyk, U.: Pile foundations. In: Smoltczyk, U. (ed.) Geotechnical Engineering Handbook, Ernst & Sohn, Berlin (2002)
17. DM-7.02: Foundations and Earth Structures. Naval Facilities Engineering Command, Washington, DC (1986)
18. EM 1110-2-2906: Design of Pile Foundations. U.S. Army Corps of Engineers, Washington, DC (1991)
19. Canadian foundation engineering manual, 4th edn, Canadian Geotechnical Society (2006)
20. Vesic, A.S.: Design of Pile Foundations. Synthesis of Highway Practice No. 42, Transportation Research Board, Washington, DC (1977)
21. Kraft, L.M.: Performance of axially loaded pipe piles in sand. J. Geotech. Eng. Div. ASCE **117**(2), 272–296 (1991)
22. Poulos, H.G., Carter, J.P., Small, J.C.: Foundations and retaining structures—research and practice. In: XV International Conference Soil Mechanics and Foundation Engineering, vol. 4, pp. 2527–2606, Istanbul (2001)
23. Fleming, K., Weltman, A., Randolph, M., Elson, K.: Piling engineering, 3rd edn. Taylor & Francis, London & New York (2009)
24. Tomlinson, M., Woodward, J.: Pile design and construction practice, 6th edn. CRC Press, Florida (2015)
25. Brown, D.A., Turner, J.P., Castelli, R.J., Loehr, E.J.: Drilled shafts: Construction procedures and design methods. Federal Highway Administration, Washington, DC (2018)
26. API Recommended Practice 2GEO: Geotechnical and Foundation Design Considerations. American Petroleum Institute, Washington, DC (2011)
27. Jardine, R., Chow, F., Overy, R., Standing, J.: ICP Design Methods for Driven Piles in Sands and Clays. Thomas Telford, London (2005)
28. BS 8004: Code of Practice for Foundations. British Standards Institution (2015)
29. AASHTO LRFD Bridge Design Specifications. American Association of State Highway and Transportation Officials (2014)
30. Bangladesh National Building Code, Housing and Building Research Institute (2015)
31. Viggiani, C., Mandolini, A., Russo, G.: Piles and Pile Foundations. Spon Press, London and New York (2012)

32. Fellenius, B.H.: Basics of Foundation Design, Electronic edition, www.fellenius.net (2009)
33. BS EN 1997-1 Eurocode 7: Geotechnical Design—Part 1: General Rules. British Standards Institution (2004)
34. German Geotechnical Society: Recommendations on Piling. Ernst & Sohn, Berlin (2014)
35. Salgado, R.: The Engineering of Foundations. McGraw-Hill Education (2006)
36. Coduto, D.P., Kitch, W.A., Yeung, M.R.: Foundation Design Principles and Practices, 3rd edn. Person (2016)
37. AS 2159: Piling—Design and Installation. Standards Australia, Sydney (2009)

# Appraisal of Innovative Finned-Pile Foundations to Resist Lateral Loads

**Pankaj Bariker and Sreevalsa Kolathayar**

## 1 Introduction

In modern developing countries like India, more preference is given to infrastructure development to enhance the requirement for the corresponding needs, which leads to its contribution toward schemes like Make in India, Digital India Program, Urban Infrastructure Development, Samudra Setu, etc. These schemes utilize the large capital to achieve the desired goal, which indirectly uses high raised buildings in onshore construction and offshore constructions for utilizing renewable resources like wind turbines, jetties, etc., for its economy.

As these platforms transmit larger loads to the subsoil, pile foundations are generally adopted. These structures have to resist larger lateral loads throughout their lifetime. Though pile foundations resist lateral loads up to an extent, they cannot carry larger lateral loads effectively with smaller length and diameter. Hence, primary preference must be given for applying an innovative pile foundation to resist these lateral loads economically with a considerably smaller diameter and length piles.

This paper gives an overview of the previous literature, its outcomes, and recommendations for adopting one such innovative pile foundations, i.e., finned-pile foundations.

Regular pile converts to the finned-piled foundation when they are stiffened at the pile head employing plates, as shown in Fig. 1, which increases the foundation system's lateral load-carrying capacity.

P. Bariker (✉) · S. Kolathayar
National Institute of Technology Karnataka, Surathkal, Mangalore, India



C. N. V. Satyanarayana Reddy et al. (eds.), *Ground Characterization and Foundations*, Lecture Notes in Civil Engineering 167,
https://doi.org/10.1007/978-981-16-3383-6_61

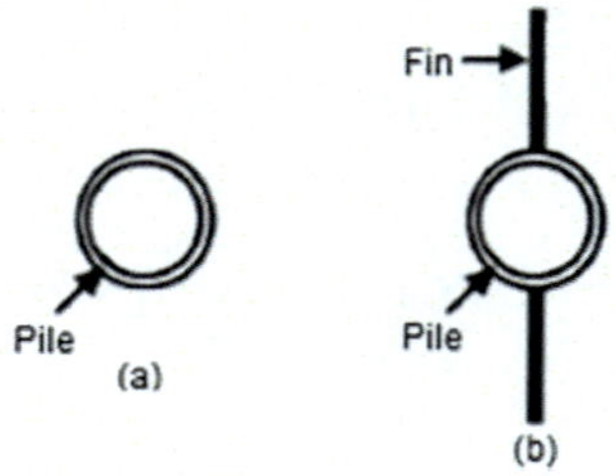


**Fig. 1** Plan view of **a** Regular pile and **b** Finned pile (Redrawn by authors after Ahmed [1])

## 2 Experimental Studies on Finned-Pile Foundations

Reese et al. [2] attempted to analyze the laterally loaded piles in the sand at the field with two piles of 601 mm dia piles for both static and cyclic loadings, and the results were presented in terms of *p–y* curves. The lateral load is applied to the full-sized piles that are open driven, and suitable instrumentation was done to record the pile head deflection and pile heal rotation along with BM, along the length of the pile. The lateral soil resistance was developed, as shown in Fig. 2. A computer program was developed to compare results with measured field values. Finally, it was concluded that a close agreement was found between the results obtained at the program's site developed.

Christos and Michael [3] made an experimental investigation over the possible effects of lateral loading over axial pile displacements, stresses, and influence of axial loads on lateral pile responses. The model tests were performed over the closed-ended aluminum piles of outer dia 19 and 1.5 mm thick. They were pushed into the soft clay bed up to 500 mm. Both axial and lateral loads were applied at increments, and responses were noted by strain gauges attached along the pile's length. The results were compared with the same set of nonlinear finite-element analysis. Finally, it is concluded that the lateral load increases the axial deformation and is dependent on the magnitude and location of the lateral load on the pile. Also, the applied lateral load reduces the axial pile stress to a smaller extent. The nonlinear finite-element analysis successfully studies the interaction between axial and lateral pile response. Still, this interaction cannot be studied by conventional elastic half-space and subgrade reaction methods.

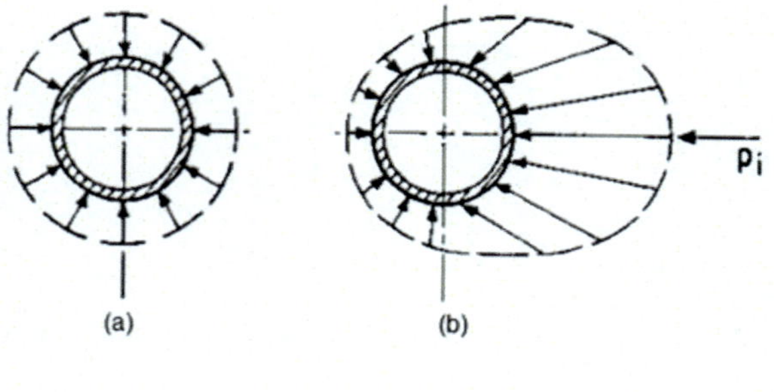


**Fig. 2** Lateral soil resistance developed **a** Before the application of lateral load **b** After applying lateral load (Redrawn by authors after Reese et al. [2])

Sastry and Meyerhof [4] studied the behavior and effect of effective embedment depth of flexible piles in layered soils for lateral and inclined loads by both model tests and field studies. Layered soils consisting of loose soil overlain by compacted sands, a PVC pile of 1.25 m length, 73 mm diameter, and 7.6 mm thick are pushed into it. They are graduated with minor equipment to measure the bending moment, lateral deflection, and lateral soil resistance at the suitable points all along the length of the pile with the spacing of 145 mm. These tests were carried out for eccentricity to embedment depth ratios (e/D) of 0.13, 0.33, and ∞ load inclinations of 0°, 15°, and 30° w.r.t vertical. The load is applied to failure in about 10–12 increments, and these values are used to validated modal tests. It is seen that the embedment depth helps in finding out the BM, lateral deflections, and lateral resistance; pile flexibility will not have its influence over axial capacity. Finally, it is concluded that the soil's lateral resistance is dependent on the stiffness of the pile and the variation of horizontal soil modulus along the pile length.

Prasad and Chari [5] attempted to verify the applicability of the previously available works of literature [3, 6, 7, 8] defining the ultimate lateral capacity of the pile which is based on the soil pressure distribution along pile length as shown in Fig. 3. Their study attempted to study this for rigid steel model piles of outer diameter, length, and thickness of 102 mm, 1135 mm, and 5.6 mm, respectively. That is driven into well-graded dry sand along with suitable arrangements made to record lateral soil resistance. Finally, a simplified method is proposed to predict soil pressure distribution and ultimate lateral capacity of the rigid piles, and a comparison was made almost the previously published literature.

Zhang et al. [9] attempted to improve the ultimate lateral resistance of piles in sands over the several available previous methods and its distribution are as shown in Fig. 4, which were difficult in selecting the suitable approximate method for designing laterally loaded piles since each method will yield different resistance values. By

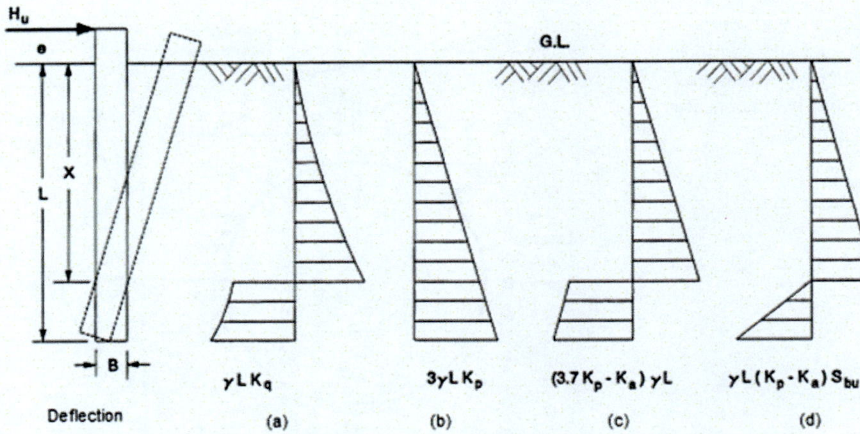


**Fig. 3** Lateral soil pressure distribution assumed by **a** Hensen, **b** Broms, **c** Petrasovits et al., and **d** Meyerhof et al., (redrawn by authors after Prasad et al. [5])

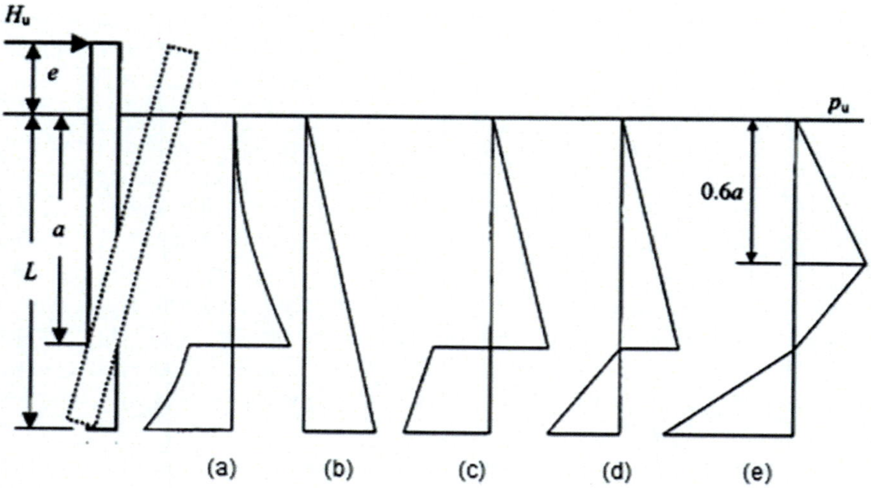


**Fig. 4** Lateral soil pressure distribution as per **a** Hensen, **b** Broms, **c** Petrasovits et al., **d** Meyerhof et al., and **e** Zhang et al., (redrawn by authors after Zhang et al. [9])

conducting model rigid pile tests, this work tried to analyze the distribution of lateral soil resistance all around the pile's c/s. Finally, proposed a method by bifurcating the lateral soil resistance in Frontal soil resistance and side shear resistance, as shown in Fig. 5, which is also satisfied for flexible piles and proposed formulae for estimating the same are as given in Table 1.

Peng [12] made an extraordinary work on laterally loaded finned piles by conducting model tests on scaled-down piles embedded in the sand for both static and cyclic loadings. An effective setup to conduct the cyclic 1D test for laterally loaded

**Fig. 5** Frontal soil resistance and side shear resistance distribution. (redrawn by authors after Zhang et al. [9])

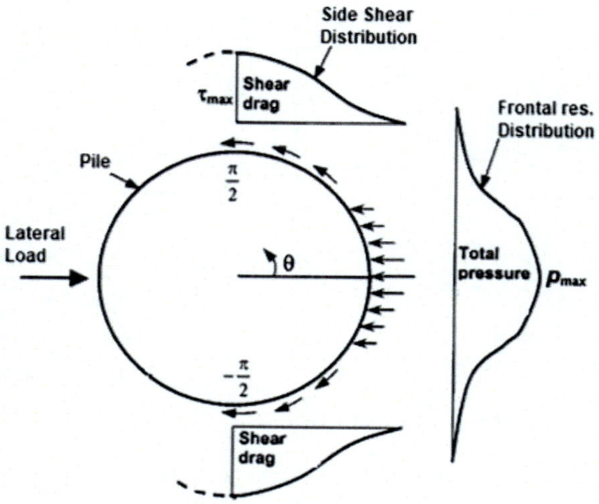

**Table 1** Proposed formulas for estimating lateral soil resistance

| Resistance developed | Ahead of pile | Behind pile |
|---|---|---|
| Frontal soil resistance | $\eta BK_p^2\gamma(0.6a)$ | $1.7\{\eta BK_p^2\gamma(0.6a)\}$ |
| Side shear resistance | $\beta BK\gamma \tan\delta\ (0.6a)$ | $1.7\{\beta BK\gamma \tan\delta\ (0.6a)\}$ |
| Where η and β are the shape factors obtained from Fleming et al. [10] and API [11] | | |

piles for varying fin dimensions was developed and tested for varying magnitude, frequency, which were later compared to monopile results to check its effectivity in carrying lateral loads. 3D FEM analysis software LUSAS was used to depict pile-soil response and soil reaction around fin. The relationship between maximum load and deflections are plotted as *p–y* curves.

It is stated that the ultimate lateral resistance is generally considered at the point on the *p–y* curve whose slope will be lesser than 0.05. From a serviceability point of view, the load is considered at the lateral deformation of 10% of the pile diameter. An increase in load by finned piles is shown in Fig. 6. Ultimate lateral capacity for both the cases, i.e., ultimate limit state and serviceability limit state, is almost the same.

Finally, it was found that the pile head deformations are reduced by the provision of fins at the pile head. To achieve efficiency in adoption, recommended optimum fin dimensions as wind width and length nearly equal to the diameter of pile and half the pile length, respectively, that helped future scope over finned-pile study in resisting lateral loads.

Peng et al. [13] studied the increase in piles' resistance subjected to cyclic lateral loads. Tests were done on small-scale finned and monopile piles to evaluate the effect of fin length over displacement for cyclically loaded piles. The tests were conducted for varying magnitude, load direction, and frequency to simulate the actual offshore

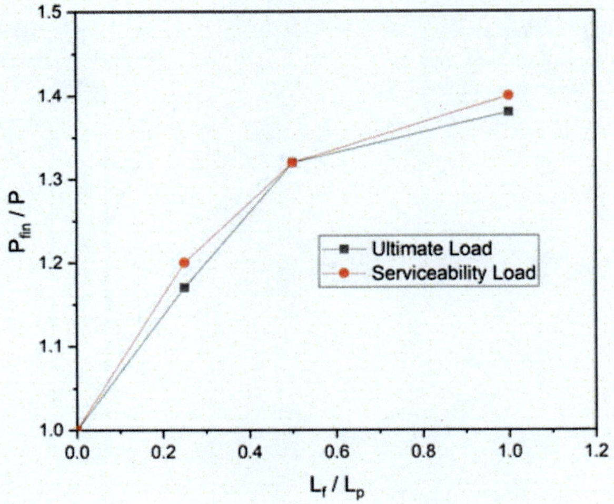


**Fig. 6** Increase in lateral load with an increase in fin length for both the cases (redrawn by authors after Peng [12])

structure loading for varying types of fin tip and length of fin. It is found that the provision of fins reduces the displacement to piles as an alternative for efficiency. The type of pile tip will not have any deterministic effect on lateral displacement. Hence, it concluded that up to half the piles' length lateral displacement would be reduced by about 50% after ten thousand loadings monopiles if the fins were provided.

Rudolph and Grabe [14] attempted to study the cyclic loading effect from varying directions over laterally loaded piles with wings, i.e., finned piles. A series of large-scale in-situ tests were carried out on two tubular steel pipes (each for monopile and wingedpile testing) equipped with two wings, i.e., finned piles to economize the real offshore condition. The cyclic loading is applied by hydraulic cylinders such that loading direction will vary by 90° after each loading, which is load controlled. Piles were equipped with sensors to note the deflection all along the pile length.

It is concluded that on proving wings to the piles below the sea bed, the loading will activate the wings with an enlarged diameter than the pile, hence reducing pile drifting. Since the soil resistance developed will be greater than the pile alone, it will increase the activated soil cross-section from circular to star-shaped. The lateral deformation of the pile gets reduced by adopting winged piles. The varying cyclic loading directions will not have any influence on lateral soil resistance.

Ahmed [1] attempted to study the numerical study using the FEM software PLAXIS 3D package and validate the same using small-scale modal tests in evaluating the response of laterally loaded finned piles embedded in the sand. Modal tests were carried out for two different sand densities corresponding to the loose and medium-dense case. Investigations were performed for varying fin dimensions (length and width) $L_f/L_p$ values of 0.1–0.5 and $W_f/D_p$ values of 0.5–2.0 for varying fin shapes, i.e., rectangular and triangular fins as shown in Fig. 7.

It was concluded from both experimental and numerical analyses that the finned piles provide higher lateral resistance along with reduced lateral displacement than regular piles. Fin length affects the resistance developed than that of fin-width. From

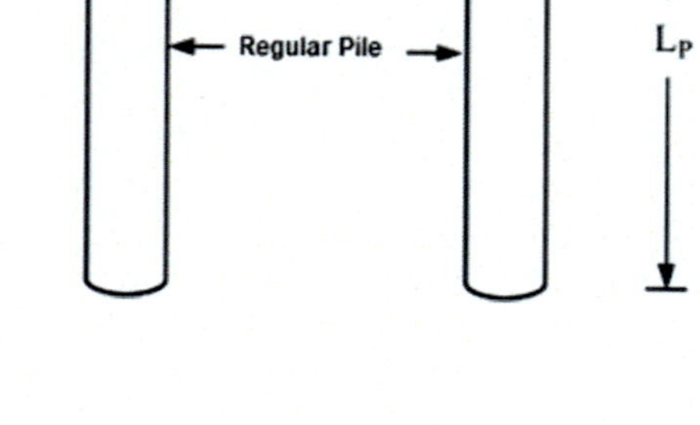


**Fig. 7** Finned-pile description (redrawn by authors after Ahmed [1])

an economical point of view, optimum fin dimensions were $L_f/L_p = 0.4$ and $W_f/D_p = 1.0$, whereas rectangular fins as the effective fin shape.

Murphy et al. [15] conducted a series of field tests to investigate winged-monopile behavior. Tests were performed at two sites with sand deposits whose soil properties were estimated with CPT and MASW tests. It was noted that the groundwater table was lower below the pile tip; hence, it did not influence pile testing. The piles installed were displacement concrete piles over which static loads are applied in increments using a hydraulic jack. The results were demonstrated as *p–y* curves, from which it was found that the Ultimate lateral Resistance was found to increase by 16 and 36% at two sites, respectively, compared to the reference pile. Even peak loads were also mobilized within lateral displacement of 10% of pile dia or lesser. The higher resistance in the second site is due to the higher initial stiffness of the sand. The ultimate serviceability load (for a displacement of 1.2% pile dia, as per Lloyd [16]) for both sites was found to be increased by 111% and 117%, respectively. Also, the lateral displacement was reduced by about 45 and 60%. The wing efficiency was greater at lower loads and decreases as the pile proceeds toward failure.

Bariker et al. [17] conducted series of small-scale model test to investigate the effectiveness of varying fin parameters like the number of fins (two and four finned), fin location along pile length (pile head, mid-length, and pile tip), fin orientation for loading (90° and 45° fin orientation), for long-flexible, intermediate and short-rigid piles, embedded in the dry sand of relative density corresponding to loose and medium-dense case, in sustaining the static-lateral loads and also an attempt is made to study the optimum fin dimension, i.e., fin length ($L_f$) and fin width ($W_f$) in resisting load effectively.

The results of lateral load response were recorded as *p–y* curves, increase in load-carrying capacity was denoted by fin efficiency as stated by Peng et al. [13] stating that the fin will effectively help in reducing the lateral deformation and increasing the load-carrying capacity, and pile head fin location was found to be optimum with fin length relative to pile length ($L_f/L_p$) of 0.4 and fin width relative to pile dia ($W_f/D_p$) of 1.0.

## 3 Numerical Studies on Finned-Pile Foundations

Many researchers like Peng [12], Hussien et al. [18], Marcelo and Jose [19], Wenjun and Ga [20], Zhang et al. [21], Zhi et al. [22] made it possible to use available software packages in modeling and recommended the key points for the regular pile problems subjected to lateral loading. It is not easy to model finned piles in software because fewer efforts were made on the same. Few people made it possible to effectively build a finned-pile problem in software packages, taking the previous literature (as specified above) as a platform to use it effectively.

Peng et al. [23] attempted to improve his work in fin piles on two-dimensional numerical analysis software LUSAS [12], which had the problem of explaining the difficulty to explain soil–pile theory because of complicated fin geometry in the

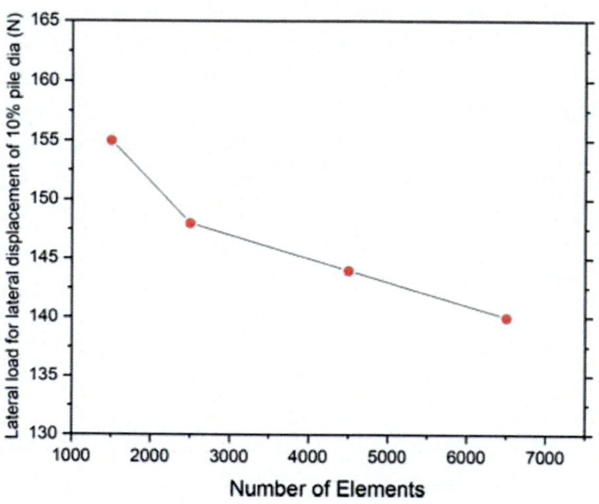


**Fig. 8** Mesh sensitivity map (redrawn by authors after Peng et al. [23])

system, by three-dimensional analysis of laterally loaded fin piles using PLAXIS-3D program to study the effectiveness of the small-scale model and full-scale studies using the pile material as steel for varying fin location on the pile, varying loading directions, length, and diameter of pile later the same results are validated with the experimental results. A close agreement was found between the results obtained also an attempt was made to check the mesh sensitivity on the laterally loaded fin piles as shown in Fig. 8, which explains that with an increase in the number of elements thereby increasing the mesh numbers or decreasing the mesh size, the results get finer and accurate.

Babu and Viswanadham [24] attempted to study the effect of fins over resisting the lateral loads effectively by performing the numerical analysis in ABAQUS software, performing three-dimensional analysis. Investigations were performed for varying fins (straight, diagonal, and star fins), location of fins (top, middle, and bottom), the direction of loading, and fin dimensions (Length and width of fins) embedded in medium-dense sand.

It was concluded that the fins placed at top resist larger lateral loads than the other two positions and increased in the number of wings in fin geometry resist larger loads, and the fin is found to be significant when fin length is 0.5 of pile length. It also concluded that the fin piles exhibit lesser bending moment than the regular piles.

Mohammad et al. [25] attempted to investigate the effectiveness of fin in enhancing the pile's lateral load-carrying capacity by conducting the numerical analysis in ABAQUS software, which were later validated from small-scale model tests. A series of investigations were made to study the effect of pile dia, fins, and aspect ratio of fins in resisting lateral loads. The results found that the aspect ratio of fin had largely influenced the fin efficiency in resisting loads. But, on comparing the influence of the length and width of fins alone, the fin's length was found to influence the design than the width of the fin because that provides higher lateral soil resistance.

An attempt was also made to study the optimum fin dimension. It was found that the fin pile of 40 mm diameter was found to resist 29% more load than the 40 mm diameter regular pile. When results of different fin diameters were made, it was found that the fin pile of 40 mm diameter to carry the same load as by 50 mm regular pile. Since fin length had influenced the design than fin-width, it was concluded that the material usage reduced by 19% resulting in the same resistance.

## 4 Conclusions

This paper reviewed the works of finned piles in resisting the lateral loads. It is found that the involvement of fin with pile is a tedious task to analyze either by basic soil-pile theory or two-dimensional analysis complicated fin geometry and its influence over resisting lateral loads. It has various parameters controlling the design of theory for finned piles like fin-type (two finned, four-straight finned, four-diagonal finned), fin location (pile head, mid-length, and pile tip), fin-length, and fin-width, the direction of loading along with pile material, pile length, the soil surrounding the pile, and nature of the load. Also, it is not easy to analyze these influencing factors by conducting small-scale experimental tests. Full-scale in-situ tests as human-made loss like experimental errors will largely influence the obtained results, which is one reason for not developing the recommendations and design methods for finned-piles testing.

It is necessary to predict the same by conducting the numerical analysis by various software packages available in the market validated by experimental results, which gives better results compared to experimental results, i.e., they do not include experimental errors. But, they fail if mistakes are made while defining the model and boundary conditions, leading to a false result. To boycott these failures, it is required to develop an advanced method to solve these influencing factors without faults to overcome experimental errors and analyze errors. This can only be tackled by modern methods like artificial neural networks (ANN) that can consider each of the influencing factors defined above.

Developing new methods can either be convolutional neural network (CNN), which generates somewhat close results, considering only the important or highly influencing factors. If accurate results are expected, recurrent neural network (RNN) must be developed considering all the influencing factors defined above and all the recommendations provided by the above-discussed literature, providing the strong basement for developing the modern methods.

## References


1. Ahmed, M.N.: Experimental and theoretical studies of laterally loaded finned piles in sand. Can. Geotech. J. **51**, 381–393 (2014). https://doi.org/10.1139/cgj-2013-0012
2. Reese, L., Coax, W., & Koop, F.: Analysis of laterally loaded piles in sand. In: Sixth Annual Offshore Technology Conference, pp. OTC 2080: 473–480. American Institute of Mining, Houston, Tex. (1974)
3. Christos, A., Michael, G.: Interaction of axial and lateral pile responses. J. Geotech. Eng. ASCE **119**(4), 793–798 (1993). https://doi.org/10.1061/(ASCE)0733-9410(1993)119:4(793)
4. Sastry, V., Meyerhof, G.: Behaviour of flexible piles in layered sands under eccentric and inclined loads. Can. Geotech. J. **31**(4), 513–520 (1994). https://doi.org/10.1139/t94-060
5. Prasad, Y.V., Chari, T.: Lateral capacity of model rigid piles in cohesionless soils. Soils Found **39**(2), 21–29 (1999)
6. Brinch Hansen, J.: The Ultimate Resistance of Rigid Piles Against Transversal Forces, pp. 5–9. Danish Geotechnical Institute, Copenhagen (1961)
7. Broms, B.: Lateral resistance of piles in cohesionless soils. ASCE J. Soil Mech. Found. Eng. **90**, 123–156 (1964)
8. Meyerhof, G., Mathur, S., Valsangkar, A.: Lateral resistance and deflection of rigid walls and piles in layered soils. Can. Geotech. J. **18**, 159–170 (1981)
9. Zhang, L., Silva, F., Grismala, R.: Ultimate lateral resistance to piles in cohesionless soils. ASCE, J. Geotech. Geoenviron. Eng. **131**(1), 78–83 (2005). https://doi.org/10.1061/(ASCE)1090-0241(2005)131:1(78)
10. Fleming, W.G.K., Weltman, A.J., Randolph, M.F., Elson, W.K.: Piling engineering. Surrey University Press, London (1992)
11. American Petroleum Institute (API): Recommended practice for planning, designing and constructing fixed offshore platforms. API recommended practice 2A (RP2A), 19th edn. Washington, D.C. (1991)
12. Peng, J.R.: The Behavior of Finned Piles in Sand under Lateral Loading. Newcastle University. Université Libanaise, London (2005)
13. Peng, J., Clarke, B., Rouainia, M.: Increasing the resistance of piles subject to cyclic lateral loading. ASCE, J. Geotech. Geoenviron. Eng. **137**(10), 977–982 (2011). https://doi.org/10.1061/(ASCE)GT.1943-5606.0000504
14. Rudolph, C., Grabe, J.: Laterally loaded piles with wings—Insitu testing with cyclic loading from varying directions. In: Proceedings of the ASME 2013 32nd International Conference on Ocean, Offshore and Arctic Engineering OMAE 2013, pp. 1–7. ASME, Nantes France (2013)
15. Murphy, G., Doherty, P., Cadogan, D., Gavin, K.: Field experiments on instrumented winged monopiles. In: Proceedings Institution of Civil Engineers, pp. 227–239. ICE Proceedings, London (2016)
16. Lloyd, G.: Guideline for the Certification of Offshore Wind turbines. Google scholar, Hamburg (2005)
17. Bariker, P., Rajesh, K.S., Raju, K.V.S.B.: A study on lateral resistance of finned piles in sands. In: Haldar, S., Patra, S., Ghanekar, R. (eds.) Advances in Offshore Geotechnics. Lecture Notes in Civil Engineering, vol. 92. Springer, Singapore. https://doi.org/10.1007/978-981-15-6832-9_18 (2020)
18. Hussien, M., Tetsu, T., Susumu, I., Mourad, K.: On the influence of vertical loads on the lateral response of pile foundation. J. Comput. Geotech. **55**, 392–403. Retrieved from https://doi.org/10.1016/j.compgeo.2013.09.022 (2014)
19. Marcelo, S., & Jose, M. R.: Evaluation of models for laterally loaded piles. J. Comput. Geotech. **48**, 316–320. Retrieved from https://doi.org/10.1016/j.compgeo.2012.07.011 (2013)
20. Wenjun, L., Ga, Z.: New p–y curve model considering vertical loading for piles of offshore wind turbine in sand. J. Ocean Eng. **203**, 107228. Retrieved from https://doi.org/10.1016/j.oceaneng.2020.107228 (2020)

21. Zhang, L., Gong, X., Yang, Z., Yu, J.-l.: Elastoplastic solutions for single piles under combined vertical and lateral loads. J. Cent. South Univ. Technol., 216–222. https://doi.org/10.1007/s11771-011-0682-x (2011)
22. Zhi, Y.A., Yuan, F.-C., Xiao, B.-J.: Behavior of laterally and vertically loaded piles in multi-layered transversely isotropic soils. J. Appl. Math. Model. **51**, 561–573. Retrieved from https://doi.org/10.1016/j.apm.2017.06.039 (2017)
23. Peng, J., Rouainia, M., Clarke, B.: Finite element analysis of laterally loaded fin piles. J. Comput. Geotech. **88**, 1239–1247 (2010). https://doi.org/10.1016/j.compstruc.2010.07.002
24. Babu, K., Viswanadham, B.: Numerical studies on lateral load response of fin piles. Int. J. Geomech. Geoeng. 1–14 (2018). https://doi.org/10.1080/17486025.2018.1535718
25. Mohammad, H., Faezeh, H., Farid, F., Milad, R.: Numerical finite element analysis of laterally loaded fin pile in sandy soil. Innov. Infrastruct. Solutions, 1–14. Retrieved from https://doi.org/10.1007/s41062-019-0200-9 (2019)

# Short Piles for a Solar Power Plant in Western Rajasthan

**Mohit Jhalani, Jitendra Kumar, Ravi Sundaram, and Sanjay Gupta**

## 1 Introduction

### *1.1 Solar Power Generation*

Solar power has emerged as a major alternative and clean source of energy in India to augment power generation. Solar energy is the most readily available source of non-polluting renewable energy resources. India is moving towards an ambitious target of making renewable energy generation at par with thermal plants.

Sunlight can be utilized in two ways viz. direct conversion into electricity through solar photovoltaic (PV) cells or indirect conversion through generating high temperatures by concentrating collectors and thereby run a steam turbine in line with a conventional thermal power plant.

In solar photovoltaic (PV) plants, thousands of solar panels are installed which are usually supported on single short piles. Being lightly loaded, the pullout and lateral capacities of the piles under wind loading are the critical loading condition.

M. Jhalani · J. Kumar
Department of Engineering, NTPC Ltd., Noida 201301, India
e-mail: mohitjhalani@ntpc.co.in

J. Kumar
e-mail: jitendrakumar07@ntpc.co.in

R. Sundaram (✉) · S. Gupta
Cengrs Geotechnica Pvt. Ltd., Noida 201309, India
e-mail: ravi@cengrs.com

S. Gupta
e-mail: sanjay@cengrs.com



C. N. V. Satyanarayana Reddy et al. (eds.), *Ground Characterization and Foundations*, Lecture Notes in Civil Engineering 167,
https://doi.org/10.1007/978-981-16-3383-6_62

**Fig. 1** Vicinity map

## 1.2 *Project Details*

The paper presents case study of a solar PV plant in Jodhpur district of western Rajasthan. The project area covers about 600 ha. The area is blessed with good sunlight over most parts, and the number of clear sunny days in a year is quite high. About 300 MW of power generation is planned.

A map of India showing the project location is illustrated in Fig. 1.

# 2 General Site Conditions

## 2.1 *Regional Geology*

A large tract of western and south-western Rajasthan and Sindh, 640 km long and 160 km wide, constitutes the Thar Desert [1]. The aeolian accumulations of the Thar is a wide expanse of windblown sand and bare rock stretching from the west of the Aravalis to the basin of the Indus and from the southern confines of Punjab to the basin of the Sutlej.

The sands cover an irregular rocky floor, but occasionally local prominences and ridges rise above the levels of the sand. Over the greater part of the area, the sands are piled up into dunes. Dune sands are of recent age rocks younger than the Marwar Supergroup occurs in northwestern part of Jodhpur district. These cover a small area and include the Badhaura sandstone and Bap boulder bed of Permo-Carboniferous age, Lathi sandstone of Jusassic age, and Kapurdi sandstone of Eocene age.

**Fig. 2** Boreholes in progress. Photo on left illustrates soil boring in progress using mechanized shell and auger. Photo on right illustrates rock coring being done using hydraulic rotary rig

## 2.2 Geotechnical Investigation

The project area was investigated by over 100 boreholes in addition to trial pits, electrical resistivity tests, etc. The boreholes were drilled to 5–7 m depth through soil and rock. Figure 2 shows typical boreholes in progress at site.

## 2.3 Stratigraphy and Geotechnical Characterization

At the project site, dune sands of the Thar Desert are underlain by sandstone. In general, the sands are poorly graded with an insignificant proportion of coarse and medium-sized sand grains. The stratigraphy at the site may be divided into two generalized strata as given below:

Stratum I Fine sand and silty fine sand.
Stratum II Sandstone.

The depth to rock varies across the site from less than 1 m to more than 7 m depth. Groundwater was not encountered to the depths investigated and is expected to be fairly deep. Contours of depth to rock are presented in Fig. 3.

In general, rock is met at shallow depth on the eastern side and outcrops in the southeastern portion. In the central and western parts, the depth of rock exceeds 3–4 m.

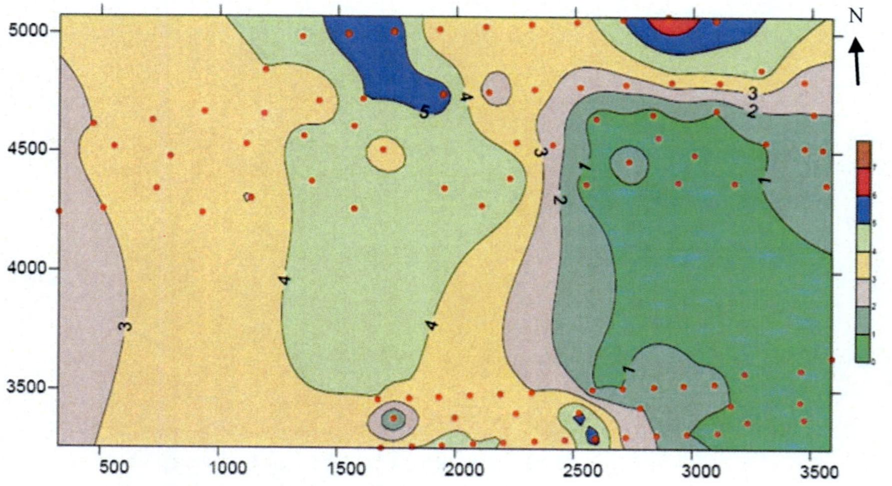


**Fig. 3** Contours of depth to rock

Alam Singh et al. [2] classify dune sand as a meta-stable or collapsible soil that goes through radical re-arrangement of particles and loss in volume upon wetting with or without load application. The SPT values and relative density of the soil are a function of the overburden.

Gupta and Sundaram [3] found that the trend of SPT values is an important feature of aeolian depositions. The soils with *N*-values less than about 10–12 may be treated as unstable portions of the dune.

Typical boreholes are illustrated in Figs. 4 and 5.

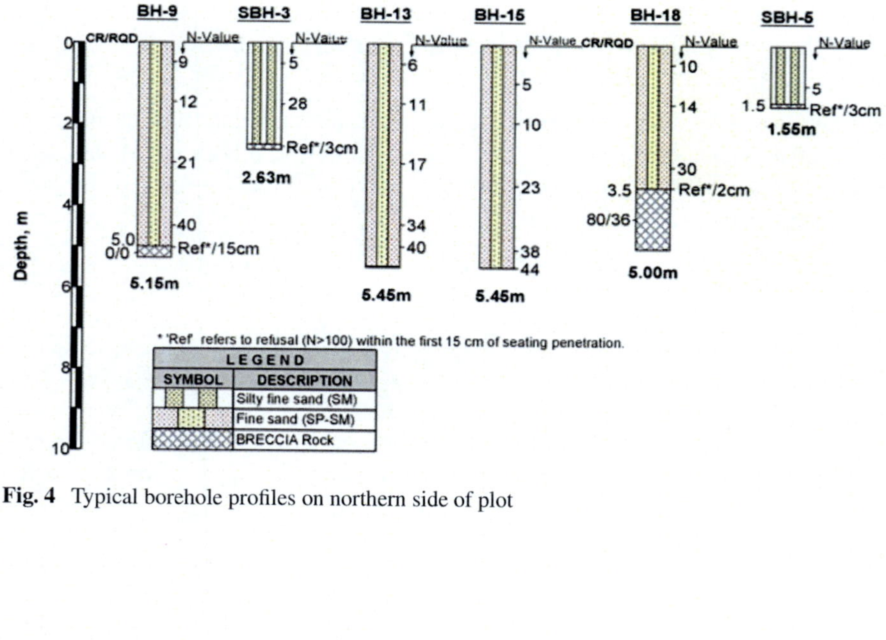


**Fig. 4** Typical borehole profiles on northern side of plot

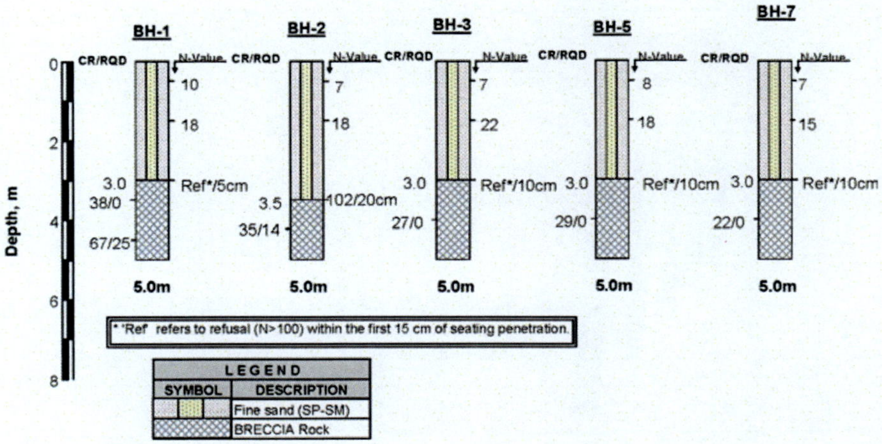


Fig. 5 Typical borehole profiles in southern side of plot

# 3 Foundation Type and Design

## *3.1 Solar PV Module Mounting Structure*

Solar PV modules are mounted on a structure that is called module mounting structure (MMS). This MMS is an arrangement of rafters and purlins that transfers the load through bracings and column posts to the foundation. The column post is generally embedded in the foundation. A typical isometric view showing different elements of MMS is placed in Fig. 6.

The module surface area is large compared to its weight; therefore the dead weight of structure is comparatively less. In view of large surface area and angle of module, wind load plays a predominant role in designing the structure and in turn foundation.

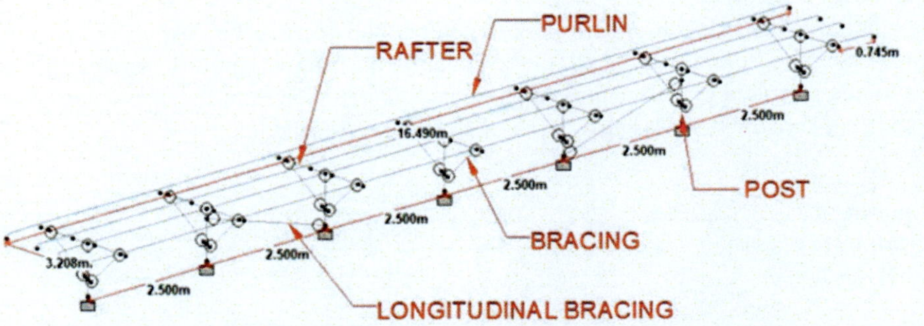


Fig. 6 Typical isometric view of MMS

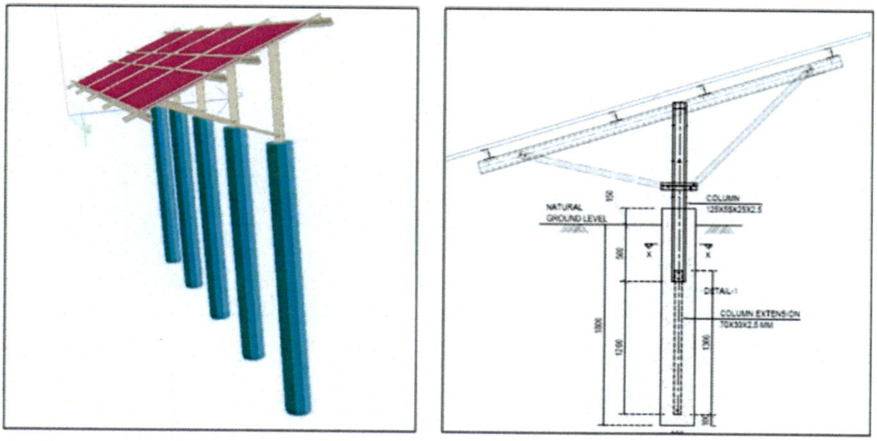


**Fig. 7** MMS table with foundation: 3D rendered view on left photo and sectional view on right

Due to wind loads, the dominant reactions obtained at the top of foundation are uplift and lateral. This necessitates design of foundation for pullout and lateral load resistance. A typical sketch showing the MMS table and foundation is placed in Fig. 7.

## 3.2 Foundation Type

Based upon the structural analysis considering the site-specific loads, the obtained reactions at each column location were in the range of 1.5–2.25 T in case of uplift and 0.8–1.5 T in case of lateral load. The load variations were due to the different table sizes/arrangements considered to maximize generation. The columns were generally spaced at 2.5 m apart below a table. The load variation range for foundation design is placed in Table 1.

Different foundation systems were explored considering the cost economics as well the speed of construction. After evaluation of different options, bored cast in-situ pile appeared to be the most economical foundation system for such loads. The speed of construction was also in line with the project schedule.

**Table 1** Range of foundation loads for pile design

| S. No. | Type of loading, m | Load range, T |
|---|---|---|
| 1 | Pullout | 1.5–2.25 |
| 2 | Lateral | 0.8–1.5 |
| 3 | Vertical | 1.0–1.5 |

### 3.3 Foundation Design

Based upon the geotechnical investigation carried out across the site, entire area was divided into two parts: one where sand was deposited up to the depth of investigation and the other where rock was encountered after few meters of loose sandy deposit. For both cases bored cast in-situ pile was found suitable.

Design of pile against the pull-out capacity was done based on the recommendations of IS:2911 Part-1 Sec-2-2010 [4]. The design was done considering factor of safety of 2 as the capacity was required to be validated through pull-out test at site.

For the design of pile against lateral load, IS:2911 Part-1, Section 2-2010 recommendations could not be used, as due to fewer loads only short pile was required. Therefore, short pile was designed considering the Broms' approach [7].

Based on different loading conditions and different geotechnical parameters, dimensions of pile were designed. The pile details are presented in Table 2.

The paper discusses the pullout and lateral load–displacement behavior for 2.65–2.8 long piles bearing on the dune sands.

## 4 Pile Testing and Results

Bored piles of 350 mm diameter were installed with about 200 mm size channel section embedded in it. Pile lengths ranged from 2.65 to 2.8 m. Since pullout and lateral loading are the critical loading conditions, sufficient number of vertical pullout and lateral load tests were performed to confirm the safe capacities.

Nearly 150,000 piles were installed at the site. About 140 initial piles were tested for compression, pullout and lateral loading to 2.5 times the design load.

**Table 2** Details of bored cast in-situ piles used at site

| Diameter of pile, mm | Length of pile, m | Approximate number of piles | Remarks |
|---|---|---|---|
| 350 | 1.8 | 20,000 | Minimum 1 m rock socket |
| 350 | 2.5 | 15,000 | Sandy soil |
| 350 | 2.65 | 65,000 | Sandy soil |
| 350 | 2.8 | 40,000 | Minimum 1 m rock socket |
| 350 | 3.1 | 10,000 | Sandy soil |

## 4.1 Pullout Tests

Pile pullout tests were performed in accordance with IS: 2911 Part 4-2013 [4]. A hydraulic jack placed over a girder was used to apply the pullout force. Reaction was obtained from supports resting on the ground. A dial gauge was placed with reference to a stable datum bar. Figure 8 presents a schematic of the test setup and a photograph of the pullout test in progress.

Typical load–displacement curves are illustrated in Fig. 9.

**Interpretation of Safe Pullout Capacity from Load–Displacement Graph.** The safe pullout capacity of pile was considered as the lower value obtained from the following two criteria given in IS: 2911 Part 4-2013 [5]:

i. Two-thirds of the load at which pile deflection attains a value of 12 mm; and
ii. 50% of the load at which the load–displacement curve shows a clear break.

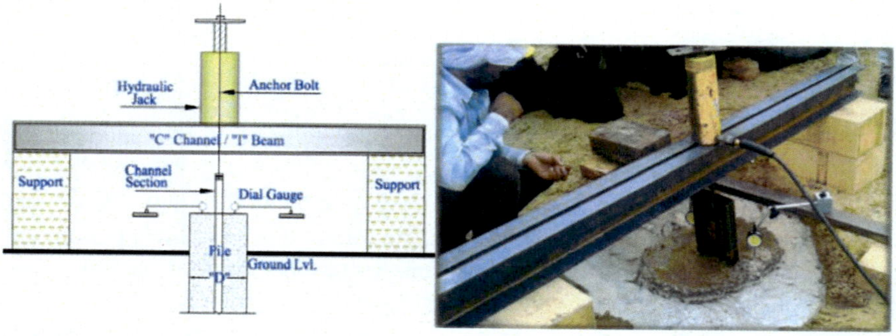


**Fig. 8** Test arrangement for pile pullout test. Sketch on left is a schematic of the test setup. Photo on right shows test in progress

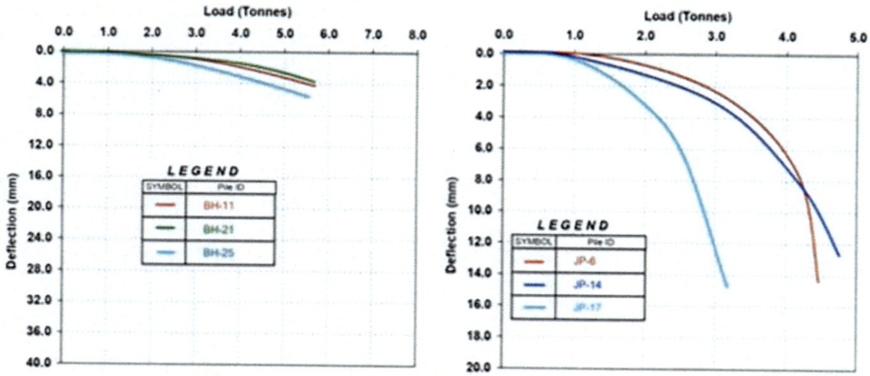


**Fig. 9** Pullout test—typical load–displacement curves

**Extrapolation of Load–Displacement Curve to Assess Safe Pile Capacity.** Many of the tests were not carried out to failure and were stopped on reaching twice the design load. Therefore an attempt was made to mathematically extrapolate the load–displacement curve assuming a hyperbolic correlation.

Kaniraj and Samantha [6] advocated the hyperbolic model to extrapolate results of vertical compression load tests on bored piles. The authors propose that a similar hyperbolic correlation may be used to extrapolate results of vertical pullout tests.

The basic premise is that the load–displacement curve follows a hyperbolic function expressed as follows:

$$\frac{S'}{Q'} = a + bS' \tag{1}$$

where

$S' = s/D$, ratio of measured displacement, $s$ to pile diameter, $D$

$Q' = Q/Q_r$, ratio of applied load $Q$ (corresponding to deflection $s$) to a reference load $Q_r$.

$a$ and $b$ are constants.

Hence, plot of $S'/Q'$ versus $S'$ shall be a straight line. To check the reliability of the correlation proposed in Eq. 1, eight tests which had been carried out to failure (displacement of more than 12–14 mm) were analyzed. A very good match was found between the actual load–displacement curve and the predicted hyperbolic curve with error of less than 3–8%.

The concept was then used to extrapolate the results of tests which had not been carried out till failure. For a good correlation, piles that had experienced deflection of at least 4 mm were used since for lesser deflections, the displacement may be primarily elastic and soil-movement may be small. Typical plot of $S'/Q'$ versus $S'$ for one test is illustrated in Fig. 10.

Typical extrapolated load-settlement curves for three tests are presented in Fig. 11. Table 3 presents the summary of the analysis.

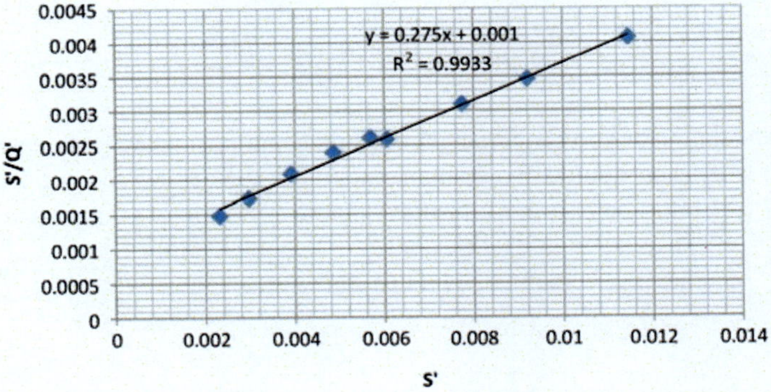


**Fig. 10** Extrapolation of load–displacement plot using hyperbolic model—of $S'/Q'$ versus $S'$ for one test and the trendline. Please note that the error is small with $R^2$ value close to 1

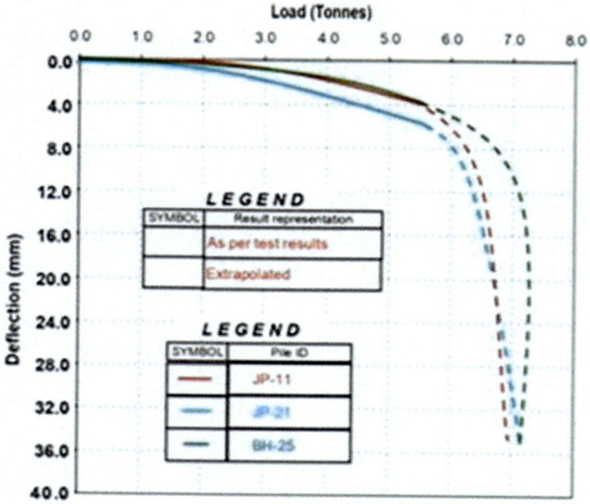


**Fig. 11** Typical load–displacement curves extrapolated using hyperbolic model. The dotted lines show the extrapolated curve beyond the test data

**Table 3** Extrapolation of load–displacement plots using hyperbolic model

| Pile No. | Maximum load applied, tonnes | Measured settlement, *s* mm | Extrapolated ultimate pullout capacity, tonnes | Interpreted safe pullout capacity, tonnes | Error in computed *s* versus measured *s*, % |
|---|---|---|---|---|---|
| JP-11 | 5.64 | 4.01 | 7.02 | 2.82 | 0–5, max 9 |
| JP-21 | 5.64 | 5.55 | 7.26 | 3.76 | 0–6, max10 |
| V5-P3 | 8.10 | 4.61 | 8.82 | 4.20 | 0–4, max 9 |
| V5-P11 | 8.10 | 3.95 | 8.84 | 4.40 | 0 –3, max 7 |
| V5-P18 | 8.10 | 4.50 | 8.73 | 4.40 | 0–7, max 9 |
| V6-P7 | 8.10 | 5.75 | 9.54 | 4.50 | 0–7 max10 |
| V6-P15 | 8.70 | 6.32 | 13.90 | 4.80 | 0–3 max 5 |
| V6-P17 | 8.10 | 5.93 | 19.80 | 5.80 | 0–3 max 6 |
| V6-P18 | 8.10 | 6.11 | 10.63 | 4.80 | 0–5, max 8 |

Back-calculation of Soil Properties. The ultimate pullout capacity of piles installed in granular soils is computed in accordance with IS: 2911 Part 1 Section 2-2010 [4] using Eq. 2 (skin friction between soil and concrete + weight of pile) as given below:

$$Q = \sigma_v k \tan\delta \, A_s L + W \tag{2}$$

where

$Q$ ultimate pullout capacity of pile.
$\sigma_v$ overburden pressure.

$k$ coefficient of earth pressure.
$\delta$ angle of friction between soil and concrete (usually taken as equal to angle of internal friction of sand, $\phi$, determined by laboratory tests).
$A_s$ surface area of pile per $m$ length.
$L$ length of pile.
$W$ weight of pile.

Substituting the value of ultimate capacity determined by the pullout test and $\phi$ value determined in the laboratory in Eq. 2 above, the value of earth pressure coefficient, $k$ may be computed.

It was observed that the value of $k$ appears to increase with increase in the SPT $N$ value. To assess the trends, Fig. 12 presents the variation of $k$ versus field $N$ value which shows the general trend of $k$ with increase of $N$ and the bandwidth of $\pm$ 10% in which scatter of the data is observed.

Reviewing the trends, the authors propose the following correlation (see Table 4) between $N$ and $k$, conservatively considering the lower end of the bandwidth: The lower end of the bandwidth of $-10\%$ has been proposed so as to ensure that the piles are likely to pass under the test load.

Linear interpolation may be done for intermediate $N$ values. The authors suggest $k$ value of 0.7 for $N \leq 5$. Further, the value of $k$ may be restricted to 1.3 for $N \geq 30$.

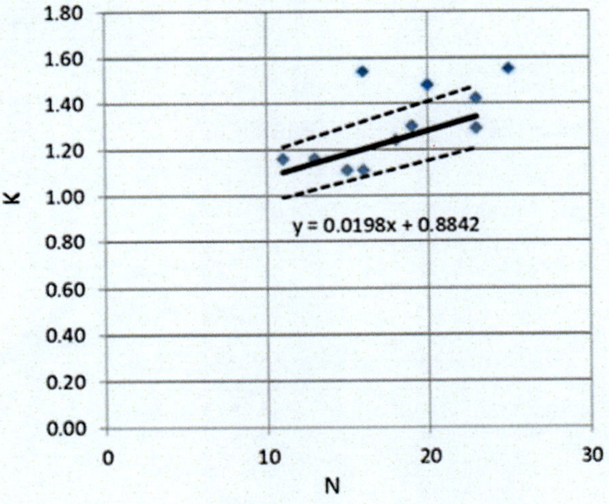


**Fig. 12** Trend of field $N$-value versus $k$ for dune sands. The continuous line presents the trend-line and the dotted lines present the bandwidth which is $\pm$ 10%

**Table 4** Correlation between SPT $N$ value and earth pressure coefficient $k$ for pile pullout capacity computation for dune sands

| Field SPTN-value | $k$ for pullout loading |
|---|---|
| 11 | 1.0 |
| 15 | 1.06 |
| 20 | 1.15 |
| 25 | 1.24 |

## 4.2 Lateral Load Tests

Lateral load tests were performed in accordance with IS: 2911 Part 4-2013 [4]. A hydraulic jack is placed between two piled to apply the horizontal force. One of the piles was used to take reaction force. A dial gauge was placed with reference to a stable datum bar. Figure 13 presents a schematic of the test setup and a photograph of the lateral load test in progress.

Typical load–displacement curves are illustrated in Fig. 14.

**Interpretation of Safe Lateral Capacity from Load–Displacement Graph.** The safe pullout capacity of pile was considered as the lower value obtained from the following two criteria given in IS: 2911 Part 4-2013 [5]:

i. Load at which the deflection attains a value of 5 mm, and
ii. Two-thirds of the load at which pile deflection attains a value of 12 mm.

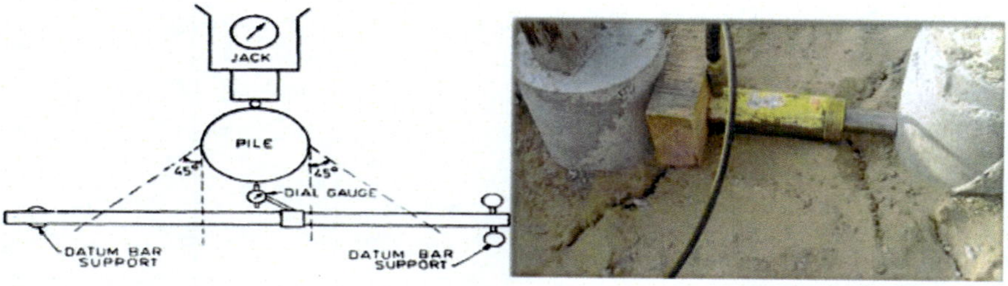


**Fig. 13** Test arrangement for lateral load test on pile. Sketch on left is a schematic of the test setup. Photo on right shows test in progress

**Fig. 14** Typical Load–Displacement curves for lateral load tests

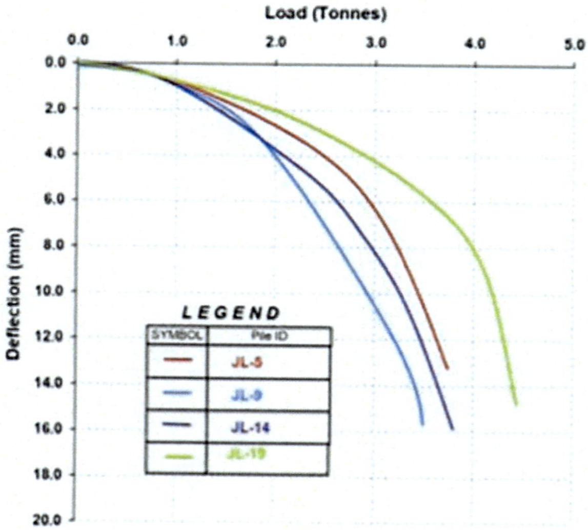

**Computation of Lateral Capacity**. Annex-C of IS: 2911 Part 1 Section 2-2010 [4] gives the procedure for analysis of lateral capacity of long piles. The standard defines a stiffness factor, $T$ for granular soils as follows:

$$T = \sqrt[5]{\frac{EI}{\eta_{\mathrm{h}}}}$$

where

E modulus of elasticity of concrete.
I moment of inertia of pile cross-section.
$\eta_{\mathrm{h}}$ modulus of subgrade reaction of soil.

While piles longer than "4 T" are considered as long pile, piles shorter than 2 T are treated as short piles.

Since the piles of 2.65–2.8 m length classify as short piles, the procedure proposed by Broms [7] has been used. Broms gives graphical plots to compute the ultimate lateral capacity and deflection of free-head and fixed-head (restrained) short piles based on the value of $\eta_{\mathrm{h}}$, $\phi$ and other properties of granular soils. The simplified solution is based on the assumption of shear failure in soil.

The simplified solution is based on the assumption of shear failure in soil. Broms's solution for the ultimate load resistance ($P_{\mathrm{u}}$) and corresponding deflections for short piles in granular soil is illustrated in Fig. 15.

A factor of safety of 3 is applied to the ultimate lateral capacity. As per the project specifications, the safe lateral load is computed for a permissible pile-head deflection of 5 mm for the free-head case. The modulus of horizontal subgrade reaction has been taken from Table 3, Annex-C of IS: 2911 Part 1 Section 2-2010 [4].

**Back-calculation of modulus of horizontal subgrade reaction** $\eta_{\mathrm{h}}$. The lateral capacity of the free-head piles was calculated using the modulus of horizontal

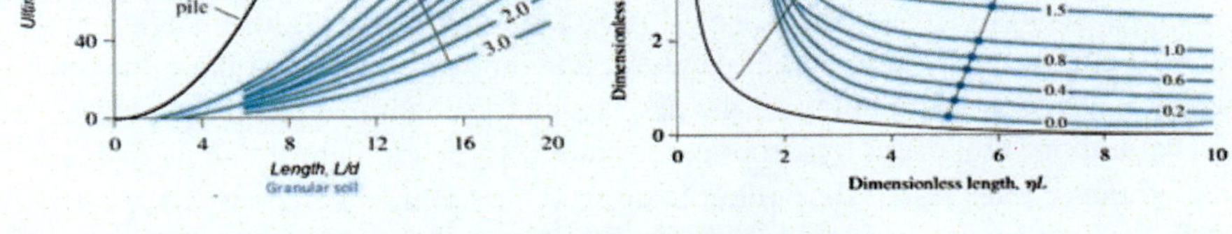


**Fig. 15** Broms' solution for laterally loaded piles in cohesionless soils

**Table 5** Safe lateral capacities and modulus of horizontal subgrade reaction $\eta_h$ computed on basis of IS: 2911 Part 1 Sec 2 and interpreted from lateral load tests

| Pile No | Field $N$-value | $\phi°$ from lab test | Safe lateral capacity | | $\eta_h$, MN/m$^3$ | |
|---|---|---|---|---|---|---|
| | | | Calculated (Broms) | Test result | IS: 2911 Part 1 Sec 2-2010 | Back calculated |
| JL5 | 13 | 30 | 0.63 | 1.83 | 3.1 | 11.0 |
| JL9 | 11 | 30 | 0.56 | 1.59 | 2.7 | 9.0 |
| JL11 | 12 | 30 | 0.60 | 2.01 | 2.9 | 11.9 |
| JL14 | 15 | 31 | 0.71 | 1.73 | 3.5 | 9.7 |
| JL17 | 15 | 31 | 0.71 | 2.14 | 3.5 | 13.5 |
| JL24 | 15 | 31 | 0.71 | 2.09 | 3.5 | 12.3 |
| JL26 | 17 | 31 | 0.79 | 1.61 | 3.9 | 9.1 |
| JL23 | 12 | 30 | 0.60 | 1.74 | 2.9 | 9.7 |

subgrade reaction $\eta_h$ values as per IS: 2911 Part 1 Section 2-2010. It was observed that the safe horizontal capacity from the load test was substantially higher than the computed capacity. The $\eta_h$ values were then changed in Broms' solution so as to match the safe capacities determined from the test.

Table 5 presents the results of the analysis for eight typical 2.65 m long piles that were tested to failure. Several additional tests in the project area tested to similar range of test loads experienced less than 4–5 mm deflection suggesting that the $\eta_h$ values would probably be higher than the values given in Table.

It is evident that the $\eta_h$ values back-calculated from the load tests are about 3–4 times more than the values given in the IS code. There is a need to analyze further data from more lateral load tests to conclusively develop a correlation to assess the trend of $\eta_h$ with other soil characteristics.

## 5 Conclusions

Short piles are used for supporting the solar panels of photo-voltaic solar power plants. These panels are lightweight structures with very small downward load. But the pullout and lateral loading is significant due to the wind loads.

The paper presents a case study of a 300 MW solar power plant in dune sands of western Rajasthan. About 150,000 RCC bored-cast piles of 350 mm diameter and 1.8–3.1 m length were installed. Due to the large number of piles installed, even a 10–20 cm change in pile length can have a significant impact on the piling cost. Hence, it is necessary to optimize the pile design to economize the project cost.

The authors propose a hyperbolic correlation to extrapolate the load–displacement graph of piles tested for pullout loading in case where the test was not carried out to failure. Analyzing the ultimate pullout loads, a correlation has been developed between SPT $N$-value and earth pressure coefficient, $k$ for computing the skin friction between pile and concrete. The authors also highlight the need to update the coefficient of horizontal subgrade reaction $\eta_h$ for sands given in Annex-C of IS: 2911 Part 1 Sec 2-2010.

## References


1. Krishnan, M.S.: Geology of India and Burma. CBS Publishers, New Delhi (1986)
2. Alam Singh, Punmia, B.C., Ohri, M.L.: Regional Deposits—Desert Soils, Indian Contributions to Geotechnical Engineering. A Commemorative Volume Released on the Occasion of XI ICSMFE, San Francisco, Sarita Prakashan, Meerut, pp 44–53 (1985)
3. Gupta, S., Sundaram, R: Geotechnical characterization of dune sands of the Thar Desert. In: Proceedings, 13th Asian Regional Conference on Geotechnical Engineering (13ARC), Kolkatta, 2.1-12IN
4. IS: 2911: Design and Construction of Pile Foundations—Code of Practice, Part 1 Section 2 Bored Cast-in-situ Piles. Bureau of Indian Standards (2010)
5. IS: 2911: Design and Construction of Pile Foundations—Code of Practice, Part 4 Load Test on Piles. Bureau of Indian Standards (2013)
6. Kaniraj, S.R., Samantha, S.: Interpretation of Safe Load from Pile Loading Test. In: Proceedings, Seminar on Piles, Indian Geotechnical Society Delhi, pp 97–105 (1996)
7. Broms, B.: The lateral resistance of piles in cohesionless soils. J. Soil Mech. Div. Am. Soc. Civil Eng. **90**(SM3), 123–156 (1964)

# Performance of Barrette Foundations in Sandy Soil Subjected to Vertical and Lateral Loading

**Sohail Qureshi, Sunil Dighe, S. W. Thakare, and A. I. Dhatrak**

## 1 Introduction

Barrette piles are cast-in-situ reinforced concrete pile which is formed in a deep trench excavated under bentonite or polymer slurry by diaphragm walling equipment. These are generally rectangular piles; however, various other shapes may also be adopted. Barrettes provide resistance to horizontal stress and to bending moments better than circular piles of the same cross-section. Barrettes provide an alternative to large diameter bored and cast-in-place piles or drilled shafts. In real situations, Barrette foundation may be used for high-rise buildings, skyscrapers, elevated expressways, towers, overpass bridges, underground car parks, etc.

The barrette pile's having considerably larger bearing capacity compared to that of the conventional circular pile is a crucial advantage in meeting the increasing demand for loading from tall buildings. Besides, the time required for trenching of barrette piles is less than that required for the drilling method employed for circular piles. Therefore, barrette piles have become increasingly common in the construction of high-rise buildings. Its cross-section and construction method differ from those of conventional circular bored piles, resulting in different bearing behaviours for the two types of piles.

S. Qureshi (✉) · S. Dighe · S. W. Thakare · A. I. Dhatrak
Government College of Engineering Amravati, Amravati, Maharashtra, India

Department of Civil Engineering, Sant Gadge Baba Amravati University, Amravati, Maharashtra, India

 
C. N. V. Satyanarayana Reddy et al. (eds.), *Ground Characterization and Foundations*, Lecture Notes in Civil Engineering 167,
https://doi.org/10.1007/978-981-16-3383-6_63

## 2 Literature Review

El Wakil et al. [1] has experimentally presented behaviour of laterally loaded small scale rectangular barrette in sand. 28 model tests were carried out to study the effect of density, aspect ratio, eccentricity and loading directions. As the aspect ratio increased, the performance of the barrette pile also increased. Relative density has a significant effect on the performance of barrette. Lateral resistance of barrette was greater when loaded along the major axis as compared to the minor axis. Submaneewong et al. [3] conducted static pile load tests to verify the lateral load capacities of T-shape barrette and bored piles. For the design approach, the back analysis suggests that input soil stiffness for T-shape barrette should be about 3 times of empirically calculated value to predict deflection values before concrete cracking similar to those of actual measurement by finite element software. Finite element analysis overestimated the later capacity ignoring the concrete cracking. Ramaswamy et al. [4] had presented a paper on the construction of barrettes for high-rise foundation. According to researchers, by adopting barrette, the designer would be able to increase the inertia and bending moment resistance in the required direction without having to increase the concrete area by properly orientating the barrette. Researcher specified the safe bearing load for the different shape of barrette with general sizes. Charles et al. [5] had presented field studies of well-instrumented barrette in Hong Kong. Zhang et al. [6] presented the behaviour of large section rectangular barrette subjected to lateral load. The load capacity was highest along the major axis and decreased as deviated from major axis to minor axis.

## 3 System Development

For the study of vertical and laterally loaded barrettes in sand, laboratory model tests were conducted on barrettes of various shapes. The test tank was made of steel frame with an inside dimension of 0.90 m (length), 0.50 m (width), and 0.65 m (depth). The bottom and vertical edges of the box were stiffened using angle sections to avoid lateral yielding during soil placement and loading of the barrette pile model.

### 3.1 *Test Setup for Vertical Loading*

The setup for vertical loading on barrette consisted of loading frame, test tank, hydraulic jack, proving ring and dial gauges. Hydraulic jack of 10 ton capacity was used to apply static loads to the pile. The settlements were measured with the help of two dial gauges with 25 mm travel and least count of 0.01 mm. The dial gauges were fixed to the magnetic bases which in turn were fixed to the sides of the tank. Proving ring of 5 kN capacity was used to precisely record the load applied

**Fig.1** Test setup used for vertical loading on Barrette piles

on the barrettes. Proving ring was fixed to the bottom plunger to transfer load from hydraulic jack to the model barrettes. The test setup is shown in Fig. 1.

## *3.2 Test Setup for Lateral Loading*

The setup for lateral loading on barrette consisted of inextensible steel wire, test tank, dial gauges, loading pan and dead weight. The test setup for lateral loading on barrette piles is shown in Fig. 2. The static lateral loads were applied to the pile caps utilizing dead weights placed on a loading pan connected to an inextensible steel-string over a pulley supported by a loading frame. Pullies of mild steel having diameter 3 cm were used in the set up to support the mild steel wires for application of lateral loads.

## *3.3 Materials*

### Model Piles

The model barrette piles were made from aluminium. The c/s area of the barrette piles was kept similar to the c/s area of the circular pile of diameter 20 mm (i.e. 314 $mm^2$). Length of all piles was kept same and equal to 300 mm (Table 1).

**Fig. 2** The test setup used in experimental investigations for applying lateral loads

**Soil**

For the model tests, uniformly graded dry cohesionless sand was used as the foundation soil. This sand is available in Vidarbha region of Maharashtra. Table 2 gives the properties of sand as determined in the laboratory as per relevant IS codal procedures.

## *3.4 Experimental Procedure for Vertical Loading*

The piles were placed at the centre of the tank vertically in the sand bed with the help of a special guide tool. The relative density of soil was maintained at 40% (medium dense) by sand rainfall technique. Initially, the tank was filled up to the pile tip level and then the model test pile is lowered in the test tank. This procedure of pile installation was assumed to simulate the stress conditions similar around cast in situ piles in the field. The pile head was screwed with the pile cap made of steel plate of 3 mm thickness. The sand was filled into the tank up to 250 mm height of pile and top 50 mm of the pile was kept projecting above soil bed. The top surface of the formed sand bed was levelled with the help of thin sharpened steel plate.

The vertical compressive load to the piles was applied with the help of hydraulic jack fitted to the loading frame. The proving ring was attached to the bottom of the jack. One metal rod was connected to the bottom threads of proving ring and rested directly on the groove made on the pile cap. Dial gauges were placed on the pile cap to measure the vertical displacement. Loads were applied in increments and settlements of the piles were measured with the help of dial gauges. Each load increment was kept constant till the rate of settlement became less than 0.1 mm per

**Table 1** Shapes and dimensions of model piles used in experimental investigation

| S. No. | Cross-section of pile | Description | Dimensions |
|---|---|---|---|
| 1 | 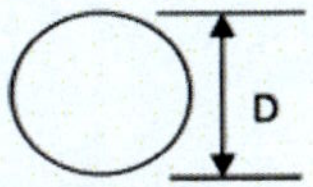  | Circular | Diameter<br>$D = 20$ mm<br>$A = 314$ mm$^2$ |
| 2 | 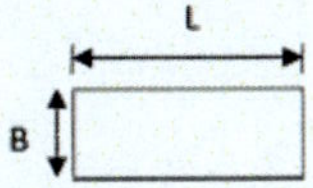  | Rectangular-1<br>Aspect ratio $(L/B) = 2$ | Length<br>$(L) = 25$ mm<br>Width<br>$(B) = 12.5$ mm |
| 3 | 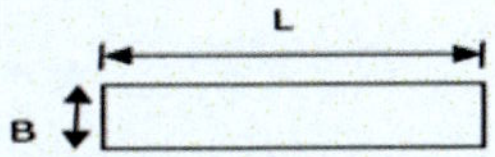  | Rectangular-2<br>Aspect ratio $(L/B) = 3$ | Length<br>$(L) = 31$ mm<br>Width<br>$(B) = 10$ mm |
| 4 | 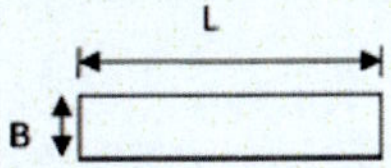  | Rectangular-3<br>Aspect ratio<br>$(L/B) = 4$ | Length<br>$(L) = 36$ mm<br>Width<br>$(B) = 9$ mm |
| 5 | 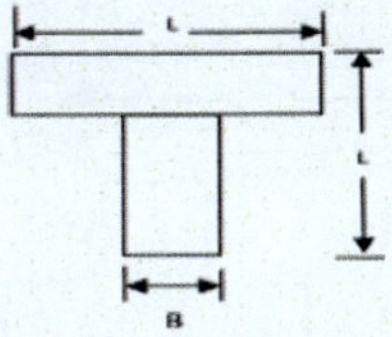 | T-Shape | Length<br>$(L) = 24.8$ mm<br>Width<br>$(B) = 7.4$ mm |

(continued)

**Table 1** (continued)

| S. No. | Cross-section of pile | Description | Dimensions |
|---|---|---|---|
| 6 | 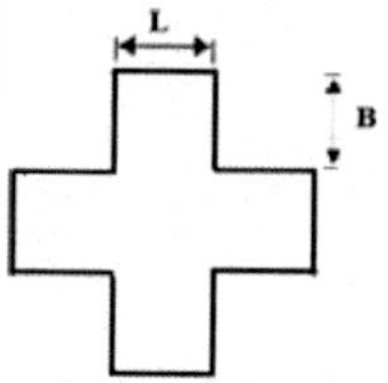 | Cruciform | Length<br>($L$) = 8 mm<br>Width<br>($B$) = 8 mm |
| 7 | 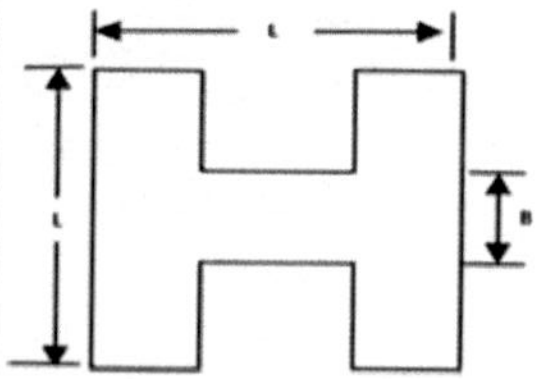 | H-Shape | Length<br>($L$) = 25 mm<br>Width<br>($B$) = 5 mm |

**Table 2** Properties of sand

| S. No. | Properties of sand | Values |
|---|---|---|
| 1 | Specific gravity | 2.67 |
| 2 | IS classification | SP (Medium Sand) |
| 3 | $\gamma_{max}$ (kN/m$^3$) | 17.5 |
| 4 | $\gamma_{min}$ (kN/m$^3$) | 15.7 |
| 5 | Angle of internal friction ($\phi$) | 38 |
| 6 | Cohesion (kN/m$^2$) | 0 |

30 min. Application of load increments was continued till failure of the model pile which was indicated by the rapid settlement of the pile.

### 3.5 Experimental Procedure for Lateral Loading

The preparation of sand bed and placement of barrette pile was similar to that of the vertical load test. Lateral loads were applied in increments and lateral displacements of the pile head were measured with the help of two dial gauges. Each load increment was kept constant until the rate of displacement became less than 0.1 mm/30 min. Application of load increments was continued till failure of the model pile which was indicated by rapid lateral displacement of the pile. The results of the tests were plotted in the form of lateral displacement versus lateral load curves.

### 3.6 Parameters Selected for the Study

The parameters selected for the experimental study are listed in Table 3.

## 4 Test Results and Discussions

### 4.1 Results of Vertical and Lateral Loading

The pile load tests were carried out on conventional circular and barrette piles and load–displacement curves were plotted to study the performance of barrette pile. The curves plotted in Figs. 3, 4, 5, 6, 7 and 8 shows the performance of circular and barrette pile subjected to vertical and lateral loads in medium dense cohesionless sand (Figs. 9 and 10).

The ultimate vertical and lateral load capacity is determined from the load–displacement curves.

**Table 3** Parameters selected for the experimental study

| S. No. | Details of parameter | Constant parameters | Varied parameters |
|---|---|---|---|
| 1 | Height of pile | 300 mm | – |
| 2 | The diameter of circular pile | 20 mm | – |
| 3 | Slenderness ratio (*L*/*D*) | 15 | – |
| 4 | Relative density of sand | 40% | – |
| 5 | Shapes of Barrette piles | – | (i) Rectangular<br>(ii) T- shape<br>(iii) Cruciform<br>(iv) H-shape |
| 6 | The aspect ratio of rectangular barrette pile | – | 2, 3, 4 |
| 7 | Type of loading | – | (i) Vertical loading<br>(ii) Lateral loading |
| 8 | Loading direction for lateral loads | - | Rectangular Barrette<br>(i) Along major axis<br>(ii) Along minor axis<br>Barrettes of other shapes<br>(i) Along the web<br>(ii) Across the web |

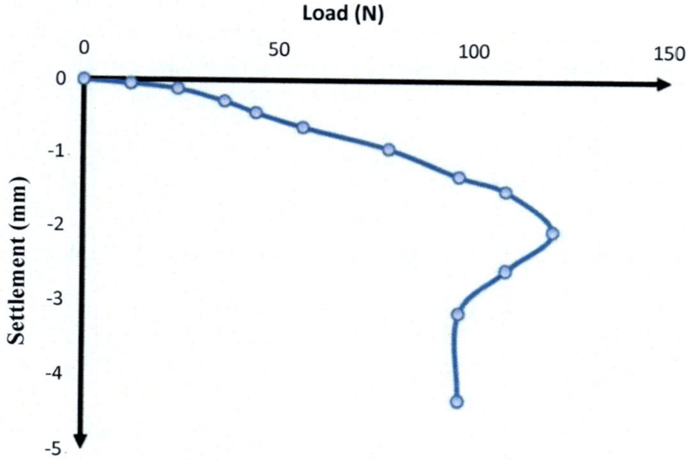


**Fig. 3** Load-settlement curve for circular pile

The ultimate vertical load capacities of the piles are shown from Tables 4, 5 and 6.

The ultimate lateral load capacities of the piles determined from the load-settlement curves are tabulated in Tables 7, 8 and 9.

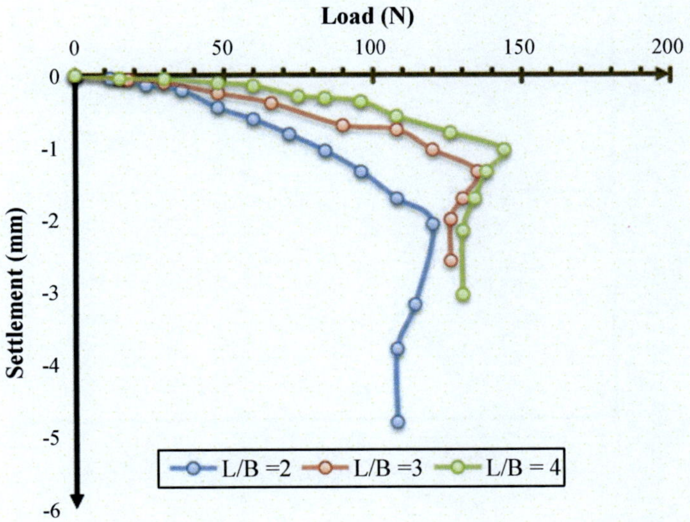


**Fig. 4** Load-settlement curves for rectangular barrettes

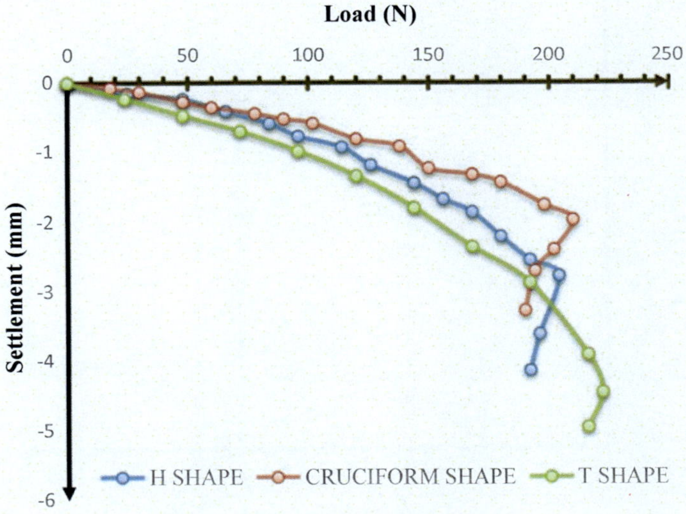


**Fig. 5** Load-settlement curves for H-shaped, cruciform and T-shaped barrette

The percentage increase in the ultimate vertical loading capacity of barrette compared to the conventional circular pile is shown in Fig. 11.

The results show the vertical loading capacities of the rectangular barrettes are higher (0–20%) as compared to the conventional circular pile and the vertical capacities of different shapes of barrette are higher (70–85%) when compared to the circular pile of the same cross-sectional area.

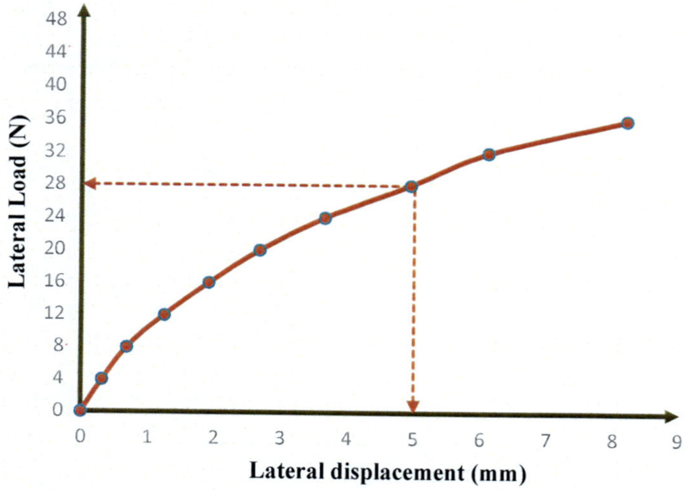


**Fig. 6** Load–displacement curve for circular pile

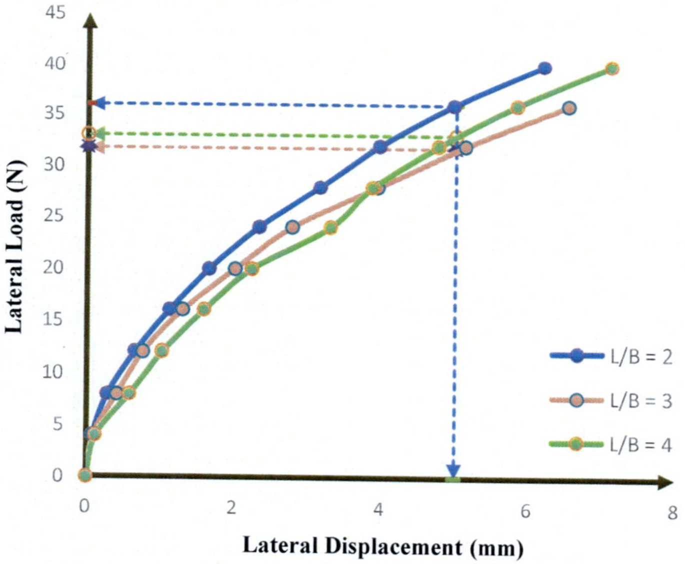


**Fig. 7** Lateral load versus lateral displacement curves for rectangular barrette pile with different aspect ratios (Loading along the minor axis)

The percentage increase in the ultimate lateral load capacities of rectangular barrettes of different aspect ratios, compared to the conventional circular pile is shown in Fig. 12.

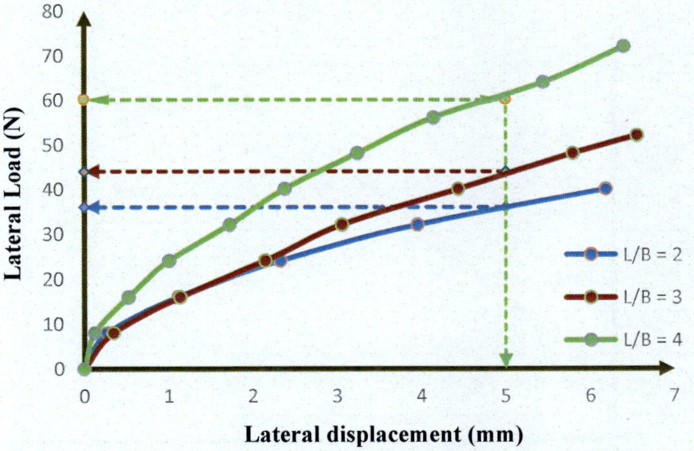


**Fig. 8** Lateral load versus lateral displacement curves for rectangular barrette pile with different aspect ratios (Loading along the major axis)

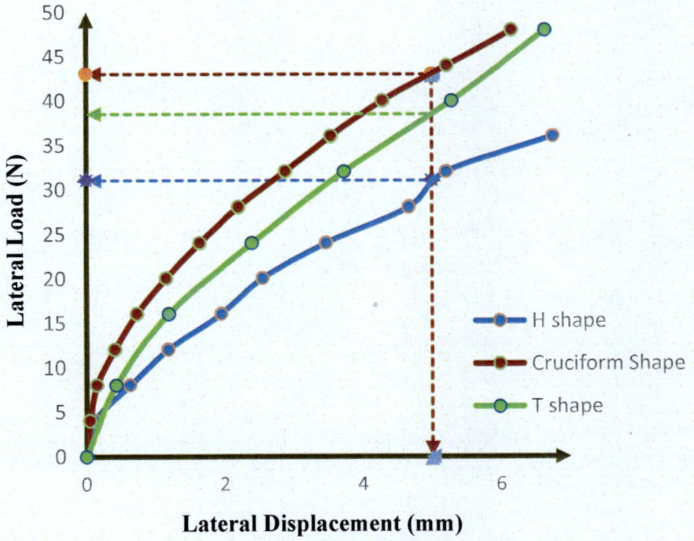


**Fig. 9** Lateral load versus lateral displacement curves for barrette piles of various shapes along Web

Thus, it can be seen that lateral load capacities of Barrette are much higher when the lateral load is applied along the major axis. Also, the lateral load capacity increases with an increase in the aspect ratio.

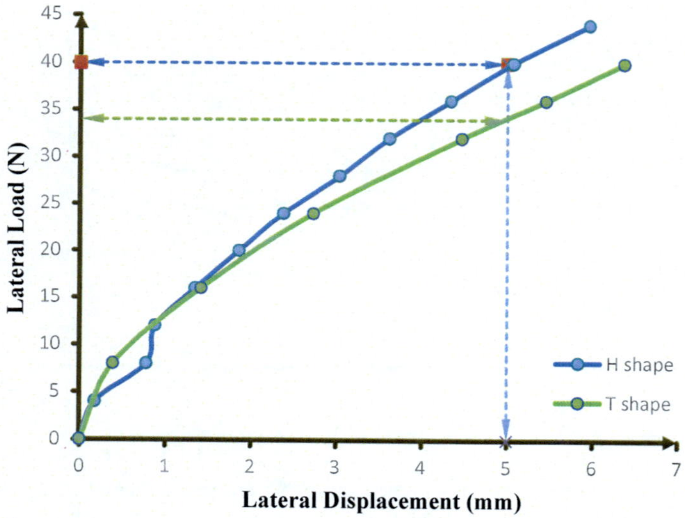


**Fig. 10** Lateral load versus lateral displacement curves for barrette piles of various shapes across the web

**Table 4** Ultimate vertical loading capacity of conventional circular piles

| Shape of pile | Ultimate vertical capacity (N) |
|---|---|
| Circular | 120 |

**Table 5** Ultimate vertical loading capacities of rectangular barrettes of different aspect ratio

| Shape of pile | Ultimate vertical capacity (N) |
|---|---|
| Rectangular ($L/B = 2$) | 120 |
| Rectangular ($L/B = 3$) | 134 |
| Rectangular ($L/B = 4$) | 142 |

**Table 6** Ultimate vertical loading capacities of barrette pile of various shapes

| Shape of pile | Ultimate vertical capacity (N) |
|---|---|
| H-shape | 205 |
| Cruciform | 210 |
| T-shape | 220 |

**Table 7** Ultimate lateral load capacity of conventional circular piles

| Shape of pile | Ultimate bearing capacity (N) |
|---|---|
| Circular | 28 |

**Table 8** Ultimate lateral load capacities of rectangular barrettes of different aspect ratio

| Aspect ratio | Ultimate lateral load capacity (N) | |
|---|---|---|
| | Along minor axis | Along major axis |
| 2 | 36 | 36 |
| 3 | 32 | 44 |
| 4 | 33 | 60 |

**Table 9** Ultimate lateral load capacities of barrette pile of various shapes

| Shape of pile | Ultimate safe lateral capacity (N) | |
|---|---|---|
| | Along web | Across web |
| T-shape | 39 | 34 |
| Cruciform shape | 43 | 43 |
| H-shape | 31 | 40 |

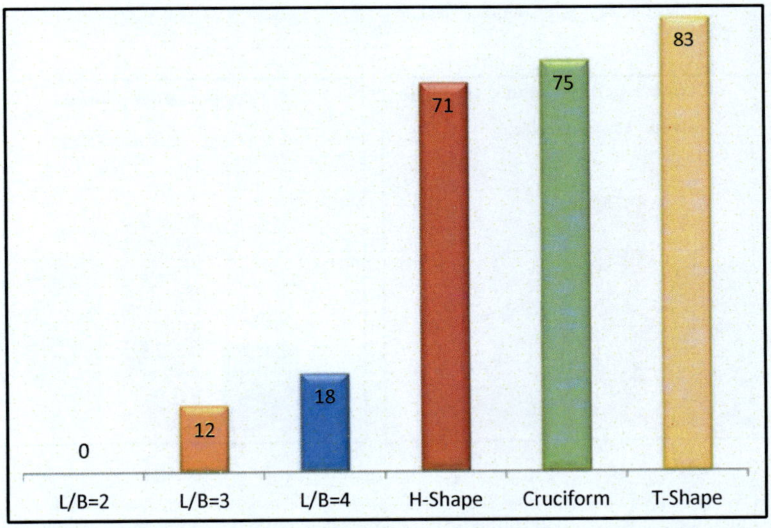


**Fig. 11** Percentage increase in ultimate vertical loading capacity of barrette pile

The percentage increase in the ultimate lateral load capacities of T-shape, Cruciform-shape, and H-shape barrettes, compared to the conventional circular pile is shown in Fig. 13.

From the above results, it is observed that the ultimate lateral load capacities of barrettes of various shapes are higher as compared to that of the circular pile. Also, the lateral load capacities of the cruciform shape barrette are higher as compared to those of barrettes of other shapes.

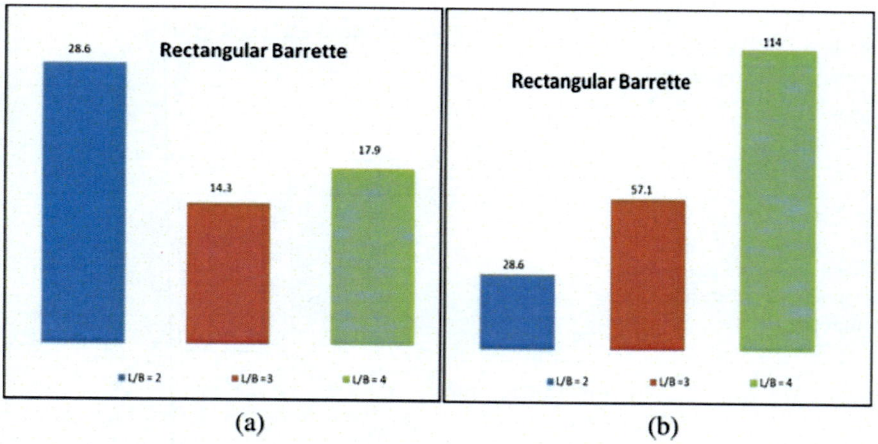


**Fig. 12** Percentage increase in the ultimate lateral load capacities **a** Rectangular barrette pile (minor axis) **b** Rectangular barrette pile (major axis)

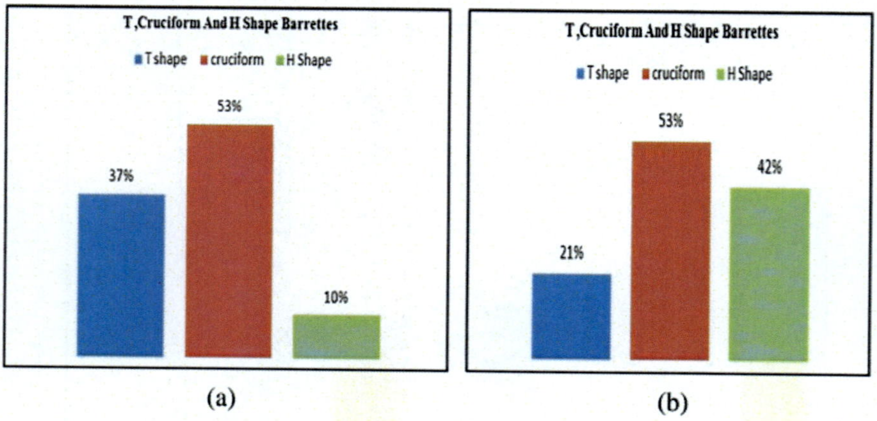


**Fig. 13** Percentage increase in the ultimate lateral load capacities **a** Cruciform-shape and H-shape barrettes (along with web) **b** Cruciform-shape and H-shape barrettes (across the web)

## 5 Conclusions

The capacities of Barrette piles of various shapes in sand and subjected to vertical loads and lateral loads were investigated using laboratory model tests. Based on the results, the effect of various parameters influencing the performance of barrette was investigated. Within the framework of the present investigation, the following broad conclusions are drawn.

1. The ultimate vertical load and lateral load capacities of Barrette Piles are higher as compared to those of circular pile of the same cross-sectional area.

2. With higher surface area, barrettes offer higher skin friction which results in increased ultimate vertical loading.
3. In the case of rectangular barrettes, the ultimate vertical capacity increases with an increase in the aspect ratio of the pile
4. The ultimate vertical load capacity of T-shaped barrette pile is much higher than other shapes of barrette pile and 83% higher than that of the circular pile
5. The ultimate lateral resistance of the rectangular barrette along the major axis is greater than the minor axis.
6. The ultimate lateral resistance of the rectangular barrette increases with an increase in the aspect ratio of the pile when the load is applied along the major axis.
7. The ultimate lateral load capacity of rectangular shape (along major axis) with an aspect ratio of 4 is much higher than other shapes of the barrette pile.
8. The lateral load capacities of the cruciform shape barrette are higher as compared to those of barrettes of other shapes.

## References


1. Wakil, A.Z., Nazir, A.K.: Behaviour of laterally loaded small scale barrettes in sand. Ain Shams Eng. J. Egypt (11 Jan 2013), 343–350 (2013). https://doi.org/10.1016/j.asej.2012.10.011
2. Thakare, S.W., Kumari A., Dhatrak, A.I.: Lateral and uplift capacities of barrette pile in sandy soil. In: Indian Geotechnical Conference, Indian Institute of Science Bengaluru (2018)
3. Submaneewong, C., Teparaksa, W.: Performance of T-shape barrette pile against lateral force in Bangkok. J. Southeast Asian Geotech Soc (Dec 2009), 247–255
4. Salem, D., Ramaswamy, F., ASCE, Pertusier E.M.: Construction of barrettes for high-rise foundations. J. Constr. Eng. Manag. 112(4) (Dec 1996) https://doi.org/10.1061/(ASCE)0733-9364 (1986)
5. Charles, W.W., Rigby, D.B., Ng, S.W.L., Lei G.H.: Field Studies of Well Instrumented Barrette in Hong Kong. J. Geotech. Geoenviron. Eng. ASCE 126:60–73 (2000)
6. Zhang, L.M.: Behavior of laterally loaded large-section barrettes. J. Geotech. Geoenviron. Eng. **129** (7), (July 1, 2003) (ASCE) https://doi.org/10.1061/10900241(2003)1297(639)
7. Dhatrak, A.I., Yaldarkar, P.S., Thakare, S.W.: Analyses of shell footing in layered sandy soil. Indian Geotech Conf (2018, Th02_045)
8. Bhoyar, S.E., Dhatrak, A.I., Thakare, S.W.: Performance of upright triangular shell footing on reinforced sand bed. Int. J. Tech. Innov. Mod. Eng. Sci. **4** (12) (Dec 2018)
9. IS:2911 (Part4)-1985: Indian Standard Code of Practice for Design and Construction of Pile Foundation Part 4 Load Test on Piles. Bureau of Indian Standards (Sept 1985)
10. Dhatrak, A.I., Ghawde, M., Thakare, S.W.: Experimental study on Belled Wedge Pile for different loadings in cohesionless soil. Indian Geotech. Conf. Indian Instit. Sci. Bengaluru 1–7 13 (2018)
11. IS:2911 (Part1/ Sec 2)-2010: Indian Standard Code of Practice for Design and Construction of Pile Foundation Part 1 Concrete piles Section 2 Bored Cast In-situ Concrete Piles. Bureau of Indian Standards (2010)

# Performance of Helical and Square Plate Anchors in Cohesionless Soil

A. I. Dhatrak, L. S. Amrutkar, S. W. Thakare, and R. R. Badnakhe

## 1 Introduction

Anchors are foundation systems that are designed primarily to resist uplift load known as tensile load. Generally, all type of foundation can be designed for uplift load and many are used to resist compressive load, lateral load, or other loading combinations. The anchor is a small-scale version and more traditional foundation types. Soil anchors are foundation systems used to transmit forces from the structure to the ground, in order to resist overturning moments and pull out forces which can threaten a structural stability. Some factors such as shear strength and unit weight of the soil surrounding an anchor can improve the anchor strength.

Now a days, helical anchors are being extensively used in the construction of pipelines, foundation of transmission tower, and braced excavations. The choice of foundation systems has an important role in the design of many structures, to ensure that they support any vertical or horizontal loads. Structures such as tunnels, seawalls, transmission towers, buried pipelines, and retaining walls are subjected to pull out forces and overturning moments. The soil anchor can also be used for tieback resistance in waterfront structures and also against thermal stresses.

Choudhary et al. [1] carried out the experimental study on the load-carrying mechanism of vertical plate anchors. Hanna et al. [2] carried out an experimental investigations on the pullout capacity of inclined shallow strip plate anchors in sand. Mittal et al. [3] conducted experimental investigations on single, double, and triple helical screw anchors under the influence of vertical compressive load. Ghosh et al. [4]

A. I. Dhatrak (✉)
Government College of Engineering, Amravati 444604, India

L. S. Amrutkar · R. R. Badnakhe
Civil Engineering Department, Government College of Engineering, Amravati 444604, India

S. W. Thakare
Department of Civil Engineering, Government College of Engineering, Amravati 444604, India

 
C. N. V. Satyanarayana Reddy et al. (eds.), *Ground Characterization and Foundations*,
Lecture Notes in Civil Engineering 167,
https://doi.org/10.1007/978-981-16-3383-6_64

carried out numerical analysis on ultimate pullout capacity of helical plate anchor in homogeneous soil to study the behavior of helical plate anchor with respect to spacing between helical plates (Sp/D), changing number of helical plates, diameter and embedded depth (H/D) of helical plates. The ultimate pullout capacity for different anchor configurations obtained from the finite element analysis was compared with the values reported by Wang et al. [5]. Their values compared reasonably well with the experimental and numerical results proposed by Wang et al. [5]. They concluded from their study that the uplift capacity of helical anchor increased with increase in Sp/D and H/D ratio.

## 2 Experimental Investigations

In the present investigations, the uplift and lateral capacities of helical and square plate anchor resting in cohesionless soil deposit with different plate configurations were determined experimentally. Different types of anchor model were cast for experimental study, where mainly the number of plates, the depth of upper- and lowermost plates, and the ratio of spacing between the plates to the diameter of the plates were varied, and ultimate uplift capacity and lateral capacity of each anchor were determined.

### 2.1 *Soil*

The sand bed was prepared for experimental work using Kanhan sand (cohesionless, dry and clean) supplied from Bhandara region (Maharashtra). The test sand was angular and of uniform yellow color, with small proportion of black flint stones. The particle size of sand decided for the test was passing through 2 mm IS sieve. Various laboratory tests were performed to determine the physical and index properties of sand (Table 1).

**Table 1** Properties of sand used for experimental investigations

| S. No. | Property of sand | Values |
|---|---|---|
| 1 | Specific gravity (Gs) | 2.67 |
| 2 | IS classification | SP (Medium sand) |
| 3 | $\gamma_{max}$ (kN/m$^3$) | 16.73 |
| 4 | $\gamma_{min}$ (kN/m$^3$) | 15.45 |
| 5 | Angle of internal friction ($\phi$) | 36° |
| 6 | Cohesion (kN/m$^2$) | 0 |

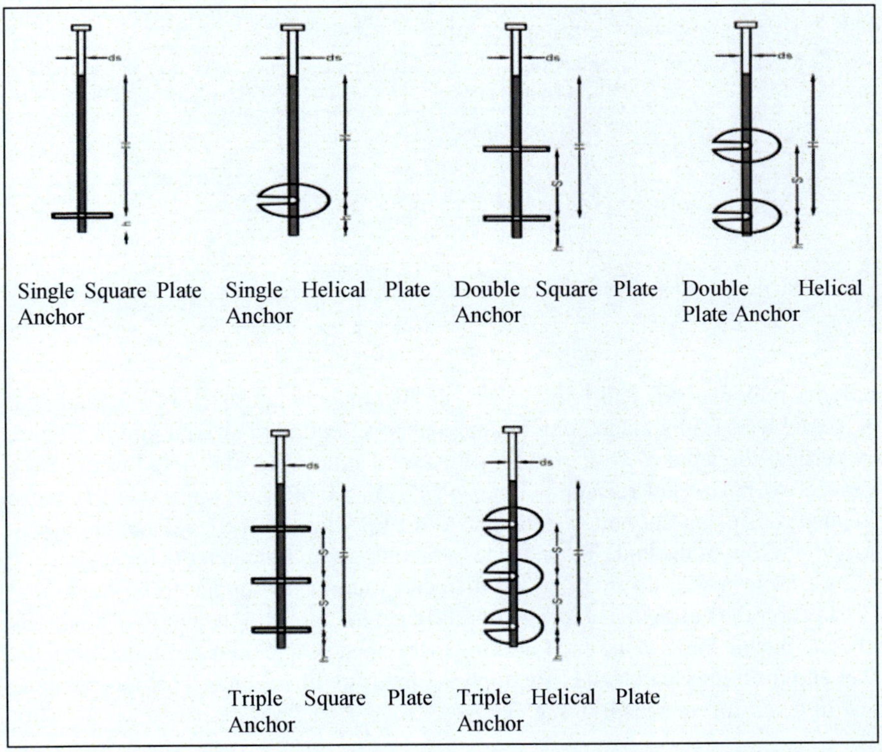


**Fig. 1** Model anchors

## 2.2 *Model Anchors*

Model square and helical plate anchors were made from mild steel. For the comparison purpose, the area of both plate anchors were kept equal, which was 1963 mm2. Accordingly, the size of square plate model anchor was 44.3 mm × 44.3 mm and that of helical plate model anchor was 50 mm dia. Length of both types of model anchor was kept equal to 250 mm. In both types of model anchors, the number of plates was varied as one, two, and three. The model anchors and parameters which were varied for experimental study are listed in Fig. 1 and Table 2, respectively.

The different anchor configurations were considered for the study whose specifications and photographic presentation are given in Table 3.

## 2.3 *Test Set up*

The tests were performed in a tank made of 3 mm thick M.S. sheets having internal dimensions 1000 × 1000 × 1000 mm ($L \times W \times H$). The loading frame used for

**Table 2** Details of parameters varied during experimental study

| Models for each type of anchor | Number of helical/square plates | Embedment ratio (λ = H/D) | Spacing between the plates (Sp/D) |
|---|---|---|---|
| A | 1 | 3.5 | – |
| B | 1 | 4 | – |
| C | 1 | 4.5 | – |
| D | 2 | – | 1 |
| E | 2 | – | 1.5 |
| F | 3 | – | 1.5 |

applying pullout loads and lateral loads on the anchors consisted of a rectangular horizontal base frame made up of rectangular pipe section of size 25 mm × 50 mm. Six vertical threaded rods (25 mm dia.) made of mild steel were fixed to this base frame. Two horizontal rectangular pipes (25 mm × 50 mm) were fixed to these threaded rods using nut bolt system, one at the middle over the tank, and another at the end, in front of the tank. Theses pipes were having an arrangement for supporting pulleys. Pulleys of mild steel having diameter 6 cm were supported by horizontal bars, to carry the non-extensible wire of mild steel for application of pullout load. The wire was having knots at the end for fixing to the anchors and hook arrangement at the other end for fixing load hanger for applying loads to the anchors. The photographic view of test setup as shown in Fig. 2.

## 2.4 Test Procedure

The anchors were placed at the center of the tank vertically in the sand bed and the relative density of soil was maintained at 55% (medium dense) by sand rainfall technique. The relative density achieved was confirmed by collecting samples in small containers placed at different locations in the tank and finding the density of sand at the time of filling. The static uplift load test and lateral load test were conducted on a model anchors to evaluate the uplift and lateral capacity of anchors. In load test, load was applied by adding iron weights in the pan attached to the string. Each load increment was maintained constant until the anchor displacement was stabilized. The anchor displacement was measured with the help of dial gauges. Loading was continued further, in increments, until failure occurred. Photographic representation of preparation of sand bed is shown in Fig. 3.

**Table 3** Details of model plate anchor used in experimental investigation

| S. No. | Description | Shape/cross section | Photographic view |
|---|---|---|---|
| 1 | Single square plate anchor | 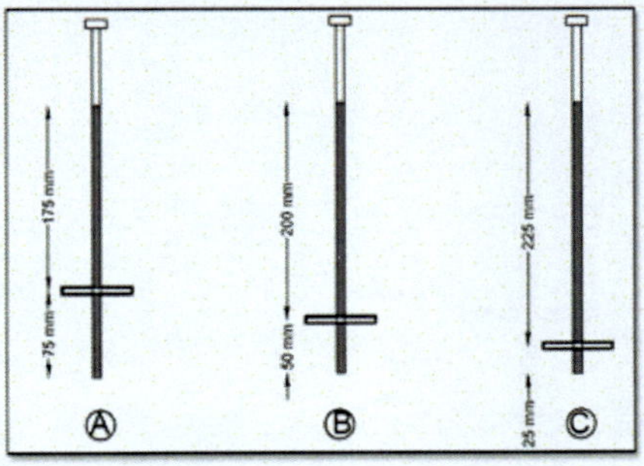  |  |
| 2 | Single helical plate anchor | 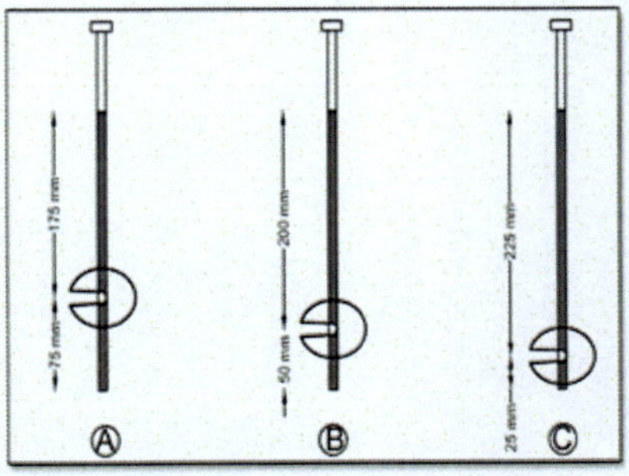  | 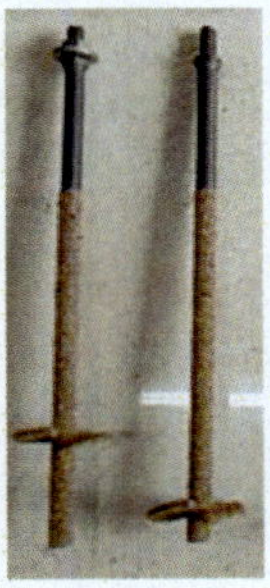 |
| 3 | Double square plate anchor | 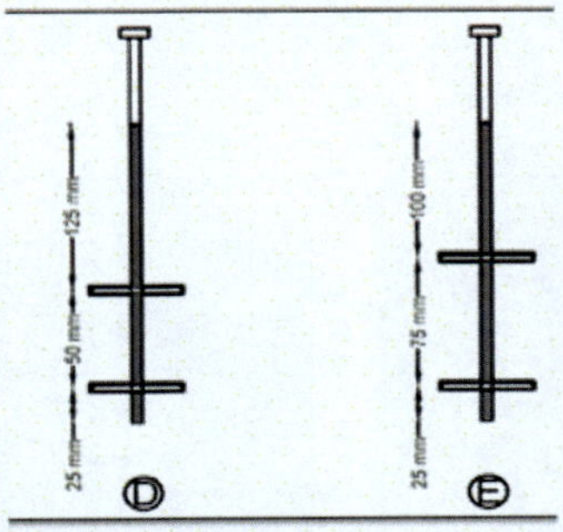  |  |
| 4 | Double helical plate anchor | 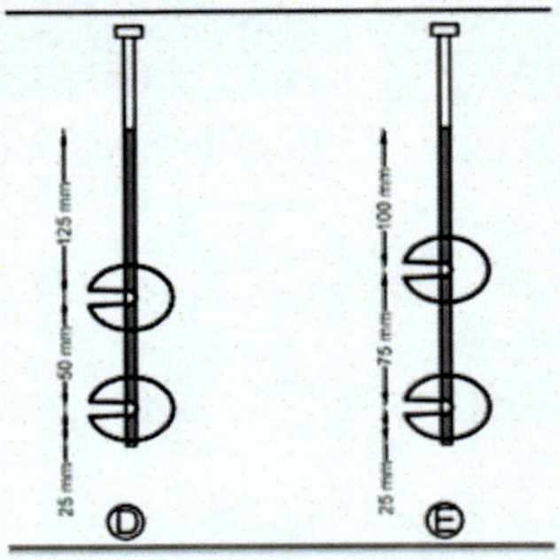  |  |

(continued)

**Table 3** (continued)

| S. No. | Description | Shape/cross section | Photographic view |
|---|---|---|---|
| 5 | Triple square plate anchor | 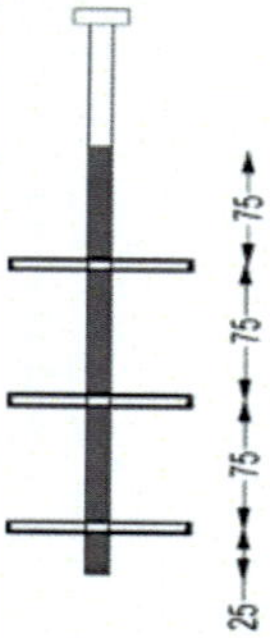  |  |
| 6 | Triple helical plate anchor | 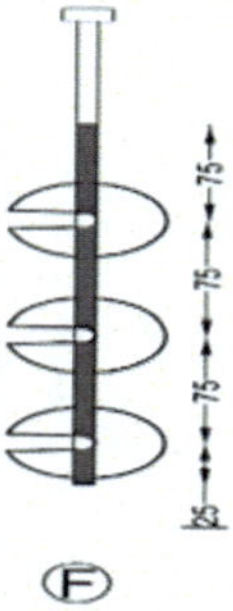  |  |

Experimental Setup used for uplift Loading

Experimental Setup used for Lateral Loading

**Fig. 2** Experimental setup

**Fig. 3** Preparation of sand bed using rainfall techniques

## 3 Results and Discussion

The experimental investigation was conducted on single plate, double plate, and triple plate square and helical anchors under uplift and lateral loading conditions for different H/D ratio viz., 3.5, 4.0, and 4.5 and different Sp/D ratio viz., 1.0 and 1.5. In present study, failure load was considered as the load at which the displacement equal to 5% of the plate diameter occurred. Based on experimental results, the load verses displacement curves for uplift and lateral loading were drawn as shown in Figs. 4, 5, 6, and 7.

From these figures, ultimate uplift and lateral capacities (Pu and PL) for each set of configuration of the square and helical anchors were obtained and given in Tables 4 and 5, respectively.

During experiments, it was observed that in case of helical plate anchors, even when the displacement was large, the shear failure area was typically near the expanded section of anchors and did not reach to the surface, indicating that local shear failure occurred. In case of square plate anchors; however, shear failure area developed up to the surface, indicating that general shear failure occurred.

From the results, it was observed that the square and helical anchors having three plates provided higher uplift capacity as compared to single and double plate anchors, also the uplift and lateral capacity increased with increase in embedment depth ($H/D$), and the spacing between the plates ($S_p/D$).

**Fig. 4** Uplift load–vertical displacement curves for single plate anchors for different embedded depth ratio

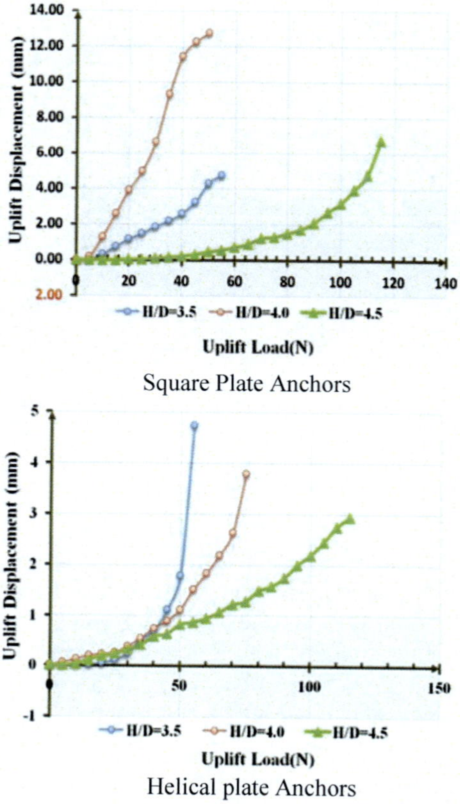


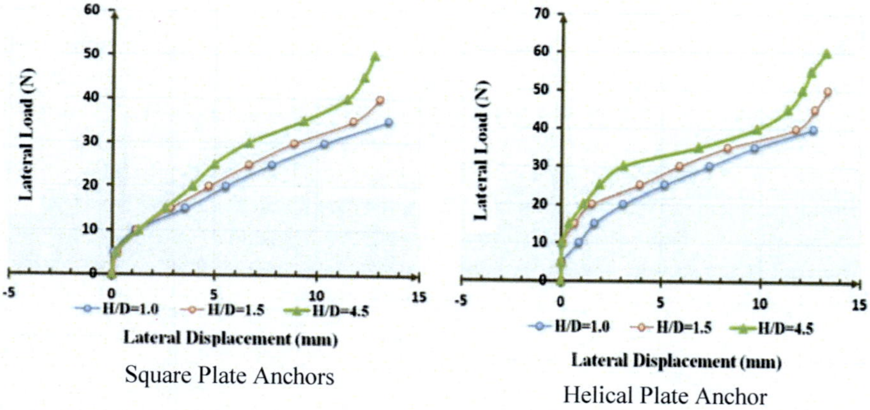


**Fig. 5** Lateral load–displacement curves for single plate anchor for different embedded depth ratio

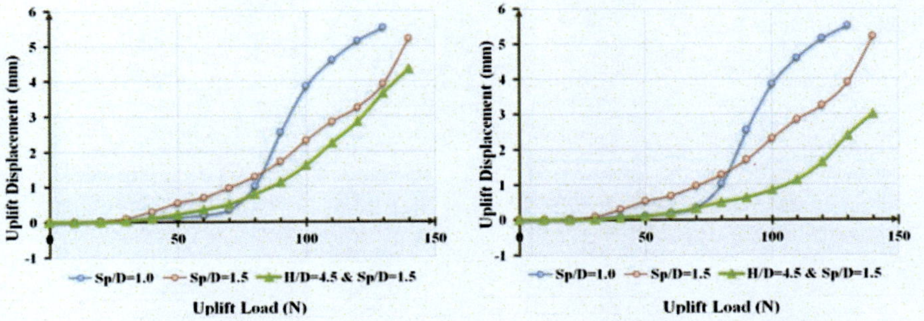


Square Plate Anchors Helical Plate Anchor

**Fig. 6** Uplift load–displacement curves for double and triple plate anchors for different spacing ratio

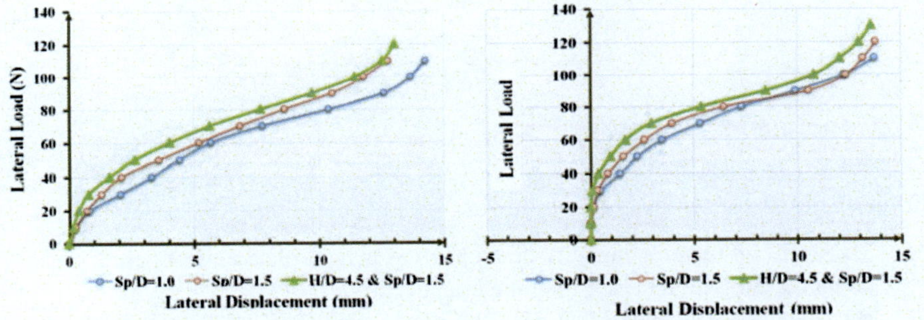


Square Plate Anchors Helical Plate Anchor

**Fig. 7** Lateral load–displacement curves for double and triple plate anchors for different spacing ratio

**Table 4** Ultimate uplift capacity for different anchor configurations

| S. No. | Types of anchors | Configuration | Ultimate uplift capacity (N) |
|---|---|---|---|
| 1 | Square plate anchors | A | 34 |
| 2 | | B | 53 |
| 3 | | C | 92 |
| 4 | | D | 98 |
| 5 | | E | 103 |
| 6 | | F | 112 |
| 7 | Helical plate anchors | A | 50 |
| 8 | | B | 69 |
| 9 | | C | 107 |
| 10 | | D | 114 |
| 11 | | E | 120 |
| 12 | | F | 131 |

**Table 5** Ultimate lateral capacity for different anchor configurations

| S. No. | Types of anchors | Configuration | Ultimate lateral capacity (N) |
|---|---|---|---|
| 1 | Square plate anchors | A | 13 |
| 2 | | B | 14 |
| 3 | | C | 16 |
| 4 | | D | 19 |
| 5 | | E | 22 |
| 6 | | F | 26 |
| 7 | Helical plate anchors | A | 20.5 |
| 8 | | B | 24 |
| 9 | | C | 30 |
| 10 | | D | 29 |
| 11 | | E | 31 |
| 12 | | F | 35 |

## 4 Conclusions

From the results of present study following broad conclusions are drawn:

1. The uplift and lateral capacity of square and helical plate anchors are found to be increased with increase in embedment ratio ($H/D$).
2. By changing the spacing between the plates with ($S_p/D$) ratio to 1.0 and 1.5, there is further increment in the uplift and lateral capacity in case of double and triple plate anchors.
3. Helical anchors are more efficient than square anchors as they are providing more uplift and lateral load-carrying capacity.
4. The ultimate uplift capacity of single plate helical anchor is up to 23% greater than that of single square plate anchor.
5. The ultimate lateral capacity of single plate helical anchor is up to 40% greater than that of single square plate anchor.
6. The ultimate uplift capacity of double and triple plate helical anchor is up to 15% greater than that of double and triple square plate anchor.
7. The ultimate lateral capacity of double and triple plate helical anchors is up to 30% greater than that of double and triple square plate anchor.

## References

1. Choudhary, A., Dash S.: Load-carrying mechanism of vertical plate anchors in sand. Int. J. Geomech. © ASCE. ISSN 1532-3641
2. Hanna, A., Ayadat, T.: Pullout capacity of inclined shallow single anchor plate in sand. Indian Geotech. J. https://doi.org/10.1007/s40098-014-0113-7

3. Mittal, S., Mukherjee, S.: Behaviour of group of helical screw anchors under compressive loads. Department of Civil Engineering, Indian Institute of Technology Roorkee, Springer (2015)
4. Ghosh, P., Samal, S.: Ultimate pullout capacity of isolated helical anchor using finite element analysis. Springer Nature Singapore Pte Ltd, https://doi.org/10.1007/978-981-13-0562-7_26
5. Wang, D., Merifield, R.S., Gaudin, C.: Uplift behaviour of helical anchors in clay. Can. Geotech. J. **50** (6), 575–584. https://doi.org/10.1139/cgj-2012-0350 (2013)

# Performance of the Helical Pile Foundation in Cohesionless Soil

**Anant I. Dhatrak, Himanshu Rajesh Varma, Minal M. Dhage, and Sanjay W. Thakare**

## 1 Introduction

A typical helical pile consists of a one or more pitched helix plates attached to either square or circular pile. Helical piles have been used in construction applications for more than 150 years. The first recorded use of helical piles was in 1836 by Alexander Mitchell when he used helical piles to underpin the Maplin Sands Lighthouse in England. The helical piles are installed into the ground by applying torque to its driving head. Helical piles have been used in various sites to provide high compressive, uplift and lateral capacities for static and dynamic loads. Their current application includes commercial building, solar farms, wind turbines, machine foundation, offshore structures and bridges. The various advantages of helical piles are as follows: ease in installation, can be driven in case of high groundwater table, it provides high compression and uplift capacity, and immediate loading can be applied after driving of the helical pile.

The compression capacity of helical piles in sand and clay was investigated by Elsherbiny and El Naggar [1] by means of field testing and numerical analysis [1]. A full-scaled study was carried out on helical pile to determine the ultimate bearing capacity and interaction between soil and pile. Bearing reduction factor was proposed for the vertical capacity of the helical pile. Sark et al. [2] conducted the full-scale axial compression and tension (uplift) testing on large capacity helical piles installed in

A. I. Dhatrak
Government College of Engineering, Amravati, Maharashtra, India

H. R. Varma (✉) · M. M. Dhage
Civil Department Government College of Engineering, M.Tech Geotechnical Engineering, Amravati, Maharashtra, India

S. W. Thakare
Civil Engineering Department, Government College of Engineering Amravati, Amravati, Maharashtra, India



C. N. V. Satyanarayana Reddy et al. (eds.), *Ground Characterization and Foundations*, Lecture Notes in Civil Engineering 167,
https://doi.org/10.1007/978-981-16-3383-6_65

cohesionless soils. George et al. [3] conducted a detailed investigation on a helical pile installed in cohesionless soil by displacement method. Laboratory experiments and numerical analysis were conducted on models to study the various factors influencing the axial bearing capacities of helical piles. The piles installed by the displacement method exhibited a higher ultimate capacity and distinct failure pattern compared to piles installed by the non-displacement method. Various studies on Helical pile were also done by other authors [7–10].

## 2 Methodology

The behavior of the helical pile was analyzed using finite element software MIDAS GTS NX. The geometry of 3D model of the helical pile and section of the helical pile considered for analysis is shown in Fig. 1. The steel material is used for the hollow circular pile and loose sand for soil stratum. The properties assigned to pile and soil are shown in Tables 1 and 2, respectively.

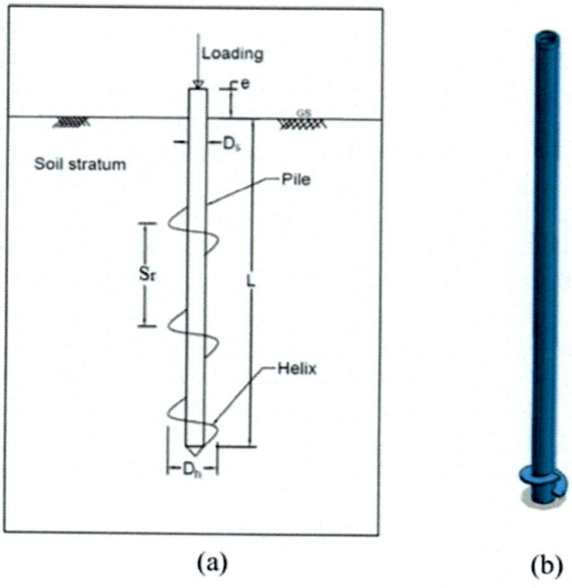


**Fig. 1** Typical helical pile model **a** used for analysis, **b** pile model developed in FEM software

**Table 1** Properties assigned to helical pile for analysis

| S. No. | Properties | Symbol | Values | Units |
|---|---|---|---|---|
| 1 | Young's modulus | $E$ | $2.1 \times 10^8$ | kN/m$^2$ |
| 2 | Density | $\rho$ | 75.5 | kN/m$^3$ |
| 3 | Poisson's ratio | $\upsilon$ | 0.15 | |

**Table 2** Properties assigned to soil layer

| Properties | Unit weight | Young's modulus | Poisson's ratio | Angle of internal friction | Cohesion |
|---|---|---|---|---|---|
| Symbols | $\gamma$ | $E$ | $\upsilon$ | Ø | $c$ |
| Unit | kN/m$^3$ | kPa | – | Degree | kPa |
| Loose sand | 16.12 | 20,133 | 0.3 | 28° | 0 |

## 3 Numerical Analysis

The plan size of the model was more than 15 times the size of the maximum helix blade used in the analysis. The soil below the tip of the pile is sufficiently deep such that its effect was minimum. The sides of soil model were restrained in the $x$-direction and $y$-directions, and the bottom boundary was restrained in all directions.

Soil was modeled using 10-node tetrahedral elements with a standard Mohr–Coulomb constitutive relationship. The pile shaft and the helix were modeled as linear isotropic elastic material. An interface element was created around the shaft and the helix plate to account the interaction between the soil and the helical pile. Mesh convergence criteria were used for selecting the mesh size for soil, pile shaft and helix blade. Figure 2 shows the helical pile embedded in the soil layer after mesh generation. The nonlinear analysis was carried out after mesh generation considering different pile geometry such as diameter of helix, inter-helix spacing, number of helix

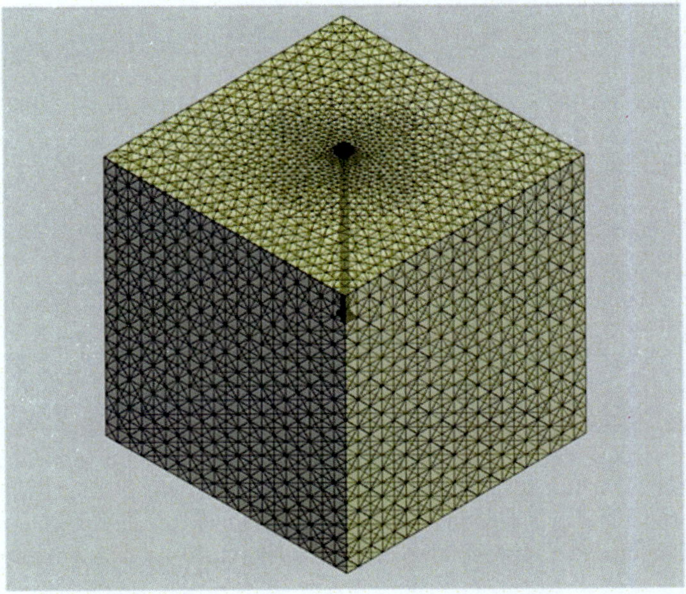

**Fig. 2** Model of the helical pile after mesh generation

**Table 3** Constant parameter used for numerical analysis

| S. No. | Type | Parameter | Values (m) |
|---|---|---|---|
| 1 | Dimensions for pile | Length | 11.75 |
| | | Diameter of pile | 0.5 |
| 2 | Soil | Loose | |

**Table 4** Varying parameters used for numerical analysis

| S. No. | Parameter | Details of parameters |
|---|---|---|
| 1 | Type of loading | i. Axial loading<br>ii. Uplift loading<br>iii. Lateral loading |
| 2 | Helix diameter ($D_h$) | $1.5D_s$, $2D_s$, $2.5D_s$ |
| 3 | Number of helix | 1, 2 and 3 |
| 4 | Inter-helix spacing ($S_r$) | $2D_h$, $2.5D_h$, $3D_h$ and $3.5D_h$ |
| 5 | Inter helical pile spacing ($S_p$) | $1.5D_h$, $2D_h$, $2.5D_h$, $3D_h$ and $3.5D_h$ |
| 6 | Number of piles | Single pile<br>Pile group with 3, 4 and 6 piles |

plate and number of piles in group. The different constant and varying parameters used for the analysis are shown in Tables 3 and 4, respectively.

where

- $D_h$ Diameter of helix.
- $D_s$ Diameter of shaft.
- $S_r$ Spacing between helix plate.
- $S_p$ Inter helical pile spacing.

# 4 Results and Discussion

## *4.1 Effect of Helix Diameter Ratio*

Initially, the analysis was carried out for conventional pile and single helical pile subjected to vertical, uplift and lateral loading in loose cohesionless soil with different helix diameter. A failure settlement criterion of 5% of the helix diameter was considered for vertical and uplift loading, whereas 12 mm displacement at pile head was considered for lateral loading condition. For conventional pile, a settlement criterion of 10% of shaft diameter was considered. The ultimate capacities of conventional and single helical pile, considering different helix diameters as determined from the load settlement curves, are presented in Table 5.

**Table 5** Ultimate capacities of helical pile with different helix diameter

| Loading condition | Ultimate capacities of pile (kN) | | | |
|---|---|---|---|---|
| | Conventional pile | Helix diameter ratio $D_h/D_s$ | | |
| | $D_h/D_s = 1$ | 1.5 | 2 | 2.5 |
| Vertical | 963 | 1147 | 1703 | 2485 |
| Uplift | 307 | 915 | 1468 | 2150 |
| Lateral | 200 | 265 | 270 | 220 |

The variation in ultimate vertical, uplift and lateral capacity of the helical pile with different helix diameter ratio ($D_h/D_s$) is shown in Fig. 3 (a), (b) and (c), respectively. Figure 3a and b shows that the ultimate vertical and uplift capacities of the helical pile are much higher than that of conventional pile and increase with the increase in helix diameter. But in the case of lateral loading condition, the ultimate lateral capacity increases up to helix diameter ratio ($D_h/D_s$) = 2 and decreases with further

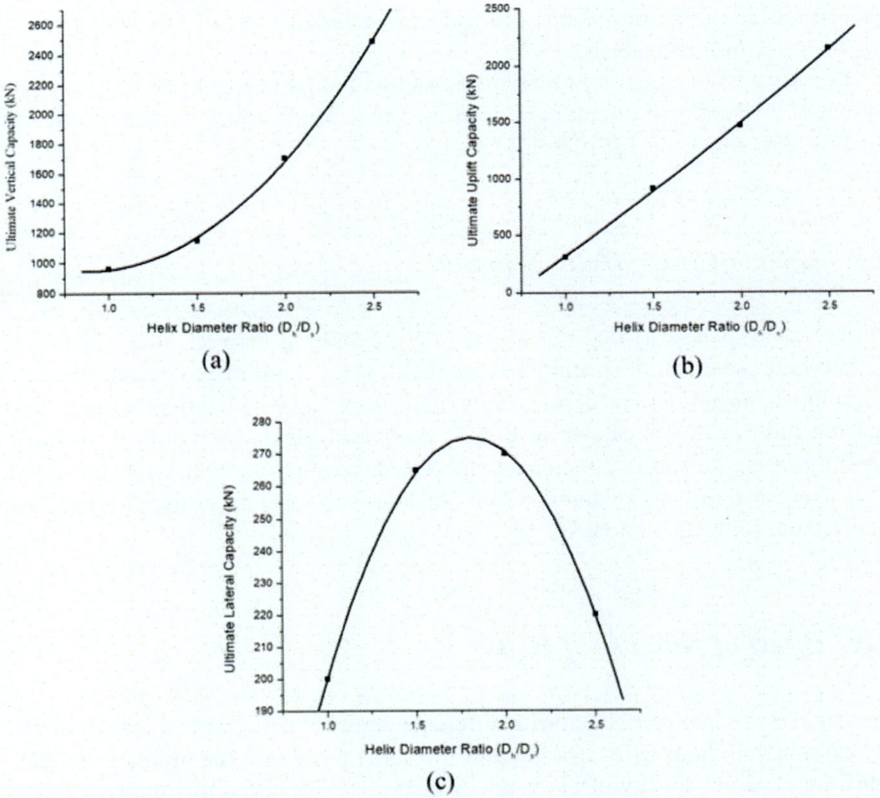


**Fig. 3** Variation in ultimate capacity with respect to the helix diamater ratio ($D_h/D_s$) (**a**) vertical, (**b**) uplift and (**c**) lateral loading condition

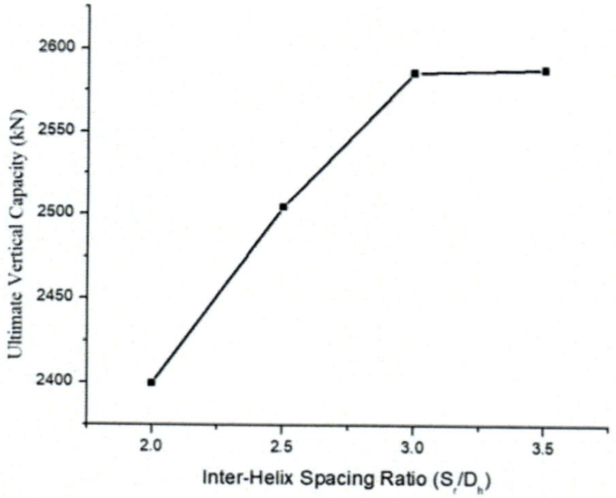


**Fig. 4** Variation in the ultimate vertical capacity with different inter-helix spacing

increase in helix diameter ratio. Helix diameter greater than $2D_s$ induces instability and assists the lateral movement of the pile. So, helix diameter of $2D_s$ is adopted as optimum for further analysis.

The percentage increase in ultimate capacity of single helical pile as compared to that of conventional circular pile is in the range of 20–160% in case of vertical loading, 200–600% in case of uplift loading.

## 4.2 *Effect of Inter-Helix Spacing*

Further analysis was carried out by varying the spacing between two helices for determining the effect of spacing between helix plate on ultimate vertical capacity. Variation in the ultimate vertical capacity of the helical pile with different inter-helix spacing ratio ($S_r/D_h$) is shown in Fig. 4. It is observed that the ultimate vertical capacity of the helical pile increases with the increasing inter-helix spacing up to ($S_r/D_h$) = 3 and further remains constant. So, inter-helix spacing ratio = 3 is adopted as optimum for further analysis.

## 4.3 *Effect of Number of Helix*

Further analysis was carried out with different number of helix plate in case of single helical pile with inter-helix spacing ratio = 3. Table 6 shows the ultimate vertical, uplift and lateral capacity of the single helical pile with different number of helix plates viz., 1, 2 and 3 helix. From the results, it is observed that the ultimate vertical and uplift capacity of the pile increase as the number of helix plate increases and the

**Table 6** Ultimate capacity with different number of helix plate

| Loading condition | Ultimate capacities of pile (kN) | | | |
|---|---|---|---|---|
| | Number of helix plate | | | |
| | 0 | 1 | 2 | 3 |
| Vertical | 963 | 1703 | 2580 | 3526 |
| Uplift | 307 | 1468 | 2364 | 3086 |
| Lateral | 200 | 270 | 212 | 273 |

ultimate lateral capacity is much higher with 3 helix plates. The percentage increase in lateral capacity of single helical pile is significant only in case of pile with three helices.

As the number of helices increases, about 95–100% increment in ultimate vertical capacity, 300–350% increment in uplift capacity and about 35% increment in ultimate lateral capacity is observed.

## 4.4 *Effect of Inter Helical Pile Spacing ($S_p/D_h$):*

The analysis was carried out on a group of three helical piles. The variation in the ultimate vertical capacity of the helical pile with different inter helical pile spacing ratio is shown in Fig. 5.

From the above results, it is observed that the ultimate vertical capacity of the group of helical pile increases with the increasing inter helical pile spacing ratio up to $S_p/D_h = 3$ and remains constant with further increase in spacing ratio.

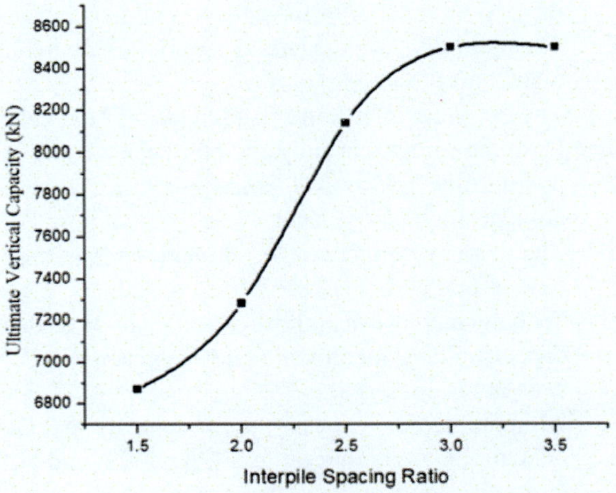


**Fig. 5** Variation in ultimate vertical capacity with respect to the inter helical pile spacing ratio

**Table 7** Ultimate capacity of piles in group

| Loading condition | Ultimate capacities of pile (kN) | | | | | |
|---|---|---|---|---|---|---|
| | Number of conventional piles in group | | | Number of helical piles in group | | |
| | 3 | 4 | 6 | 3 | 4 | 6 |
| Vertical | 2630 | 3516 | 5304 | 8490 | 10,383 | 13,520 |
| Uplift | 904 | 1330 | 1806 | 7365 | 8980 | 12,237 |
| Lateral | 459 | 1330 | 847 | 722 | 8980 | 1323 |

## *4.5 Performance of Group of Helical Pile*

The analysis was carried on pile groups consisting of conventional as well as the helical piles subjected to different types of load. Number of piles in the group was varied as 3, 4 and 6. For the helical pile group analysis, three helix plate configurations were adopted. Table 7 presents the ultimate capacities obtained from the analysis of group of pile subjected to vertical, uplift and lateral loading condition.

Percentage increase in ultimate vertical capacity of helical pile group as compared to that of conventional circular pile group is in the range of 150–200%, whereas 500% in case of ultimate uplift capacity and about 50% in case of ultimate lateral capacity.

# 5 Conclusions

Following broad conclusions are drawn from the analysis of helical piles.

1. The ultimate vertical and uplift capacity of the helical pile increases with increase in number of the helices and diameter of the helix.
2. The optimum ultimate lateral capacity of the helical pile is achieved at the helix diameter ratio = 2.
3. The optimum inter-helix spacing ratio of the helical pile is equal to 3.
4. Ultimate vertical capacity and uplift capacity of the single helical pile are higher up to 160% and 600%, respectively, as compared to those of conventional circular pile in loose sand.
5. The percentage increase in ultimate lateral capacity is significant only in the case of pile with 3 helices.
6. With the increase in number of helix, about 100% increment in ultimate vertical capacity, 350% increment in uplift capacity and about 35% increment in lateral capacity is observed.
7. Ultimate vertical capacity, uplift capacity and lateral capacity of the group of helical pile are higher up to 200%, 500% and 50%, respectively, as compared to group of conventional circular pile.

## References

1. Elsherbiny, Z.H., El Naggar, M.H.: Axial compressive capacity of helical piles from field tests and numerical study. Can. Geotech. J. **50**, 1191–1203. https://doi.org/10.1139/cgj-2012-0487 (2015)
2. Sark, M.: Installation and Performance Characteristics of High Capacity Helical Piles in Cohesionless Soils. J. Deep Found. Inst **5** (1). https://doi.org/10.1179/dfi.2011.004 (2011)
3. George, B.E., Banerjee, S., Gandhi, S.R.: Helical piles installed in cohesionless soil by displacement method. Int. J. Geomech. © ASCE (2019). https://doi.org/10.1061/(ASCE)GM.1943-5622.0001457 (2017)
4. IS:2911 (Part-IV) 1985: Indian Standard Code of Practice for Design and Construction of Pile Foundation. Bureau of Indian Standards. New Delhi, India (1985)
5. ASTM D 1143—07: Standard Test Methods for Deep Foundations under Static Axial Compressive Load, ASTM. Annual Book of ASTM Standards (2007)
6. ASTM D 3689: Standard Test Methods for Deep Foundations under Static Axial Tensile Load. ASTM, Annual Book of ASTM Standards (2007)
7. Darin Willis: How to Design Helical Piles per the 2009 International Building Code. Ram Jacked Systems Distributions, LCC
8. Fahmy, A., El Naggar, M.K.: Axial performance of helical tapered piles in sand. Geotech. Geol. Eng. **35** (4), 1–28. https://doi.org/10.1007/s10706-017-0192-1 (2007)
9. Ghaly, A., Hanna, A.: Response of anchors to variations in displacement-based loading. Can. Geotech. J. **40** (3), 694–701. https://doi.org/10.1139/t03-003 (2003)
10. Ghaly, A., Hanna, A., Hanna, M.: Installation torque of screw anchors in dry sand. Soils Found. **31** (2), 77–92. https://doi.org/10.3208/sandf1972.31.2_77 (1991)
11. Dhatrak, A.I., Maithily Ghawde and Thakare, S.W.: Experimental study on Belled Wedge Pile for different loadings in cohesionless soil. In: Indian Geotechnical Conference, Indian Institute of Science Bengaluru, pp. 1–7 (2018)
12. Tale, N.G., Dhatrak, A.I., Thakare, S.W.: Numerical analysis of spin fin pile under different loading conditions. Int. J. Tech. Innov. Mod. Eng. Sci. (IJTIMES) **5** (05), 659–664. e-ISSN: 2455-2585 (2019)
13. Thakare, S.W., Wankhade, P.P, Dhatrak, A.I.: Experimental investigations on performance of spin fin pile under different loading modes. Int. J. Tech. Innov. Mod. Eng. Sci. (IJTIMES) **5** (05), 501–506. e-ISSN: 2455-2585 (2019)

# Analysis of Pile Group and Piled Raft as a Foundation System

**Shreyansh Kumar Golchha, Jay Kumar Shukla, and Nitin H. Joshi**

## 1 Introduction

In a traditional foundation design, a shallow foundation is used to carry the structural load. Nowadays in most high-rise buildings, the combination of both piles and raft are used since only raft cannot satisfy design requirements. So, by addition of piles with raft load carrying capacity also increases and also reduces settlement. Piled Raft foundation have been considered as one of the most economical foundation systems and are widely used to support high-rise structures. The analysis of a piled raft foundation is a complex process as there are many parameters which influences the behavior of the foundation. As there is very little literature available to analyze the exact behavior of piled raft foundation. An optimized design may be defined as a design achieving maximum economy of the solution, that is minimum cost for the installation of the foundation, while maintaining satisfactory performance [1]. The traditional design practice for pile foundations is based on the assumption that the piles are free-standing, and that the whole external load is carried by the piles, in which any footing contribution being ignored. This approach is over conservative, since the raft or pile cap is actually in direct contact with the soil, and thus carries a significant proportion of the load [2]. Pile is mostly installed in groups, it can be divided into two main types based on the connectivity of the pile cap with the underlying soil i.e. pile group and piled raft. In pile group, pile cap is not in contact with the underlying soil whereas in piled raft, the raft is in contact with the underlying soil [3]. For both types of pile groups, it is expected that piles interact and influence each other's capacity. In case of raft in contact with underlying soil,

S. K. Golchha (✉) · N. H. Joshi
Applied Mechanics Department, MSU, Vadodara, Gujarat 390002, India

J. K. Shukla
GEO DYNAMICS, Vadodara, Gujarat 390023, India

 
C. N. V. Satyanarayana Reddy et al. (eds.), *Ground Characterization and Foundations*, Lecture Notes in Civil Engineering 167,
https://doi.org/10.1007/978-981-16-3383-6_66

the interaction between the raft-soil and the pile group depends upon several factors like the geometry of the raft, type of soil, pile length, and spacing between piles [4].

## 2 Model Validation

In this research work, parametric study has been carried out using the software FLAC3D. Here, analysis is based on the finite difference method. Model verification was carried out for the model shown in Figs. 1 and 2. Here, there are total nine identical piles with the properties as shown in Fig. 1. Also, the raft geometry and

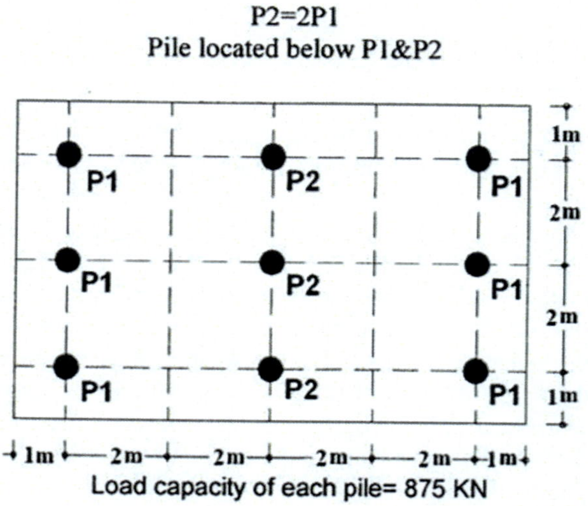


**Fig. 1** Pile and load configuration for model validation [5]

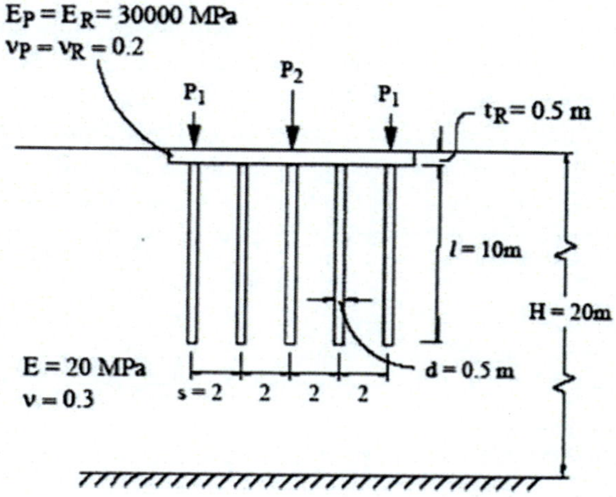


**Fig. 2** Model condition and material properties for model validation [5]

properties are shown in Figs. 1 and 2. At the center piles load applied is double that of the load applied at the side piles. After the analysis, the results obtained are shown in Figs. 3, 4 and 5. Load-settlement relationship is obtained of the model shown in Figs. 2 and 3 with the help of FLAC3D software. Variation of maximum settlement,

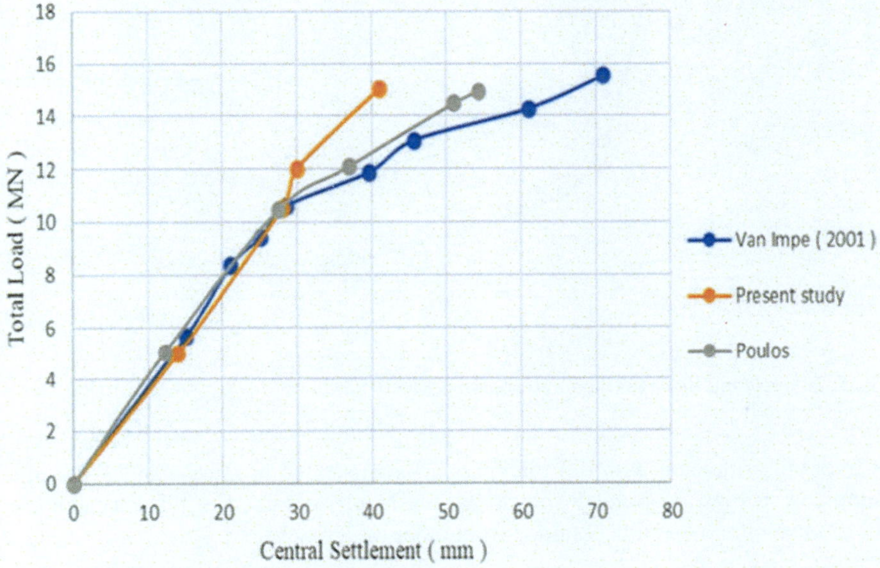


**Fig. 3** Load-settlement analysis for model validation

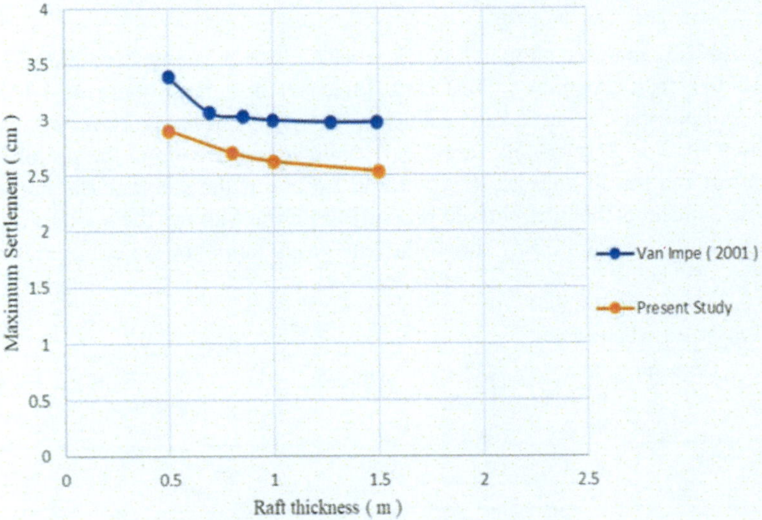


**Fig. 4** Effect of raft thickness on maximum settlement (Comparison with other methods)

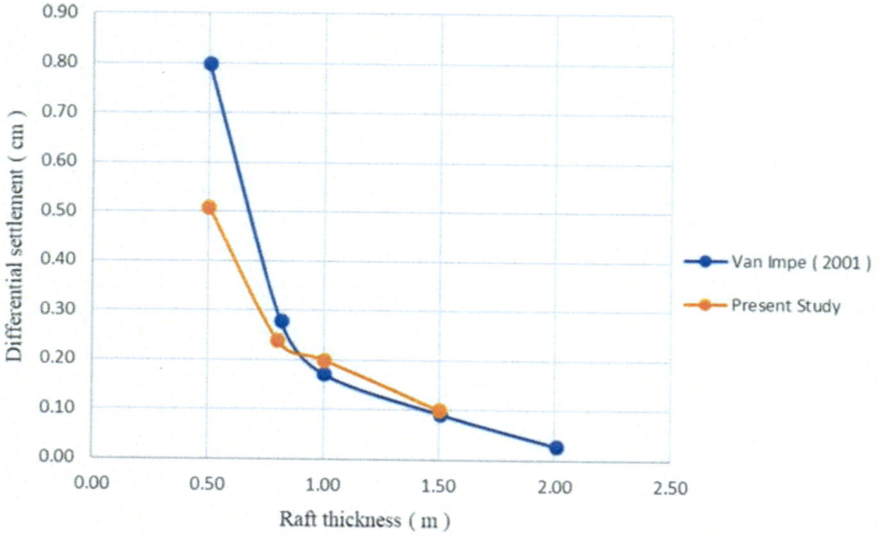


**Fig. 5** Effect of raft thickness on differential settlement

differential settlement with raft thickness are represented in Figs. 4 and 5. Results computed in Figs. 3, 4 and 5 shows that the results obtained in this research are acceptable with the results obtained from other researchers.

## 3 Geometry Model and Material Properties

In Pile cap/Raft material property is taken as $\mu = 0.2$, Young Modulus (kPa) = 29,580,000, Density (kg/m$^3$) = 2500. Here, Piled raft top surface, there are 4 different thickness of pile cap/raft are taken i.e. 800, 1000, 1200, 1500 mm. The configuration for 4 piles are 2 × 2, for 6 piles are 2 × 3, for 10 piles are along the periphery of pile cap/raft and for 20 piles are 4 × 5. Table 1 presents the soil material properties and Table 2 presents the pile material properties. Figure 6 shows the piled raft model (4 x 5 configuration) and Fig. 7 shows the Pile group model (4 x 5 configuration).

**Table 1** Soil material properties

| Soil | Depth, m | Young's modulus, kPa | Cohesion kN/m$^2$ | Friction (°) | $\mu$ | Submerged density, kg/m$^3$ | Dilation |
|---|---|---|---|---|---|---|---|
| Layer-1, top | 20 | 12,300 | 6 | 30 | 0.334 | 840 | 0 |
| Layer-2, middle | 30 | 16,800 | 5 | 30 | 0.334 | 830 | 0 |

**Table 2** Pile material properties

| Pile property | Diameter (m) | Length (m) | $\mu$ | Young modulus (kPa) | Area ($m^2$) | Perimeter (m) | Density ($kg/m^3$) |
|---|---|---|---|---|---|---|---|
| Values | 0.8 | 35 | 0.2 | 29,580,000 | 0.5024 | 2.51 | 2500 |

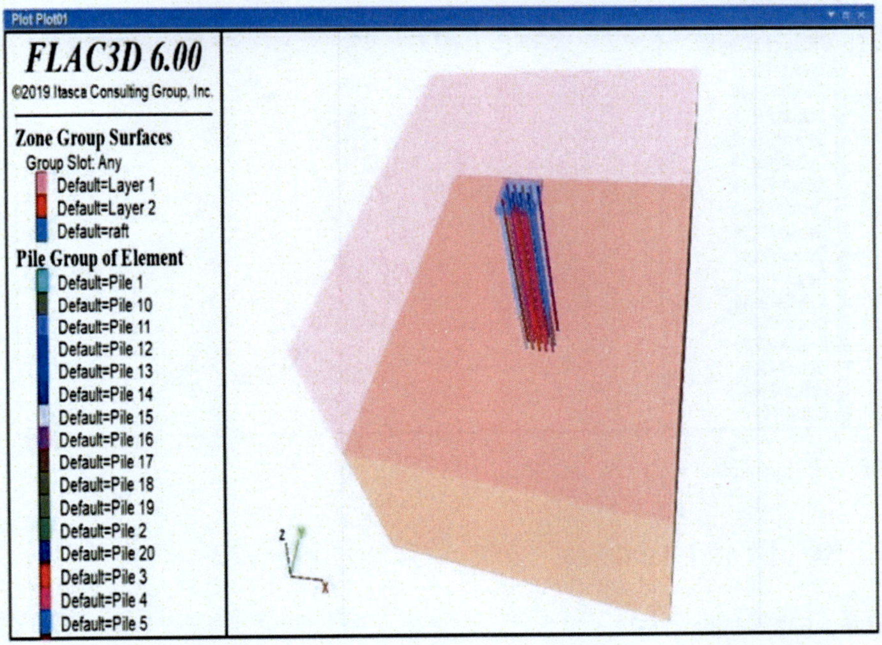


**Fig. 6** Piled raft model (4 x 5 configuration)

## *3.1 Method of Analysis*

Here, the analysis is done by the FLAC 3D software which is based on finite difference method. To carry out numerical modeling and simulation includes grid generation, assigning material properties, specifying models, boundaries and initial conditions, stepping to equilibrium and interpretation of results. There are many ways to create zones such as brick, cylinder-shell, cylinder, pyramid, wedge, etc. in FLAC3D. In this the zone is of brick type. Soil block is grouped into two layers i.e. 'Layer1', 'Layer2'. Here, Mohr-Coulomb model is applied to the soil.

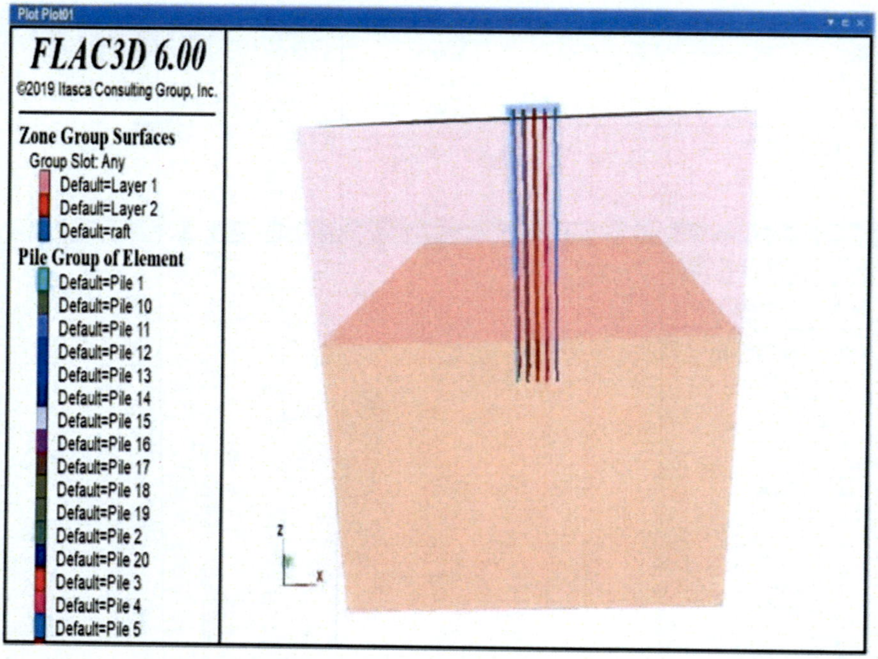


**Fig. 7** Pile group model (4 x 5 configuration)

## 4 Results and Analysis

Abbreviation used

| | |
|---|---|
| PG | Pile Group |
| PR | Piled Raft |
| pl | point load at the center of the pile cap/raft |
| udl | uniformly distributed load on the pile cap/raft |
| PT | Pile cap Thickness |
| RT | Raft Thickness. |

Notation used in the graph

- Pile Group
  _____: point load at the center of the pile cap/raft.
- Piled Raft
  _ _ _ _ _: uniformly distributed load on the pile cap/raft.

## 4.1 Pile Cap/Raft Thickness Variation of 4 Pile Configuration

Point load applied at the center of the pile cap/raft = 8000 KN Uniformly distributed load applied on the pile cap/raft = 500 KPa Pile diameter = 800 mm.

Area of pile cap/raft = 16 m$^2$.

Here, in Fig. 8b it shows that as the pile cap/raft thickness increases the differential settlement decreases. By comparing pile group and piled raft, it shows that differential settlement is less in case of piled raft compared to pile group. Also, in case of point load differential settlement is more compared to uniformly distributed load. There is a negligible differential settlement for pile cap/raft thickness of 1200 and 1500 mm. From Fig. 8c it is concluded that as pile cap/raft thickness increases, the maximum bending moment of the pile increases. The increase in the maximum bending moment

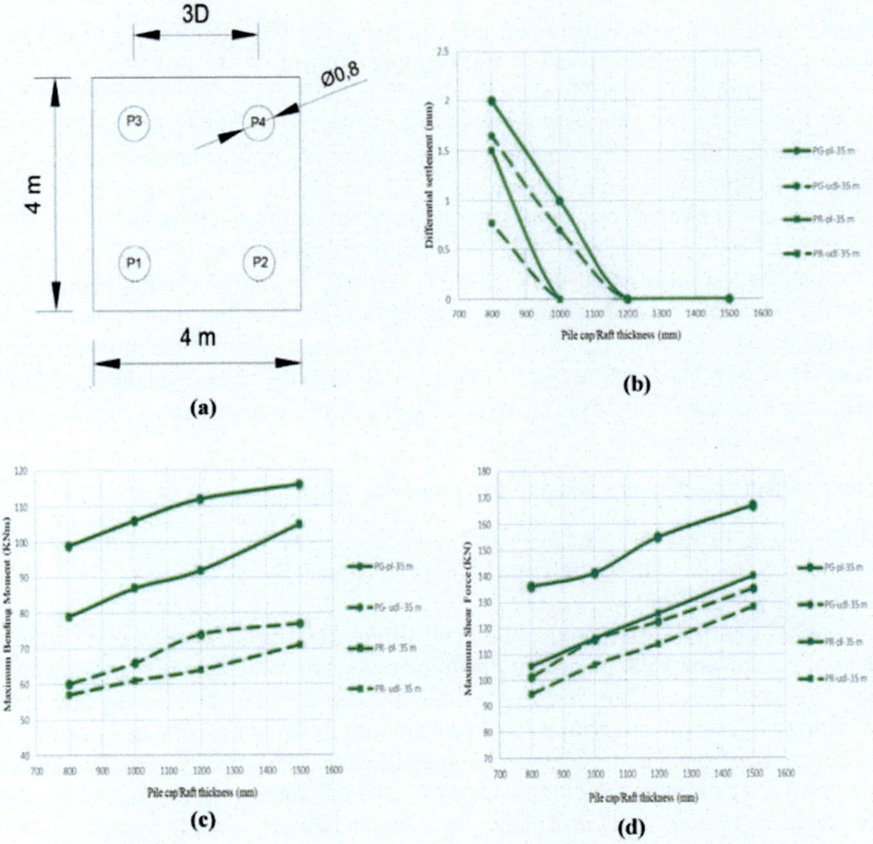


**Fig. 8** **a** Top view of 4 pile configuration of both pile group and piled raft, **b** Effect of pile cap/raft thickness on differential settlement (for 4 pile configuration), **c** Effect of pile cap/raft thickness on maximum bending moment (for 4 pile configuration), **d** Effect of pile cap/raft thickness on maximum shear force (for 4 pile configuration)

is more in point load application as compared to uniformly distributed load. Also, in case of pile group it shows more maximum bending moment of the pile in comparison to piled raft. The variations for the point load is more compared to uniformly distributed load as pile cap/raft thickness increases. From Fig. 8d shows that as pile cap/raft thickness increases maximum shear force of the pile also increases. Here, the maximum value is up to 167 KN which is for pile group of 35 m pile length and pile cap/raft thickness 1500 mm subjected to point load at the center of the pile group.

### 4.2 Pile Cap/Raft Thickness Variation for 6 Pile Configuration

Point load applied at the center of the pile cap/raft = 12,000 kN Uniformly distributed load applied on the pile cap/raft = 500 kPa Pile diameter = 800 mm.

Area of pile cap/raft = 24 $m^2$.

In Fig. 9b, as the pile cap/raft thickness increases the difference in differential settlement decreases for point load and uniformly distributed load. Here, the maximum differential settlement occurs is 3.8 mm for the pile group of pile length 35 m and pile cap/raft thickness 800 mm subjected to point load. Figure 9c shows that maximum bending moment of the pile is nearly consistent for pile group and piled raft subjected to uniformly distributed load. Also, the maximum bending moment of pile for pile group is more compared to piled raft. For pile group and piled raft of pile length 35 m subjected to point load the increase in maximum bending moment is more. From Fig. 9d it is seen that as Pile cap/Raft thickness increases the maximum shear force increases slightly. The maximum shear force is more in case of pile group compared to the piled raft.

**Pile cap/Raft thickness variation for 10 pile configuration**

Point load applied at the center of the pile cap/raft = 18,000 KN Uniformly distributed load applied on the pile cap/raft = 375 kPa Pile diameter = 800 mm.

Area of pile cap/raft = 48 $m^2$.

In Fig. 10b, the differential settlement difference is less between pile cap/raft thickness 1200 and 1500 mm. Here, the maximum value of differential settlement is 5.9 mm of pile group of pile length 35 m and pile cap thickness 800 mm subjected to point load. There is a steep decrease in differential settlement as pile cap/raft thickness increases from 800 to 1200 mm. Figure 10c shows that as pile cap/raft thickness increases the maximum bending moment increases slightly for pile group and piled raft subjected to uniformly distributed load. The maximum bending moment difference between pile group and piled raft for pile length 35 m subjected to point load varies in a non-uniform way as pile cap/raft increases. Here, the range of maximum bending moment for pile group and piled raft subjected to point load varies between 991.2 and 830.4 KN whereas for pile group and piled raft subjected to uniformly

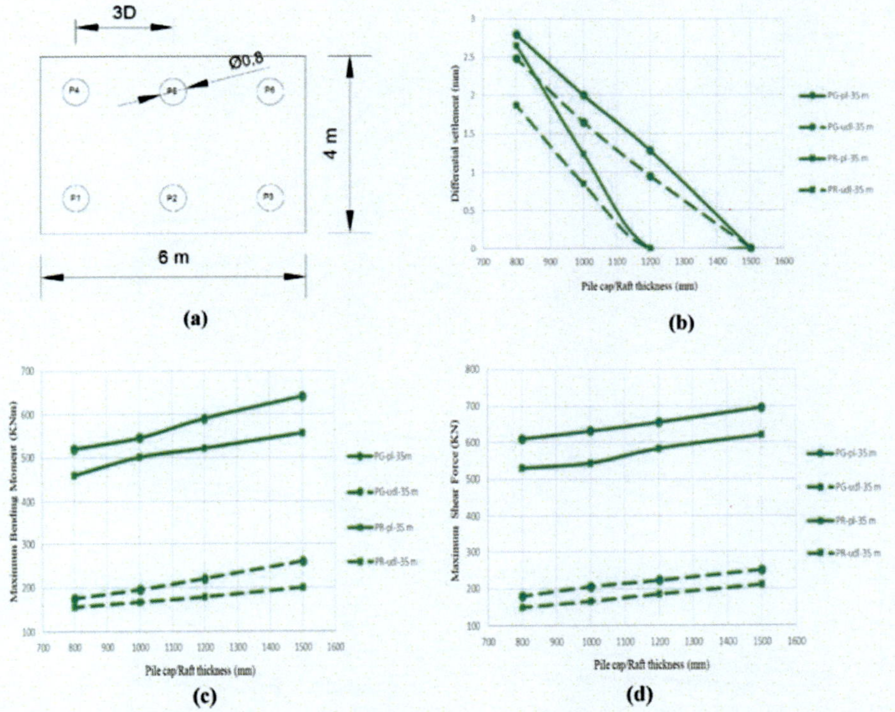


**Fig. 9** **a** Top view of 6 pile configuration of both pile group and piled raft, **b** Effect of pile cap/raft thickness on differential settlement (for 6 pile configuration), **c** Effect of pile cap/raft thickness on maximum bending moment (for 6 pile configuration), **d** Effect of pile cap/raft thickness on maximum shear force (for 6 pile configuration)

distributed load varies from 385.2 and 266 KN. In Fig. 10d, as pile cap/raft thickness increases there is a small increase in the maximum shear force. Also, the maximum shear force of pile group is more in comparison to pile draft.

### Raft thickness variation for 20 pile configuration

Point load applied at the center of the pile cap/raft = 32,000 KN Uniformly distributed load applied on the pile cap/raft = 400 kPa Pile diameter = 800 mm.

Area of pile cap/raft = 80 m$^2$ **(b)**.

In Fig. 11b, the maximum differential settlement goes up to 11.6 mm for the pile group of pile length 35 m and pile cap thickness 800 mm subjected to point load. As the pile cap/raft thickness increases the differential settlement decreases. In Fig. 11c, as pile cap/raft thickness increases the maximum bending moment increases slightly. The maximum bending moment for pile group is more compared to piled raft. Also, the maximum bending moment is more for point load at the center of the pile cap/raft in comparison to uniformly distributed load on the pile cap/raft. In Fig. 11d, the maximum shear force difference between pile group and piled raft subjected to uniformly distributed load is nearly constant. Here, the maximum shear

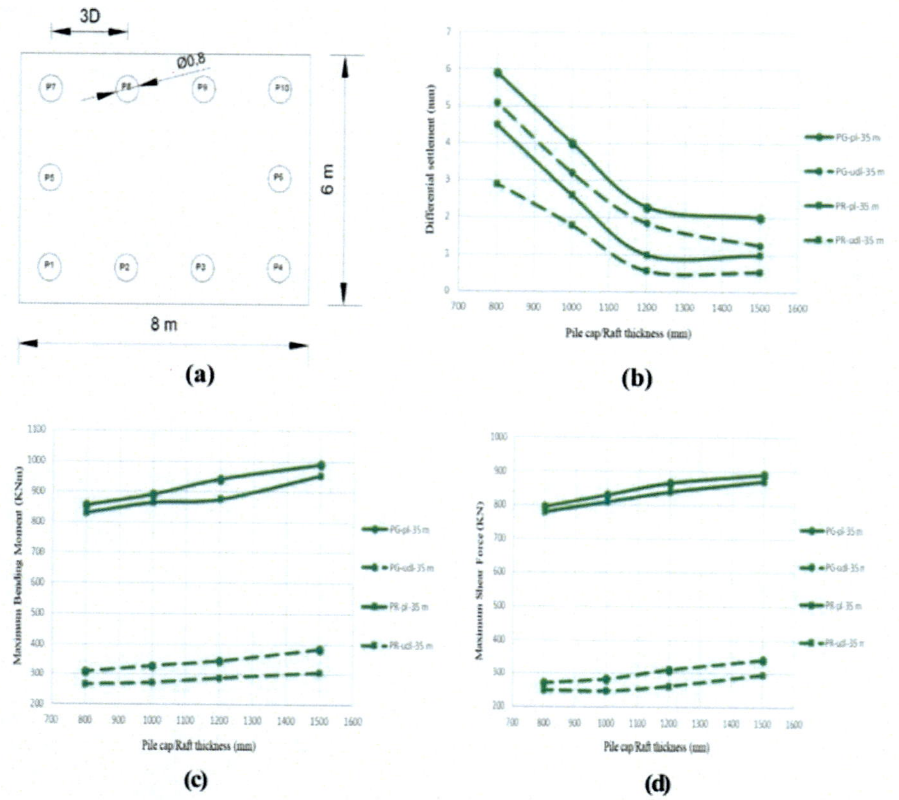


**Fig. 10** **a** Top view of of10 pile configuration of both pile group and piled raft, **b** Effect of pile cap/raft thickness on differential settlement (for 10 pile configuration), **c** Effect of pile cap/raft thickness on maximum bending moment (for 10 pile configuration), **d** Effect of pile cap/raft thickness on maximum shear force (for 10 pile configuration)

force varies from 382.44 KN for piled raft subjected to uniformly distributed load to 1290.59 KN for pile group subjected to point load. Also, the maximum shear force is more for pile group compared to pile draft.

## 5 Conclusions

1. As the pile cap/raft thickness increases, the differential settlement decreases i.e. more the pile cap/raft thickness less the differential settlement. It has been noted that point load produces more differential settlement than uniformly distributed load.
2. It has been concluded that maximum bending moment of pile increases with the increase in pile cap/raft thickness for both pile group and piled raft. The

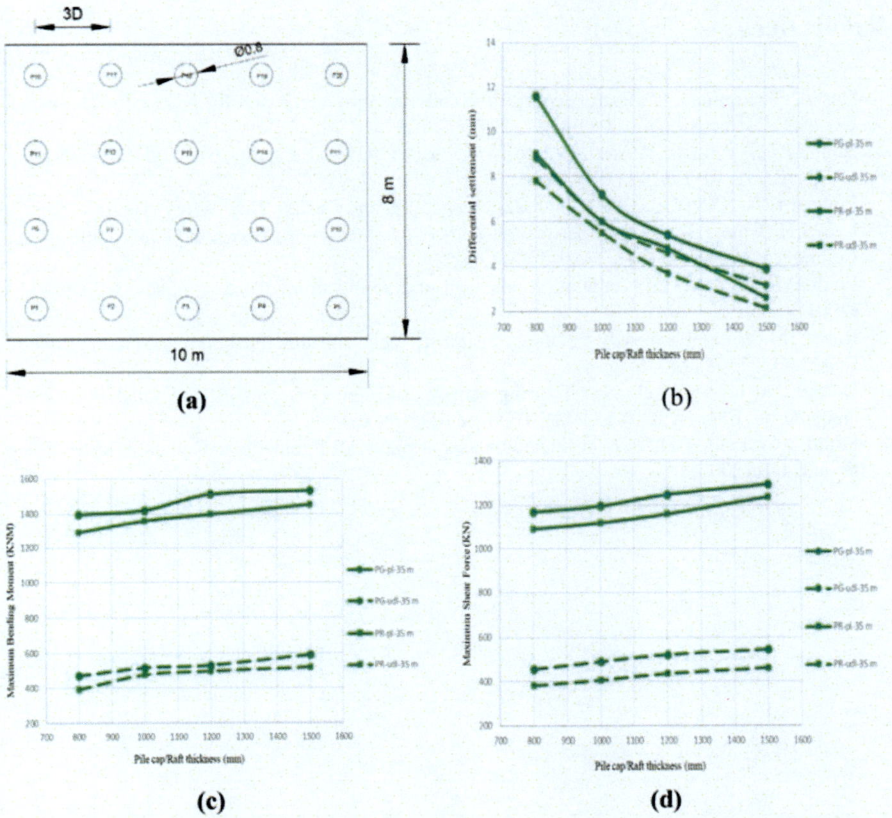


**Fig. 11** **a** Top view of 20 pile configuration of both pile group and piled raft, **b** Effect of pile cap/raft thickness on differential settlement (for 20 pile configuration), **c** Effect of pile cap/raft thickness on maximum bending moment (for 20 pile configuration), **d** Effect of pile cap/raft thickness on maximum shear force (for 20 pile configuration)

increase in maximum bending moment of pile is less in case of pile group and piled raft subjected to uniformly distributed load as compared to pile group and piled raft subjected load point load.

3. The maximum bending moment and maximum shear force is more in case of pile group as compared to piled raft for both point load and uniformly distributed load.
4. There has been increase in a maximum shear force of pile as pile cap/raft thickness of pile group and piled raft increases. Also, the maximum shear force due to point load is more in case of pile group and piled raft subjected to uniformly distributed load.
5. It has been noted that as pile cap/raft thickness increases the percentage of load carried by the piles is almost uniform for both pile group and piled raft subjected to point load as well as uniformly distributed load.

## References

1. Russo, G., Viggiani, C.: Factors controlling soil-structure interaction for piled rafts. Darmstadt Geotech. Darmstadt Univ. Technol. **4**, 297–322 (1998)
2. Long, P.D., Vietnam, V.W.: Piled raft—A cost-effective foundation method for high-rises. Geotech. Eng. **41**(1), 149 (2010)
3. Poulos, H.G., Davis, E.H.: Pile Foundation Analysis and Design. New Wiley (1980)
4. Lee, S.H., Chung, K.C.: An experimental study of the interaction of vertically loaded pile groups in sand. Can. Geoteh. J. **42**, 1485–1493 (2005)
5. Poulos, H.G.: Piled raft foundations: Design and applications. Géotechnique **51**(2), 95–113 (2001)
6. Rabiei, M., Choobbasti, A.J.: Piled raft design strategies for high rise buildings. Geotech. Geol. Eng. **34**(1), 75–85 (2016)
7. De Sanctis, L., Mandolini, A.: Bearing capacity of piled rafts on soft clay soils. J. Geotech. Geoenviron. Eng. (ASCE) **132**(12), 1600–1610 (2006)
8. Small, J.C., Poulos, H.G.: Non-linear analysis of piled raft foundations. Contemp. Issues Deep Found, 1–9 (2007)

# Experimental Investigation on Performance of Helical Pile in Cohesionless Soil

**A. I. Dhatrak, M. M. Dhage, H. R. Varma, and S. W. Thakare**

## 1 Introduction

Helical piles were invented by Alexander Mitchell in nineteenth century. This pile type was first used as the foundation for Maplin Sands Lighthouse in England. Helical pile is made up of steel shaft either a solid square shaft or a circular pipe with one or more helices welded to it with required pitch distance. They are installed into the soil by applying a torque to the upper end of the shaft. Helical piles are used to resist uplift, compressive forces and overturning moment. They are used in the construction of structures such as telecommunication and power transmission towers, machine foundations, residential and commercial buildings, buried pipelines and bridges. The advantages of helical pile over conventional concrete and steel piles are that it is lightweight, it has high vertical and uplift capacity, quick installation, and it can be installed in limited access condition. Helical piles do not produce loud noise during installation process. In case of high groundwater level, helical piles save dewatering or pumping of construction site.

A. I. Dhatrak
Dean Academics, Government College of Engineering, Amravati, Maharashtra, India

M. M. Dhage (✉) · H. R. Varma
Civil Engineering Department, Government College of Engineering, M.Tech Geotechnical Engineering, Amravati, Maharashtra, India

S. W. Thakare
Civil Engineering Department, Government College of Engineering, Jalgaon, Maharashtra, India

 
C. N. V. Satyanarayana Reddy et al. (eds.), *Ground Characterization and Foundations*, Lecture Notes in Civil Engineering 167,
https://doi.org/10.1007/978-981-16-3383-6_67

## 2 Literature Review

George et al. [1] conducted a detailed investigation on a helical pile installed in cohesionless soil by displacement method. Laboratory experiments were conducted on models to study the various factors influencing the axial bearing capacities of helical piles. To augment the investigation, finite-element analyses were carried out using PLAXIS 3D AE software and compared with experimental results. The piles installed by the displacement method exhibited a higher ultimate capacity and distinct failure pattern. Abdrabbo et al. [2] conducted experimental investigation on helical piles with different helices diameters, numbers and spacing subjected to the horizontal loadings. The effect of these parameters was monitored, and comparative study between helical piles and piles without helices was accomplished. They found that the most beneficial helical depth ratio was between 1/3 and 1/2. Farhad Nabizadeh et al. [3] conducted field test on the piles with various numbers of helices to investigate the behaviour of helical piles in sand and silty clay soils. Furthermore, the effect of post grouting on the strength of these piles was assessed. In sandy soil, pile resistance increased due to the penetration of grouting, but in clayey soil because of the compaction grouting, there was less increase. Sakr [4] conducted the full-scale axial compression and tension testing on large capacity helical piles installed in cohesionless soils. He concluded that helical piles with relatively large diameter up to 508 mm were successfully installed into dense to very dense soils.

## 3 Experimental Investigations

To study the performance of helical pile under vertical loads, uplift loads and lateral loads, laboratory model tests were conducted on helical piles with different helix diameter.

### 3.1 *Test Setup*

The test setup for vertical load tests on helical pile consisted of test tank, hydraulic jack, loading frame, dial gauges and proving ring. The M.S. tank with an inside dimension of 0.65 m (length), 0.65 m (width) and 0.65 m (depth) was used for the experimental investigation. The hydraulic jack was used for loading. The settlements were measured with the help of two dial gauges with 25 mm travel and least count of 0.01 mm. The dial gauges with magnetic base were fixed to the sides of the tank. A proving ring (5 kN capacity) was used to precisely record load applied on the pile. The model pile load test on conventional pile was conducted according to IS:2911-(part 4) 1985 [5] and the vertical load capacity was evaluated. Ultimate vertical capacity of helical pile was considered as the load at which the displacement equal to 5% of helix diameter ($D_h$). The test setup is shown in Fig. 1.

The lateral load test was conducted according to IS: 2911-(part 4) 1985, and lateral load on the piles was applied with the help of pulley and non-extensible wire system as shown in Fig. 2. A dial gauge of 0.01 mm sensitivity, horizontally fixed on side edge of tank, was used to measure the lateral displacement of pile at the top. Standard weights were used for loading.

**Fig. 1** Test setup used for vertical loading on helical piles

**Fig. 2** Test setup for lateral loading on helical pile

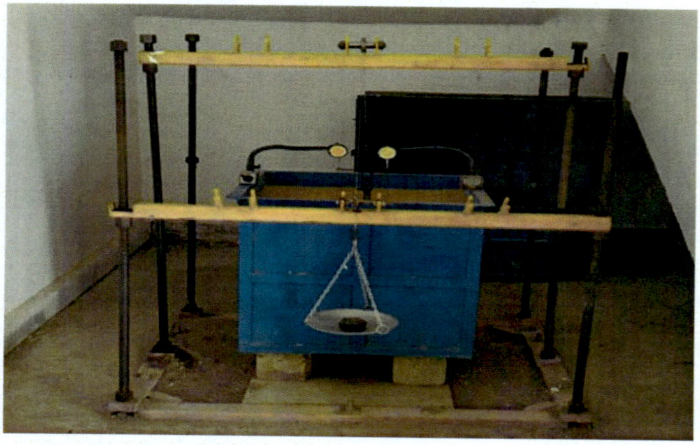

**Fig. 3** Test setup for uplift loading on helical pile

The uplift load tests were conducted on piles in accordance of IS: 2911-(part 4) 1985. The uplift load was applied in number of increments by adding standard weights through a loading arrangement as shown in Fig. 3. Pile head movement was measured for each load increment. When the pile head movement was stopped, the next load increment was applied. The procedure was repeated till the failure displacement of pile was recorded for the sudden pull out of pile occurred.

## 3.2 Materials

The model helical piles used for experimental investigation were made from mild steel. Piles were 300 mm in length and 12 mm in diameter. Helical piles with different

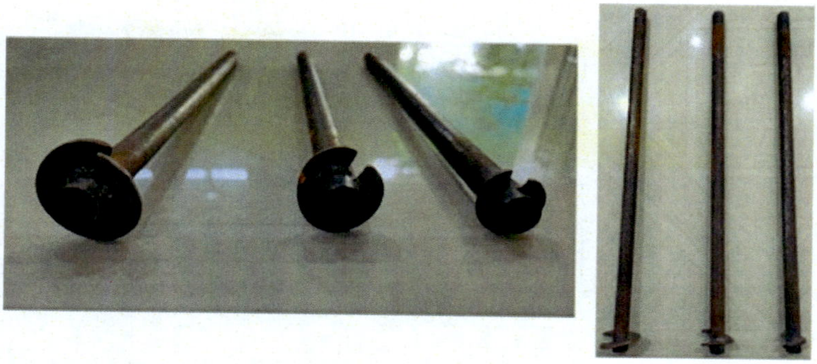

**Fig. 4** Model helical pile used in experimental investigation

**Table 1** Properties of sand

| S. No. | Properties of sand | Values |
|---|---|---|
| 1 | Specific gravity | 2.65 |
| 2 | $\gamma_{max}$ (kN/m$^3$) | 17.94 |
| 3 | $\gamma_{min}$ (kN/m$^3$) | 15.45 |
| 4 | Angle of internal friction ($\phi$) | 36° |
| 5 | Cohesion (kN/m$^2$) | 0 |
| 6 | IS classification | SP (medium sand) |

helical blade diameter are as shown in Fig. 4. For the model tests, uniformly graded dry cohesionless sand was used. Properties of sand are as shown in Table 1.

### *3.3 Sand Bed Preparation and Installation of Model Pile*

The relative density of sand bed was maintained by using sand raining technique. The height of fall to achieve the desired relative density was determined prior by performing a series of trials with different height of fall. In this study, the height of fall was selected as 24 cm, and the corresponding relative density was maintained at 55%. Initially, tank was filled up to the pile tip, and the pile was placed vertically at the centre of tank by using special guide. The non-displacement method of pile installation was used. Remaining tank was filled by using same technique. Pile was kept vertical during this process. The top surface of the sand bed was levelled and checked by spirit level.

### *3.4 Test Program*

Experimental tests were carried out on model helical pile and conventional circular pile embedded in cohesionless soil. Parameters selected for study are presented in Table 2.

## 4 Test Results and Discussion

The pile load tests were carried out on conventional circular pile and helical piles with different helix diameter ratio ($D_h/D_s$), and load-settlement curves were plotted to study the performance of helical pile. Figures 5 and 6 show the performance of circular and helical pile subjected to vertical loads in medium dense sand.

**Table 2** Parameters selected for the study

| S. No. | Details of parameters | Constant parameters | Varying parameters |
|---|---|---|---|
| 1 | Length of pile | 300 mm | – |
| 2 | Diameter of pile ($D_s$) | 12 mm | – |
| 3 | Slenderness ratio ($L/D$) | 25 | – |
| 4 | Type of soil | Sand | – |
| 5 | Density of sand | 55% | – |
| 6 | Type of loading | | Vertical, lateral, uplift loading |
| 7 | Helix diameter ratio ($D_h/D_s$) | | 2, 2.5 and 3 |

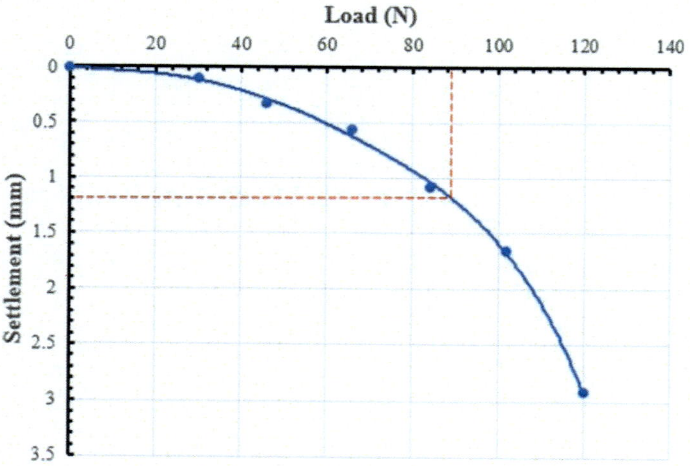


**Fig. 5** Load-settlement curve for circular pile

Figures 7 and 8 show the performance of circular and helical pile subjected to lateral loads in medium dense sand. For ultimate lateral capacity, the failure criteria taken as the load at which displacement is equal to 5 mm.

Figures 9 and 10 show the performance of circular and helical pile subjected to uplift loads in medium dense sand.

The ultimate vertical, lateral and uplift load capacities are determined from the load-settlement/displacement curves. The ultimate capacities of the piles are shown in Tables 3 and 4.

The percentage increase in ultimate vertical capacity of helical pile with helix diameter ratio 2, 2.5 and 3 is 31%, 75% and 99%, respectively, as compared to conventional circular pile. The percentage increase in ultimate uplift capacity of

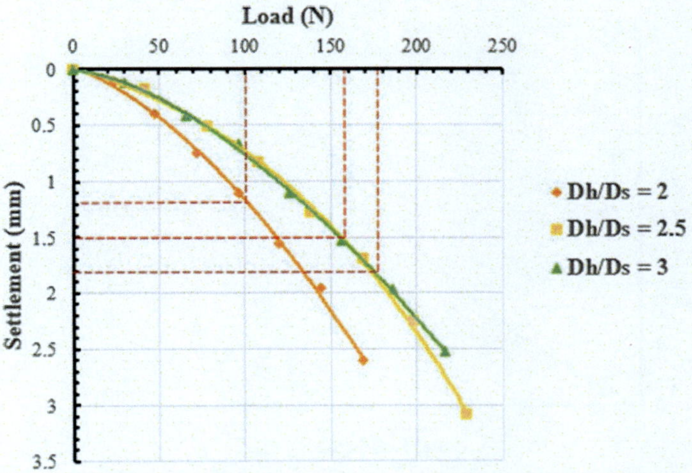


**Fig. 6** Load-settlement curve for helical piles with different helix diameter ratio

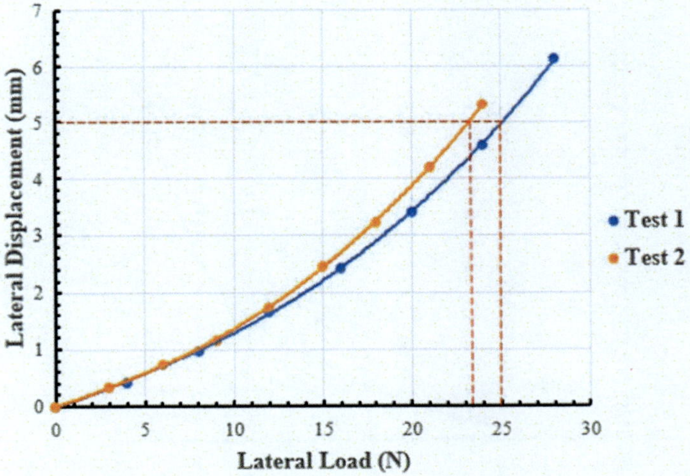


**Fig. 7** Lateral load versus displacement curve for circular pile

helical pile with helix diameter ratio 2, 2.5 and 3 is 18%, 27% and 45%, respectively, as compared to conventional circular pile. The percentage increase in ultimate vertical capacity of helical pile with helix diameter ratio 2, 2.5 and 3 is 16%, 8% and 4%, respectively, as compared to conventional circular pile.

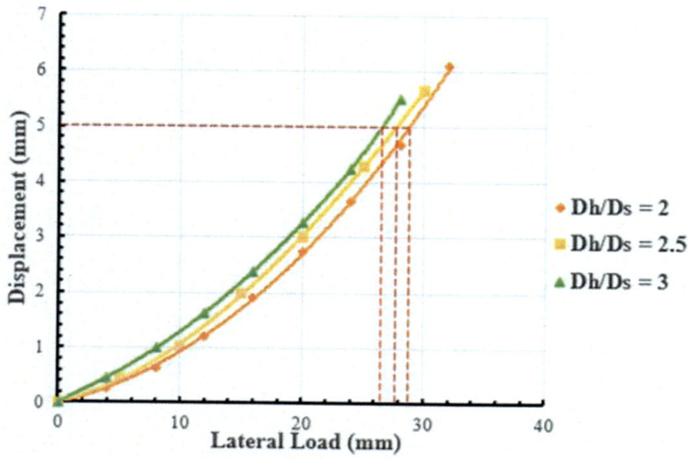


**Fig. 8** Lateral load versus displacement curve for helical piles with different helix diameter ratio ($D_h/D_s$)

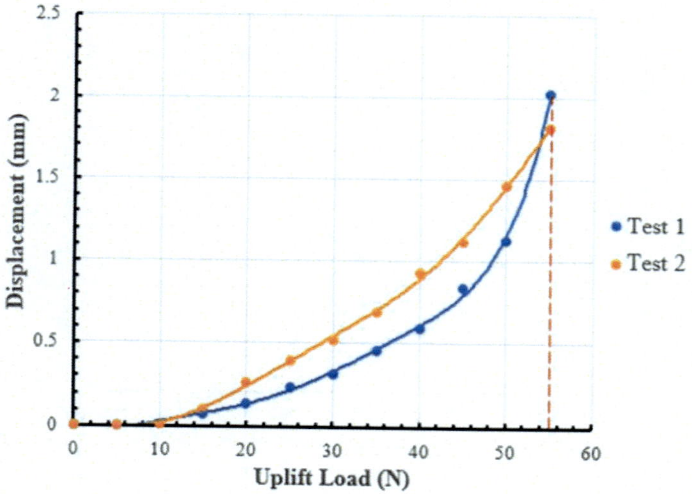


**Fig. 9** Uplift load versus displacement curve for circular pile

## 5 Conclusions

From the results of present study, following conclusions are drawn:

1. The ultimate vertical capacity and uplift capacity of helical pile increases with increase in diameter of helical blade.
2. The ultimate lateral load capacity of helical pile decreases with increase in helix diameter ratio.

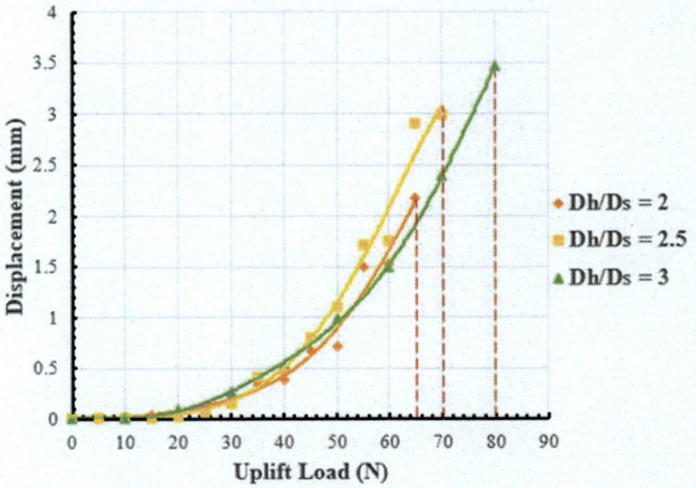


**Fig. 10** Uplift load versus displacement curve for helical pile with different helix diameter ratio

**Table 3** Ultimate capacity of conventional circular piles

| Type of pile | Ultimate vertical capacity ($N$) | Ultimate lateral capacity ($N$) | Ultimate uplift capacity ($N$) |
|---|---|---|---|
| Circular | 88 | 24 | 55 |

**Table 4** Ultimate capacities of helical pile with different helix diameter ratio

| Helix diameter ratio ($D_h/D_s$) | Ultimate vertical capacity ($N$) | Ultimate lateral capacity ($N$) | Ultimate uplift capacity ($N$) |
|---|---|---|---|
| 2 | 101 | 29 | 65 |
| 2.5 | 154 | 27 | 70 |
| 3 | 175 | 26 | 80 |

3. Up to 99% increment in vertical capacity is observed with varying helix diameter ratio as compared to the conventional circular pile.
4. Up to 45% increment in uplift capacity is observed with varying helix diameter ratio as compared to the conventional circular pile.
5. Maximum increment of 21% in lateral capacity is observed in helical pile with helix diameter ratio equals to 2.

## References

1. George, B.F., Banerjee, S., Gandhi, S.R.: Helical piles installed in cohesionless soil by displacement method. Int. J. Geomech. ASCE **19**(7) (2019)
2. Abdrabbo, F.M., El Wakil, A.Z.: Laterally loaded helical piles in sand. Alex. Eng. J. **55**(4), 3239–3245 (2016)
3. Nabizadeh, F., Chaoobbasti, A.J.: Field study of capacity helical piles in sand and Silty clay. Transp. Infrastruct. Geotechnol. **4**(1), 3–17 (2016)
4. Sakr, M.: Installation and performance characteristics of high capacity helical piles in cohesionless soils. DFI J. **5**(1) (2011)
5. Bureau of Indian Standard.: Code of practice for design and construction of pile foundations—load test on piles. IS 2911-Part 4. Bureau of Indian Standard, New Delhi, India

# Instrumented Pile Load Tests in Southern India

**Sujan Kulkarni, Ravikiran Vaidya, and P. Sriram**

## 1 Introduction

A conventional vertical static load test on a pile provides limited information as one monitors load and displacement only at pile top. Such testing does not provide any quantitative information on the load-transfer mechanism (magnitude of the toe resistance and the distribution of shaft resistance). Similarly, conventional lateral load test only provides the load–deflection curve for the top of the pile and pile deflection along the length as well as point of fixity is unknown. Yet, this information is what the consultant often needs in order to verify his design. Therefore, more and more frequently, the conventional test arrangement is expanded to include instrumentation to obtain the required information.

This paper presents a case study for the instrumented tests performed in Kochi for a Test pile at Kochi Metro Rail Project. Geo Dynamics in association with Kochi Metro Rail Corporation (KMRL) performed state-of-the-art instrumentation studies during vertical as well as lateral load tests. Embedment-type strain gages were installed in the pile during pile casting to perform instrumentation study. An inclinometer casing was also installed in order to monitor the deflection of pile along the length during lateral load test. The pile was a test pile (mono pile) with diameter of 2 m and length of 50 m. A crosshole sonic logging test was also performed before the load tests, and it was concluded that the pile has major defect from 44 m to pile toe.

S. Kulkarni (✉) · R. Vaidya
Geo Dynamics, Vadodara, Gujarat 390023, India
e-mail: sujan@geodynamics.net

P. Sriram
DMRC, Kochi Metro Rail Corporation, Kochi, India

 
C. N. V. Satyanarayana Reddy et al. (eds.), *Ground Characterization and Foundations*, Lecture Notes in Civil Engineering 167,
https://doi.org/10.1007/978-981-16-3383-6_68

## 2 Subsurface Conditions

During boring for the pile, samples at every meter were collected. Those samples were used for visual classification and presented as Table 1.

**Table 1** Subsurface conditions

| Sample depth (m) | Classification | Soil type |
|---|---|---|
| 0–6 | No samples collected (fill?) | Fill |
| 6–12 | Dark gray marine clay/plastic silt | CH/MH |
| 12–17 | Olive gray marine clay/plastic silt | |
| 18 | Olive gray silt (less plastic) | ML |
| 19 | Reddish yellow gray mix sandy silt | |
| 20–22 | Reddish brown spotted brownish yellow mix, sandy silt, some clay | |
| 22 | Reddish brown mix with yellow gray sandy clay/clayey sand | SC/CL |
| 23 | Brownish yellow mix with reddish brown plastic silt with some sand | MH |
| 24 | Dark gray clay with some organics | CL |
| 25 | Dark gray clay with some more organics | |
| 26–28 | Yellowish gray fat clay | CH |
| 28 | Brownish yellow sandy silt | ML |
| 29 | Brownish yellow reddish gray sandy silt | |
| 30 | Yellowish gray/brownish yellow clayey sand | SC |
| 31 | Medium to coarse brown sand | SP-SM |
| 32 | White sand trace clay | |
| 33 | Brownish gray clayey sand | SC |
| 34 | Gray medium sand | SP-SM |
| 35 | Plastic silt olive gray | MH |
| 36 | Brownish gray medium sand | SP-SM |
| 37 | Light gray sand | |
| 38 | Gray sandy clay | CL |
| 39 | Gray sand with trace clay | SP-SC |
| 40–42 | Dark gray clay trace organic | CL |
| 42 | Gray sand with trace clay | SP-CL |
| 43 | Dark gray peat | PT |
| 44–46 | Dark gray clay trace peat | CL |
| 46 | Brownish gray sandy clay | |
| 47 | Dark brown clayey sand | SC |
| 48–50 | Brownish gray silt | ML |

## 3 Vertical Load Test

A cyclic load test was planned and the design load on the pile was 850 tons while the test load was 1275 tons. Combination of anchor piles and kentledge was used to provide reaction during testing. The instrumentation consisted of 56 embedment type vibrating wire strain gages. It was planned to install gages at every stratigraphy change (however not greater than 3 m). The strain gages were installed in sets of 2 and 4 alternatively. Whenever two gages were installed they were at 180° and whenever four strain gages were installed they were located at 90°. Photograph of the strain gage installation is presented below as Fig. 1.

A cyclic load test was performed on the test pile in order to evaluate the load–settlement behavior of the pile. The load test was directed and carried out by DMRC as per their method statement. Four jacks, each having 500 tons capacity, were used for loading. The pile was loaded to an initial load of 170 tons (first cycle) and was unloaded to zero load. The next load cycles were 339.1 tons, 508.6 tons, 678.4 tons, 848 tons, 1017.4 tons, 1186.9 tons, and 1277.4 tons. After each load increment, corresponding settlements were measured and pile was unloaded to zero load. The pile was then again loaded to 1277.4 tons and the same load was maintained for 24 h after which the testing was terminated. The failure criterion was considered to be settlement of 12 mm. Load–settlement plot of the pile is presented as Fig. 2.

The maximum settlement was observed to be 12.08 mm when the test load of 1277.4 tons was maintained for 24 h. Once the pile was unloaded to zero load then the net settlement of the pile was observed to be 5.6 mm. The elastic recovery was

**Fig. 1** Strain gage installation in progress

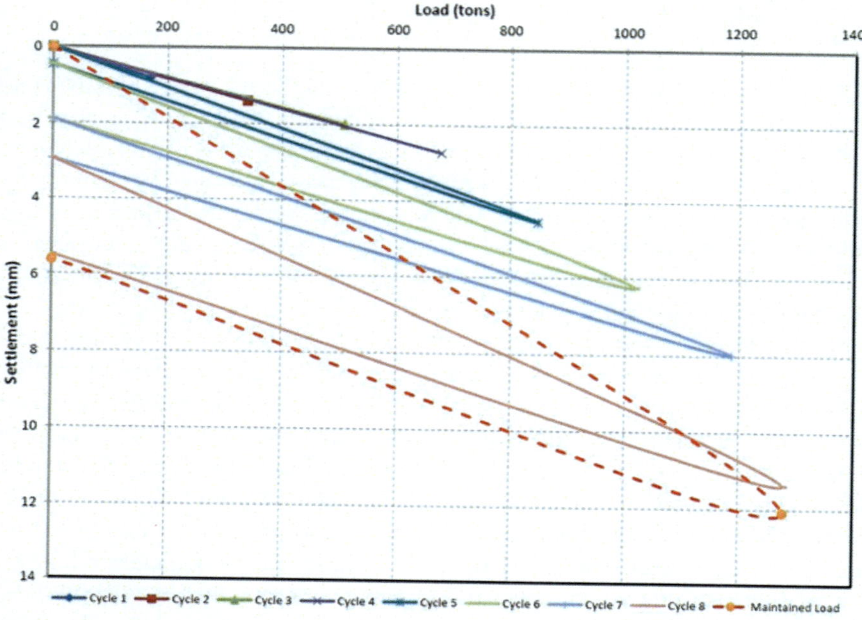


**Fig. 2** Load–settlement plot—vertical load test

around 6.5 mm. Since pile had major defect from 44 m to pile toe, the load carried by the pile was due to frictional resistance only.

As stated above, 56 vibrating wire strain gages were installed in the pile. At the time of testing 9 gages were not functional. It is possible that while cage lowering operations or during concreting these gages and/or cables might have damaged. However, other gages provided reasonable data for the analyses and interpretation. Strain gages located below 44 m indicated unusual readings due to presence of defect. Other strain gages provided consistent readings implying that the pile shaft is under compression. Photograph shows the data collection in progress (Fig. 3).

The strain gage readings were used further to calculate the load transfer at each level (presented below as Fig. 4). The load transfer was estimated based on state-of-the-art published literature by renowned professor [1, 2]. For each load increment, load transfer up to 6.7 m was similar indicating not much resistance offered by the surfacial soil. However, after this level, there was significant decrease in load transfer indicating high amount of skin friction. Strain data obtained from the gages located beyond 44 m were not used for load transferred calculation due to their unusual behavior.

The skin friction provided by the soil is calculated as difference in load transferred to the pile and presented as Fig. 5.

For the maximum test load, the skin friction was estimated to be around 89% while end bearing was only 11%. Note that this 11% also includes skin friction from 41 m to pile toe.

**Fig. 3** Instrumented vertical load test—data collection in progress

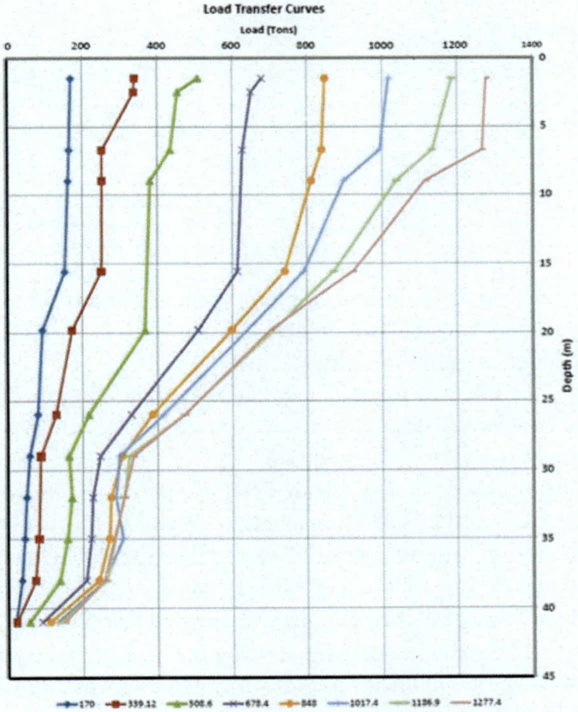


**Fig. 4** Load-transfer curves for each loading cycle

# 4 Lateral Load Test

Upon completion vertical load test, an instrumented cyclic lateral load test was performed on the same pile. The design lateral load on the pile was 45 tons while the test load was 67.5 tons. Pile deflection along the length of the pile was measured

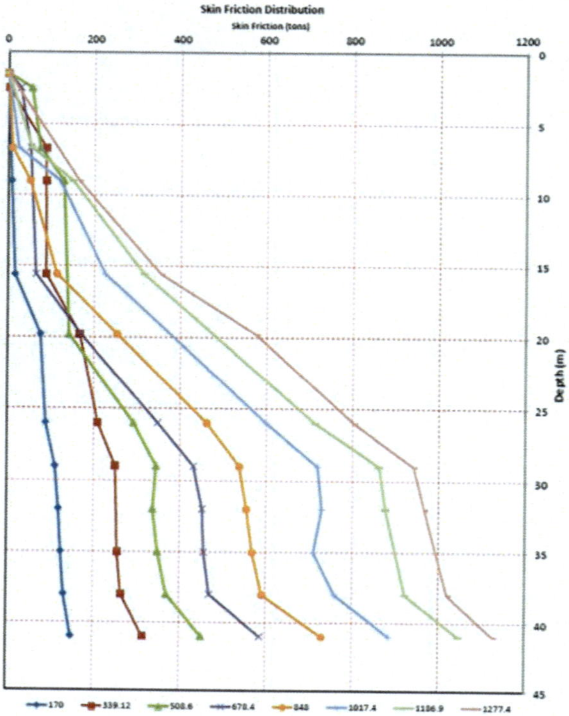


**Fig. 5** Skin friction distribution for each loading cycle

by means of inclinometer. Inclinometer casing was installed during pile concreting. The depth of the inclinometer casing was 15 m which was determined based on theoretical point of fixity (12 m).

The pile was loaded to an initial load of 9.24 tons (first cycle) and was unloaded to zero load. The next load cycles were 18.48 tons, 27.72 tons, 36.96 tons, 46.2 tons, 55.44 tons, 64.68 tons, and 67.76 tons. After each load increment, corresponding displacements were measured and pile was unloaded to zero load. The pile was then again loaded to 67.76 tons and the same load was maintained for 24 h after which the testing was terminated. Before maintaining the load, all the dial gages were reset to zero. The failure criterion was considered to be displacement of 12 mm. Load displacement plot of the pile is presented below as Fig. 6.

The maximum settlement was observed by dial gages to be 8.54 mm when the load was maintained for 24 h. Once the pile was unloaded to zero load, then the net displacement of the pile was observed to be 3.6 mm. The elastic recovery was around 4.94 mm.

Inclinometer readings were taken after each loading cycle as shown in Fig. 7. A typical load deflection along the depth is also presented as Fig. 8. Generally good agreement was observed between dial gage readings and deflection at the top observed by inclinometer. Based on the data collected by inclinometer, it can be

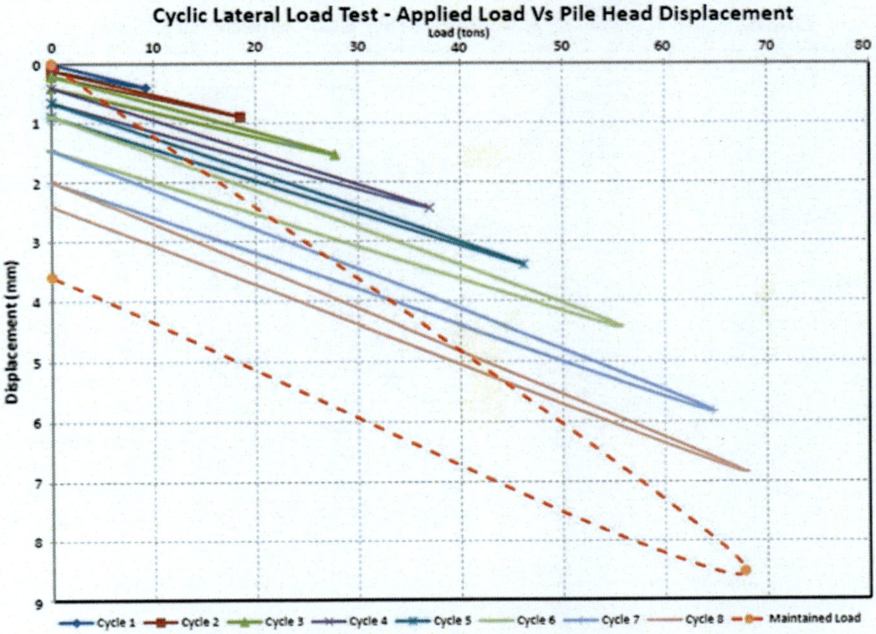


**Fig. 6** Load–displacement plot—lateral load test

**Fig. 7** Inclinometer reading being taken during lateral load test

**Fig. 8** Load deflection graph for each loading cycle

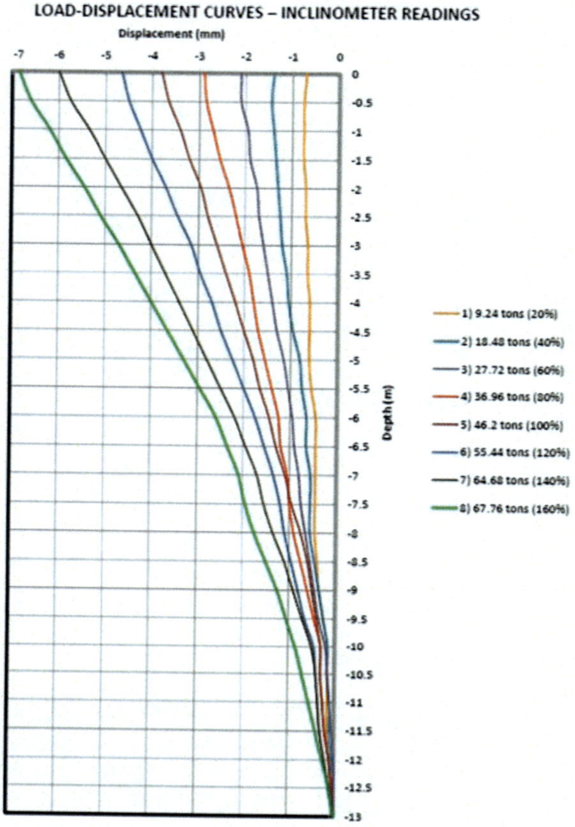


inferred that the pile undergone some deflection even at 12 m and below indicating that the actual point of fixity is somewhat lower than the theoretical calculations.

## 5 Conclusions

1. Based on CSL results, the pile has major defect form 44 m to pile toe, and this was verified by the strain gage readings at corresponding levels.
2. Strain gage readings were used to compute the skin friction distribution and amount end bearing mobilized. 89% of the load was resisted by skin friction (up to 41 m) while only 11% was resisted by end bearing (which also includes friction from 41 m to pile toe).
3. An instrumented lateral load test was performed on a test pile, and this report presents the results of the inclinometer readings. The pile was loaded to a maximum load of 67.76 tons and the maximum settlement was observed to be around 8.5 mm by inclinometer (8.54 mm as per dial gages).

4. Inclinometer readings were used to verify theoretical point of fixity, and based on the data, it can be concluded that the actual point of fixity is somewhat lower than the theoretical calculations.

## References

1. Fellenius, B.H., Brusey, W.G., Pepe, F.: Soil setup, variable concrete modulus, and residual load for tapered instrumented piles in sand. In: American Society of Civil Engineers, ASCE, Specialty Conference on Performance Confirmation of Constructed Geotechnical Facilities. University of Massachusetts, Amherst, April 9–12, 16 p (2000)
2. Fellenius, B.H.: From strain measurements to load in an instrumented pile. Geotech. News Mag. **19**(1), 35–38 (2001)

# Bidirectional Static Load Test (BDSLT) on a Versatile Barrette Foundation to 18000 tonnes

**Anil Cherian**

## 1 Introduction

Barrettes are large rectangular piles that are constructed either by the use of a device with rotating cutter heads. The excavation for a barrette is performed under bentonite or a polymer that keeps the hole open as for a conventional drilled pile. Once the barrette is excavated and the bentonite de-sanded, a steel cage is lowered into the hole, and then, concrete is trimmed into the base of the hole, displacing the bentonite. Barrettes can be constructed in L, T, H, or cruciform shapes in the plan if so desired by cutting the rectangular hole several times to form the shape. Barrette foundations have been used for many years and are treated as rectangular piles in foundation engineering applications. In the Middle East, a fast-growing construction hub, excavated rectangular barrettes, and large diameter bored piles are commonly adopted as the foundation units in high-rise buildings and infrastructure projects [1]. The founding strata for these units are usually medium to hard rocks, relying on shaft resistance, are designed. Nevertheless, pile design parameters must be geotechnically and structurally verified by using preliminary pile loading tests before they are used in the final design. Deprived of performing such load tests on-site, unit shaft resistance and settlement cannot be identified and not normally permitted by the local authorities. In the last two decades, many full-scale compression loading tests on instrumented barrettes and bored piles have been performed using BDSLT in the United Arab Emirates (UAE) to verify the design parameters. The bidirectional static load test (BDSLT) has been around since the early 1970s. The first commercial development came about in the early 1980s in Brazil and about a decade later in the USA [2–4]. Due to the many advantages over the conventional top-down load test, the bidirectional static loading test (BDSLT) using the hydraulically driven jacks is becoming an increasingly popular way to determine the ultimate capacity of deep

A. Cherian (✉)
Strainstall Middle East LLC (James Fisher and Sons plc, UK), Dubai, United Arab Emirates



C. N. V. Satyanarayana Reddy et al. (eds.), *Ground Characterization and Foundations*, Lecture Notes in Civil Engineering 167,
https://doi.org/10.1007/978-981-16-3383-6_69

foundations. The high capacity sacrificial jack is installed within the foundation unit at the chosen location, where it is typically halfway down the pile capacity length of the foundation [5].

This method is internationally accepted and referred to in several international standards [6–8]. BDSLT has been employed on several barrette projects in the Middle East and other Asian countries and has been developed into an efficient and cost-effective method. With an increase in demand for the foundations units that utilize barrette construction for their foundation design in the UAE, it is evident that BDSLT will play a vital role in future barrette foundation developments. This article discusses the application of BDSLT on an instrumented versatile deep barrette foundation to identify the settlement, load distribution, and unit skin friction, and thereby to verify the foundation design for value engineering at La Maison Residential high-rise building, Dubai, UAE.

## 2 Geological Conditions

The geology of the United Arab Emirates, and the Arabian Gulf area, has been substantially influenced by the deposition of marine sediments associated with numerous sea level changes during relatively recent geological time. With the exception of mountainous regions shared with Oman in the north-east, the country is relatively low-lying, with near surface geology dominated by Quaternary to late Pleistocene age, mobile aeolian dune sands, and sabkha/evaporites deposits. The geologically stable Arabian Plate is separated from the unstable Iranian fold belt by the Arabian Gulf. It is believed that a tilting of the entire Arabian Plate occurred during the early Permian period, resulting in uplift in southern Yemen, and depression to the north-east. Crustal deformations and igneous intrusions occurred in the north-east as a result of this movement. During the Plio-Pleistocene epoch, tectonic movements near the folding of the Iranian Zagros Range probably contributed to the formation of both the Arabian Gulf depression and the mountainous regions shared by the United Arab Emirates and Oman in the northeast. The near surface geology of the Dubai region is dominated by Aeolian dune sand deposits of the Holocene to Pleistocene age. These deposits typically comprise fine-grained silty calcareous sand, which is commonly dense and variably cemented beneath a shallow, loose, normally consolidated mobile layer. The degree of cementation generally increases with depth from variably cemented sand grades to predominantly calcareous sandstone. Very silty, gypsiferous sabkha and evaporate layers occur occasionally within the Aeolian sand deposits [9]. Although surficial sabkha deposits are found throughout the coastal belt of the Arabian Gulf, and far inland in the western and southern parts of the United Arab Emirates, they are not particularly common in the Dubai region. These superficial deposits were underlain by alternating beds of siliceous calcarenite, calcareous sandstone, siltstone, and conglomerates [10].

The Barrette test location is positioned in Dubai Business Bay, about 2.0 km south-east of the famous Burj Khalifa tower. The site is of rectangular shape featuring an approximate area of 90 m by 60 m. The proposed tower footprint is covering a 30 m by 50 m area. The local geology is characterized by the presence of the Barzaman Formation, which is encountered at depths greater than 23 m. The Barzaman formations include reddish-brown conglomerates, brecciated dolomitic calcisitites, and breccias with clasts of coarse gravel and cobble size of limestone [10]. The cementing material in the calcisiltic breccias is relatively weaker, a greenish-grey in colour and is probably due to the presence of palygorskite clay mineral. The Barzaman formation is overlain by the reddish-brown sandstones, which are extremely weak to weak with localized medium beds of calcilutite breccia. The sandstones are fine to medium sand size with a cementing material that imparts a very inconsistent strength to the rock. The reddish-brown sandstones are overlain by a brown to light brown Calcarenite (Ghayathi Formation). The Calcarenite is locally thinly laminated, fine to medium-grained. Localized medium beds of imperfectly laminated or massive Calcarenite with fine to medium clasts are also encountered [5]. The general geotechnical parameter used for the foundation design is provided in Table 1.

**Table 1** The general geotechnical parameters

| Strata | Depth (mDMD) | SPT N | Allowable unit skin friction (kPa) |
|---|---|---|---|
| Medium dense silty fine sand with some cemented pieces | 0.00 to −18.00 | 20–50 | – |
| Calcarenite/Calcareous Sandstone, slightly to moderately weathered interbedded with cemented calcareous sand | −18.00 to −27.50 | – | 100–250 |
| Weak to moderately weak brown Conglomerate, slightly to moderately weathered | −27.50 to −33.00 | – | 170–210 |
| Calcisiltite/Calcareous Siltstone moderately weathered interbedded with cemented calcareous silt | −33.00 to −40.00 | – | 100–140 |
| Weak to moderately weak light brown Calcareous Siltstone interbedded with cemented calcareous silt | −40.00 to −45.00 | | 100–130 |
| Weak to moderately weak light brown Calcisiltite interbedded with cemented calcareous silt | −45.00 to −70.00 | | 100–150 |

# 3 Methodology

Barrette excavation was carried out under the piling contractor's work plan as approved by the engineer/contractor. Upon reaching the final toe elevation, the pile bottom was cleaned and approved by the Engineer/Contractor for concrete placement. The hydraulic cell assembly, related hydraulic supply, and instrumentation were lowered into the pile attached to the steel cage. The steel cage was fabricated in several pieces and spliced together over the bored hole. The number of cages was kept to a minimum to speed up the installation process. The first section of the reinforcing cage containing the hydraulic cell assembly was lowered into the borehole. The second cage section was then lowered vertically into position and spliced to the top of the first cage (Fig. 1). After the entire reinforcing cage was lowered into the shaft, without any steel casing as the working platform level was about 1.0–1.50 m above the cutoff level during concrete placement. Concrete placement commences utilizing a suitable size tremie pipe of sufficient length to extend beyond the hydraulic cell assembly to the toe of the pile. Cutouts of sufficient sizes were provided in the hydraulic cell steel bearing plates to accommodate the tremie pipe. A funnel was also constructed between the opening in the top plate of the hydraulic cell assembly and the main vertical rebar to guide the concrete tremie pipe through the steel bearing plates. The funnel also serves as a means of preventing the tremie pipe from accidentally hitting the hydraulic fittings on the cell top by forming a physical barrier apart from serving as a guide. Further, protection for the hydraulic hoses was in the form of foam shields and protection bars leading from the hydraulic fittings to the cell top the cage vertical rebar which protects the hoses from the effects of flowing concrete. The concrete was placed up to the designed cutoff level as per the concreting procedures. Reinforcing steel or steel angle iron was welded between the top and bottom bearing plates before the lifting process. These temporary supports were cut out when the cage was lowered into the hole. Tell-tale tubes were installed to measure the cell top and bottom movements. The hydraulic cell is attached to the reinforcement steel cage to ensure its location and depth are located precisely. The size of the barrette was 2.80 m × 1.20 m, and one 350 mm diameter tremie pipe was used for the inflow of class C75/20 (OPC + 36%GGBS + 6%MS) concrete.

After the concrete reaches a minimum required strength, the test may be started (Fig. 2). As the load is applied to the hydraulic cell, it begins working in two directions: upward against upper skin friction and downward against lower skin friction and base resistance. BDSLT is considered to be complete after reaching the ultimate capacity above or below the hydraulic cell or upon reaching the maximum capacity of the hydraulic cells. Instrumentation includes hydraulic cell expansion using tell-tale rods and displacement transducers; pile movement uses displacement transducers; skin friction, strain, and load transfer at different levels use vibrating wire concrete embedded strain gauges. The load increments were applied as specified in the loading schedule, and each successive load increment was held constant by adjusting the hydraulic jack pressure until the settlement criteria were met. Data acquisition of all installed instruments was connected to a data logger to a laptop

**Fig. 1** Barrette installation

computer allowing the data to be recorded and stored automatically at stipulated intervals and displayed in real-time.

A total of three bidirectional static load tests was carried out on the proposed High-rise Residential Tower project on Plot no. BB-B04-001 at Business Bay, Dubai. Tests were carried out from 27th December 2016 to the 6th of January 2017. The barrette size was 2.80 × 1.20 m with a maximum length of 48.60 m below the finished level. All barrettes were tested to a maximum load of more than three (3) times the expected working load to verify their capacity (Table 2). Tests were performed in sacrificial barrettes using a hydraulic jack assembly comprising of nine 900 tonne capacity bidirectional jacks, each jack can achieve an additional capacity of 15–20% during over jacking. Eight levels of Geokon vibrating wire-concrete embedment 4200 type

**Fig. 2** Barrette load test set up

**Table 2** Instrumentation details of test barrettes

| Test Barrette details | GB1 | GB2 | GB3 | Rock profile |
|---|---|---|---|---|
| Diameter (mm) | 2.80 × 1.20 | 2.80 × 1.20 | 2.80 × 1.20 | Calcarenite/Calcareous Sandstone (−1800 to −27.50)<br>Conglomerate (−27.50 to −33.00)<br>Calcisiltite/Calcareous Siltstone (−33.00 to −40.00)<br>Calcareous Siltstone (−40.00 to −45.00)<br>Calcisiltite (−45.00 to −70.00) |
| Length(m) | 45.125 | 47.825 | 48.60 | |
| Cutoff level (mDMD) | −24.875 | −22.175 | −20.375 | |
| Toe level (mDMD) | −70.00 | −70.00 | −68.975 | |
| Jack position (mDMD) | −48.30 | −46.90 | −45.40 | |
| Working load (kN) | 54,000 | 54,000 | 54,000 | |
| Test load (kN) | 162,000 | 162,000 | 162,000 | |
| Maximum achieved load (kN) | 180,320 | 183,110 | 183,120 | |
| Strain gauge levels (mDMD) | −27.0, −33.10, −39.2, −45.3, −51.3, −57.2, −63.1, and −69.0 | −24.3, −30.8, −37.3, −43.9, −49.9, −56.2, −62.5, and −69.0 | −22.5, −29.1, −35.7, −42.4, −48.4, −54.9, −61.4, and −67.9 | |

strain gauges comprising four units at each level were also installed on the test pile to measure strains at nominated locations. The data obtained from the site was analysed using an equivalent top-loading method to identify the elastic settlement [5, 11].

# 4 Results and Discussion

The data obtained from all the three tests were analysed, and results are presented. Table 2 gives the details of barrettes with hydraulic jack position and strain gauge levels. Table 3 summarizes load and settlement obtained for the three barrette load tests performed.

## *4.1 FEM Modelling*

Generally, a geotechnical analysis counts to ensure that the subsoil can stand the load transmitted by the supporting system to ensure a proper foundation design. Soil-structure interaction phenomena were found to have a significant impact on the design. The analysis is required for foundation elements when soil-structure interaction and multistage loading types are considered. The axial capacities of barrettes can be parametrically verified for different soil materials. The three-dimensional finite element programme, MIDAS GTS-NX was chosen to analyse the barrette capacity using settlement and unit skin friction parameters, obtained from the load tests. Finite element analyses were carried out to support the structural design and to obtain settlement and capacity estimates. Results on a 47 m barrette, when applying the revised soil parameters, indicated a total vertical settlement of 6.5 mm under the working load of 54,000 kN and a total settlement of 15.70 mm at 200% of the working load which compares well with the load settlement curve developed from load tests (Table

**Table 3** Settlement summary of Bidirectional static load tests

| Barrette no | Working load (kN) | Test load (kN) | Achieved load (kN) |
|---|---|---|---|
| GB1 | 54,000 | 162,000 | 180320 |
| | Elastic settlement (mm) | | |
| | 6.40 | 19.20 | 24.40 |
| GB2 | 54,000 | 162,000 | 183110 |
| | Elastic settlement (mm) | | |
| | 6.20 | 20.80 | 25.60 |
| GB3 | 54,000 | 162,000 | 183120 |
| | Elastic settlement (mm) | | |
| | 6.10 | 20.30 | 25.70 |

**Table 4** Settlement analysis of a single barrette element

| Barrette | | | Total vertical settlement (mm) | | |
|---|---|---|---|---|---|
| Method | Dimensions (m × m) | Length (m) | 100% working load | 150% working load | 200% working load |
| Load test | 2.80 × 1.20 | 47 | 6.20 | 9.70 | 13.90 |
| FEM | 2.80 × 1.20 | 47 | 6.50 | 11.10 | 15.70 |

4). It should be noted that induced settlement at the working load is well within the acceptable limit of 1.5% of the equivalent diameter as specified from BS 8004 [12].

The above results show that the results of the load test are in good agreement with the design values. The results of the single barrette model will be further utilized to modify the ground parameters to match the actual results in the Barrette group modelling in further stages of the group analysis to estimate total settlement under the group behaviour. The three barrette tests showed similar results with settlement ranging from 6.10 to 6.40 mm at the working load of 54,000 kN and up to 19.20–20.80 mm at three times the working load. Settlement results of a 47 m deep, 2.80 × 1.20 m barrette indicate a total settlement of 6.5 mm at the working load of 54,000 kN that is in good agreement with the 6.20 mm induced from the bidirectional static load tests.

## 4.2 *Barrette Capacity*

Based on the available results from barrette load tests, the ultimate skin friction provided [13] was revisited to match the results of 100% loading conditions. Mobilized skin friction for the three tests at 100, 200, and 300% of the working load (Table 5) and theoretical values are presented (using the highest cutoff level barrette no.GB3, −20.375 m DMD) in Fig. 3. Based on the analysis, modified capacity for a 2.80 × 1.20 m barrette with depth is obtained (Fig. 4). A comparison between the revised (from barrette load test results) and preliminary barrette capacity in compression is presented in Fig. 5. The test results show a good agreement with the theoretical and design parameters.

In order to monitor foundation behaviour, strain gauges were installed in eight levels of barrettes to compare the calculated and actual behaviour of the foundation. Preliminary load tests, performed on three barrettes, were established in order to achieve skin friction up to 3.0 times larger than the estimated serviceability values. During the assessment of the load test results, we considered that design skin friction may be increased from values reported before pile testing and the revised recommended values of allowable shaft friction are required. The side resistances determined from the load tests are higher than the initial design values adopted, leading to possible optimizations. The initial theoretical allowable unit skin friction was used for the preliminary barrette design (Fig. 3). Based on the three load test results, the

**Table 5** Mobilized unit skin friction from BDSLT

| Strain gauge levels (from top to bottom) | GB1 | | | GB2 | | | GB3 | | |
|---|---|---|---|---|---|---|---|---|---|
| | Unit skin friction (kPa) | | | | | | | | |
| | 100% Load | 200% Load | 300% Load | 100% Load | 200% Load | 300% Load | 100% Load | 200% Load | 300% Load |
| 1–2 | 64 | 106 | 195 | 70 | 134 | 151 | 38 | 161 | 253 |
| 2–3 | 162 | 275 | 361 | 117 | 263 | 438 | 128 | 239 | 400 |
| 3–4 | 179 | 417 | 655 | 182 | 352 | 540 | 185 | 340 | 494 |
| 4 to Jack | 283 | 573 | 812 | 299 | 553 | 826 | 325 | 564 | 737 |
| 5 to Jack | 325 | 603 | 896 | 405 | 649 | 896 | 317 | 559 | 727 |
| 5–6 | 192 | 425 | 590 | 201 | 421 | 636 | 222 | 353 | 464 |
| 6–7 | 172 | 298 | 441 | 83 | 211 | 363 | 116 | 276 | 404 |
| 7–8 | 42 | 109 | 211 | 40 | 90 | 122 | 20 | 119 | 287 |

maximum mobilized skin friction value calculated from the strain gauge readings is in the range of around 253–896 kPa. The unit skin friction values are increasing linearly and do not show any evidence of developing geotechnical failure. For all of the barrettes, the unit shaft resistance was mobilized at an average value of settlement equal to about 20 mm. This indicates that the barrettes can be still loaded to mobilize ultimate skin friction resistance along the complete shaft length. It can be concluded that the load tests can appropriately represent the characteristics of soil strata and the side resistances determined are much larger than the design values adopted. Hence, based on the theoretical load test results, revised theoretical values were derived and these values were used for the execution of barrettes in the site. Based on the above result, for a permanent compression load of 54,000 kN (Table 6) for the barrettes and assuming a cutoff level at −23.00 m DMD, a barrette length of 42 m is found to be sufficient for the foundation design. This corresponds to a total length reduction of 11% after the interpretation of barrette load test results.

## 5 Conclusions

Load test results and analysis indicate that the barrette capacities can be further optimized maintaining a sufficient factor of safety. The outcomes indicate that the barrette design can be optimized in length up to 11%, reducing the current barrette length from 47 to 42 m deep. Preliminary settlement results of a 47 m deep, 2.80 × 1.20 m size barrette indicate a total settlement of 6.5 mm at the working load of 54,000 kN that is in good agreement with the 6.20 mm induced from the bidirectional static load tests. It is identified that test barrettes can be used for foundation testing to virtually any high loads. The results obtained from the testing have to lead to a reassessment of the original pile design to benefit future stakeholders in designing

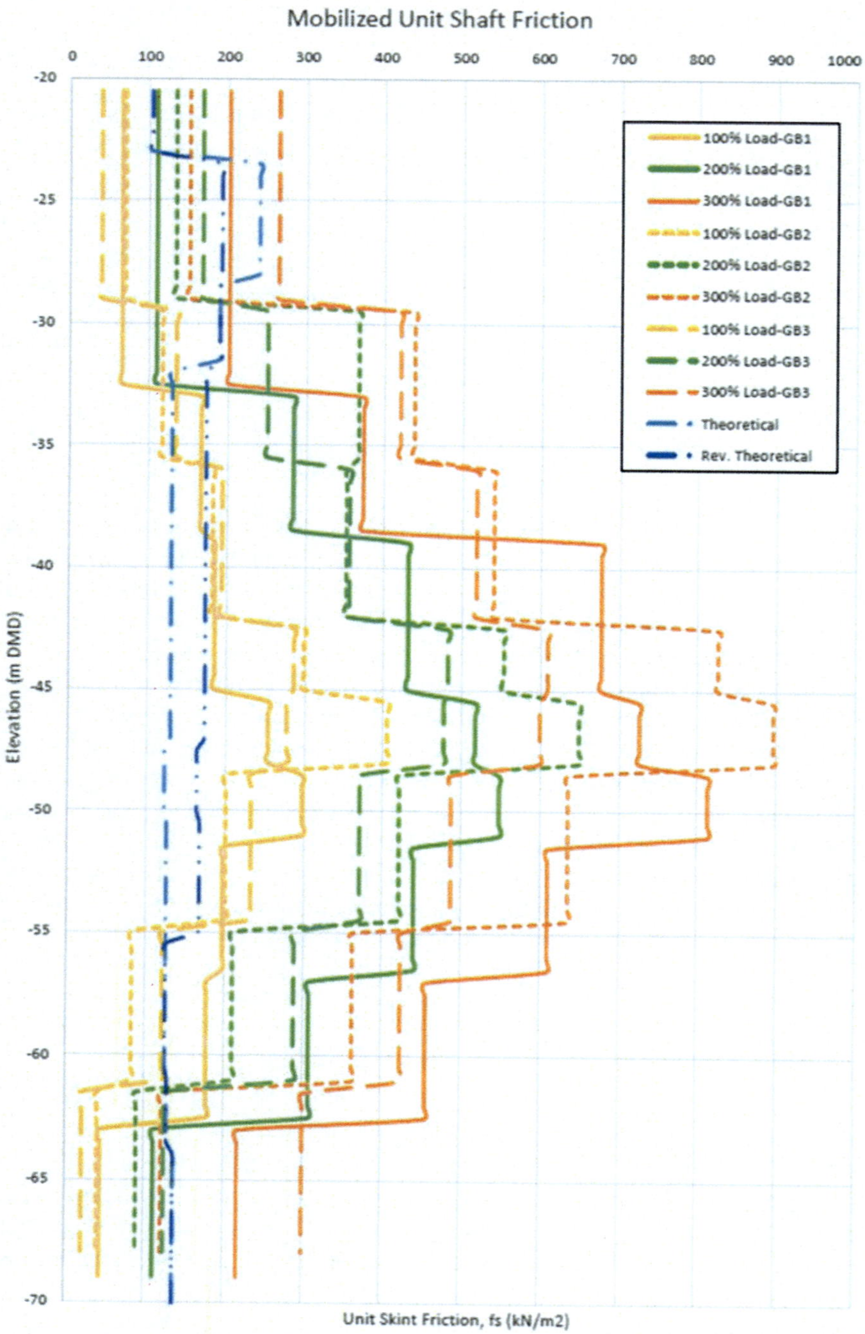


**Fig. 3** Mobilized skin friction versus load

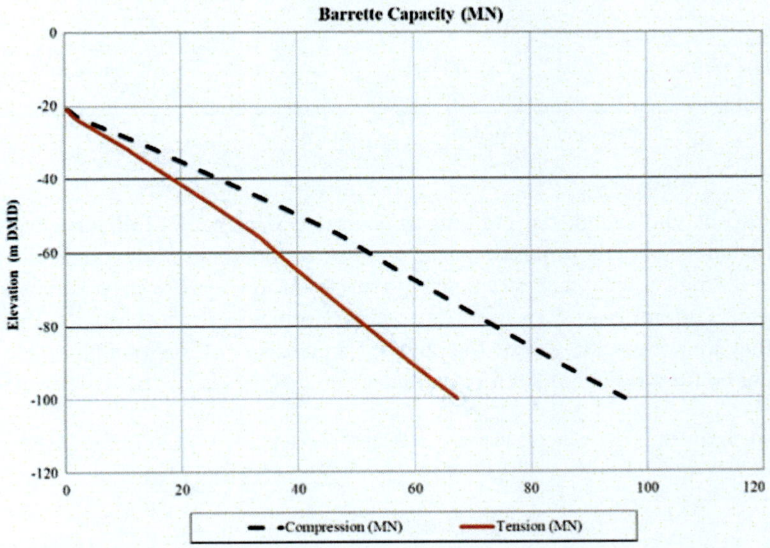


**Fig. 4** Revised barrette capacity based on BDSLT results

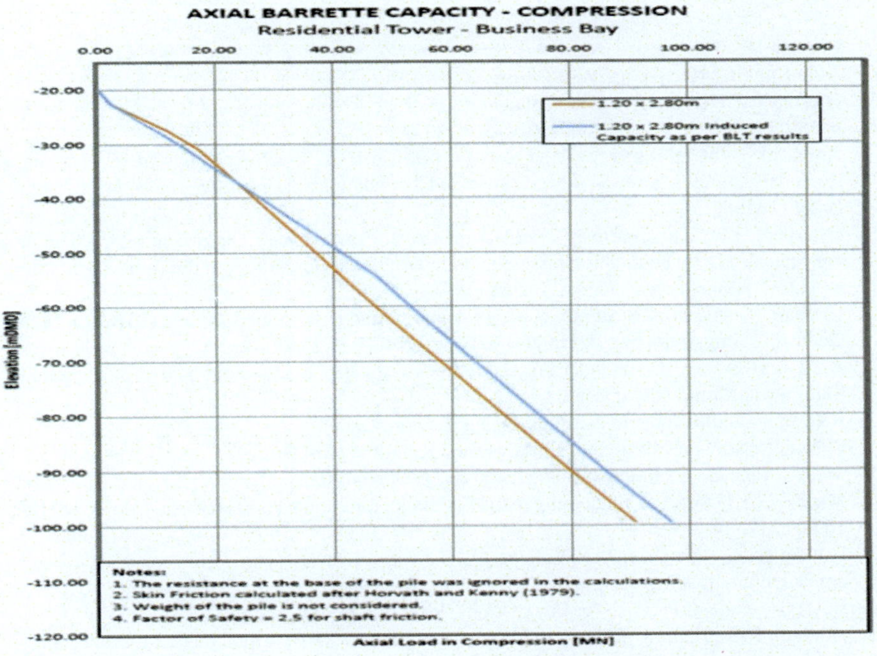


**Fig. 5** A comparison between load test and preliminary barrette capacity

**Table 6** Designed barrette after BDSLT

| Structural element | | Cutoff level (mDMD) | Toe level (mDMD) | Compression load (kN) |
|---|---|---|---|---|
| Barrette | 2.80 m × 1.20 m | −23.00 | −65.00 | 54,000 |

economically viable high-rise building projects. BDSLT enables full-scale testing of the foundation element proposed for the working foundations, allowing the designer to quantify the geotechnical parameters precisely. Moreover, the large dimension of the barrette often permits a variety of loading arrangements to virtually any high loads that are not constrained by the physical dimensions of the foundation element as would be the case in a piled foundation.

**Acknowledgements** The author is thankful to Strainstall Middle East LLC, Dubai, UAE for their support and inspiration. Also grateful to anonymous reviewers for their suggestions.

## References

1. Cherian, A.: The construction industry in the perspective of an economic boost of the United Arab Emirates (UAE). Int. Res. J. Eng. Technol. **7**(8), 270–276 (2020)
2. Gibson, G.L., Devenny, D.W.: Concrete to bedrock testing by jacking from the bottom of a borehole. Can. Geotech. J. **10**(2), 304–306 (1973)
3. Amir, J.M.: Interpretation of load tests on piles in rock. In: Proceedings of the 7th Asian Regional Conference on Soil Mechanics and Foundation Engineering, Haifa, pp. 235–238, Israel (1983)
4. Osterberg, J.O.: The Osterberg load test method for drilled shaft and driven piles—The first ten years. In: Deep Foundation Institute, 7th International Conference and Exhibition on Piling and Deep Foundations, Vienna, Austria (1998)
5. Cherian, A.: Assessment of pile capacity using bidirectional static load test (BDSLT). Indian Geotech. J. https://doi.org/10.1007/s40098-020-00447-x (2020)
6. ATM D8169/8169-18: Standard Test Methods for Deep Foundations under Bi-Directional Static Axial Compressive Load, USA (2018)
7. ICE: Manual of Geotechnical Engineering, London (2012)
8. FPS: Federation of Piling Specialists, United Kingdom (2006)
9. Materal Lab: Geotechnical Investigation report, Dubai (2012)
10. Macklin, S., Gaba, A.: Engineering in the Barzaman formation, coastal Dubai, UAE. Civil Eng. **162** (6), 18–24 (2009)
11. Hoyoung, S., Rozbeh, B.M., William, D.L.: Assessment of methods for construction of an equivalent top loading curve from O-cell test data. Soils Found. **56**(5), 889–903 (2016)
12. BS 8004: Code of Practice for Foundations, Section 6.8.2.1, London (2015)
13. Horvath, R.G., Kenney, T.C.: Shaft resistance of rock socketed drilled piers. In: Proceeding of Symposium on Deep Foundations, ASCE, New York (1979)

# Numerical Analysis of Load-Carrying Capacity of Fibre-Reinforced Polymer Piles

**A. I. Dhatrak, Swetha N. Bhadke, and S. W. Thakare**

## 1 Introduction

Fibre-reinforced polymer (FRP) is a composite material which contains fibre and resin that provide high tensile strength. This material is also known as fibre-reinforced plastics or advanced composite materials (ACMs). Fibre-reinforced materials are lightweight, have high specific strength, durability and are resistance to chemicals and corrosive environments. FRP is well known for its high ratio of strength to weight and high ratio of longitudinal/transversal shear modulus. It can be categorized into Aramid FRP (AFRP), Basalt FRP (BFRP), Carbon FRP (CFRP) and Glass FRP (GFRP) depending on the materials of fibre.

There are wide utilizations of piles whereby some issues occur when piles are located in harsh environments especially in marine or coastal conditions. The piles with traditional materials could be destroyed due to corrosion of steel, deterioration of timber and degradation of concrete as shown in Fig. 1. FRP piles can be used as an alternative construction material to these piles which gets damaged in the aggressive environment. FRP composite piles are composed of a hollow FRP pipe that is filled before installation with an expanding concrete as shown in Fig. 2 and is coated with a durable corrosion-resistant coating layer. The hollow pipe is produced from unsaturated polyester or epoxy resins with reinforcement rovings (E-glass) and appropriate filler material to form a rigid material. E-glass is incorporated as continuous rovings and is set in resin under pressure during the fabrication process. The FRP composite

A. I. Dhatrak
Government College of Engineering, Amravati, Maharashtra, India

S. N. Bhadke (✉)
M.Tech Geotechnical Engineering, Civil Department Government College of Engineering, Amravati, Maharashtra, India

S. W. Thakare
Civil Engineering Department, Government College of Engineering, Amravati, Maharashtra, India

 
C. N. V. Satyanarayana Reddy et al. (eds.), *Ground Characterization and Foundations*, Lecture Notes in Civil Engineering 167,
https://doi.org/10.1007/978-981-16-3383-6_70

(a) (b) (c)

**Fig. 1** Traditional piles in aggressive environment **a** corrosion of steel, **b** deterioration of timber and **c** degradation of concrete [1]

(a) (b)

**Fig. 2** FRP composite pile **a** hollow FRP pipe and **b** concrete-filled FRP pile [1]

pile can be used in marine environments in a variety of applications, viz. fender piling, dauphins, light structural piling and bridge pier protection (Fig. 3). FRP piles can also be used as deep foundations in marine front structures such as jetties, boardwalks and seaside buildings. The load-bearing capacity of FRP piles is governed by combination of end bearing and friction resistance.

Shaia et al. [3] carried experimental test on CFRP and GFRP piles subjected to the axial and lateral load. A long-term study was undertaken to assess the effect of FRP degradation on the pile capacity where the piles were exposed to acidic and alkaline soil environments for 180 days. Numerical analysis was carried out using ABAQUS software to simulate and validate the experimental results of FRP

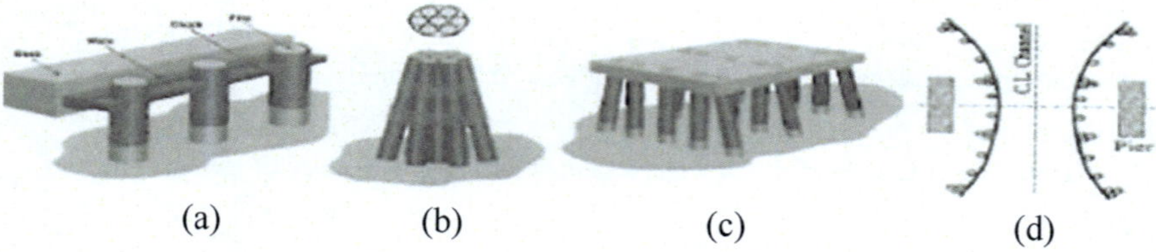

(a) (b) (c) (d)

**Fig. 3** Different applications of FRP composite piles **a** fender pile, **b** dauphin, **c** light structural piles and **d** bridge pier protection [2] pile model. Valez [4] conducted laboratory investigation on small-scale hollow FRP piles in undisturbed clay samples and compared results with the steel piles. Their findings showed that the FRP piles in soft clays can be used as an alternative solution to traditional steel piles. Nemr et al. [5] investigated the FRP pile behaviour in sandy soil using finite element modelling under static loading condition. Their findings show that as the stiffness of the pile increases the settlement decreases. Giraldo et al. [6] conducted small-scale field test on hollow fibre-reinforced polymer (FRP) piles in the 'Canadian Geotechnical Research Site' located at Gloucester, Ontario. The load transfer behaviour of FRP piles manufactured using glass and carbon fibres was studied and compared to that of traditional steel piles. Fibre orientation had major role in determining the pile capacity.

## 2 Methodology

In the present study, FRP pile behaviour was investigated using finite element modelling under static loading condition using the MIDAS GTS NX software. Figure 4 shows the typical FRP pile used in analysis and pile model developed in MIDAS GTS NX. The numerical model represents the cylindrical tube pile filled with concrete embedded in marine layered soil bed of width 40d, where d is the diameter of pile with water table at the ground level. The tubes were modelled as orthotropic unidirectional laminates in three direction (direction 1 being the fibre direction and direction 2, 3 being the perpendicular to the fibre). Table 1 provides the mechanical properties of CFRP, and GFRP laminates in the hoop and longitudinal direction. For both CFRP and GFRP type, the layer of composite tubes was modelled using unidirectional mechanical properties in longitudinal orientation (0°). The concrete pile was modelled as a linearly elastic perfectly plastic model. Table 2 shows the properties assigned to concrete pile for analysis. The characteristics and

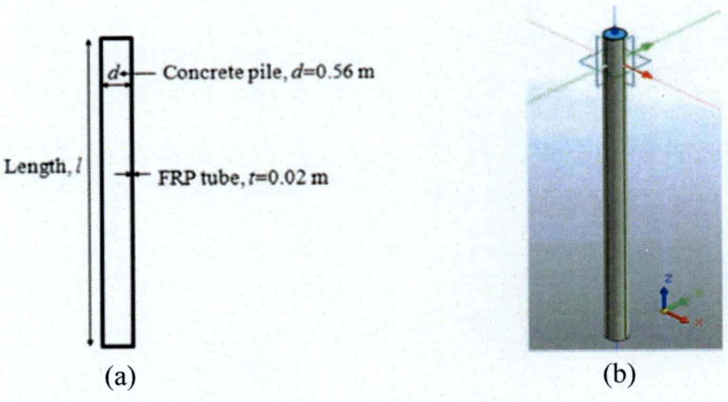


**Fig. 4** FRP pile used in analysis **a** schematic diagram of FRP piles and **b** FRP pile model developed in MIDAS GTS NX

**Table 1** Properties assigned to FRP pile for analysis [3]

| Property | Symbol | Unit | CFRP | GFRP |
|---|---|---|---|---|
| Density | $\rho$ | g/cm$^3$ | 1.6 | 2.0 |
| Longitudinal modulus[a] | $E_{11}$ | MPa | 135,000 | 50,000 |
| Transverse in-plane modulus[a] | $E_{22}$ | MPa | 10,000 | 40,000 |
| Transverse out-plane modulus[a] | $E_{33}$ | MPa | 10,000 | 8500 |
| In-plane shear modulus[b] | $G_{12}$ | MPa | 5000 | 4300 |
| Out-of-plane shear modulus[b] | $G_{23}$ | MPa | 1900 | 3500 |
| Out-of-plane shear modulus[b] | $G_{13}$ | MPa | 5000 | 4300 |
| Major in-plane Poisson's ratio[b] | $\upsilon_{12}$ | – | 0.3 | 0.27 |
| Poisson's ratio[b] | $\upsilon_{23}$ | – | 0.3 | 0.27 |

Where *CFRP*—carbon fibre-reinforced polymer, *GFRP*—glass fibre-reinforced polymer, [a]provided by manufacture, [b]provided by Wu [7]

**Table 2** Properties assigned to concrete pile for analysis [3]

| Material | Model | Young's modulus, $E$ (Mpa) | Poisson's ratio, $\upsilon$ | Unit weight, $\gamma$ (kN/m$^3$) |
|---|---|---|---|---|
| Concrete pile | Elastic | 28,000 | 0.22 | 25 |

thickness of various layers had been selected from the Chennai port trust (India) with the actual soil profile of the marine environment. The soil layers were modelled as elastic perfectly plastic model based on the Mohr–Coulomb failure criterion. Table 3 shows the properties assigned to soil layers for analysis. The results of CFRP and

**Table 3** Properties assigned to soil for analysis [8]

| Depth, m | Unit weight, $\gamma$ (kN/m$^3$) | Poisson's ratio, $\upsilon$ | Effective cohesion, $c$ (kN/m$^2$) | $\phi°$ | Modulus of elasticity, $E$ (MPa) |
|---|---|---|---|---|---|
| 0–14 (Loose silty sand) | 12 | 0.25 | – | 28 | 8000 |
| 14–20 (Medium dense silty sand) | 15 | 0.3 | – | 30 | 17,000 |
| 20–33 (Dense silty sand) | 19 | 0.35 | – | 38 | 50,000 |
| 33–35 (Slightly weathered to fresh clay) | 21 | 0.4 | 10 | – | 52,500 |
| 35–40 (Highly to moderately weathered granite) | 22 | 0.4 | 33.3 | – | 120,000 |

GFRP piles tube filled with concrete were compared to that of conventional concrete piles and conclusions were drawn.

## 3 Numerical Analysis

The model developed was auto-connected by using Boolean operation. Separate meshing was provided for pile and soil model. Mesh was generated by using tetrahedron elements produced by the auto-mesh generation function. Figure 5 shows geometry of FRP pile embedded in soil after mesh generation. Boundary condition for geometry was defined by restraining the sides of soil model in the *x*-direction and y-direction and the bottom boundary in all directions. An interface element was generated around the FRP pile to account the soil–pile interaction. The analysis was then carried out by selecting the nonlinear analysis case.

Thus, the analysis carried out aims to investigate the load settlement response of FRP pile and conventional concrete pile embedded in marine layered soil bed. The various parameters considered for the study were type of FRP pile material, type of loadings, number of piles in the group and slenderness ratio. The parameters of pile which were kept constant are given in Table 4. Table 5 shows various parameters varied during analysis.

**Fig. 5** Geometry of FRP pile embedded in soil after mesh generation in MIDAS GTS NX

**Table 4** Constant dimensions of FRP pile

| S. No. | Parameter | Values |
|---|---|---|
| 1 | Dimensions for pile | Diameter of pile = 0.6 m |
| | | Thickness of FRP tube, $t = 0.02$ m |

**Table 5** Details of parametric study for FRP pile

| S. No. | Parameter | Values |
|---|---|---|
| 1 | Slenderness ratio (*l*/*d*) (*d* constant) | 11/0.6, 16/0.6, 21/0.6, 26/0.6, 31/0.6 |
| 2 | Type of FRP pile material | Glass fibre-reinforced polymer piles (GFRP)<br>Carbon fibre-reinforced polymer piles (CFRP) |
| 3 | Type of loadings | Axial loading<br>Lateral loading<br>Uplift loading |
| 4 | Number of piles (*N*) | (i) Single pile<br>(ii) Pile group with 5 and 7 |
| 5 | Spacing between piles | 3d |

# 4 Results and Discussions

## *4.1 Performance of Single FRP Pile*

Initially, the analysis was conducted on single conventional circular pile and FRP pile subjected to vertical load, uplift load and lateral load in marine layered soil bed. The ultimate load capacity for conventional pile and FRP pile were considered as per the provision of IS: 2911 (Part-4) 1985. The ultimate capacities as determined from the load settlement curves are presented in Table 6. The results obtained from the analysis were discussed, and the effect of *l*/*d* ratio on performance of single GFRP and CFRP pile was studied. Figure 6 shows the variation in ultimate load capacity, ultimate uplift capacity and ultimate lateral load capacity for conventional, GFRP and CFRP pile with respect to the *l*/*d* ratios.

The percentage increase in ultimate vertical load, uplift load and lateral load compared with conventional circular pile is given in Table 7. The percentage increase in ultimate vertical, uplift and lateral load capacities of single GFRP and CFRP pile with slenderness ratio is shown in Fig. 7 in the form of bar charts.

From the above results, it is observed that the ultimate vertical, uplift and lateral load capacities of single GFRP as well as CFRP pile are higher than that of conventional concrete pile. Ultimate vertical, uplift and lateral capacities of GFRP and CFRP pile increase rapidly with increase in *l*/*d* ratio up to 43.33; with further increase in *l*/*d* ratio, there is no significant increase in these capacities. Therefore, analysis of group of FRP piles for ultimate load capacities was carried out by keeping *l*/*d* ratio of 43.33 as constant parameter. Also, the percentage increase in ultimate load capacities of

**Table 6** Ultimate load capacities of conventional and FRP pile with different slenderness ratio

| Loading condition | Type of pile | Ultimate load capacities (kN) | | | | |
|---|---|---|---|---|---|---|
| | | Slenderness ratio ($l/d$) | | | | |
| | | $l/d = 18.33$ | $l/d = 26.66$ | $l/d = 35$ | $l/d = 43.33$ | $l/d = 51.66$ |
| Vertical | Conventional concrete pile | 990 | 2300 | 4000 | 4900 | 5100 |
| | GFRP | 1300 | 2800 | 5200 | 6400 | 7600 |
| | CFRP | 1600 | 3200 | 5600 | 7200 | 8000 |
| Uplift | Conventional concrete pile | 625 | 1200 | 1900 | 2800 | 3150 |
| | GFRP | 700 | 1300 | 2195 | 3250 | 3500 |
| | CFRP | 820 | 1410 | 2500 | 3820 | 4200 |
| Lateral | Conventional concrete pile | 140 | 160 | 180 | 190 | 210 |
| | GFRP | 180 | 200 | 250 | 275 | 300 |
| | CFRP | 190 | 230 | 260 | 280 | 310 |

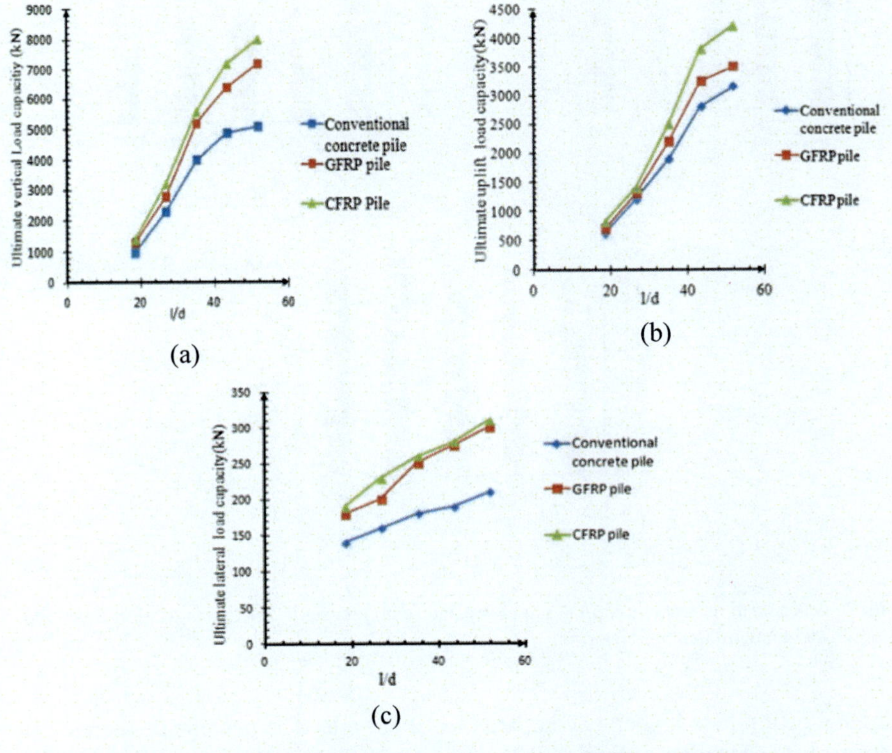


**Fig. 6** Variation in ultimate load capacities of piles with respect to $l/d$ ratio **a** vertical, **b** uplift and **c** lateral loading condition

**Table 7** Percentage increase in ultimate load capacities of FRP pile with different slenderness ratio

| Loading condition | Type of pile | Percentage increase in pile capacity (%) | | | | |
|---|---|---|---|---|---|---|
| | | Slenderness ratio ($l/d$) | | | | |
| | | $l/d = 18.33$ | $l/d = 26.66$ | $l/d = 35$ | $l/d = 43.33$ | $l/d = 51.66$ |
| Vertical | GFRP | 31 | 22 | 30 | 31 | 41 |
| | CFRP | 41 | 39 | 40 | 47 | 57 |
| Uplift | GFRP | 12 | 8 | 16 | 16 | 11 |
| | CFRP | 31 | 18 | 32 | 36 | 33 |
| Lateral | GFRP | 29 | 25 | 39 | 45 | 43 |
| | CFRP | 36 | 44 | 44 | 47 | 48 |

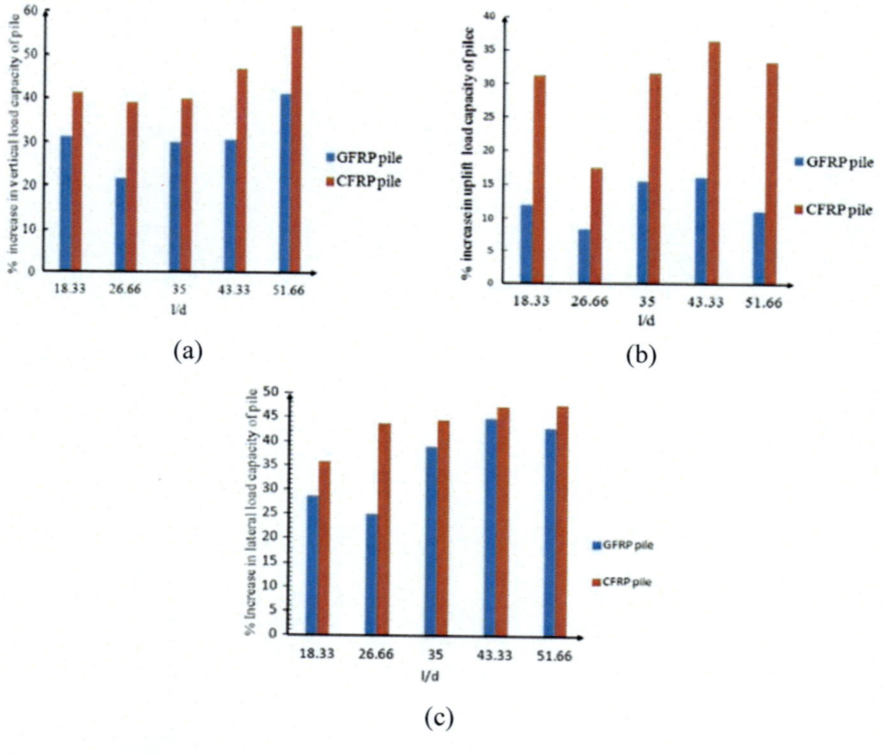


**Fig. 7** Percentage increase in ultimate load capacity with respect to $l/d$ ratio variation **a** vertical, **b** uplift and **c** lateral loading condition

single CFRP pile is higher as compared to GFRP pile. The percentage increase in vertical load capacities of single FRP piles is higher as compared to uplift and lateral load capacities.

**Table 8** Ultimate load capacities of conventional, GFRP and CFRP pile group for vertical, uplift and lateral loading

| Loading conditions | Type of pile | Ultimate load capacities (kN) | |
|---|---|---|---|
| | | No. of piles in group | |
| | | 5 (square arrangement with pile at centre) | 7 (hexagonal arrangement with pile at centre) |
| Vertical | Conventional concrete pile | 29,000 | 36,000 |
| | GFRP | 39,150 | 51,000 |
| | CFRP | 46,666 | 64,000 |
| Uplift | Conventional concrete pile | 7988 | 11,000 |
| | GFRP | 10,050 | 16,500 |
| | CFRP | 14,000 | 20,500 |
| Lateral | Conventional concrete pile | 700 | 1100 |
| | GFRP | 1022 | 1705 |
| | CFRP | 1302 | 2100 |

## 4.2 *Performance of Pile Groups*

The analyses were conducted on pile groups consisting of conventional as well as FRP piles subjected to vertical, uplift and lateral load. The analyses were performed consisting of group of five piles in square arrangement with pile at centre and seven piles in hexagonal arrangement with pile at centre for conventional, GFRP and CFRP pile and their ultimate load capacities were determined. The ultimate capacities for the pile group as determined from the load settlement curves are presented in Table 8.

The percentage increase in ultimate vertical load, uplift load and lateral load compared with conventional circular pile group is given in Table 9.

The percentage increase in ultimate vertical, uplift and lateral load capacities of group of piles with respect to number of piles in group is shown in Fig. 8, respectively, in the form of bar chart.

From the above results, it is observed that the ultimate vertical, uplift and lateral load capacity of FRP pile group is higher than that of conventional concrete pile. Also, the ultimate vertical, uplift and lateral load capacity of CFRP pile group is higher than GFRP pile group. The ultimate vertical, uplift and lateral load capacities of group of seven piles in hexagonal arrangement with pile at centre is higher as compared to group of five piles in square arrangement with pile at centre. The percentage increase in ultimate load capacities of CFRP pile group is higher as compared to GFRP pile group. The percentage increase in lateral load capacities of FRP pile group is higher as compared to vertical and uplift load capacities.

**Table 9** Percentage increase in ultimate load capacities of FRP pile groups compared with conventional circular pile group

| Loading condition | Type of pile | Percentage increase in pile capacities (%) | |
|---|---|---|---|
| | | No. of piles in group | |
| | | 5 (square arrangement with pile at centre) | 7 (hexagonal arrangement with pile at centre) |
| Vertical | GFRP | 35 | 42 |
| | CFRP | 61 | 78 |
| Uplift | GFRP | 26 | 50 |
| | CFRP | 75 | 86 |
| Lateral | GFRP | 46 | 55 |
| | CFRP | 86 | 91 |

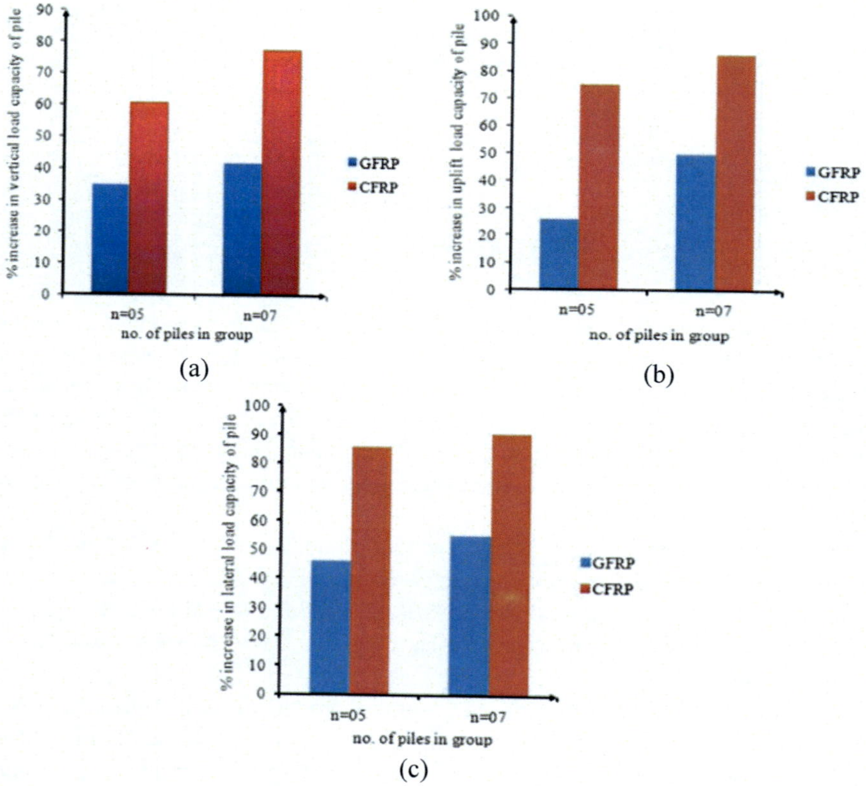


**Fig. 8** Percentage increase in ultimate load capacities of FRP pile groups **a** vertical, **b** uplift and **c** lateral loading condition

## 5 Conclusions

In present study, FRP pile foundation has been analysed in marine layered soil of Chennai port trust (India) using MIDAS GTS NX software. From the present study, following broad conclusions are drawn.

1. The ultimate vertical, uplift and lateral load capacity of single FRP pile is higher than that of conventional concrete pile. The percentage increase in ultimate load capacities of single CFRP pile is higher as compared to GFRP pile. Also, the percentage increase in vertical load capacities of single FRP piles is higher as compared to uplift and lateral load capacities.
2. Ultimate vertical, uplift and lateral capacities of GFRP and CFRP pile increase with the *l*/*d* ratio up to 45. With further increase in *l*/*d* ratio, there is no further significant increase in these capacities.
3. The ultimate vertical, uplift and lateral load capacity of FRP pile group are higher than that of conventional concrete pile group. The capacities of CFRP pile group are higher than GFRP pile group.
4. The percentage increase in ultimate load capacities of CFRP pile group is higher as compared to GFRP pile group.
5. The percentage increase in ultimate load capacities of group of seven piles is higher as compared to group of five piles.
6. The percentage increase in lateral load capacities of FRP pile group is higher as compared to vertical and uplift load capacities.

## References

1. Juran, I., Komornik U.: Behavior of Fiber-Reinforced Polymer (FRP) Composite Piles Under Vertical Loads, FHWA-HRT-04-107 (2006)
2. Zhou, J., Erwin, Oh.: A review of geotechnical application of fibre reinforced polymer materials. In: Eighth International Conference on Fibre Reinforced Polymer (FRP) Composites in Civil Engineering, pp. 856–861 (2016)
3. Shaia, H., Abuel-Naga, H.: Behaviour of Fibre Reinforced Polymer Composite Piles: Experimental and Numerical Study. Ph.D. Thesis, Department of Civil Engineering, Manchester University (2013)
4. Valez, J.: Experimental Study of Hollow Fibre Reinforced Polymer Piles in Soft Clay. Department of Civil Engineering, Carleton University, Thesis (2013)
5. Nemr, M., Hekal, G.: Finite element modeling of confined concrete piles with FRP tubes in sandy soil under static loading. In: The Sixth International Conference on SEMC, pp. 2122–2127 (2016)
6. Giraldo, J., Rayhani, T.: Axial and lateral load transfer of fibre reinforced polymer (FRP) Piles in soft clay. Int. J. Geotech. Eng. **11**(2) (2016)
7. Wu, S.: Finite Element Analysis and Theoretical Study on FRP Wrap Concrete Cylinders. Department of Civil Engineering, Manchester University, Masters (2011)
8. Premalatha, P.V., Kumar, S., Baskar, K.: Influence of change in pile diameter at various locations of a pile group in a Berthing Structure. Indian J. Geo Marine Sci. (2017)

# Engineering Performance of the Foundation of Thanjavur Brihadeeswarar Temple

**K. Arunkumar and K. Premalatha**

## 1 Introduction

The Indian heritage is respected and held in high esteem throughout the world. One of the major features of this rich heritage includes the ancient religious monuments. Ancient era was the witness when religious practices flourished the most and temples became the world's storehouse of knowledge and culture. Ancient temple is representing the history of culture, religion, science and technology. India has many splendid temples that have found a place in the World Heritage list. These include Sun temple at Konark, Khajuraho temple, Ajanta Caves, Brihadeeswarar Temple and Sanchi Stupas. These temples are resistant to most natural calamities due to their geometry, material and construction technique. The foundation is considered to be the major contributing factor for their structural stability. Their work could be carried out in many different aspects, taking into account in particular: the kind of foundation and the object placed upon it, the material, the shape and dimensions of the foundation, bearing capacity and settlement of foundation and geological substructure conditions. Stone foundations were mostly used for ancient structures and public utility structures. Various types of stones were used, depending on the function of specific elements of a building. Granite, limestone, sandstone, sandstone volcanic tuff, and marble were used for this type of foundation. Research on ancient foundation techniques can bring about a huge impact on today's foundation technologies. Most of the studies carried out so far concentrate more on superstructures like architecture, material, arrangement, order, pattern, science behind the distribution of spaces. But there has been noticeably less study on the foundations of the historical structures, especially Indian temples, since most of these temples are still living and it is very difficult to access or get permission to excavate to understand the

K. Arunkumar (✉) · K. Premalatha
Division of Soil Mechanics and Foundation Engineering, Department of Civil Engineering, College of Engineering Guindy, Anna University, Chennai 600025, India

 
C. N. V. Satyanarayana Reddy et al. (eds.), *Ground Characterization and Foundations*, Lecture Notes in Civil Engineering 167,
https://doi.org/10.1007/978-981-16-3383-6_71

nature of the foundations. There is no study about Thanjavur Brihadeeswarar temple foundation. Since the present study is about static load behaviour of foundation of Thanjavur Brihadeeswarar temple Garphagriha tower.

The Thanjavur temple was constructed by the great Raja Raja Cholan who ruled Tamil Nadu in the period 1010 CE with help of architect Kunjalamallan. It took nearly 7 years to construct the above temple. The Brihadeeswarar temple plan and development, utilize the axial and symmetrical geometry rules. The temple complex is a rectangle that is almost two stacked squares, covering 240.79 m east to west, and 121.92 m north to south. The height of the temple is about 59.75–65.85 m (As per IGNCA = 59.82 m). The entire structure is made of the granites by using buzzle technology which means force applied at the top to prevent the movement of rock without any binder. The main Vimana (Shikhara) is a massive 16 stories tower of which 13 are tapering squares [1, 2]. It dominates the main quadrangle. It sits above a 30.18 m (99.0 ft.) sided square [3]. During thousand-year celebration function in 2010 Archaeology Department have drilled a borehole for up to 100 feet distance in which they have got sand only [4]. This sand obtained was different from soil nearby area. The location coordinates of this temple are 10°46′ 58″ N 79°7′ 54″ E [5]. The aerial photographic view of the Thanjavur Brihadeeswarar temple is shown in Fig. 1. The inner and outer view of the grabhagriha tower is shown in Fig. 2.

**Fig. 1** Aerial view of the Thanjavur Big temple

**Fig. 2** Outer and Inner view of Thanjavur temple Garbhagriha Tower

## 2 Literature Review

Tan et al. [6] studied about visual modelling of Chinese temple construction for student education. This study concludes that temple construction site present with incompetent soil, to increase the strength without excavating deeper depth. To excavate the soil at shallow depth, fill with dense sand. To install the stones around the perimeter and insides also. This method of installation gives higher strength without excavating deeper depth. Daka et al. [7] studied about the sandbox technique which was used to build the foundation for two major temples in Telangana viz. Ramappa temple and Thousand pillar temple. As part of this study, experiments were carried out on sand (dry and wet) and wet red soil materials by building a model which mimics the sandbox technique. In this context, piezoelectric knock sensors are used to capture the vibrations. This study determined the dampening of vibrations for sand (dry and wet) and wet Red soil for various types of loads. Wethyavivorn et al. [8] studied about the model verification of Thai historic masonry monuments. Two dominant styles of Thai historic masonry monuments, bell-shaped and corn-shaped, were investigated. Both styles of monuments were analyzed under their self-weight using homogenized material properties. Based on the results it was concluded that the mass of underlying subsoil foundation which must be included in the model is at least 99.8% of the mass of the monument. Iwasaki et al. [9] studied about the authenticity of the soils and foundation of heritage structure of Bayon temple, Angkor, Cambodia. The heritage structure of Bayon in Angkor stands upon shallow direct foundation. Based on this study it was concluded that the original ground was excavated 2–3 m from the surface and filled back with compacted sandy soil and the manmade fill shows very high bearing strength due to special characteristics of the grain size distribution of sandy soil. Sargunan [3] studied the "THE BIG TEMPLE, THANJAVUR" [3]. The architects conducted load tests on small sections of the rock

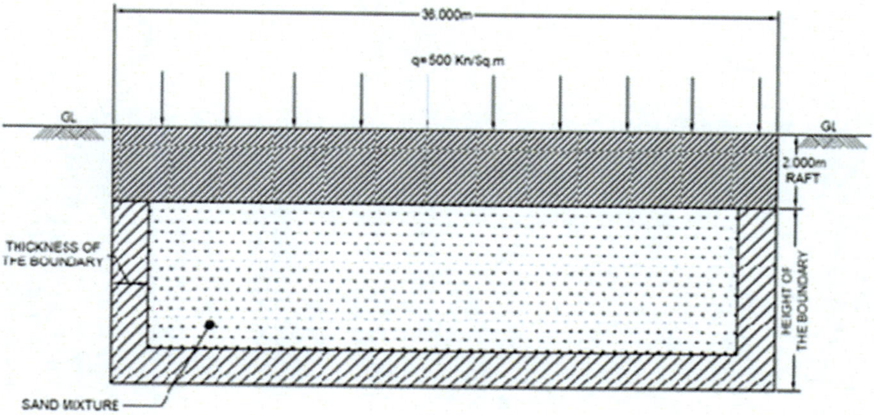


**Fig. 3** Cross-section view of foundation model

(at foundation level). The test plot was around 1.2 m × 1.2 m and four stone blocks are stacked up layer by layer at 0.6 m × 0.6 m × 0.6 m. Based on this study it was concluded that the tower was designed for zero settlement.

## 3 Methodology

The present study is focused on the behaviour of Thanjavur temple garbhagriha tower foundation for static load conditions. The various factors influencing the performance of ancient monuments are identified based on the detailed review of literatures. The influence of these parameters on the behaviour of the foundation is to be investigated by numerical analysis. The numerical analysis is performed by using PLAXIS 3D software. Subsoil profile details of this place collected from the State Ground and Surface water Resources Data centre, Tharamani, Chennai. The soil interface material is modelled as a 10 noded tetrahedral element. Soil is modelled by using the Mohr–Coulomb model. The various parameters are to be considered for the numerical study and they are (a) Without the boundary (b) Thickness of the boundary c) Height of the boundary. The cross-section view of the foundation model [3] shown in Fig. 3.

## 4 Validation of the Software

To verify the FEM model, the data reported by Desai and Desai [10] is selected. The soil and foundation details (Table 1) as reported by the author are given in Table 1. The author modelled the foundation in PLAXIS 2D modelling software. For the validation, the soil layer and the foundation model are simulated using PLAXIS 3D.

**Table 1** Soil and footing parameters [10]

| Parameter | Value |
|---|---|
| Material model | Mohr–Coulomb |
| Material type | Drained |
| Unsaturated unit weight (kN/m$^3$) | 16 |
| Saturated unit weight (kN/m$^3$) | 17.12 |
| Young's Modulus ($E$) (kN/m$^2$) | 40,000 |
| Poisson's ratio ($\mu$) | 0.3 |
| Cohesion ($c$) (kN/m$^2$) | 0 |
| Angle of internal friction ($\Phi$) (degree) | 30 |
| Width of the footing ($B$) (mm) | 2000 |
| Thickness ($d$) (m) | 0.353 |

In 3D modelling, boundary dimension of 50 m × 50 m × 20 m is used. The output details obtained from the PLAXIS 3D modelling are given in the form of settlement contour in Fig. 5. The settlement contour is given by Desai and Desai [10] is presented in Fig. 4. From the results, the settlement of the foundation is found to be 73 mm, which is almost comparable with the settlement data reported by Desai and Desai [10]. The settlement results (Table 2) are comparable and hence the modelling and analysis are valid.

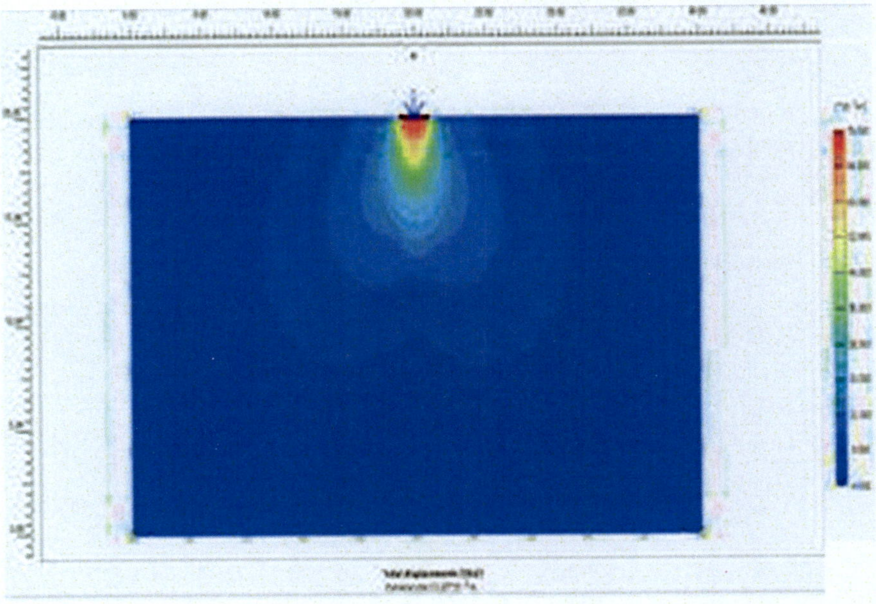

**Fig. 4** Output from PLAXIS 2D—settlement contour [10]

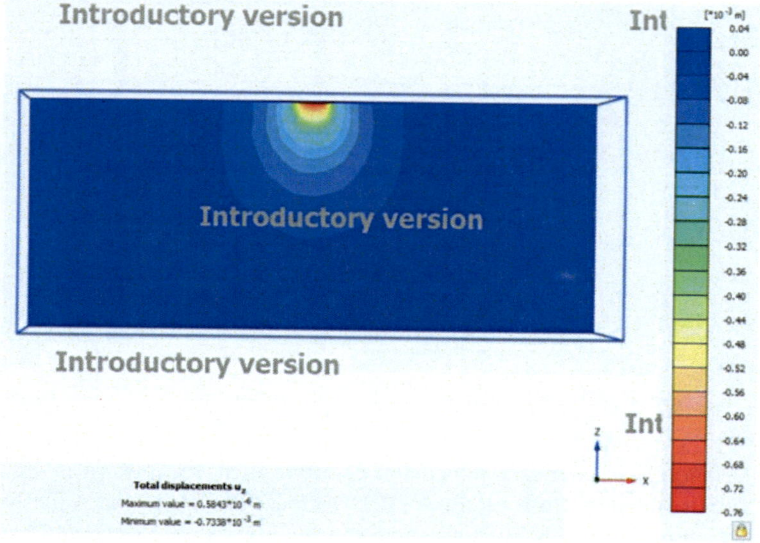


**Fig. 5** Output from PLAXIS 3D—settlement contour

**Table 2** Comparison of settlement results

| Methods (mm) | Vertical settlement (mm) |
|---|---|
| PLAXIS 2D [10] | 80 |
| PLAXIS 3D (validation analysis) | 74 |

The properties of materials (Table 3) used in the numerical analysis are tabulated below. The properties of the materials are taken from the corresponding IS codes [11–14].

**Table 3** Properties of materials

| Properties | Laterites | Sandstone | Granite | Sand mixture |
|---|---|---|---|---|
| Density ($\gamma$) (kN/m$^3$) | 20 | 21 | 27 | 20 |
| Young's modulus (E) (kPa) | $1 \times 10^6$ | $30 \times 10^6$ | $(40–70) \times 10^6$ | $100 \times 10^3$ |
| Poisson's ratio ($\mu$) | 0.3 | 0.2 | 0.2 | 0.3 |
| Angle of internal friction ($\Phi$) (degree) | 30 | 30 | 51 | 40 |
| Cohesion © (kPa) | 5000 | $30 \times 10^6$ | 55,000 | |

## 5 Results and Discussion

Displacement and Vertical stress values for the different boundary conditions are determined by considering the Mohr–Coulomb model by using PLAXIS 3D finite element software. A model of 200 m × 200 m × 80 m is created in the PLAXIS in which the foundation is formed to get the displacement and vertical stress values. The results obtained are plotted in the form of graph to show the displacement and vertical stress variations based on boundary conditions. Height of the boundary Vs Displacement for various boundary condition is shown in Fig. 6.

The displacement value of 2 m raft is more than 3 m thickness of the boundary. Because of the boundary thickness increases, displacement distributed along the sides of confinement. Displacement profile for 3 m boundary thickness and 6 m height of the boundary is shown in Fig. 7.

Height of the boundary versus Vertical stress for various boundary conditions are shown in Fig. 8.

The most of the stresses due to applied load are taken by the boundary of the sandbox because of the confinement (see Fig. 9). The maximum stress values are obtained at these coordinates of the boundary of the sandbox. Due to the boundary i.e. by providing sandbox, the stresses transferred are less than without boundary. Vertical stress for 3 m boundary thickness and 7 m height of the boundary is shown in Fig. 9.

This temple foundation constructed by granite stone. The young's modulus of granite stone is varying between 40 and 70 kPa. The young's modulus value increases, settlement and stress values decreased. But young's modulus values varying between

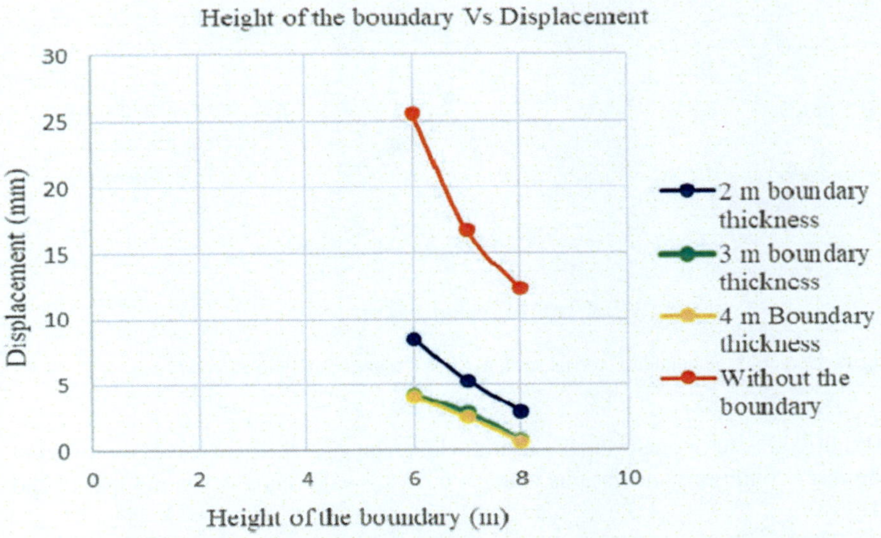


**Fig. 6** Height of the boundary versus displacement for various boundary conditions

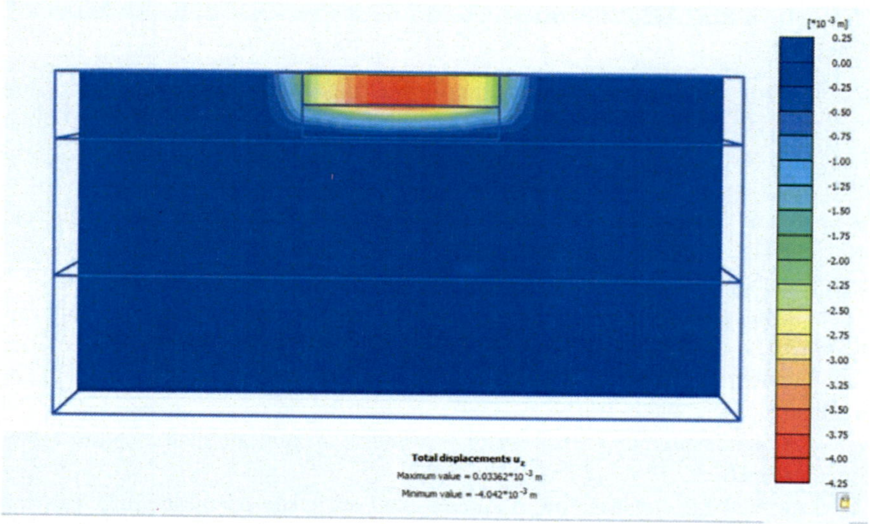


**Fig. 7** Displacement profile for 3 m boundary thickness and 6 m height of the boundary

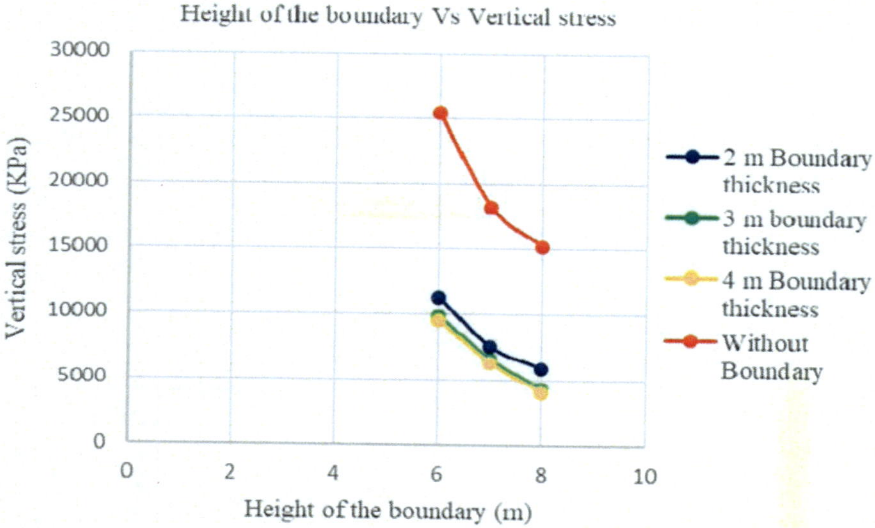


**Fig. 8** Height of the boundary versus vertical stress for various boundary conditions

60 and 70 kPa, settlement values varying is low. The displacement values for different boundary conditions and various young's modulus values are shown in Figs. 10 and 11.

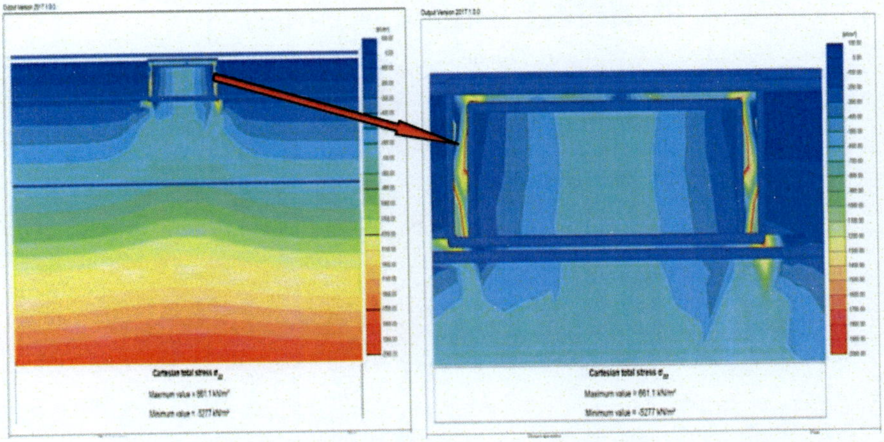

**Fig. 9** Vertical stress for 3 m boundary thickness and 7 m height of the boundary

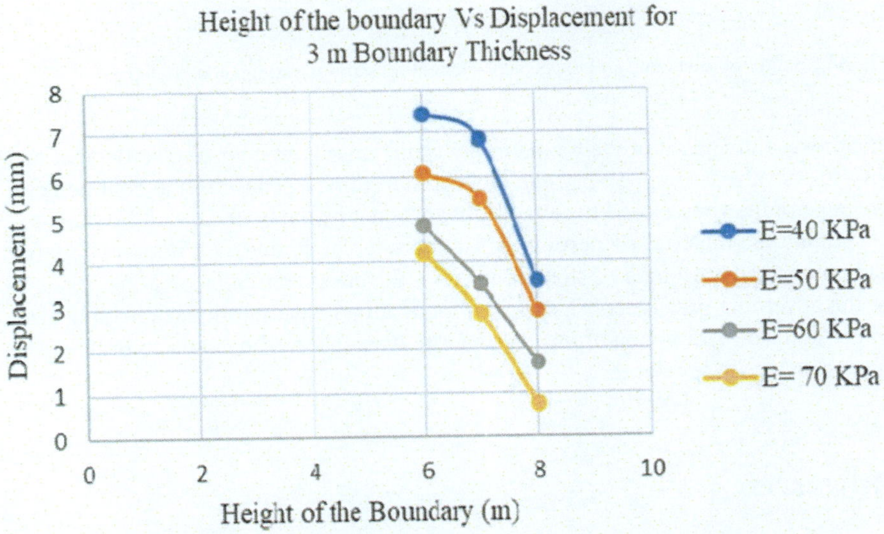


**Fig. 10** Height of the boundary versus displacement for various young's modulus (E)

## 6 Conclusions

Based on the results, it was concluded that the Thanjavur Brihadeeswarar temple tower foundation was designed for zero settlement. The foundation designed without the boundary, the settlement value was more than the permissible limit. But the foundation designed with boundary, the settlement value was decreased. The shallow foundation 36 m square raft and thickness of the footing is 2 m with 3 m boundary

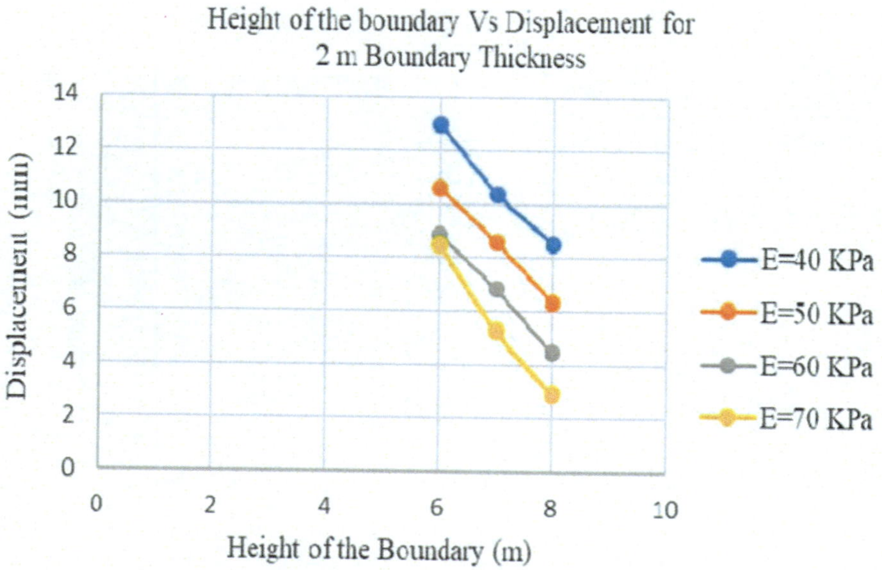


**Fig. 11** Height of the boundary versus displacement for various young's modulus (E)

thickness, 8 m Height of the boundary gives the approximately zero settlement and low-stress values compare than the other dimensions. Because of the confinement, the most of the stresses due to self-weight are taken by the boundary of the sandbox.

Among the material properties Elastic modulus of stone ($E$), Friction angle ($\phi$) and compressive strength of stone joints ($f_c$), the elastic modulus of stone seems to be the governing parameter. This type of foundation resists earthquake loading also. In earthquake loading conditions, the energy was observed by sand to transfer the superstructure in low level.

## References

1. Brihadeeswarar Temple-Architecture Home page. https://en.wikipedia.org/wiki/BrihadesswararTemple-Architecture. Last accessed 18 July 2020
2. Architecture of Tamil Nadu Home Page. https://en.wikipedia.org/wiki/ArchitectureofTamilNadu. Last accessed 18 July 2020
3. Sargunan, A.: The Big Temples, Thanjavur. Personal Communication, Trichy (2009)
4. www.youtube.com/SecretsofThanjavurTemple. Last accessed 18 July 2020
5. Thanjavur temple Home page. https://en.wikipedia.org/wiki/thanjavurtemple. Last accessed 18 July 2020
6. Adrian hadipriono Tan, Shilun Hao, Jiyuan Li, Lei Zhao, and Fabian Hadipriono Tan 'Visual Modeling of Chinese Temple Construction for Student Education', Proceedings of 122nd ASEE Annual Conference and Exposition, pp.1–23 (2015)

7. Daka, T., Udatha, L., Pasupuleti, V.D.K., Kalapatapu, P., Raja-ram, B.: Anciens Sandbox technique: an experimental study using piezoelectric sensors. Euro-Mediterranean Conf. 173–184 (2018)
8. Wethyavivorn, B., Surit, S., Apichatyakul, C., Lerdsivanon, N.: Model verification of Thai historic masonry monuments. J. Perform. Construct. Facil. **30**(1) (2016)
9. Iwasaki, Y., Fukuda, M., McCarthy, R., Ishizuka, M., Nakagawa, T.: Authenticity of the soils and foundation of heritage structure of Bayon Temple, Angkor, Cambodia. In: International Symposium on Geotechnical Aspects of Heritage Structures ISGHS Chennai 2019, IIT Madras, Chennai, India (2019)
10. Dr. MutalikDesai, V.G., Desai, A.: Analysis of strip footing using PLAXIS 2D. Int. J. Res. Sci. Inn. (IJRSI) **IV**(VI), 33–38 (2017)
11. IS-3316: Specification of Structural Granite. Bureau of Indian Standards, New Delhi (R-2003) (1974)
12. Tripathi, A.K., Kant, R.: Investigation of geo-mechanical properties of rock samples. Int. Refereed J. Eng. Sci. **8**, 28–36
13. Goodman, R.: Introduction to rock mechanics. Wiley India Pvt. Ltd., New Delhi
14. IS-3620: Indian Standard Specifications for Laterite Stone Block Masonry. Bureau of Indian Standards, New Delhi (1974)

# Re-evaluation of Failure of Silo Tower Foundations

**J. Y. V. Shiva Bhushan, Asif Ali Mohammed, and Madhav Madhira**

## 1 Introduction

Silos are large structures that are commonly used for storing food grains, petrol, fertilizers, etc., for long periods. It is usual practice to design silos with circular or ring footings founded at depth from the ground surface based on requirement of bearing capacity. Skempton [10] proposed a method to estimate ultimate bearing capacity ($q_u$) of foundations on homogeneous soil as

$$q_u = cN_c + p \tag{1}$$

where $c$ is the average shear strength of the soil to a depth below the foundation equal to two-thirds of the diameter, $N_c$—bearing capacity factor, $p$ ($= \gamma D_f$)—overburden pressure at the level of the footing, $\gamma$—the unit weight of soil and $D_f$—depth of footing. Meyerhof [7] considers the average shear strength of soil as cohesion, '$c$' in Eq. (1). Normally consolidated alluvial or marine soils, when exposed to atmosphere, are subjected to seasonal changes [5] and form a crust or desiccated layer because of diurnal heating and cooling, lowering of water table, wetting, drying, etc. Desiccation causes an increase in unit weight and a significant increase in undrained strength of near-surface layers.

Undrained shear strength varies linearly with depth for a young normally consolidated deposit (Fig. 1a). With time, the deposit may gain in shear strength due to ageing (Fig. 1b). The strength profile of an aged deposit with a crust/stiff layer on top of NC soil may be represented as in Fig. 1c. Solutions for bearing capacity of circular footing resting on non-homogeneous aged clays with shear strength profile shown in Fig. 1b and c are given by Davis and Booker [4], as in Eqs. (2) and (3), respectively.

J. Y. V. Shiva Bhushan · A. A. Mohammed (✉) · M. Madhira
VNR Vignana Jyothi Institute of Engineering and Technology, Hyderabad 500090, India



C. N. V. Satyanarayana Reddy et al. (eds.), *Ground Characterization and Foundations*,
Lecture Notes in Civil Engineering 167,
https://doi.org/10.1007/978-981-16-3383-6_72

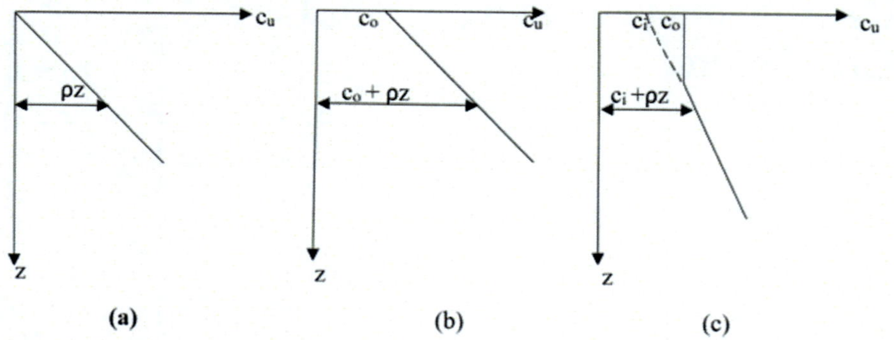


**Fig. 1** Shear strength variation with depth of **a** normally consolidated clay, **b** aged deposits and **c** normally consolidated clay with crust (after Davis and Booker [4])

$$q_{\mathrm{uf}} = 1.2F_{\mathrm{R}}\left[c_0 N_{\mathrm{c}} + \frac{\rho D}{4}\right] \tag{2}$$

$$q_{\mathrm{uf}} = 1.2F_{\mathrm{RC}}\left[c_0 N_{\mathrm{c}} + \frac{\rho D}{4}\right] \tag{3}$$

where $N_c = (2 + \pi) = 5.14$; $\rho$—the rate of increase of shear strength with depth, $c_o$—undrained shear strength at ground level, $D$—width of the footing and $F_R$ and $F_{RC} = f(\rho D/c_o)$—correction factors for roughness of the footing for footing resting on clays without and with crust are given in Figs. 2and 3, respectively, [4]. The intercept, $c_i$, of the linear profile at the top, shown in Fig. 1c, is less than $c_o$ and can be zero for normally consolidated deposits with no crust.

**Fig. 2** Correction factor, $F_R$, for rough footing (after Davis and Booker [4])

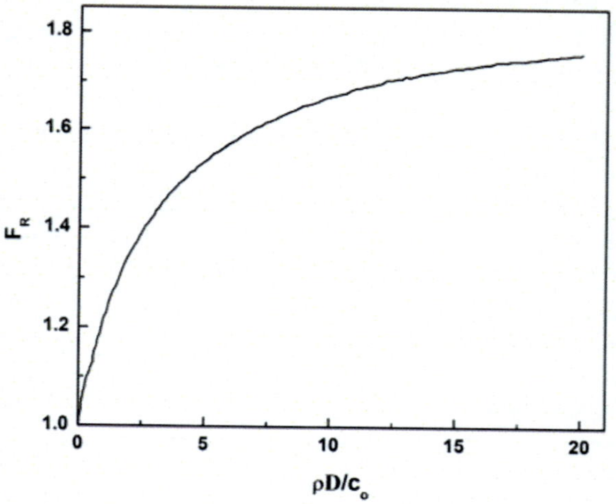

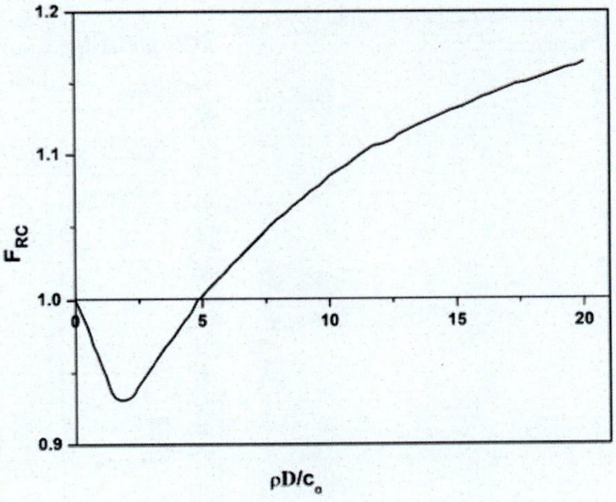


**Fig. 3** Correction factor, $F_{RC}$, for rough footing with crust (after Davis and Booker [4])

## 2 Case Histories of Collapsed Silos

Failures of several silo structures in Canada in 1970s were presented by Bozozuk [1–3]. The following are the details of the silos with their soil profiles and reported causes of failure.

### *2.1 Vankleek Hill Silo*

This silo tower 6 m in dia. and 21 m high was constructed in May, 1970. The soil profile is shown in Fig. 4. Organic soil 0.3 m thick overlies desiccated silty clay of 3 m thickness. Undrained strength in this layer decreases from 100 kPa at 1.8 m to nearly 12 kPa at 4.0 m illustrating the reducing effect of desiccation with depth. Normally consolidated silty clay with some black mottling, with undrained strength increasing linearly with depth exists from 3.4 to 10.3 m. The average plasticity index and natural water content were, respectively, 36% and 55%. The average bearing pressure due to weight of superstructure and foundation was evaluated as 47 kPa. Ring footing 1 m thick and 1.2 m deep was constructed, and a concrete apron/raft of diameter 21 m was provided as foundation. The average pressures due to combined load of structure and silage as per owner and calculated from data were 166 kPa and 150 kPa, respectively. The non-uniform loading of silage into silos and overlapping of pressure bulbs of adjacent silos resulted in tilting of the structures (Fig. 5). The tilting of silo ultimately caused its failure.

**Fig. 4** Shear strength profile at Vankleek Hill tower silo

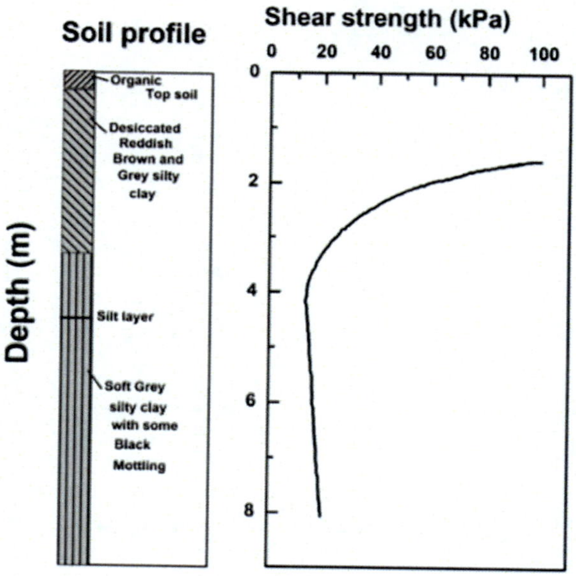


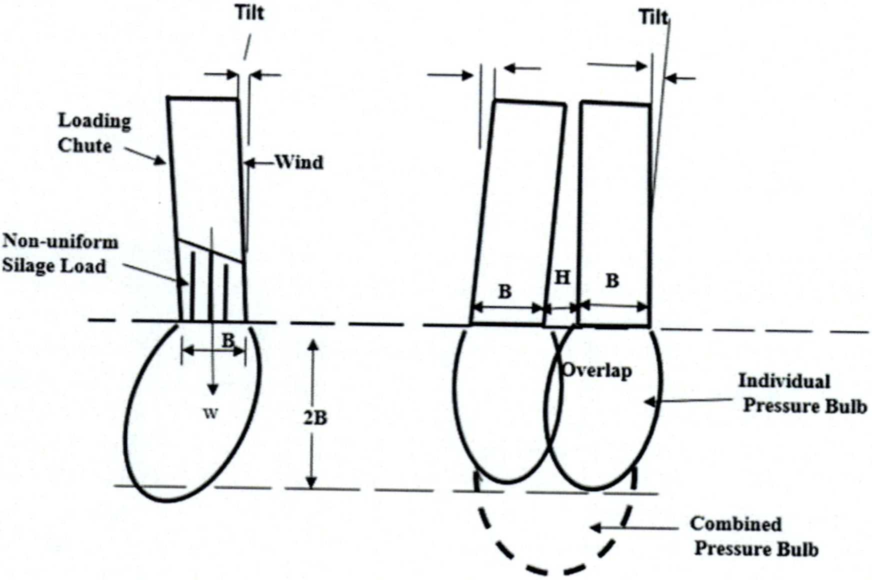


**Fig. 5** Non-uniform loading and overlapping pressure bulbs (Redrawn after Bozozuk [2])

**Fig. 6** Shear strength profile at Richmond silo

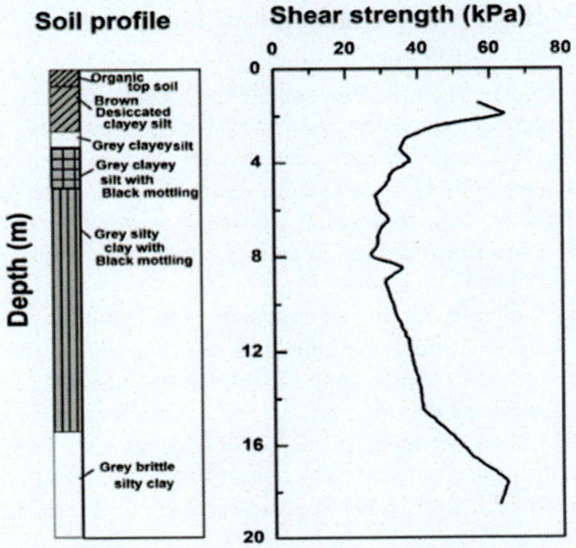


## 2.2 *Richmond Silo*

A concrete tower silo of 9.14 m diameter and 32.3 m height was constructed on normally consolidated marine deposits of the Champlain Sea in August, 1975. The soil and strength profiles are shown in Fig. 6. Brown desiccated clayey silt exists below the top organic soil, up to 2.4 m at which depth groundwater level was met with. Grey clayey silt follows to a depth of 5 m and extends with black mottling up to a depth of 15.5 m. Further down up to 20.0 m, thick brittle grey silty clay was observed.

The liquid limit and plasticity index of soil in the desiccated layer were 40 and 20, respectively. Undrained strength decreases from about 60 kPa at 2.0 m to 25 kPa at 5.0 m and increases linearly with further increase in depth (Fig. 6). Ring footing with inner and outer diameters of 7.62 m and 11.89 m, respectively, was provided. The thickness of the cast-in-place non-reinforced footing was 0.61 m. The silo with a combined pressure of 203.5 kPa, due to silage and dead load overturned and failed within a month of filling.

In both these cases, the actual shear strength variations, i.e. decrease with depth in the zone of desiccation and further a linear increase with depth of clays, have not been considered. Instead, average strength to a depth on 0.67B was used [1] to estimate the ultimate bearing capacities of the silo foundations.

## 3 Problem Definition and Formulation

Foundations for the above silos rest on a desiccated layer overlying normally consolidated soil. However, Bozozuk [1] used Skempton's [10] solution (Eq. 1) to estimate the ultimate bearing capacity of the foundations, which is strictly valid for only homogenous deposits with an average undrained strength corresponding to a depth of 0.67B from the footing level. Based on these case histories of silos, bearing capacity of foundations resting on desiccated layer overlying normally consolidated soil is formulated as follows.

A circular footing of diameter, $D$, at depth, $D_f$, from the ground, rests on soil whose shear strength profile is as shown in Fig. 7. The undrained shear strengths of soil at the base of footing ($D_f$) and at the interface of desiccated soil and normally consolidated clay ($z_o$) are $c_{uB}$ and $c_{uo}$, respectively. Desiccation strength ratio, $\mu_B$ ($= c_{uB}/c_{uo}$), is defined as a ratio of strength at the footing level to that at the bottom of the desiccated layer. The footing rests within the desiccated layer of thickness, $z_o$. Normally consolidated soil whose strength increases linearly with depth lies beneath the desiccated layer. The rate of increase of shear strength with depth in the normally consolidated deposit is denoted by $\rho$.

The shear strengths, $c_{uz}$, of soil at a depth, $z$, within desiccated and normally consolidated layers are given in Eqs. (4) and (5), respectively.

$$z < z_0,\ c_{uz} = \frac{c_{uB}(z_o - z) + c_{u0}(z - D_f)}{(z_o - D_f)} \tag{4}$$

$$z > z_0, c_{uz} = c_{u0} + \rho(z - z_0) \tag{5}$$

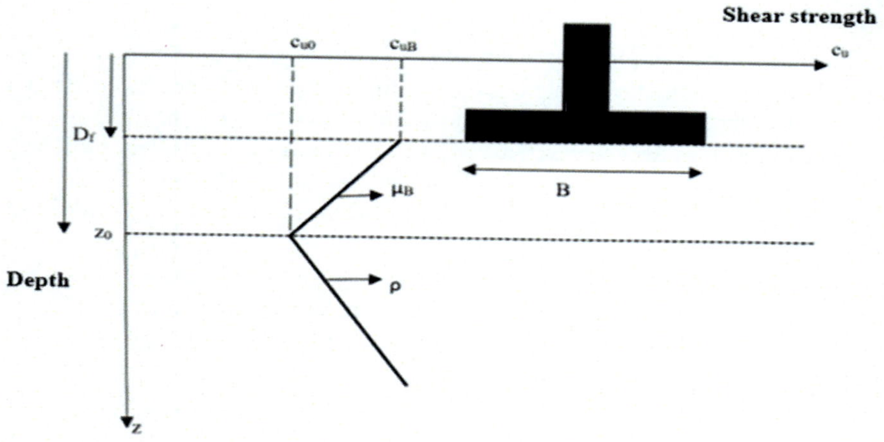


**Fig. 7** Undrained shear strength profile

Footing resting on strong desiccated soil overlying normally consolidated soil with undrained strength increasing with depth is modelled as a two-layered system similar to Meyerhof [8] and Meyerhof and Hanna [6, 9] but with the proviso that the strength in the desiccated layer is not constant but decreases with depth.

The mass of soil below the base of the footing within the desiccated layer (Fig. 8a) subjected to punching shear is considered. The various forces acting on the soil element below the footing up to the desiccated layer are shown in Fig. 8b. The forces acting on the elemental mass of thickness ($z_o - D_f$) are the adhesive force $C_a$, between the soil mass and the adjacent soil in desiccated clay, ultimate bearing capacities $q_{uf}$ and $q_{uc}$ acting on top and bottom of the soil mass, respectively. The self-weight, $W$, of cylindrical soil element acts vertically downwards.

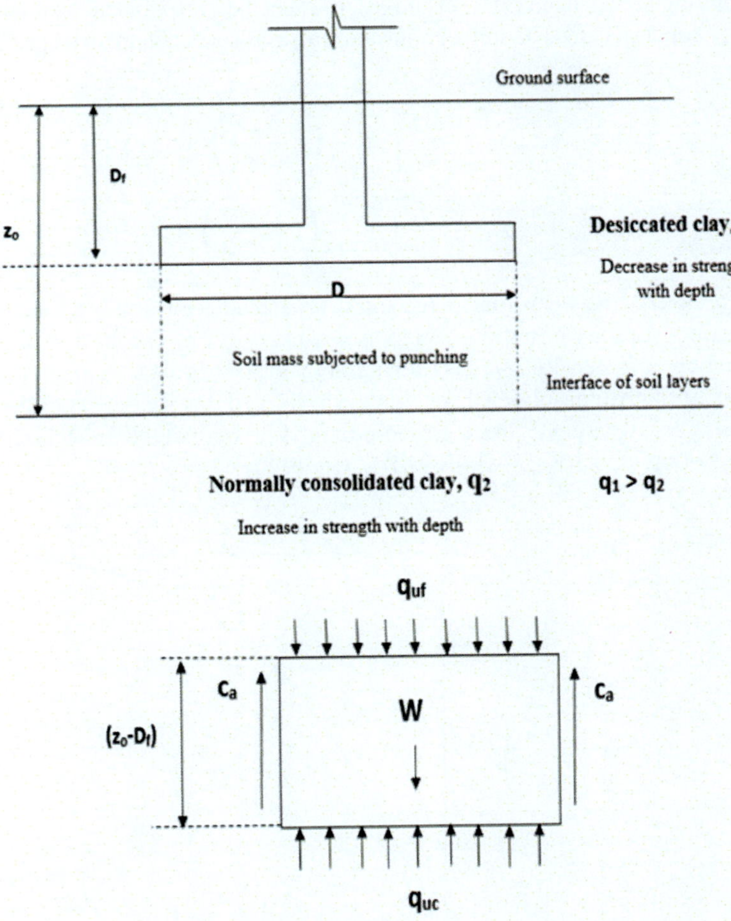


**Fig. 8** Free body diagram of soil mass within desiccation layer

Ultimate bearing capacity, $q_{uf}$, of circular footing for NC soil with desiccated layer is obtained from equilibrium conditions.

$$q_{uf}\left(\frac{\pi D^2}{4}\right) = q_{uc}\left(\frac{\pi D^2}{4}\right) + c_a(z_0 - D_f)(\pi D) - \gamma(z_0 - D_f)\left(\frac{\pi D^2}{4}\right)$$

$$q_{uf} = q_{uc} + \frac{4}{D}c_a(z_0 - D_f) - \gamma(z_0 - D_f) \quad (6)$$

where $q_{uc} = 1.2F_R\left[c_{u0}N_c + \frac{\rho B}{4}\right]$ is the ultimate bearing capacity of a circular footing on non-homogeneous ground with strength increasing with depth [4] with $\rho$ the rate of increase of strength with depth as defined earlier. Simplifying Eq. (6) and substituting for $q_{uc}$ from Eqs. (2) and (3), ultimate bearing capacities of circular footing, $q_{uf}$, for desiccated NC soil without and with a crust are obtained, respectively, as

$$q_{uf} = 1.2F_R\left[c_{uo}N_c + \frac{\rho D}{4}\right] + \frac{z_o}{D}\left(1 - \frac{D_f}{z_o}\right)(4c_a - \gamma D) \quad (7)$$

$$q_{uf} = 1.2F_{RC}\left[c_{uo}N_c + \frac{\rho D}{4}\right] + \frac{z_o}{D}\left(1 - \frac{D_f}{z_o}\right)(4c_a - \gamma D) \quad (8)$$

The adhesion, $c_a$, between considered soil mass and adjacent soil is a function of ratio of bearing capacity of bottom and top soils, i.e. $q_2/q_1$, $c_1$ is cohesion of upper layer. Variation of $c_a/c_1$ with $q_2/q_1$ is shown in Fig. 9 [9]. Ultimate bearing capacity of the lower normally consolidated soil, $q_2$, is $c_{uo}\,N_C{}^*$ and that for the top desiccated layer $q_1$, as $c_1\,N_C$, where $c_1$ is the average undrained strength of the desiccated layer below the footing, $= 0.5(c_{uB} + c_{u0}) = 0.5c_{uo}\,(\rho_B + 1)$.

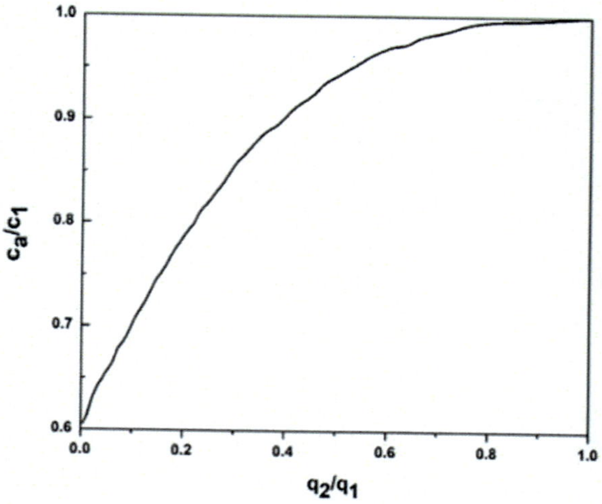


**Fig. 9** Variation of $\frac{c_a}{c_1}$ with $\frac{q_2}{q_1}$ (after Meyerhof et al. [9])

The footing is at depth $D_f$ in the desiccated layer of total thickness, $z_0$. Considering surcharge stress, $q_0 = \gamma D_f$, Eqs. (7) and (8) become

$$q_u = q_{uf} + \gamma D_f$$

$$q_u = 1.2F_R\left[c_{uo}N_c + \frac{\rho D}{4}\right] + \frac{z_o}{D}\left(1 - \frac{D_f}{z_o}\right)(4c_a - \gamma D) + \gamma D_f \quad (9)$$

$$q_u = 1.2F_{RC}\left[c_{uo}N_c + \frac{\rho D}{4}\right] + \frac{z_o}{D}\left(1 - \frac{D_f}{z_o}\right)(4c_a - \gamma D) + \gamma D_f \quad (10)$$

Gross ultimate bearing capacity of the circular footing resting on desiccated layer overlying normally consolidated soil without and with a crust on top of NC clay given in Eqs. (9) and (10), respectively, is derived by incorporating Davis and Booker's [4] solution into Meyerhof's [9] punching failure approach for two-layered soil. Bearing capacity is normalized with undrained strength, $c_{u0}$, and this ratio of gross ultimate bearing capacity to undrained strength is defined as normalized bearing capacity (NBC), and Eq. (10) is written as

$$\text{NBC} = \frac{q_u}{c_{uo}} = 1.2F_R\left[5.14 + \frac{\rho D}{4c_{uo}}\right] + \frac{z_o}{D}\left(1 - \frac{D_f}{z_o}\right)\left(\frac{4c_a - \gamma D}{c_{uo}}\right) + \frac{\gamma D_f}{c_{uo}} \quad (11)$$

Variations of NBC with $\rho D/c_{uo}$, for different $D_f/z_0$ and for $z_o = 0.5D$ are shown in Fig. 10. NBC increases with the normalized rate of increase of strength, $\rho D/c_{uo}$ of NC layer but decreases with increase in the depth of the footing as the strength of desiccated layer decreases with depth. For a footing at mid-depth of the desiccated layer, $D_f/z_0$ equal to 0.5, NBC increases from 9.6 for homogenous deposit ($\rho B/c_{uo} = 0$ strength constant with depth) to 24.9 for NC layer with $\rho D/c_{uo}$ equal to 20. For

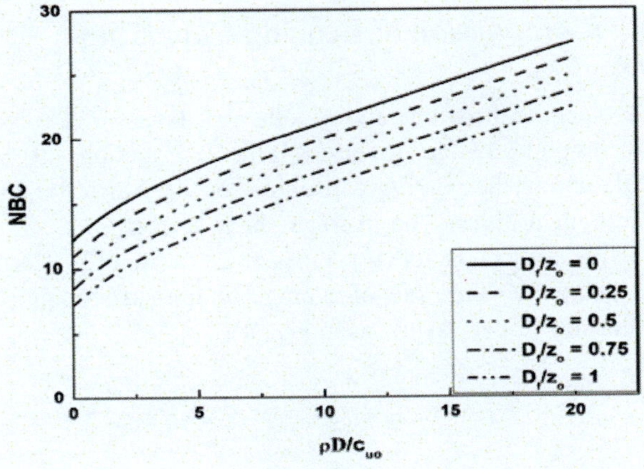


**Fig. 10** NBC versus $\rho D/c_{uo}$—effect of $D_f/z_o$ for $z_o = 0.5D$

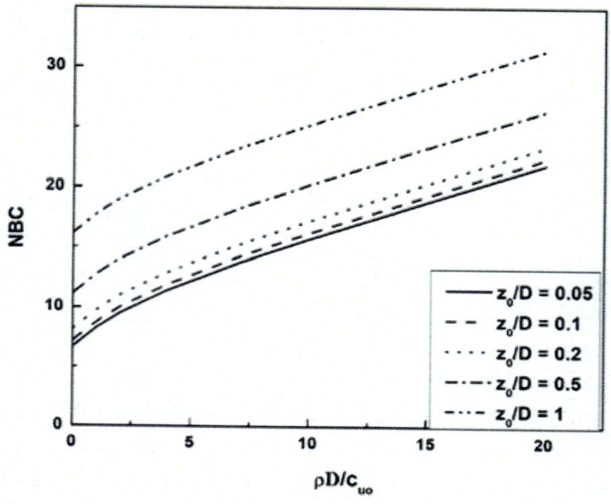


**Fig. 11** NBC versus $\rho D/c_{uo}$—effect of $z_0/D$ for $D_f = 0$

a moderate rate of increase of strength with depth, i.e. $\rho D/c_{uo}$ equal to 5.0, NBC decreases from 17.8 for a footing at the surface to 12.8 for the footing at the bottom of the desiccated layer, a decrease of about 28%.

Variations of NBC with $\rho D/c_{uo}$, for different $z_0/D$ and $D_f = 0$, are shown in Fig. 11. NBC increases with the normalized rate of increase of strength, $\rho D/c_{uo}$, of NC layer and with increase in the thickness of desiccated layer. For a footing at the surface, i.e. on top of the desiccated layer, $z_0/D$ equal to 0.5, NBC increases from 11.2 for homogenous deposit ($\rho D/c_{uo}$, strength constant with depth) to 26.5 for NC layer with $\rho D/c_{uo}$ equal to 20. For a moderate rate of increase of strength with depth, i.e. $\rho D/c_{uo}$ equal to 5.0, NBC increases from 12.2 for a thin desiccated layer, i.e. $z_0$ = 5% of size of footing to 21.5 for $z_0$ = 100% of size of footing, an increase of about 76% is observed.

## 4 Comparison of Bearing Capacities

Geotechnical parameters of soils and details of the footings for the Vankleek and Richmond silos are given in Table 1. Equations (9) and (10) are the utilized for calculating the bearing capacities for these silo foundations considering the given strength profiles. The average shear strength, $c_{u\,\mathrm{avg}}$, is estimated from Skempton [10] bearing capacity (Eq. 1), as given in Eq. (12). Normalized shear strength, NSS, is defined as the ratio of average undrained strength to undrained strength at the interface of two layers, as in Eq. (13).

$$c_{u\,avg} = \frac{q_{\mathrm{u}}}{N_{\mathrm{c}}^{*}} \tag{12}$$

**Table 1** Parameters of soils and foundations at silo locations

| Parameter | Vankleek Hill silo | Richmond silo |
|---|---|---|
| Diameter of footing $D$, m | 8 | 11.89 |
| Depth of footing $D_f$, m | 1.52 | 0.61 |
| Unit weight of soil $\gamma$, kN/m$^3$ | 15 | 15 |
| Thickness of desiccated layer $z_0$, m | 3.7 | 2.4 |
| Rate of increase of shear strength, $\rho$ | 1.35 | 1 |
| Strength reduction ratio $\mu_B$ | 6.25 | 2 |
| Undrained cohesion at the interface of desiccated and non-homogeneous layers $c_{u0}$, kPa | 16 | 30 |
| Roughness correction factor $F_R$ | 1.06 | 1.08 |
| Bearing capacity factor $N_c{}^*$ | 6.6 | 6.6 |
| Undrained cohesion of desiccated soil $c_1$, kPa | 58 | 45 |
| Ratio of bearing capacities $q_2/q_1$ | 0.28 | 0.67 |
| Ratio of adhesion to cohesion in desiccated clay $c_a/c_1$ | 0.84 | 0.98 |
| Adhesion between footing and soil in desiccation $c_a$, kPa | 48.72 | 44.1 |

$$\text{NSS} = \frac{c_{u\,avg}}{c_{uo}} = \frac{q_u}{c_{uo} N_c^*} = \frac{\text{NBC}}{N_c^*} \tag{13}$$

where $q_u$ is the ultimate bearing capacities obtained from Eqs. (8) and (9), $N_c^*$ is Bearing capacity factor (taken as 6.6 as per Skempton [10]) and $c_{uo}$ is undrained cohesion at the interface of desiccated and non-homogeneous layers.

Variation of NSS with $\rho D/c_{uo}$ for different $z_0$, $D_f = 0$ and $N_c = 6.6$ are shown in Fig. 12. NSS increases with the normalized rate of increase of strength, $\rho D/c_{uo}$, of

**Fig. 12** NSS versus $\rho B/c_{uo}$—effect of $z_0/D$

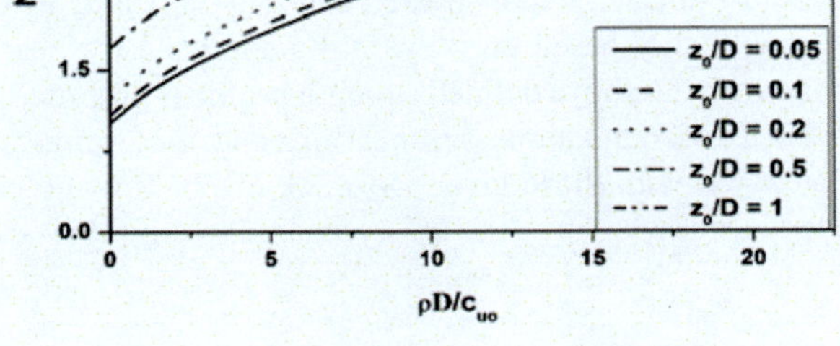

**Table 2** Comparison of average shear strengths of soils and bearing capacities of foundations of silos

| S. No. | Name of the silo | Shear strength $c_{u\,avg}$, (kPa) | | Ultimate bearing capacity $q_u$, (kPa) | |
|---|---|---|---|---|---|
| | | Bozozuk | Proposed | Bozozuk | Proposed |
| 1 | Vankleek | 24.4 | 22.9 | 205 | 151.3 |
| 2 | Richmond | 36.5 | 30.5 | 250 | 201.2 |

NC layer and increases with increase in thickness of desiccated layer. For a footing at the surface, i.e. on top of the desiccated layer, $z_0/D$ equal to 0.5, NSS increases from 1.7 for homogenous deposit ($\rho D/c_{uo} = 0$, strength constant with depth) to 4 for NC layer with $\rho D/c_{uo}$ equal to 20. For a moderate rate of increase of strength with depth, i.e. $\rho D/c_{uo}$ equal to 5.0, NBC increases from 1.8 for a minimal thickness of desiccated layer, i.e. $z_0 = 5\%$ of size of footing to 3.3 for $z_0 = 100\%$ of size of footing, an increase of about 83% is observed.

Bearing capacities of the silo foundations as estimated and reported by Bozozuk [1, 2] based on Skempton [10] equation are compared with those estimated by the present approach. From Table 2, it can be observed that Bozozuk had overestimated the bearing capacities and these high values might have led to the failure of the silos. Bearing capacities estimated by Bozozuk [1, 2] are more and on the unconservative side being 205 and 250 kPa compared to the values 151.3 and 201.2 kPa estimated by considering the actual undrained strength profiles, i.e. considering both the effect of desiccation and the increase in undrained strength with depth for normally consolidated soil.

The bearing capacities from the proposed equation and by Bozozuk [1, 2] differs by 26% and 20% in cases of Vankleek and Richmond silo, respectively. The proposed rigorous theory incorporating the true variation of undrained strength with depth provided a less ultimate bearing capacity of silo foundations.

## 5 Conclusions

Failure of silos reported by Bozozuk [1, 2] is examined in the light of developments in geotechnical engineering, in particular the linear increase in undrained strength with depth of normally consolidated soil and increased strength in the desiccated layer. A new theory based on Meyerhof's approach for bearing capacity of two-layered soils with the above defined strength profile is proposed for the estimation of bearing capacity of foundations. The bearing capacities estimated by the proposed theory are compared with those given Bozozuk [1, 2]. Bearing capacities from the proposed approach are less by 20–26% compared to the values estimated by Bozozuk [1, 2].

## References

1. Bozozuk, M.: Foundation failure of the Vanleek hill tower silo. Proc. Speciality Conf. Performs Earth Earth-Support. Struct. **1**, 885–902 (1972)
2. Bozozuk, M.: Bearing capacity of clays for tower silos. Can. Agric. Eng. **16**, 13–17 (1974)
3. Bozozuk, M.: Problems with concrete tower silos. Can. Agric. Eng. **21**, 69–77 (1979)
4. Davis, E.H., Booker, J.R.: The effect of increasing strength with depth on the bearing capacity of clays. Geotechnique **23**, 551–563 (1973). https://doi.org/10.1680/geot.1973.23.4.551
5. Day, R.W.: Desiccation theory for soft cohesive soils. J. Geotech. Eng. **122**, 943–947 (1996)
6. Hanna, A.M., Meyerhof, G.G.: Design charts for ultimate bearing capacity of foundations on sand overlying soft clay. Can. Geotech. J **17**, 300–303 (1980). https://doi.org/10.1139/t80-030
7. Meyerhof, G.G.: The ultimate bearing capacity of foundations. Geotechnique **2**, 301–322 (1951)
8. Meyerhof, G.G.: Ultimate bearing capacity of footings on sand layer overlying clay. Can. Geotech. J. **11**, 223–229 (1974)
9. Meyerhof, G.G., Hanna, A.M.: Ultimate bearing capacity of foundations on layered soils under inclined load. Can. Geotech. J. **15**, 565–572 (1978). https://doi.org/10.1139/t78-060
10. Skempton, A.W.: The bearing capacity of clays. In: Proceedings Building Research Congress, London, pp. 180–189 (1951)